INSTRUMENTAL
METHODS
OF ANALYSIS

Fifth Edition

HOBART H. WILLARD Professor Emeritus of Chemistry
University of Michigan

LYNNE L. MERRITT, JR. Professor of Chemistry
Indiana University

JOHN A. DEAN Professor of Chemistry
University of Tennessee at Knoxville

D. VAN NOSTRAND COMPANY

New York Cincinnati Toronto London Melbourne

D. Van Nostrand Company Regional Offices:
New York Cincinnati Millbrae

D. Van Nostrand Company International Offices:
London Toronto Melbourne

Published by D. Van Nostrand Company
450 West 33rd Street, New York, N.Y. 10001

Published simultaneously in Canada by
Van Nostrand Reinhold Ltd.
10 9 8 7 6 5 4 3 2

Preface

The Fifth Edition continues to survey modern instrumental methods of chemical analysis. Most of the chapters have been extensively revised and some have been completely rewritten.

Changes in order of presentation now place molecular fluorescence and phosphorescence methods after ultraviolet and visible absorption methods, Raman spectroscopy after infrared spectroscopy, and flame emission and atomic absorption spectrometry before emission spectroscopy. This arrangement is more logical than the order of presentation in the earlier editions.

Among the new topics treated in this edition are: turbidimetry and nephelometry, the vacuum ultraviolet, reflectance measurements, Fourier transform infrared, laser-Raman spectroscopy, Mössbauer spectroscopy, interfacing gas chromatography with mass spectrometry, and all classes of selective ion electrodes. Atomic absorption has been expanded and integrated with flame emission methods. Classical polarography has been absorbed within an enlarged chapter on voltammetry, polarography, and related techniques. Emphasis continues to be placed on structural identification of compounds through infrared and Raman spectra, nuclear magnetic resonance and electron spin resonance spectroscopy, ultraviolet absorption spectra, and mass spectrometry.

Individual chapters are designed, in general, to stand alone. Consequently, the order of presentation is not critical. Instructors will be able to select material for several levels of achievement. References to the literature and collateral readings are included in each chapter. The book should also be suitable as a reference manual.

Numerous examples are incorporated within the text, including those illustrating mathematical operations. These introduce the student to the unit of measurement and reduce, and possibly eliminate, the dependence upon additional problem books. There are 390 numerical problems; answers to virtually all are given separately at the end of the text. Many of these problems contain data that would be obtained in the laboratory experiments and are thus of particular value for schools unable to furnish equipment for specific areas of instrumentation, for supplementing experiments when laboratory periods are limited in number, or for self-study.

Experiments have been selected to illustrate the principles discussed in the theoretical portions of each chapter. Some experiments are described in considerable detail and thus are suitable for use by less experienced undergraduate students. Others are merely sketched outlines or suggestions for work to give instructors in advanced courses flexibility in eliciting from students a degree of independence and originality in the outline and execution of experimental work.

Because some confusion may arise over the meanings of abbreviations and the uses of symbols, particularly the overlapping uses of certain symbols in the diverse techniques covered in this book, separate listings of abbreviations and symbols are included in pages xii to xix. Whenever available, recommendations of concerned nomenclature commissions have been followed. In addition, the Appendices provide a fairly comprehensive tabulation of standard-reduction potentials in aqueous solution, polarographic half-wave potentials and diffusion-current constants, acid dissociation constants, formation constants of some metal complexes, flame emission and atomic absorption spectra, and a conversion table involving values of absorbance for percent absorption. A four-place table of common logarithms, a table of 1971 atomic weights, and a periodic chart of the elements facilitate computations and provide ready reference data.

The authors remain greatly indebted to the manufacturers who have so generously furnished schematic diagrams, photographs, and technical information of their instruments. Thanks are expressed also to many colleagues who have kindly helped with suggestions and improvements.

HOBART H. WILLARD
LYNNE L. MERRITT, JR.
JOHN A. DEAN

Contents

List of Experiments

Experiment

Experiment

Experiment

Abbreviations

absorption	Abs
alpha particle	α
alternating current (adj.)	ac
ampere	A
angstrom	Å
atmosphere	atm
atomic weight	at. wt
attenuated total reflectance	ATR
barn	b
beta particle	β
boiling point	bp
calorie	cal
capacitance	C
conductance	$1/R$
coulomb	C, Q
counts per minute (second)	c/m 1, cpm (c/s)
cubic centimeter	cm^3
curie	Ci
cycles per second (hertz)	Hz
day	d
decibel	db
degree Celsius	°C
degree Kelvin	°K
deuteron	d
diameter	diam
differential scanning calorimeter	DSC
differential thermal analysis	DTA
direct current (adj.)	dc
disintegrations per minute (second)	dpm, d/m; dps, d/s
dropping mercury electrode	dme
dyne	dyn
electromotive force	emf
electron	e^-, e
electron paramagnetic resonance	epr

electron spin resonance	esr
electron volt	eV
equivalent weight	equiv wt
ethyl	Et
ethylenediamine	en
ethylenediamine-N, N, N', N'-tetraacetic acid	EDTA
(the anion)	Y^{4-}
exempli gratia (for example)	e.g.
exponential	exp
farad	F, f
formal (concentration)	F
frequency	f
gamma radiation	γ
gas (physical state)	g
gauss	G
gram	g
hertz	Hz
hour	hr, h
ibidem (in the same place)	Ibid.
id est (that is)	i.e.
inch.	in.
indicator	ind
inductance	L
infrared	ir
inside diameter	i.d.
joule	J
kilo- (prefix)	k-
kilocalorie	kcal
liquid (physical state)	liq, l
liter	liter (alone), l (with prefixes)
logarithm (common)	log
logarithm (natural)	ln
maximum	max
meg- (prefix)	M-
melting point	mp
meter	m
methyl	Me
micro- (prefix)	μ-
micrometer (micron)	μm
milli- (prefix)	m-
milliequivalent	mequiv
milliliter	ml
millimole	mM
minimum	min

minute	min, m
molar	M
mole	mol
molecular weight	mol wt
nano- (prefix)	n-
nanometer (millimicron)	nm
Naperian base	e
negative	neg
neutron	n
normal (concentration)	N
normal hydrogen electrode	NHE, SHE
nuclear magnetic resonance	nmr
ohm	Ω
optical speed	f/number
outside diameter	o.d.
oxidant	ox
page(s)	p. (pp.)
parts per billion, volume	ng/ml
parts per billion, weight	ng/g
parts per million, volume	μg/ml
parts per million, weight	μg/g
percent	%
phenyl	ϕ, Ph
pico- (prefix)	p-
positive	pos
potential	E
positron	β^+
proton	p
proton magnetic resonance	pmr
quantum (energy)	$h\nu$
radiofrequency	rf
reciprocal ohm	mho (Ω^{-1})
reductant	red
reference	ref
resistance	R
revolutions per minute	rpm
saturated calomel electrode	SCE
second	sec, s
solid (physical state)	s
specific gravity	sp gr
standard hydrogen electrode	SHE, NHE
standard temperature and pressure	STP
temperature	temp, T
thermal gravimetric analysis	TGA

torr (mm of mercury)	torr
tritium	t, ^{3}H
ultraviolet	uv
vacuum	vac
vacuum tube voltmeter	VTVM
versus	vs.
volt	V
volume	vol, V, v
volume per volume	v/v
volume per weight	v/w
watt	W
wavenumber	cm^{-1}
wavenumber difference (Raman)	Δcm^{-1}
year	yr, y

Symbols

A	absorbance; activity (radiochemistry); area; atomic weight
A_{nm}	transition probability of spontaneous emission ($m \rightarrow n$ energy level)
a	specific absorptivity
a_i	hyperfine coupling constant (esr)
a_x	activity of species x
B	source brightness
B_{mn}	transition probability of absorption ($n \rightarrow m$ energy level)
B_{nm}	transition probability of induced or stimulated emission ($m \rightarrow n$ energy level)
b	distance, optical path length, thickness
C	concentration; capacitance
C_M	concentration of solute in mobile phase
C_S	concentration of solute in stationary phase
c	velocity of light
D	dielectric constant; diffusion coefficient
D_{MO}	dissociation energy (of metal oxide)
d	diameter, distance, or spacing
d_f	thickness of liquid film
d_p	particle diameter
E	electrode potential; potential of a half-reaction; energy
$E°$	standard electrode potential
$E_{1/2}$	half-wave potential
E_i	ionization potential; energy of electronic state
E_j	junction potential; energy of electronic state
e	electronic charge; Naperian base (logarithms)
F	faraday; fluorescence
F_c	volume flow rate of gas
F_T	total flux transmitting power
f	focal length; fractional abundance
f_{nm}	oscillator strength ($n \rightarrow m$ energy level)
f_x	activity coefficient of species x
f/number	effective aperture ratio
G	high-frequency conductance
$\Delta G°$	Gibbs free energy

g	spectroscopic splitting factor; statistical weights of particular energy levels
H	magnetic field strength, plate height (chromatography)
ΔH	enthalpy change; peak-to-peak separation (esr)
h	height; Planck constant
I	radiant intensity; spin quantum number
I_d	diffusion current constant
I_ν	emission line intensity
i	angle of incidence; current
i_d	diffusion current
i_{lim}	limiting current
i_r	residual current
J	spin-spin coupling constant
j	compressibility factor (gas chromatography)
K_a	acid dissociation constant
K_d	partition coefficient
K_f	formation constant
K_i	ionization constant (gaseous state)
K_{sp}	solubility product
K_w	ion product of water
k	Boltzmann constant; partition ratio and capacity factor (chromatography); force constant (ir); general constant
k_ν	absorption coefficient (optical)
L	length or distance; lightness (color), inductance
M_s	angular momentum quantum number
m	mass of mercury (dme); order number (optical); metastable
m^+	ionized mass fragment
m/e	mass-to-charge ratio
N	noise; plate number (chromatography); total number of something
N_A	Avogadro number
N_j, N_m	number of species in excited energy state
N_n, N_0	number of species in ground energy state
n, n_D	refractive index (at D sodium line)
n	number of electrons transferred in an electrode reaction; unshared p-electrons
P	pressure; radiant power
P_M	parent mass peak
P_0	incident radiant power
p	pressure; type of electron; depolarization ratio (Raman)
p-	(prefix) negative logarithm of, pico-
Q	flow rate; heat capacity
R	gas constant; resolving power
R.I.	retention index (Kovats)
r	radius; counting rate; resolution (recorders); angle of diffraction
r_D	specific refraction

S	electron spin; saturation factor (radiochemistry)
S_1	first excited singlet state
S_0	ground electronic state
ΔS	entropy
S/N	signal-to-noise ratio
T	temperature; transmittance
T_1	spin-lattice relaxation; first excited triplet state
T_c	column temperature
t	time; prism base length
$t_{1/2}$	half-life
t_R	retention time
V	volume; voltage
V_g°	specific retention volume at 0°C
V_M	volume of mobile phase
V_N	net retention volume
V_R	retention volume
V_R'	adjusted retention volume
v	volume; velocity
W	weight; zone width at base line (chromatography)
$W_{1/2}$	zone width at $\frac{1}{2}$ peak height
W_f	flux
W_L	weight of liquid phase
w	effective aperture width
X_C	capacitive reactance
X_L	inductive reactance
Z	atomic number
z	valence
z_+, z_-	ionic charge
α	degree of ionization; relative retention ratio
$[\alpha]$	specific rotation
β	blaze angle; buffer value; volumetric phase ratio
β_N	Bohr magneton
γ	activity coefficient; emulsion characteristic (photography); ratio of specific heats at constant pressure and constant volume; surface tension
Δ	(prefix) symbol for finite change
δ	chemical shift (nmr); thickness of diffusion layer
∂	(prefix) partial derivative
ϵ	molar absorptivity
ϵ_{max}	molar absorptivity at wavelength of an absorption maximum
η	viscosity
η_D	refractive index (D line of sodium)
θ	cell constant (conductance)
$[\theta]$	molecular ellipticity
κ	specific conductance

Λ	equivalent conductance
Λ_∞	equivalent conductance at infinite dilution
λ	decay constant (radioactivity); wavelength
λ_+, λ_-	ionic conductance
λ_{max}	wavelength of an absorption maximum
μ	ionic strength; linear absorption coefficient; magnetic moment
μ_m	mass absorption coefficient
μ/ρ	mass absorption coefficient
ν	frequency; designation of vibrational levels
$\bar{\nu}$	wavenumber
$\Delta\nu_D$	Doppler broadening
$\Delta\nu_L$	Lorentz broadening
π	pi (3.1416 ...); type of electron or bond
ρ	density; resistivity
σ	capture cross section; shielding constant (nmr); standard deviation
σ_{hkl}	reciprocal lattice vectors
τ	chemical shift (nmr); mean emission lifetime; resolving time; time constant
υ	velocity
Φ	neutron flux
ϕ	quantum efficiency
χ	Pauling electronegativity
ω	chopping frequency; overpotential
ω_c	angular velocity
[ ]	molar concentration of

1

Electronics: Passive Elements and Basic Electrical Measurements

This chapter should be a review for those who have had a course in physics covering ac and dc circuits or a course in electrical measurements and may be omitted if these subjects are still well in mind. The material in this section is not intended to be a complete discussion. The following units of measurement and definitions will be useful.

UNITS

Coulomb The coulomb (Q) is the unit of electrical charge or quantity of electricity. Because one electron carries a charge of 1.602×10^{-19} coulomb, the coulomb represents 6.24×10^{18} electrons.

Faraday The faraday (F) is the charge carried by one equivalent weight of an ion and is equal to 96,487 Q.

Ampere The common unit of current (symbol i) is the ampere (abbreviated A) which corresponds to a charge flow rate of one coulomb per second.

Ohm The unit of resistance (symbol R) is the ohm (Ω). It is that resistance through which a difference of potential of one volt will produce a current of one ampere.*

Conductance The reciprocal of the resistance, $1/R$, is known as the conductance and is measured in reciprocal ohms, Ω^{-1}, often written as mho.

Volt The volt (V), the unit of electrical potential or pressure, is the potential developed across a resistance of one ohm when carrying a current of one ampere. The abbreviation for potential is E when potential is measured in volts, V.

Watt The unit of electrical power, the watt (W), is the time rate of work equal to one joule per second. It is equal to the product of current times potential; thus:

$$W = Ei = i^2 R \qquad (1\text{-}1)$$

*See *Nat. Bur. Std. Circ.* 524, p. 103 (Aug. 1953) for references to publications describing these measurements.

Decibel The decibel (db) is often used to express the ratio of two power levels and is defined as $10 \log P_1/P_2$.

dc CIRCUITS

Fundamental Laws and Principles: Ohm's Law and Kirchhoff's Laws

Perhaps the most important law in electrical measurements, Ohm's law, concerns the relationship between current, potential, and resistance in a circuit containing pure resistance only. The law can be simply stated thus:

$$E = iR \tag{1-2}$$

The following two laws, first expressed by Kirchhoff, are very useful in analyzing circuits: (a) The sum of the currents about any single point of an electrical circuit is zero, and (b) the sum of the voltages around a closed electrical circuit is zero.

Analysis of Simple dc Circuits

With the preceding laws one may determine the currents and voltages in a simple or complex circuit consisting of batteries and resistances. As examples of such analyses, consider the circuits in Fig. 1-1.

For the circuit shown in Fig. 1-1(a), application of Kirchhoff's second law starting clockwise from point *a* gives

$$-i_1 R_1 - i_1 R_2 + E = 0 \tag{1-3}$$

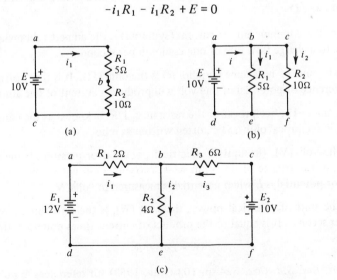

Fig. 1-1. Circuits containing only batteries and resistances.

or

$$i_1(R_1 + R_2) = E \tag{1-4}$$

which leads to a value of 2/3 A for the current i_1 after substitution of the values of R_1, R_2, and E in Eq. 1-4. Furthermore, Eq. 1-4 shows that resistances in series such as R_1 and R_2 are additive. Thus the rule for resistances in series can be stated by Eq. 1-5:

$$R = R_1 + R_2 + \cdots \tag{1-5}$$

By application of Kirchhoff's second law to the circuit of Fig. 1-1(b), one can arrive at three equations for the three possible closed paths, but of these only two, any two, are independent. Thus, for the two paths $a\text{-}b\text{-}e\text{-}d\text{-}a$ and $a\text{-}c\text{-}f\text{-}d\text{-}a$, the equations are

$$-i_1 R_1 + E = 0 \tag{1-6}$$

$$-i_2 R_2 + E = 0 \tag{1-7}$$

To continue the analysis, one must apply Kirchhoff's first law. Selecting point b, one finds

$$i - i_1 - i_2 = 0 \tag{1-8}$$

or

$$i = i_1 + i_2 \tag{1-9}$$

Substituting Eqs. 1–6 and 1–7 in Eq. 1-9 yields

$$i = \frac{E}{R_1} + \frac{E}{R_2} = E\left(\frac{1}{R_1} + \frac{1}{R_2}\right) \tag{1-10}$$

When the values of R_1 and R_2 are substituted in these equations, one finds that $i_1 = 2$ A, $i_2 = 1$ A, and $i = 3$ A. Furthermore, Eq. 1-10 indicates that the rule for adding resistances in parallel is

$$\frac{1}{R} = \frac{1}{R_1} + \frac{1}{R_2} + \cdots \tag{1-11}$$

The analysis of a more complex circuit, such as that shown in Fig. 1-1(c), follows the same rules and leads to the following equations:

$$12 - 2i_1 - 4i_2 = 0 \tag{1-12}$$

$$10 - 6i_3 - 4i_2 = 0 \tag{1-13}$$

and

$$i_1 - i_2 + i_3 = 0 \tag{1-14}$$

Solving these three equations for the three currents leads to the values: $i_1 = 20/11$ A, $i_2 = 23/11$ A, $i_3 = 3/11$ A. A negative answer for a current would mean that the actual current was flowing in the opposite direction to the direction arbitrarily chosen at the start of the analysis.

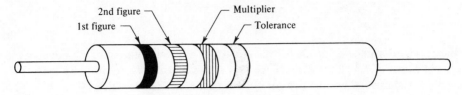

2nd figure

1st figure

Multiplier

Tolerance

Fig. 1-2. Color code of resistors.

Designation of Values of Commercial Resistors

Most commercial resistors are marked with a color code which indicates their resistance value and the tolerance of this value. The code is indicated in Fig. 1-2 and Table 1-1. The resistance value is given by the relationship

$$R = (\text{1st figure} \times 10 + \text{2nd figure}) \times 10^{\text{multiplier}} \pm \text{tolerance }\% \qquad (1\text{-}15)$$

For the tolerance band, silver represents 10% and gold 5%. Resistors more precise than 5% are known as *precision resistors*, and the value is stamped on the resistor. A resistor with the colors yellow, violet, green, and silver would have a resistance of 4,700,000 Ω (or 4.7 MΩ), within ±10%.

The Potentiometer

The potentiometer is widely used for the accurate measurement of electrical potentials. It employs the *null-balance principle*, by which the potential to be measured is balanced by an equal potential but in the opposing sense, so that no current is drawn from the circuit whose potential is to be measured. This potentiometric compensation technique forms the basis of many electrical and electronic measuring circuits and automatic recording devices. The simplified circuit is shown in Fig. 1-3.

A battery of constant emf, larger than any to be measured, is connected across a uniform wire $\overline{AB}$ of high resistance. The emf to be measured, such as a titration cell, is connected to A, with the polarity such that the negative pole of the battery corresponds to the negative pole of the unknown emf. The positive pole of the unknown is connected to the galvanometer G, then through a key which can be closed momentarily by tapping,

TABLE 1-1 Meaning of Colors in Resistor Color-Code

Black	0	Yellow	4	Violet	7
Brown	1	Green	5	Gray	8
Red	2	Blue	6	White	9
Orange	3				

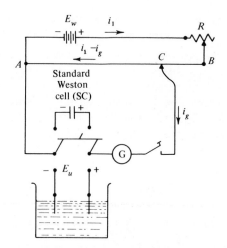

Fig. 1-3. The potentiometer.

and finally to the sliding contactor C, which can be moved across $\overline{AB}$. By repeated trials, the position of C is so adjusted that no current flows through the galvanometer G. Using Kirchhoff's laws, we can derive the condition for balance, that is, zero current through the galvanometer.

Let the currents i_1 and i_g be as shown on Fig. 1-3, and let E_w and E_u be the potentials of the working cell and the unknown, respectively. Let R_g be the resistance of the galvanometer, R_{AC} the resistance of the $\overline{AC}$ section of the wire, R_{CB} the resistance of the $\overline{BC}$ section of the wire, and R_R the resistance of the variable, calibrating resistor R. then

$$E_u + i_g R_g - (i_1 - i_g) R_{AC} = 0 \qquad (1\text{-}16)$$

Solving Eq. 1-16 for i_g and equating this to zero gives Eq. 1-17 and the condition for balance, Eq. 1-18.

$$i_g = \frac{i_1 R_{AC} - E_u}{R_g + R_{AC}} = 0 \qquad (1\text{-}17)$$

or

$$E_u = i_1 R_{AC} \qquad (1\text{-}18)$$

Thus the potential of the unknown is exactly equal to the potential developed across the $\overline{AC}$ section of the slidewire.

If the double-throw double-pole switch is now thrown so that the Weston Standard cell SC, with an exactly known potential E_{SC}, replaces the unknown E_u, and the contactor C is readjusted until a new point of balance is reached at C', an analysis of the circuit would give:

$$E_{SC} = i_1 R_{AC'} \qquad (1\text{-}19)$$

Elimination of i_1 from Eq. 1-18 and 1-19 yields:

$$E_u = \frac{R_{AC}}{R_{AC'}} E_{SC} \qquad (1\text{-}20)$$

Because the slidewire has a uniform resistance along its length, the resistances are directly proportional to length, and Eq. 1-20 can be rewritten in terms of length as:

$$E_u = \frac{\overline{AC}}{\overline{AC'}} E_{SC}$$ (1-21)

By placing a variable resistance R between one of the ends of the slidewire and the battery, it is possible to vary i_1 so that $E_{SC}/R_{AC'}$ in Eq. 1-19 is some simple number, and thus the slidewire can be made direct-reading in millivolts per millimeter or some other simple function of voltage per unit of length.

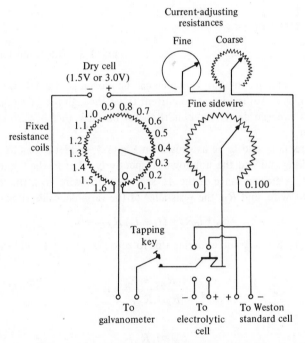

Fig. 1-4. Simplified circuit diagram of a typical commercial potentiometer.

A typical commercial potentiometer is shown in Fig. 1-5. In this instrument the variable resistance, R, of Fig. 1-3 is replaced by two resistances in series: the fine and coarse current-adjusting resistances. Likewise the slidewire, $\overline{AB}$ of Fig. 1-3, is replaced by 16 fixed 10-Ω resistors and one slidewire of 10 Ω resistance graduated into 100 or 1000 divisions. The current can be adjusted by the standard cell and current-adjusting resistors so that each 10-Ω resistor corresponds to 0.1 V and a division on the slidewire corresponds to 1 mV or 0.1 mV. In this type of potentiometer, connections are made with heavy-gauge copper wire and soldered contact points. Sliding contacts are made with metal leaves brushing over large-area surfaces in order to reduce extraneous contact resistances to a minimum.

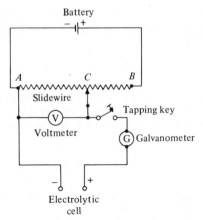

Fig. 1-5. A simple, functional potentiometer. Voltmeter, 0–3 V (10-kΩ resistance); galvanometer, 10^{-6} A/mm; slidewire, 100-Ω resistance; battery, 1.5–3.0 V dry cell.

A very simple and less accurate potentiometer can be constructed as shown in Fig. 1-5. In this case the slidewire need not be uniform in resistance or graduated because the voltage across it is read by the voltmeter when the galvanometer indicates that the potential exactly balances that of the electrolytic cell. Precision is limited by the precision of the voltmeter reading.

As the balance-point detector for general purposes, a simple pointer galvanometer whose sensitivity is of the order of 0.2 μA/mm is satisfactory. For more accurate work a suitably damped mirror galvanometer (Fig. 1-6) should be used. In the latter type the optical lever arm is made very long, yet the entire instrument is kept to a reasonable size by using mirrors to give a folded optical path. When the potential to be measured has a

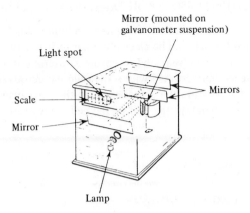

Fig. 1-6. Optical design principle of a high-sensitivity, portable galvanometer based upon a folded optical lever arm of long total path length. (Courtesy of Rubicon Instrument Co.)

very high resistance, greater than 1 MΩ, the ordinary D'Arsonval galvanometer is not sensitive enough to serve as the null-balance detector. One must resort to a high-impedance vacuum tube voltmeter (VTVM).

Example 1-1

Suppose that the true potential between an indicator and a reference electrode in a given solution is 1.0 V, and the internal resistance of the cell is 100 MΩ. If the voltage measuring device has a resistance of 10^9 Ω, the current that will flow in the circuit is

$$i = \frac{E}{R} = \frac{1.0}{10^8 + 10^9} = 9.0 \times 10^{-10} \text{ A}$$

This current, flowing through the cell resistance, generates a voltage difference across it equal to

$$iR_{cell} = (9 \times 10^{-10})(1 \times 10^8) = 0.09 \text{ V}$$

The voltage is opposite in polarity to the cell emf; hence, the net voltage apparent to the voltmeter is only 0.91 V. Thus, the measured voltage is 9% in error due to the current that flowed during the measurement. In order to reduce the error to the order of 0.1%, the resistance of the voltmeter would have to be at least as great as 10^{12} Ω, so that the current being drawn from the electrode would not be greater than 10^{-12} A.

Standard Weston Cells

The accurate measurement of a cell emf requires the use of a standard cell whose emf is precisely known. For this purpose the Weston cell is almost universally used. A diagram of this cell is shown in Fig. 1-7. It may be represented as follows:

$$\text{Cd (Hg)} \mid \text{CdSO}_4 \cdot 8/3\text{H}_2\text{O}_{\text{sat'd}}, \text{Hg}_2\text{SO}_{4 \text{ sat'd}} \mid \text{Hg}$$

The cadmium electrode is a saturated, two-phase cadmium amalgam containing about 10–12% of cadmium by weight. The electrolyte above it is saturated cadmium sulfate solution in contact with undissolved solid to ensure saturation at all temperatures. The other electrode is pure mercury covered with some solid mercurous sulfate.

At any constant temperature the activities of all the participants in the electrode reaction are constant, and therefore the emf of the cell is invariant. The emf varies very

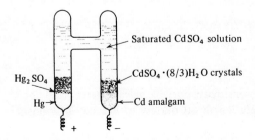

Saturated CdSO₄ solution

Hg₂SO₄

CdSO₄·(8/3)H₂O crystals

Hg

Cd amalgam

Fig. 1-7. Weston standard cell (saturated type).

slightly with temperature due, primarily, to the changing solubility of cadmium sulfate. Between $0°$ and $40°C$ the emf of the saturated Weston cell is given by the equation

$$E_t = 1.018300 - 0.0000406\,(t - 20) - 0.00000095\,(t - 20)^2 \qquad (1\text{-}22)$$

The temperature coefficient is thus approximately -0.04 mV/$°C$.

In most instruments an unsaturated Weston cell is used. Such a cell is made by saturating the electrolyte at about $4°C$ with cadmium sulfate and by using no excess salt so that the cell is unsaturated at ordinary temperatures. These cells have a much lower temperature coefficient, less than 0.01 mV/$°C$. The emf varies from cell to cell, usually in the range of 1.0181 to 1.0191 V, and the exact value for any given cell is determined by comparison with a saturated Weston cell. The emf of unsaturated cells usually decreases very slightly with time, usually not more than 0.1 mV per year unless the cell has been damaged by drawing excessive currents from it.

ac CIRCUITS

Fundamental Laws and Principles

An alternating current is a current in which the direction of the flow of electrons reverses periodically. The brief discussion which follows will be limited to pure sine waves, which can be described by:

$$i = I_P \sin 2\pi ft \qquad (1\text{-}23)$$

where I_P represents the peak or maximum current, i is the instantaneous current at time t, and f is the frequency in hertz (Hz). A second sine wave might have the same frequency as the first but differ in the times at which the current starts from zero. Such waves are said to be "out of phase," and the different waves can be expressed by similar equations, such as

$$i' = I_P' \sin (2\pi ft + \theta) \qquad (1\text{-}24)$$

where θ is the *phase angle* between the wave and the reference wave. Two sine waves with a phase difference of $90°$ are shown in Fig. 1-8.

If an ac generator is connected to a pure resistance, R, the voltage at any instant across the resistor is given by Ohm's law, $e = iR$, and thus the equation for the instantaneous voltage, e, will have the same form as that for the current, i. For example, the voltage across a resistor of 10 Ω connected across an ac generator with current output

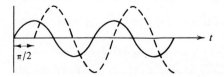

Fig. 1-8. Two sine waves with different amplitudes and $90°$ phase difference.

given by Eq. 1-23 would be

$$e = 10I_P \sin 2\pi ft \qquad (1\text{-}25)$$

The current and voltage are said to be "in phase."

The average voltage or current in a sinusoidal wave is found by integrating current times the time over one-half cycle and dividing by the time for one-half cycle π, as in Eq. 1-26.

$$I_{av} = \frac{1}{\pi} \int_0^\pi I_P \sin 2\pi ft \ \partial t \qquad (1\text{-}26)$$

$$= \frac{2I_P}{\pi} = 0.637 \ I_P \qquad (1\text{-}27)$$

Another value characteristic of sinusoidal waves needed in power calculations is the root-mean-square value, rms. This value is given by Eqs. 1-28 and 1-29.

$$I_{rms} = \left(\frac{1}{\pi} \int_0^\pi I_P^2 \sin^2 2\pi ft \ \partial t \right)^{1/2} \qquad (1\text{-}28)$$

$$= \frac{I_P}{2} \sqrt{2} = 0.707 \ I_P \qquad (1\text{-}29)$$

Capacitors

A capacitor consists of two sheets of conducting metal separated by a thin film of a dielectric material. The dielectric material commonly consists of paper, mica, plastics of various kinds, oil, ceramics, or a thin film of the metallic oxide formed by electrolytic action on the metal sheet. Electrolytic capacitors must always be used in a circuit where the polarity remains in one direction. If the polarity is reversed, the oxide film will be reduced and the capacitor will be destroyed.

If a capacitor is connected across the terminals of a battery, electrons will start to flow from the negative terminal of the battery to one of the metal foils. Electrons on the other metal foil are repelled and flow to the positive terminal of the battery. This flow of current gradually slows down and eventually ceases when the potential across the capacitor terminals equals the potential of the battery. The capacitance, C, of the capacitor is given by Eq. 1-30.

$$C = \frac{Q}{e} \qquad (1\text{-}30)$$

Where C is capacitance in farads, Q is the charge in coulombs, and e is the voltage across the capacitor.

Capacitors of various types are available in ranges from a few picofarads ($pF = 10^{-12}$ farad), to several thousand microfarads ($\mu F = 10^{-6}$ farad). Electrolytic capacitors generally have the highest capacitances. However, at the same time, they also have the lowest *breakdown voltages*, the voltage at which the dielectric breaks down and the capacitor becomes ruined. The larger the area of the metal foil, the thinner the dielectric; the higher the dielectric constant of the dielectric, the higher the capacitance.

Capacitors connected in parallel are additive, that is,

$$C = C_1 + C_2 + C_3 + \cdots \tag{1-31}$$

and capacitors connected in a series follow Eq. 1-32:

$$\frac{1}{C} = \frac{1}{C_1} + \frac{1}{C_2} + \frac{1}{C_3} + \cdots \tag{1-32}$$

If one considers the events that occur as a capacitor is charged as described above, one notices that when the charging cycle begins the current will be at a maximum and the voltage will be zero at this point. When the voltage reaches a maximum, the current has ceased. In other words, the potential across a capacitor lags behind the current by 90° or $\pi/2$. Thus, if the current is given by Eq. 1-23, the potential across a capacitor will be given by:

$$e = E_P \sin\left(2\pi f t - \frac{\pi}{2}\right) \tag{1-33}$$

The relationship between I_P and E_P for a capacitor can be obtained if one remembers that the current is the time rate of flow of charge.

$$i = \frac{\partial Q}{\partial t} \tag{1-34}$$

From Eqs. 1-30, 1-33, and 1-34,

$$i = \frac{\partial Q}{\partial t} = C\frac{\partial e}{\partial t} = C\frac{\partial}{\partial t}\left[E_P \sin\left(2\pi f t - \frac{\pi}{2}\right)\right] \tag{1-35}$$

$$= 2\pi f C\, E_P \cos\left(2\pi f t - \frac{\pi}{2}\right) \tag{1-36}$$

The maximum value of i, I_P, is then clearly given by Eq. 1-37, because the maximum value of the cosine is 1.

$$I_P = 2\pi f C E_P \tag{1-37}$$

The equivalent of the resistance R, or E/i in a purely resistive circuit, is in a capacitive circuit given by the ratio of E_P/I_P, and is called the *capacitive reactance*, X_C;

$$X_C = \frac{E_P}{I_P} = \frac{1}{2\pi f C} \tag{1-38}$$

The capacitive reactance is a measure of the impedance to the flow of charge by the capacitor and, like resistance, is measured in ohms. The capacitive reactance, or impedance, decreases with increasing frequency, and it is infinite when the frequency is zero. Thus capacitors offer infinite resistance to direct currents (except for negligible leakage resistances) and can be used to isolate direct from alternating currents.

Inductors

An inductor is a loop or coil of wire which may be wound around an iron core to increase the magnetic flux through the coil. An inductor resists a change in current passing through it, because a conductor that carries a current generates a magnetic field around the wire. A varying magnetic field, in turn, induces a potential in any conductor in that field in such a way as to oppose the change in the magnetic field. Thus when a current starts to pass through the wires of an inductor, a magnetic field is generated which induces a counter-potential in the wires. This counter-potential tends to impede, or hinder, the flow of current. When the current is steady, the magnetic field does not vary and no counter-potential is induced.

The inductance, L, of a coil is given by:

$$e = L \frac{\partial i}{\partial t} \tag{1-39}$$

where L represents the inductance in henrys (H), e is the counter-potential or emf in volts, and i is the current in amperes. Inductors are available in the range of several hundred henrys down to microhenrys. It should be pointed out that whereas capacitors can be made with almost negligible leakage—that is, with extremely high resistances—inductors cannot be made with zero resistances, but always have appreciable resistances, and therefore they are not "pure" inductances.

Inductors connected in series are additive, that is,

$$L = L_1 + L_2 + L_3 + \cdots \tag{1-40}$$

whereas inductors connected in parallel behave according to Eq. 1-41:

$$\frac{1}{L} = \frac{1}{L_1} + \frac{1}{L_2} + \frac{1}{L_2} + \cdots \tag{1-41}$$

From Eqs. 1-39 and 1-23, the relationship in Eq. 1-43 can be derived:

$$e = L \frac{\partial i}{\partial t} = L \frac{\partial}{\partial t} (I_P \sin 2\pi f t) \tag{1-42}$$

$$= 2\pi f L I_P \cos 2\pi f t = 2\pi f L I_P \sin \left(2\pi f t + \frac{\pi}{2} \right) \tag{1-43}$$

This shows that the potential leads the current by 90° or $\pi/2$ in an inductor. The value of the peak potential, E_P, is also obtained as

$$E_P = 2\pi f L I_P \tag{1-44}$$

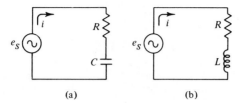

(a)   (b)

Fig. 1-9. Simple *RC* and *RL* circuits.

In a manner analogous to that used for capacitors and resistors, the ratio of E_P/I_P is called the *inductive reactance* and is given by:

$$\frac{E_P}{I_P} = 2\pi f L = X_L \tag{1-45}$$

Inductive reactance, X_L, is frequency-dependent and is greater with higher frequencies.

Analysis of Simple ac Series Circuits

Simple ac circuits containing resistors, capacitors, and inductors can often be understood by using vector diagrams. As an example, consider the circuit shown in Fig. 1-9(a). If the current is represented by a horizontal vector **i**, the potential across the resistor will be a parallel vector of length **i**R and the potential across the capacitor will be a vector at right angles to (90° behind) the current vector and of length **i**X$_C$. The potential vectors then form a diagram as shown in Fig. 1-10. Note that in such diagrams the vectors are conceived as rotating in a counter-clockwise fashion and that vector **i**X$_C$ is 90° behind vector **i**R. The vector **i**Z represents the potential of the source given by the product of the current, *i*, and the total impedance, *Z*, of the circuit. Because *i* can be eliminated from each quantity represented in Fig. 1-10, an exactly similar diagram for impedances can be drawn as in Fig. 1-11. It is apparent from the diagram that

$$Z = \sqrt{R^2 + X_C^2} \tag{1-46}$$

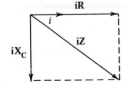

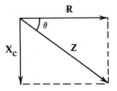

Fig. 1-10. Potential vectors for a series *RC* circuit.
Fig. 1-11. Impedance vectors for a series *RC* circuit.

and that

$$\tan \theta = \frac{X_C}{R} \qquad (1\text{-}47)$$

A similar set of diagrams can be constructed for the series RL circuit shown in Fig. 1-9(b), except that the potential in the inductor leads the current by $90°$. The diagrams are shown in Figs. 1-12 and 1-13.

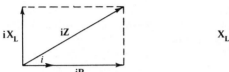 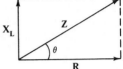

Fig. 1-12. Potential vectors for a series RL circuit.
Fig. 1-13. Impedance vectors for a series RL circuit.

For a series RL circuit, the impedance, Z, is given by Eq. 1-48, and the phase angle, θ, is given by Eq. 1-49.

$$Z = \sqrt{R^2 + X_L^2} \qquad (1\text{-}48)$$

$$\tan \theta = \frac{X_L}{R} \qquad (1\text{-}49)$$

A series circuit containing resistance, inductance, and capacitance is represented in Fig. 1-14, and the corresponding vector diagram for the impedances is shown in Fig. 1-15.

It is apparent that the impedance and the phase angle for a series RLC circuit is given by the Eqs. 1-50 and 1-51, respectively.

$$Z = \sqrt{R^2 + (X_L - X_C)^2} \qquad (1\text{-}50)$$

$$\tan \theta = \frac{X_L - X_C}{R} \qquad (1\text{-}51)$$

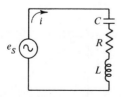 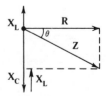

Fig. 1-14. A simple series RLC circuit.
Fig. 1-15. Impedance vectors for a series RLC circuit.

Note that when $X_L = X_C$ the impedance becomes merely R. For this to be true, the frequency must be given by Eq. 1-53.

$$X_L = 2\pi f L = \frac{1}{2\pi f C} = X_C \qquad (1\text{-}52)$$

$$f = \frac{1}{2\pi \sqrt{C}} \qquad (1\text{-}53)$$

Such a circuit is known as a *series resonant circuit*. It can be used as a selective filter for the resonant frequency given by Eq. 1-53.

Analysis of Simple ac Parallel Circuits

Simple parallel RC and RL circuits and an RLC resonant circuit are shown in Fig. 1-16. In parallel circuits, the potential across each unit must be the same. This leads to the following relationships between the currents, potentials, and impedances for the circuit shown in Fig. 1-16(a):

$$e_S = e_C = e_R \qquad (1\text{-}54)$$

$$i_C = \frac{e_S}{X_C} \quad \text{and} \quad i_R = \frac{e_S}{R} \qquad (1\text{-}55)$$

$$i = \sqrt{\left(\frac{e_S}{X_C}\right)^2 + \left(\frac{e_S}{R}\right)^2} = e_S \sqrt{\left(\frac{1}{X_C}\right)^2 + \left(\frac{1}{R}\right)^2} \qquad (1\text{-}56)$$

The vector relationships between the currents for the circuits diagramed in Fig. 1-16 are shown in Fig. 1-17. The circuit shown in Fig. 1-16(c) is more complex, and the current diagram is an approximation to the actual situation. However, this circuit is very interesting. It can be seen from the current diagram that when the vertical component of i_L is equal (but of opposite sense) to i_C, then i will be in phase with e_S. Furthermore, it is the effect of the resistance R which makes i_L not exactly parallel with i_C. If this resistance is small, then the resultant i is small and the circuit will possess a high impedance. Ideally, the maximum impedance occurs when i_C is equal to i_L, or in other words, when

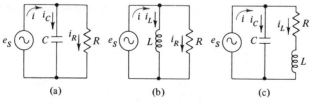

(a)   (b)   (c)

Fig. 1-16. Parallel RC, RL, and RLC resonant circuits.

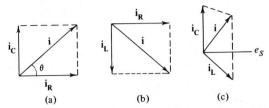

(a)   (b)   (c)

Fig. 1-17. Current vector diagrams for parallel RC, RL, and RLC circuits.

X_L is equal to X_C, because e_S is the same for both branches. This condition of resonance results when the frequency is given by Eq. 1-53. Note that the parallel resonant circuit shows a high impedance to the resonant frequency, whereas the series resonant circuit shows a low impedance.

THE WHEATSTONE BRIDGE

The value of an unknown resistance R_x can be accurately determined by comparison with known resistors R_1, R_2, and R_3 by the Wheatstone bridge method, which employs the arrangement shown in Fig. 1-18. The variable resistor R_3 is adjusted until a balance is obtained as indicated by a zero reading on the galvanometer. When the galvanometer indicates that no current is flowing through it, that is, when i_g is zero, the following relationships can be derived by using Kirchhoff's laws:

$$i_1 R_1 + i_2 R_3 \quad \text{and} \quad i_1 R_2 = i_2 R_x \tag{1-57}$$

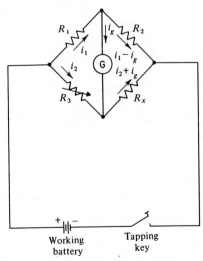

Fig. 1-18. Wheatstone bridge for measuring resistances.

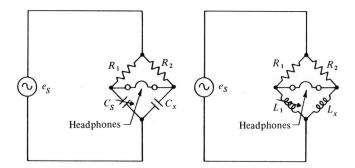

Fig. 1-19. Bridge method for measuring capacitance or inductance.

$$R_x = R_3 \cdot \frac{R_2}{R_1} \tag{1-58}$$

By a very similar procedure, unknown capacitances or inductances can be measured. The setups are shown in Fig. 1-19. An ac source (frequently 1000 Hz) replaces the dc source in Fig. 1-17, and an ac detector, such as a set of headphones, replaces the galvanometer. When balance is obtained in the bridges, the following equations are satisfied:

$$C_x = C_S \left(\frac{R_1}{R_2} \right) \quad \text{or} \quad L_x = L_1 \left(\frac{R_2}{R_1} \right) \tag{1-59}$$

PROBLEMS

1. How many watts of power will be dissipated by each of the resistors in Fig. 1-1?

2. Suppose that the potential of an unknown cell to be measured on a potentiometer such as that shown in Fig. 1-3 is of the order of 1 V. The galvanometer resistance is 1000 Ω; the resistance of the slide wire is 1000 Ω per meter of length and R has been so adjusted that the potentiometer slidewire is direct-reading in millivolts per millimeter. What current flows in the slidewire at balance? What current would have to be detectable by the galvanometer if an imbalance of 0.001 V is to be detectable?

3. Calculate the impedance and the phase angle for a series RLC circuit in which $R = 1000 \ \Omega$, $C = 0.1 \ \mu f$, and $L = 1$ H if the frequency of the current source is 1000 Hz.

4. What is the resonance frequency of the circuit in Problem 3?

5. Assume that an inductance of 10 H is available. It is desired to construct a circuit with a very high impedance at 60 Hz. What must be the value of the capacitance in the parallel LC circuit?

BIBLIOGRAPHY

Brophy, J. J., *Basic Electronics for Scientists*, McGraw-Hill, New York, 1966.
Malmstadt, H. V., C. G. Enke, and E. C. Toren, Jr., *Electronics for Scientists*, W. A. Benjamin, Menlo Park, Calif., 1962.
Smith, A. W., *Electrical Measurements in Theory and Application*, McGraw-Hill, New York, 1943.
Suprynowicz, V. A., *Introduction to Electronics*, Addison-Wesley, Reading, Mass., 1966.

2

Electronics: Vacuum Tubes, Semiconductors, and Associated Circuits

In this chapter the treatment of circuits will be limited to those that are simplest and most widely used in analytical instrumentation. A brief discussion of the characteristics of nonlinear electronic devices, vacuum tubes, and semiconductors is followed by some simple circuits, such as rectifiers and amplifiers, and by a description of a cathode-ray tube. Operational amplifiers are introduced to give the student some ability in reading block diagrams and to indicate the many uses to which these commercially available, compact units can be put.

ELECTRONIC TUBES

Diodes

A diode consists of a plate, or anode, and a cathode enclosed in a high vacuum. A diagram of a typical diode is shown in Fig. 2-1.

The cathodes are generally of two types, those heated directly and those heated indirectly. For the directly heated cathode, a tungsten or thoriated tungsten wire is heated to 2200° or 1300°C, respectively, by the passage of a current, usually dc. For the indirectly heated cathode, a nickel foil covered with a film of alkaline earth oxides is heated to 700°–900°C by a filament close to but not touching it. The filament is made of resistance wire and may be heated by the passage of either ac or dc current through it. Because the filament does not touch the actual cathode, use of ac current has no effect on the cathode potential.

The amount of electron emission by a cathode depends upon the material of which it is made and its temperature. The current density J_0, in amperes per square centimeter, is closely represented by Eq. 2-1,

$$J_0 = AT^2 e^{-w/kT} \qquad (2\text{-}1)$$

where A is a constant depending on the cathode material (1×10^2 for oxide-coated tungsten and 6×10^5 for tungsten), T is the temperature, k the Boltzmann coefficient, and w the work function in ergs (about 1 for oxide coatings and 4.5 for tungsten).

19

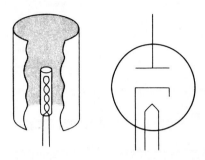

Fig. 2-1. Diode, indirectly heated. Envelope of glass or metal not shown.

The anode is usually a cylinder that surrounds the cathode. When electrons strike the anode or plate, heat is generated and, if the temperature becomes too high, gases may be evolved that may then ionize and bombard the cathode, thus destroying it. Anodes are, therefore, usually blackened to increase heat radiation, increased in size, and, in high-power tubes, made of some special material such as platinum or tantalum.

The *plate characteristic* of a typical diode is shown in Fig. 2-2. When the plate of a diode is at zero potential or is negative with respect to the cathode, only electrons with appreciable kinetic energy can reach the plate. Over region A the number of electrons reaching the plate is nearly proportional to the applied voltage; at the end of region B, all electrons emitted by the cathode are being collected by the plate. Further increase of voltage, as in region C, succeeds in collecting only a few extra electrons that are emitted from the cathode due to the Schottky effect, a decrease in work function caused by the strong field.

Unless the potential between the plate and cathode is high (regions B and C), the electrons emitted by the cathode are not all attracted to the plate, and the excess electrons form a cloud around the cathode. This "space charge" repels electrons emitted by the cathode, and some electrons return to the cathode. A plot of potential as a function of distance between the cathode and plate as well as the field strength is shown in Fig. 2-3. Note the minimum in the potential curve that occurs near the cathode where the "space charge" is greatest.

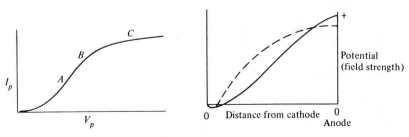

Fig. 2-2. Plate characteristic of a diode.

Fig. 2-3. Potential as a function of distance from cathode. (Field strength shown as dotted line.)

A triode has a cathode and plate similar to those in a diode. In addition, it has a control grid, an open mesh of wires or a screen, which is placed between the cathode and plate but closer to the cathode. In normal operation very little current flows to the control grid since it is usually kept negative with respect to the cathode. Grid current will be drawn, however, if the grid becomes positive and, in such a case, a grid of larger wires will be needed to prevent damage.

A circuit, such as that shown in Fig. 2-4, may be used to measure the characteristics of a triode. A typical set of curves for a triode is shown in Fig. 2-5.

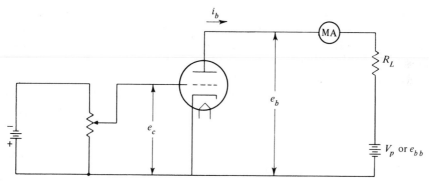

Fig. 2-4. Circuit for measuring characteristics of a triode with voltages and currents designated.

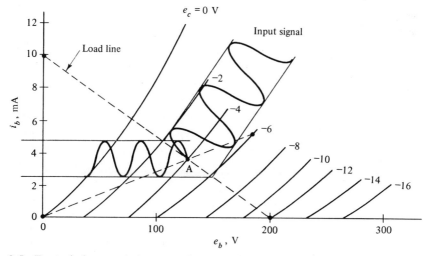

Fig. 2-5. Typical characteristic curves for a triode, with load line for $R_L = 20,000 \ \Omega$ and operating point A for $R_K = 1000 \ \Omega$ shown. An input signal of 4 V is shown along with the corresponding output current.

The constants associated with any given triode are generally three in number and are defined in Eqs. 2-2, 2-3, and 2-4.

$$\text{plate resistance} = r_p = \left|\frac{\partial e_b}{\partial i_b}\right|_{e_c} \tag{2-2}$$

$$\text{amplification factor} = \mu = \left|\frac{\partial e_b}{\partial e_c}\right|_{i_b} \tag{2-3}$$

$$\text{transconductance} = g_m = \left|\frac{\partial i_b}{\partial e_c}\right|_{e_b} \tag{2-4}$$

Pentodes

A pentode contains two grids in addition to the control grid of a triode. These grids are the screen grid, a mesh of wires closer to the plate than the control grid, and a suppressor grid, a mesh of wires very close to the plate (i.e., between the plate and the screen grid). The screen grid is maintained at a positive potential and serves to screen the plate from the effect of the control grid. The suppressor grid is usually connected to the cathode and is, therefore, negative with respect to the plate and serves to return to the plate any secondary electrons knocked out of it by the electrons from the cathode as they impinge on the plate. A pentode has quite different characteristic curves than does a triode (Fig. 2-6) and shows high-amplification factors, up to 2000, and plate resistances that may exceed 1 megohm (MΩ). The corresponding values for a triode are usually about 2 orders of magnitude less.

SEMICONDUCTORS

n-type and p-type

When a crystal of a semiconductor, such as germanium or silicon, is grown from a melt that contains a small amount of an impurity element of a higher group in the periodic table, an n-type crystal is obtained. Such crystals have an excess of electrons compared with those in a pure crystal. On the other hand, if the impurity element is of a periodic group lower than the main element, the semiconducting crystal that results is said to be a p-type. Such crystals contain a deficiency of electrons or, equivalently, an excess of holes.

Diodes

A semiconductor diode is prepared by arranging an n-type and p-type junction in a crystal (Fig. 2-7). There are various ways of producing such a junction, for example, a bit

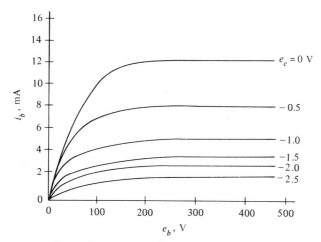

Fig. 2-6. Typical pentode characteristics.

of one type of impurity (dopant) and then a bit of the other type of impurity may be added to the crystal as it is being drawn from a melt of the pure material.

At the *p-n* junction, some holes and electrons combine in a narrow region around the junction. The *n*-type material becomes positively charged as holes move into it and, conversely, the *p*-type material becomes negatively charged as electrons move in. Thus a barrier field is created that repels further movement of electrons or holes. If a potential is now applied to the diode such that the *p*-type material is made positive with respect to the *n*-type, the barrier potential is opposed by the applied potential and electrons will flow across the barrier toward the *p*-type material; holes will flow in the opposite direction, that is, a "forward current" flows. If the applied potential is in the opposite direction, it enhances the barrier potential and no or practically no electrons and holes traverse the barrier. That is, there is only a very small "reverse current."

The characteristics of a semiconductor diode are shown in Fig. 2-8. Note that there is a change in scale on both axes on either side of zero. The forward-biased current is appreciable, whereas the reversed-biased current is very small until the Zener limit is reached, when breakdown occurs and a large current flows. A semiconductor diode, then, operates in much the same fashion as a vacuum tube diode and can be used as a switching device to control the flow of current depending upon the polarity of the applied voltage. Semiconductors normally operate at considerably lower voltages than vacuum tubes. They consume very little power compared to vacuum tubes, which generate much useless heat. They are usually more reliable because they do not have filaments that eventually disintegrate and burn out.

Fig. 2-7. Semiconductor diode with schematic representation.

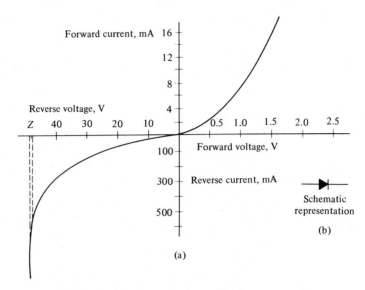

Fig. 2-8. Characteristics of a semiconductor diode.

Transistors

A transistor is a three-layer unit in which an emitter and collector of one type of semiconducting material are separated from each other by a thin slice (0.02 mm or less) of the other type of semiconducting material. Thus transistors are either *n-p-n* or *p-n-p* types.

An *n-p-n* type transistor is shown in Fig. 2-9. When the emitter is biased in the forward direction, that is, when the emitter is negative with respect to the base, electrons pass from emitter to base and, because the base is so thin, most of these electrons (95–99 percent) pass on into the collector region. From 1 to 5 percent of the electrons combine with holes in the *p* region, the base, and form the base current. At the same time, the collector is biased in the reverse direction, that is, positive with respect to the base.

Although the current is nearly the same in the emitter and collector circuits, reference

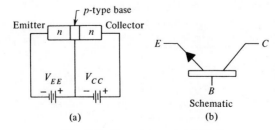

Fig. 2-9. Transistor, *n-p-n* type.

to Fig. 2-8 will show that a small change of voltage in the *forward* region causes a great change in the current which would require a large voltage change in the *reverse-biased* region. In other words, the emitter current flows in a circuit with a low resistance, and the collector current flows in a high-resistance circuit. If a signal is present in the emitter circuit and a load resistance of, say, 10,000 Ω is added to the collector circuit, a high-voltage signal will appear across the load resistance. Since the currents in the two circuits are approximately equal but the output voltage is much higher, *power* or *voltage* amplification has been achieved.

A *p-n-p* transistor is shown in Fig. 2-10. When this type of transistor is biased in the sense shown in the figure, similar considerations to those just discussed apply.

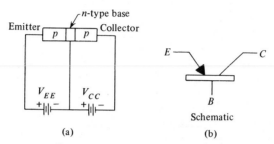

Fig. 2-10. Transistor, *p-n-p* type.

Common-Base Collector Characteristics

The circuit shown in Fig. 2-9 is known as a *common-base* circuit because both V_{EE} and V_{CC} are connected to one common point, the base. When I_C, the collector current, is plotted against the collector-base voltage V_{CB} for fixed values of the emitter current I_E, characteristic curves, such as those in Fig. 2-11, are obtained.

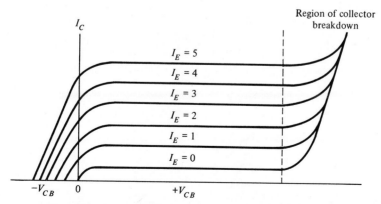

Fig. 2-11. Common-base transistor circuit characteristics.

Common-Emitter Collector Characteristics

If the emitter section of the transistor is in common with the base and collector voltage sources as shown in Fig. 2-12, a common-emitter circuit results. A set of typical collector characteristic curves for this circuit are shown in Fig. 2-13.

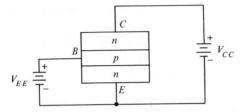

Fig. 2-12. Common-emitter circuit.

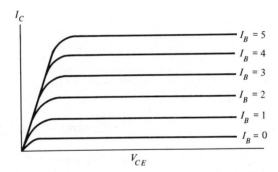

Fig. 2-13. Common-emitter transistor circuit characteristics.

Common-Collector Collector Characteristics

The third type of transistor circuit has the collector of the transistor connected to both the input and output potential sources. This circuit is often called an *emitter-follower* circuit. It has a high input impedance, a low output impedance, and excellent linearity; thus it is often used in impedance-matching circuits where the output impedance should be low.

Field-Effect Transistors

If a bar of one type of semiconductor material is "gated" by applying a region of opposite type material on the top and bottom surfaces of the bar, as shown in Fig. 2-14, the channel width through which holes or electrons flow can be varied by applying a potential between the source and gate. If a reverse bias is applied, the space charge is in-

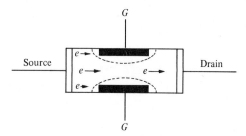

Fig. 2-14. Field-effect transistor.

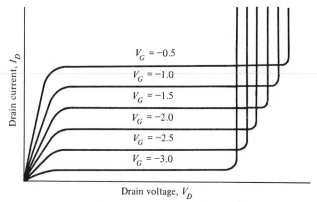

Fig. 2-15. Characteristic curves for a field-effect transistor.

creased and the channel width for carrier flow is decreased. Thus the gate voltage controls the current. A set of typical characteristic curves is shown in Fig. 2-15.

TYPICAL ELECTRONIC CIRCUITS

Some simple electronic circuits that are often met as component parts of more complicated circuits will be discussed below.

Rectifier Circuits

Since either solid-state or vacuum tube diodes conduct current only in one direction, they can be used as rectifiers in ac circuits. A half-wave rectifier circuit and the corresponding voltages and circuits in it are shown in Fig. 2-16. The flow of electrons, which is opposite in direction to the conventional "current flow," is shown by an arrow.

A full-wave rectifier can be made by using two diodes. Such a circuit with the associated currents and voltages is shown in Fig. 2-17.

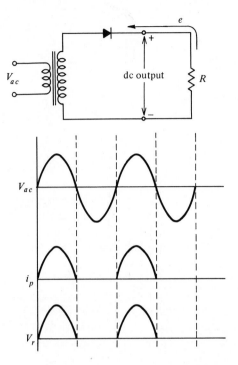

Fig. 2-16. A half-wave rectifier circuit.

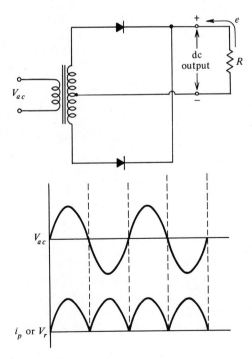

Fig. 2-17. Full-wave rectifier circuit.

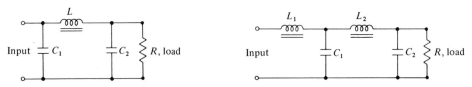

Fig. 2-18. Capacitor-input filter.

Fig. 2-19. Choke-input filter.

Filters Associated with Rectification

When a rectifier is used to produce a dc voltage from an ac source, a filter circuit is usually necessary to smooth out the pulsating direct current that would be obtained directly from the rectifier (see Figs. 2-16 and 2-17). Filter circuits are generally classified as either capacitor- or inductor- (choke) input filters depending upon the first element used after the rectifier. An example of a capacitor-input filter is shown in Fig. 2-18 and a choke-input filter in Fig. 2-19. The action of a filter circuit can be understood by consideration of Fig. 2-20, which shows the effect of placing a capacitor such as C_1 of Fig. 2-18 across the terminals of a full-wave rectifier. The dashed line indicates the voltage across C_1, which is seen to be more constant than the voltage across a rectifier used alone, as indicated by the solid line. An inductance further smooths out the voltage, since inductances show high resistances to the ripple frequency (twice the line frequency), which still exists in the output of the rectifier-capacitor circuit. The inductance presents only a very low resistance to the dc component of the output.

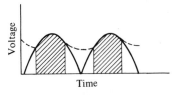

Fig. 2-20. Effect of a capacitor, such as C_1 of Fig. 2-18, on output voltage of full-wave rectifier. Dotted line is voltage across capacitor. Solid line is rectifier output voltage. Shaded area is charging interval.

The two types of filter circuits have somewhat different characteristics and therefore different uses. The rectifier current flows continuously in the choke-input filter and only intermittently in the capacitor-input variety. For the same load current, the peak diode current is higher in the capacitor-input circuit. The voltage regulation (the variation of voltage output with variation of the current drawn from the circuit) is better with a choke-input filter. At very small current drains, however, the choke loses its effectiveness. The average rectified voltage is higher with a capacitor-input filter.

Regulated Power Supplies

For many uses, the dc voltage output of a rectifier and filter circuit must be more closely controlled or regulated. Two simple methods of control are shown in Figs. 2-21 and 2-22.

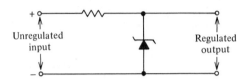

Fig. 2-21. Shunt-regulated circuit using Zener diode.

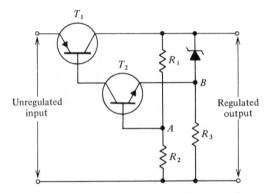

Fig. 2-22. Series-regulated circuit.

A Zener diode as used in Fig. 2-21 is essentially a diode operated in the Z breakdown region of Fig. 2-8, where it can be seen that the voltage is nearly independent of the current. The equivalent in tube circuits is to use a glow discharge tube, that is, a diode containing a gas under reduced pressure. When the glow discharge is started by application of a sufficiently high voltage to ionize the gas, the voltage across the tube remains fairly constant regardless of the current passing through the tube (up to a certain limit).

The operation of the more elaborate regulator circuit shown in Fig. 2-22 can be understood by considering what would happen should the voltage across the output terminals suddenly tend to decrease. Since the voltage across the Zener diode is constant, the voltage across resistor R_3 would decrease. Since R_1, R_2, R_3, and the Zener diode are so chosen that the normal operating voltage between AB is slightly positive, a decrease in voltage at point B will further increase the base-emitter voltage of T_2 and thereby increase the base current. The collector current will then increase by an amount equal to the change in base current multiplied by the current gain of the transistor. Current will now flow into the base of the transistor T_1 and produce a lowering of the emitter-collector voltage of T_1. The output voltage increases, then, since the drop in voltage across T_1 has decreased. This brings the output voltage back to its regulated value.

Amplifiers

A simple triode tube amplifier and the designations commonly used for the voltages in the circuit are illustrated in Fig. 2-23.

The actual operation of the triode can be understood by drawing a load line and establishing the operating point on the tube characteristic curves. The load line is given by Eq. 2-5,

$$E_{bb} = i_b R_L + E_B + i_b R_K \qquad (2\text{-}5)$$

which can be derived by considering the plate current flowing in the output circuit. The two points on the extremities of the load line are given by the conditions of Eqs. 2-6 and 2-7.

If $E_B = 0$,

$$i_b(R_L + R_K) = E_{bb} \qquad (2\text{-}6)$$

If $i_b = 0$,

$$E_B = E_{bb} \qquad (2\text{-}7)$$

Such a line where $R_L = 20{,}000 \ \Omega$, $R_K = 1000 \ \Omega$, and $E_{bb} = 200 \ V$ is shown in Fig. 2-5. The normal operating point, that is, the condition of the tube with no input signal, depends on the value of e_c and will be zero if a cathode resistor and a grid bias E_{cc} are not

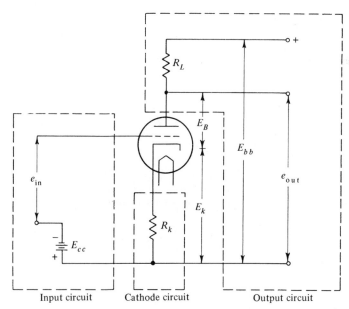

Fig. 2-23. Typical triode amplifier circuit.

used. If a cathode resistor is used, one can determine the operating point by noting the intersection of the load line with the cathode bias line drawn according to Eq. 2-8.

$$-e_c = i_b R_K \tag{2-8}$$

The amplification properties of the tube can be understood by assuming that the input signal of magnitude 4 V shown in Fig. 2-5 is caused by a current flowing through an input resistor of 10,000 Ω. The signal current would then have been

$$i_{signal} = \frac{4}{10,000} = 0.4 \text{ mA} \tag{2-9}$$

The output current is of the order of 2 mA and the output signal would have a total magnitude of

$$e = (0.002)(20,000) = 40 \text{ V} \tag{2-10}$$

The actual output signal given by

$$e_{out} = E_{bb} - i_b R_L \tag{2-11}$$

would vary by plus and minus 20 V around the value given by considering the normal operating point of the tube in this circuit, or, approximately,

$$e_{out} = 200 - (0.004)(20,000) = 120 \text{ V} \tag{2-12}$$

An exact treatment of the amplifying action of a vacuum tube can be made by noting that an input signal on the cathode of value $e_{in} = e_c$ would cause the tube to produce a voltage e_b equal to $-\mu e_c$. This voltage would, in effect, be that produced by the plate current i_b flowing through the plate resistance r_p and a load resistor R_L connected in series. Thus,

$$i_b(R_L + r_p) = -\mu e_c \tag{2-13}$$

The output signal, however, is

$$e_{out} = i_b R_L = \frac{-\mu e_c R_L}{R_L + r_p} \tag{2-14}$$

and the gain A of the circuit would be given by

$$A = \frac{e_{out}}{e_{in}} = \frac{-\mu e_c R_L}{e_c(R_L + r_p)} = \frac{-\mu R_L}{R_L + r_p} \tag{2-15}$$

The negative sign indicates that the output signal is $180°$ out of phase with the input signal. Since $\mu = g_m r_p$, Eq. 2-15 can also be written as

$$A = \frac{-g_m r_p R_L}{r_p + R_L} \tag{2-16}$$

When r_p is very large compared with R_L, as it usually is in pentode circuits, the gain is approximately that given by Eq. 2-17.

$$A \approx -g_m R_L \tag{2-17}$$

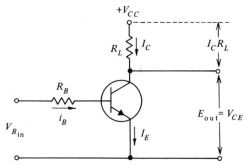

Fig. 2-24. A typical common-emitter transistor amplifier circuit.

Analysis of transistor amplifier circuits is much more complex. It can be done graphically much as in Fig. 2-5 by superimposing a load line on the I versus V curves for the transistor circuit. A typical common-emitter amplifier circuit is shown in Fig. 2-24. The related curves with a load line based upon an R_L value of 1000 Ω and V_{CC} = 10 V are shown in Fig. 2-25. By varying I_B the operating point can be shifted almost at will to any point along the load line. The load line is given by Eq. 2-18.

$$V_{CE} = V_{CC} - I_C R_L \qquad (2\text{-}18)$$

It can be seen that a small change in I_B will result in a large change in I_C and, thus, current amplification. The ratio I_C/I_B is known as the *dc current gain* or the *dc forward current transfer ratio* or the *dc beta*. The usual symbols are given in Eq. 2-19.

$$\beta_N = h_{FE} = \frac{I_C}{I_B} \qquad (2\text{-}19)$$

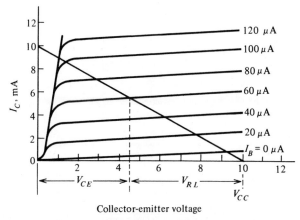

Collector-emitter voltage

Fig. 2-25. Characteristic curves for circuit of Fig. 2-24.

Inverse Feedback Principle

Equations 2-13 through 2-16 were developed for an amplifier circuit similar to that shown in Fig. 2-23 but without a cathode resistor R_K. If such a cathode resistor is added, the following equations indicate the behavior of the amplifier.

Let a signal ∂e_s be impressed on the grid of the tube. The corresponding change in plate current is called ∂i_b. This current, flowing through the cathode resistor, would cause the cathode to become more positive by $R_K \partial i_b$ V, and the signal would therefore be decreased by this amount. This is known as *inverse feedback*. The actual change in the grid voltage with respect to ground potential would be

$$\partial e_c = \partial e_s - R_K \partial i_b \tag{2-20}$$

Substitution of Eq. 2-20 into Eq. 2-4 gives

$$\partial i_b = g_m \partial e_c = g_m \partial e_s - g_m R_K \partial i_b \tag{2-21}$$

thus

$$\partial i_b = \frac{g_m \partial e_s}{1 + g_m R_K} \tag{2-22}$$

Let us call g' the effective transconductance of this circuit with inverse feedback.

$$g' = \frac{\partial i_b}{\partial e_s} = g_m \left(\frac{1}{1 + g_m R_K} \right) \tag{2-23}$$

We see from Eq. 2-23 that the introduction of the cathode resistor R_K reduces the effect of the tube constant g_m on the circuit by the factor $1/(1 + g_m R_K)$. If $R_K \gg 1$, then

$$g' = \frac{g_m}{g_m R_K} = \frac{1}{R_K} \tag{2-24}$$

In this case the circuit would be independent of the tube constants. This is the advantage of inverse feedback: It makes circuits more stable, that is, less dependent on the charac-

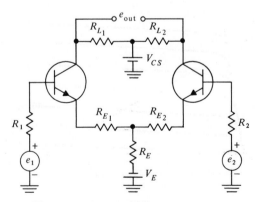

Fig. 2-26. Simple difference amplifier.

teristics and vagaries of the tubes (or transistors) at the expense, however, of some loss in amplification. Similar principles and equations apply also to transistor circuits.

Difference Amplifiers

A special type of amplifier circuit frequently used in instruments for analytical purposes is the difference amplifier, an example of which is shown in Fig. 2-26.

It can be shown that the output signal in such an amplifier is directly proportional to the difference in the input signals and is greatly amplified.

Operational Amplifiers

High-gain dc amplifiers with inverse feedback for stabilization and with high-input impedances and low-output impedances and with good high-frequency responses are now commercially available. Such amplifiers find many uses in instrumental design and are known as operational amplifiers. An operational amplifier has two input and two output terminals and is often indicated in schematic circuit drawings as a triangle, as shown in Fig. 2-27(a) and, in simplified form, in Fig. 2-27(b).

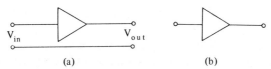

(a)   (b)

Fig. 2-27. Schematic representation of operational amplifiers.

The operational amplifier is really a high-gain difference amplifier. Very frequently the common input-output terminal of Fig. 2-27(a) is connected to ground, and an input impedance Z_1 and a feedback impedance Z_2 are added, as in Fig. 2-28.

Since the input impedance of the operational amplifier is so high, i_c is practically zero and can be neglected. Thus $i_1 = i_2$. Furthermore, the high gain of the operational ampli-

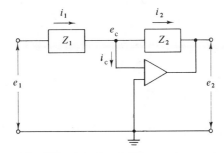

Fig. 2-28. Operational amplifier with impedances.

fier ensures that the potential e_c must be very nearly equal to that of the other input terminal—in this case, ground potential. The point e_c is said to be a *virtual ground*. Under these conditions,

$$i_1 = \frac{1}{Z_1}(e_1 - e_c) = \frac{e_1}{Z_1} = i_2 = \frac{1}{Z_2}(e_c - e_2) = \frac{-e_2}{Z_2} \tag{2-25}$$

Therefore,

$$-e_2 = \frac{Z_2}{Z_1} e_1 \tag{2-26}$$

The operational amplifier can be used in a large number of ways. One method is to furnish an output voltage which is proportional to the sum of several currents. Such a circuit is shown in Fig. 2-29. Since e_c must be zero and since the total current flows through R_F, there being practically no current through the amplifier,

$$e_0 = -(i_1 + i_2 + i_3)R_F \tag{2-27}$$

Voltages can be added by the circuit shown in Fig. 2-30. Each voltage can be multiplied by a different factor if desired. In this case,

$$e_0 = -R_F \left(\frac{e_1}{R_1} + \frac{e_2}{R_2} + \frac{e_3}{R_3}\right) \tag{2-28}$$

If the feedback impedance is a capacitance instead of a resistance, as in Fig. 2-31, a current integrator results.

The point e_c must remain at ground potential and all of the current must pass through the capacitor. Thus the output potential e_0 always equals the potential of the capacitor. Since capacitance $C = Q/E$, where Q is charge and E is potential, and since the charge is the time integral of the current, or $Q = \displaystyle\int_0^t i\,\partial t$, it follows that

$$E = \frac{1}{C} \int_0^t i\,\partial t = -e_0 \tag{2-29}$$

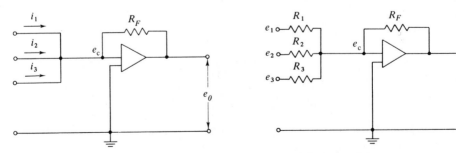

Fig. 2-29. Operational amplifier circuit to measure the sum of three currents.
Fig. 2-30. Operational amplifier circuit to measure the sum of three voltages.

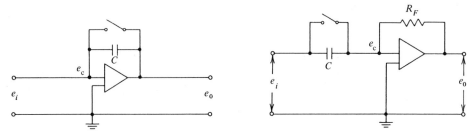

Fig. 2-31. Operational amplifier integrating circuit.

Fig. 2-32. Operational amplifier differentiating circuit.

By placing a resistance in the input circuit, we can integrate potentials rather than currents. The discharge switch is necessary to reset the circuit.

A differentiating circuit for potentials can be obtained by using the circuit of Fig. 2-32. In this case, the output e_0 is given by Eq. 2-30.

$$e_0 = -R_F C \frac{\partial e_i}{\partial t} \tag{2-30}$$

Many other examples of circuits using operational amplifiers could be given, but these may serve to indicate the wide applicability of these amplifiers. It should be understood that several circuits can be combined, and complicated equations can therefore be solved by these devices.

IMPEDANCE MATCHING AND MAXIMUM POWER TRANSFER

One electronic circuit will often not furnish sufficient power to operate the final load, and two or more circuits may have to be connected together. Furthermore, in many electronic circuits it is necessary to transfer the maximum amount of power from the final circuit to a load such as an antenna, loudspeaker, or recording device. It is therefore of interest to determine the conditions for the transfer of a maximum amount of power from the output circuit of one device to the input circuit of a second device.

Thévenin's theorem is a very useful device for representing the output impedance of a circuit. This theorem states that the output of any two-terminal network containing a finite number of voltage sources and impedances can be replaced by a series combination of a voltage source, V_{eq}, and an impedance, R_{eq}. The impedance, R_{eq}, is that which would be obtained across the output terminals when the voltage sources are replaced by their internal impedances. The voltage, V_{eq}, is the open-circuit voltage between the two terminals. In the general case, voltages and impedances will be functions of frequency and so a complex function may result. However, for simplicity we may neglect the interelectrode capacitances of vacuum tubes except at high frequencies and we may take the plate resistance of the tube as its output impedance.

The Thévenin equivalent circuit for a simple vacuum tube circuit connected to a load resistance, R_L, may be represented by Fig. 2-33.

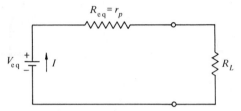

Fig. 2-33. Thévenin equivalent circuit for a simple vacuum-tube amplifier.

The power in the load resistor is

$$P_L = I^2 R_L \tag{2-31}$$

According to Ohm's law

$$I = \frac{V_{eq}}{R_L + r_p} \tag{2-32}$$

therefore

$$P_L = I^2 R_L = \left(\frac{V_{eq}}{R_L + r_p}\right)^2 R_L \tag{2-33}$$

To find the value of R_L that will permit the maximum amount of power to be transferred from the tube circuit to the load, differentiate Eq. 2-33 with respect to R_L and equate the result to zero. Thus

$$\frac{dP_L}{dR_L} = \frac{V_{eq}^2}{(R_L + r_p)^2} - \frac{2R_L V_{eq}^2}{(R_L + r_p)^3} = 0 \tag{2-34}$$

$$\frac{V_{eq}^2}{(R_L + r_p)^2} = \frac{2R_L V_{eq}^2}{(R_L + r_p)^3} \tag{2-35}$$

$$1 = \frac{2R_L}{R_L + r_p} \tag{2-36}$$

$$r_p = R_L \tag{2-37}$$

This proves that the maximum power is delivered to the load resistor when it is equal to the plate resistance of the tube or, in general, the internal impedance of the circuit delivering the power. Under such conditions the load is said to be matched to the impedance of the source. Half of the total power is lost in the internal impedance of the circuit.

When it is required to furnish maximum power to a load with a small impedance, special circuits may be needed since the internal resistance of vacuum tubes is usually quite high. A special circuit known as a cathode follower circuit may be used, Fig. 2-34. The amplification of a cathode follower circuit is less than one, but its output impedance is relatively low.

For impedance-matching transistor circuits, the common-collector circuit is useful. A typical emitter-follower circuit using this configuration is shown in Fig. 2-35. This circuit has a high input impedance, a low output impedance, and excellent linearity.

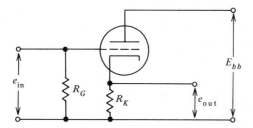

Fig. 2-34. Cathode-follower circuit.

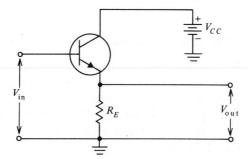

Fig. 2-35. Emitter-follower circuit.

Transformers may also be used to match impedances of circuits. Impedances are transformed by transformers in proportion to the square of the ratio of turns on the primary and secondary windings. Of course, transformers can only be used for ac current circuits.

CATHODE-RAY TUBES

A cathode-ray tube, shown in schematic diagram in Fig. 2-36, forms the display unit in many modern electrical devices and especially in the oscilloscope. The oscilloscope is widely used as a major test instrument to display electrical signals, that is, voltage as a function of time.

The grid is normally negative with respect to the cathode, the more negative the grid the less intense the beam. The focusing anode is positive and is adjusted to give a sharp spot when the electrons strike the phosphorescent screen. The accelerating anodes are still more positive and serve, as the name indicates, to accelerate the electron beam further, the direction of the beam being determined by the sign and magnitude of the voltages on the horizontal and vertical deflecting plates. The electrons are further accelerated by application of a high positive charge to the inside of the tube near the screen. The screen is covered with a phosphor, and tubes may be obtained with different types of phosphors, those with long or short persistence and with various colors. Besides the cathode ray tube

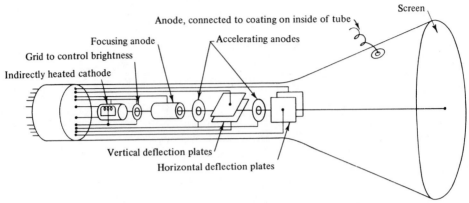

Fig. 2-36. Schematic diagram of a cathode-ray tube.

with electrostatic focusing and deflection illustrated here, tubes are also available with magnetic focusing or magnetic deflection, or both. In most uses, a voltage varying in a sawtooth fashion with respect to time is applied to the horizontal plates so that a "sweep" linear with time and recurring at regular intervals is obtained. The horizontal plates are connected to the signal to be displayed.

PROBLEMS

1. Trace or copy Fig. 2-5. Draw a load line and a cathode bias line for the circuit shown in Fig. 2-23 with $E_{CC} = 3.0$ V, $R_K = 500$ Ω, $E_{bb} = 250$ V, and $R_L = 30,000$ Ω. What is the normal operating point of the circuit?

2. In the circuit of the above problem, what would be the magnitude of the output signal in volts for an input of 4 V? What would be the normal operating output voltage?

3. Calculate the gain of a triode amplifier with $g_m = 3100\ \mu\Omega^{-1}$; $r_p = 6250$ Ω and $R_L = 100,000$ Ω. Calculate the gain of a pentode amplifier with $g_m = 4500\ \mu\Omega^{-1}$, $r_p = 1.5$ MΩ, and $R_L = 0.47$ MΩ.

4. Indicate by a block diagram a circuit which could be used to solve the equation $e_0 =$ constant $\int_0^t (0.1e_1 + 0.2e_2)\, \partial t.$

BIBLIOGRAPHY

Bair, E. J., *Introduction to Chemical Instrumentation*, McGraw-Hill, New York, 1962.
Brophy, J. J., *Basic Electronics for Scientists*, McGraw-Hill, New York, 1966.
Elmore, N. C., and M. Sands, *Electronics*, McGraw-Hill, New York, 1949.

Hill, R. W., *Electronics in Engineering*, McGraw-Hill, New York, 1949.

Malmstadt, H. V., and C. G. Enke, *Digital Electronics for Scientists*, Benjamin, Menlo Park, Calif., 1969.

Malmstadt, H. V., C. G. Enke, and E. C. Toren, Jr., *Electronics for Scientists*, Benjamin, Menlo Park, Calif., 1962.

Müller, R. H., R. L. Garman, and M. E. Droz, *Experimental Electronics*, Prentice-Hall, Englewood Cliffs, N.J., 1945.

Phillips, A. B., *Transistor Engineering and Introduction to Integrated Semiconductor Circuits*, McGraw-Hill, New York, 1962.

Reilley, C. N., and D. T. Sawyer, *Experiments for Instrumental Methods*, *Part V*, McGraw-Hill, New York, 1961.

Seely, S., *Electron-Tube Circuits*, 2nd ed., McGraw-Hill, New York, 1958.

Suprynowicz, V. A., *Introduction to Electronics*, Addison-Wesley, Reading, Mass., 1966.

Uzunoglu, V., *Semiconductor Network Analysis and Design*, McGraw-Hill, New York, 1964.

3

Ultraviolet and Visible Absorption Instrumentation

Changes in the electronic configuration and energy of molecules produce spectra in the visible and ultraviolet region of the electromagnetic spectrum. For our purposes, the visible and ultraviolet region will be defined as radiation associated with absorption in the range 200-800 nm, which is the spectral range of a conventional ultraviolet/visible spectrophotometer. In 1941, upon the introduction of the Beckman DU spectrophotometer with its quartz optics and ultraviolet accessory unit, the chemist was able for the first time to obtain reliable absorption spectra conveniently and within a reasonable time. With the advent of automatic recording and many improvements in instrumentation, there are now a wide variety of excellent commercially available spectrophotometers that are capable of meeting all the requirements of the analytical chemist. In this chapter certain characteristic design and operational features of ultraviolet/visible spectrophotometers will be considered.

Absorption measurements involve determination of the reduction in power suffered by a beam of radiation as a consequence of passing through the absorbing medium. The wavelength at which an absorbance maximum is found depends on the magnitude of the energy involved for a specific electronic transition. Absorption spectra often serve for qualitative identification. A wide range of quantitative applications exist, as will be discussed in Chapter 4. Although this technique has been used largely to determine trace quantities of constituents, it may also be applied to identify a substance constituting the major part of the sample.

TERMINOLOGY AND BASIC COMPONENTS

Instruments designed for measuring the emission or the absorption of radiant energy from substances have various names: photometers, spectrometers, and spectrophotometers. The following definitions will be employed in this text:

Photometer An instrument that furnishes the ratio, or some function of the ratio, of radiant power of two electromagnetic beams.

Spectrometer, optical An instrument with an entrance slit, a dispersing device, and one or more exit slits with which measurements are made at selected wavelengths within the spectral range, or by scanning over the range. The quantity detected is a function of radiant power.

Spectrophotometer A spectrometer with associated equipment, so that it furnishes the ratio, or a function of the ratio, of the radiant power of two electromagnetic beams as a function of spectral wavelength. These two beams may be separated in time, space, or both.

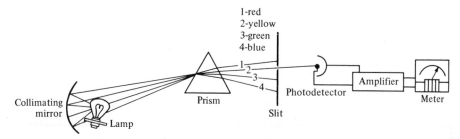

Fig. 3-1. Schematic of a spectrophotometer.

Basic components Figure 3-1 illustrates how a spectrophotometer works. All absorption instruments must contain a source of radiation, and each spectral region has its own requirements. (In emission spectrometers the sample serves as its own source.) All spectrometers include some way of discriminating between different radiation frequencies by dispersion of the radiation with a prism or grating into a spectrum of wavelengths. In the visible region every color is observed from violet to red in a continuous rainbow; violet merges into blue, blue merges into green, and so on. The dispersed radiation is swept past a slit until the desired emission line or band of color falls on the slit opening and reaches the sample. The slit, often adjustable to allow any bandwidth to be chosen, blocks off all but a narrow band of radiation. The sample absorbs a portion of the light; the remainder is transmitted through the sample and strikes a detector where it is changed into an extremely small electrical signal. The signal is then sent to an amplifier. By greatly increasing the strength of the minute signal from the detector, the amplifier eliminates the need for delicate parts, permitting a rugged meter to indicate the amount of light passing through the sample. From an engineering standpoint, it is desirable that the system be detector-limited; that is, the limiting factor should be the noise generated by the detector. By making a series of such measurements and using different wavelengths of light, the analyst can construct a curve showing the exact location and degree of the absorption of the sample over a wide range of wavelengths—the unmistakable "fingerprint" of the sample.

RADIANT ENERGY SOURCES

The principal requirements of the light source are that it emits a sufficient intensity of radiation in the desired spectral region and generates this radiation in a reproducible and controlled fashion. For molecular measurements, the resolution of monochromators is normally adequate to permit continuous sources to be used. Generally, brightness is not a problem, although in design considerations it must be remembered that flux density of radiant energy varies inversely as the square of the distance from the source. Two types of sources are in common use: incandescent sources and luminous gas sources.

Incandescent Sources

For work in the visible and near-infrared regions, the most common source is the tungsten, or tungsten-iodide, incandescent lamp. Figure 3-2 shows the spectral distribution of the radiation from an incandescent solid (acting as a blackbody radiator) at different temperatures. Displacement of the wavelength maximum toward shorter wavelengths at higher temperatures is obvious. Unfortunately, the tungsten lamp emits the major portion of its energy in the near-infrared; only about 15% of the radiant energy falls within the visible region at an operating temperature of about 2850°K. In size and shape the tungsten lamp varies from a 100-W projection lamp to a 6- or 12-V automobile headlamp with coiled filament, or a small flashlight bulb. Lamps with prefocus bases are useful with respect to easy replacement in an optical system. Often a heat-absorbing filter is inserted between the lamp and sample holder in filter photometers in order to absorb most of the infrared radiation without seriously diminishing radiant energy at other wavelengths.

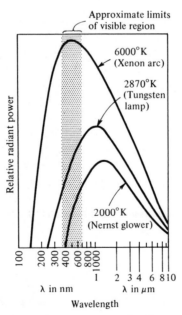

Fig. 3-2. Spectral distribution curves of radiant energy sources.

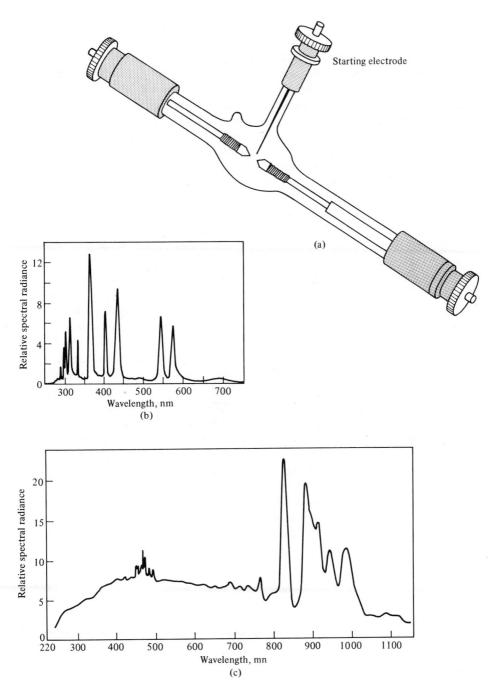

Fig. 3-3. (a) Illustration of mercury arc lamp. (b) Spectral radiance of mercury lamp.
(c) Spectral radiant energy distribution of xenon arc lamp.

Luminous Gas Sources[4]

Work in the ultraviolet region is done mainly with hydrogen or deuterium discharge lamps operated under low pressure and dc conditions. They provide continuous ultraviolet emission down to 1650 Å with an optical window of fused silica. At wavelengths longer than about 3600 Å, the hydrogen discharge has emission lines superimposed on the continuum so that routine measurements above this wavelength are usually made with an incandescent source. A vital feature of a discharge lamp is a mechanical aperture between the cathode and the anode that constricts the discharge. Normally the anode is placed close to the aperture, which creates an intensely radiating ball of light about 0.10 to 0.15 cm in diameter on the cathode side of the opening. The lamp is imaged at the entrance slit of the spectrophotometer.

Low-pressure mercury lamps have sharp line emissions with little background or continuum present. The lamp contains a starting electrode in addition to the two current-carrying electrodes. When the voltage is applied, a discharge initiates in the argon between the starting electrode and the main electrode adjacent to it. This causes the heating and ionization of the gas in the rest of the tube, and a discharge forms between the two main electrodes. This produces further heating, and mercury present in the tube vaporizes. The lamp output increases continuously for several minutes as the mercury vapor pressure builds up, until a stable operating condition is reached. The lamps may be cooled by an air blast.

The highest luminous intensities are obtained if a gas tube is operated under conditions corresponding to the formation of an arc discharge. This is achieved by employing a high gas pressure in a high-brightness mercury or xenon lamp. During operation the arc is compressed in the narrow gap between the electrodes (Fig. 3-3) and becomes extremely bright. Medium and high-pressure mercury lamps produce abundant energy throughout the ultraviolet and midvisible region; this continuum provides excitation energy for those compounds measured fluorometrically. The spectral output of a high-pressure, xenon-arc lamp produces brightness qualities with the equivalent color temperatures of about 6000°K. The visible radiation has the same white appearance as sunlight. The useful output extends in a fairly continuous fashion from 3000 Å to 1.3 μm, with several radiation peaks in the near-infrared region of 0.8 to 1.1 μm. Arc flicker sets the limit of short-term stability, and is about 0.3%. The long-term stability is a drift of 1% per hour and is limited by electrode wear and arc wander.

Source Stability

High, short-term stability of an incandescent source is required for single-beam photometers. When a detector is illuminated, the photocurrent i can be expressed as a function of voltage V applied across the lamp terminals by the equation

$$i = kV^x \tag{3-1}$$

The exponent x has a value between 3 and 4 for tungsten lamps. To reproduce the photocurrent within 0.2%, the attainable spectrophotometric precision, the lamp voltage must be regulated within a few thousandths of a volt. To attain such close voltage regulation, the lamp must be operated from a storage battery under continuous charge, or from a well-regulated power supply. The effects of voltage fluctuations from an ordinary house main can be cancelled by properly designed double-beam spectrophotometers.

PHOTOSENSITIVE DETECTORS

A detector of radiant energy should have a linear response in the spectral region used and a sensitivity sufficient for the particular task at hand. A *barrier-layer photocell* is the simplest and requires little additional equipment, but its response is difficult to amplify. Its use is restricted generally to instruments with an optical system that permits a wide band of radiant energy to strike the detector. Instruments that restrict the bandwidth of the radiant energy reaching the detectors will need to employ phototubes and amplifier units to boost the output signal. The *electron multiplier phototube* provides the maximum signal and permits use of extremely narrow slits.

Human Eye

The human eye is limited in response to visible light, and in this region, its response is most acute for green light. It suffers from numerous frailties, including fatigue, slowness of response, and a tendency to respond more readily to dominant colors. Perhaps its more serious limitation is its inability to determine the radiant power level except by matching with a reference. To a human being color is a subjective, psychological judgement.

Barrier-Layer or Photovoltaic Cells

In photovoltaic cells electrons are excited across the potential barrier between suitable pairs of materials that are sensitive to visible radiation. A cell consists of a plate of metal upon which a thin layer of a semiconductor has been deposited. Very frequently selenium is deposited upon an iron base. A very thin transparent layer of silver is sputtered over the selenium to act as the collector electrode. The iron base acts as the second electrode. The construction is shown in Fig. 3-4.

Radiant energy falling upon the semiconductor surface excites electrons at the silver-

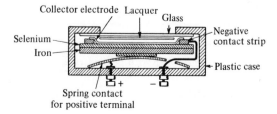

Fig. 3-4. Construction of a barrier-layer cell. (Courtesy of General Electric Co.)

selenium interface that are released and pass to the collector electrode. A hypothetical barrier region appears to exist near the interface across which electrons pass easily from the semiconductor to the collector electrode, whereas a moderate resistance opposes the electron flow in the reverse direction. Consequently, this cell generates its own electromotive force and no external power supply need be applied to observe a photocurrent. If the cell is connected to a galvanometer, a current will flow if the resistance in the external circuit, including the resistance of the coil of the meter, is relatively small. This cell yields photocurrents as high as 0.08 μA per μW. When the external resistance is about 400 Ω or less, the photocurrent is very nearly proportional to the radiant power of the incident light beam at low levels of illumination (Fig. 3-5).

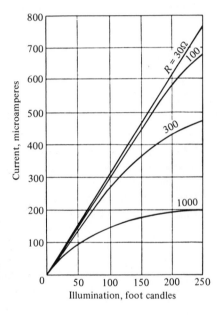

Fig. 3-5. Relationship between illumination and output for a typical barrier-layer cell.

The spectral response of a selenium cell with a glass protective cover adequately covers the visible region, with its sensitivity greatest for the green through yellow wavelengths (Fig. 3-6). Because of the low impedence of the cell, the output current cannot be amplified unless a regenerative feedback type of amplifier is used. Consequently, this

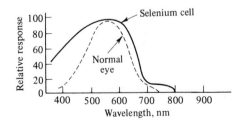

Fig. 3-6. Spectral response of selenium barrier-layer cell with protective glass cover. Response of human eye shown for comparison.

type of detector finds use mainly in filter photometers where fairly high levels of illumination exist and where there is no need to amplify the signal.

Barrier-layer cells show fatigue effects. Upon illumination, the photocurrent rises to a value several percent above the apparent equilibrium value and then falls off gradually with time. This difficulty can be overcome by the use of a gravity-controlled light shutter that permits light to strike the cell only while readings are being taken with a cuvette inserted in the sample holder. The effect is more pronounced at high levels of illumination and with improper circuit resistances. The selenium cell also has a high temperature coefficient, and if readings are taken before the body of the instrument has attained ambient temperature after turning on the light source, erroneous results are likely to occur. Its modulation ability is also poor; that is, the barrier-layer cell fails to respond immediately to changes in levels of illumination such as would occur if the light beam were to be interrupted 15-60 times per second by a mechanical chopper.

Photoemissive Tubes

The vacuum phototube comprises a negative electrode, often hemihedral in shape and coated with a light-sensitive layer, and an anode that is either an axially centered wire or a rectangular wire that frames the cathode. These electrodes are sealed within an evacuated glass envelope (Fig. 3-7). The anode is maintained at a positive potential by means of a power supply. Electrons will be ejected from the cathode surface if the energy of the radiation quanta exceeds the *work function* of the cathode material. The electrons pass over to the anode where they are collected and return via the external circuit.

The spectral sensitivity of a photocathode depends on the composition of the enve-

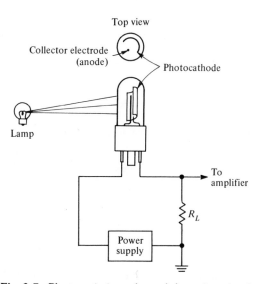

Fig. 3-7. Photoemissive tube and detection circuit.

lope and on the nature of the substance coating the cathode; the latter can be varied by using different alkali metals or by variation in the preparation of the surface. For example, the Ag-O-Cs type cathode surface is made in the following manner: A silver-plated nickel cathode of proper dimensions is carefully oxidized by means of a glow discharge to form a layer of silver oxide; then a layer of cesium metal is distilled onto the silver oxide surface; and finally the tube is subjected to a baking process, which results in the formation of some cesium oxide on the surface through interaction of cesium metal with the silver oxide layer, and the formation of some metallic silver throughout the silver oxide layer. The composite semiconductor coating consists of cesium metal admixed with cesium oxide, silver oxide, and silver metal.

Photoemission involves three steps: (1) Absorption of a photon resulting in the transfer of energy from photon to electron; (2) motion of the electron toward the material-vacuum interface; and (3) escape of the electron over the potential barrier at the surface into the vacuum of the tube. The energy required is just present at the *threshold wavelength*. Light absorption is efficient if the photon energy exceeds E_G, the energy of the band gap between the valence band and the conduction band. Losses by transmission and reflection reduce the quantum efficiency so that the photocathode surface must not be transparent nor highly reflective. In the second step, the photoelectrons may lose energy by collision with other electrons or with the lattice (phonon scattering); the latter is the predominant loss mechanism in semiconductors, and therefore the escape depth is much greater than in metals. The threshold wavelength is given by the value of $E_G + E_A$, where E_A is the electron affinity of the semiconductor surface. Remarkable improvements in the photoemission from semiconductors have been obtained through deliberate modification of the energy-band structure, mainly to reduce the electron affinity and thus to permit the escape of electrons that have been excited into the conduction band at greater depths within the material.

There are two types of photocathodes: (1) The *opaque* photocathode where the light is incident on a thick photoemissive material and the electrons are emitted from the same side as that struck by the radiant energy, and (2) the *semitransparent* surface where the photoemissive material is deposited on planar or spherical-section windows so that the electrons are emitted from the side of the photocathode opposite the incident radiation. Spectral-response curves for photocathodes commonly used are shown in Fig. 3-8; the ordinate scale is amperes per incident watt. The spectral response is modified by the window material. Therefore, the short-wavelength limit is generally characteristic of the window rather than the photocathode. Window inserts of fused silica extend the range into the far ultraviolet.

Photoemissive tubes require an external power supply to maintain a fairly high voltage between the cathode and anode. As the voltage is increased, the point is reached where all the photoelectrons are swept to the anode as soon as they are released, and a *saturation photocurrent* is obtained. The potential necessary to achieve saturation increases with an increase in the intensity of the radiant energy, as shown in Fig. 3-9. An excessively high voltage is undesirable because it contributes to *leakage (dark) currents* without any gain in response. A high vacuum within the tube avoids scattering of the photoelectrons by collisions with gas molecules.

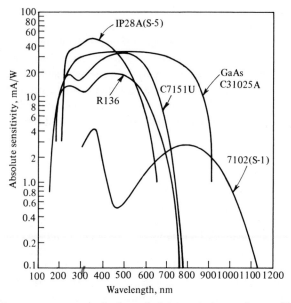

Fig. 3-8. Spectral response curves of selected photoemissive surfaces. (Courtesy of Radio Corporation of America.)

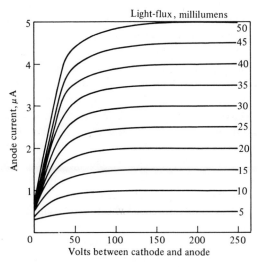

Fig. 3-9. Typical anode characteristics of vacuum photoemissive tube; light source is a tungsten-filament lamp operated at $2870°K$.

A typical circuit for use with a vacuum phototube is shown in Fig. 3-7. The resistor may be the grid load resistor for the amplifier circuit. Care must be exercised that the resulting iR drop across the load resistor will not lower the voltage across the phototube to a value below the minimal anode potential for saturation photocurrents. A vacuum phototube behaves as a high-impedance current generator and in critical applications should be used in conjunction with an electrometer. Strict proportionality between saturation photocurrent and light intensity is a fundamental law of photoelectricity, but its realization in practice demands a carefully designed phototube and attendant circuits.

Photomultiplier Tubes

The electron multiplier phototube, or photomultiplier tube as it is commonly called, is a combination of a photodiode and an electron multiplying amplifier. It is extremely sensitive, as well as extremely fast in response. The tube combines photoemission with multiple cascade stages of secondary electron emission. It achieves a large amplification of primary photocurrent within the envelope of the phototube and linear response is retained. As shown in Fig. 3-10 for two popular configurations, the tube is constructed so that the primary photoelectrons from the cathode are accelerated by an electric field so as to strike a small area on the first dynode. Each dynode is an electrode whose surface is of Be-Cu, Cs-Sb, or similar material. The impinging electrons strike with sufficient energy to dislodge and eject two to five *secondary electrons*. These secondary electrons are accelerated to the second dynode by an additional positive potential and more electrons are dislodged. This process is repeated at the successive dynodes, of which there may be 9 to 16 stages. The overall rounded shape of the dynode converges the electrons in one dimension, whereas the field-forming ridges at or near the dynode ends converge the electrons in the second dimension. In the final step the anode collects secondary electrons from the last dynode. The simplest anode structure is a gridlike collector in which secondary electrons from the next-to-last dynode pass through the grid to the last dynode. Secondary electrons leaving the last dynode are then collected on the anode.

Each initial photoelectron resulting from the absorption of a quantum on the photocathode gives rise to a burst of about 10^6 electrons at the anode, that is, 4^9 where an average of four secondary electrons are assumed to be released per stage and there are nine stages. The transit time between absorption of the photon and the arrival of the shower of electrons is typically in the range of 10 to 100 nsec. The rise time of a sharp pulse is in the range of 2 to 20 nsec, times short enough to allow a photomultiplier to be used as a quantum counter at very low light levels of approximately 10^6 quanta per sec (cf. use in scintillation counters). At higher light levels the steady current output is measured by amplifying the voltage produced when the current flows through a load resistor (100 kΩ to 100 MΩ).

Successive stages are operated at voltages increasing in equal steps of 75-150 V, the actual voltage depending upon the particular tube. This requires a very stable, high-voltage power supply with a chain of resistors between the high voltage and ground terminal to provide the correct voltage steps for the dynodes. The current through the dynode

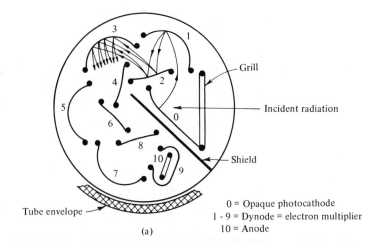

0 = Opaque photocathode
1 - 9 = Dynode = electron multiplier
10 = Anode

(a)

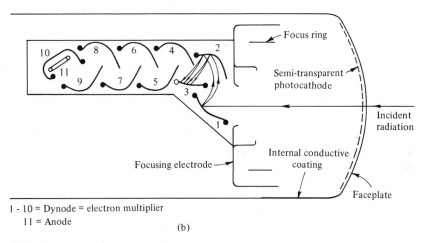

1 - 10 = Dynode = electron multiplier
11 = Anode

(b)

Fig. 3-10. Photomultiplier design. (a) The circular-cage multiplier structure in a "side-on" tube, and (b) the linear-multiplier structure in a "head-on" tube. (Courtesy of Radio Corporation of America.)

chain should be at least 10 times, and preferably 100 times, the maximum value of anode current that is likely to flow, because the electrons ejected from the dynodes by the secondary emission process have to be replaced from the power supply via the dynode-resistor chain. A typical circuit is shown in Fig. 3-11. Current amplification may be controlled, or the output signal may be modulated, by the adjustment of the voltage applied to a single or to two consecutive central dynodes. Amplification can be pushed to the limits set by the *shot-effect noise*—the inherent dark current due to thermionic emission and other random noise signals, which is approximately 6×10^{-12} A.

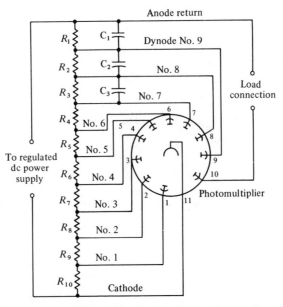

Fig. 3-11. Typical voltage-divider arrangement for a photomultiplier tube. R_1 through R_{10} = 20,000 to 1,000,000 Ω. Power supply adjustable between 500 and 1250 V.

The dark current may be decreased by lowering the temperature at which the tube is operated. The further the peak response pushes into the red region of the spectrum, the greater is the need for the tube to be cooled. By taking the entire tube and its potted resistor network down to liquid nitrogen temperature, a typical red-sensitive tube can be upgraded by a factor close to 10^5, its noise equivalent power decreasing to 10^{-16} W, and its useful range extended to 1.2 μm. A blue-sensitive tube would be upgraded by a factor of 100. A heater submerged in a Dewar of liquid nitrogen forces dry nitrogen gas at a temperature just above that of the liquid around the entire tube to achieve uniform, shock-free cooling.

Photomultiplier fatigue, leading to signal drift, can be a limitation unless overcome by proper circuit design and experimental conditions. When stability is of prime consideration, the use of average anode currents of 1 μA or less is recommended. Adequate light shielding should be provided to prevent extraneous light from reaching any part of the photomultiplier tube. Extensive information on photomultiplier tubes is available.[7]

At wavelengths below 2000 Å, it is easiest to use a blue-sensitive photomultiplier in conjunction with a thin film of sodium salicylate phosphor deposited on the vacuum side of an exit window from the light source or on the outside of the tube itself, if it is included within the vacuum chamber. Such a film has a fairly uniform quantum efficiency of 60-90% between 300 and 3000 Å and a fluorescence output that peaks at 4200 Å, the wavelength of maximum sensitivity of an S-11 photocathode. When extremely weak light signals are to be detected in the vacuum ultraviolet, the "solar-blind" photomultipliers with lithium fluoride windows and photocathodes of CsI or Cs_2Te are more sensitive.

DISPERSING DEVICES

Spectrophotometric methods call for the isolation of more or less narrow wavebands of radiant energy. Additionally, in an emission mode, one must be able to select the most favorable ratio between background and analytical line radiation. The important characteristics of a dispersing device are its *bandpass* (the range of wavelengths at which the transmittance is one-half the peak transmittance), the *transmittance* at the nominal wavelength, and the *nominal wavelength*. These are illustrated in Fig. 3-12.

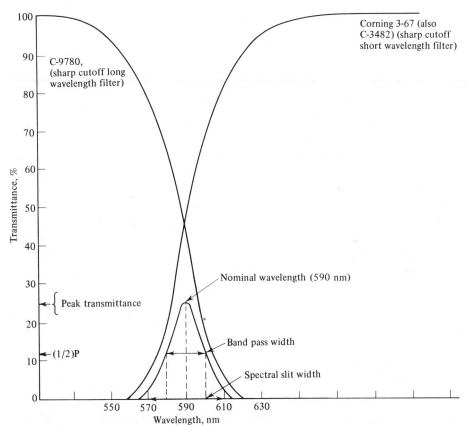

Fig. 3-12. Spectral transmittance characteristics of a composite glass absorption filter and its components.

Less expensive instruments may use an interference filter to isolate the radiant energy. Better isolation of spectral energy can be obtained with a prism or grating as dispersing medium. These dispersing devices, in conjunction with entrance and exit slits, suitable baffles, and mirrors, form a *monochromator*.

Filters

The *tinted glass absorption filter* consists of a solid sheet of glass that has been colored by a pigment which is either dissolved or dispersed in the glass. Composite filters are constructed from sets of unit filters. One series consists of sharp *cutoff filters* that pass long wavelengths, the red and yellow series; the other comprises long-wavelength cutoff filters, the blue and green series. The transmittance of the individual cutoff filters and the combined transmittance for a filter with a nominal wavelength at 5900 Å are shown in Fig. 3-12. Glass absorption filters have a wide bandpass of 350-500 Å at one-half maximum transmittance, and their peak transmittance is only 5-20%, decreasing with improved spectral isolation. Combination filters are employed only in inexpensive photometers. However, cutoff filters enjoy wide use as blocking filters to suppress unwanted spectral orders from gratings and interference filters.

Interference filters employ thin metallic or dielectric layers to produce interference phenomena at desired wavelengths, thus permitting rejection of unwanted radiation by selective reflection. Their construction follows: A semitransparent metal film is deposited on a plate of glass. Next, a thin layer of some dielectric material, such as MgF_2, is evaporated on top of this, and then the dielectric layer is in turn coated with a film of metal. Finally another plate of glass is placed over the films for mechanical protection. The completed filter is shown in Fig. 3-13. A portion of the light incident upon the face of the filter is reflected back and forth between the metal films. Constructive interference between the different pairs of superposed light rays occurs only when the path difference is exactly one wavelength, or a multiple thereof. Since the path difference is now in the dielectric of refractive index n, the wavelengths of maximum transmission for normal incidence are given by

$$\lambda = \frac{2nb}{m} \tag{3-2}$$

where b is the thickness of the dielectric spaces and m is the order number. For example, a dielectric layer of $n = 1.35$ that is 185 nm thick will provide a first order filter of 5000

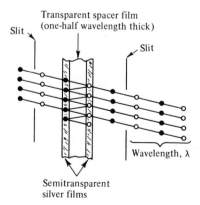

Slit

Transparent spacer film
(one-half wavelength thick)

Slit

Wavelength, λ

Semitransparent
silver films

Fig. 3-13. Schematic of an interference filter and path of light rays through the filter.

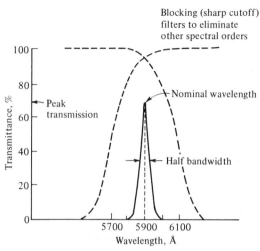

Fig. 3-14. Transmission of an interference filter and its cutoff (blocking) filters (dashed lines).

Å peak wavelength. This filter also passes bands centered at 2500 and 1670 Å. Unwanted transmission bands can be eliminated by using an appropriate sharp cutoff absorption filter for the protecting glass cover, as indicated in Fig. 3-14. Interference filters have a bandpass of 100-150 Å and peak transmittance of 40-60%.

Multilayer interference filters consist of layers of nonabsorbing material of alternately high and low indices of refraction deposited on an optical base, followed by a transparent spacer film, and then by more layers of similar material.[1] These filters have transmissions of 50-70% and bandwidths of 10-50 Å.

By depositing a wedge-shaped layer of dielectric between the semireflecting metallic layers, a *continuously variable transmission interference filter* is obtained. At each point along the base of this filter a different wavelength band will be transmitted. In use the interference wedge is moved past a slit to select wavelengths (see Fig. 4-12). The circular variable filter is similar in construction. Simple monochromators have been devised with these dispersing devices.

Prisms

The action of a prism depends on the refraction of light by the prism material. The dispersive power depends on the variation of the index of refraction with wavelength. A light ray entering a prism at an angle of incidence i will be bent toward the normal (vertical to the prism face) and, at the prism-air interface, it is bent away from the vertical (see Fig. 3-15). To minimize astigmatism of a prism and achieve best definition, the prism should be illuminated by parallel light with the slit parallel to the prism edge and oriented so that the light rays pass through a plane parallel to the prism base. The rays should pass through the prism symmetrically so that the incident and emergent beams form equal

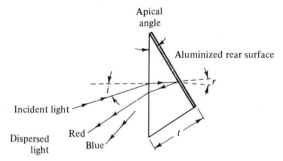

Fig. 3-15. The prism as a dispersing medium. Littrow-type mounting; i is angle of incidence, r is angle of refraction, t is base width of prism, and apical angle is $30°$.

angles to the faces; the prism is then used at minimum deviation. The entrance slit is projected onto the exit slit in a series of images ranged next to each other, caused by light of shorter wavelengths being more strongly bent than light of longer wavelengths. A nonlinear wavelength scale results.

Flint glass provides about threefold better dispersion than fused silica, and is the material of choice for the near-infrared-visible region of the spectrum. Fused silica is required for work in the ultraviolet region.

Gratings

A grating consists of a large number of parallel, equally spaced grooves ruled upon a highly polished surface. The light incident on each groove is diffracted or spread out over a range of angles, and in certain directions reinforcement or constructive interference

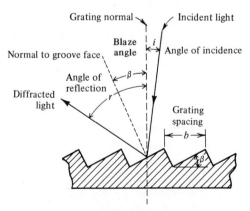

Fig. 3-16. Cross-section diagram of a diffraction grating showing the "angles" of a single groove, which are microscopic in size on an actual grating.

occurs, as stated in the grating formula:

$$b(\sin i \pm \sin r) = m\lambda \qquad (3\text{-}3)$$

where b is the distance between adjacent grooves, i is the angle of incidence, r is the angle of diffraction, and m is the order number (Fig. 3-16). The larger the number of grooves per millimeter, the greater the angle of diffraction for a given order of a particular wavelength. A positive sign applies where incoming and emergent beams are on the same side of the grating normal.

Example 3-1

Let us apply the grating equation to find the primary angle at which light of 3000 Å will be diffracted at normal incidence $(i = 0°)$, by a grating ruled 1180 grooves/mm, in the first order.

$$\sin r = \frac{m\lambda}{b - \sin i}$$

$$= \frac{1 \times 3000 \times 10^{-8}\ cm}{[1/11{,}800\ cm]} - 0$$

$$= 0.354$$

The angle having this sine is $20.73°$.

Gratings give rise to overlapping orders (Fig. 3-17). On the other hand, the grating has a nearly constant dispersion throughout the spectrum and, consequently, a linear scale for wavelength equal to m/b. The smaller the b-value, the more widely spread will be the spectrum. The second-order spectrum has twice the dispersion of the first order, the third three times, and so on. In general, it is good practice to utilize the first-order spectrum whenever possible to avoid overlapping orders in the visible and ultraviolet. Overlapping can be reasonably overcome, however, in orders up to the third with proper cutoff filters. The zeroth order (or direct image) arises where there is zero phase difference between the energy diffracted by consecutive grooves. It is the direction in which a mirrorlike reflection would occur if the grating surface were replaced by a mirror. All wavelengths coincide in this direction, and we get an undispersed image of the radiation source.

Two factors limit the number of grooves per millimeter that can be used for a given wavelength. Neither i nor r can be greater than about $65°$, and the groove aspect as "seen" by the light cannot be substantially less than the wavelength of the light, otherwise the grating acts as a mirror, reflecting light rather than dispersing it. Standard gratings for the visible–ultraviolet region will have approximately 590, 1180, or 1770 grooves/mm ruled on an area varying from 25 × 25 mm to 102 × 102 mm. The ruled area of a grating should be large enough to intercept all the incident light even when the grating is turned to its extreme angular position. Any smaller area will decrease the useful light in the spectrum by increasing that going into the zeroth order as well as increasing that light which misses the grating altogether.

When the grating is ruled, the *groove (or blaze) angle*, β, can be adjusted so that most

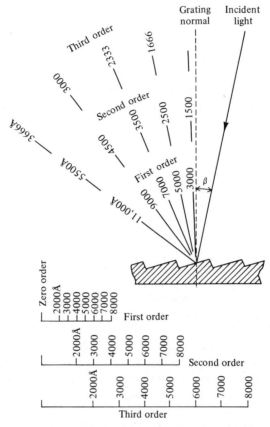

Fig. 3-17. Overlapping orders of spectra from a reflection grating.

of the light intensity will be concentrated in the wavelength region of greatest interest. The *blaze wavelength*, λ_β, is defined as that wavelength for which the angle of reflectance from the groove face and the angle of diffraction from the grating are identical. This value is usually specified as the first-order wavelength although the grating is also blazed for the second-order of half the wavelength, and so on. It is customary to calculate and specify the blaze angle and blaze wavelength assuming that the angle of the incident light is the same as the angle of diffraction. In this case the grating equation reduces to:

$$m\lambda_\beta = 2b \sin \beta \qquad (3\text{-}4)$$

The spectral region wherein the intensity is greater than one-half the intensity at the blaze wavelength in its first order extends approximately from two-thirds its blaze wavelength to twice the blaze wavelength. In the second and higher orders, the grating intensity curve tends to be more balanced and extends from $\frac{2}{3}\lambda_\beta$ to $\frac{3}{2}\lambda_\beta$. A grating looks like a good mirror to wavelengths longer than the groove spacing.

Monochromator Designs

The two most popular monochromator designs are the *Ebert* (or a modification) and the *Littrow*. Both are compact designs.

In the Ebert mounting, focusing is achieved by off-center reflections from a large concave mirror which collimates the entrant light before it strikes the plane grating, and intercepts the dispersed beam and focuses it on the exit slit. In the side-by-side Czerny-Turner mounting (Fig. 3-18), two smaller concave mirrors replace the single large mirror. These mountings are *stigmatic*, that is, vertical and horizontal foci are the same, and also *achromatic*, so that the rays of all wavelengths are brought to focus at the exit slit without changing the slit-to-mirror distance. Entrance and exit beams are stationary. Wavelength is readily adjusted without affecting focus by rotating the grating on its axis. A sine-bar drive produces a direct wavelength readout. Fastie suggested an "under-over" design in which the entrant rays pass below the grating whereas the emergent rays pass above.

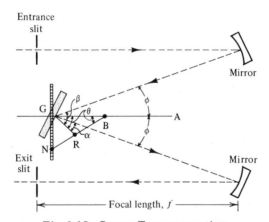

Fig. 3-18. Czerny-Turner mounting.

The Littrow mounting shown in Fig. 3-19, will accommodate a $30°$ prism backed by an aluminized surface, a $60°$ prism and separate Littrow mirror, or a grating. The source illuminates a condensing mirror that brings the reflected beam to a focus on the plane of the entrance slit of the monochromator. The image of the entrance slit is collimated by the parabolic mirror and directed onto the dispersing device. Then the refracted or diffracted beam is sent back to the same collimating mirror, but at a different height, and the beam is then projected and focused onto the exit slit that selects a portion of the dispersed spectrum for transmission through the sample and on to the detector. The upper and lower portions of the same slit assembly are used as entrance and exit slits, thus providing perfect correspondence of slit widths. The slit system is continuously adjustable. In the $30°$ arrangement the prism is rotated by means of a mount connected to the wavelength scroll. The Littrow (plane) mirror behind the $60°$ prism reflects the light beam and

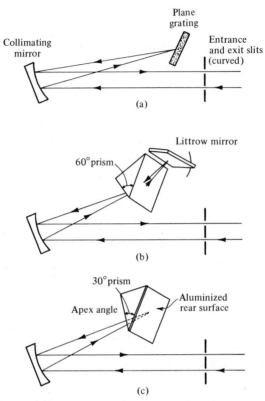

Fig. 3-19. Littrow mounting: (a) with plane grating, (b) 60° prism with Littrow mirror, and (c) 30° prism with aluminized backing.

returns it through the prism a second time, thus doubling the dispersion; the mirror is turned through a small angle to obtain the different wavelengths at the exit slit. With prisms, the wavelength scale must be calibrated relative to the refractive index of the prism material. The Littrow mount is stigmatic.

Associated Equipment

The transmittance of materials suitable for windows, lenses, and sample containers is a critical factor. The absorbance of any material should be less than 0.2 at the wavelength of use. Ordinary silicate glasses transmit satisfactorily from 3500 Å to 3.0 μm. Special Corex glass extends the ultraviolet range to about 3000 Å. Quartz or fused silica must be used for work below 3000 Å; the limit for quartz is about 2100 Å. If the monochromator is flushed with nitrogen or argon to eliminate absorption by oxygen, useful measurements can be taken down to 1650 Å with fused silica.

Beam reduction is accomplished by condensers that function as simple microscopes.

With these, it is possible to reduce the beam size by a factor of at least 25 with a loss in energy of less than 2. It then is possible to achieve reliable spectra of small crystals, biological cells, or subcellular particulates as small as 2 μm in diameter.

To minimize light losses, front-surfaced mirrors are used in place of lenses to focus or collimate light beams. The entrance slit is located at the focal length of the collimator. The diameter of the collimator is chosen as required to illuminate completely the entire area of the dispersing device. Upon returning from the dispersing device, the beam is still parallel, and so it is refocused by a second collimator on the exit slit as a series of images of the entrance slit, again at the focal length of the collimator mirror. Mirrors are aluminized on their front surfaces to avoid image distortions that occur from the multiple reflections from a back-surface mirror. Thin film coatings, generally SiO in the visible and MgF$_2$ in the ultraviolet, are used to improve reflection from mirrors and to protect the metal film on the surface. Parabolic mirrors are free of *chromatic aberrations*, which arise in lenses because different wavelengths of light focus at different distances from a single lens due to the variation of refractive index with wavelength.

Beam Splitters

Achromatic beam-splitting plates, also called *dichroic mirrors* (Fig. 3-20), are used when an incident beam of light must be split spectrally. They are semitransparent mirrors, typically used at an angle of 45°, on one surface of which is deposited a coating so that light is reflected at any desired value, the remainder of the incident light being transmitted. With white illumination the two separated beams are complementary in color.

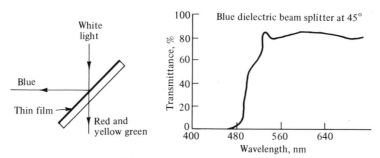

Fig. 3-20. Beam splitter.

Fiber Optics

Fiber optic bundles are composed of numerous strands of light-transmitting glass or plastic with polished ends. A single fiber will transmit light, but a bundle of fibers will transmit both light and images. Individual fibers with an index of refraction n_1 are clad with a material (index of refraction n_2) that eliminates light leakage and traps light within the fiber for *total internal reflection*; that is, light that enters one end of a fiber at the proper angle will be reflected from the walls as it traverses the length of the fiber and will emerge from the other end at the same angle. When $n_1 > n_2$, total internal reflection is possible, so that from Snell's law:

$$n_3 \sin \theta = \sqrt{n_1^2 - n_2^2} \qquad (3\text{-}5)$$

where n_3 is the refractive index of the medium through which light approaches the end of the fiber. The sine of the half-angle of acceptance of the incoming light at the fiber end is defined as the *numerical aperture*, and the angle is the maximum at which a light ray can be transmitted through the fiber. The commercial glasses normally used for preparing fiber optics transmit light in the 4000-9000 Å range; for infrared transmission, arsenic trisulfide fibers are available; and for the ultraviolet, fused quartz is used.

Slits and Irises

Basically the purpose of the entrance slit is to provide a narrow source of light so that after dispersion and refocusing in the plane of the exit slit, the amount of overlapping of the monochromatic images is limited. The exit slit then selects a narrow band of the dispersed spectrum for observation by the detector. In practice it is desirable to have the entrance and exit slits of equal width because it can be shown that for conditions of a given resolving power, maximum radiant power is passed by the spectrometer when this is true. Under these conditions the width of a monochromatic image of the entrance slit is such that it will just be passed by the exit slit.

For perfect image formation, the monochromator must be equipped with bilaterally opening slits with curved jaws, where the radius of curvature of the jaws equals one-half the distance between the entrance and exit slits. At full height, the slits retain resolution while maximizing photoelectric light-gathering power. Resolution degradation becomes noticeable for straight slits when slit height exceeds 4 mm. With prism monochromators the slit jaws are adjustable and operated by direct pressure on a steel spring. Fixed slit openings, 10, 25, 50, or 100 μm in width, are employed with grating mounts.

An adjustable iris diaphragm may be placed in the light path at a point in the beam where there is no imaging of the source, for example, immediately before or behind the filter or one of the lenses. If set at about two-thirds open, the diaphragm can be adjusted to compensate for changes in sensitivity during prolonged operation in emission instruments. A calibrated iris diaphragm can serve as the readout scale; when operated by a logarithmic cam, the readings will be in absorbance units (see Fig. 3-23).

Recording System

In recording spectrophotometers there is always some time lag between the recorded reading and the actual value. Scanning speeds should be selected to ensure that the detecting system can follow the signal from narrow emission lines or absorption bands. By definition, the *time constant* of an electronic detection system is the time required for the measuring device to reach 63% $(100 - 100/e)$ of its maximum scale value when a step-function signal is received at the detector. Roughly four time constants are required for one full peak height (98%); this is often called the *response time.*

Assuming that the recorder pen traverse speed is faster than the time constant of the electronics producing the signal being recorded, the maximum scanning speed needed to produce a faithful record of a spectrum is equal to the *bandpass*, in angstroms, divided by the response time, in seconds. Increasing the amplification involves an increase in noise, which is proportional to the square root of the amplification. This requires a longer response time to dampen out fluctuations. Thus the response time should be increased in proportion to the square root of the change in amplification. For example, when one switches from a 1X to a 10X scale (10 times amplification increase), the response time would be increased by about 3 and, correspondingly, the scanning speed should be reduced.

MONOCHROMATOR PERFORMANCE

The performance of a monochromator involves three interrelated factors: resolution, light-gathering power, and purity of light output. The resolution depends on the dispersion and on the perfection of the image formation, whereas the purity is determined mainly by the amount of stray or scattered light. Large dispersion and high resolving power in monochromators are necessary to measure accurately emission spectra with discrete lines or sharp absorption bands, whereas emission bands and usual broad absorption bands show up with instruments of medium dispersion.

Dispersion

Dispersion is a measure of the linear spread between two spectral lines, $dx/d\lambda$, in the plane of the exit slit of the monochromator. It is common practice to specify it as the *reciprocal linear dispersion*; when expressed in angstroms per millimeter, it is the difference in the wavelengths of the two lines divided by the observed separation of these lines. The lower the number, the better the dispersion. Also, as the number becomes lower it becomes possible to achieve a given bandpass with wider slits, thus admitting more light. The relation between linear dispersion and the *angular dispersion*, $d\theta/d\lambda$, is given by

$$\frac{dx}{d\lambda} = f\frac{d\theta}{d\lambda} \tag{3-6}$$

where f is the focal length of the focusing mirror or lens.

For a grating spectrometer, the reciprocal linear dispersion may be found by differentiating Eq. 3-3, with the angle of incidence constant, and combining the result with Eq. 3-6 to give

$$\frac{d\lambda}{dx} = \frac{b\cos r}{mf} \tag{3-7}$$

or since cos r will be virtually constant (r is about $6°$), approximately

$$\frac{d\lambda}{dx} = \frac{b}{mf} \tag{3-8}$$

For example, with a grating monochromator with a reciprocal linear dispersion of 16 Å/mm, two spectrum lines separated in wavelength by 6 Å would be 0.38 mm apart in the plane of the exit slit. Stated another way, using slits 0.100 mm in width, the bandpass would be 1.6 Å, which equals the physical slit width, in millimeters, multiplied by the reciprocal linear dispersion of the monochromator.

Dispersion for a prism is a function of wavelength. For a medium quartz prism monochromator of focal length 600 mm, typical values of reciprocal linear dispersion would be 6 Å/mm at 2300 Å, 10.4 Å/mm at 2700 Å, 15.6 Å/mm at 3100 Å, 29 Å/mm at 3700 Å, 54 Å/mm at 4500 Å, and 120 Å/mm at 6000 Å.

Resolution

The *resolution* (or *separation power*) of a spectrophotometer is its ability to distinguish adjacent absorption bands or two very close spectral lines as separate entities. Resolution is determined by size and dispersing characteristics of the prism or grating, the optical system of which the prism or grating is a part, and the slit width of the monochromator. It is also a function of the recording system at the scan speed in recording spectrophotometers.

The fundamental resolution of any dispersing device is

$$\frac{\lambda}{d\lambda} = Rs = w\frac{d\theta}{d\lambda} \tag{3-9}$$

and is limited only by w, the effective aperture width, and the angular dispersion. When narrow entrance and exit slits are used, spectral features which are quite close together can be resolved. However, sufficient light must reach the detector to enable the spectral features to be distinguished above the general background (noise) signal, thus the light-gathering power is also important. The resolving power of an actual instrument is generally poorer than the theoretical maximum value because of optical aberrations and other deleterious effects. Also, just what is the criterion for calling two features resolved? The *Rayleigh criterion* suggests a 19% valley between two equally intense lines. Perhaps one would prefer a valley that just attains the baseline between two lines without any stipulation about line intensities. Such a definition would require essentially a bandpass, $\Delta\lambda$, equal to twice the bandwidth of a spectral line at one-half maximum intensity. In separating absorption peaks, too little resolution depresses the peak height, limits sensitivity, and invites uncertainties due to increased peak height dependence on minor slit-width changes. Too much resolution can result in absorbance uncertainties because of unnecessary noise superimposed upon the scan. To achieve maximum sensitivity and to measure an absorption band free from instrumental distortion, it is important to work at slit widths that provide spectral bandwidths of 0.1 or less of the natural bandwidth. At 0.1 of the natural bandwidth the recorded peak height will be within 0.5% of the true height of the band.

For a grating, the *effective aperture width* is simply the width of an individual ruling, b, multiplied by the total number of rulings, N, and by cos r; that is, bN cos r. Because $d\theta/d\lambda = m/b$ cos r,

$$Rs = mN \qquad (3\text{-}10)$$

For example, a grating ruled with 600 grooves/mm and 50 mm in width has a resolving power in the first order of 30,000. At the sodium wavelength of 5890 Å, the smallest wavelength interval resolved will be $\Delta\lambda$ = 5890/30,000 = 0.2 Å. It is the product, mN, that is significant, so that a 50-mm grating of 1200 grooves/mm, where N = 60,000, will have no better resolution in the first order that a 50-mm grating of 600 grooves/mm, where N is only 30,000, when the latter is used in the second order. The advantage of a finer ruling is merely that it permits a higher resolution in the first order.

The resolving power of a prism,

$$Rs = t\,\frac{dn}{d\lambda} = \frac{d}{f}\cdot\frac{dx}{d\lambda} \qquad (3\text{-}11)$$

is limited by the base length of the prism, t, and the dispersive power of the prism material. The latter is not constant for a prism but increases from long wavelengths to shorter wavelengths. This requires a knowledge of the refractive index of the dispersing material and its rate of change as a function of wavelength, or the linear dispersion as a function of wavelength. A graph supplying this information should be provided with each instrument by the vendor.

Optical Speed

Photoelectric and photographic speed must be distinguished from each other. A photoelectric detector responds to the total flux impinging on its surface; a photographic plate responds to the flux falling on a single silver halide grain. Tall slits increase the light-gathering power photoelectrically, but have no effect on the photographic speed.

The *aperture ratio*, or *f*/number, is used to designate the optical speed of a spectrometer. The *f*/number is given by *f*/*d*, the focal length divided by the diameter of the objective. For example, *f*/8 might represent a system involving a 7.5-cm diameter collimating mirror of focal length 60 cm. Because the ruled area of a grating presents a rectangular aperture to the light beam, the effective aperture ratio equates the useful rectangular area to an equivalent circular aperture. The effective aperture area becomes

$$2\left(\frac{hw\,\cos i}{\pi}\right)^{1/2} \qquad (3\text{-}12)$$

where h is the height, w is the width, and i is the angle of incidence. For a 25 $\times$ 25 mm grating and a focal length of 250 mm, the effective aperture ratio is *f*/8.8. The smaller the *f*/number, the greater the solid angle of energy from the source put through the spectrometer optics, and hence the greater the total energy flux. Assuming the same sensitivity detector, a smaller f/number instrument will have more total energy to work with at low transmissions than one with a larger f/number.

The total flux transmitting power, F_T, for a monochromator is given by [3]:

$$F_T = \frac{4BTsLhw \cos i}{\pi f^2}$$

(3-13)

where B is the source brightness, T is the effective transmission of the optical elements, s is the slit width, L is the slit length, and the other terms have been defined previously. Upon comparing two grating spectrometers, one of which is twice the size of the other, the larger instrument will have twice the flux transmitting power of the smaller, if the slit widths are not altered. However, the grating also plays a role by virtue of its dispersion. The flux transmitting power of an instrument is proportional to the linear dispersion because the slit width is involved. For an equivalent bandwidth and neglecting resolution limiting effects, if the dispersion is increased by a factor of 2, the slit width can also be increased by a factor of 2, thereby increasing the flux transmitting power by the same amount. The slit length should be made as long as possible, and the area of the grating should be as large as possible. The focal length should be reduced. Unfortunately, it is not possible to increase the slit height and grating area and simultaneously decrease the focal length beyond certain limits without encountering aberrations that limit the resolving power of the spectrometer.

Spectral Purity and Stray Light

In a well-designed monochromator, stray radiant energy resulting from reflections from optical and mechanical members is minimized by matte blackening of all internal mechanical parts, by avoiding reflecting edges as far as possible (i.e., mounting mirrors within matte black frames), and a careful finish of all optical surfaces. Some radiant energy, caused by nonspecular scattering by the optical elements, will remain. This unwanted radiant energy can be reduced through the use of a second monochromator or a filter in combination with a monochromator. Unfortunately, any process of monochromatization is accompanied by a reduction of the radiant power, and the more complex the monochromator the greater the burden upon the measuring system. The effect of stray light and apparent deviation from Beer's law is discussed in Chapter 4.

INSTRUMENTS

Instruments for absorption photometry may be classified as *visual comparators, filter photometers*, and *spectrophotometers*. One also recognizes the constructional differences between single-beam and double-beam light paths, and whether the photoelectric instrument is direct reading or employs a balanced circuit. Special features include double monochromation and dual-wavelength monochromators. In the final selection of an instrument, one should consider the initial cost, maintenance, flexibility of operation, and what characteristics of resolution, wavelength range, accuracy, and additional features need to be met. No single instrument will meet all needs.

The sample is inserted somewhere between the source and the detector. In infrared spectrophotometers the holder is located in front of the dispersing device so that all of the light from the source passes through the sample. Light scattering by the sample will not be serious because most of the scattered light will be rejected by the monochromator. However, fluorescence and photosensitivity are more serious problems in the ultraviolet-visible, and so the sample is placed after the dispersing device and immediately before the detector.

Visual Comparators

The simplest types of color comparators use side-by-side viewing of light from a common source, such as diffused daylight, through a pair of tubes containing the unknown and the standard, respectively. A series of standard solutions are prepared in tubes of constant depth and diameter. The sample solution is transferred to a duplicate tube and diluted to volume. When the color is matched, the test solution has the same concentration as the standard. When not matched exactly, the color can be adjudged to lie between two standard solutions and, perhaps, the position in the interval estimated. Semipermanent artificial color standards are available in the form of tinted glasses (often mounted in a comparative color wheel) for a limited number of color reactions.

Filter Photometers

Relatively inexpensive instruments employing filters are adequate for a large proportion of the methods used, especially for absorbing systems with broad absorption bands. Actually, it is usually necessary to sacrifice spectral purity in order to obtain sufficient sensitivity for measurement with a rugged galvanometer with an instrument employing photovoltaic cells.

A single-beam, direct-reading, filter photometer is illustrated in Fig. 3-21. The optical path is simply from the light source, through the filter and sample holder, and to the detector. Light from the tungsten-filament lamp in the reflector is defined in area by fixed apertures in the sample holder and restricted to a desired band of wavelengths by an absorption or interference filter. After passing through the container for the solution, the light strikes the surface of a barrier-layer cell, the output of which is measured by the deflection of a light-spot galvanometer. The lamp is energized by a 6- or 12-V storage battery or by the output of a constant-voltage transformer. The location of the apertures in the sample container determines the minimum volume of solution on which a measurement can be made.

To operate a filter photometer of this type, the reference material (a solvent blank or reference solution) is positioned in the light path and the instrument is adjusted to read 0% transmittance when no light passes to the detector (shutter closed or light off) and

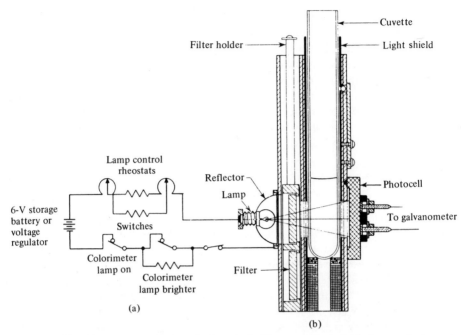

Fig. 3-21. (a) Schematic optical and electrical diagram of a single-beam photometer. (b) Evelyn photoelectric colorimeter. (Courtesy of Rubicon Company.)

100% with the shutter open and light on. The 100% adjustment is accomplished usually in one of three ways: (1) by a diaphragm somewhere in the light beam; (2) by a rheostat in the source circuit to alter the lamp brightness; or (3) by adjusting electrically the galvanometer pointer by means of a resistance in the detector–galvanometer circuit or by a potentiometer "bucking" circuit. After these adjustments have been made, the sample is placed in the light path and the absorbance or transmittance is read and related to the concentration either through the use of calibration curves or appropriate algebraic methods.

Double-beam photometers employing barrier-layer cells fall into two categories. In a *bridge-potentiometer arrangement* (Fig. 3-22), the null-balance galvanometer may be considered as receiving the photocurrent from each photocell through a universal shunt. Each shunt is a low-resistance (about 400 Ω), linearly wound potentiometer. The beam of filtered light is divided, part passing through the solution in the cuvette before falling on the measuring photocell, and the other part passing directly onto the reference photocell. The opposing currents through the galvanometer may be written

$$(i_g)_m = kTP \frac{a}{R_1 + r_g} \quad \text{and} \quad (i_g)_{\text{ref}} = kP \frac{x}{R_2 + r_g} \tag{3-14}$$

where T is the transmittance of the cuvette and solution, P is the radiant power of the filtered light, and a and x are the contactor positions on R_1 and R_2, respectively. At balance these currents are equal; hence, since $R_1 = R_2$, then $x = Ta$. In other words, the

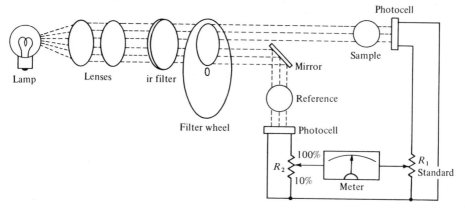

Fig. 3-22. Schematic diagram of the Electrophotometer II, a filter photometer with a double-beam, bridge-potentiometer circuit. (Courtesy of Fisher Scientific Co.)

potentiometer setting on R_2, namely x, is directly proportional to the transmittance. The slidewire scale is calibrated usually in 100 linear scale divisions, or in logarithmic units, or both.

To operate this type of double-beam photometer, the null-balance galvanometer is adjusted mechanically to position the needle or light spot at midscale with the lamp off. Then, with the lamp on, the reference solution in the light path, and the measuring slide-wire set at 100, balance is restored either by adjusting the contactor on slidewire R_1 (Fig. 3-22), adjusting the intensity of the reference light beam by means of a diaphragm, or insertion of a neutral-density wedge or a series of fixed apertures plus rotation of the

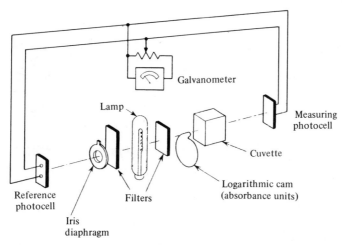

Fig. 3-23. Schematic optical and electrical diagram of Hilger-Spekker absorptiometer. (Courtesy of Hilger & Watts, Ltd.)

reference photocell about an axis perpendicular to the light beam through an angle of 60°. Subsequently, these adjustments remain unchanged while standards and unknowns are introduced and the contactor on the slidewire R_2, in series with the reference photocell, is adjusted to obtain the scale reading. Figure 3-23 illustrates a purely *optical arrangement*. A cam-shaped diaphragm, with an associated logarithmic scale, in the measuring circuit is adjusted to allow more light to reach the detector to compensate for the absorbance of the specimen. Adjustment to zero absorbance with a blank in the light path is accomplished by an iris diaphragm before the reference photocell.

All the double-beam circuits selected for illustration compensate for normal variations in the lamp supply voltage, enabling the lamp to be operated from ordinary alternating mains. However, not all double-beam circuits compensate for variations in the intensity of the source. Compensation, or lack of compensation, has been discussed in the literature.[2,5]

Single-Beam Spectrophotometers

In a single-beam spectrophotometer there is only one light path from the source to the detector (Fig. 3-24). The instrument is usually operated at a fixed wavelength and is primarily employed for the quantitative determination of the concentration of a single component when a large number of samples are to be analyzed. An obvious requirement of single-beam spectrophotometers is a high degree of stability both of the light source and the detector system.

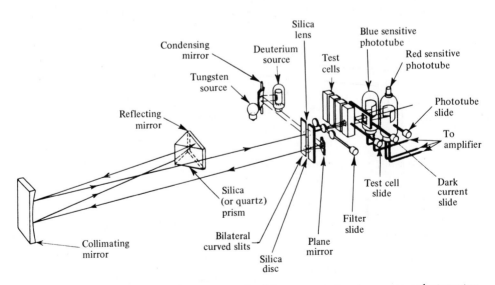

Fig. 3-24. Schematic optical arrangement of a Littrow-mount, prism spectrophotometer. (Courtesy of Beckman Instruments, Inc.)

Direct-Reading, Double-Beam Spectrophotometers

An inexpensive and popular grating spectrophotometer is illustrated in Fig. 3-25. A small replica grating provides the dispersion and, in conjunction with fixed slits, provides a bandpass of 20 nm. A difference amplifier is used (Fig. 3-26). Ejection of electrons

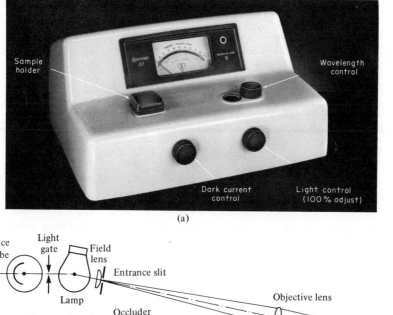

(a)

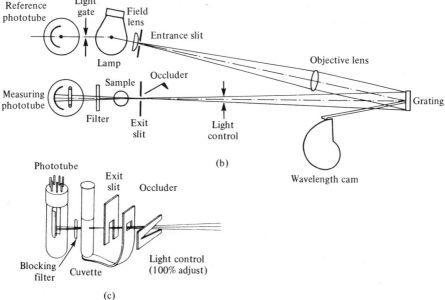

Fig. 3-25. Bausch & Lomb Spectronic 20 colorimeter. (a) View of instrument with controls labeled. (b) Schematic optical diagram. (c) Detail of photocell. (Courtesy of Bausch & Lomb.)

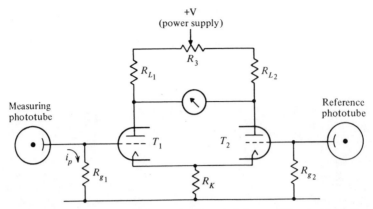

Fig. 3.26. Schematic diagram of the cathode-coupled amplifier used in the Bausch & Lomb Spectronic 20 colorimeter. $R_{g,1} = R_{g,2} = 10$ MΩ; $R_{L,1} = R_{L,2} = 0.1$ MΩ; $R_3 = 50$ kΩ. (Courtesy of Bausch & Lomb Optical Co.)

from the photocathode causes a corresponding number to be drawn from the grid of tube T_1, making it more positive and increasing the output voltage of the tube. The meter in the bridge circuit, which is graduated in linear scale divisions from 0 to 100% transmittance, indicates the intensity of the signal on grid 1 and thus the light incident on the phototube. Whenever the sample container is removed from the instrument, an occluder falls into the light beam so that with the phototube dark, the amplifier control (dark current) is adjusted to bring the meter needle to zero on the transmittance scale. This is accomplished by varying the position of the tap on resistor R_3, which alters the relative value of the plate voltage furnished each tube from the power supply. The instrument is balanced at 100% transmittance by means of a variable V-shaped slit (light control) in the dispersed light beam. The range of the spectrophotometer is from 340 to 650 nm with a blue-sensitive phototube, and can be extended to 950 nm by the addition of a red-blocking filter and substitution of a red-sensitive phototube. A second phototube, located to the rear of the light source, serves as a reference to monitor fluctuations in the source; its output is placed across resistor, $R_{g,2}$.

Double-Beam Recording Spectrometers

This type of spectrophotometer features a continuous change in wavelength and an automatic comparison of light intensities of sample and reference material; the ratio of the latter is the transmittance of the sample, which is plotted as a function of wavelength. The automatic operation eliminates many time-consuming adjustments and provides a rapid spectrogram. These instruments are well suited for qualitative analysis where complex curves must be obtained over a large spectral range.

In the *double beam-in-time* arrangement, energy from a dispersed source passes through the exit slit of the monochromator and is alternated between reference and sample com-

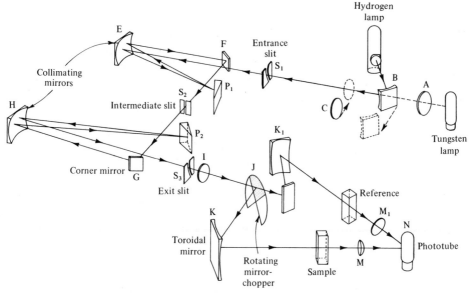

Fig. 3-27. Schematic optical diagram of a double beam-in-time spectrophotometer with double monochromation—Cary Model 16. (Courtesy of Cary Instruments, a Varian subsidiary.)

partments at a rate of 60 Hz (or other fixed frequency). These two beams alternately strike a single detector where their optical energy is converted to an electrical signal (Fig. 3-27). The output of the detector is an alternating signal whose amplitude is proportional to the difference in intensities in the two channels. The reference signal is maintained constant by an automatic slit servo system which widens or closes the slits to achieve a 100% transmittance baseline. The switching between reference and sample paths is done by a rotating half-sector mirror system, or by a vibrating mirror assembly, or by a stack of thin horizontal glass plates that are silvered along their edges and alternately oriented, or by a prismatic beam splitter. A chopper is placed either before the entrance slit or after the exit slit of the monochromator. For a prism monochromator, a control voltage that varies with wavelength according to the prism dispersion curve, actuates the slit servo amplifier to keep the monochromator at a constant preselected bandwidth. With grating monochromators, a sine-bar linkage is used to solve the grating equation.

In the second arrangement, called *double beam-in-space*, two separate light paths are created by a beam splitter and mirrors. Each beam is reflected by appropriate mirrors through the sample and reference compartments. Unabsorbed radiation falls on individual detectors (Fig. 3-28). The readout is based on measurement of the ratio of intensities in the two channels. If desired, the individual signals can be added, subtracted, or treated in any manner suited to a research problem, giving versatility in data presentation. *Electrical ratioing*, as this type of measurement is called, makes no attempt to achieve a physical equalization of the sample and reference beam intensities. Light chopping is done on the source side of the sample and reference beams.

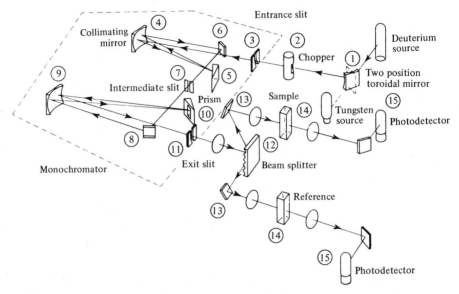

Fig. 3-28. Schematic optical diagram of a dual prism, double beam-in-space spectrophotometer—Cary Model 14. (Courtesy of Cary Instruments, a Varian subsidiary.)

Double Monochromation

In double monochromation two dispersing systems are used in series with an intervening slit. White light impinges on the entrance slit. At the intermediate slit the light has been dispersed into an order spectrum. This intermediate slit is the exit slit of the first monochromator and the entrance slit of the second monochromator wherein additional dispersion occurs. Two arrangements are shown in Figs. 3-27 and 3-28. Double monochromation drastically diminishes stray light, which is always a problem in precise measurements.

Double- (Dual-) Wavelength Spectrophotometer[6]

The double-wavelength spectrophotometer provides information from two wavelengths per unit time by passing radiation of two different wavelengths through the same sample before reaching the detector. The optical layout of a typical instrument is shown in Fig. 3-29. Light from the source passes into a Czerny-Turner monochromator through slit, S_1, to form an image of the mask on gratings G_1 and G_2. The dispersed radiation of both is focused by the collimator mirror, M_3, split, and chopped. Bilateral optical attenuators O_1 and O_2, situated at the pupil image of each beam, are used to vary the intensity of radiation of each beam continuously to compensate for intensity differences. Mirrors M_{7-1} and M_{7-2} converge the two beams through a single cuvette in the sample compartment. The time separating sample, reference, and zero signals are compared.

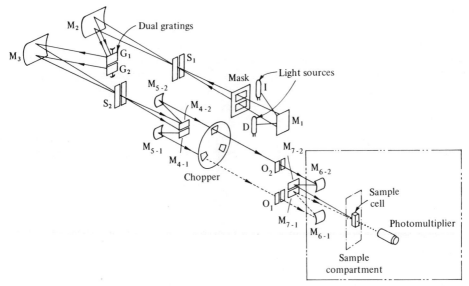

Fig. 3-29. Schematic optical diagram of a double- (dual-) wavelength spectrophotometer. (Courtesy of Perkin-Elmer Corporation.)

Types of double-wavelength measurements include the following:

1. Sample and reference beams are set at different wavelengths (nonscanning). Two wavelengths are chosen such that the absorbance of the interfering component at both wavelengths is identical or their absorbance difference is exactly zero. Of course, the sample wavelength should be a sensitive measure of the test substance. This mode is used in analytical situations where the interfering substance highly overlaps the test substance or when the effect of turbidity must be minimized.
2. Sample and reference beams are set at different wavelengths (nonscanning), and the output at each wavelength is measured independently. This mode is particularly suitable in reaction kinetics where absorbance changes of two species can be monitored simultaneously. When compounds of protein or nucleic acid origin are studied, the relative absorbances at 254 and 280 nm are often used. Protein compounds absorb more at 280 nm than at 254 nm; nucleic acid compounds do just the opposite.
3. The sample beam scans and the reference beam is fixed. This mode is used to determine the spectral characteristics of a highly turbid sample. Light is scattered in a zigzag manner when passing through highly turbid samples; both wavelengths are scattered to the same extent and subject to the same path length in the cuvette.
4. Derivative absorption measurements, made by scanning with the two monochromators operating with a fixed small-wavelength difference between them, is useful in analytical situations where the test substance in a two-component system occurs on the side of the absorption band of the interfering component. If the interference is approximated by a straight line, quantitative analysis of the test substance can be accomplished.

Data Acquisition Systems

Fully automated and semiautomated systems are designed to improve the accuracy and reproducibility of repetitive spectrophotometric determinations. A special cuvette is inserted into the sample compartment. With the use of a vacuum probe pump or positive displacement plunger, a sample is drawn by means of a stainless steel or plastic probe into the cuvette. One, two, or three optional rinses precede the final filling, each followed by emptying the solution from the sample cell into a waste jar. In the manual sample presentation, the operator places the cuvette sampling probe into a sample and presses the fill button on the spectrophotometer to initiate a programmed cycle—the rinsing, reading, and draining sequence. The reading signal is recorded on a strip-chart recorder to resemble a bar graph, or presented in digital form, or presented on a visual readout meter for manual recording. The rigidity of the cuvette minimizes misalignment of cell holders and variations among cuvettes.

Accessories for completely programmed operation enable a series of samples inserted on a turntable to be measured automatically and the readings presented in digital form on tape. This is an excellent timesaver for repetitive analyses and eliminates operator mistakes in obtaining and recording readings.

LABORATORY WORK

Experiment 3-1   Determination of Spectral Response of Photosensitive Detectors

A direct-reading spectrometer will provide everything required. The following directions apply specifically to a Bausch & Lomb Spectronic 20 spectrometer.

Adjust the instrument, photocell dark, with the zero control in the normal manner. Insert into the sample holder a test tube filled with water. Rotate the 100% control until the meter needle reads near midscale. Now rotate the wavelength control until a maximum transmittance reading is achieved (readjusting the 100% control if necessary, to keep the meter needle on scale).

At the wavelength of maximum transmittance, carefully balance the instrument at 100% transmittance. Without changing any controls, henceforth, change the wavelength in intervals of 20 nm and record the corresponding transmittance reading (actually the relative response reading).

Repeat the series of operation with other phototubes. Plot the results on graph paper, plotting wavelength as abscissa. These results represent the overall response of the spectrometer—phototube, emissivity of light source, and intensity diffracted by the grating, each of which is a function of wavelength. Since the main difference lies for the most part with the tungsten source, it would be possible to estimate the relative response of the phototube to light of constant intensity by correcting each observed transmittance reading for the relative radiant power of the tungsten lamp.

Experiment 3-2   Determination of Effective Slit Width of Spectrometers

An atomic line source—a mercury lamp or a flame emission attachment—and a spectrometer, either grating or prism, is required.

Insert the atomic line source in place of the usual continuous light source. Adjust the wavelength scale to the vicinity of one known emission line.

Adjust the slit aperture to a given width, if using a prism instrument; this term will be fixed for a grating instrument.

Carefully adjust the dark current control, then rotate the wavelength drum until a signal is detected. Change the wavelength setting by a small increment, and measure the signal again. Repeat until the signal has risen to a maximum and diminished to zero.

Repeat the operation for additional slit apertures and for different nominal wavelengths.

Prepare your results as a series of graphs, each normalized to the same maximum reading at the nominal wavelength. From the known dispersion of the instrument, or from a graph of dispersion versus wavelength for prism instruments, compute the effective slit width at each nominal wavelength, and compare with the experimental results.

PROBLEMS

1. Show that the violet of the third-order spectrum overlaps the red of the second-order spectrum.

2. Assuming that the limits of the visible spectrum are approximately 380 nm and 700 nm, find the angular breadth of the first-order visible spectrum produced by a plane grating having 780 grooves/mm with the light incident normally on the grating.

3. For each individual plane reflection grating, supply the missing information.

Grating	Grooves/mm	Blaze Wavelength, Å	Blaze Angle, °
A		5000	26.4
B	1180		8.1
C	1180	3000	
D		4000	13.7
E	1180		17.2
F	1180	6000	
G		5000	8.5
H	590		17.2
I		1.0 μm	8.5
J	197	2.6 μm	
K		5.0 μm	21.6
L	74	10.0 μm	

4. Compare the resolution obtainable with a 15.2-cm grating of 1180 grooves/mm to another grating of the same size but with 295 grooves/mm when each grating is used in any of the first four orders.

5. Find the reciprocal linear dispersion, the speed, and the flux transmitting power of the two Ebert monochromators for the following cases: (a) Both have gratings of 590 grooves/mm, 65% efficiency for the 2537 Å line of mercury, ruled areas of 52 × 52 mm which are limited apertures. For 2537 Å, the incident angle will be about 9°. M_2 has a focal length of 1000 mm; M_1 of 500 mm. (b) Assume that M_2 has a grating of 104 × 104 mm, that the aberrations of the two instruments remain the same, and that M_2 has a grating with 1180 grooves/mm and M_1 has a grating with 590 grooves/mm.

6. A description of one commercial monochromator follows: Diffraction grating with 600 grooves/mm, ruled area 30 × 32 mm. Focal length of collimating mirror, 330 mm. Blaze in first order, 500 nm. Dispersion at exit slit (nm/mm), 5 in first order. (a) What is the theoretical resolving power? (b) What is the resolvable wavelength difference? (c) Could the two emission lines of hydrogen, H-alpha (656.28 nm) and D-alpha (656.10 nm), be resolved?

7. A green spectrum line of wavelength 5300 Å is observed as a close doublet. What is the wavelength separation between these two lines if they are just resolved in the third order of a grating ruled 780 grooves/mm, ruled area 64 × 64 mm?

8. The sodium yellow line 5893 Å is actually a doublet of 6 Å peak separation. (a) What is the minimum number of lines that a grating must have to resolve this doublet in the first three orders? (b) What must be the spectral bandwidth to achieve baseline resolution? (c) What slit width is necessary if the monochromator has a reciprocal linear dispersion of 16 Å/mm in the first order?

9. A difficult problem in emission spectroscopy involves the resolution of the lines of gallium (4032.98 Å), potassium (4044.14, 4047.20 Å), and manganese (4030.76, 4033.07, 4034.49 Å) from each other. (a) What is the minimum number of lines that a grating must have to resolve the closest pair of lines? (b) What slit width is necessary if the monochromator has a reciprocal linear dispersion of 16 Å/mm in the first order? (c) Just to resolve the lines of one element from another, what slit width is necessary? Usual fixed slit widths are 100, 50, 25, and 10 μm.

10. Calculate the thickness of dielectric spacer required for the production of interference filters whose nominal wavelength peaks at: (a) 4227 Å, (b) 4551 Å, (c) 5900 Å, and (d) 7670 Å. Assume that the dielectric material is magnesium fluoride (n = 1.38).

11. Assuming ideal bandpass characteristics, would it be feasible to employ an interference filter with a bandpass of 100 Å to isolate the calcium emission line at 4227 Å from the potassium emission line at 4044 Å?

12. An Ebert spectrometer has a mirror of 500 mm focal length and 150 mm in diameter. The plane grating is ruled 1180 grooves/mm over a 52 × 52 mm area. (a) Calculate

the effective aperture ratio, assuming $i = 9°$. (b) What is the theoretical resolution in the first order? (c) What is the reciprocal linear dispersion?

13. If the number of secondary electrons produced at each dynode is 2, what is the total gain for a 9-stage multiplier phototube?

BIBLIOGRAPHY

Bauman, R. P., *Absorption Spectroscopy*, Wiley, New York, 1962.

Clark, G. L., Ed., *The Encyclopedia of Spectroscopy*, Van Nostrand Reinhold, New York, 1960.

Jenkins, F. A. and H. E. White, *Fundamentals of Optics*, 3rd ed., McGraw-Hill, New York, 1957.

Lothian, G. F., *Absorption Spectrophotometry*, 2nd ed., Hilger and Watts, London, 1958.

Lott, P. F., "Recent Instrumentation for UV-Visible Spectrophotometry," *J. Chem. Educ.*, **45**, A89, A169, A273 (1968).

Mellon, M. G., Ed., *Analytical Absorption Spectroscopy*, Wiley, New York, 1950.

Strobel, H. A., *Chemical Instrumentation*, Addison-Wesley, Reading, Mass., 1960.

West, W., *Physical Methods of Organic Chemistry*, 3rd ed., Vol. 1, Part III, A. Weissberger, Ed., Wiley-Interscience, New York, 1960.

LITERATURE CITED

1. Baumeister, P. and G. Pincus, "Optical Interference Coatings," *Sci. Am.*, pp. 59–75 (December 1970).

2. Brice, B. A., *Rev. Sci. Instr.*, **8**, 279 (1937).

3. Jarrell, R. F., in *The Encyclopedia of Spectroscopy*, G. L. Clark, Ed., Van Nostrand Reinhold, New York, 1960, pp. 247–250.

4. Lewin, S. Z., *J. Chem. Educ.*, **42**, A165 (1965).

5. Müller, R. H., *Ind. Eng. Chem., Anal. Ed.*, **11**, 1 (1939).

6. Porro, T. J., "Double-Wavelength Spectroscopy," *Anal. Chem.*, **44** (4), 93A (1972).

7. RCA Corporation, *RCA Photomultiplier Manual—Theory, Design, Application*, Harrison, N.J., 1970.

4

Ultraviolet and Visible
Absorption Methods

THE ELECTROMAGNETIC SPECTRUM

Electromagnetic radiation can be considered as consisting of waves of energy. For each wave the distance from crest to crest (or trough to trough) is called the wavelength, λ. The product of the wavelength and the frequency, ν (the number of cycles per second, in units of hertz, Hz), is the speed of light, c (essentially the speed of light in a vacuum):

$$c = \lambda \nu \tag{4-1}$$

In any material medium the speed of propagation is smaller than this and is given by $nc = 2.9979 \times 10^{10}$ (cm sec^{-1}), where n is the refractive index of the medium.

Radiation is absorbed or emitted only in discrete packets called *photons*. The energy of the photons is proportional to the frequency of the radiation:

$$E = h\nu \tag{4-2}$$

The *intensity* of a beam of radiation is characterized by its radiant power, P or I, which is proportional to the number of photons per second that are propagated in the beam. A beam carrying radiation of only one discrete wavelength is said to be *monochromatic;* a *polychromatic* beam contains radiation of several wavelengths.

The various regions in the electromagnetic spectrum are displayed in Fig. 4-1, along with the nature of the changes brought about by the radiation. Unfortunately, individual areas of spectroscopy were developed rather independently of each other, which has led to the lack of a consistent set of units and of the recognition of similarities between the areas. Visible light represents only a very small part of the electromagnetic spectrum; it is generally considered to extend from 3800 to 7800 Å (380 to 780 nm).

Absorption occurs when a quantum of radiant energy coincides with an allowed transition to a higher energy state of the atom or molecule under study. In the ultraviolet (2100 to 3800 Å), the absorption of light in atoms is caused by orbital shell electron transitions; in molecules, it is caused by electronic transitions associated with "clouds" of electrons taking part in certain types of bonds. Absorption resulting from inner orbital shell electron transitions requires higher energy quanta and thus is measurable only in the vacuum ultraviolet (10 to 2100 Å). Absorption causes objects to be dark and when it varies with wavelength, to have color.

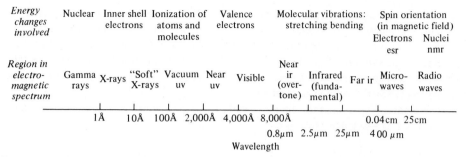

Fig. 4-1. Schematic diagram of electromagnetic spectrum. Note that wavelength scale is nonlinear.

FUNDAMENTAL LAWS OF PHOTOMETRY

The time rate at which energy is transported in a beam of radiant energy is denoted by the symbol P_0 for the incident beam, and by P for the quantity remaining unabsorbed after passage through a sample or container. The ratio of radiant power transmitted by the sample to the radiant power incident on the same is the transmittance, T:

$$T = \frac{P}{P_0} \tag{4-3}$$

The logarithm (base 10) of the reciprocal of the transmittance is the *absorbance, A*:

$$A = -\log_{10} T = \log_{10}\left(\frac{1}{T}\right) \tag{4-4}$$

It is implied that compensation has been made for reflectance losses, solvent absorption loss, refractive effects (if present), and that attenuation by scattering is negligible. Taking the ratio of the intensity transmitted by the sample to the intensity transmitted by the solvent is an exact method for correcting for reflection losses.

The fundamental law underlying the practice of absorption photometry is called *Beer's law* although we know that others made contributions to its formulation. When a beam of monochromatic light, previously rendered plane-parallel, enters an absorbing medium at right angles to the plane, parallel surfaces of the medium, the rate of decrease in radiant power with the length of light path (cuvette interior) b, or with the concentration of absorbing material C (in grams per liter) will follow an exponential progression:

$$T = 10^{-A} = 10^{-abC} \tag{4-5}$$

If a certain thickness of material absorbs half the radiant energy, then an equal thickness of material which follows will absorb half of the remainder, and so on. Doubling the concentration of an absorbing material has the same effect as doubling the path length.

The absorbance is given by

$$A = abC \tag{4-6}$$

where a is the *absorptivity* of the component of interest in solution. The absorptivity is a constant dependent upon the wavelength of the radiation and the nature of the absorbing material. The product of absorptivity and molecular weight is called *molar absorptivity* and is given the symbol ϵ.* The sensitivity of a colorimetric method is governed basically by the molar absorptivity of the color species formed, but can be increased by developing the color in as small a volume as possible.

The term $A_{1\,cm}^{1\%}$ represents the absorbance of a 1-cm layer of solution that contains 1% by weight of absorbing substance.

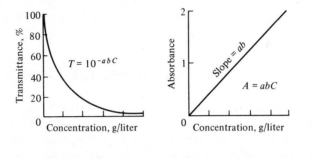

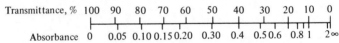

Fig. 4-2. Representation of Beer's law and comparison between scale in absorbance and transmittance.

A plot of absorbance versus concentration will be a straight line passing through the origin, as shown in Fig. 4-2. This is much more convenient than the relationship between transmittance and concentration. For this reason readout meters on spectrophotometers are also calibrated to read absorbance, although the instrument actually measures the light that is transmitted.

Example 4-1

Over what concentration range could analyses be performed for an iron(II) chelate which possesses a molar absorptivity of 12,000 if it is desired to confine the transmittance readings within the range from 0.200 to 0.650? Assume an optical length of 1.00 cm.

Answer The concentration could range from $C_1 = A_1/\epsilon b$ to $C_2 = A_2/\epsilon b$. Conversion of the transmittance readings to corresponding absorbance units gives

$$A_1 = -\log 0.200 = 0.699$$

$$A_2 = -\log 0.650 = 0.187$$

*Molar absorptivity is given in units of liter mole^{-1} cm^{-1}. These units, however, are rarely stated after molar absorptivity values.

from which

$$C_1 = \frac{0.699}{(12{,}000)\,(1.00)} = 5.83 \times 10^{-5} \text{ mole/liter}$$

$$C_2 = \frac{0.186}{(12{,}000)\,(1.00)} = 1.55 \times 10^{-5} \text{ mole/liter}$$

For iron, the limits would extend from 1.04 to 3.26 mg/liter.

The molar absorptivity is governed largely by the probability of the electronic transition and the polarity of the excited state. In order for interaction to take place, a photon must obviously strike a molecule approximately within the space of the molecular dimensions, and the *transition probability*, g, will be the proportion of target hits which lead to absorption. Thus,

$$-\frac{dP}{P} = \frac{1}{3} g C N_A \, A \, \frac{db}{1000} \qquad (4\text{-}7)$$

where N_A is the Avogadro number, A is the cross-sectional target area, and $1/3$ is a statistical factor to allow for random orientation. On integration and insertion of numerical constants,

$$\frac{\log\,(P_0/P)}{bC} = \epsilon = (0.87 \times 10^{20}) g A \qquad (4\text{-}8)$$

For many organic molecules the cross-sectional area is about 10 Å, so that for a transition of unit probability, $\epsilon \simeq 10^5$. The highest molar absorptivities observed are of this order. Absorption with $\epsilon > 10^4$ is considered high-intensity absorption.

Sources of Error

It is necessary to establish the relationship between the absorbance and concentration, and to determine the range over which this relationship may be considered linear in calculations. In most analyses where the absorption band (or emission line) is completely resolved, there will be a linear relationship between the measured absorbance and the concentration. In analyses where the absorption band is not completely resolved, or where the state of the absorbing component changes with concentration, the relationship between absorbance and concentration may be nonlinear. Such a curve is still useful in quantitative analysis, but the concentration of such a material must be read from a standard curve that must be verified at frequent intervals.

Discrepancies are usually found when the absorbing solute dissociates or associates in solution, since the nature of the species in solution will vary with the concentration. Absorbance readings taken at an *isosbestic point** usually circumvent difficulties when the

---

*The term is applied to any wavelength where the molar absorptivity is the same for two materials that are interconvertible without regard to the equilibrium position of the reaction between them.

absorbing species is part of an equilibrium system. Indicator systems of weak acids or bases are shown in Fig. 4-19. Potassium dichromate solutions involve the chromate ion-dichromate ion equilibrium:

$$2CrO_4^{2-} + 2H^+ \rightleftharpoons Cr_2O_7^{2-} + H_2O$$

$$(\lambda_{max}, 375 \text{ nm}) \quad (\lambda_{max}, 350, 450 \text{ nm}) \tag{4-9}$$

At relatively small hydrogen-ion concentrations, the equilibrium is shifted by increasing the pH, upon dilution, or by changing total chromium concentration. In solutions of high hydrogen-ion concentration, Beer's law is obeyed.

Temperature often shifts ionic equilibria and, in addition, an increase in temperature shifts absorption bands to longer wavelengths.

In the derivation of Beer's law the use of a beam of monochromatic radiation is implied. However, most spectrophotometers and all filter photometers employ a finite group of frequencies. The wider the bandwidth of radiation passed by the filter or other dispersing device, the greater will be the apparent deviation of a system from adherence to Beer's law. Often the deviation becomes evident at higher concentrations on an absorbance versus concentration plot when the curve bends towards the concentration axis. Fundamentally, this departure arises because in all photometers it is the radiant power of the component wavelengths which are additive (or nearly so), whereas Beer's law requires that the logarithms be additive. Only when the absorption curve is essentially flat over the spectral bandwidths employed can Beer's law be expected to apply. Lack of adherence to Beer's law in the negative direction is always undesirable because of the rather large increase in relative error of reported concentrations.

Kortum and Seiler[13] point out that Beer's law is only a limiting law and should be expected to apply only at low concentrations. It is not a or ϵ, which is constant, independent of concentration, but $an/(n^2 + 2)^2$, where n is the refractive index of the solution. At low concentrations ($< 10^{-3} M$), n is practically constant, but at higher concentrations the refractive index may vary appreciably. This effect may be encountered in high-absorbance differential spectrophotometry.

Stray light which enters the detector is another source of error. Although corrections may be made for it, estimation of stray radiation is always subject to some uncertainty.[19] Because the photometer measures the ratio of the light which has passed through the sample to that incident upon it, absorption of stray light by the sample will reduce the numerator and make the ratio too small; the absorbance value, therefore, will be too large. If the sample does not absorb the stray radiation, both the numerator and the denominator will be increased by the same amount; the ratio will be increased and the absorbance value reduced. The latter case is the one most frequently encountered in analytical chemistry. Elimination by a double monochromator or by suitable filters is the only way to minimize this error. Scattered light from suspensions, or fluorescence, may also cause deviations from Beer's law.

The passage of part of the incident radiation more than once through the sample, caused by repeated reflections by the front and back cuvette surfaces, results in too high an absorbance value. This is most likely to occur in measuring solutions of very low absorbance. A convergence error arises from rays passing through the sample at an angle and thus traveling a greater distance through the sample than those traveling parallel to the optical axis. This error can be particularly important when the beam is strongly convergent as in beam-condensing units and micro-illuminators.

PRESENTATION OF SPECTRAL DATA

Molecular absorption spectra are displayed in several ways, as shown in Fig. 4-3. Increasing values of molar absorptivity, absorbance, or transmittance as ordinate may be plotted on a linear or logarithmic scale against increasing values of wavelength (in nanometers or angstrom units) or decreasing values of frequency (in cm^{-1}). The transmittance scale is not very useful except, perhaps, for characterization of filters and solvents. The frequency scale possesses the theoretical advantage of being a linear scale, and from a graph of frequency versus wavelength, an integrated absorption value of the oscillator strength can be calculated from the area under the absorption band. A plot of absorbance versus wavelength is generally the most useful presentation. Because the path length is fixed experimentally, and because the molar absorptivity is a function of wavelength, the concentration becomes directly proportional to absorbance at any specific wavelength.

To facilitate comparison of spectra in qualitative identification, it is advantageous to plot the logarithm of absorbance versus wavelength. Each curve is displaced along the log A axis by the amount of log bC, thus the sample concentration and cell path appear only as a fixed displacement of the recorded curve. The curve of log A versus wavelength can be recorded directly with an auxiliary recorder having a linear logarithmic scale. Reference files of these recordings facilitate the identification of unknown compounds. A log

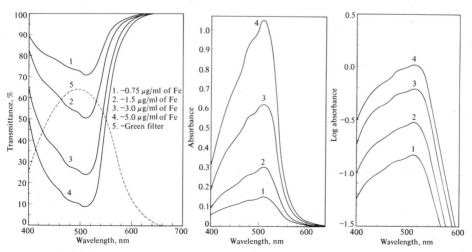

Fig. 4-3. Presentation of spectral data for the iron(II)-1,10-phenanthroline system (1-cm cuvettes). Curve 5 is the spectral transmittance plot for a green filter which could be used in a filter photometer in measuring this colored system.

A recording is also a valuable tool in studies of the kinetics of first-order reactions. With time-drive accessories, one can obtain at a fixed wavelength curves of concentration versus time whose slope is equal to the reaction rate constant.

QUANTITATIVE METHODOLOGY

The basic procedure of most quantitative absorption methods consists in comparing the extent of absorption of radiant energy at a particular wavelength by a solution of the test material with a series of standard solutions. Work with visual comparators, although requiring simple equipment, is subject to the vagaries of the human eye, in particular fatigue and unavoidable low sensitivity under 450 and above 675 nm. Precision is always less than that attainable with photoelectric instruments except, perhaps, when dealing with very weak colors. Filter photometers are suitable for many routine methods that do not involve complex spectra. Precise work is done with a spectrophotometer which is able to employ narrow bandwidths of radiant energy and which can handle absorption spectra in the ultraviolet region if equipped with fused silica optics.

The limitations of many colorimetric procedures lie in the chemical reactions upon which these procedures are based rather than upon the instruments available. Many instances arise when a specimen does not possess suitable chromogenic properties; sometimes it may be converted to an absorbing species or be made to react with an absorbing reagent. Formation of metal–organic complexes is well known; for example, since the iron(II) ion is very weakly colored, a complexing agent, 1,10-phenanthroline, is added to form an ion-association species which is suitable for the determination of very small amounts of iron. A few moments of reflection will bring to mind a number of possibilities among organic compounds. For example, although alcohols possess no absorption spectra between 200 and 1000 nm, treatment of an alcohol with phenyl isocyanate yields the corresponding phenyl alkyl carbamate which absorbs at about 280 nm. Semicarbazones display maxima which are shifted to longer wavelengths by 30 to 40 nm, with an average increase in molar absorptivity of 10,000 compared with the original carbonyl compound. Conversely, the strong absorption of anthracene can be eliminated by a Diels-Alder reaction with 1,2-dicyanoethylene.

Although very few reactions are specific for a particular substance, many reactions are quite selective, or can be rendered selective through the introduction of masking agents, control of pH, use of solvent extraction techniques, adjustment of oxidation state, or by prior removal of interferents.[5] Both the color-developing reagent and the absorbing product must be stable for a reasonable period of time. It is often necessary to specify that the color comparisons be made within a definite period of time, and it is always advisable to prepare standards and unknowns on a definite time schedule. When extraneous color bodies are present, the standards should match the composition of the sample solution. Adherence to Beer's law is desirable, for then the absorbance is directly proportional to concentration and only a few points are required to establish the calibration curve. In any event the standard curve should be checked at frequent intervals. High ionic strength of the medium, appreciable temperature variations, and use of polychromatic radiation can cause departure of the calibration curve from linearity.

Cutoff

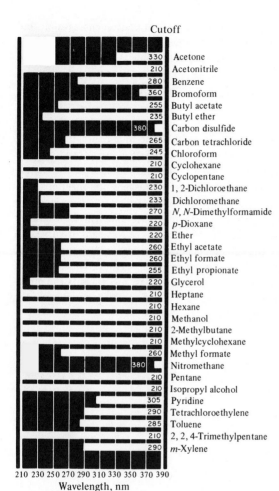

	Cutoff	
	330	Acetone
	210	Acetonitrile
	280	Benzene
	360	Bromoform
	255	Butyl acetate
	235	Butyl ether
	380	Carbon disulfide
	265	Carbon tetrachloride
	245	Chloroform
	210	Cyclohexane
	210	Cyclopentane
	230	1, 2-Dichloroethane
	233	Dichloromethane
	270	N, N-Dimethylformamide
	220	p-Dioxane
	220	Ether
	260	Ethyl acetate
	260	Ethyl formate
	255	Ethyl propionate
	220	Glycerol
	210	Heptane
	210	Hexane
	210	Methanol
	210	2-Methylbutane
	210	Methylcyclohexane
	260	Methyl formate
	380	Nitromethane
	210	Pentane
	210	Isopropyl alcohol
	305	Pyridine
	290	Tetrachloroethylene
	285	Toluene
	210	2, 2, 4-Trimethylpentane
	290	m-Xylene

210 230 250 270 290 310 330 350 370 390
Wavelength, nm

Fig. 4-4. Transmittance of selected solvents in the ultraviolet region. The cutoff point in the ultraviolet region is the wavelength at which the absorbance approaches unity using a 1-cm cell path with water as the reference. (Courtesy of Matheson, Coleman & Bell.)

Water is a common solvent for many inorganic substances. The ultraviolet transmittance characteristics and cutoff wavelengths of a number of "spectro-quality solvents" for organic substances are shown in Fig. 4-4. Cyclohexane is a desirable solvent for aromatic compounds. When a more polar solvent is required, 95% ethanol is a good choice.

Choice of Wavelength

The wavelength selected for a particular assay is chosen so that the material of interest will absorb light at this wavelength, and so that the absorption will be as little affected as possible by interfering substances or variations in the procedure. If the material has a characteristic visual color, its complementary color may indicate the proper wavelength region. When filter photometers are to be employed, the proper filter can be selected during the course of preparing the calibration curve. A series of standard solutions is prepared, including a blank. Using one filter at a time, a series of calibration curves is plotted in terms of absorbance versus concentration. The filter which permits closest adherence to linearity over the widest absorbance interval and which yields the largest slope (but with a small or zero intercept) will constitute the best choice. If a spectrophotometer is available, the wavelength of maximum absorbance is quickly ascertained from the absorbance-wavelength curve.

Dilemmas arise in practice. Consider the absorption curves shown in Fig. 4-5. The unreacted reagent and the metal complex absorb strongly in the blue region of the spectrum and, although this region would yield the steepest calibration curve (and greatest sensitivity in analysis), the choice of wavelength must be on the shoulder of the metal-complex absorption curve at about 500 nm where the unreacted reagent no longer absorbs. If the unreacted reagent could somehow be removed from the system, the wavelength at 420 nm could be used. When no linear portion free from interference can be found, it may be necessary to work on a curved portion of the absorbance-wavelength curve. In this case Beer's law will not hold unless the isolated spectral band is quite narrow. There is no real objection to operation on a steep part of an absorption curve, provided that the usual standard calibration curve is obtained, except that with most instruments the reproducibility of the absorbance readings will be poor unless a fixed wavelength setting of the monochromator is maintained or unless a filter is used.

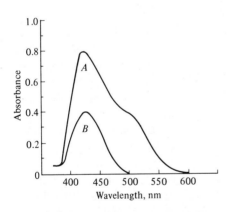

Fig. 4-5. Spectral data for A, the cobalt complex, and B, 1-nitroso-2-naphthol-3,6-sulfonic acid—reagent blank.

With spectrometers generally, and filter photometers in particular, the spectral response of the detector and the spectral distribution curve of the light source, as well as the absorption curve of the sample and any filter employed, must be considered. For example, Tobias acid has an absorption band around 350 nm. A red color is developed upon coupling to β-naphthol. The red substance exhibits a maximum at 500 nm. Molar absorptivities are equal. Whether one should utilize the self color or the developed color is answered by considering other interferences and whether an ultraviolet spectrophotometer is available. As another example, the silica-molybdenum blue complex exhibits an absorption maximum at 820 nm and an absorption plateau around 630 nm, whose absorbance is about one-half the value observed at the absorption maximum. With instruments equipped with a red-sensitive phototube, the choice is obviously the maximum at 820 nm. But with filter photometers equipped with a barrier-layer cell, the measurements should be made with a filter whose nominal wavelength lies close to the plateau around 630 nm.

For systems sensitive to pH, and for which an isosbestic point can be found, measurements at the wavelength of the latter are advised when the pH is not, or cannot, be controlled.

The purity of the radiant energy used in absorption measurements may profoundly affect the results, especially the absorption curves with narrow absorption bands. The importance of good resolution is shown by the tracings of a portion of the didymium glass absorption curve (Fig. 4-6). As long as the actual width of absorption bands (natural bandwidth) is considerably greater than the wavelength interval included by the image of the exit slit, molar absorptivity will not be affected. On the other hand, if a measurement is made at a sharp maximum using a slit width which accepts wavelengths on either side of the maximum, the measured absorbance will be smaller than the true value, and the converse effect will be found at a minimum or in the vicinity of a shoulder.

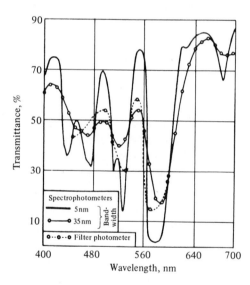

Fig. 4-6. Resolution of the absorption spectrum of a didymium glass at different band widths.

Simultaneous Photometric Determinations

When no region can be found free from overlapping spectra of two different absorbing groups, it is still possible to devise a method based on measurements at two wavelengths. Two dissimilar absorbing groups must necessarily have different powers of light absorption at some point or points in the absorption spectrum. If, therefore, measurements are made on each solution at two such points, a pair of simultaneous equations may be obtained from which the two unknown concentrations may be determined. For best

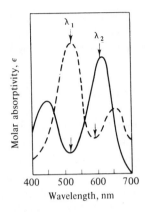

Fig. 4-7. Simultaneous spectrophotometric analysis of a two-component system. Selection of analytical wavelengths indicated by arrows.

precision it is desirable to select two points on the wavelength scale where the ratio of the molar absorptivities are maxima. For the system illustrated in Fig. 4-7, $(\epsilon_1/\epsilon_2)_{\lambda_1}$ and $(\epsilon_2/\epsilon_1)_{\lambda_2}$ are maxima. Neither of these wavelengths need necessarily coincide with an absorption maximum for either component. Next, it is necessary to calculate the molar absorptivity for each component using a particular set of sample containers and the instrument to be employed. Since absorbance is directly proportional to the product of molar absorptivity and concentration, if the light path remains constant, two simultaneous equations may be written,

$$C_1(\epsilon_1)_{\lambda_1} + C_2(\epsilon_2)_{\lambda_1} = A_{\lambda_1} \tag{4-10}$$

$$C_1(\epsilon_1)_{\lambda_2} + C_2(\epsilon_2)_{\lambda_2} = A_{\lambda_2} \tag{4-11}$$

and solved for the concentration of each component. Simultaneous determinations rest on the assumption that the substances concerned contribute additively to the total absorbance at an analytical wavelength. This assumption should be tested with known mixtures of the test materials.

Relative Concentration Error

There is interest in the conditions under which the maximum precision in the unknown concentration can be achieved. This precision can often be maximized by judicious choice of such experimental parameters as sample thickness, concentration, or gas pressure. Hence a quantitative analysis should be conducted at that value (or within that range) of transmittance for which a given error or uncertainty in transmittance, ΔT, will cause the least uncertainty in concentration, ΔC. Instruments with thermal detectors (thermocouple, bolometer, or Golay cell) will show optimum precision in one range, and spectrometers with photoemissive detectors will perform best in a somewhat different range. Uncertainty of the transmittance setting of most instruments will be on the order of 0.01 to 0.002 of the total scale; the latter value is considered a practical limit in ordinary work. The uncertainty may be evaluated experimentally by taking a standard solution and measuring its transmittance repeatedly. Each measurement should include the operations of emptying, refilling, and repositioning the cuvette. Usually ΔT is taken as twice the average deviation of the replicate readings in order to include the uncertainty involved in setting the scale to 0% and 100% T.

Assume first that the noise is constant at any value of T; this situation corresponds to an instrument with a thermal detector which is noise limited, or instruments whose amplifier noise is greater than the noise from the shot effect in the photoemissive tube. Rewriting Beer's law,

$$C = - \frac{\log T}{ab} \tag{4-12}$$

and differentiating,

$$\frac{dC}{dT} = \frac{-0.4343}{T(ab)} \tag{4-13}$$

Replacing the term ab by its equivalent, and rearranging, gives

$$\frac{dC/C}{dT} = \frac{0.4343}{T \log T} \simeq \frac{\Delta C/C}{\Delta T} \tag{4-14}$$

The optimum transmittance occurs at $1/e$ or 36.8% which may be shown by differentiating Eq. 4-14 and setting the derivative equal to zero, and the minimum error is

$$\frac{dC}{C} = -e \, dT = -2.72 \, dT \tag{4-15}$$

Thus, a 0.1% error in T produces a 0.27% error in C. As shown in Fig. 4-8, the plot of relative concentration error as a function of absorbance is rather flat between 0.2 and 0.7 so that careful adjustment to 0.43 is of little value.

When shot noise predominates, the net noise includes contributions from both reference and sample channels and is given by

$$N = \sqrt{N_{\text{ref}}^2 + N_{\text{sple}}^2} \tag{4-16}$$

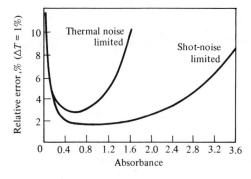

Fig. 4-8. Relative   concentration   error, $\Delta C/C$, in percent, for a constant transmittance error.

In the sample channel the noise varies with the square root of the light level, which decreases with the transmittance of the sample; hence,

$$N_{\text{sple}} = N_1 \sqrt{T} \tag{4-17}$$

where

$$N_1 = N_{\text{ref}} \sqrt{2} = N_{\text{sple}} \sqrt{2} \tag{4-18}$$

when the setting at 100% T is made. In the reference channel where $T = 1.00$ in all cases, the noise appears across the measurement slidewire and is attenuated proportional to the contactor position which is a direct measure of T:

$$N_{\text{ref}} = N_1 T \tag{4-19}$$

The net noise, in units of transmittance, is equal to the transmittance uncertainty:

$$N = N_1 \sqrt{T^2 + T} = \Delta T \tag{4-20}$$

Differentiating Beer's law, written as $2.303\, A = -\ln T$,

$$dA = -0.4343 \frac{dT}{T} \tag{4-21}$$

The relative concentration error is then given by

$$\frac{dC}{C} = \frac{dA}{A} = \frac{-0.4343\, N_1}{A \sqrt{1 + 1/T}} \tag{4-22}$$

The minimum error occurs approximately at an absorbance of $2/e$ or 0.868 ($T = 0.135$). The plot of relative concentration error as a function of transmittance is quite flat from 0.01 to 0.50. A shot-noise limited instrument has the tremendous advantage of providing better precision over a much wider transmittance range. By working at higher absorbance, the effects of errors in cell matching and window cleanliness are less significant.

Differential Spectrophotometry

The greater the incident radiant energy, the greater will be the measured intensity for a given value of concentration and hence absorbance, and thus the smaller will be the relative concentration error. Under normal operating procedures the maximum value of P_0 is limited by the length of the potentiometer slidewire, since one end corresponds to zero intensity and the other end to the full intensity at zero concentration. By setting the full-scale deflection with a strong absorber in the light beam, the region of interest can be fitted onto the scale. The absorber can be a sample slightly less concentrated than that to be measured, or a neutral-density filter or a wire screen.

To compensate for the lesser amount of transmitted energy reaching the detector, the photometer is brought to a balance with the absorbance scale set at zero by increasing the slit width, or by increasing the source intensity, or by increasing the amplifier-phototube output (without, however, increasing the noise). Not all instruments possess these pro-

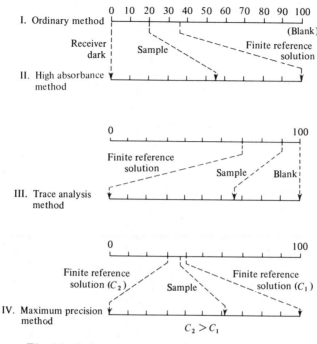

Fig. 4-9. Relative absorbance spectrophotometry.

visions for reserve sensitivity. Under these conditions, the actual scale is effectively lengthened, although there is no change in its physical dimensions. Thus, if a sample lies in the 0–36% transmittance range, this range is made to cover 0–100% transmittance by using a standard with 36% transmittance to make the 100% setting (Fig. 4-9).

In differential spectrophotometry the relative error curve is altered and becomes finite at zero absorbance (see Fig. 4-10). The relative concentration error becomes dependent in a pronounced manner on the actual absorbance of the comparison standard, and is given by

$$\frac{dC/C}{dT} = \frac{0.4343}{(TA)_{\text{sple}} + A_{\text{ref}}} \qquad (4\text{-}23)$$

where A_{ref} is the absorbance of the reference solution used for setting the instrument to zero absorbance. As the actual concentration of reference absorber increases, the position of minimum error gradually shifts to zero absorbance. This amounts to comparing standard and unknown at the same scale reading, namely zero absorbance.

Example 4-2

Table 4-1 shows the sequence and results for differential measurements of nickel(II) ion at 720 nm.[3] Each weight of nickel, in increments of 0.2 g per 100 ml, was read against the lower weight beginning with pure solvent; the absorbance

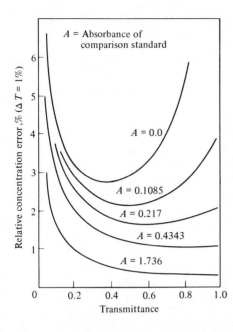

Fig. 4.10. A plot of the relative error coefficient. (After C. F. Hiskey, *Anal. Chem.*, **21**, 1440 (1949). Courtesy of *Analytical Chemistry*.)

readings are shown in column 4. Assuming that all segments of the calibration curve are linear over the range covered, the slopes are given in column 5. The product of the slope and the weight in the reference standard are given in column 6. These increase rapidly to a maximum, then decrease slowly. There is always a gain in precision until the deviation from Beer's law gives a slope which falls off more rapidly than the concentration of the reference standard increases; this occurs when the weight of the reference standard exceeds 0.8 g per 100 ml.

The optical path length of the cuvettes must be known with a precision equal to the best precision expected for the differential method, or else one cuvette must be used for

TABLE 4-1   Differential Spectrophotometry

Nickel, g/100 ml		Slit Aperture, mm	Absorbance	Slope Beer's Law Plot	Product of Slope and Weight Nickel to Set Scale Zero
To Set Scale Zero	To Obtain Reading				
0.0000	0.2002	0.034	0.735	3.67	—
0.2002	0.4002	0.078	0.714	3.57	0.715
0.4002	0.6006	0.187	0.711	3.55	1.42
0.6006	0.8000	0.439	0.695	3.47	2.09
0.8000	1.0003	0.864	0.561	2.80	2.24
1.0003	1.2000	1.26	0.405	2.03	2.03
1.2000	1.4000	1.52	0.341	1.71	2.05
1.4000	1.6005	1.69	0.278	1.39	1.95

all absorbance measurements. To minimize volumetric errors, aliquots should be taken by weight. For interesting discussions of the differential method and a rigorous treatment of the error function, the reader is referred to papers by Bastian,[1,2] Hiskey,[9,10] and Reilley and Crawford.[4,17]

In the *trace analysis procedure*,[4,17] the absorbance scale is set to zero with the blank in the usual manner, but the zero energy is "faked" by placing in the beam a standard or screen which actually transmits a significant fraction of the incident radiation, but less than that of the most concentrated unknown solution. The effective length of the trans-mittance scale is again increased, this time favoring low absorbing systems. For example, if a sample series falls in the 70-100% transmittance range, this range can be made to cover 0-100 scale units by using the standard with 70% transmittance to make the zero transmittance setting. It is now possible to increase the gain, or the slit width may be increased, to move the P_0 level to the 100% transmittance position. The error at the low transmittance end of the scale now becomes finite. The relative concentration error becomes

$$\frac{dC/C}{dT} = \frac{0.4343(1 - T_0)}{T \log T} \tag{4-24}$$

where T_0 is the actual transmittance of the reference solution used for setting the zero transmittance reading. There is a limitation to the trace analysis method which is related to the balancing voltage attainable from the "zero" or "dark current" control. In using this method, one is forced to construct a calibration curve, since the increase in sensitivity was achieved at the expense of a positive deviation from Beer's law.

These ideas can be carried one step further in the *maximum precision* method. Both ends of the reading scale are defined by using finite reference solutions closely spaced around a transmittance value somewhere on the flat portion of the relative concentration error curve. For a component present at a level of $C \pm x$ g/liter, reference solutions for the ends of the reading scale should contain the desired component in the concentration of $C + 2x$ and $C - 2x$ for setting the 0 and 100% transmittance readings, respectively. The relative concentration error is then given by

$$\frac{dC/C}{dT} = \frac{0.4343(T_{ref} - T_0)}{T \log T} \tag{4-25}$$

Photometric Titrations[7,8]

The change in absorbance of a solution may be used to follow the change in concentration of a light-absorbing constituent during a titration. The absorbance is linearly proportional to the concentration of absorbing constituent rather than logarithmically as in potentiometric methods. This means that in a titration in which the titrant, the reactant, or a reaction product absorbs, the plot of absorbance versus titrant will consist, if the reaction be complete, of two straight lines intersecting at the end point—similar to amperometric and conductometric titrations. For reactions that are appreciably incomplete, extrapolation of the two linear segments of the titration curve establishes the intersection and

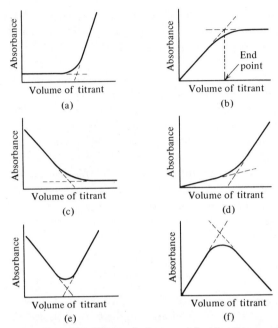

Fig. 4-11. Possible shapes of photometric titration curves.

end-point volume. Possible shapes of photometric titration curves are shown in Fig. 4-11. Curve (a), for example, is typical of the titration where the titrant alone absorbs, as in the titration of arsenic(III) with bromate-bromide, where the absorbance readings are taken at the wavelength where the bromine absorbs. Curve (b) is characteristic of systems where the product of the reaction absorbs, as in the titration of copper(II) with EDTA. When the analyte is converted to a nonabsorbing product—for example, titration of p-toluidine in butanol with perchloric acid at 290 nm—curve (c) results. When a colored analyte is converted to a colorless product by a colored titrant—for example, bromination of a red dyestuff—curves similar to (e) are obtained. Curves (d) and (f) might represent the successive addition of ligands to form two successive complexes of different absorptivity.

Photometric titrations have several distinct advantages over a direct photometric determination. The presence of other absorbing species at the analytical wavelength, as in curve (a) of Fig. 4-11, does not necessarily cause interference, since only the change in absorbance is significant. However, the absorbance of nontitratable components (color or turbidity) must not be intense because, if so, the absorbance readings will be limited to the undesirable upper end of the scale unless the slit width or amplifier gain can be increased. Only a single absorber needs to be present from among the reactant, the titrant, or the reaction products. This extends photometric methods to a large number of nonabsorbing constituents. Precision of 0.5% or better is attainable because a number of pieces of information are pooled in constructing the segments of the titration curve.

The analytical wavelength is selected on the basis of two considerations: (1) avoidance of interference by other absorbing substances and (2) need for a molar absorptivity which will cause the change in absorbance during the titration to fall within a convenient range. Often the chosen wavelength lies well apart from an absorption maximum.

Volume change is seldom negligible, and straight lines are obtained only if correction is made. This is done simply by multiplying the measured absorbance by the factor $(V + v)/V$ where V is the volume initially and v is the volume of titrant added up to any point. If the correction is not made, the lines are curved down toward the volume axis and erroneous intersections are obtained. Use of a micro-syringe and relatively concentrated titrant is desirable. Stray-light error also affects the linearity of the titration curve. The upper limit of concentration permissible is found by delivering from the buret into a

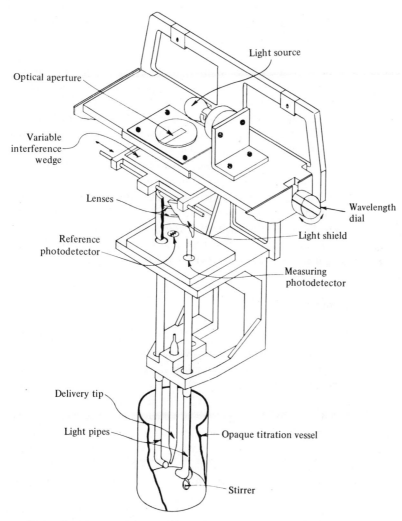

Fig. 4-12. The Fisher Titralyzer. (Courtesy of Fisher Scientific Co.)

beaker of transparent liquid measured portions of a colored substance known to obey Beer's law. After correcting for dilution, the plot of absorbance versus concentration will be a straight line up to the absorbance value where the stray-light error becomes detectable.

Areas of particular applicability are for solutions so dilute that the indicator blank is excessive by other methods, or when the color change is not sharp due perhaps to titration reactions which are incomplete in the vicinity of the equivalence point, or when extraneous colored materials are present in the sample. Ordinarily there is no difficulty in working in solutions of either high or low ionic strength or in nonaqueous solvents. One of the attractive features of photometric titrations is the ease with which the sensitivity of measurements can be changed, simply by changing the wavelength or the length of the light path. When self-indicating systems are lacking, an indicator can be deliberately added, but in relatively large amount to provide a sufficient linear segment on the titration curve beyond the equivalence point.

All one needs to carry out photometric titrations in the visible region is a light source, a series of narrow bandpass filters, a titration vessel (which can be an ordinary beaker), a receptor, and a buret or other titrant delivery unit. The entire assembly is housed in a light-tight compartment. Photometers or spectrophotometers with provision for inclusion of a suitable titration vessel from 5- to 100-ml capacity are suitable. It is imperative that the titration vessel remain stationary throughout the titration. By the use of Vycor beakers and an appropriate spectrophotometer, titrations may be conducted in the ultraviolet region. Provision for magnetic stirring from underneath or some type of overhead stirrer is desirable; otherwise, manual agitation after the addition of each increment of titrant is necessary. For the transmission of photometric end points, the ends of two fiber optic light pipes can be located facing each other across an area below the buret tip in the titrating vessel. One pipe conducts dispersed light to the sample, and the other pipe carries the transmitted light to the photodetector (Fig. 4-12).

Indirect Spectrophotometry

In indirect spectrophotometric methods of the bleaching type, the element in question bleaches or lowers the absorbance of a given reagent. Since the absorbance in the direct spectrophotometric method cannot go much above 1.4 without incurring a rapidly increasing relative concentration error, the precision is limited. There are no such limitations on the absorbance which can be bleached. The relative concentration error is now given by $dC/C = dA/(A_2 - A_1)$. As long as one is willing to increase the concentration of the element doing the bleaching and the reagent being bleached (A_2), and the absorbance of the reaction products is low, the denominator ($A_2 - A_1$) can be made very small. Here A_1 is the residual absorbance after the reaction is completed. Thus, if one bleaches 99% of the reagent and measures the residual reagent with a precision of 1%, the overall precision is also about 1%.

Indirect procedures increase the number of useful reagents. The strain is taken off the photometry and the errors lie mainly in the reproducibility of the chemical reactions involved. The advantage over photometric titrations is that an end point is not sought, so that errors in locating the end point are eliminated. There is no instrumental limit on how high the absorbance of the unbleached solution is because it is not necessary to make any spectrophotometric measurements on it. This is an important advantage over differential spectrophotometry.

REFLECTANCE MEASUREMENTS

Reflection occurs whenever a ray of light encounters a boundary between two media. The light reflected from the first surface of contact is called the *specular* (gloss, sheen) component. These encounters are repeated over and over again in granular or fibrous structures where a light beam will encounter a new interface every few millionths of a centimeter. These repeated encounters result in thorough diffusion such that the surface tends to appear uniformly bright in all directions—this reflected light is called the *diffuse* component and it is responsible for color where color exists. Particle size plays an important role. Whiteness and reflectance gain as the diameter is reduced to about one-half of the wavelength of light. At very small particle diameters ($< \lambda/4$), scattering takes over and diffuse reflection falls off.

Color[11]

Color may be represented in terms of tristimulus values for red, blue, and green spectral colors (CIE system of XYZ coordinates); of green, amber, and blue reflectances (GAB system); of R_d, a, and b or its closely related system: L, a_L, and b_L. With these scales it is possible to represent colors by position in a three-coordinate system (Fig. 4-13). The diffuse reflectance term, R_d, is the percentage of light reflected by the sample relative to that reflected by a magnesium oxide standard; it is equal to the value Y (CIE system) and G (GAB system). The term L measures lightness and corresponds closely to visual estimates of this quantity: $L = 10\sqrt{Y} = 10\sqrt{R_d}$. In the rectangular diagram, a and b are the chromaticity coordinates; the vertical axis is the L value. If a sample has zero value for a and b, it is black, gray, or white, depending on the value of L. A plus value for a indicates

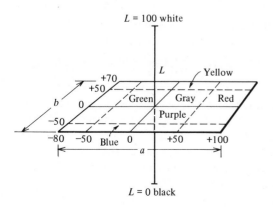

Fig. 4-13. The L, a, b system in which the colors of the specimens may be specified and visually interpreted. (Courtesy of Hunter Associates Laboratory, Inc.)

redness; a minus value, greenness. A plus value for b indicates yellowness; a minus value, blueness. When the information being sought is color difference rather than the colors themselves, the color difference is computed as follows:

$$\text{color difference} = \sqrt{(\Delta L)^2 + (\Delta a)^2 + (\Delta b)^2} \qquad (4\text{-}26)$$

Reflectance curves of color materials may look somewhat like a transmittance curve (Fig. 4-14). A perfectly white sample would reflect all of the light, and its curve would be a horizontal line at 100%. A theoretically perfect black sample would absorb all of the light and its curve would be a horizontal line at zero. A neutral gray would give a horizontal line, its position on the plot depending upon the depth of shade. The brighter a color, the more nearly vertical the reflectance band, while duller shades will produce flatter curves (approaching a gray).

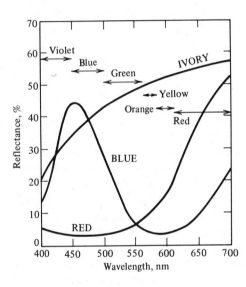

Fig. 4-14. Spectral reflectance curves for selected colors.

Gloss

Gloss, or *specular reflection*, may be defined as the degree to which a surface possesses the light-reflecting property of a perfect mirror. A perfectly diffusing surface, even under unidirectional illumination, has a constant luminance regardless of the angle from which it is viewed—this corresponds to a *matte* or zero gloss (Fig. 4-15). Surfaces with intermediate properties between a matte and a perfect mirror can have their characteristics measured with a *goniophotometer*. The gloss of any surface increases as the angle of incidence increases; that is, as the incident ray departs from a line perpendicular to the surface. Gloss classification over a wide range from high to low is best made with the incidence angle and aperture of a 60° instrument. Goniophotometric curves for gloss, analogous to spectrophotometric curves for color, have maxima in the direction of mirror reflection; the narrower and higher the peak, the higher the gloss.

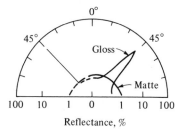

100   10   1   0   1   10   100

Reflectance, %

Fig. 4-15. Goniophotometric curves. (Courtesy of Hunter Associates Laboratory, Inc.)

Instruments[20]

In diffuse reflectance the radiation reflected from the surface of a sample is detected and recorded as a function of wavelength. Instruments fall into three categories: *reflectometer*, *spectroreflectometer*, or *colorimeter*. A reflectometer employs a series of filters to obtain approximately monochromatic radiation. The spectroreflectometer embodies a monochromator. The colorimeter employs three filter-photocell combinations, each with its own readout dial calibrated for the color coordinate values of the system selected. Attachments for making reflectance measurements are available for many standard spectrophotometers.

Optical arrangements for reflectance measurements are of three types: (1) integrating sphere type; (2) annular, ellipsoidal mirror type; and (3) reflection type. In the integrating sphere type (Fig. 4-16), monochromatic radiation enters the sphere through a side aperture, strikes the sample, the reflected radiation is collected by the sphere which is coated on the interior with magnesium oxide, and then measured by a photodetector. Sample and reference material are pressed against the measuring (and reference) aperture by means of a spring-loaded disk. The specular component of the reflected radiation may be rejected by using light traps of black velveteen located at an angle of 90° to the sample and reference materials.

In the second type, the sample is illuminated at a zero angle of incidence, the reflected radiation from the sample is collected by an annular, ellipsoidal mirror, and then detected by the photodetector. A screen attenuator is sometimes located in the reference beam to permit the reference beam to be standardized against a reference sample.

The third type is a simple arrangement where the reflected radiation is detected by a photodetector without any type of collector device being used. Radiation from the slit and field lens strikes the sample surface at an angle of 45° to the perpendicular. The reflected radiation at 90° to the surface of the sample is measured. Reflected light is distributed to three phototubes, each having a different tristimulus filter in front of it.

In a *gloss meter*, light from a prefocused incandescent source falls on the sample at a specific angle. Light specularly reflected at an equal but opposite angle falls on a photocell, and the intensity is indicated on a meter. A multi-angle glossmeter consists of five prefocused systems at angles of 20°, 45°, 60°, 75°, and 85°. Small angles serve best to differentiate among surfaces of high gloss, and large angles best among surfaces of low gloss. Instruments with 45° geometry meet special requirements of the plastics and ceramics industries; the 75° geometry meets the needs of the book and paper industry.

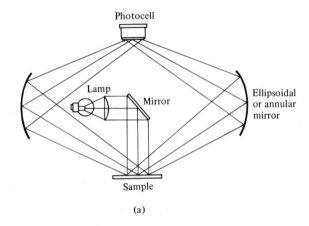

(a)

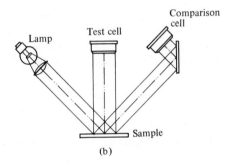

(b)

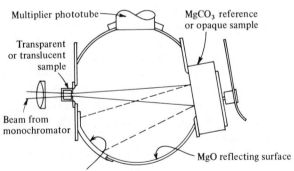

Specular component may be either:
1. Rejected by absorption in black velveteen
or
2. Accepted by reflection from white MgO

(c)

Fig. 4-16. Types of reflectance instruments. (a) Annular, ellipsoidal mirror type. (b) Reflection type. (c) Integrating sphere type.

With a goniophotometer, the reflected light is measured as a function of a variable angle of incidence.

One method of image-gloss evaluation involves viewing the specular reflection at 13.5° and through a series of *Landolt rings*—a circle with a small gap in it—ranging in diameter from 0.00356 to 31.8 mm, and reporting the smallest pattern in which the orientation of the gap can be distinguished visually.

Haze, defined as percentage of transmitted light which in passing through the specimen deviates from the incident beam by forward scattering by more than 2.5°, is measured be separating the scattered light from the directly transmitted light. This is done by passing the light beam through an integrating sphere to a light trap (0°) or tilting the sphere and using the walls for reflectance. Nonuniform samples are rotated.

TURBIDIMETRY AND NEPHELOMETRY

Turbidimetric and nephelometric methods of analysis are based upon the phenomenon whereby light, passing through a medium with dispersed particles of a different refractive index than the medium, is attenuated in intensity by scattering. In turbidimetry, measurement is made of the intensity of light transmitted through the medium, that is, of unscattered light. In nephelometry, the intensity of the scattered light is measured, usually, but not necessarily, at right angle to the incident light beam. Turbidimetric methods are similar to colorimetric methods in that both involve measurement of the intensity of light transmitted through a medium. Both employ similar apparatus. Nephelometric methods are similar to fluorimetric methods (see Chapter 5); however, the scattering is elastic in nephelometry so that both incident and scattered light are of the same wavelength in a nephelometric determination.

The intensity of light scattered at any particular angle is a function of the concentration of scattering particles, of their size, of their shape, of the wavelength of light, and of the difference in refractive indices of the particle and the medium. For a particle which is small compared to the wavelength of the light, the scatter depends upon the area intercepting the beam, and so is proportional to the square of the effective radius of the particle. The total scatter from a number of particles, assuming no multiple scattering interactions, is simply the sum of the individual scatterings, and the absorbance of the system is directly proportional to the concentration of particles. The intensity of scatter is inversely proportional to the fourth power of the wavelength of the light. Thus, blue light is scattered to a greater extent than is red light. The angular distribution of the scattered radiation is symmetrical in the forward and backward directions.

As the particle size increases, more of the radiation is scattered into the forward direction. Interference occurs among the several portions of the scattered light beam, and the resulting angular dependence of the scattered light is indicative of the shape and size of the particle. The relationship between any measurable indication of scattered light intensity and concentration of particles is not simple. Analytical determinations are therefore usually empirical.

Virtually any colorimeter or spectrophotometer can be used for turbidimetry and, if provision is made for viewing at right angles, for nephelometry as well. The choice of wavelength is generally not critical; however, if the sample solution contains a colored solute in addition to the turbid component, a wavelength region of minimum absorption is preferable. If the scattering component itself absorbs characteristically, a wavelength region of maximum absorption should be used so as to utilize the combined absorption and scattering effects. Care must be exercised that light scattered from the sample is not rescattered from the walls and housing in such a way that some of it gets to the detector and interferes with the turbidimetric reading. Sample cuvettes with black side walls will absorb the scattered light.

For the measurement of very small amounts of turbidity (transmittance greater than 90%), the nephelometric method is the one of choice. If the liquid is entirely free of particles, no scattered light will reach the photodetectors, and the indicating meter will read zero. Increasing turbidity gives an increase in meter reading. A very strong light source can be used which permits a high degree of sensitivity. The detection limit is set by the noise level of the detector-amplifier system. Because it is difficult to prepare a standard of very low turbidity, a mechanical reflectance standard, either a polished glass or stainless steel rod, is often inserted into the nephelometer in front of the photo-detectors. For intermediate ranges of turbidity, certified standards are available which consist of suspensions of an inorganic salt in a highly viscous polyethylene polymer, sealed into glass tubes.

The nephelometric method can also be used for high turbidities if the surface scatter approach is used. When a very narrow beam of light strikes the surface of the sample at a very low angle, part of the beam is reflected by the water surface and escapes to a light trap. The remaining portion enters the sample at approximately a 45° angle. If particles of turbidity are present, light scattering will occur and some of the scattered light will reach the detector located slightly above the surface of the sample.

The Parr turbidimeter is an extinction type of instrument which consists of a cylinder to contain the turbid suspension, a lamp filament of fixed intensity at the base, and an adjustable plunger through which visual observation is made. Measurement is made of the depth of turbid medium necessary to extinguish the image of the lamp filament. The Hellige turbidimeter is also a variable depth type of instrument using visual detection. A combination of vertical and horizontal illumination of the sample and a split ocular permit the eye to function merely to compare the intensities of two images simultaneously appearing in the ocular. Adjustment is made of a slit in the path of the direct, or vertical, illumination of the sample, and the calibration curve consists of a plot of this slit opening versus concentration.

Applications of nephelometry and turbidimetry are widely varied. Some determinations involve systems which are turbid prior to entering the analytical laboratory, such as in the determination of suspended material in waters. Measurement of the clarity of beverages and pharmaceuticals is typical of the simple kind of nephelometric determination. This is essentially an appearance measurement designed to evaluate the amount of haze, or cloudiness, present in a sample. Clarity and sparkle are important characteristics of product quality; the presence of suspended materials, even in amounts so small as to be invisible at the bottling point, will after bottling and storage ultimately result in an unsightly and unpalatable sediment.   The suitability of industrial process waters, and clarity of boiler feed waters and condensates, are typical examples of determinations in everyday use.

Many applications are possible in which a turbidity is developed from the test sample under controlled conditions.   Accuracy is frequently limited by the nonreproducibility of the physical form of the precipitate or by the instability of the precipitate once it is formed.   The precipitation reaction should be one that takes place rapidly, and the precipitated particles should be low in solubility and small in particle size.   Surfactants and other protective colloids are helpful, not only to make the process more reproducible but also to make the particles small by enhancing nucleation at the expense of crystal growth.

The variation of light scattering with size is utilized in the *light-scattering photometer*.   An individual particle passing rapidly through a light beam sends a pulse of light to a photomultiplier which produces an electrical pulse which is particle-size dependent.   Suitable electronic equipment is used to size and count the electrical pulses which are used to determine the aerosol size distribution.   The instrument must be calibrated with standards of known particle size.   In the *Science Spectrum* instrument, a scattering sample contained in a cuvette is illuminated by a laser beam, while a collimated detector with $2°$ resolution circumferentially scans the light scattered by the illuminated volume.   A signal proportional to the intensity of the scattered light is recorded as a function of the angle formed between the incident and scattered beams for subsequent examination and analysis.   Information can be obtained from individual microparticles, such as single bacteria, or aerosol and latex particles of 100 to 5000 nm diameter, which can be levitated and automatically held stationary in the beam by means of electrical fields.   Average size, size distribution, shape, and other characteristics of particle suspensions in their natural state can be determined.

CORRELATION OF ELECTRONIC ABSORPTION SPECTRA WITH MOLECULAR STRUCTURE

When molecules interact with radiant energy in the visible and ultraviolet region, the absorption of energy consists in displacing an outer electron in the molecule.   Rotational and vibrational modes will be found combined with electronic transitions.   Broadly, the spectrum is a function of the whole structure of a substance rather than of specific bonds. No unique electronic spectrum will be found; this is a poor region for product identification by the "fingerprint" method.   Information obtained from this region should be used in conjunction with other evidence to confirm the identity of a compound, for example, previous history of a compound, its synthesis, auxiliary chemical tests, and other spectroscopic methods.   On the other hand, electronic absorption often has a very large magnitude.   Molar absorptivity values frequently exceed 10,000, whereas in the infrared they rarely exceed 1,000.   Thus, dilute solutions are adequate in visible-ultraviolet spectrophotometry.

Structural Features

We will consider only those molecules capable of absorption within the wavelength region from 185 to 800 nm. Compounds with only single bonds involving σ-valency electrons exhibit absorption spectra only below 150 nm and will be discussed only in interaction with other kinds. In covalently saturated compounds containing heteroatoms—for example, nitrogen, oxygen, sulfur, and halogen—unshared p-electrons are present in addition to σ-electrons. Excitation promotes a p-orbital electron into an antibonding σ orbit, that is, an $n \rightarrow \sigma^*$ transition, such as occurs in ethers, amines, sulfides, and alkyl halides. In unsaturated compounds absorption results in the displacement of π-electrons. Molecules containing single absorbing groups, called *chromophores*, undergo transitions at approximately the wavelengths indicated in Table 4-2.

Molecules with two or more insulated chromophores will absorb light of nearly the same wavelength as a molecule containing only a single chromophore of a particular type,

TABLE 4-2 Electronic Absorption Bands for Single Chromophores

Chromophore	System	λ_{max}, nm	ϵ_{max}	λ_{max}, nm	ϵ_{max}
Nitrile	—C≡N	160	—		
Acetylide	—C≡C—	175–180	6000		
Sulfone	—SO$_2$—	180	—		
Ether	—O—	185	1000		
Oxime	—NOH	190	5000		
Azido	>C=N—	190	5000		
Ethylene	—C=C—	190	8000		
Ketone	>C=O	195	1000	270–300	18–30
Thioether	—S—	195	4600	215	1600
Amine	—NH$_2$	195	2800		
Thiol	—SH	195	1400		
Disulfide	—S—S—	195	5500	255	400
Thioketone	>C=S	205	Strong	~495	Weak
Ester	—COOR	205	50		
Bromide	—Br	208	300		
Carboxyl	—COOH	200–210	50–70		
Aldehyde	—CHO	210	Strong	280–300	11–18
Sulfoxide	>S→O	210	1500		
Azoxy	—N=N(O)—	217	7250	274(sh)	40
Nitro	—NO$_2$	210	Strong		
Azide	—N$_3$	220	150	287	20
Nitrite	—ONO	220–230	1000–2000	300–400	10
Nitrosoamine	—N—NO	233	7400	350	90
Thionoacetate	—OCSCH$_3$	241	8100	369	18
Dithioacetal	—S—C—S—	235	540		
Isothiocyanate	—N=C=S	245	1000		
Thiocyanate	—S—C≡N	248	50		
Nitrate	—ONO$_2$	270	12		
Azo	—N=N—	285–400	3–25		
Nitroso	—N=O	302	100		

but the intensity of the absorption will be proportional to the number of that type of chromophore present in the molecule. Appreciable interaction between chromophores does not occur unless they are linked to each other directly; interposition of a single methylene group, or *meta*-orientation about an aromatic ring, is sufficient to insulate chromophores almost completely from each other. However, certain combinations of functional groups afford chromophoric systems which give rise to characteristic absorption bands.

The unsubstituted conjugated *diene*, heteroannular or open chain, absorbs nominally near 214 nm unless both double bonds are contained within the same ring. A homoannular diene absorbs at 253 nm for 1,3-cyclohexadiene, at 228 nm for cyclopentadiene, and at 241 nm for 1,3-cycloheptadiene. An alkyl or O-alkyl group displaces the absorption band to the red about 5 nm, as does an exocyclic bond, a Cl- or a Br-substituent. A S-alkyl or another double bond in conjugation displaces the band 30 nm to the red. A red shift of 60 nm is associated with a N-alkyl group. Change of solvent has little effect. Heteroannular and acylic dienes display molar absorptivities in the 8,000–20,000 range, whereas homoannular dienes display values in the 5,000–8,000 range.

α,β-Unsaturated carbonyl compounds absorb in a similar spectral range to conjugated dienes. From the parent system, absorbing at 215 nm,

$$\overset{\displaystyle O}{\overset{\displaystyle \|}{-C}}-\overset{\alpha}{C}=\overset{\beta}{C}-\overset{\gamma}{C}=\overset{\delta}{C}$$

the wavelength shifts to the red by 10 nm for an α-alkyl substituent, 12 nm for a β-alkyl substituent, 18 nm for γ- or δ-alkyl substituent, 39 nm if a homoannular enone system is present, and 30 nm for an exocyclic double bond or for a double bond extending the conjugation. Subtract 10 nm for a 5-membered ring ketone, 5 nm for aldehydes, and 20 nm for carboxylic acids and esters. Although the influence of other substituents is not so predictable, typical red shifts in ethanol medium (in nm) are: —OH, α(35), β(30), δ(50); —OCOCH$_3$(6); —OCH$_3$, α(35), β(30), γ(17), δ(31); —S-alkyl, β(85); —Cl, α(15), β(12); —Br, α(25), β(30); and —N(alkyl)$_2$, β(95). Enones can be distinguished from the dienes since only the enones display significant solvent effects which take the form of a blue shift of the carbonyl band in water of 8 nm, and a red shift of 11 nm for aliphatic hydrocarbons and 5 nm for dioxane (with ethanol as reference). The molar absorptivity of cisoid enones is usually less than 10,000 whereas it is greater than 10,000 for transoid enones.

In an inert solvent, benzene shows absorption bands at 180–185 nm ($\epsilon = 40,000$), 193–204 nm ($\epsilon = 5,000$), and 230–270 nm (B-band, $\epsilon = 250$). In polynuclear aromatic systems all the bands are displaced to longer wavelengths and become more intense. Ring substitution affords red shifts and intensification of the spectrum. The absorption of the B-band moves to longer wavelengths in cases where the new substituent is electron donating or capable of conjugation. With electron-withdrawing substituents, practically no change in the maximum position is observed. The vibrational fine structure of the benzenoid spectrum tends to disappear on substitution, and if the spectrum is determined in solvents more polar than aliphatic hydrocarbons. Extremely useful and identifying solvent shifts are observed in the spectra of phenol between neutral and alkaline media, and in the aniline spectra between neutral and acidic media.

When electronically complementary groups are situated *para* to each other in disubstituted benzenes, there is a pronounced red shift in the main absorption band due to the extension of the chromophore from the electron-donating group through the ring to the electron-withdrawing group. When the *para* groups are not complementary, or when the groups are situated *ortho* or *meta* to each other, the observed spectrum is usually similar to that of the separate, noninteracting chromophores.

Multiple bonded oximes, nitriles, nitro and azo compounds show no or only weak absorption. If conjugated, the transitions and absorption wavelengths resemble those of the dienes and enones.

Stereochemical Effects

In predicting the wavelength of maximum absorption, allowance must be made for the likely shape of the molecule and for any unusual strain. Distortion of the chromophore may lead to red or blue shifts, depending on the nature of the distortion. In extreme cases conjugation is almost completely inhibited through rotation of the system out of planarity, and the molecule absorbs as two distinct entities. Steric interactions can also cause a change in the absorption intensity of the main conjugation band. In *cis-trans* isomerism, generally the *cis*-configuration of a conjugated compound absorbs with lower intensity due to shorter distance between the ends of the chromophore.

A great deal of "negative" information may be deduced regarding molecular structures. If a compound is highly transparent throughout the region from 220 to 800 nm, it contains no conjugated unsaturated or benzenoid system, no aldehyde or keto group, no nitro group, and no bromine or iodine. If the screening indicates the presence of chromophores, the wavelength(s) of maximum absorbance are ascertained and tables are searched for known chromophores. Further information may be deduced from the shape, intensity, and detailed location of the bands. Finally, the absorption spectrum is compared with the spectra in standard compilations of ultraviolet-visible spectra.[14,18] Oftentimes structural details can be inferred from the close resemblance of a compound's spectrum with that of a compound of known and related structure—for example, in petroleum ether, the spectra of toluene and chlorobenzene are similar.

VACUUM ULTRAVIOLET

For our purposes the discussion will be confined to wavelengths from 584 to 2100 Å. Since atmospheric oxygen absorbs nearly all radiation at wavelengths shorter than 2100 Å, it is necessary to carry out all investigations in vacuum, or in one of the few gases transparent in portions of the region.

Several emission sources are needed to cover the entire region for absorption spectrophotometry; each source exhibits continua over only narrow spectral ranges:

hydrogen (or deuterium) lamp above 1650 Å; xenon lamp from 2000 to 1500 Å; krypton, 1500 to 1300 Å; argon, 1300 to 1100 Å; and a windowless helium lamp, 1100 to 584 Å. These continua are emitted after excitation by a continuous electric discharge through the gas under moderately low pressure. One arrangement places the sources in a radial array. A rotatable diagonal mirror (MgF_2 coated) deflects the energy from each source to the entrance slit of the monochromator. Since the reflectivity of coated mirrors decreases rapidly below 1200 Å, the argon and helium sources are located directly in front of the entrance slit and the diagonal mirror is moved out of the path.

Among the several optical configuration of monochromators,[6,15] the simplest is the normal incidence mounting which utilizes a concave grating which is self-focusing and does not require ancillary optics. The Seya-Namioka mounting utilizes the unique properties of a fixed 70°15' angle of arc between the entrance and exit beams; only slight defocusing occurs over the vacuum ultraviolet. It is the only mounting capable of operation in the farthest vacuum ultraviolet with the exception of the costly grazing incidence spectrometer. Plane grating systems have a wavelength cutoff at 1050 Å due to the three reflecting surfaces in the Czerny-Turner mounting. The optical arrangement in the McPherson instrument (Fig. 4-17) employs an unconventional configuration which is achieved through the use of 0.3-m focal length aspheric mirrors in a crossed optical system with an effective aperture ratio of $f/5.3$. This design lends itself to a short focal length for high flux density. It allows maintenance of normal incidence stigmatic focus, but permits a wide dihedral slit-entrance angle. For wavelengths below 1050 Å, the grazing incidence principle must be used (Fig. 4-18). The grating moves on the Rowland circle and the exit slit is transposed on the circle. Angles of incidence range from 82° to 88°.

If photographic detection is used, special emulsions are necessary. These consist of AgBr grains at the surface of a very thin layer of gelatin film, or ordinary emulsions are coated with fluorescent (terphenyl) lacquer or else dipped into sodium salicylate for sensitization just prior to use. A photomultiplier tube, blue-sensitive and covered with a thin film of sodium salicylate, responds with a constant quantum efficiency over the

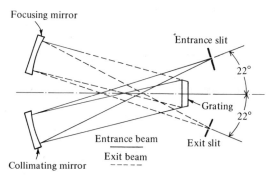

Fig. 4-17. Optical diagram of McPherson 0.3-m focal length monochromator for vacuum ultraviolet region. (Courtesy of McPherson Instrument Co.)

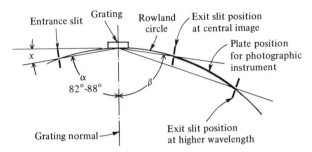

Fig. 4-18. Optical diagram of a grazing incidence spectrometer.

range from 2000 to 1000 Å. In an alternate arrangement the photomultiplier tube can be mounted behind a Vycor window coated with sodium salicylate; Vycor passes most of the blue fluorescent band of salicylate but absorbs shorter wavelengths. The windowless beryllium–copper electrostatic detector possesses a high work function; it responds only to very short wavelengths of radiation. As such, it is solar-blind and is usable below 1000 Å where a serious stray light problem exists with other detectors.

The choice of transparent material for cell windows and lenses is limited to LiF, CaF_2, and synthetic sapphire, which are usable to 1150, 1250, and 1400 Å, respectively. When working with liquid samples, one is usually obliged to work with very short optical-path cuvettes because of the high absorbance of solutes and solvents. The sandwich-type cells familiar to the infrared spectroscopists are more useful.

With the exception of the rare gases, hydrogen, nitrogen, and saturated hydro-carbons, virtually all of the inorganic gases and organic vapors absorb in the 1700–2200 Å region. The absorption spectra are often very detailed. In the case of liquids the specificity of vacuum ultraviolet spectra is about the same as that for the near ultraviolet liquid spectra. All olefins absorb around 1800 Å, ketones at 1950 Å, organic iodides around 2000 Å, organic bromides around 1850 Å, and inorganic iodides at 1950 and 2250 Å. References to vacuum ultraviolet spectra of many compounds are given by Kaye.[12] Solutions can be studied to about 1720 Å without undue difficulty. Water and *n*-heptane are among the most transparent liquids. The hydrocarbon is purified by passage through a column of activated silica. Oxygen normally dissolved in liquids should be removed. Highly purified perfluorinated solvents can be used to about 1565 Å. Argon, nitrogen, and hydrogen may be compressed and cooled to the liquid state in which state they serve as solvents for work to 1300 Å.

The band at 1940 Å for acetone has been applied to its analysis in acetylene. Analysis of any unsaturate in a nonabsorbing matrix is possible with a sensitivity of a few parts per million; fatty acid solutions have been studied to about 1750 Å and their absorptivities have been correlated with the number of double bonds in the particular acid.

In emission spectroscopy (Chapter 13) the main use of the vacuum ultraviolet region has been confined to the emission lines for carbon (1657 Å), phosphorus (1783 Å), sulfur (1820 Å), arsenic (1890 Å), mercury (1849 Å), and the halogens. Sometimes the monochromator, detector, and arc-spark stand are flushed with dry nitrogen, which is transparent above 1600 Å.

LABORATORY WORK

Operation of Photometers

For an instrument which produces a response R, directly proportional to the radiant power P falling on the detector, as do most photometers, the response is given by

$$R = k_\lambda k_\alpha P + k'$$

where k_λ is a proportionality constant which depends on the slit width and response of the detector, k_α is a second constant which reflects the gain in the amplifier control (or sensitivity setting), and k' is the term which reflects the current that flows in the receiver and amplifier circuit when the receiver is dark. In the ordinary method of operation, with the receiver dark to represent infinite concentration of sample, a dark current or zero suppressor control is adjusted to balance the circuit with the reading scale at zero. For instruments with barrier-layer cells, this involves merely the mechanical alignment of the galvanometer index at the zero mark for direct-reading instruments or at midscale for null-balance photometers. Next, with the solvent or a reference solution in the light beam and the reading scale set at 100% transmittance (zero absorbance) to represent zero concentration, the slit width control or amplifier control, or both, are adjusted to balance the circuit once again. Finally, the series of standards are inserted successively into the light beam and the scale reading is noted. From these readings a calibration curve can be constructed.

Selected Colorimetric Methods

Experiment 4-1   Determination of Iron with 1,10-Phenanthroline

Standard iron solution   Dissolve 0.7022 g of reagent grade $(NH_4)_2 Fe(SO_4)_2 \cdot 6H_2O$ in 100 ml of distilled water, add 3 ml of 18 M H_2SO_4, and dilute to 1 liter. One milliliter contains 0.100 mg of iron(II).

For a standard iron(III) solution, dissolve 0.864 g $NH_4 Fe(SO_4)_2 \cdot 12H_2O$ water, add 3 ml of H_2SO_4, and dilute to 1 liter. Alternatively, the ferrous ammonium sulfate solution can be carefully oxidized with $KMnO_4$ added dropwise until a slight pink coloration remains after stirring well, before diluting to 1 liter. One milliliter contains 0.100 mg of iron.

1,10-Phenanthroline   Dissolve 0.25 g of the monohydrate in 100 ml of water, warming if necessary.

Hydroxylamine hydrochloride, 10% (w/v)   Dissolve 10 g in 100 ml of water.

Sodium acetate, 2 M Dissolve 17 g of sodium acetate in 100 ml of water.

Procedure   Take an aliquot portion of the unknown solution containing 0.1 to 0.5 mg of iron and transfer it to a 100-ml volumetric flask. Determine by the use of a similar aliquot portion containing a few drops of bromophenol blue, the volume of sodium acetate solution required to bring the pH to 4.0 (or use a pH meter). Add the same volume of acetate solution to the original aliquot and then 5 ml each of the hydroxylamine hydrochloride and 1,10-phenanthroline solutions. Dilute to the mark, mix well, and measure the absorbance after 10 min in the region 460–520 nm.

Both iron(II) and iron(III) can be determined simultaneously. Only the iron(II) complex absorbs at 515 nm, and both complexes have identical absorption at 396 nm, the amount being additive. The solution should be buffered at pH 3.9.

Experiment 4-2　Determination of Manganese as Permanganate

Standard manganese solution Reduce 200 ml of 0.0100 M KMnO$_4$ solution with a little sodium sulfite after the addition of 1 ml of 18 M H$_2$SO$_4$. Remove the SO$_2$ by boiling and dilute to 1 liter. One milliliter contains 0.110 mg of Mn.

Procedure Weigh out accurately a suitable quantity of a steel sample (0.1–0.2 g for average steels) into a conical flask, dissolve it in about 40 ml of water and 10 ml of 15 M HNO$_3$, and boil for 2–3 min after any violent bubbling subsides to expel oxides of nitrogen. Cool the solution, and add slowly about 1 g of ammonium peroxydisulfate [(NH$_4$)$_2$S$_2$O$_8$]. Boil gently for 10 min to oxidize carbon compounds and to destroy the excess peroxydisulfate. Dilute to about 100 ml with water, add 10 ml of 85% H$_3$PO$_4$, and then add 0.5 g of potassium periodate. Boil gently for 3 min and keep hot for 10 min. Cool to room temperature and transfer to a 250-ml volumetric flask; dilute to the mark with distilled water. Measure the absorbance at either 525 or 545 nm (Fig. 4-19).

When ready for measurement, the solution should not contain more than 2 mg of

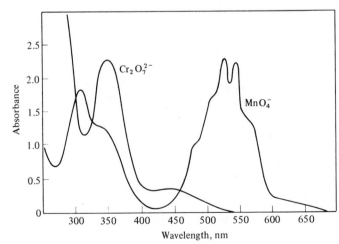

Fig. 4-19. Spectral absorption curves for permanganate ion and dichromate ion (each 0.001 M) in 1 M sulfuric acid.

Mn per 100 ml.   Standard iron(III) solution can be added to aqueous standard so-lutions of manganese to match the iron content in steel samples. The interference of chromium(VI) may be removed by decolorizing one portion by the dropwise addition of potassium nitrite solution until the permanganate color just disappears. The decolorized solution is then used as the reference solution.

Experiment 4-3   Determination of Titanium as Its Peroxide

Standard titanium solution   Fuse 0.2500 g of TiO_2 with 3–4 g of potassium pyro-sulfate in a platinum or a porcelain crucible. Dissolve the melt in 50 ml of hot $4 N\ H_2SO_4$ and dilute to 250 ml with the same acid. One milliliter contains 1.00 mg of TiO_2.

Procedure   Take an aliquot of the standard or unknown solution such that the final concentration is 1–8 mg of TiO_2 in 100 ml. Add sufficient H_2SO_4 to adjust the final acidity to 1.5–3.5 N. Add 10 ml of 3% hydrogen peroxide solution and dilute the solu-tion to 100 ml in a volumetric flask. Measure the absorbance at 400–420 nm (see Fig. 4-20).

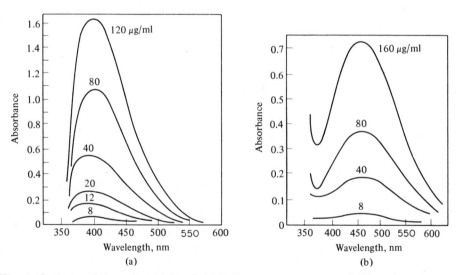

Fig. 4-20. Spectral absorption curves for (a) titanium-peroxide complex and (b) vanadium-peroxide complex, each 1 M perchloric acid solution.

Experiment 4-4   Comparison of Dichromate Solutions

Standard chromium(VI) solution   Dissolve 2.823 g of $K_2Cr_2O_7$ in water and dilute to 1 liter. One milliliter contains 1.00 mg of Cr.

Procedure   Take an aliquot of the standard or unknown solution such that the final concentration of chromium is 2-10 mg per 100 ml. Add 1.5 ml of 18 M H_2SO_4 and dilute to 100 ml in a volumetric flask. Measure the absorbance at 440 nm.

Experiment 4-5   Determination of Cobalt with Nitroso-R-Salt

Standard cobalt solution   Dissolve 0.404 g of $CoCl_2 \cdot 6H_2O$ in water and dilute to 1 liter. One milliliter contains 0.100 mg of Co.

Nitroso-R-salt, 1% (w/v)   Dissolve 1 g of reagent in 100 ml of water.

Procedure   Take an aliquot that contains 0.1-0.4 mg of cobalt. Evaporate almost to dryness, add 1 ml of 15 M HNO_3, and continue the evaporation just to dryness. Dissolve the residue in 20 ml of water containing 1 ml each of 6 M HCl and 1 M HNO_3. Boil for a few minutes. Cool to room temperature, add 2 ml of 1% nitroso-R-salt solution, and 2 g of hydrated sodium acetate. The pH of the solution should be close to 5.5 (check with a pH meter or test with bromocresol green indicator on a drop of the solution). Boil for 1 min, add 1 ml of 12 M HCl, and boil again for 1 min. Cool, and dilute to 100 ml in a volumetric flask. Measure the absorbance at 500 nm (see Fig. 4-5).

Experiment 4-6   Determination of the Dissociation Constant of Indicators

Indicator solution, 0.04% (w/v)   Dissolve 0.1 g of the indicator (sulfonphthalein dyes) in water, adding 1-3 ml of 0.1 M NaOH, if necessary. Dilute to 250 ml.

Buffer solutions   Any standard buffer series may be used. It is desirable to maintain the ionic strength constant. In the following directions, 0.1 M sodium acetate solution is employed. Dissolve 2.30 g of sodium acetate in water and dilute to 250 ml.

Procedure   Transfer exactly 10.0 ml of 0.1 M sodium acetate solution to a 100-ml volumetric flask. Add exactly 5.00 ml of 0.04% bromocresol green solution and dilute to the mark with distilled water. Mix thoroughly. Determine the absorption spectrum of the sodium acetate solution of bromocresol green (the "alkaline" color).

Pour the contents of the cuvette and the volumetric flask into a 250-ml beaker. Add precisely 2.00 ml of an acetic acid solution (0.25 M in acetic acid and 0.1 M in KCl). Mix well with a stirring rod, and measure the pH with a pH meter. Measure the absorbance at the wavelength of maximum absorbance of the "alkaline" form of the indicator. Repeat this procedure for additional 2.00-ml increments of acetic acid solution. Determine the absorption spectrum for the buffer solution which is equimolar in acetate and acetic acid. After five such measurements have been made, add 1.0 ml of 3 M HCl and again determine the entire spectrum of the "acid" form of the indicator.

Correct all absorbance readings for the effect of dilution by multiplying each observed value by the factor $(100 + V)/100$, where V is the volume of acetic acid (and HCl) added. Plot the three absorption spectra on a single sheet of graph paper.

Another curve is plotted from the absorbance readings at the wavelength of maximum absorbance vs. the pH value for each buffer solution. Draw a smooth line through the points. The "acid" and "alkaline" solutions are assumed to represent the limiting values of absorbance. Determine the value of pK_a in these ways: (1) from the inflection point of the absorbance-pH curve, and (2) by insertion of appropriate absorption readings into Eq. 20-12 and averaging the values obtained.

Other indicators in the sulfonphthalein series may be studied. These include bromophenol blue, bromocresol purple, bromothymol blue, phenol red, and cresol red. Other compounds that are suitable include vanillin ($\epsilon = 25,900$), $4 \times 10^{-5}M$ at 347 nm; p-bromophenol, 1 mg/100 ml in hexane, at 244 nm; and o-chlorotoluic acid, $2 \times 10^{-5}M$, at 280 nm.

Use at least five buffer solutions whose pH values lie within the range $pK_{ind} \pm 1$. Also prepare the fully acid and fully alkaline solutions. Figure 4-21 illustrates some typical results.

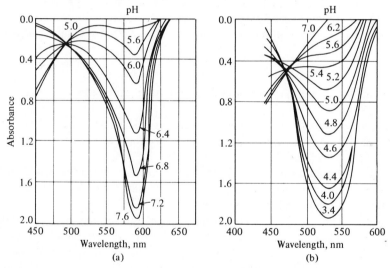

Fig. 4-21. Spectrophotometric absorption curves for (a) bromocresol purple (b) and methyl red.

A study of ionic strength effects merely involves assigning a fixed ionic strength to each of several individuals working with the same indicator. See R. W. Ramette, *J. Chem. Education*, **40**, 252 (1963).

Experiment 4-7   Simultaneous Determination of Binary Mixtures

Each student will be assigned a pair of solutes and the type of sample to be analyzed. Suggested systems include the following:

1. Manganese, as permanganate, 0.0004–0.00008 M, plus chromium as dichromate, 0.0004–0.0017 M, in 0.5 M H_2SO_4. Analytical wavelengths are 440 and 525 nm. See Fig. 4-19.
2. Titanium, 0.8–8.0 mg of TiO_2 per 100 ml, plus vanadium, 1.0–20 mg as V per 100 ml, as their peroxide complexes. Analytical wavelengths are 400 and 460 nm. See Fig. 4-20.
3. Iron(II) and iron(III) as their 1,10-phenanthroline complexes at pH 3.9. Analytical wavelengths are 396 and 515 nm.

Procedure Measure the absorbance of 3–5 standard solutions of each solute, whose concentrations span the interval suggested, at both analytical wavelengths. Compute the molar absorptivities from each individual measurement and determine the average value of the molar absorptivity and its associated deviation at the 95% confidence level.

Test the assumption that each solute contributes additively to the total absorbance at each analytical wavelength. Combine known aliquots of the individual solutions prepared for the preceding step, and measure the absorbance at each analytical wavelength. Calculate the amount of each solute present.

Determine the concentration of each solute in an unknown sample and report the results.

Experiment 4-8   High-Absorbance Differential Spectrophotometry

Each student will be assigned a single solute or colorimetric system. The range of concentrations to be studied and the particular spectrophotometer to be used will be designated.

Prepare a series of solutions whose absorbance readings extend from 0.2 to 2.0 (or approximately three times the maximum strength usually recommended) and separated from each other by about 0.2 absorbance unit.

Proceed in the ordinary way to set the instrument scale reading to zero transmittance when the photocell is darkened, and to 100 scale divisions (zero absorbance) with pure solvent in the sample container. Measure the absorbance and transmittance of each solution. Prepare a graph of absorbance vs. concentration.

Remove the solvent from its cuvette and substitute in it some of the solution which had the 0.2-absorbance value. Measure the transmittance to ascertain the ratio T_2/T_1. If not unity, this preliminary value can be used to correct subsequent readings for difference in path length of the two cuvettes: $-\log (T_2/T_1) = ab_1 C [\beta - 1]$, where $\beta = b_2/b_1$. Place the comparison standard (i.e., solution with 0.2-absorbance value originally) in the light beam, with the reading scale set at 100 scale units, and balance the instrument by increasing the slit opening or by increasing the gain control. Measure the absorbance for the more concentrated solutions. Plot the results on the same graph.

Repeat the procedure with progressively more concentrated solutions being used as the comparison standard for setting the reading scale at 100 until it becomes impossible to establish the balance of the instrument.

For each series of readings, calculate the relative concentration error (assuming $\Delta T = 0.004$). Plot the relative concentration error vs. transmittance for each series as a family of curves on a single graph (see Fig. 4-10).

Experiment 4-9   Low-Absorbance Differential Spectrophotometry

Each student will be assigned a specific system and instrument.

Prepare a series of solutions whose transmittance values range from 0.368 to 0.950 and are separated from each other by about 0.100 transmittance.

Proceed in the ordinary way to set the instrument scale reading to zero transmittance when the photocell is darkened, and to 100 scale units with pure solvent in the sample container. Measure the absorbance and transmittance of each solution. Prepare a graph of absorbance vs. concentration; also of transmittance vs. concentration.

Insert the most concentrated solution into the light beam and, with the reading scale set at zero transmittance, adjust the dark current or zero control. Then adjust the 100 setting with the solvent in the light beam. Recheck the zero setting and readjust if necessary. It may be necessary to repeat each setting, in sequence, several times until neither the zero nor the 100 setting change when going from one to the other. Measure the transmittance of the other solutions. Plot the results on the same graph of transmittance vs. concentration. Calculate the relative concentration error for each measurement. Plot these results as relative concentration error vs. transmittance. Assume $\Delta T = 0.004$.

If sufficient zero-suppression is available, repeat the cycle of balancing operations with a less concentrated reference solution.

Experiment 4-10   Maximum Precision Spectrophotometry

Each student will be assigned a specific system and instrument.

Prepare a set of five solutions whose transmittance readings are approximately 0.200, 0.300, 0.350, 0.400, and 0.500. Measure the transmittance in the ordinary manner, using pure solvent for the setting at 100% transmittance.

Insert the least concentrated standard (about 50% T) and set the reading scale to 100; the most concentrated standard (about 20% T) is used to set the reading scale to zero transmittance. Readjust the 100 reading, if necessary. Recheck the zero reading and readjust if necessary. It may be necessary to repeat the cycle a few times until neither the zero nor the 100 setting change when going from one to the other. Measure the transmittance of the other solutions. Plot the results as transmittance vs. concentration.

Calculate the relative concentration error for each reading. Plot these results as relative concentration error vs. transmittance. Assume $\Delta T = 0.004$.

If sufficient zero-suppression and gain control are available, the balancing operation can be repeated with the solutions of 30% and 40% transmittance serving as comparison standards.

Experiment 4-11 Photometric Titrations

Operating technique Set the spectrophotometer to the analytical wavelength or insert the proper filter into the photometer. Adjust the dark current to zero. Place the sample to be titrated into the light beam and set the instrument to read zero absorbance (if the reactant is colorless) or to some other starting value (if the reactant is colored) that lies within the range of linearity of the instrument. If the initial absorbance reading is too large, readjust the concentration of reactant.

Place the buret so that the tip extends into the solution. Turn on the stirrer and adjust the stirring rate so that the vortex does not obstruct the light path. Commence the titration by adding an increment from the buret, waiting until the absorbance reading is constant, and recording that value. More increments are then added and the absorbance noted. For exploratory work, 0.2- to 0.5-ml increments are desirable when using a 10-ml buret. Once the shape of the titration curve is known, it is usually necessary to take only three to four points on each side of the end point.

A plot of absorbance vs. milliliters of titrant is then made and the best straight lines are drawn between points taken well before and after the equivalence point. The intersection of the linear segments is taken as the end point. Correct all absorbance readings for the change in volume at each point on the titration curve.

Suggested Systems

Acid-base titrations Mixture of acetic acid with *p*-nitrophenol, each 10^{-3} *M*, titrated with 0.1 *M* NaOH at 400–420 nm.

A solution of any strong acid, 10^{-3} to 10^{-4} *M*, titrated with 0.01 *M* NaOH in the presence of thymol blue ($pK_a = 9.0$) at 615 nm.

Oxidation-reduction titrations A solution of arsenic(III), 10^{-3} *M*, with ceric sulfate, 0.01 *M*, at 320 nm, with a trace (10^{-5} *M*) of osmium tetroxide as catalyst.

Complexometric titrations Copper(II), 0.04 *M*, with 0.1 *M* EDTA at pH 4.0 and a wavelength of 745 nm.

Calcium plus copper(II), each 0.02–0.06 *M*, with 0.1 *M* EDTA in a 1 *M* ammonia-ammonium chloride buffer adjusted to pH 9.0 ± 0.2 and at a wavelength of 745 nm.

EDTA, 0.002 *M*, titrated with 0.02 *M* zinc solution in an approximately 1 *M* ammonia–ammonium chloride buffer adjusted to pH 9 and with 0.002 g of Superchrome Black TS indicator present. Wavelength is 550 nm.

Iron(III), 0.05 *M*, titrated with 0.1 *M* EDTA in an acetic acid solution (1.0 *M*) adjusted to pH 1.7–2.3 with HCl. Salicylic acid (4 ml of a 1% solution is added per 100 ml total volume) serves as indicator; the analytical wavelength of the iron-salicylic acid complex is at 525 nm.

Experiment 4-12   Composition of Complexes: Mole-Ratio Method, Method
of Continuous Variations, and Slope-Ratio Method

A. Mole-Ratio Method

In this method the concentration of metal ion is held fixed and the concentration of
the reagent is increased stepwise. On the graph of absorbance vs. moles of reagent added,
the intersection of the extrapolated linear segments determines the ratio: moles of
reagent/moles of metal.

Procedure   Transfer exactly 4 ml of the standard metal solution to 11 separate
50-ml volumetric flasks. Add any appropriate buffer solution and other necessary reductants. Add exactly 4, 6, 8, 10, 12, 14, 16, 18, 20, 22, and 24 ml of the standard
reagent solution to each flask. Dilute to the mark with distilled water and mix. After
any suggested waiting period, measure the absorbance at the designated wavelength.

B. Method of Continuous Variations

In this method the sum of the molar concentrations of the two reactants is kept
constant as their ratio is varied. The abscissa of the extrapolated peak will correspond to
the ratio present in the complex.

Procedure   Transfer exactly 0, 2, 4, 6, 8, 10, 12, 14, 16, 18, and 20 ml of the metal
solution to separate 50-ml volumetric flasks. Add 5 ml of the appropriate buffer solution
and any necessary reductants. To the above flasks, arranged in serial order, add these
amounts of the reagent solution, also in serial order: 20, 18, 16, 14, 12, 10, 8, 6, 4, 2,
and 0 ml. Dilute to the mark and mix. After the suggested waiting period, measure the
absorbance at the designated wavelength.

C. Slope-Ratio Method

In this method two series of solutions are prepared. In the first series various
amounts of metal ion are added to a large excess of the reagent, while in the second
series different quantities of reagent are added to a large excess of metal ion. The absorbance of the solutions in each series is measured and plotted vs. the concentration of
the variable component. The combining ratio of the components in the complex is equal
to the ratio of the slopes of the two straight lines.

Procedure   Transfer exactly 30 ml of the standard metal solution to five separate
50-ml volumetric flasks. Add appropriate amounts of buffer solution and any necessary
reductant. Add 1, 2, 3, 4, and 5 ml of the reagent solution, respectively, to the flasks,
dilute to volume, and mix. After any waiting period, measure the absorbance at the
designated wavelength. In the second series, use 30 ml of the reagent solution, and add
1, 2, 3, 4, and 5 ml, respectively, of the metal solution.

Suggested Systems

1. Iron(III) nitrate, 0.002 M (plus 0.01 M HNO_3), and potassium thiocyanate, 0.002 M (plus 0.015 M HCl), measured at 480 nm. [See W. R. Carmody, *J. Chem. Educ.*, **41**, 615 (1964).] Suitable only for method of continuous variations; ϵ = 6300.

2. Iron(II) ammonium sulfate, 0.0005 M; 1,10-phenanthroline, 0.0005 M; acetic acid–sodium acetate buffer (total acetate = 0.1 M), pH 5.0 (use 10 ml for each solution); hydroxylamine hydrochloride, 5% (w/v), use 1 ml for each solution; measure at 510 nm after standing 10 min; ϵ = 12,000.

Experiment 4-13   Identification of Complex Ions in Solution

If two or more compounds can be formed from components A and B, the mole ratio at which the absorbance is a maximum should vary with the wavelength of the light used for the measurement. It should then be possible to determine whether or not more than one compound is formed by scanning the absorption spectra of solutions containing a fixed number of moles of metal ion and integral mole ratios of complexing material.

Procedure   Prepare a 0.08 M nickel solution (21.03 g $NiSO_4 \cdot 6H_2O$ per liter) and a 0.08 M ethylenediamine solution (4.81 g/liter).

Into five 100-ml volumetric flasks, transfer 20.0 ml of the nickel solution. Add these volumes of ethylenediamine: none, 20.0, 40.0, 60.0, and 80.0 ml. Dilute to the mark with distilled water. Mix well. Scan the absorption spectra from 400 to 700 nm and note the number of absorption maxima and the wavelength of each maximum. Use 4-cm cell path lengths.

Three distinct wavelength maxima should be noted. These occur at 622 nm for a 1 : 1 mole ratio, at 578 nm for a 1 : 2 mole ratio, and at 545 nm for a 1 : 3 mole ratio.

Verification for the existance of three coordination species of nickel ion with ethylenediamine can be obtained from the method of continuous variations (Experiment 4-12, part B). To 100-ml volumetric flasks, add these volumes of 0.08 M nickel solution, and fill to the mark with 0.08 M ethylenediamine solution: 100, 80, 75, 60, 56, 54, 52, 50, 48, 44, 40, 36, 34, 32, 30, 28, 26, 24, 22, 20, 15, 10, and 5 ml. Measure the absorbance (in 4-cm cuvettes) at 622 nm for the solutions containing 100 to 44 ml of nickel; at 578 nm for solutions containing 50 to 24 ml of nickel; and at 545 nm for solutions containing 32 to 5 ml of nickel.

From the absorbance of the nickel-only solution at 622 nm, calculate the proportional absorbance for the other solutions due to nickel only present. Subtract the respective absorbance due to nickel only from the total absorbance measured, for each mixture, and plot the net absorbance versus the volume of nickel solution taken. At the maximum, the ratio: volume of ethylenediamine/volume of nickel, gives the mole ratio.

In a similar manner, estimate the contribution to the overall absorbance at 578 nm from the contributions of the 1 : 1 coordination species. Plot the net absorbance vs. the volume of nickel solution taken. At the maximum, the ratio: volume of ethylenediamine/volume of nickel, gives the mole ratio of the second species.

Proceed in a similar manner to ascertain the mole ratio of nickel and ethylenediamine at 545 nm.

PROBLEMS

1. A Duboscq colorimeter has a 50.0-mm scale for measuring the depth of each solution. The colors for a known and an unknown solution matched when the known solution was set at 40.0 mm and that for the unknown at 33.0 mm. Calculate how much iron must be in 50 ml of unknown solution if the known solution contained 2.00 mg of iron per 100 ml of solution.

2. The molar absorptivity of a particular solute is 2.1×10^4. Calculate the transmittance through a cuvette with a 5.00-cm light path for a 2.00×10^{-6} M solution.

3. With a certain photoelectric photometer using a 510-nm filter and 2-cm cells, the reading on a linear scale for P_0 was 85.4. With a 1.00×10^{-4} M solution of a colored substance, the value of P was 20.3. Calculate the molar absorptivity of the colored substance.

4. A substance is known to have a molar absorptivity of 14,000 at its wavelength of maximum absorption. With 1-cm cells, calculate what molarity of this substance could be measured in a spectrophotometer if the absorbance reading is to be 0.850.

5. The simultaneous determination of titanium and vanadium, each as their peroxide complex, can be done in steel. When 1.000-g samples of steel were dissolved, colors developed, and diluted to 50 ml exactly, the presence of 1.00 mg of Ti gave an absorbance of 0.269 at 400 nm and 0.134 at 460 nm. Under similar conditions, 1.00 mg of V gave an absorbance of 0.057 at 400 nm and 0.091 at 460 nm (see Fig. 4-18). For each of the following samples, 1.000 g in weight and ultimately diluted to 50 ml, calculate the percent titanium and vanadium from these absorbance readings:

Sample	A_{400}	A_{460}	Sample	A_{400}	A_{460}
1	0.172	0.116	5	0.902	0.570
2	0.366	0.430	6	0.600	0.660
3	0.370	0.298	7	0.393	0.215
4	0.640	0.436	8	0.206	0.130
			9	0.323	0.177

6. Develop equations for the spectrophotometric determination of the concentration of three substances present in a solution if the substances have the following specific absorptivity a at the following wavelengths. Express concentration in milligrams per milliliter. Let the cell thickness b be 1 cm.

Substance	400 nm	Wavelength 500 nm	600 nm
A	0	0	1.00
B	2.00	0.05	0
C	0.60	1.80	0

7. The absorption spectra for tyrosine and tryptophan in 0.1 N NaOH are shown in illustration (a) below. Select appropriate wavelengths for the simultaneous determination of each component in mixtures.

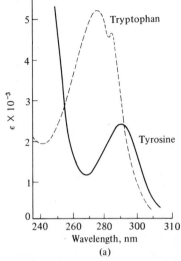

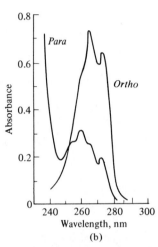

Problem 4-7. (a) Absorption spectra of tyrosine and tryptophan in intact proteins in 0.1 N NaOH. (b) Absorption spectra of p- and o-toluene sulfonamides (0.1 g/liter) in 0.1 N HCl.

8. Toluene sulfonamides in 0.1 N HCl have the absorption spectra shown in the figure for Problem 4–7(b). Select appropriate wavelengths for the determination of the *ortho* and the *para* fraction.

9. A mixture of sodium acetate and o-chloroaniline solution, 10 ml each, were titrated in glacial acetic acid at 312 nm with a 0.1010 N HClO$_4$ solution. Sodium acetate does not absorb in the ultraviolet portion of the spectrum, but it is a stronger base than o-chloroaniline. These results were obtained (corrected for dilution).

Volume of Titrant, ml	Absorbance	Volume of Titrant, ml	Absorbance
0.00	0.68	8.25	0.37
1.00	0.68	8.50	0.32
2.00	0.68	8.75	0.26
3.00	0.68	9.00	0.20
4.00	0.67	9.25	0.14
5.00	0.66	9.50	0.09
6.00	0.63	10.50	0.02
7.00	0.56	11.00	0.02
8.00	0.42	11.50	0.02

Plot the results and calculate the concentration of the original aliquots of sodium acetate and of o-chloroaniline.

10. Mixtures of two amines can be determined if their rates of acetylation are sufficiently different and possess separate absorption bands. The following titration of aniline and 2-naphthylamine was carried out at the wavelength of maximum absorption of 2-naphthylamine, the amine with the slower rate of acetylation. Ascertain the volume of acetic anhydride consumed in reaching each end point (all volumes corrected for dilution).

Volume of Titrant, ml	Absorbance	Volume of Titrant, ml	Absorbance
0.150	0.460	1.500	0.233
0.300	0.460	1.600	0.192
0.450	0.460	3.000	0.040
1.200	0.361	3.250	0.034
1.300	0.318	3.500	0.034
1.400	0.275	4.000	0.034

11. In a photometric titration of magnesium with 0.00130 M EDTA at 222 nm, the following procedure was employed. All reagents except the magnesium-containing solution were placed in the titration cell, and the slit width was adjusted to give zero absorbance. The following readings were observed after additions of the standard EDTA:

Absorbance	EDTA Added, ml
0.000	0.00
0.014	0.10
0.200	0.40
0.429	0.60
0.657	0.80
0.906	1.00

At this point, the magnesium solution was added and the absorbance fell to zero. The titration was continued with the following results:

Absorbance	EDTA Added, ml
0.000	1.00
0.020	1.50
0.065	2.00
0.160	2.50
0.240	2.80
0.360	3.00
0.580	3.20
0.803	3.40
1.000	3.60
1.220	3.80

Plot the results, explain the curves obtained, and calculate the number of micrograms of magnesium present, assuming that the reagent has exactly the concentration stated above.

12. A compound is known to contain four carbon atoms, one bromine atom, and one double bond. Ultraviolet spectral data is featureless above 210 nm. Write the structure.

13. Mesityl oxide exists in two isomeric forms: $CH_3-C(CH_3)=CH-CO-CH_3$ and $CH_2=C(CH_3)-CH_2-CO-CH_3$. One exhibits an absorption maximum at 235 nm with a molar absorptivity of 12,000; the other shows no high-intensity absorption beyond 220 nm. Identify the isomers.

14. Pyrethrolone contains a five-membered carbon ring, three ethylenic bonds, and one keto group; λ_{max} = 227 nm and ϵ_{max} = 28,000. Upon catalytic hydrogenation of the ethylenic groups, λ_{max} = 230 nm and ϵ_{max} = 12,000. Write the structural formula for pyrethrolone.

15. Assign the structures shown to the respective isomer on the basis of this information: the α-isomer shows a peak at 228 nm (ϵ = 14,000) while the β-isomer has a band at 296 nm (ϵ = 11,000).

structure I   structure II

16. From infrared spectral information, it is known that a compound whose empirical formula is $C_9H_{10}O$ possesses a benzene ring, a carbonyl group, a methylene group, and a methyl group. The placement of the carbonyl group is uncertain; however, the ultraviolet absorption spectrum shows these bands (with log ϵ_{max} values): 245 nm (4.1); 280 nm (3.1); and 320 nm (1.9). Write the structure of the compound.

17. Explain how the ultraviolet spectrum can be used to decide between the following isomeric systems:

(a)   (b)

(c)   (d)

18. The main ultraviolet absorption bands in light petroleum for three biphenyls occur as follows:

$$\lambda_{max} \ 259.5 \text{ nm } (\epsilon = 740), \text{ inflection at } 228 \text{ nm } (\epsilon = 6,000)$$

$$\lambda_{max} \ 250.5 \text{ nm } (\epsilon = 16,100)$$

$$\lambda_{max} \ 254.5 \text{ nm } (\epsilon = 21,000)$$

Assign these bands to compounds a, b, and c.

19. At a wavelength of 356 nm, the molar absorptivity of the phenolic alkaloid numismine in 0.1 M HCl is 400; in 0.2 M NaOH, 17,100. Determined in pH 9.50 buffer, the molar absorptivity is 9,800. Calculate the pK_a.

20. Calculate the dissociation constant of 2-nitro-4-chlorophenol in water at 25°C from this data taken at 427 nm. The absorbance of the undissociated phenol, measured in 0.01 M HCl, was 0.062. The absorbance of the fully ionized phenol, measured in 0.01 M NaOH, was 0.855. In a buffer solution of pH 6.22, the absorbance was 0.356.

21. Graph the data and determine the acid dissociation constant for these materials.

	p-Nitrophenol			Papaverine (Cation Form)	
	$\epsilon \times 10^{-3}$			$\epsilon \times 10^{-4}$	
pH	3170 Å	4070 Å	pH	2390 Å	2510 Å
3.0		0.33	2.0	3.36	5.90
4.0	9.72	0.33	3.0	3.36	5.90
5.0	9.72	0.50	4.0	3.39	5.83
6.0	9.03	1.66	5.0	3.48	5.63
6.2	8.61	2.28	5.6	3.86	5.19
6.4	8.19	3.99	5.8	3.93	4.91
6.6	7.36	5.14	6.0	4.30	4.61
6.8	6.39	7.22	6.2	4.61	4.15
7.0	5.55	9.16	6.4	4.86	3.71
7.2	4.45	11.65	6.6	5.22	3.30
7.4	3.61	13.40	6.8	5.46	2.77
7.6	2.92	15.00	7.0	5.66	2.51
7.8	2.08	16.90	7.4	6.03	2.00
8.0	1.81	17.50	8.0	6.27	1.63
9.0	1.39	18.33	11.0	6.43	1.56
10.0	1.39	18.33	12.0	6.44	1.56

22. Determine the acid dissociation constant for each indicator from the absorbance, at the wavelength of maximum absorbance, measured as a function of pH. Ionic strength was 0.05.

Bromophenol Blue		Methyl Red		Bromocresol Purple	
λ_{max} = 592 nm		λ_{max} = 530 nm		λ_{max} = 591 nm	
Absorbance	pH	Absorbance	pH	Absorbance	pH
0.00	2.00	2.00	3.20	0.00	4.00
0.18	3.00	1.78	4.00	0.24	5.40
0.58	3.60	1.40	4.60	0.66	6.00
0.98	4.00	0.92	5.00	0.87	6.20
1.43	4.40	0.48	5.40	1.13	6.40
1.75	5.00	0.16	6.00	1.37	6.60
2.10	7.00	0.00	7.00	1.72	7.00
				2.00	8.00

23. A series of chromium(III) nitrate solutions were measured according to the ordinary method with 0% T set with the phototube darkened and 100% T set with the pure solvent. From the results obtained at 550 nm, (a) calculate the relative concentration error for each measurement, assuming ΔT = 0.004; and (b) plot the results as relative concentration error vs. absorbance.

Concentration, M	Absorbance	Concentration, M	Absorbance
Blank	0	0.0300	0.357
0.0050	0.060	0.0400	0.476
0.0100	0.119	0.0500	0.595
0.0150	0.179	0.0600	0.714
0.0200	0.238	0.0800	0.952
0.0250	0.298	0.1000	1.190
		0.1100	1.309

24. From the series of solutions used in Problem 23, the 0.0500 M solution was used to set the 100% T reading. These results were obtained:

Concentration, M	Transmittance	Concentration, M	Transmittance
0.0500	1.000	0.110	0.230
0.0600	0.775	0.120	0.162
0.0700	0.584	0.130	0.120
0.0800	0.453	0.140	0.097
0.0900	0.380	0.150	0.074
0.1000	0.295		

(a) Calculate the relative concentration error for each measurement, assuming ΔT = 0.004. (b) Plot the results as relative concentration error vs. transmittance.

25. In using the low absorbance method a 0.100 M solution of chromium(III) nitrate was used as the standard with which the scale was set at 0% T by means of the zero-suppressor (dark-current) control. Pure solvent was used for setting the 100% T point. These results were obtained:

Concentration, M	Transmittance	Concentration, M	Transmittance
Blank	1.000	0.0500	0.197
0.0100	0.745	0.0600	0.130
0.0200	0.546	0.0700	0.077
0.0300	0.420	0.0800	0.037
0.0400	0.286	0.0900	0.012

(a) Graph the results as absorbance vs. concentration. (b) Calculate the relative concentration error for each measurement, assuming $\Delta T = 0.004$. (c) Plot the results as relative concentration error vs. transmittance.

26. With the maximum precision method, a 0.0500 M chromium(III) nitrate solution was used to set the 100% T point, and a 0.100 M solution was used to set the 0% T point. These results were obtained:

Concentration, M	Transmittance	Concentration, M	Transmittance
0.0500	1.000	0.0800	0.247
0.0600	0.695	0.0900	0.128
0.0700	0.437	0.100	0.000

(a) Graph these results as absorbance vs. concentration. (b) Calculate the relative concentration error for each measurement, assuming $\Delta T = 0.004$. (c) Plot the results as relative concentration error vs. transmittance.

27. In the determination of acetone in biological fluids, the following calibration curve between absorbance and concentration was obtained.

Acetone Standards, mg/100 ml	Absorbance
Reagent blank	0.045
0.5	0.057
1.0	0.069
2.0	0.092
4.0	0.137
6.0	0.182
8.0	0.229

(a) Calculate the molar absorptivity and the specific absorptivity. (b) Estimate the relative concentration error, and the increase in precision that would be obtained for each measurement if the solution containing 8.0 mg/100 ml were used for setting the zero transmittance reading. (c) Samples of blood and urine from a normal subject and from a ketotic patient were analyzed with these results: normal blood, $A = 0.068$; ketotic blood, $A = 0.189$; normal urine, 0.097; ketotic urine, $A = 0.198$ (1/25 dilution). Calculate the acetone concentration in each sample assuming absorbance values of the samples are net values after correction for the blank.

28. The absorbance of a series of phosphate solutions, using the phosphovanadomolybdate complex at 420 nm are as follows when referred to a 5.0-mg phosphate solution as reference standard (actual $A = 1.075$):

mg P_2O_5/100 ml	Absorbance
5.0	0.000
5.2	0.046
5.4	0.092
5.6	0.138
5.8	0.184
6.0	0.230
6.2	0.276

(a) Determine the increase in precision for the measurement of the 5.2-mg phosphate solution as compared with the normal method. (b) Estimate the relative concentration error for the 5.6-mg phosphate solution when the 6.0-mg phosphate solution is used to set the zero transmittance reading in addition to the 5.0-mg phosphate solution being used for the 100% transmittance reading.

29. A series of solutions is prepared in which the amount of iron(II) is held constant at 2.00 ml of 7.12×10^{-4} M, while the volume of 7.12×10^{-4} M, 1,10-phenanthroline is varied. After dilution to 25 ml, absorbance data for these solutions in 1.00-cm cuvettes at 510 nm is as follows:

1,10-Phenanthroline, ml	Absorbance
2.00	0.240
3.00	0.360
4.00	0.480
5.00	0.593
6.00	0.700
8.00	0.720
10.00	0.720
12.00	0.720

(a) Evaluate the composition of the complex. (b) Estimate the value of the formation constant of the complex.

30. The method of continuous variation was used to investigate the species responsible for the absorption at 510 nm when the indicated volumes of 6.72×10^{-4} M iron(II) solution were mixed with sufficient 6.72×10^{-4} M 1,10-phenanthroline to equal a total volume of 10.00 ml, after which the entire system was diluted to 25 ml. Cuvettes were 1.00 cm.

Iron(II), ml	Absorbance	Iron(II), ml	Absorbance
0.00	0.000	5.00	0.565
1.00	0.340	6.00	0.450
1.50	0.510	7.00	0.335
2.00	0.680	8.00	0.223
3.00	0.794	9.00	0.108
4.00	0.680	10.00	0.000

(a) Elucidate the composition of the ion-association complex. (b) Calculate the molar absorptivity of the complex.

31. Evaluate the composition of the iron(II)–1,10-phenanthroline complex with an absorption peak at 510 nm on the basis of the following absorbance data obtained, after dilution to 25 ml, in 1.00-cm cuvettes.

Iron(II) Constant at 5.00 ml of 7.00×10^{-4} M		Ligand Constant at 10.00 ml of 2.10×10^{-3} M	
7.00×10^{-4} M Ligand, ml	A_{510} nm	7.00×10^{-4} M Iron(II), ml	A_{510} nm
1.00	0.117	0.50	0.177
2.00	0.235	1.00	0.352
3.00	0.352	1.50	0.530
4.00	0.470	2.00	0.706
5.00	0.585	2.50	0.883

32. The dissociation of the complex between thorium and quercetin can be expressed as $ThQ_2 \rightleftharpoons Th + 2Q$ (omitting formal charges). For a solution that was 2.30×10^{-5} M in thorium and contained a large excess of quercetin, sufficient to ensure that all of the thorium is present as the complex, the absorbance was 0.780. When the same amount of thorium is mixed with a stoichiometric amount of quercetin, the absorbance was 0.520. Calculate (a) the degree of dissociation and (b) the value of the formation constant of the complex.

33. Estimate the relative concentration error due to stray light from this data:

Absorbance with Stray Light at:		
Zero	0.5%	5.0%
0.000	0.000	0.000
0.301	0.299	0.280
0.602	0.595	0.541
0.903	0.890	0.772
1.203	1.17	0.963

34. Suggest a method for monitoring low concentrations of phenols in waste water by differential ultraviolet photometry.

BIBLIOGRAPHY

Bauman, R. P., *Absorption Spectroscopy*, Wiley, New York, 1962.

Brode, W. R., *Chemical Spectroscopy*, Wiley, New York, 1943.

Cheng, K. L., *Advan. Anal. Chem. Instr.*, **9**, 321 (1971).

Forbes, W. R., in *Interpretive Spectroscopy*, S. K. Freeman, Ed., Van Nostrand Reinhold, New York, 1965, Chapter 1.

Mellon, M. G., *Analytical Absorption Spectroscopy*, Wiley, New York, 1950.

Silverstein, R. M., and G. C. Bassler, *Spectrometric Identification of Organic Compounds*, 2nd ed., Wiley, New York, 1967.

West, W., Ed., *Chemical Applications of Spectroscopy*, Vol. IX, Part 1, 2nd rev. ed., Wiley, New York, 1968.

LITERATURE CITED

1. Bastian, R., *Anal. Chem.*, **21**, 972 (1949).
2. Bastian, R., R. Weberling, and F. Palilla, *Anal. Chem.*, **22**, 160 (1950).
3. Bastian, R., *Anal. Chem.*, **23**, 580 (1951).
4. Crawford, M., *Anal. Chem.*, **31**, 343 (1959).
5. Dean, J. A., *Chemical Separation Methods*, D. Van Nostrand, New York, 1969.
6. Gilmore, J., *Anal. Chem.*, **38** (10) 27A (1966).
7. Goddu, R. F. and D. N. Hume, *Anal. Chem.*, **26**, 1679, 1740 (1954).
8. Headridge, J. B., *Photometric Titrations*, Pergamon, New York, 1961.
9. Hiskey, C. F., *Anal. Chem.*, **21**, 1440 (1949).
10. Hiskey, C. F., J. Rabinowitz, and J. G. Young, *Anal. Chem.*, **22**, 1464 (1950).
11. Hunter, R. S., *Off. Dig. Feder. Soc. Paint Technol.*, **35**, 350 (1963).
12. Kaye, W. I., *Appl. Spectr.*, **15**, 130 (1961).
13. Kortum, G. and M. Seiler, *Angew. Chem.*, **52**, 687 (1939).
14. Lang, L., *Absorption Spectra in the Ultraviolet and Visible Region*, Academic Press, New York, 1961-.
15. Milazzo, G. and G. Cecchetti, *Appl. Spectrosc.*, **23**, 197 (1969).
16. Poulson, R. E., *Appl. Opt.*, **3**, 99 (1964).
17. Reilley, C. N. and C. M. Crawford, *Anal. Chem.*, **27**, 716 (1955).
18. Sadtler Research Laboratories, *Ultraviolet Reference Spectra*, Philadelphia.
19. Slavin, W., *Anal. Chem.*, **35**, 561 (1963).
20. Wendlandt, W. W., *J. Chem. Educ.*, **45**, A861, A947 (1968).

Molecular Fluorescence
and Phosphorescence
Methods

The photoluminescent methods of fluorescence and phosphorescence are closely related to molecular absorption spectrophotometry. After molecules have absorbed radiant energy and been excited to a higher electronic state, they must lose their excess energy in order to return to the ground electronic state. *Fluorescence* is the immediate emission (in the order of 10^{-8} sec) of light from a molecule after it has absorbed radiation, as opposed to *phosphorescence* which is the delayed release of the absorbed energy. There is also a theoretical distinction; fluorescence arises from a singlet–singlet transition whereas a triplet–singlet transition is involved in phosphorescence.

FUNDAMENTAL MECHANISM

Electronic states of most organic molecules can be grouped into two broad categories of singlet and triplet states. In a singlet state all of the electrons in the molecule have their spins paired, whereas in a triplet state one set of electron spins are unpaired. A schematic energy-level diagram for a diatomic molecule is shown in Fig. 5-1. The abscissa represents the internuclear distance between the two atoms; the ordinate represents potential energy. For a more complex molecule, the potential energy has to be represented by a surface in a polydimensional space. The two-dimensional representation is then obtained by a suitable cross section through a surface. The values of the internuclear distance at the minima of the different curves represent the equilibrium configurations of the vibrating atoms in the corresponding electronic states.

The absorption of appropriate radiant energy by a molecule raises the molecule from a vibrational level in the ground state to one of many vibrational levels in one of the excited electronic levels, usually the first excited singlet state, S_1. The absorption step occurs within 10^{-15} sec. A number of vibrational levels of the excited state are populated immediately following absorption. However, molecules in a higher vibrational level of the excited singlet state quickly return to the lowest vibrational level of the excited state by transferring their excess energy to other molecules through collisions, as well as by partitioning the excess energy to other possible modes of vibration or rotation within the excited molecule.

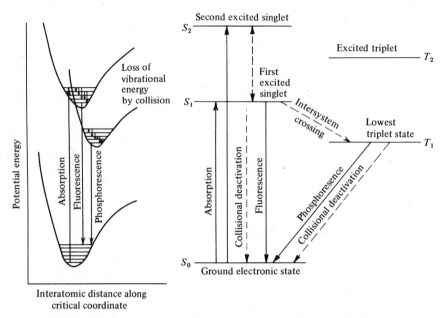

Fig. 5-1. Schematic energy-level diagram for a diatomic molecule.

Fluorescence results from the spontaneous radiative transition that occurs when molecules return to the ground electronic level.[1] This radiative process ($S_1 \rightarrow S_0$) has a short lifetime (about 10^{-8} sec) so that in many molecules it can compete effectively with other processes capable of removing the excitation energy, such as *internal conversion* or *intersystem crossing*. Of course, if the absorption process leads to an electronic state in which the energy exceeds the bond strength of one of the solute's linkages, then excitation energy is lost by molecular dissociation before fluorescence can take place.

If the potential energy curve of the singlet state crosses that of the triplet state,[7] some singlet-excited molecules may pass over to the lowest triplet state via an intersystem crossing which involves vibrational coupling between the excited singlet state, S_1, and the triplet state, T_1 (Fig. 5-1). Although singlet-triplet transitions are forbidden processes, there is some probability that the internal conversion from the excited singlet to the triplet state will occur, because the energy of the lowest vibrational level of the triplet state is lower than that of the singlet state. The probability of intersystem crossing is greater when the potential energy curves cross at the lowest point on the excited singlet curve. Once indirect occupation of the triplet state has been achieved, further vibrational energy will be lost by radiationless transitions to the vicinity of the zero-point vibrational level of the triplet state. Return from here to a vibrational level of the ground state constitutes phosphorescence emission, again a transition which will occur with low probability since spin reversal must once more occur. Consequently, the triplet state persists for a relatively long average lifetime and the rate of phosphorescence emission is very slow (10^{-2} to 100 sec). *Spin-orbit coupling*, which is a magnetic perturbation capable of flipping spins, is believed to be the main source of phosphorescence transitions back to the ground singlet state. While some deactivation of excited-state molecules may occur as a result of collisions with solvent molecules before the energy is reemitted as fluorescence, this effect becomes preponderantly important in phosphorescence due to the long lifetimes of $S_1 \to T_1$ and $T_1 \to S_0$ transitions. Phosphorescence is virtually unknown for dissolved molecules and can only be observed on any scale when the phosphor is frozen into glasses at low temperature so that collisional deactivation is prevented or at least severely restricted.

In both fluorescence and phosphorescence the lower energy photon is emitted in an arbitrary direction and at wavelengths longer than the excitation wavelength. Since raising a molecule to its excited state is a matter of total energy required, the emitted photoluminescence is at the same wavelength regardless of the wavelength of the absorbed energy. With the exception of X-ray fluorescence (Chapter 10), most of the work lies in the wavelength region between 2000 and 8000 Å. Fluorescence and phosphorescence each provide two kinds of spectra for identification—excitation and emission. The fact that fluorescence measurements can be made under a wider range of conditions than phosphorescence measurements accounts for the more extensive use of fluorescence spectroscopy.

EXCITATION AND EMISSION SPECTRA

The *excitation spectrum* is represented by a plot of the exciting wavelength against the luminescent intensity observed at the wavelength of a luminescence maximum. In practice, this spectrum is obtained by setting the emission spectrometer to an emission wavelength of the sample and then scanning a range of wavelengths with the excitation spectrometer. By adjusting the excitation spectrometer to an excitation wavelength for a sample and causing the emission spectrometer to scan, the *emission spectrum* of the sample will be recorded.

The emission and excitation spectra taken from a sample of anthracene are shown in Fig. 5-2. Curve *A* in bold outline from 3500 to 5000 Å is the emission spectrum; curve *B* in light outline from 2200 to 3900 Å is the excitation spectrum of anthracene. The wavelength positions of each band in the absorption spectrum are the same as for the bands in the excitation spectrum. This means that the excitation spectrum would provide the absorption characteristics of a sample but at concentrations much lower than it would be possible to use if working in absorption spectrophotometry.

The frequency of the wave produced by the vibrationless transition characterizes the transition from the zero-point level of the excited electronic state to the zero-point level of the ground state. It will be the highest frequency band of the phosphorescence spectrum, but it will occur at lower frequencies than a fluorescence band originating from a corresponding transition. There is a gap between the highest phosphorescent and the lowest absorption frequency (see Fig. 5-9). The vibrational emission pattern represents the spacings of the vibrational levels of the ground electronic state of the emitting molecule.

When a scanning spectrometer is used at constant slit width, the excitation and emission spectra will be a function of the source emission, band width of the mono-

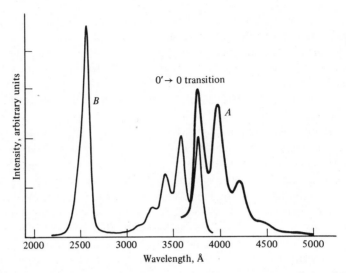

Fig. 5-2. Emission spectrum, *A*, of anthracene shows mirror image relationship with excitation spectrum, *B*.

chromator, and detector characteristics. For purely routine work, uncorrected spectra are usually adequate, but it is desirable to correct the spectra if the information is to be of maximum value to other workers. An energy-recording, or better, a quantum-recording spectrofluorophotometer circumvents the tedious point-by-point corrections.[4,5,6]

INSTRUMENTATION FOR FLUORESCENCE MEASUREMENTS

Commercial instruments range from simple filter fluorometers to highly sophisticated spectrophotofluorometers.[3] Each will contain four principal components: (1) a source of excitation energy, (2) a sample cuvette, (3) a detector to measure the photoluminescence, and (4) a pair of filters or monochromators for selecting the excitation and emission wavelengths.

A *filter fluorometer* (sometimes spelled fluorimeter), shown in Fig. 5-3, is set up so that the source light passes through the primary (excitation) filter and an aperture plate, directly to the sample cuvette. The aperture plate controls the amount of light incident on the sample; other devices that are used include a V-shaped variable slit or a guillotine shutter. Primary filters are selected to pass only specific excitation wavelengths; these are often one of the main spectral lines of a mercury vapor lamp which occur at 303, 313.5, 334, 365, 405, 436, 546, and 577/579 nm. Fluorescent light, generated when the excitation light strikes the sample, passes out of the cuvette through the secondary filter, which is selected to pass only wavelengths corresponding to maximum fluorescent emission and to reject any scattered excitation light. The fluorescent emission strikes the detector, usually a photomultiplier tube, which transforms the light into electrical energy which

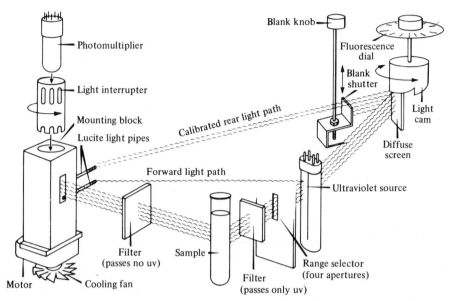

Photomultiplier

Light interrupter

Mounting block

Lucite light pipes

Calibrated rear light path

Forward light path

Filter
(passes no uv)

Sample

Motor   Cooling fan

Filter
(passes only uv)

Range selector
(four apertures)

Ultraviolet source

Blank knob

Fluorescence
dial

Blank
shutter

Light
cam

Diffuse
screen

Fig. 5-3. Details of the Turner model 110 fluorometer optical system. (Courtesy of Turner Instrument Co.)

may then be amplified and displayed on a readout meter or recorded. A single detector with a light interrupter is used in the optical bridge circuit illustrated to measure the light differential between the fluorescent emission and a standard calibrated (rear light path) beam. Rotation of the reading (fluorescence) dial and connected diffuse screen adjusts the calibration beam to equal the sample emission intensity. When balanced, the photomultiplier detects no difference in signal intensity and thus is a null-balance instrument. This circuit arrangement cancels out variations in line voltage, light source, and photomultiplier sensitivity.

Problems involved in filter selection include "cross-talk" in which the longer wavelengths that are transmitted through the primary filter also pass through the shorter wavelength region of the secondary filter, resulting in a high blank, and the difficulty of finding filters with the nominal wavelength matching the sample excitation characteristics. Secondary filters are usually the short-wavelength cutoff type (see Fig. 3-12).

A true fluorescence spectrometer possesses a pair of monochromators, one providing the narrow band selection of exciting radiation and the other the narrow band selection of emission radiation for passage to the detector. Figure 5-4 is a schematic functional diagram of the instrument optics. As described in Chapter 3, a double monochromator can be substituted for each single unit. The primary light source is a xenon arc (usually 150 W, which emits a continuous spectrum from 2000 to 8000 Å) whose output passes into the first (excitation) monochromator that disperses the light and provides monochromatic radiation to excite the sample. The excited sample becomes the source for the second (emission) monochromator. A fluorescence spectrophotometer can provide two kinds of spectra, excitation and emission, from any fluorescent sample. An optical filter

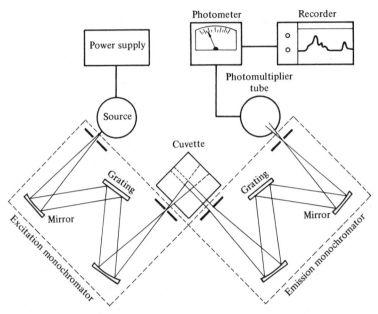

Fig. 5-4. Schematic diagram of a fluorescence spectrophotometer.

may be inserted into the light path between the cuvette and the excitation mono-chromator if it is desired to reject second-order excitation from a grating monochromator when excitation wavelengths longer than 4000 Å are being used.

Fluorescence measurements usually are made by reference to some arbitrarily chosen standard. The standard is placed in the instrument and the circuit balanced with the reading scale at any chosen setting. Without readjusting any circuit components, the standard is replaced by standard solutions of the test material and the fluorescence of each recorded. Finally, the fluorescence of the solvent and cuvette alone is measured to establish the true zero concentration. Some fluorometers are equipped with a zero-adjust circuit. A plot of scale (fluorescence) readings against concentration of the reference solutions furnishes the fluorescence–concentration curve. The initial standard solution often represents the highest concentration of test material anticipated. In order to adjust the instrument quickly to some definite setting for routine analyses, it is convenient to use reference glasses or fluorescence standards. Some of the most commonly used fluorescence standards are quinine bisulfate in $0.1N\,H_2SO_4$, tryptophan in water, and anthracene in cyclohexane, ethanol, or benzene.

To adapt a spectrofluorometer to record energy, two control loops are added to the instrument.[6] A constant proportion of the energy emerging from the excitation mono-chromator is directed to a reference thermocouple whose response is essentially constant with wavelength. The thermocouple signal controls a continuously variable slit servo mechanism on the excitation monochromator. This servo loop keeps the thermocouple signal constant and also keeps the output energy from the excitation monochromator constant, thus correcting for variations of source brightness as a function of wavelength.

A second servo, attached to the slits of the fluorescence monochromator, is operated from an adjustable electrical cam which is set to compensate for the variation of the spectral response of the detector and the other wavelength-dependent variables. The cam adjustment is made by putting a neutral scattering material in the sample position and using the excitation system as a source of constant energy.

To record the spectrum on a quantum rather than on an energy basis, the thermocouple is replaced with a screen of a material known to have constant quantum efficiency and which absorbs all the incident radiation.[5] From this screen, a photomultiplier detects the fluorescence signal which is proportional to the quanta in the beam. If this is allowed to control the slit servo, spectra result in which the ordinate is in units of quanta.

The cuvettes used in fluorometers are made of glass, if the exciting wavelengths are above 3200 Å, or fused silica for work at shorter wavelengths. All cells should be checked for fluorescence of their own and rejected if they have such. Cuvettes that have become scratched through usage will scatter light excessively and should be discarded.

Figure 5-5 highlights two viewing modes. At the top is the 37° viewing mode in which both the exciting and the resulting emitted radiation interact with only the front surface of the horizontally placed sample. This mode is generally useful for all types of samples, but is essential for the measurement of opaque samples, turbid solutions, concentrated solutions, or thin-layer (paper) chromatograms. The mode in the center illustrates the optical path of the 90° viewing mode, generally limited to dilute solutions or gases, or for making measurements of fluorescence polarization. In the classic 90° viewing mode, the excitation radiation passes through a fairly long solution path so that the output radiation may be affected by the absorbance of the sample. Use of a front surface type of sampling mode tends to minimize this effect. In Fig. 5-6 note the differences in both curve shape and wavelength of maximum intensity between the two sampling modes. The shift to higher wavelengths shown in the 90° mode is caused by the reabsorption of the yellow fluorescence in the blue region.

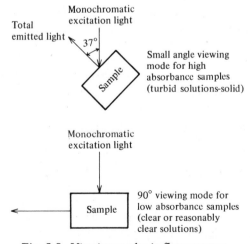

Fig. 5-5. Viewing modes in fluorescence.

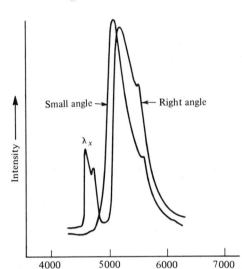

Fig. 5-6. Fluorescence emission spectrum of fluorescein taken at two sampling modes.

INSTRUMENTATION FOR PHOSPHORESCENCE MEASUREMENTS

Instrumentation for phosphorescence investigations is identical to that described for fluorescence measurements with the addition of a rotating slotted disk and provision for immersion of the sample in a Dewar for liquid nitrogen temperatures. Excitation light from a xenon lamp, after dispersion by the excitation monochromator, is admitted to the sample via a fixed slit system and a set of slotted disks with equally spaced ports rotating around the sample cuvette. The slots are so arranged that openings in one are in line with uncut portions of the second disk. Thus the sample is alternately exposed to the excitation light and then darkened; while it is dark the phosphorescence is passed to the emission monochromator and thence to the detector system. The cuvette is a small Dewar made of fused silica and silvered, except in the region where the optical path traverses the Dewar. The solvent frequently used is a mixture of diethyl ether, isopentane, and ethanol in a volume ratio of 5 : 5 : 2, respectively. It is referred to as the *EPA mixed solvent*. When cooled to liquid nitrogen temperature, it gives a clear transparent glass.

The resolution time of the instrument is the length of time between the cutoff of each pulse of excitation light admitted to the sample and the clearing of the optical path by the second disk to allow the phosphorescence emission to enter the emission monochromator. The time is a function of the motor speed, the size and spacing of the ports, and the relative radial positions of the ports to each other. Decay curves can be recorded if the detector circuit is equipped with an oscillograph. These curves follow the decay in the intensity of phosphorescence with time after removal of the excitation light.

PHOTOLUMINESCENCE INTENSITY AS RELATED TO CONCENTRATION

The relationship of concentration and fluorescence (or phosphorescence) may be derived from Beer's Law. The fraction of light transmitted is

$$\frac{P}{P_0} = e^{-\epsilon bC} \tag{5-1}$$

and the corresponding fraction of light absorbed is

$$1 - \frac{P}{P_0} = 1 - e^{-\epsilon bC} \tag{5-2}$$

Rearranging, the amount of light absorbed is

$$P_0 - P = P_0(1 - e^{-\epsilon bC}) \tag{5-3}$$

The total fluorescence intensity is proportional to the quanta of light absorbed and to the fluorescence efficiency, ϕ, which is the ratio of quanta absorbed to quanta emitted.

$$F = (P_0 - P)\phi = P_0\phi(1 - e^{-\epsilon bC}) \tag{5-4}$$

For very dilute solutions in which not over 2% of the total excitation energy is absorbed and the term ϵbC is not greater than 0.05 (0.01 in phosphorescence), Eq. 5-4 reduces to

$$F = k\phi P_0 \epsilon bC \tag{5-5}$$

where the term k has been introduced to handle instrumental artifacts and the geometry factor because fluorescence is emitted in all directions but is viewed only through a limited aperture. Therefore, the maximum concentration that will give a linear calibration curve will be $C_{max} = 0.05/\epsilon b$, where ϵ is the molar absorptivity at the wavelength of excitation.

When the concentration increases above C_{max}, the plot of fluorescence intensity against concentration becomes nonlinear, although a calibration curve can be constructed to extend the useful range at least another order of magnitude above C_{max}. Eventually, when absorption of incident radiation is almost complete, Eq. 5-5 becomes: $F = k\phi P_0$; that is, the detector signal is independent of fluorescer concentration. Before this situation prevails the fluorescence becomes noticeably concentrated where the incident radiation enters the sample. Although the exciting radiation does penetrate the sample, it is not evenly distributed along the sample length.

Equation 5-5 suggests a possible way to increase sensitivity. Any decrease in concentration, C, may be balanced against a corresponding increase in the excitation radiation, P_0. This approach is limited by the emission and interference characteristics of impurities and solvent which become proportionally more significant as concentration is decreased; also photodecomposition must not occur. There is a fundamental difference between absorption spectrophotometry, in which the detection limit is set by the minimum detectable difference in intensity between the incident and transmitted radiation, and fluorescence where the detection limit is set essentially by the noise level of the detector–amplifier circuit. Fluorometric techniques can readily result in the determination of material at very low concentrations and are 1,000 to 10,000 times more sensitive than absorption techniques. There is also the additional specificity inherent in fluorescence because it has spectral requirements for both emission and absorption. The following situations are possible: (1) Each compound may absorb at identical wavelengths, but only one may fluoresce. (2) Each compound may absorb at the same wavelength but fluoresce at different wavelengths. (3) If each compound absorbs and emits at identical wavelengths, it is often possible to quench the fluorescence of one by the use of selected reagents. Phosphorescence is more specific than fluorescence because time-resolved techniques will distinguish phosphorescent from fluorescent emissions and, perhaps, will distinguish between several phosphorescent emissions if lifetimes differ significantly.

The disadvantage of fluorescence is that it is strongly temperature dependent, and the intensity of fluorescence depends on the excitation intensity. Neither of these is true for spectrophotometry. Hercules has compared fluorescence with ultraviolet spectrophotometry.[2]

CORRELATION OF PHOTOLUMINESCENCE WITH MOLECULAR STRUCTURE[8]

Fluorescence may be expected generally in molecules that are aromatic or contain multiple-conjugated double bonds with a high degree of resonance stability, although substituents may alter greatly the degree of fluorescence. A substituent which delocalizes the π-electrons, such as NH_2, OH, F, OCH_3, $NHCH_3$, and $N(CH_3)_3$ groups, is likely to increase the fluorescence, whereas those containing Cl, Br, I, $NHCOCH_3$, NO_2, or COOH are apt to decrease or completely quench the fluorescence. Polycyclic aromatic systems in which the number of π-electrons available is greater than in benzene, and the derivatives of such systems, are usually much more fluorescent than benzene and its derivatives. The fluorescence of heterocyclic systems depends upon the nature of the hetero atom and upon substituents. Heterocyclic systems containing doubly bound nitrogen tend to be nonfluorescent unless there are substituents present which counteract the localization of electrons on the nitrogen. Thus, pyridine is nonfluorescent whereas 3-hydroxypyridine is fluorescent because of the effect of the electron-donating OH group. Ring closure is conducive to the fluorescence of aromatic compounds. Compare, for example, fluorescein, a fluorescing molecule, with phenolphthalein, which is nonfluorescent. The bridging-oxygen induces rigidity in fluorescein. Metal chelate formation often promotes fluorescence. Also, fluorescing structures are coplanar with respect to the chromophore, whereas nonfluorescing structures exhibit a molecular crowding which interferes with coplanarity.

Introduction of paramagnetic metal ions, such as copper(II) and nickel(II), gives rise to phosphorescence but not fluorescence in metal complexes. By contrast, magnesium and zinc compounds show strong fluorescence but no phosphorescence. Generally only those cations which are diamagnetic when coordinated and are nonreducible will form fluorescent complexes. The transition metals with unfilled outer d orbitals will quench fluorescence completely. On the other hand, whereas paramagnetic species quench fluorescence, they strongly promote intersystem crossing so that at low temperature, those cations will be observed to promote phosphorescence.

Phosphorescence lifetimes are also affected by molecular structure. Unsubstituted cyclic and polycyclic hydrocarbons and their derivatives containing CH_3, NH_2, OH, COOH, and OCH_3 substituents have lifetimes in the range of 5 to 10 sec for most benzene derivatives, and 1 to 4 sec for many naphthalene derivatives. The nitro group diminishes the intensity of phosphorescence and the lifetime of the triplet state to about 0.2 sec. Aldehydic and ketonic carbonyl groups diminish the lifetime to about 0.001 sec. Introduction of bulky substituents which force a planar configuration to become nonplanar markedly shortens lifetimes.

Fluorescence intensity and wavelength often vary with solvent. Solvents capable of operating strong van der Waal's binding forces with the excited state molecule will prolong the lifetime of a collisional encounter and favor deactivation. Solvents which possess molecular substituents such as Br, I, NO_2, or N=N groups are undesirable because the strong magnetic fields which surround their bulky atomic cores promote spin decoupling of electrons and triplet-state formation which give rise to a marked quenching of fluorescence although these solvents may promote phosphorescence. Indole illustrates the wavelength shifts that may occur. It shows the same maximum excitation at 2850 Å in each solvent, but the wavelength of maximum fluorescence is 2970 Å in cyclohexane, 3050 Å in benzene, 3100 Å in dioxane, 3300 Å in ethanol, and 3500 Å in water. Fluorescence of chlorophyll is remarkably enhanced upon the addition of a polar solvent to dry nonpolar solvents. Pyridoxine fluoresces at 3350 Å in dioxane and at 4000 Å in water, due to ionization. Substances fluoresce more brightly in a rigid state or in viscous solutions. The probability of intermolecular energy transfer between the fluorescer and other molecules tends to be reduced at low temperature and in a medium of high viscosity in which the rotational relaxation time of the fluorescer is much longer than the lifetime of the excited state.

Changes in the pH of the system, if it affects the charge status of the chromophore, may have an influence on fluorescence. Both phenol and anisole fluoresce at 3000 and 3100 Å at pH 7, but at pH 12 phenol is converted into the nonfluorescent phenoxide ion whereas anisole remains unchanged. Similarly, aniline fluoresces at pH 7 and 12, but the

TABLE 5-1 Typical Fluorescence Indicators

Name of Indicator	pH Range	Color Change
3,6-Dihydroxyphthalimide	0.0– 2.5	Colorless to yellow-green
Erythrosin B	2.5– 4.0	Colorless to green
Chromotropic acid	3.0– 4.5	Colorless to blue
Fluorescein	4.0– 6.0	Colorless to green
β-Naphthoquinoline	4.4– 6.3	Blue to colorless
Umbelliferone	6.5– 8.0	Faint blue to bright blue
o-Coumaric acid	7.2– 9.0	Colorless to green
Naphthol AS	8.2–10.3	Colorless to yellow-green

anilinium ion is nonfluorescent at pH 2. In fact, some substances are so sensitive to pH that they can be used as indicators in acid–base titrations (Table 5-1). The merit of such indicators is that they can be employed in turbid or intensely colored systems. The solution to be titrated is placed in a dark box, illuminated with a mercury lamp equipped with a black glass envelope, and the progress of the titration observed visually through a viewing port.

LABORATORY WORK

Experiment 5-1   Determination of Quinine

Standard solution of quinine Dissolve 0.100 g of quinine bisulfate in 0.1 N H_2SO_4 solution and dilute to 1 liter with additional acid solution. Dilute 10.0 ml of the foregoing solution to 1 liter with 0.1 N acid solution. The resulting solution contains 0.00100 mg/ml of quinine bisulfate.

Dilute sulfuric acid, 0.1 N Add 3 ml of concentrated sulfuric acid to 100 ml of water and dilute to 1 liter in a graduated cylinder.

Procedure Pipet 10.0, 25.0, 35.0, and 50.0 ml of the dilute standard quinine solution into a set of 100-ml volumetric flasks. Dilute to the mark with 0.1 N H_2SO_4 solution.

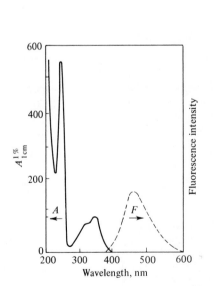
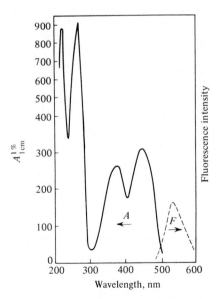

Fig. 5-7. Absorption and fluorescence spectra of quinine bisulfate in 0.1 N H_2SO_4.
Fig. 5-8. Absorption and fluorescence spectra of riboflavin in water.

Treat the unknown sample similarly. Measure the fluorescence and prepare a calibration curve.

The fluorescence of quinine is constant in the concentration interval from 0.01 to 0.2 N H_2SO_4. Proper primary and secondary filters, or wavelengths for excitation and fluorescence measurement, can be ascertained from Fig. 5-7.

Experiment 5-2　Determination (Simplified) of Riboflavin

Standard riboflavin solution　Dissolve 10.0 mg of riboflavin in 1 liter of 1% acetic acid solution. Keep the solution in a cool, dark place. This solution contains 10.0 μg/ml of riboflavin.

Procedure　Prepare a series of standard riboflavin solutions, the strongest of which does not contain more than 1.0 μg/ml of riboflavin. Prepare a calibration curve. Appropriate excitation and fluorescence wavelengths can be deduced from the spectra in Fig. 5-8. One suggested set of filters is Corning 3–73 plus 5–58 ($\frac{1}{2}$ standard width) in the primary beam and Corning 3–69 in the secondary beam.

Experiment 5-3　Selection of a Proper Pair of Filters

Prepare a series of standard fluorescence solutions. Insert a primary and a secondary filter into the fluorometer. Balance the fluorometer at about 50% of the normal full-scale reading when the strongest standard solution is employed. Preliminary selection of a primary filter is made by observing visually the fluorescence of the strongest standard solution when different primary filters are inserted in the exciting beam. Likewise, secondary filters can be selected by holding them between the eye and the top of the cuvette when the sample is irradiated with the exciting beam.

Measure the fluorescence reading of each of the remaining solutions in the series. Include a blank in the series. Plot your results.

Now remove the primary filter and replace with another. Without changing any instrument settings, measure the fluorescence of each member of the standard series of solutions, including the blank. Plot your results. Repeat these operations until all available primary filters have been used.

From the preceding results, select the primary filter which yields the steepest slope (and perhaps the smallest reading for the blank), and change the secondary filters, one at a time. For each secondary filter measure the fluorescence of the series of standard solutions. Plot your results.

On the basis of the complete sets of data, choose the optimum pair of filters. Factors to be considered include linearity of the calibration curve, slope of the curve, and minimal fluorescence of the blank.

Experiment 5-4   Excited State Dissociation

Prepare a series of 10^{-6} M solutions of 2-naphthol with pH varying from 1 to 12. To avoid interferences, buffers should not be used. Dilute 1 ml of a stock solution of 2-naphthol (10^{-4} M) to 100 ml with a solution which has been adjusted to the approximate pH with sodium hydroxide or sulfuric acid. Suggested solutions are one each of pH 1.0, 2.0, 2.3, 2.5, 2.8, 3.0, 3.3, 4.0, 6.0, 8.0, 8.5, 9.0, 9.3, 9.5, 9.8, 10.0, 10.5, 11.0, and 12.0. The pH of the resultant solutions must be measured with a pH meter.

Measure the ionic fluorescence intensity at 429 nm using quartz cuvettes in an instrument equipped with a source capable of exciting molecular 2-naphthol at 330 nm and below. Graph the results and estimate the two pK_a values.

If a spectrofluorometer is available, one can obtain simultaneously both the molecular and the ionic fluorescence intensities as a function of pH. The former is observed at 359 nm. The absorption spectra should also be obtained for both species (pH 1 and 12).

For additional discussion and presentation of results, see the article by D. W. Ellis, *J. Chem. Educ.*, **43**, 259 (1966).

PROBLEMS

1. A 1.00-g sample of a cereal product was extracted with acid and treated so as to isolate the riboflavin plus a small amount of extraneous material. The riboflavin was oxidized by the addition of a small amount of $KMnO_4$, the excess of which was removed by H_2O_2. The solution was transferred to a 50-ml volumetric flask and diluted to the mark. A 25-ml portion was transferred to the sample holder and the fluorescence measured. Initially the fluorometer had been adjusted to read 100 scale divisions with a solution of quinine bisulfate. The solution read 6.0 scale divisions. A small amount of solid sodium dithionite was added to the cuvette to convert the oxidized riboflavin back to riboflavin. The solution now read 55 scale divisions. The sample was discarded and replaced in the same cuvette by 24 ml of the oxidized sample plus 1 ml of a standard solution of riboflavin which contains 0.500 $\mu g/ml$ of riboflavin. A small amount of solid sodium dithionite was added. The solution read 92 scale divisions. Calculate the micrograms of riboflavin per gram of cereal.

2. Solutions of varying amounts of aluminum were prepared, 8-quinolinol was added, and the complex was extracted with chloroform. The chloroform extracts were all diluted to 50 ml and compared in a fluorometer. These readings were obtained:

Aluminum $\mu g/50$ ml	Fluorometer Reading	Aluminum $\mu g/50$ ml	Fluorometer Reading
2	10	12	53
4	19	14	60
6	28	16	66
8	37	18	71
10	45		

Plot the fluorometer reading vs. the aluminum concentration. Does the calibration curve follow the relationship predicted by Eq. 5-5? Over what concentration range is the approximation of Eq. 5-5 valid?

3. From the information given in Fig. 5-2, select the optimum wavelength for excitation and for fluorescence emission. Do the same for the quinine and riboflavin systems, Figs. 5-7 and 5-8.

4. Repeat Problem 3, but assuming this time that only a filter fluorometer is available. Select the optimum primary and secondary filters.

5. From the appropriate spectra for naphthalene (Fig. 5-9) and phenanthrene (Fig. 5-10), devise a method of analysis for each component assuming that a fluorescence spectrometer and a phosphorescence spectrometer is available.

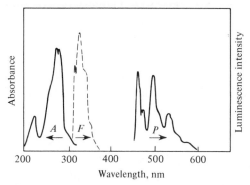

Fig. 5-9. Absorption, fluorescence emission, and phosphorescence emission spectra of naphthalene at 77°K in EPA.

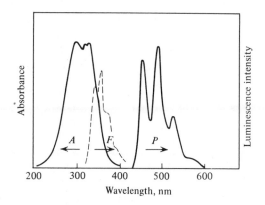

Fig. 5-10. Absorption, fluorescence emission, and phosphorescence emission spectra of phenanthrene at 77°K in EPA.

6. The phosphorescence emission spectrum of 4-nitrobiphenyl ($\tau = 0.080$ sec), benzaldehyde ($\tau = 0.006$ sec), and benzophenone ($\tau = 0.006$ sec) are given in Fig. 5-11. Outline a scheme of analysis for each component based on considerations of wavelength
of phosphorescence and the phosphorescence lifetimes.

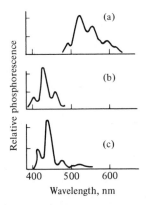

Fig. 5-11. Phosphorescence emission spectra for (a)
4-nitrobiphenyl, (b) benzaldehyde, (c) benzophenone,
each at 77°K in EPA.

7. From the phosphorescence plus fluorescence spectrum and the phosphorescence
spectrum alone, the following peaks were obtained for trivalent metal chelates of
dibenzoyl methane. Construct approximate energy-level diagrams for the fluorescence
and phosphorescence transitions to the vibrational levels in the ground electronic
state.

	Metal Ion		
	Al	Sc	Y
λ_{ex} for F, Å	4175	4250	4290
λ_{ex} for P, Å	4785	4850	4910
Phosphorescence, Å	4783	4840	4910
	4950	5000	5080
	5120	5150	5250
	5308	5370	5480
	5480	5560	5800
	5755	5735	
Phosphorescence plus	4171	4340	4283
Fluorescence, Å	4400	4500	4520
	4676	4840	4890
	4810	5150	5080
	4950	5200	5250
	5120	5370	5480
	5303	ca 5560	ca 5690
	ca 5500		
	ca 5720		

8. From the spectra for naphthalene at 77°K, construct an approximate energy-level diagram for the fluorescence and phosphorescence transitions to the vibrational levels in the ground electronic state.

Phosphorescence peaks (λ_{ex} = 310 nm): 475, 490, 515, 565 nm

Fluorescence peaks (λ_{ex} = 274 nm): 310, 320, 335, 355 nm

BIBLIOGRAPHY

Guilbault, G. G., Ed., *Fluorescence*, Marcel Dekker, New York, 1967.

Hercules, D. M., Ed., *Fluorescence and Phosphorescence Analysis*, Wiley-Interscience, New York, 1966.

Parker, C. A., *Photoluminescence in Solutions*, American Elsevier, New York, 1968.

Passwater, R. A., *Guide to Fluorescence Literature*, Plenum, New York, 1967.

Radley, J. A. and J. Grant, *Fluorescence Analysis in Ultraviolet Light*, 4th ed., Van Nostrand Reinhold, New York, 1953.

Udenfriend, S., *Fluorescence Assay in Biology and Medicine*, Academic, New York, 1964.

Winefordner, J. D., S. G. Schulman, and T. C. O'Haver, *Luminescence Spectrometry in Analytical Chemistry*, Wiley-Interscience, New York, 1972.

Zander, M., *Phosphorimetry*, Academic, New York, 1968.

LITERATURE CITED

1. Flygare, W. H., *Accounts Chem. Res.*, **1**, 121 (1968).
2. Hercules, D. M., *Anal. Chem.*, **38** (12) 39A (1966).
3. Lott, P. F., *J. Chem. Educ.*, **41**, A327, A421 (1964).
4. Mehluish, W. H., *J. Phys. Chem.*, **64**, 762 (1960).
5. Parker, C. A. and W. T. Rees, *Analyst*, **85**, 587 (1960).
6. Slavin, W., R. W. Mooney, and D. T. Palumbo, *J. Opt. Soc. Am.*, **51**, 93 (1961).
7. Turro, N. J., *J. Chem. Educ.*, **46**, 2 (1969).
8. Williams, R. T. and J. W. Bridges, *J. Clin. Pathol.*, **17**, 371 (1964).

Infrared Spectroscopy

The infrared region of the electromagnetic spectrum extends from the red end of the visible spectrum to the microwaves; that is, the region includes radiation at wavelengths between 0.7 and 500 μm or, in wavenumbers, between 14,000 cm^{-1} and 20 cm^{-1}. The spectral range of greatest use is the mid-infrared region, which covers the frequency range from 200 cm^{-1} to 4000 cm^{-1} (50 to 2.5 μm). Infrared spectroscopy involves the twisting, bending, rotating, and vibrational motions of atoms in a molecule. Upon interaction with infrared radiation, portions of the incident radiation are absorbed at particular wavelengths. The multiplicity of vibrations occurring simultaneously produces a highly complex absorption spectrum, which is uniquely characteristic of the functional groups comprising the molecule and of the overall configuration of the atoms as well.

MOLECULAR VIBRATIONS

Atoms or atomic groups in molecules are in continuous motion with respect to each other. The possible vibrational modes in a polyatomic molecule can be visualized from a mechanical model of the system, shown schematically in Fig. 6-1. Atomic masses are represented by balls, their weight being proportional to the corresponding atomic weight, and arranged in accordance with the actual space geometry of the molecule. Mechanical springs, with forces that are proportional to the bonding forces of the chemical links, connect and keep the balls in positions of balance. If the model is suspended in space and struck a blow, the balls will appear to undergo random chaotic motions. However, if the vibrating model is observed with a *stroboscopic* light of variable frequency, certain light frequencies will be found at which the balls appear to remain stationary. These represent the specific vibrational frequencies for these motions.

For infrared absorption to occur, two major conditions must be fulfilled. First, the energy of the radiation must coincide with the energy difference between the excited and ground states of the molecule. Radiant energy will then be absorbed by the molecule, increasing its natural vibration. Second, the vibration must entail a change in the electrical dipole moment, a restriction which distinguishes infrared from Raman spectroscopy.

Stretching vibrations involve changes in the frequency of the vibration of bonded atoms along the bond axis. The vibrational modes for a methylene group are illustrated in Fig. 6-2. In a symmetrical group such as methylene there are identical vibrational frequencies. For example, the asymmetric vibration (b) occurs in the plane of the paper

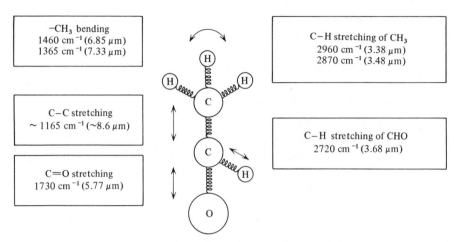

Fig. 6-1. Vibrations and characteristic frequencies of acetaldehyde.

and also in the plane at right angles to the paper. In space these two are indistinguishable and said to be one *doubly degenerate* vibration. In the symmetric stretching mode (a) there will be no change in the dipole moment as the two hydrogen atoms move equal distances in opposite directions from the carbon atom, and the vibration will be infrared inactive. In (b), however, there will be a change in the dipole moment, since during these vibrations the centers of highest positive charge (hydrogen) and negative charge (carbon)

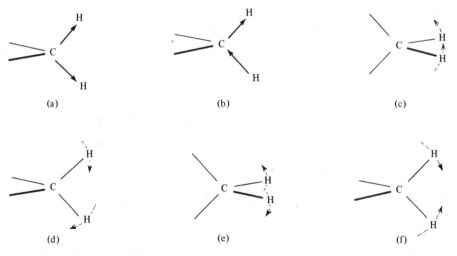

Fig. 6-2. Vibrational modes of the H—C—H group. (a) Symmetrical stretching, (b) asymmetrical stretching, (c) wagging or out-of-plane bending, (d) rocking or asymmetrical in-plane bending, (e) twisting or out-of-plane bending, (f) scissoring or symmetrical in-plane bending.

will move in such a way that the electrical center of the group is displaced from the carbon atom. These vibrations will be observed in the infrared spectrum of the methylene group.

When a three-atom system is part of a larger molecule, it is possible to have bending or deformation vibrations. These are vibrations which imply movement of atoms out from the bonding axis. Four types can be distinguished:

1. *Deformation or scissoring* The two atoms connected to a central atom move toward and away from each other with deformation of the valence angle.

2. *Rocking or in-plane bending* The structural unit swings back and forth in the symmetry plane of the molecule.

3. *Wagging or out-of-plane bending* The structural unit swings back and forth in a plane perpendicular to the molecule's symmetry plane.

4. *Twisting* The structural unit rotates back and forth around the bond which joins it to the rest of the molecule.

Splitting of bending vibrations due to in-plane and out-of-plane vibrations is found with larger groups joined by a central atom. An example is the doublet produced by the *gem*-dimethyl group. Bending motions produce absorption at lower frequencies than fundamental stretching modes.

Molecules composed of several atoms vibrate not only according to the frequencies of the bonds, but also at overtones of these frequencies. When one bond vibrates, the rest of the molecule is involved. The *harmonic (overtone)* vibrations possess a frequency which represents approximately integral multiples of the fundamental frequency. A *combination band* is the sum, or the difference between, the frequencies of two or more fundamental or harmonic vibrations. The uniqueness of an infrared spectrum arises largely from these bands which are characteristic of the whole molecule. The intensities of overtone and combination bands are usually about one-hundredth of those of fundamental bands.

The intensity of an infrared absorption band is proportional to the square of the rate of change of dipole moment with respect to the displacement of the atoms. In some cases, the magnitude of the change in dipole moment may be quite small, producing only weak absorption bands, as in the relatively nonpolar $C \equiv N$ group. By contrast, the large permanent dipole moment of the $C = O$ group causes strong absorption bands, which is often the most distinctive feature of an infrared spectrum. If no dipole moment is created, as in the $C = C$ bond (when located symmetrically in the molecule) undergoing stretching vibration, then no radiation is absorbed and the vibrational mode is said to be *infrared inactive*. Fortunately, an infrared inactive mode will usually give a strong Raman signal.

As defined by quantum laws, the vibrations are not random events but can occur only at specific frequencies governed by the atomic masses and strengths of the chemical bonds. Mathematically, this can be expressed as

$$\bar{\nu} = \frac{1}{2\pi c} \sqrt{\frac{k}{\mu}} \tag{6-1}$$

where $\bar{\nu}$ is the frequency of the vibration, c is the velocity of light, k is the force constant, and μ is the reduced mass of the atoms involved. The frequency is greater the smaller the mass of the vibrating nuclei and the greater the force restoring the nuclei to the equilibrium position. Motions involving hydrogen atoms are found at much higher frequencies than are motions involving heavier atoms. For multiple bond linkage, the force constants of double and triple bonds are roughly two and three times those of the single bonds, and the absorption position becomes approximately two and three times higher in frequency. Interaction with neighbors may alter these values, as will resonating structures, hydrogen bonds, and ring strain.

Example 6-1

Calculate the fundamental frequency expected in the infrared absorption spectrum for the C—O stretching frequency. The value of the force constant is 5.0×10^5 dyn cm^{-1}.

Answer From Eq. 5-1,

$$\bar{\nu} \text{ (in cm}^{-1}) = \frac{1}{(2)(3.14)(3 \times 10^{10})} \sqrt{\frac{(5 \times 10^5)(12 + 16)(6.023 \times 10^{23})}{(12)(16)}}$$

$$= 1110 \text{ cm}^{-1}$$

INSTRUMENTATION

It is convenient to divide the infrared region into three segments with the dividing points based on instrumental capabilities (Table 6-1). Different radiation sources, optical systems, and detectors are needed for the different regions. The standard infrared spectrophotometer is an instrument covering the range from 4000 to 650 cm^{-1} (2.5 to 15.4 μm). Although many prism instruments are still in use, there has been an almost complete transition to filter-grating and prism-grating spectrophotometers. Grating instruments offer higher resolution that permits separation of closely spaced absorption bands, more accurate measurements of band positions and intensities, and higher scanning speeds for a given resolution and noise level. Modern spectrophotometers generally have attachments that permit speed suppression, scale expansion, repetitive scanning, and automatic control of slit, period, and gain. Accessories such as beam condensers, reflectance units, polarizers, and micro cells can usually be added to extend versatility or accuracy.

Spectrophotometers for the infrared region are composed of the same basic components as instruments in the ultraviolet–visible region, although the sources, the detectors, and the materials used in the fabrication of the optical components are different, except in the near-infrared. Radiation from a source emitting in the infrared region is chopped at a low frequency, often 10–26 times per second, and is passed alternately through the sample and the reference before entering the monochromator. This minimizes the effect of stray radiation, a serious problem in most of the infrared region. Temperature and relative humidity in the room housing the instrument must be controlled.

TABLE 6-1 Components of Infrared Spectrophotometers

	Region of Electromagnetic Spectrum		
	Near-infrared	Mid-infrared	Far-infrared
Wavenumber, cm^{-1}	12,500 4000	200	10
Wavelength, μm	0.8 2.5	50	1000
Source of Radiation	Tungsten filament lamp	Nernst glower, Globar, or coil of Nichrome wire	High-pressure mercury-arc lamp
Optical System	One or two quartz prisms or prism-grating double monochromator	Two to four plane diffraction gratings with either a fore-prism monochromator or infrared filters	Double-beam grating instruments for use to 700 μm; interferometric spectrometers for use to 1000 μm.
Detector	Lead sulfide photoconductive	Thermocouple, bolometer, Golay	Golay, pyroelectric

Radiation Sources

In the region beyond 5000 cm^{-1}, blackbody sources without envelopes commonly are used. The same spectral characteristics cited for the tungsten incandescent lamp apply to these as well (see Fig. 3-2). Unfortunately, the emission maximum lies in the near-infrared. A fraction of the shorter wavelength radiation will be present as stray light, and this will be particularly serious for long-wavelength measurements.

A close-wound *Nichrome helix* can be raised to incandescence by resistive heating. A black oxide forms on the wire which gives acceptable emissivity. Temperatures up to 1100°C can be reached. This source is recommended where reliability is essential, such as in nondispersive process analyzers and inexpensive spectrophotometers. Although simple and rugged, this source is less intense than other infrared sources.

A hotter, and therefore brighter, source is the *Nernst glower*, which has an operating temperature as high as 1500°C. Nernst glowers are constructed from yttria-stabilized zirconium oxide in the form of hollow rods 2 mm in diameter and 30 mm in length. The ends are cemented to short ceramic tubes to facilitate mounting; short platinum leads provide power connections. Nernst glowers are fragile. They have a negative coefficient of resistance and must be preheated to be conductive. Therefore, auxiliary heaters must be provided as well as a ballast system to prevent overheating. A glower must be protected from drafts, but at the same time adequate ventilation is needed to remove surplus heat and evaporated oxides and binder. Radiation intensity is approximately twice that of Nichrome and Globar sources except in the near-infrared.

The *Globar*, a rod of silicon carbide 4 mm in diameter and 50 mm in length, possesses characteristics intermediate between heated wire coils and the Nernst glower. It is self-starting and has an operating temperature near 1300°C. The temperature coefficient of resistance is positive and may be conveniently controlled with a variable transformer. Its resistance increases with the length of time used so that provision must be made for increasing the voltage across the unit. Its electrodes must be water cooled.

In the very far infrared, beyond 200 cm^{-1}, blackbody-type sources lose effectiveness since their radiation decreases with the fourth power of wavelength. High-pressure mercury arcs, with an extra quartz jacket to reduce thermal loss, give intense radiation in this region. Output is similar to that from blackbody sources, but additional radiation is emitted from a plasma which enhances the long-wavelength output.

Detectors[2]

At the short-wavelength end, below about 1.2 μm, the preferred detection methods are the same as those used for visible and ultraviolet radiation. The detectors used at longer wavelengths can generally be classified into two broad groups: (1) *quantum detectors*, which depend on internal photoconductive effects resulting from the transition of an electron from one valence band to a conduction band within the semiconductor receptor, and (2) *thermal detectors*, in which the radiation produces a heating effect that alters some physical property of the detector. Quantum detectors are faster and more sensitive, but severely restricted with respect to range of wavelengths to which they can respond. Thermal detectors are usable over a wide range of wavelengths, actually over the entire spectral region in which the light-absorbing part of the detector can be regarded as black, but they suffer from relatively low sensitivity and slow response. The response time sets an upper limit to the frequency at which the radiation can usefully be chopped; the total mass represented by receiver, absorbing material, and temperature-sensing element must heat or cool during each half-cycle of chopping frequency.

The basic forms of thermal radiation detectors are the *radiation thermocouple*, the *Golay detector*, and the *bolometer*. The thermocouple, which is most widely used of all infrared detectors, is usually fabricated with a small piece of blackened gold foil, to absorb the radiation, welded to the tips of two wire leads made of dissimilar metals chosen to give a large thermoelectric emf. Lead materials may be semiconductors, one having a large positive thermoelectric power with respect to gold and the other a large negative power. The entire assembly is mounted in an evacuated enclosure with an infrared-transmitting window so that conductive heat losses are minimized. Typical thermocouple detectors would have a sensitive area of 0.5 mm^2, a response time of 40 msec, a dc resistance between 10 and 200 Ω, a signal voltage of 0.1–0.2 μV, and noise-equivalent power of 10^{-10} W at a chopping frequency of 5 Hz. To prevent the faint signals from being lost in the stray (noise) signals picked up by the lead wires, a preamplifier is located as close to the detector as possible. The "cold" junction of the thermocouple actually consists of heavy copper lugs in contact with the thermocouple wires. Since the detector only needs to respond to chopped radiation to give an ac output, only changes of temperature are significant, hence the actual temperature of the cold junction is unimportant. Receiver size is chosen to match a reduced image of the spectrometer's exit slit.

The Golay detector is pneumatic in principle. The unit consists of a small metal cylinder, closed by a rigid blackened metal plate (2-mm square) at one end and by a flexible silvered diaphragm at the other end. The chamber is filled with xenon. Radiation passes through a small infrared-transmitting window and is absorbed by the blackened plate. Heat, conducted to the gas, causes it to expand and deform the flexible diaphragm. To amplify distortions of the mirror-surface of the flexible diaphragm, light from a lamp inside the detector housing is focused upon the mirror which reflects the light beam onto a vacuum phototube. Motion of the diaphragm moves the light beam across the phototube surface and changes its output. In an alternative arrangement, the rigid diaphragm is used as one plate of a dynamic condenser; a perforated diaphragm a slight distance away serves as the second plate. The distortion of the solid diaphragm relative to the fixed plate alters the plate separation and hence the capacity. With either arrangement the alternating output corresponds to the chopping frequency. Response time is on the order of 20 msec, corresponding to a chopping frequency of 15 Hz. The Golay detector has a sensitivity similar to that of a thermocouple. The angular aperture is $60°$ and so the unit must be used with a system of condensing mirrors to concentrate the radiation. The detector is significantly superior for the far infrared beyond $50 \, \mu m$.

A bolometer is a miniature resistance thermometer usually constructed of metal or semiconductor. The resistance of a metal increases with temperature about 0.35% per degree Celsius, whereas that of a semiconductor decreases about 7% per degree Celsius. A small flake of lightly doped germanium or silicon, cooled with liquid helium, constitutes a very effective bolometer. Two sensing elements, as identical as possible, mounted close to each other with one shielded from the radiant energy, form two arms of a Wheatstone bridge. For a thin slice of germanium cooled to $1.2°K$, a signal of about 10^{-12} W can be detected, which is two orders of magnitude smaller than what can be detected with a Golay cell. Another detector in the far infrared is made with a very pure piece of InSb. It is generally considered to be an electronic bolometer, where the electron gas can be heated for a short time without coupling to the lattice. Its sensitivity is comparable to that of the germanium bolometer, but the time constant is 10^{-7} sec, considerably shorter.

The *photoconductive effect*, occurring in semiconducting materials, provides a class of quantum detectors that are particularly useful in the near infrared. The incident photon flux interacts with electrons in bound states in the valence band or in the trapping level and excites them into free states in the conduction band where they remain for a characteristic lifetime. A positive hole is left behind. Both electron and hole contribute to electrical conduction. At wavelengths longer than about $10 \, \mu m$, photons have insufficient energy to promote electrons across the forbidden zone between the valence and conduction bands. Semiconductors must be doped with impurity atoms to provide some intermediate energy levels. Lead sulfide detectors are sensitive to radiation below $3 \, \mu m$, lead telluride below $4 \, \mu m$, and lead selenide below $5 \, \mu m$. These long-wavelength limits are increased about 50% on cooling the detectors to $20°K$. Doped germanium and silicon detectors cooled to 4°K possess useful sensitivity at wavelengths as long as 120 μm. Response time for lead selenide and telluride is less than 10 μsec, and for doped germanium and silicon about 1 nsec. The speed of response is determined by the time required for the excited charge carriers to become immobilized through recombination. These detectors consist of a film of semiconductor, 0.1 μm in thickness, deposited on a glass or quartz base and sealed into an evacuated envelope. The limiting noise is *generation-recombination noise*. It is associated with fluctuations in the density of free charge carriers, produced either by lattice vibrations, when the detector is not cooled sufficiently, or by the random arrival of photons from the background.

The *pyroelectric detector*[8] consists of a slice of a ferroelectric material, usually a single crystal of triglycine sulfate. Radiation absorbed by the crystal is converted to heat which alters the lattice spacings. Below the Curie temperature of the crystal, a change in lattice spacings produces a change in the spontaneous electric polarization. The crystal is mounted between two parallel electrodes, one of them infrared-transparent, and both normal to the polarization axis. If the electrodes are connected to an external circuit, a current is set up to balance the polarization charge at the crystal faces. This current then produces a voltage signal across an appropriate load resistor which is applied directly to a field-effect transistor, which is an integral part of the detector package. A distinct advantage of this detector is the decrease of its sensitivity (close to that of a Golay detector) by $\omega^{-1/2}$ when the chopping frequency ω of the beam increases, instead of the ω^{-1} relation for most other thermal detectors.

Spectrophotometers

Most infrared spectrophotometers are double-beam instruments in which two equivalent beams of radiant energy are taken from the source. By means of a combined rotating mirror and light interrupter, the source is flicked alternately between the reference and sample paths. In the optical-null system, the detector responds only when the intensity of the two beams is unequal. Any imbalance is corrected for by a light attenuator (an optical wedge or shutter comb) moving in or out of the reference beam to restore balance. The recording pen is coupled to the light attenuator. Although very popular, the optical-null system has serious faults. Near zero transmittance of the sample, the reference-beam attenuator will move in to stop practically all light in the reference beam. Both beams are then blocked, no energy is passed, and the spectrometer has no way of determining how close it is to the correct transmittance value. The instrument will go dead. However, in the mid-infrared region the electrical beam-ratioing method is not an easy means of avoiding the deficiencies of the optical-null system. To a large extent it is trading optical and mechanical problems for electronic problems.

Monochromators employing prisms for dispersion utilize a Littrow 60° prism-plane mirror mount (Fig. 3-19). Mid-infrared instruments employ a sodium chloride prism for the region from 4000 to 650 cm^{-1} (2.5 to 15.4 μm), with a potassium bromide or cesium iodide prism and optics for the extension of the useful spectrum to 400 cm^{-1} (25 μm) or 270 cm^{-1} (37 μm), respectively. Quartz monochromators, designed for the ultraviolet-visible region, extend their coverage into the near-infrared (to 2500 cm^{-1} or 4 μm).

Plane-reflectance grating monochromators dominate today's instruments. To cover the wide wavelength range, several gratings with different ruling densities and associated higher order filters are necessary. This requires some complex sensing and switching mechanisms for automating the scan with acceptable accuracy. Because of the nature of the blackbody emission curve, a slit programming mechanism must be employed to give near constant energy and resolution as a function of wavelength. The principal limitation is energy. Resolution and signal-to-noise ratio are limited primarily by the emission of the blackbody source and the noise-equivalent power of the detector. Two gratings are often mounted back-to-back so that each need be used only in the first order; the gratings are changed at 2000 cm^{-1} (5.0 μm) in mid-infrared spectrometers. Grating instruments incorporate a sine-bar mechanism to drive the grating mount when a wavelength readout is desired, and a cosecant-bar drive when wavenumbers are desired. Undesired overlapping orders can be eliminated with a fore-prism or by suitable filters.

The optical arrangement for a filter-grating spectrometer is shown in Fig. 6-3. The filters are inserted near a slit or slit image when the required size of the filter is not excessive. The circular variable filter is simple in construction. It is frequently necessary to use gratings as reflectance filters when working in the far-infrared in order to remove unwanted second and higher orders from the light incident on the far-infrared grating. For this purpose small plane gratings are used which are blazed for the wavelength of the unwanted radiation. The grating acts as a mirror reflecting the wanted light into the instrument and diffracting the shorter wavelengths out of the beam; a grating "looks" like a good mirror to wavelengths longer than the groove spacing.

Probably the most elegant filter is a prism because it provides a narrow band of wavelengths with high efficiency over a relatively broad spectral range. The prism and grating must track together over consecutive grating orders. Light from the parabolic mirror enters the fore-prism where it is dispersed so that only a relatively narrow band of wavelengths is allowed to fall on the grating. The resolution of the prism can be quite low, because it need only exclude the adjacent orders, but for the higher orders, the interval gets successively narrower. Thus, it is preferable to use two gratings and confine their application to lower orders.

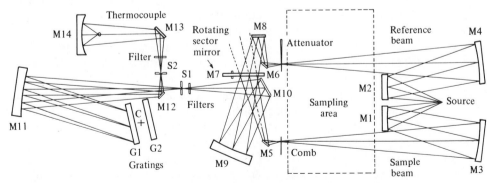

Fig. 6-3. Optical schematic of a filter-grating infrared spectrometer employing the double-beam, optical-null principle. (Courtesy of Perkin-Elmer Corp.)

In process-stream analyzers, elasticity of use has been sacrificed for economy, dependability, and extreme stability. Instruments fall into two categories: *dispersive* and *nondispersive*.

A dispersive instrument is patterned after a double-beam spectrophotometer. Radiation at two fixed wavelengths passes through a cell containing the process-stream to provide a continuous measurement of the absorption ratio. At one wavelength the material absorbs selectively, and at the other wavelength the material does not absorb or else exhibits a constant, but small, absorption. The ratio of transmittance readings is converted directly into concentration of absorber and recorded. This type of instrument can handle liquid systems as well as gas streams, and has the ability to analyze quite complex mixtures.

No prisms or gratings are used in nondispersive instruments. The total radiation from an infrared source is passed through the sample, providing more signal power. By filling one or both cells containing the detector with the pure form of the gas being determined, these analyzers show high selectivity and virtually infinite resolving power although they employ a very simple optical train. Two modes of detection are employed.

A schematic diagram of a *negative filter* type of nondispersive analyzer is shown in the upper portion of Fig. 6-4. An infrared source sends radiation through the sample

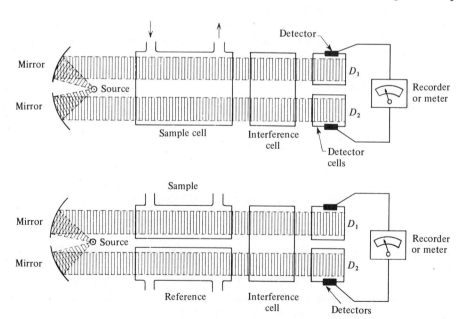

Fig. 6-4. Nondispersive types of process-stream infrared analyzers. Top, negative-filter arrangement; D_1 contains the gas being determined and D_2 contains a nonabsorbing gas. Bottom, positive-filter arrangement in which both detector cells are filled with the gas being determined.

chamber. Half of the beam is intercepted by each detector. One cell is filled with the pure form of the gas being determined (component A); the other is filled with a nonabsorbing gas. The former absorbs all the radiation in its beam which is characteristic of component A, and a thermal detector in the cell records the temperature rise. As radiation passes through a gas stream in the sample cell which is devoid of component A, the detector filled with the nonabsorbing gas absorbs none of the radiant energy. The net difference in the two signals—the maximum difference—is the 0% reading. When the process stream contains some of the component A, a proportionate amount of the characteristic radiation is absorbed in the sample cell and fails to reach the detectors, thus decreasing the signal from the detector which is filled with pure component A, but no change occurs in the signal from the other detector. Thus, as the concentration of component A in the process stream approaches 100%, the signal difference between the two detectors approaches zero.

In the *positive filter* arrangement, shown in the lower portion of Fig. 6-4, the beam of radiation from the source is split into two parallel beams. One beam passes through the reference cell and the other through the sample cell. Each detector, in this arrangement, is filled with the pure form of the gas being determined. When some of the latter is present in the sample beam, the sample detector receives less radiant energy by the amount absorbed by the sample component at its characteristic wavelength. The difference in signal from the two detectors is related to absorber concentration.

If some other absorbing component is present in the process stream, the analyzer is "desensitized" by filling an intermediate cell (in both light paths) with the pure form of the interfering gas or a sufficient concentration to remove its characteristic wavelengths adequately from both light paths. The analyzer operates on the remaining regions of the spectrum. Of course, this reduces somewhat the sensitivity of the analyzer toward the component being analyzed.

Interferometric (Fourier Transform) Spectrophotometer[5]

The basic configuration of the interferometer portion of a Fourier transform spectrometer, shown in Fig. 6-5(a), includes two plane mirrors at a right angle to each other and a beam splitter at 45° to the mirrors. Modulated light from the source is collimated and passes to the beam splitter which divides it into two equal beams for the two mirrors. An equal thickness of support material (without the semireflection coating), called the compensator, is placed in one arm of the interferometer to equalize the optical path length in both arms. When these mirrors are positioned so that the optical path lengths of the reflected and transmitted beams are equal, the two beams will be in phase when they return to the beam splitter and will constructively interfere. Displacing the movable mirror by one-quarter wavelength will bring the two beams 180° out of phase and they will destructively interfere. Continuing the movement of the mirror in either direction will cause the field to oscillate from light to dark for each quarter-wavelength movement of the mirror, corresponding to $\lambda/2$ changes. When the interferometer is illuminated by monochromatic light of wavelength λ, and the mirror is moved with a velocity v, the sig-

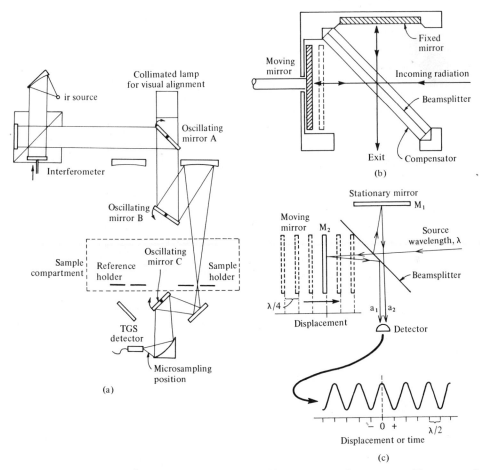

Fig. 6-5. A Fourier transform spectrometer. (a) FTS-4 configuration. (Courtesy of Block Engineering, Inc.) (b) Simplified diagram of a Michelson interferometer. (c) Mirror movement translated into cosine wave pattern.

nal from the detector has a frequency $f = 2v/\lambda$. A plot of signal vs. mirror distance is a pure cosine wave. With polychromatic light, the output signal is the sum of all the cosine waves, which is the Fourier transform of the spectrum. Each frequency is given an intensity modulation, f, which is proportional both to the frequency of the incident radiation and to the speed of the moving mirror. For example, with a constant mirror velocity of 0.5 mm/sec, radiation of 1000 cm^{-1} (10 μm and a frequency of 3×10^{14} Hz) will produce a detector signal of 50 Hz. For 5-μm radiation, the signal is 100 Hz, and so on. An appropriate inverse transformation of the interferogram will give the desired spectrum. Rather than dispersing polychromatic radiation as would a conventional dispersive spectrometer, the Fourier transform spectrometer performs a frequency transformation. Data reduction requires digital computer techniques and analog conversion devices.

To make any sense out of the intensity measurement, the displacement of the movable mirror has to be known precisely. With a constant velocity of mirror motion, the mirror should move as far and as smoothly as possible. If the velocity is precise, an electronically timed coordinate can be generated for the interferogram. Severe mechanical problems limit this approach. The interferometer itself, however, can be used to generate its own time scale. In addition to processing the incoming spectral radiation, a line from a laser source is used to produce a discrete signal which is time-locked to the mirror motion and hence to the interferogram. This is the *fringe-reference* system and is analogous to the frequency/field lock in nmr. The mirror position can be determined by measuring the laser line interferogram, counting the fringes as the mirror moves from the starting position—denoted by a burst of light from an incandescent source.

Dispersion or filtering is not required, so that energy-wasting slits are not needed, and this is a major advantage. With energy at a premium in the far-infrared, the superior light-gathering power of the interferometric spectrometer is a welcome asset for this spectral region.

In the near- and mid-infrared, germanium coated on a transparent salt, such as NaCl, KBr, or CsI, is a common beam splitter material. In far-infrared spectrometers, the beam splitter is a thin film of Mylar whose thickness must be chosen for the spectral region of interest. For example, a Mylar film 0.25 mil thick can effectively cover the range from 60 to 375 cm^{-1}.

Resolution is related to the maximum extent of mirror movement so that a 1-cm movement results in 1-cm^{-1} resolution and a 2-cm movement yields 0.5-cm^{-1} resolution. Resolution can also be doubled by doubling the measurement times, or resolution can be traded for rapid response. Because the detector of the interferometer "sees" all resolution elements throughout the entire scan time, the signal-to-noise ratio, S/N, is proportional to $\sqrt{T}$, where T is the measurement time. For example, when examining a spectrum composed of 2000 resolution elements with an observation time of 1 sec per element assumed for the desired S/N, the interferometric measurement is complete in 1 sec. Improving the S/N by a factor of 2 would require only 4 sec to complete the measurement. Comparable times for a dispersive spectrometer are 33 and 72 min, respectively. Repetitive signal-averaged scans are very feasible with an interferometer.

SAMPLE HANDLING

Infrared instrumentation has reached a remarkable degree of standardization as far as the sample compartment of various spectrometers is concerned. Sample handling itself, however, presents a number of problems in the infrared region. No rugged window material for cuvettes exists that is transparent and also inert over this region. The alkali halides are widely used, particularly sodium chloride, which is transparent at wavelengths as long as 16 μm (625 cm^{-1}). Cell windows are easily fogged by exposure to moisture and require frequent repolishing. Silver chloride is often used for moist samples, or aqueous solutions, but it is soft, easily deformed, and darkens on exposure to visible light. Teflon has only C—C and C—F absorption bands. For frequencies under 600

TABLE 6-2 Infrared Transmitting Materials

Material	Wavelength Range, μm	Wavenumber Range, cm^{-1}	Refractive Index at 2 μm
Sodium chloride	0.25 to 16	40,000 to 625	1.52
Potassium bromide	0.25 to 25	40,000 to 400	1.53
Silver chloride	0.4 to 23	25,000 to 435	2.0
Cesium bromide	1 to 37	10,000 to 270	1.67
Cesium iodide	1 to 50	10,000 to 200	1.74
Irtran-2	2 to 14	5,000 to 714	2.26
KRS-5	0.5 to 40	20,000 to 250	2.4
Sapphire	0.5 to 5.6	20,000 to 1,780	1.7
Germanium	0.5 to 16.7	20,000 to 600	4.02
Polyethylene (high density)	16 to 300	625 to 33	1.54

cm^{-1}, a polyethylene cell is useful. Infrared transmission materials are compiled in Table 6–2. Crystals of high refractive index produce strong, persistent fringes.

Gases

In the analysis of gases the usual path length is 10 cm. When this is too short to measure the spectra of minor components encountered in trace analysis, a variable-path microgas cell provides path lengths ranging from 10 cm to 10 m. The light path is folded using internal mirrors so that the light beam passes through the length of the cell as many as 40 times to attain a total length of 10 m. Further gains in sensitivity can be realized by increasing the pressure of the gas sample in the cell. Pressure broadening of absorption bands is troublesome in quantitative work, however.

Liquids and Solutions

Samples that are liquid at room temperature are usually scanned in their neat form, or in solution. The sample concentration and path length should be chosen so that the transmittance lies between 15 and 70%. For neat liquids this will represent a very thin layer, about 0.001–0.05 mm in thickness. For solutions, concentrations of 10% and cell lengths of 0.1 mm are most practical. Unfortunately, not all substances can be dissolved in a reasonable concentration in a solvent that is nonabsorbing in regions of interest. When possible, the spectrum is obtained in a 10% solution of CCl_4 in a 0.1-mm cell in the region 4000 to 1333 cm^{-1} (2.5 to 7.5 μm), and in a 10% solution of CS_2 in the region 1333 to 650 cm^{-1} (7.5 to 15.4 μm). Transparent regions of selected solvents are given in Fig. 6-6. To obtain solution spectra of polar materials which are insoluble in CCl_4 or CS_2, chloroform, methylene chloride, acetonitrile, and acetone are useful solvents. Sensitivity can be gained by going to longer path lengths if a suitably transparent solvent can

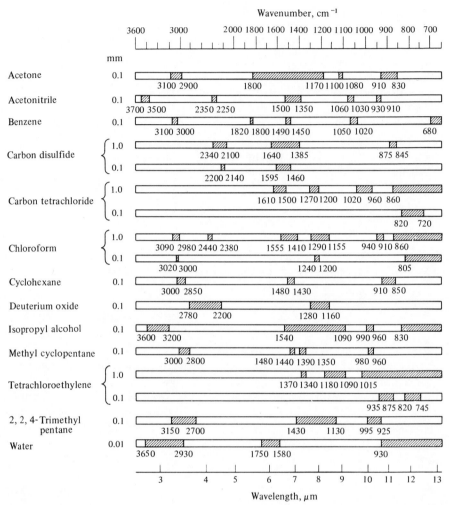

Fig. 6-6. Transmission characteristics of selected solvents. The material is considered transparent if the transmittance is 75% or greater. Solvent thickness is given in millimeters.

be found. In a double-beam spectrophotometer a reference cell of the same path length as the sample cell is filled with pure solvent and placed in the reference beam. Moderate solvent absorption, now common to both beams, will not be observed in the recorded spectrum. However, solvent transmittance should never fall under 10%.

The possible influence of a solvent on the spectrum of a solute must not be overlooked. Particular care should be exercised in the selection of a solvent for compounds which are susceptible to hydrogen-bonding effects. Hydrogen bonding through an —OH or —NH— group alters the characteristic vibrational frequency of that group; the stronger the hydrogen bonding, the greater is the lowering of the fundamental frequency. To differentiate between inter- and intramolecular hydrogen bonding, a series of spectra at different dilutions, yet having the same number of absorbing molecules in the beam, must be obtained. If, as the dilution increases, the hydrogen-bonded absorption band decreases while the unbonded absorption band increases, the bonding is intermolecular. Intramolecular bonding shows no comparable dilution effect.

Infrared solution cells are constructed with windows sealed and separated by thin gaskets of copper and lead which have been wetted with mercury. The whole assembly is securely clamped together. As the mercury penetrates the metal, the gasket expands, producing a tight seal. The cell is provided with tapered fittings to accept the needles of hypodermic syringes for filling. In demountable cells, the sample and spacer are placed on one window, covered with another window, and the entire sandwich is clamped together.

Cell Thickness

One of two methods may be used to measure the path length of infrared absorption cells: the *interference fringe* method and the *standard absorber* method. The interference fringe method is ideally suited to cells whose windows have a high polish. With the empty cell in the spectrophotometer on the sample side and no cell in the reference beam, the spectrophotometer is operated as near as possible to the 100% line. Enough spectrum is run to produce 20–50 fringes. The cell thickness, b, in centimeters, is calculated from the expression

$$b = \frac{1}{2\eta_D}\left(\frac{n}{\bar{\nu}_1 - \bar{\nu}_2}\right) \tag{6-2}$$

where n is the number of fringes (peaks or troughs) between two wavenumbers $\bar{\nu}_1$ and $\bar{\nu}_2$, and η_D is the refractive index of the sample material. If measurements are made in wavelength (micrometers), the equation is

$$b = \frac{1}{2\eta_D}\left(\frac{n\lambda_1\lambda_2}{\lambda_2 - \lambda_1}\right) \tag{6-3}$$

where λ_1 is the starting wavelength and λ_2 the finishing wavelength. The fringe method also works well for measurement of film thickness.

The standard absorber method may be used with a cell in any condition and with cavity cells whose inner faces do not have a finished polish. The 1960 cm^{-1} (5.10 μm) band of benzene may be used for calibrating cells which are less than 0.1 mm in path length, and the 845 cm^{-1} (11.8 μm) band for cells 0.1 mm or longer in path length. At the former frequency, benzene has an absorbance of 0.10 for every 0.01 mm of thickness; at 845 cm^{-1}, benzene has an absorbance of 0.24 for every 0.1 mm of thickness.

Films

Spectra of liquids not soluble in a suitable solvent are best obtained from capillary films. A large drop of the neat liquid is placed between two rock-salt plates which are then squeezed together and mounted in the spectrometer in a suitable holder. Plates need not have high polish, but must be flat to avoid distortion of the spectrum.

For polymers, resins, and amorphous solids, the sample is dissolved in any reasonably volatile solvent, the solution poured onto a rock-salt plate, and the solvent evaporated by gentle heating. If the solid is noncrystalline, a thin homogeneous film is deposited on the plate which then can be mounted and scanned directly. Sometimes polymers can be "hot pressed" onto plates.

Mulls

Powders, or solids reduced to particles, can be examined as a thin paste or mull by grinding the pulverized solid (about 1 mg) in a greasy liquid medium. The suspension is pressed into an annular groove in a demountable cell. Multiple reflections and refractions off the particles are lessened by grinding the particles to a size an order of magnitude less than the analytical wavelength and surrounding the particles by a medium whose refractive index more closely matches theirs than does air. Liquid media include mineral oil or Nujol, hexachlorobutadiene, perfluorokerosene, and chlorofluorocarbon greases (fluorolubes). The latter are used when the absorption by the mineral oil masks the presence of C—H bands. For qualitative analysis the mull technique is rapid and convenient, but quantitative data are difficult to obtain, even when an internal standard is incorporated into the mull. Polymorphic changes, degradation, and other changes may occur during grinding.

Pellet Technique

The pellet technique involves mixing the fine ground sample (1–100 μg) and potassium bromide powder, and pressing the mixture in an evacuable die at sufficient pressure (60,000–100,000 psi) to produce a transparent disk. Potassium bromide becomes quite plastic at high pressure, and will flow to form a clear disk. Grinding–mixing is conveniently done in a vibrating ball-mill (Wig-L-Bug). Other alkali halides may also be used, particularly CsI or CsBr for measurements at longer wavelengths. Good dispersion of the sample in the matrix is critical; moisture must be absent. Freeze-drying the sample is often a necessary preliminary step.

KBr wafers can be formed, without evacuation, in a Mini-Press. Two highly polished bolts are turned against each other in a steel cylinder. Pressure is applied with wrenches for about 1 min to 75 to 100 mg of powder, the bolts are removed, and the cylinder—now a cell complete with window—is installed in its slide holder in any spectrophotometer.

Quantitative analyses are readily performed since an accurate measurement can be made of the weight ratio of sample to internal standard added in each disk or wafer.

Attenuated Total Reflectance (ATR)

The scope and versatility of infrared spectroscopy as a qualitative analytical tool have been increased substantially by the *attenuated total reflectance* (also known as internal reflectance) technique.[4,11] When a beam of radiation enters a plate (or prism), it will be reflected internally if the angle of incidence at the interface between sample and plate is greater than the critical angle (which is a function of refractive index). On internal reflection, all the energy is reflected. However, the beam appears to penetrate slightly (from a fraction of a wavelength up to several wavelengths) beyond the reflecting surface, and then return. When a material is placed in contact with the reflecting surface, the beam will lose energy at those wavelengths where the material absorbs due to an interaction with the penetrating beam. This attenuated radiation, when measured and plotted as a function of wavelength, will give rise to an absorption spectrum characteristic of the material which resembles an infrared spectrum obtained in the normal manner.

Most ATR work is done by means of an accessory readily inserted in, and removed from, the sampling space of a conventional infrared spectrophotometer (Fig. 6-7). The accessory consists of a mirror system which sends the source radiation through the attachment and a second mirror system which directs the radiation into the monochromator. The width of the crystal is chosen to be equal to, or greater than, the height of the spectrometer slits (10–15 mm). The length-to-thickness ratio determines the number of reflections once the angle of incidence is selected; dimensions vary from 0.25 to 5 mm of thickness, and lengths from 1 to 10 cm. Parallelism and flatness of sampling surfaces and surface polish are critical. In the single-pass plate, light is introduced through an entrance aperture consisting of a simple bevel at one end of the plate and, after propagation via multiple internal reflection down the length of the plate, leaves by means of an exit aperture either parallel or perpendicular to the entrance aperture. The angle of the bevel determines the interior angle of incidence. This type of plate is useful for bulk materials, thin films, and surface studies. In the double-pass plate, light enters as before, propagates down the length of the plate, is totally reflected at the opposite end from a surface perpendicular to the sample surfaces, and returns to leave the plate via the exit aperture. The free end of the plate can be dipped into liquids or powders or placed in closed systems requiring only one optical window.

The apparent depth to which the radiation penetrates the sample is only a few micrometers and is independent of sample thickness. Consequently, ATR spectra can be obtained for many samples that cannot be studied by normal transmission methods. These include samples which show very strong absorptions, resist preparation in thin films, are characteristic only as thick layers, and are available on a nontransparent support. Aqueous solutions can be handled without compensating for very strong solvent absorptions. Samples containing suspended matter, such as dispersed solids or emulsions, that produce high backgrounds in transmission spectra due to scatter, give better results by ATR. Excellent contact efficiency at the sample/crystal interface is achieved when the sample is a self-adhering mobile liquid, flows slightly under modest pressure, or can be evaporated from solution. Samples that absorb only weakly, or those that do not give intimate contact with the crystal surface, such as rough-textured fabric, can be handled using multiple internal reflections, analogous to increasing the path length in transmission studies.

The appearance and intensity of an ATR spectrum will depend upon the difference of the indices of refraction between the reflection crystal and the rarer medium containing the absorber, and upon the internal angle of incidence. Thus a reflection crystal of

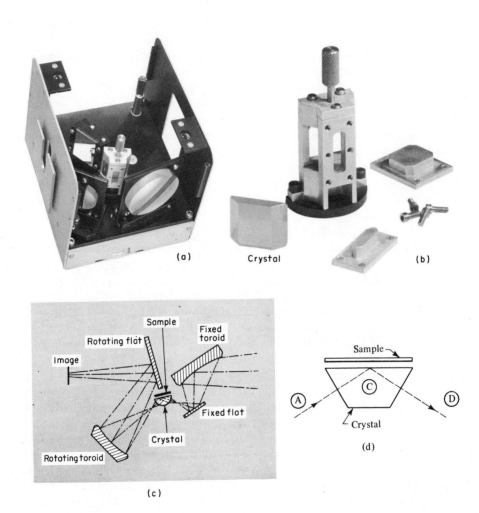

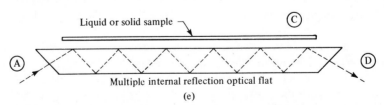

Fig. 6-7. (a) Attenuated total reflection attachment. (b) Exploded view of holder. (c) Schematic optical diagram with the crystal and sample shown enlarged as with (d) trapezoidal prism and (e) a multiple internal optical flat. (Courtesy of Barnes Engineering Co.)

relatively high index of refraction should be used. Two materials found to perform most satisfactorily for the majority of liquid and solid samples are KRS-5 and AgCl. KRS-5 is a tough and durable material with excellent transmission properties. Its index of refraction is high enough to permit well-defined spectra of nearly all organic materials, although it is soluble in basic solutions. AgCl is recommended for aqueous samples because of its insolubility and lower refractive index. An overall angle of incidence should be selected that is far enough from the average critical angle of the sample vs. reflector so that the change of the critical angle through the region of changing index of refraction (the absorption band) has a minimum effect on the shape of the ATR band. Unfortunately, when the index of refraction of the crystal is considerably greater than that of the sample so that little distortion occurs, the total absorption is reduced. With multiple reflection equipment, however, ample absorption can be obtained at angles well away from the critical angle.

Infrared Probe

Resembling a specific ion electrode, the infrared probe contains a sensitive element which is dipped into the sample (Fig. 6-8). To operate it, the user selects the proper wavelength by rotating a calibrated, circular variable filter, then adjusts the gain and slits to bring the meter to 100%. Next the probe is lowered into the sample. The meter indicates the absorbance. This value can be converted into concentration by reference to a previously prepared calibration curve. To detect the presence or absence of a particular functional group, one scans through the portion of the spectrum where the absorption bands characteristic of that group appear.

The infrared probe utilizes attenuated total reflection to obtain the absorption information. The probe crystal is made from a chemically inert material such as germanium or

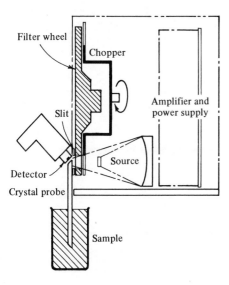

Fig. 6-8. Schematic of the infrared probe. (Courtesy of Wilks Scientific Corp.)

synthetic sapphire. The reflecting surfaces are masked so that the same area is covered by sample each time an analysis is made. A single-beam optical system is employed, chopped at 45 Hz. Since the air path is less than 5 cm, as opposed to well over a meter in conventional infrared spectrophotometers, absorption due to atmospheric water vapor and carbon dioxide is insignificant.

QUANTITATIVE ANALYSIS

The application of infrared spectroscopy as a quantitative analytical tool varies widely from one laboratory to another. However, the use of high-resolution grating instruments materially increases the scope and reliability of quantitative infrared work. Quantitative infrared analysis is based on Beer's law; apparent deviations arise from either chemical or instrumental effects. In many cases the presence of scattered radiation makes the direct application of Beer's law inaccurate, especially at high values of absorbance. Since the energy available in the useful portion of the infrared is usually quite small, it is necessary to use rather wide slit widths in the monochromator. This causes a considerable change in the apparent value of the molar absorptivity; therefore, molar absorptivity should be determined empirically.

The base-line method, shown in Fig. 6-9, involves selection of an absorption band of the substance under analysis which does not fall too close to the bands of other matrix components. The value of the incident radiant energy P_0 is obtained by drawing a straight line tangent to the spectral absorption curve at the position of the sample's absorption band. The transmittance P is measured at the point of maximum absorption. The value of $\log(P_0/P)$ is then plotted against concentration.

Many possible errors are eliminated by the base-line method. The same cell is used for all determinations. All measurements are made at points on the spectrum which are sharply defined by the spectrum itself, thus there is no dependence on wavelength settings. Use of such ratios eliminates changes in instrument sensitivity, source intensity, or changes in adjustment of the optical system.

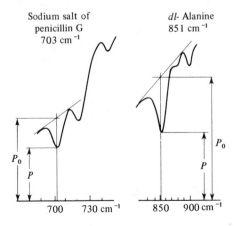

Fig. 6-9. The baseline method for calculation of the transmittance ratio in quantitative analysis.

Pellets from the disc technique can be employed in quantitative measurements. Uniform pellets of similar weight are essential, however, for quantitative analysis. Known weights of KBr are taken, plus a known quantity of the test substance from which absorbance data a calibration curve can be constructed. The discs are weighed and their thickness measured at several points on the surface with a dial micrometer. The disadvantage of measuring pellet thickness can be overcome by using an internal standard. Potassium thiocyanate makes an excellent internal standard. It should be preground, dried, and then reground, at a concentration of 0.2% by weight with dry KBr. The final mix is stored over phosphorus pentoxide. A standard calibration curve is made by mixing about 10% by weight of the test substance with the KBr–KSCN mixture and then grinding. The ratio of the thiocyanate absorption at 2125 cm^{-1} (4.70 μm) to a chosen band absorption of the test substance is plotted against percent concentration of the sample.

For quantitative measurements, the single-beam system has some fundamental characteristics that can result in greater sensitivity and better accuracy than the double-beam systems. All other things being equal, a single-beam instrument will automatically have a greater signal-to-noise ratio. There is a factor of 2 advantage in looking at one beam all the time rather than two beams half the time. Electronic switching gives another factor of 2 advantage. Thus, in any analytical situation where background noise is appreciable, the single-beam spectrometer should be superior.

CORRELATION OF INFRARED SPECTRA WITH MOLECULAR STRUCTURE

The infrared spectrum of a compound is essentially the superposition of absorption bands of specific functional groups, yet subtle interactions with the surrounding atoms of the molecule impose the stamp of individuality on the spectrum of each compound. For qualitative analysis, one of the best features of an infrared spectrum is that the absorption or the *lack of absorption* in specific frequency regions can be correlated with specific stretching and bending motions and, in some cases, with the relationship of these groups to the remainder of the molecule. Thus, by interpretation of the spectrum, it is possible to state that certain functional groups are present in the material and that certain others are absent. With this one datum, the possibilities for the unknown can be sometimes narrowed so sharply that comparison with a library of pure spectra permits identification.

Near-Infrared Region

In the near-infrared region, which meets the visible region at about 12,500 cm^{-1} (0.8 μm) and extends to about 4000 cm^{-1} (2.5 μm), are found many absorption bands resulting from harmonic overtones of fundamental bands and combination bands often associated with hydrogen atoms. Among these are the first overtones of the O—H and N—H stretching vibrations near 7140 cm^{-1} (1.4 μm) and 6667 cm^{-1} (1.5 μm),

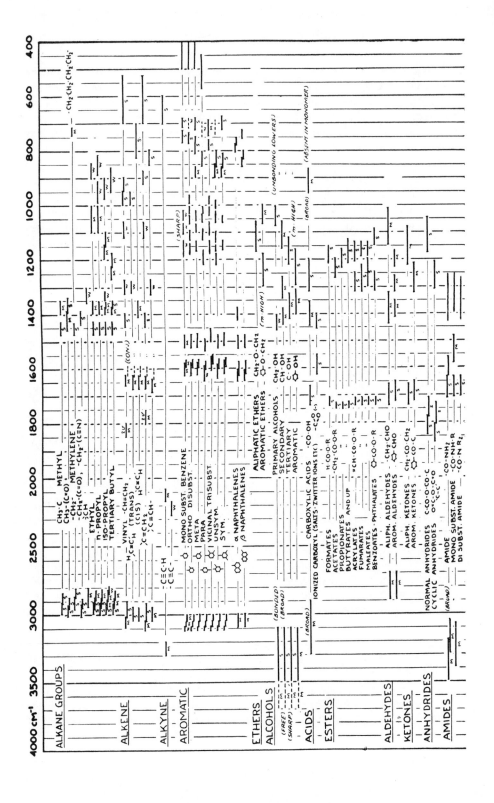

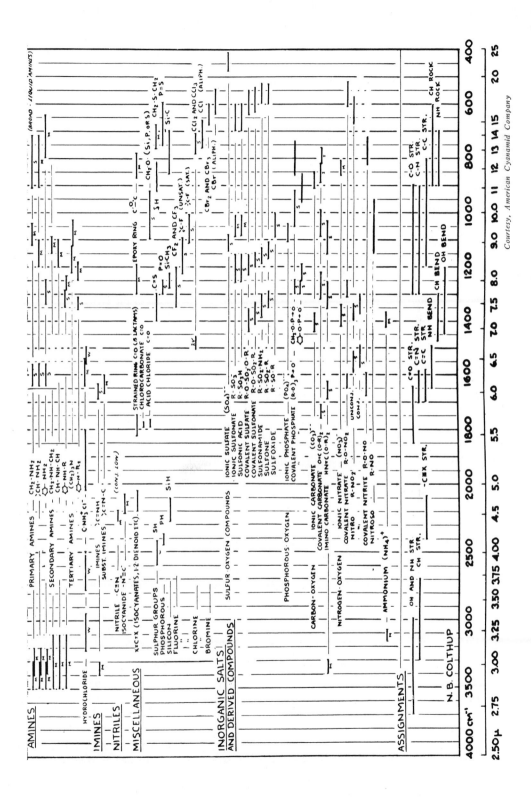

Courtesy, American Cyanamid Company

respectively, combination bands resulting from C—H stretching, and deformation vibrations of alkyl groups at 4548 cm^{-1} (2.2 μm) and 3850 cm^{-1} (2.6 μm). Thicker sample layers (0.5–10 mm) compensate for lessened molar absorptivities. The region is accessible with quartz optics, and this is coupled with greater sensitivity of near-infrared detectors and more intense light sources. The near-infrared region is often used for quantitative work.

Water has been analyzed in glycerol, hydrazine, Freon, organic films, acetone, and fuming nitric acid. Absorption bands at 2.76, 1.90, and 1.40 μm are used depending on the concentration of the test substance. Where interferences from other absorption bands are severe or where very low concentrations of water are being studied, the water can be extracted with glycerol or ethylene glycol.

Near-infrared spectrometry is a valuable tool for analyzing mixtures of aromatic amines. Primary aromatic amines are characterized by two relatively intense absorption bands near 1.97 and 1.49 μm. The band at 1.97 μm is a combination of N—H bending and stretching modes and the one at 1.49 μm is the first overtone of the symmetric N—H stretching vibration. Secondary amines exhibit an overtone band but do not absorb appreciably in the combination region. These differences in absorption provide the basis for rapid, quantitative analytical methods. The analyses are normally carried out on 1% solutions in CCl$_4$, using 10-cm cells. Background corrections can be obtained at 1.575 and 1.915 μm. Tertiary amines do not exhibit appreciable absorption at either wavelength. The overtone and combination bands of aliphatic amines are shifted to about 1.525 and 2.000 μm, respectively. Interference from the first overtone of the O—H stretching vibration at 1.40 μm is easily avoided with the high resolution available with near-infrared instruments.

Mid-Infrared Region

Many useful correlations have been found in the mid-infrared region (Fig. 6-10). This region is divided into the "group frequency" region—4000 to 1300 cm^{-1} (2.5 to 8 μm), and the "fingerprint" region—1300 to 650 cm^{-1} (8.0 to 15.4 μm). In the group frequency region the principal absorption bands may be assigned to vibration units consisting of only two atoms of a molecule, that is, units which are more or less dependent only on the functional group giving the absorption and not on the complete molecular structure. Structural influences do reveal themselves, however, as significant shifts from one compound to another. In the derivation of information from an infrared spectrum, prominent bands in this region are noted and assigned first. In the interval from 4000 to 2500 cm^{-1} (2.5 to 4.0 μm), the absorption is characteristic of hydrogen stretching vibrations with elements of mass 19 or less. When coupled with heavier masses, the frequencies overlap the triple-bond region. The intermediate frequency range, 2500 to 1540 cm^{-1} (4.0 to 6.5 μm), is often termed the *unsaturated* region. Triple bonds, and very little else, appear from 2500 to 2000 cm^{-1} (4.0 to 5.0 μm). Double-bond frequencies fall in the region from 2000 to 1540 cm^{-1} (5.0 to 6.5 μm). By judicious application of accumulated empirical data, it is possible to distinguish among C=O,

$C=C$, $C=N$, $N=O$, and $S=O$ bands. The major factors in the spectra between 1300 and 650 cm^{-1} (7.7 to 15.4 μm) are single-bond stretching frequencies and bending vibrations (skeletal frequencies) of polyatomic systems which involve motions of bonds linking a substituent group to the remainder of the molecule. This is the fingerprint region. Multiplicity is too great for assured individual identification, but collectively the absorption bands aid in identification.

Far-Infrared Region

The region between 667 and 10 cm^{-1} (15 to 1000 μm) contains the bending vibrations of carbon, nitrogen, oxygen, and fluorine with atoms heavier than mass 19, and additional bending motions in cyclic or unsaturated systems.[10] The low-frequency molecular vibrations found in the far-infrared are particularly sensitive to changes in the overall structure of the molecule. When studying the conformation of the molecule as a whole, the far-infrared bands differ often in a predictable manner for different isomeric forms of the same basic compound. The far-infrared frequencies of organometallic compounds are often sensitive to the metal ion or atom, and this, too, can be used advantageously in the study of coordination bonds. Moreover, this region is particularly well suited to the study of organometallic or inorganic compounds whose atoms are heavy and whose bonds are inclined to be weak.[3]

Structural Analysis

After the presence of a particular fundamental stretching frequency has been established, closer examination of the shape and exact position of an absorption band often yields additional information. The shape of an absorption band around 3000 cm^{-1} (3.3 μm) gives a rough idea of the CH group present. Alkyl groups have their $C-H$ stretching frequencies lower than 3000 cm^{-1}, whereas alkenes and aromatics have them slightly higher than 3000 cm^{-1}. The CH$_3$ group gives rise to an asymmetric stretching mode at 2960 cm^{-1} (3.38 μm) and a symmetric mode at 2870 cm^{-1} (3.48 μm). For $-CH_2-$ these bands occur at 2930 cm^{-1} (3.42 μm) and 2850 cm^{-1} (3.51 μm).

Next, one should examine regions where characteristic vibrations from bending motions occur. For alkanes, bands at 1460 cm^{-1} (6.85 μm) and 1380 cm^{-1} (7.25 μm) are indicative of a terminal methyl group attached to carbon exhibiting in-plane bending motions; if the latter band is split into a doublet at about 1397 and 1370 cm^{-1} (7.16 and 7.30 μm), geminal methyls are indicated. The symmetrical in-plane bending is shifted to lower frequencies when the methyl group is adjacent to $>C=O$ (1360-1350 cm^{-1}), $-S-$ (1325 cm^{-1}), and silicon (1250 cm^{-1}). The in-plane scissor motion of $-CH_2-$ at 1470 cm^{-1} (6.80 μm) indicates the presence of that group. Four or more methylene groups in a linear arrangement gives rise to a weak rocking motion at about 720 cm^{-1} (13.9 μm). Figure 6-11 illustrates the typical spectrum of an alkane.

The substitution pattern of an aromatic ring can be deduced from a series of weak but very useful bands in the region 2000 to 1670 cm^{-1} (5 to 6 μm) coupled with the

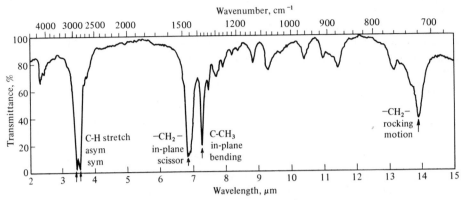

Fig. 6-11. Typical infrared spectrum of a saturated *n*-alkane.

position of the strong bands between 900 and 650 cm^{-1} (11.1 and 15.4 μm) which are due to the out-of-plane bending vibrations (Fig. 6-12). Absence of the symmetrical breathing mode at 690–710 cm^{-1} in the spectra of *para*- and *ortho*-substituted rings is helpful. The spectrum of *o*-xylene, shown in Fig. 6-13, is characteristic of aromatic systems. Ring stretching modes are observed near 1600, 1570, and 1500 cm^{-1} (6.25, 6.37, and 6.67 μm). These characteristic absorption patterns are also observed with substituted pyridines and polycyclic benzenoid aromatics.

The presence of an unsaturated C=C linkage introduces the stretching frequency at 1650 cm^{-1} (6.07 μm), shown in Fig. 6-14, and which may be weak or nonexistent if symmetrically located in the molecule. Mono- and tri-substituted olefins give rise to more intense bands than *cis*- or *trans*-disubstituted olefins. Substitution by a nitrogen or

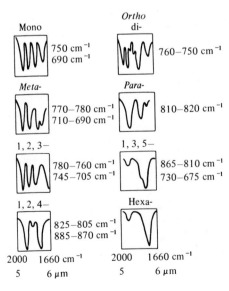

Fig. 6-12. Benzene ring substitution—pattern of combination bands between 2000 and 1670 cm^{-1} (5 to 6 μm). To the right of each curve are the approximate positions of the C—H out-of-plane bending bands between 900 and 650 cm^{-1} (11.1 to 15.4 μm). (After C. W. Young, R. B. Duvall, and N. Wright, *Anal. Chem.*, **23**, 709, (1951). Courtesy of *Analytical Chemistry*.)

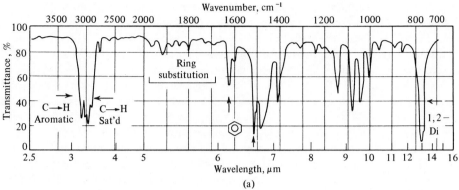

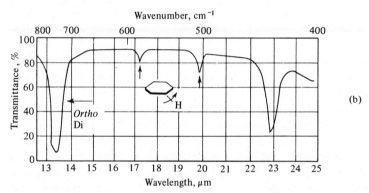

Fig. 6-13. Infrared spectrum of *o*-xylene. (a) Sodium chloride region and (b) potassium bromide region.

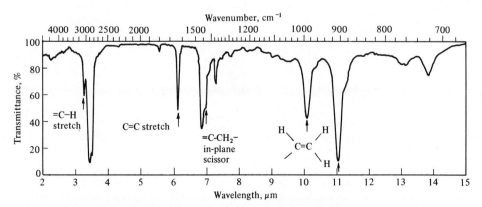

Fig. 6-14. Infrared spectrum of octene-1.

oxygen functional group greatly increases the intensity of the C=C absorption band. Conjugation with an aromatic nucleus causes a slight shift to lower frequency, but with a second C=C or C=O, the shift to lower frequency is 40 to 60 cm^{-1} with a substantial increase in intensity. The out-of-plane bending vibrations of the hydrogens on a C=C linkage are very valuable. A vinyl group gives rise to two bands at about 990 cm^{-1} (10.1 μm) and 910 cm^{-1} (11.0 μm). The =CH$_2$ (vinylidene) band appears near 895 cm^{-1} (11.2 μm) and is a very prominent feature of the spectrum. *Cis-* and *trans-*disubstituted olefins absorb near 685–730 cm^{-1} (13.7–14.6 μm) and 965 cm^{-1} (10.4 μm), respectively. The single hydrogen in a trisubstituted olefin appears near 820 cm^{-1} (12.2 μm).

In alkynes the ethynyl hydrogen appears as a needle-sharp and intense band at 3300 cm^{-1} (3.0 μm). The absorption band for —C≡C— is located in about the range from 2100 to 2140 cm^{-1} (4.67–4.76 μm) when terminal, but in the region from 2260 to 2190 cm^{-1} (4.42–4.56 μm) if nonterminal. The intensity of the latter type band decreases as the symmetry of the molecule increases; it is best identified by Raman spectroscopy. When the acetylene linkage is conjugated with a carbonyl group, however, the absorption becomes very intense.

For ethers the one important band appears near 1100 cm^{-1} (9.09 μm) and is due to the antisymmetric stretching mode of the —C—O—C— links. It is quite strong and may dominate the spectrum of a simple ether.

For alcohols the most useful absorption is that due to the stretching of the O—H bond. In the free or unassociated state, it appears as a weak but sharp band at about 3600 cm^{-1} (2.78 μm). Hydrogen bonding will greatly increase the intensity of the band and move it to lower frequencies and, if the hydrogen bonding is especially strong, the band becomes quite broad. Intermolecular hydrogen bonding is concentration dependent, whereas intramolecular hydrogen bonding is not concentration dependent. Measurements in solution under different concentrations are invaluable. The spectrum of an acid is quite distinctive in shape and breadth (Fig. 6-15) in the high-frequency region. The

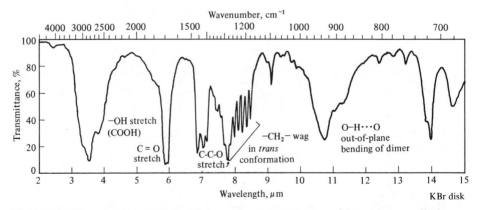

Fig. 6-15. Infrared spectrum of stearic acid. Solid spectra of long-chain *n*-alkyl compounds exhibit a series of evenly spaced bands in the region 1350 to 1180 cm^{-1} that are characteristic of the chain length: 2 × number of bands = number of —CH$_2$— groups.

TABLE 6-3 Classification of Various Types of Alcohols

	Position of C—O Bands	
Type of Alcohol	cm^{-1}	μm
Saturated tertiary Highly symmetrical secondary	1200–1125	8.30–8.90
Saturated secondary α-Unsaturated or cyclic tertiary	1125–1085	8.90–9.22
α-Unsaturated secondary Alicyclic secondary (5- or 6-membered ring) Saturated primary	1085–1050	9.22–9.52
Highly α-unsaturated tertiary Di-α-unsaturated secondary α-Unsaturated and α-branched secondary Alicyclic secondary (7- or 8-membered ring) α-Branched and/or α-unsaturated primary	< 1050	> 9.52

SOURCE: M. Gianturco in S. K. Freeman, Ed., *Interpretive Spectroscopy*, Van Nostrand Reinhold, New York, 1965, p. 56; by permission.

distinction between the several types of alcohols is often possible on the basis of the C—O stretching absorption bands, as indicated in Table 6-3. A spectrum of an alcohol is shown in Fig. 6-16.

The carbonyl group is not difficult to recognize; it is often the strongest band in the spectrum. Its exact position in the region, extending from about 1825 to 1575 cm^{-1} (5.48 to 6.35 μm), is dependent upon the double-bond character of the carbonyl group (Table 6-4). Anhydrides usually show a double absorption band. Aldehydes are distinguished from ketones by the additional C—H stretching frequency of the CHO group at about 2720 cm^{-1} (3.68 μm). In esters (Fig. 6-17), two bands related to C—O,

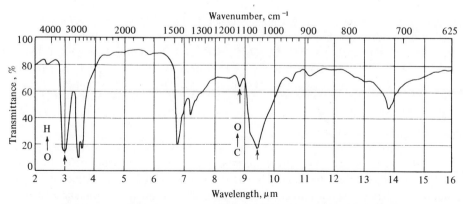

Fig. 6-16. Infrared spectrum of lauryl alcohol, $CH_3(CH_2)_{10}CH_2OH$.

TABLE 6-4 Carbonyl Absorptions[a]

Typical of	Wavenumber, cm^{-1}	Wavelength, μm
Anhydrides of carboxylic acids:		
aliphatic	1825[b]	5.48[b]
	1754	5.70
aromatic	1802[b]	5.55[b]
	1754	5.70
Chloride of carboxylic acids	1812	5.52
Carboxylic acids (monomers)	1776	5.63
Phenyl esters	1770	5.65
Vinyl esters of carboxylic acids	1770	5.65
Vinylidene esters of carboxylic acids	1764	5.67
Vinyl-type carbonates	1761	5.68
Normal carbonates	1751	5.71
Methyl esters of carboxylic acids	1748	5.72
Esters of carboxylic acids	1736	5.76
Esters of formic acid	1733	5.77
Aldehydes	1736	5.76
Ketones:		
aliphatic	1724	5.80
aromatic	1680–1645	5.95–6.08
Carboxylic acids (dimers)	1720–1700	5.81–5.88
Carbamates	1689	5.92
Amides (1°) of carboxylic acids	1718 sh	5.84 sh
	1684	5.94
Amides (2°) of carboxylic acids	1669	5.99
Amides (3°) of carboxylic acids	1667	6.00
Salts of carboxylic acids	1575	6.35

SOURCE: M. Gianturco in S. K. Freeman, Ed., *Interpretive Spectroscopy*, Van Nostrand Reinhold, New York, 1965, p. 86; by permission.
[a]All values in CCl_4.
[b]Weakens as colinearity is approached.

stretching and bending are recognizable, between 1300 and 1040 cm^{-1} (7.7 and 9.6 μm), in addition to the carbonyl band. The carboxyl group, in a sense, shows bands arising from the superposition of C=O, C—O, C—OH, and O—H vibrations (Fig. 6-15). Of five characteristic bands, three of these (2700, 1300, and 943 cm^{-1}; 3.7, 7.7, and 10.6 μm) are associated with vibrations of the carboxyl OH. They disappear when the carboxylate ion is formed. When the acid exists in the dimeric form, the O—H stretching band at 2700 cm^{-1} disappears, but the absorption band at 943 cm^{-1} due to OH out-of-plane bending of the dimer remains.

The spectrum of an amine is shown in Fig. 6-18. Of particular interest in a primary amine (or amide) are the N—H stretching vibrations at about 3500 and 3400 cm^{-1} (2.86 and 2.94 μm), the in-plane bending of N—H at 1610 cm^{-1} (6.2 μm), and the out-of-plane bending of —NH$_2$ at about 830 cm^{-1} (12.0 μm), which is broad for primary amines. By contrast, a secondary amine exhibits a single band in the high-frequency region at about 3350 cm^{-1} (2.98 μm). The high-frequency bands broaden and shift about 100 cm^{-1} to

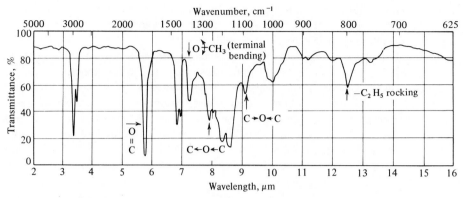

Fig. 6-17. Infrared spectrum of dimethyl-2, 5-diethyldipate,

$$CH_2CH_3 \qquad\qquad CH_2CH_3$$
$$CH_3-C-CH-CH_2-CH_2-CH-C-OCH_3$$
$$\overset{\|}{O} \qquad\qquad\qquad\qquad \overset{\|}{O}$$

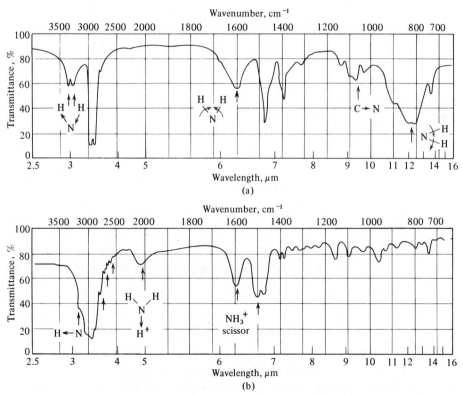

Fig. 6-18. Infrared spectrum of *n*-hexylamine. (a) Free amine and (b) hydrochloride.

lower frequency when involved in hydrogen bonding. When the amine salt is formed, these bands are markedly broadened and lie between 3030 and 2500 cm^{-1} (3.3 and 4.0 μm), resembling the —COOH bands in this region.

The nitro group is characterized by two equally strong absorption bands at about 1560 and 1350 cm^{-1} (6.41 and 7.40 μm), the asymmetric and symmetric stretching frequencies. In an N-oxide, only a single very intense band is present in the region from 1300 to 1200 cm^{-1} (7.70 to 8.33 μm). In addition there are C—N stretching and various bending vibrations whose positions should be checked in Fig. 6-10 (see also Fig. 7-6). Quite analogous bands are observed for bonds between S and O; all are intense. Stretching frequencies for SO$_2$ appear around 1400-1310 and 1230-1120 cm^{-1} (7.14-7.63 and 8.13-8.93 μm); for S=O at 1200-1040 cm^{-1} (8.33-9.62 μm); and for S—O around 900-700 cm^{-1} (11.11-14.28 μm).

Compound Identification

In many cases the interpretation of the infrared spectrum on the basis of characteristic frequencies will not be sufficient to permit positive identification of a total unknown, but, perhaps the type or class of compound can be deduced. One must resist the tendency to overinterpret a spectrum, that is, to attempt to interpret and assign all of the observed absorption bands, particularly those of moderate and weak intensity in the fingerprint region. Once the category is established, the spectrum of the unknown is compared with spectra of appropriate known compounds for an exact spectral match. If the exact compound happens not to be in the file, particular structure variations within the category may assist in suggesting possible answers and eliminating others. Several collections of spectra are available commercially.[1,6,7,9]

PROBLEMS

1. What would be the frequency of the fundamental absorption if its first overtone was observed at 1820 cm^{-1}?

2. The molecular heterotope ^{35}Cl ^{37}Cl has a fundamental band at 554 cm^{-1} in the gaseous state. Where would one expect the first and second overtones? What window material would be suitable?

3. Assuming a simple diatomic molecule, obtain the frequencies of the absorption band from the force constants given. Compare your answers with the tabulated positions in Fig. 6-10.
    (a) $k = 5.1 \times 10^5$ dynes cm^{-1} for C—H bond in ethane.
    (b) $k = 5.9 \times 10^5$ dynes cm^{-1} for C—H bond in acetylene.
    (c) $k = 4.5 \times 10^5$ dynes cm^{-1} for C—C bond in ethane.
    (d) $k = 7.6 \times 10^5$ dynes cm^{-1} for C—C bond in benzene.
    (e) $k = 17.5 \times 10^5$ dynes cm^{-1} for C≡N bond in CH$_3$CN.
    (f) $k = 12.3 \times 10^5$ dynes cm^{-1} for C=O bond in formaldehyde.

4. The apparent specific absorptivities are given for various infrared absorbers. Calculate the minimum liquid concentrations determinable (mg/ml) in 0.025-mm cells (for an absorbance reading of 0.005).

(a) $a = 900$ for $CHCl_3$ at 1216 cm^{-1}.

(b) $a = 1320$ for CH_2Cl_2 at 1259 cm^{-1}.

(c) $a = 4900$ for C_6H_6 at 1348 cm^{-1}.

(d) $a = 6080$ for $COCl_2$ at 1810 cm^{-1}.

(e) $a = 4400$ for $CH_2ClCOCl$ at 1821 cm^{-1}.

(f) $a = 1010$ for water at 1640 cm^{-1} –

5. The presence of ethylene in samples of ethane is determined by using the absorption band of ethylene at 2080 cm^{-1} (5.2 μm). A series of standards gave the following data:

% Ethylene:	0.50	1.00	2.00	3.00
Absorbance:	0.120	0.240	0.480	0.719

Calculate the percentage of ethylene in an unknown sample that had an absorbance of 0.412 when using the same cell and the same instrument.

6. The incorporation of an allyl group into one or both of the side chains of a barbiturate is always associated with the appearance of strong absorption bands at 10.1 and 10.8 μm. What alteration in these absorption bands would be expected by replacement of the hydrogen atom attached to the central carbon atom of the unsaturated allyl group by bromine?

7. Estimate the minimum concentration detectable ($A = 0.005$) in 0.05-mm cells for each of the following compounds, given their molar absorptivities:

(a) Phenol at 3600 cm^{-1}, $\epsilon = 5000$.

(b) Aniline at 3480 cm^{-1}, $\epsilon = 2000$.

(c) Acrylonitrile at 2250 cm^{-1}, $\epsilon = 590$.

(d) Acetone at 1720 cm^{-1}, $\epsilon = 8100$.

(e) Isocyanate (in polyurethane foam) monomer at 2100 cm^{-1}, $\epsilon = 17,000$.

8a. Calculate the thickness of the four cells from their interference fringes.

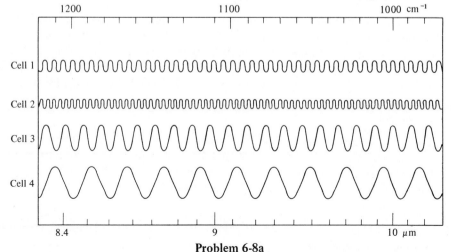

Problem 6-8a

b. Calculate the pathlength of the infrared absorption cell from the transmittance curve for benzene shown.

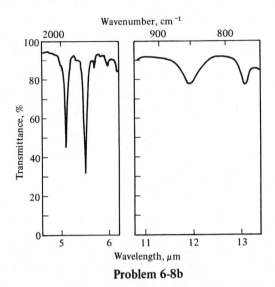

Problem 6-8b

9. Identify the particular xylene from the infrared data:
   Compound A: Absorption bands at 767 and 692 cm^{-1} (13.0 and 14.4 μm).
   Compound B: Absorption band at 792 cm^{-1} (12.6 μm).
   Compound C: Absorption band at 742 cm^{-1} (13.5 μm).

10. A bromotoluene, C_7H_7Br, has a single band at 801 cm^{-1} (12.50 μm). What is the correct structure?

11. A chlorobenzene exhibits no absorption bands between 900 and 690 cm^{-1} (11.1 and 14.5 μm). What is the probable structure?

12. An aromatic compound, C_7H_8O, has these features in its infrared spectrum: Absorption bands present at 3040 cm^{-1} (3.30 μm), 1010 cm^{-1} (9.90 μm), 3380 cm^{-1} (2.96 μm), 2940 cm^{-1} (3.4 μm), 1460 cm^{-1} (6.85 μm), and 690 cm^{-1} and 740 cm^{-1} (14.5 and 13.5 μm), whereas bands were absent at 1735 cm^{-1} (5.77 μm), 2720 cm^{-1} (3.68 μm), 1380 cm^{-1} (7.25 μm), 1182 cm^{-1} (8.45 μm). Identify the mode of each absorption band present (and absent), and write the structure of the compound.

13. The only significant absorption bands observed in the infrared spectrum were stretching of saturated C—H at 2960 and 2870 cm^{-1} (3.38 and 3.49 μm), methylene bending at 1461 cm^{-1} (6.85 μm), terminal methyl at 1380 cm^{-1} (7.25 μm), and the rocking of ethyl groups at 775 cm^{-1} (12.9 μm). Deduce the structure of this compound, C_6H_{14}.

14. A crystalline material is believed to be either a substituted hydroxylethyl cyanamide(I) or an imino oxazolidine(II):

$$N\equiv C-NH_2^+-CH-CH_2OH \qquad HN=CH-NH-\overset{\overset{\textstyle O}{\textstyle \|}}{C}-CH_2-$$
$$\text{I} \qquad\qquad\qquad\qquad \text{II}$$

Sharp bands are located at 3330 cm^{-1} (3.0 μm) and 1600 cm^{-1} (6.25 μm), but there are no bands at 2300 cm^{-1} (4.35 μm) or 3600 cm^{-1} (2.78 μm). Which structure fits the infrared data?

15. Deduce the structure of the compound whose empirical formula is C_4H_5N. Sharp, distinctive absorption bands, and virtually nothing else, occur at 3080 cm^{-1} (3.25 μm), 2960 cm^{-1} (3.38 μm), 2260 cm^{-1} (4.43 μm), 1865 cm^{-1} (5.36 μm), 1647 cm^{-1} (6.08 μm), 1418 cm^{-1} (7.05 μm), 990 cm^{-1} (10.1 μm), and 935 cm^{-1} (10.7 μm). The band at 1865 cm^{-1} is weak.

16. As far as possible, deduce the structural formula from the following information contained in the infrared spectrum. Cell thickness: 0.05 mm. Absorption bands (and their transmittance) at: 3080 cm^{-1} (0.28); 2950, 2910, 2835 cm^{-1} (0.28–0.37); 1848 cm^{-1} (0.71); 1650 cm^{-1} (0.65); 1455, 1440, 1410 cm^{-1} (0.32–0.45); 990 cm^{-1} (0.30); and 910 cm^{-1} (0.29).

17. Deduce the structure of the compound whose empirical formula is C_7H_5OCl using the following data. The infrared spectrum was obtained using a cell of thickness 0.1 mm. Absorbances are given in parentheses after each frequency: 3080 (0.07), 2810 (0.19), 2720 (0.17), 1705 (1.0), 1593 (0.27), 1573 (0.30), 1470 (0.20), 1438 (0.22), 1383 (0.22), 1279 (0.14), 1196 (0.82), 1070 (0.14), 900 (0.20), and 871 (0.20). The nmr spectrum showed a singlet at δ 9.95 and a 4-line pattern, symmetrical in appearance, with centers at δ 7.45 and δ 7.75 ($J = 7$ Hz in both cases); the number of protons was in the ratio of 1:2:2.

18. Deduce the product that gave the infrared spectrum shown, taken as a liquid film. The compound has a boiling point of about 101°C, $n_D^{20} = 1.3890$, and a molecular formula $C_6H_{12}O_2$.

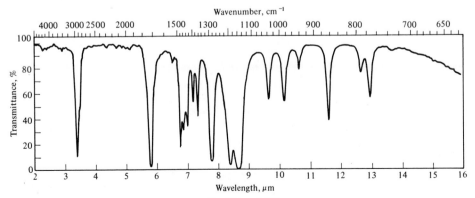

Problem 6-18

19. Deduce the structure of the compound with molecular formula $C_6H_{12}O_2$ whose infrared spectrum is shown.

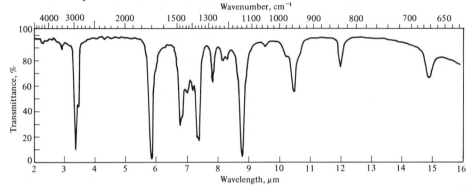

Problem 6-19

20. From the infrared spectrum shown, deduce the molecular structure. The compound has a boiling point of $69°C$.

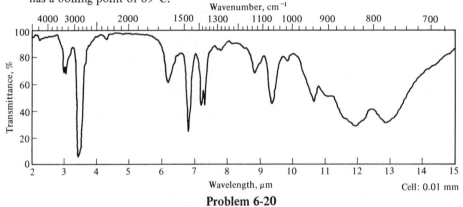

Cell: 0.01 mm

Problem 6-20

21. From the infrared spectrum shown, and the molecular weight of 131, deduce the compound.

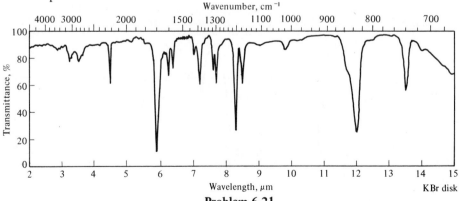

KBr disk

Problem 6-21

22. Deduce the molecular structure of the compound, molecular weight 176, whose infrared spectrum shown in the figure was taken as a liquid film.

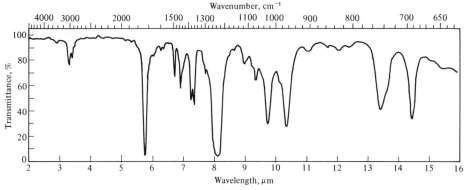

Problem 6-22

23. From the infrared spectrum of the liquid film which is shown in the figure, and the molecular weight of 118, write the structure of the compound.

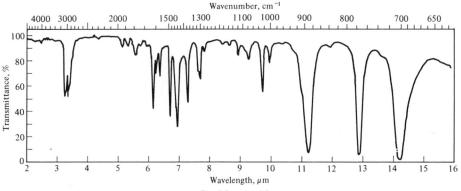

Problem 6-23

BIBLIOGRAPHY

Bauman, R. P., *Absorption Spectroscopy*, Wiley, New York, 1962.

Bellamy, L. J., *The Infrared Spectra of Complex Molecules*, 2nd ed., Wiley, New York, 1958.

Colthup, N. B., L. H. Daly, and S. E. Wiberley, *Introduction to Infrared and Raman Spectroscopy*, Academic Press, New York, 1964.

Gianturco, M., in *Interpretive Spectroscopy*, S. K. Freeman, Ed., Van Nostrand Reinhold, New York, 1965; Chapter 2.

Herzberg, G., *Molecular Spectra and Molecular Structure*, Vols. I and II, Van Nostrand Reinhold, New York, 1945.

Nakanishi, K., *Infrared Absorption Spectroscopy–Practical*, Holden-Day, San Francisco, 1962.

Silverstein, R. M., and G. C. Bassler, *Spectrometric Identification of Organic Compounds*, 2nd ed., Wiley, New York, 1967.

LITERATURE CITED

1. "ASTM-Wyandotte Index, Molecular Formula List of Compounds, Names and References to Published Infrared Spectra," *Am. Soc. Testing Materials, Spec. Tech. Publ.* 131 (1962) and 131-A (1963).
2. Ewing, G. W., *J. Chem. Educ.*, **48**, A521 (1971).
3. Ferraro, J. R., *Anal. Chem.*, **40** (4), 24A (April 1968).
4. Harrick, N. J., *Internal Reflection Spectroscopy*, Wiley, New York, 1967.
5. Low, M. J. D., *Anal. Chem.*, **41** (6), 97 A (May 1969); *J. Chem. Educ.*, **47**, A163, A255, A349, A415 (1970).
6. Nyquist, R. P. and R. O. Kagel, *Infrared Spectra of Inorganic Compounds*, Academic Press, New York, 1971.
7. Pouchert, C. J., *The Aldrich Library of Infrared Spectra*, Aldrich Chemical Co., Inc., Milwaukee, 1970.
8. Putley, E. H., "The Pyroelectric Detector" in *Semiconductors and Semimetals*, Vol. 5, R. K. Willardson and A. C. Beer, Eds., Academic Press, New York, 1970.
9. Sadtler Research Laboratories, *Catalog of Infrared Spectrograms*, 3314 Spring Garden St., Philadelphia, Pa.
10. Stewart, J. E. in *Interpretive Spectroscopy*, S. K. Freeman Ed., Van Nostrand Reinhold, New York, 1965, pp. 131–169.
11. Wilkes, P. A., Jr., "A Practical Guide to Internal Reflection Spectroscopy," *Am. Laboratory*, **4** (11), 42 (Nov. 1972).

7

Raman
Spectroscopy

When monochromatic light is scattered by molecules, a small fraction of the scattered light is observed to have a different frequency from that of the irradiating light; this is known as the *Raman effect*.* Since its discovery in 1928, the Raman effect has been important as a method for the elucidation of molecular structure, for locating various functional groups or chemical bonds in molecules, and for the quantitative analysis of complex mixtures, particularly for major components. Although Raman spectra are related to infrared absorption spectra, a Raman spectrum arises in a quite different manner and thus provides complementary information. Vibrations that are active in Raman may be inactive in the infrared, and vice versa. A unique feature of Raman scattering is that each line has a characteristic *polarization*, and polarization data provide additional information related to molecular structure.

THEORY

The Raman effect arises when a beam of intense monochromatic exciting radiation passes through a sample that contains molecules that can undergo a change in *molecular polarizability* as they vibrate. For example, in the symmetric stretching mode of carbon dioxide, there will be no change in the dipole moment as the two negative centers move equal distances in opposite directions from the positive center. The electron cloud around the molecule, however, alternately elongates and contracts, changing the polarizability accordingly. In such a case, the changing molecular polarizability will cause a modulation of the scattered light at the vibrational frequency. Hence, the induced classical oscillating dipole radiates not only at the frequency of the incident light but also at frequencies corresponding to the sum and the difference of this frequency and the molecular vibrational frequencies.

Most collisions of the incident photons with the sample molecules are elastic (Rayleigh scattering). About one in every million collisions, however, is inelastic and

*The Raman effect should not be confused with *Rayleigh* scattering in which radiation is scattered in all directions by interaction with atoms in its path; or with *Tyndall* scattering which is due to the interaction of light with small particles rather than with atoms; or with fluorescence in which a photon is actually absorbed and another photon is emitted at a later time.

189

involves a quantized exchange of energy between the scatterer and the incident photon which gives weak scattered lines that are separated from the exciting line by frequencies equal to vibrational frequencies of the scatterer. In the quantum mechanical representation of the origin of Raman lines, the incident photon elevates the scattering molecule to a quasi-excited state whose height above the initial energy level equals the energy of the exciting radiation (Fig. 7-1). This quasi-excited state then radiates light in all directions

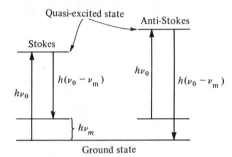

Fig. 7-1. Quantum representation of energy interchange involved in the Raman effect.

except along the line of action of the dipole, that is, the direction of the incident radiation. On the return to the ground electronic level, a vibrational quantum of energy may remain with the scatterer; if so, there is a decrease in the frequency of the re-emitted radiation. If the scattering molecule is already in an excited vibrational level of the ground electronic state, a vibrational quantum of energy may be emitted from the scatterer, leaving it in a lower vibrational level, and thus increasing the frequency of the scattered radiation. For either case, the shift in frequency of the scattered Raman radiation is proportional to the vibrational energy involved in the transition. Thus, the Raman spectrum occurs as a series of discrete frequencies, shifted symmetrically above and below the frequency of the exciting radiation and in a pattern characteristic of the molecule. The shift is independent of the frequency of the incident radiation; however, the intensity of the scattered radiation varies with the fourth power of the frequency of the incident radiation. The lines that are usually studied are those on the low-frequency side—the *Stokes* lines, which are more intense than those on the high-frequency side. By convention the positions of Raman lines are expressed as wavenumbers, but more correctly they are wavenumber differences, $\Delta \bar{\nu}$.

Some shortcomings of the Raman technique exist. The frequency of the exciting radiation must be lower than electronic energies of the molecule so that no absorption of the incident light occurs. If a substance is absorbing in the exciting region, it will reabsorb the Raman scattering and this will make it difficult to detect the Raman lines. After absorption, fluorescence may occur, and the intensity of fluorescence may be orders of magnitude stronger than the Raman effect, completely obscuring the Raman spectrum. The excitation frequency of Raman sources are selected to lie below most S–S* electronic transitions and above most fundamental vibrational frequencies. Both solid and liquid samples must be free from dust particles or the Raman spectrum may be masked by Tyndall scattering.

INSTRUMENTATION

Intense monochromatic light source, sensitive detection, and high light-gathering power, coupled with freedom from extraneous stray light, must be built into a Raman spectrometer. Many of the commercial instruments employ a He–Ne laser as the light source. A double monochromator (often $f/6$ with a focal distance of 75 to 100 cm, 0.25 cm^{-1} resolution) keeps stray light from the unshifted laser wavelength to a minimum, a special problem with powders and "dirty" samples. Scattered light is detected by a multiplier phototube with a red-sensitive response (if a He–Ne or Kr laser is the source). Visible-type optics are used throughout, and the entire spectral region is covered by a single grating. The schematic optical arrangement is shown in Fig. 7-2. High-quality

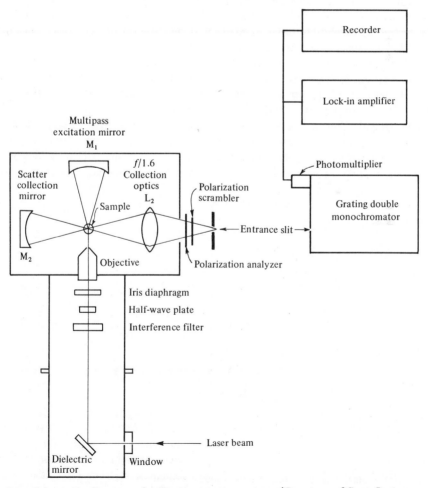

Fig. 7-2. Schematic diagram of a Raman spectrometer. (Courtesy of Spex Instruments.)

gratings are required for Raman work because imperfections in ruling that could lead to ghosts or other anomalies produce spectral artifacts in the Raman single-beam emission technique. Resolution and signal-to-noise ratio most often are limited by the effectiveness of the sample as a Raman scatterer because the Raman process is incredibly inefficient.

The laser beam is focused into the sample by a lens system. Another lens collects and focuses the scattered light into the monochromator. A laser provides an almost ideal monochromatic source of narrow linewidth. It emits radiant energy that is coherent, parallel, and polarized. A laser beam can be kept as a very slim cylinder only a few micrometers in cross section. Compared to the old mercury-arc sources, the sample image at the slit may be as much as 1000 times brighter with the laser. Sample volume can be diminished accordingly. The He–Ne laser line at 6328 Å is favorably located in the spectrum where the least amount of fluorescent problems appear in routine analyses. Although the Raman excitation efficiency decreases with the fourth power of the exciting frequency, fluorescence excitation efficiency usually decreases still faster. Also, strongly colored or photosensitive samples can frequently be examined by exciting in the red unless, of course, the sample is opaque to red.

Other laser sources offer advantages in certain applications. These include the argon laser with strong lines at 4880 and 5145 Å, and the line at 5682 Å from the krypton laser. Higher available power with these two lasers will improve Raman intensity but, at present, their cost is higher and their lifetime is shorter.

Photographic detection has given way almost entirely to photoelectric detection. Choice of phototube response depends on which laser line is used. The S-20 trialkali photocathode has about 7% quantum efficiency at 6328 Å and falls in efficiency four-fold for every 1000 cm^{-1}. Raman shifts of 3700 cm^{-1} require response to approximately

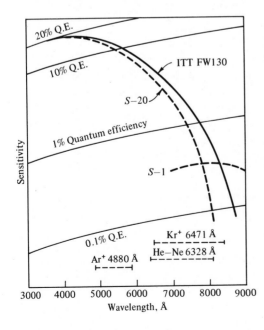

Fig. 7-3. Sensitivity of several photomultiplier tubes. The dashed horizontal lines represent the range of 3500 cm^{-1} Raman shift from three laser exciting lines.

8266 Å for the He–Ne laser. The extended red multialkali cathode and gallium arsenide photocathode designs have much higher quantum efficiency in the red region out to 9000 Å. By contrast, photomultiplier tubes are near their peak sensitivity at 4880 Å, the operating region for argon and krypton lasers. Obviously, laser selection and detector choice are interwoven, as shown in Fig. 7-3.

SAMPLE ILLUMINATION AND HANDLING

Examination of liquids is usually performed using a single pass of the laser beam either axially or transverse to the neat liquid sample contained in a glass capillary tube. If the liquid is clear, the beam focused to a diffraction-limited point in a small sample will pass through the liquid and can be reflected back again for another pass. Considerable gain in Raman intensity can be achieved beyond a single pass. It permits work with extremely small samples, in the microliter or even nanoliter range. The volume of the focused He–Ne laser beam is about 8 nl at the diffraction-limited point. When greater volumes of sample are available, the exciting radiation may be multipassed through larger cells. Gas chromatography fractions may be examined by trapping directly in glass capillary tubes; it is possible to produce Raman spectra with a sample of as little as 10 μg. Water is a weak scatterer and therefore an excellent solvent for Raman work. This has important consequences in studies of biochemical interest and in the pharmaceutical industry. Other widely used solvents are carbon disulfide, carbon tetrachloride, chloroform, and acetonitrile. Their obscuration ranges are shown in Fig. 7-4.

Gas samples can be handled with powerful laser sources and efficient multipassing or

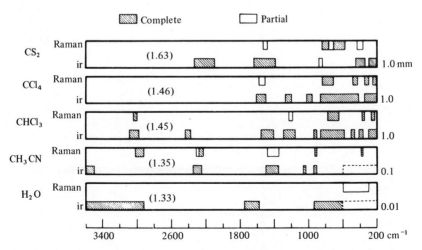

Fig. 7-4. The obscuration ranges of the most useful solvents for Raman spectrometry in solution. The infrared obscuration at the indicated path lengths is given for comparison; the refractive index of the solvent is given in parentheses.

intra-laser-cavity techniques, but are difficult to study because of their low scattering. At the moment Raman spectroscopy is not a routine technique for the study of gases.

Powders are tamped into an open-ended cavity for front surface illumination, or into a transparent glass capillary tube for transverse excitation. Forward (180°) sample illumination provides higher collection efficiency (better S/N ratio); however, Raman lines

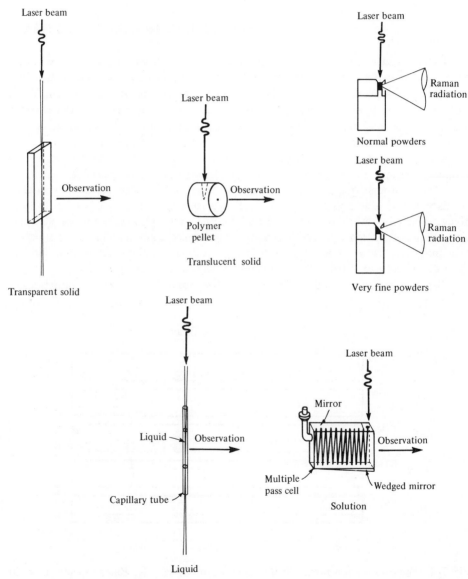

Fig. 7-5. Experimental arrangements for laser-excitation of specimens in various physical forms.

in the low-frequency region are more easily observed with right-angle viewing because the ratio of Raman to Tyndall and Rayleigh scattering is improved. For a translucent solid, the laser beam is focused into a conical cavity on the face of the sample, cut either into a cast piece or a pellet formed by compression of powder. The cavity functions as a light trap, producing multiple scattering of the incident photons via the intrinsic reflectivity and transmittance of the specimen to the laser frequency. These arrangements are shown in Fig. 7-5. A transparent solid sample preserves the directionality of the laser beam as it passes through the specimen, producing a scattering image which is collinear with the monochromator slit aperture for scattering observed at 90° to the direction of incidence. Bulk polymer samples and also single, very fine fibers may be run intact. Polymerization studies on such individual fibers often yield information about their orientation and crystalline properties.

DIAGNOSTIC STRUCTURAL ANALYSIS

In Raman spectroscopy those vibrations originating in relatively nonpolar bonds with symmetrical charge distributions and which are symmetrical in nature produce the greatest polarizability changes and are the most intense. Vibrations from $-C=C-$, $-C\equiv C-$, $-C\equiv N$, $-C=S$, $-C-S-$, $-S-S-$, $-N=N-$, and $-S-H$ bonds are readily observed. Raman lines are more characteristic than infrared absorption bands of the skeletal vibrations of finite chains and rings of saturated and unsaturated hydrocarbons.

The position of the symmetric ring stretching vibration of cyclic compounds is characteristic of the type and size of ring present in the compound. Aromatic compounds have particularly strong spectra; all have a strong ring deformation mode at 1600 ± 30 cm^{-1}. Monosubstituted compounds have an intense symmetric ring stretching vibration at about $1000 \ cm^{-1}$, a strong in-plane hydrogen bending vibration at about $1025 \ cm^{-1}$, and a weak depolarized in-plane bending vibration at about $615 \ cm^{-1}$ (Fig. 7-6). *Meta-*

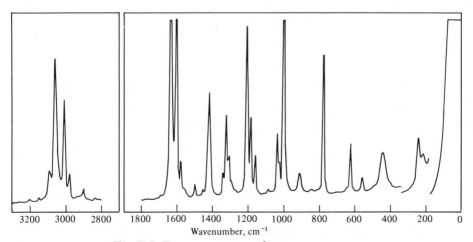

Wavenumber, cm^{-1}

Fig. 7-6. Raman spectrum of styrene monomer.

and 1,3,5-trisubstituted compounds have only the line at 1000 cm^{-1}. *Ortho*-substituted compounds have a line at 1037 cm^{-1}, and *para*-substituted compounds have a weak line at 640 cm^{-1}.

The intense Raman band near 500 cm^{-1} is characteristic of the —S—S— linkage, and the band near 650 cm^{-1} derives from —C—S— stretching (Fig. 7-7). Also, the —S—H stretching band near 2500 cm^{-1}, normally weak in the infrared, shows high intensity in the Raman.

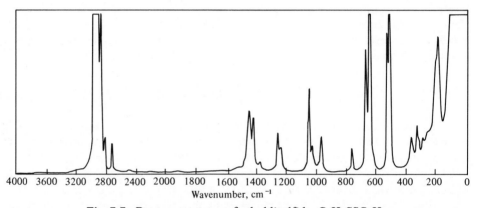

Fig. 7-7. Raman spectrum of ethyldisulfide, $C_2H_5SSC_2H_5$.

Although the —C≡N stretch appears in both the infrared and Raman spectra, it is markedly diminished in the infrared when an electronegative group such as chloride is α-substituted. The intensity is retained in the Raman spectrum.

Whenever the 3300 cm^{-1} region in the infrared is badly obscured by intense OH absorption, Raman spectroscopy is helpful because the OH band is weak whereas the NH and CH stretching frequencies still exhibit moderate intensity.

In the Raman spectrum the symmetrical methyl deformation frequency near 1380 cm^{-1} is sensitive to the environment of the methyl group. It is quite weak in alkyl compounds, but the band intensity is considerably enhanced when the methyl is attached to an aromatic ring and some types of double bonds.

Skeletal motions are very characteristic and highly useful for cyclic and aromatic rings, steroids, and long chains of methylenes. The spectrum in the region 800 to 1500 cm^{-1} is quite characteristic. In solid samples, sharpening and intensifying of certain bands appear to be a function of crystallinity.[1]

For all molecules that have a center of symmetry, a band allowed in the infrared is forbidden in the Raman and vice versa. In molecules with symmetry elements other than a center of symmetry, certain bands may be active in the Raman, infrared, both, or neither. For a complex molecule that has no symmetry save the identity element, all of the normal vibrational modes are allowed in both the infrared and Raman spectra.

Raman spectroscopy has a distinct advantage in the detection of low-frequency vibrations; the lower limit is dictated by the nature of the sample. With gases, information can be taken to within $2\ cm^{-1}$ of the exciting line. In less favorable cases, within 50–$100\ cm^{-1}$ is more typical. This corresponds to the far-infrared region where measurements are difficult. Most of the important vibrations involved in metal bonding of inorganic and organometallic compounds fall in the low-frequency region as a result of the large masses of the metal atoms. Raman spectroscopy has been applied to the analysis of strong acids and other aqueous solutions, and to the determination of the degree of dissociation of strong electrolytes and their corresponding activity coefficients.

The Raman technique has proved to be particularly valuable in the study of single crystals where the infrared technique has greater limitations on sample size and geometry. In addition, polarization data obtained from Raman spectra allow unambiguous classification of fundamentals and lattice modes into the various symmetry classes. Although Raman spectroscopy will never challenge X-ray diffraction as a tool for quantitative structural analysis, it is the preferred technique when qualitative information is sufficient, because it is faster and less expensive.

POLARIZATION MEASUREMENTS

The output of the laser is linearly polarized by passage through the Brewsters angle windows at each end of the plasma tube. This property can be put to use to determine the *depolarization ratio*, defined by

$$p = I_\perp / I_\parallel \qquad (7\text{-}1)$$

where $I_\parallel$ is the intensity of the scattered radiation polarized parallel and $I_\perp$ the intensity polarized perpendicular (with the analyzer and polarizer crossed). The parallel component is always preponderant. The depolarization ratio approaches a value of zero for highly symmetrical types of vibrations. For all nontotally symmetrical vibrations, the depolarization ratio will have a value of 0.86, the theoretical maximum, when the Raman line is said to be depolarized. Results are valuable in assigning frequencies to particular modes of vibration. A vibrational frequency that is antisymmetric or degenerate to one or more symmetry elements will be depolarized if observable, or else will be inactive.

The depolarization unit consists of an analyzer prism and a depolarization compensator in the path of the Raman scattered light. With this accessory unit the analyzer prism is set at $0°$, then at $90°$, and the ratio of the band intensities measured. Polarization characteristics of the monochromator are eliminated by the polarization compensator, perhaps a quartz wedge, which completely scrambles the polarized Raman scattering entering the spectrometer. Rotation of the plane of polarization is achieved by a half-wave retardation plate between the laser source and sample.

Depolarization is illustrated by the partial Raman spectrum of $CHCl_3$, shown in Fig. 7-8. Nonspherical symmetry exists for the molecule. The antisymmetric $C\text{—}Cl_3$ bending vibration at $261\ cm^{-1}$ and the $C\text{—}Cl$ stretching mode at $760\ cm^{-1}$ are depolarized. At $366\ cm^{-1}$ the polarized $C\text{—}Cl_3$ symmetric bending mode appears, as does the polarized symmetric $C\text{—}Cl$ stretching mode at $667\ cm^{-1}$. The $C\text{—}H$ stretching mode is not shown.

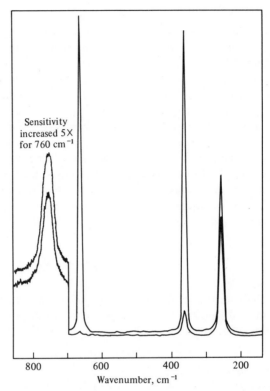

Fig. 7-8. Partial Raman spectrum of $CHCl_3$ illustrating depolarization. Lower trace is $I_\perp$, and upper trace is $I_\parallel$.

QUANTITATIVE ANALYSIS

The determination of the absolute intensities of Raman bands is even more difficult than the determination of the absolute intensity of infrared absorption bands. For this reason the intensity of a Raman line is usually measured in terms of an arbitrarily chosen reference line, usually the line of CCl_4 at 459 cm^{-1}, which is scanned before and after the spectral trace of the sample. Scattering intensities, or peak heights on the spectrum, are then converted to *scattering coefficients* by dividing the recorded height of the sample peak by the average of the heights of the dual traces of the CCl_4 peak. Both standard and sample must be recorded in cells of the same dimension.

For quantitative analysis the intensity of Raman lines is directly proportional to the number of scatterer molecules and thus to the scattering coefficient. For mixtures in which the components are all of the same molecular type, there is a direct proportionality between the scattering coefficient and the volume fraction of the compound present. For mixtures of dissimilar type, Raman shifts will vary among the various compounds, and a broad band is recorded at the position characteristic of these bond types. The area under the recorded peak can be used as a measurement of scattering intensity.

PROBLEMS

1. What instrumental factors have led to a false but widely held belief that C—H stretching vibrations are weak in Raman spectroscopy?

2. Suppose we wish to scan a Raman spectrum at the fastest possible speed. The minimum time constant of the electronics is 0.1 sec, and the described resolution (band pass) is 8 cm^{-1}. (a) What should the scan speed be? (b) If the dispersion of a double spectrometer with 1200 grooves/mm grating, working in the first order, is 5.5 Å/mm at the exit slit, what can the maximum slit opening be?

3. By what factor is the scattered intensity of a given band reduced in changing the excitation frequency from an argon laser (4,880 Å) to a neodymium-doped laser (10,650 Å)? Ignore changes in detector response, as well as grating and reflector efficiencies.

4. What are the relative intensities of a Raman line excited by a He–Ne laser at 6328 Å and one excited by an argon laser at 4880 Å?

5. When excited by the mercury line at 4358 Å, the spectral trace of benzene contains Raman lines at 606, 850, 991, 1176, 1584, 1605, 3047, and 3063 cm^{-1}. At what wavelengths will these Raman lines appear if benzene is irradiated with (a) a He—Ne laser (6328 Å), (b) an argon ion laser (4880 and 5145 Å), and (c) a krypton laser (5682 and 6471 Å)?

6. If unfiltered laser excitation from either the krypton or argon laser were used to excite the Raman spectrum of benzene (Problem 5), to what extent would each suite of Raman lines overlap?

7. For carbon disulfide, all vibrations that are Raman active are infrared inactive, and vice versa, whereas for nitrous oxide (N_2O) the vibrations are simultaneously Raman and infrared active. What can one conclude concerning the structures of N_2O and CS_2?

8. For CCl_4, four principal Raman lines appear at 218, 314, 458, and 791 cm^{-1}. None of these infrared frequencies is absorbed. The depolarization ratios are 0.86, 0.86, 0.046, and 0.83, respectively. What can be concluded about the symmetry of the molecule? Is its spatial configuration planar or tetrahedral?

9. For each unknown mixture of the trimethylbenzenes, compute the volume percent of each. The scattering coefficient of the pure compound at each analytical wave number is tabulated:

	1,2,3-		1,2,4-		1,3,5-	
	652 cm^{-1}	0.627	716 cm^{-1}	0.208	570 cm^{-1}	0.555
Mixture A		0.209		0.069		0.185
B		0.251		0.054		0.189
C		0.157		0.077		0.211

10. Deduce the structure of the compound whose spectra is shown in the accompanying figure. The molecular weight is 140.

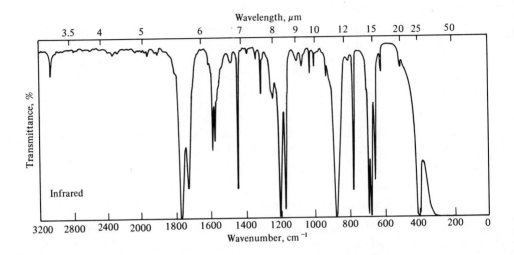

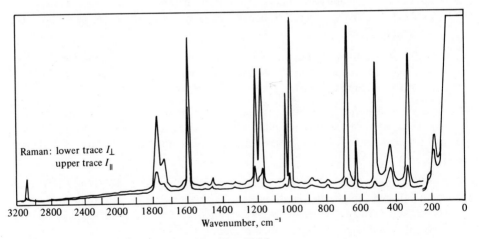

Problem 7-10

11. The molecular weight of the compound is 54. Using the spectra in the accompanying figure, determine the structure of the compound?

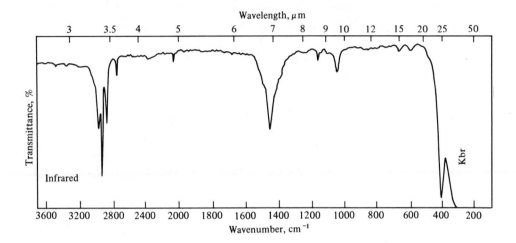

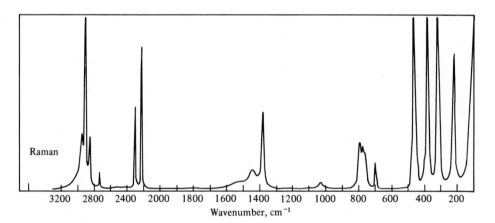

Problem 7-11

12. From the infrared and Raman spectra shown, and the molecular weight of 123, deduce the structure of the compound.

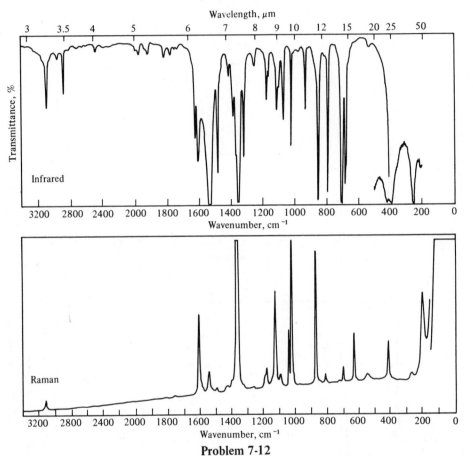

Problem 7-12

BIBLIOGRAPHY

Bulkin, B. J., "Raman Spectroscopy," *J. Chem. Educ.*, **46**, A781, A859 (1969).
Sloane, H. J., "The Technique of Raman Spectroscopy; A State-of-the-Art Comparison to Infrared," *Appl. Spectrosc.*, **25**, 430 (1971).
Tobias, R. S., "Raman Spectroscopy in Inorganic Chemistry," *J. Chem. Educ.*, **44**, 2 (1967).
Tobin, M. C., *Laser Raman Spectroscopy*, Wiley-Interscience, New York, 1971.

LITERATURE CITED

1. Sloane, H. J. in *Polymer Characterization*, C. D. Craver, Ed., Plenum Press, New York, 1971, pp. 15–36.

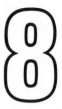

Nuclear Magnetic
Resonance Spectroscopy

The usefulness of *nuclear magnetic resonance* (*nmr*) spectroscopy in chemistry explains its rapid development since its discovery in 1946. Along with other spectroscopic methods, nmr has revolutionized the identification and characterization of molecules. Nuclear magnetic resonance spectroscopy concerns radio-frequency (rf) induced transitions between quantized energy states of magnetic nuclei that have been oriented by magnetic fields. Much information is provided by nmr because of the effects of inter- and intramolecular interactions on the values of the magnetic field strength at the nuclear sites in molecules. The spectra obtained answer many questions such as (referring to specific nuclei): Who are you?; Where are you located in the molecule?; How many of you are there?; Who and where are your neighbors?; How are you related to your neighbors? The result is often the delineation of complete sequences of groups or arrangement of atoms in the molecule.

BASIC PRINCIPLES

In addition to charge and mass, about half of the known nuclei possess spin or angular momentum. The spinning charge generates a magnetic field, and associated with the angular momentum is a magnetic moment, μ. These nuclei resemble a tiny bar magnet, the axis of which is coincident with the axis of spin. When placed in a powerful, uniform magnetic field, H_0, such nuclei are acted upon by a torque and tend to assume an allowed orientation with respect to the external field. The field aligns the spinning nuclei against the disordering tendencies of thermal processes. However, the nuclei do not align perfectly parallel (or antiparallel) to the field. Instead, their spin axes are inclined to the field and, like the top of a gyroscope, precess about the field direction—each pole of the nuclear axis sweeps out a circular path in the xy-plane (Fig. 8-1). Increasing the strength of the field only makes the nuclei precess faster. They can be made to undergo a transition to a higher energy level by applying a second, much weaker rf field, H_1, at right angles to the uniform magnetic field, H_0. When the frequency of the rotating component of this second field reaches the precession frequency, the spinning nuclei will absorb energy and flip into a higher energy level—one antiparallel to the field H_0 in the case of protons and other nuclei with a spin of $\frac{1}{2}$.

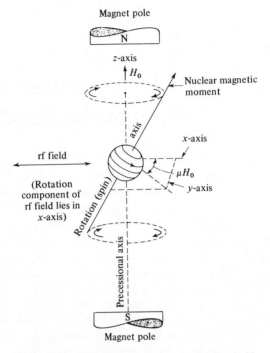

Fig. 8-1. Spinning nucleus in a magnetic field.

The resonance frequency, ν, that will effect transitions between energy levels is derived by equating the Planck quantum of energy with the energy of reorientation of a magnetic dipole:

$$\Delta E = h\nu = \frac{\mu H_0}{I} \tag{8-1}$$

where I is the spin quantum number in units of $h/2\pi$, h is Planck's constant, and E is energy. There will be $2I + 1$ possible orientations and corresponding energy levels. Nuclei with $I = \frac{1}{2}$ give the best resolved spectra because their *electric quadrupole moment** is zero. They act as though they were spherical bodies possessing a uniform charge distribution, which circulates over their surfaces. These nuclei include ^{1}H, ^{13}C, ^{19}F, and ^{31}P. Probably more than 90% of all nmr work is on proton spectra. Its natural abundance, importance in chemistry, strong magnetic moment and consequent nmr sensitivity, and spin of $\frac{1}{2}$, which implies only two energy states, account for its dominance. Conditions for ^{19}F resonance are nearly as favorable. All other nuclei give much weaker signals. Nuclei with spins of 1 or more also possess nuclear electric quadrupole moments, and so

---

*The electric quadrupole moment measures the electric charge distribution within a nucleus when it possesses nonspherical symmetry.

are readily disturbed by molecular electric field gradients. The result is a shortening of spin lifetime in a given state, and smearing-out of the nmr signal.

The tiny proton magnet is restricted to just two possible orientations in the applied field, as diagrammed in Fig. 8-2. When a sample is placed between the poles of a magnet, the protons in the sample can occupy one of two quantum states. They can either line up with the external field, the lower energy state, or against it, the higher energy state. Separation of nmr spin levels is about 10^{-3} times that of electron spin levels, but the line widths in nmr are smaller than in electron spin resonance (see Chapter 9) and of the order of 10^{-3} gauss. Consequently, more information can be supplied by nmr.

The frequency of the resonance absorption varies with the value of the applied field, as indicated by Eq. 8-1. Because the strength of the absorption signal is roughly proportional to the square of the magnetic field, larger values of field strength lead to a stronger signal. Nuclear characteristics for the more useful nuclei are listed in Table 8-1.

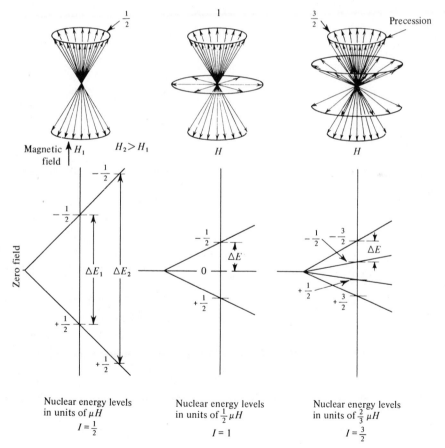

Fig. 8-2. Nuclear orientation and energy levels of nuclei in a magnetic field for different spin numbers.

TABLE 8-1   Properties of Nuclei

Isotope	Magnetic Moment, μ^a	nmr Frequency Values (in MHz) for		nmr Field Values (in kG) for	
		14.09 kG	23.49 kG	60 MHz	90 MHz
^1H	2.79268	60.000	100.000	14.092	21.06
^{13}C	0.70220	15.086	25.147	56.05	
^{19}F	2.6273	56.444	94.087	14.98	22.47
^{31}P	1.1305	24.288	40.485	34.81	52.22

^{a}In multiples of the nuclear magneton, $eh/4\pi Mc$.

Example 8-1

For a proton, $\mu = 1.41 \times 10^{-30}$ J·G^{-1}. From Eq. 8-1,

$$\nu = \frac{(1.41 \times 10^{-30} \text{ J} \cdot \text{G}^{-1})(14{,}092 \text{ G})}{(6.626 \times 10^{-34} \text{ J} \cdot \text{s})(\frac{1}{2})} = 60 \times 10^6 \text{ sec}^{-1}$$

Thus, in a magnetic field of 14,092 gauss, the protons will precess 60 million times per second, or 60 MHz—the rf resonance frequency required.

The energy difference between the two energy levels is not very large compared to thermal energies, kT, only about 10^{-2} eV or 0.01 cal. Consequently, thermal agitation diminishes the slight excess of nuclei in the lower energy state. At room temperature and a magnetic field of 14.09 kG, it is unfortunate that only 20 protons out of each 10 million serve as the effective participating population, because of the cancellation of opposing magnetic vectors for the remainder. The population ratio is given by the Boltzmann relation:

$$\frac{n_{upper}}{n_{lower}} = e^{-\mu H/IkT} \tag{8-2}$$

If an absorption signal is to persist, some mechanism must be provided for replenishing the number of nuclei in the lower energy state otherwise, in time, the rf field would equalize the populations of the energy levels; that is, the spin system would become saturated.

Energy absorbed and stored in the upper energy level can be dissipated and the nucleus returned to the lower energy level by a process called *spin-lattice relaxation*. It is brought about by interaction of the spin with the fluctuating magnetic fields produced by the random motions of neighboring nuclei. (Neighboring nuclei are called the "lattice" whether the material is crystalline, amorphous, or fluid.) In solids and viscous liquids, the relaxation time is in the order of hours, but in typical organic liquids and dilute solutions the time is in the range of 1 to 20 sec.

SAMPLE HANDLING

Nmr is a relatively insensitive analytical technique. For satisfactory detection on a routine basis, the concentration of the proton being observed must generally be 0.01 M or greater. Minimum sample size is about 2.5 mg dissolved in 50 μl. Cavity size in the sample probe can be as small as 25 μl; skillful handling technique is required. Selection of a suitable solvent is sometimes a problem. A solvent must be found which will dissolve the sample and yet not give an interfering nmr spectrum of its own. It should be chemically inert, magnetically isotropic, and preferably devoid of hydrogen atoms for proton nmr. CCl_4 is ideal. When compounds are insufficiently soluble in CCl_4, CS_2, cyclohexane, and $CDCl_3$ should be tried. Other solvents include perdeutero derivatives of acetonitrile, dimethylformamide, trifluoroacetic acid, and dimethylsulfoxide. The latter is often added in small amounts to solubilize many compounds that are poorly soluble in other solvents. Oxygen, being paramagnetic, broadens the natural line widths and should be removed if resolution better than 1 Hz is desired.

nmr SPECTROMETER (CONTINUOUS-WAVE)

Nmr instrumentation involves six basic units: (1) a magnet to separate the nuclear spin energy states, (2) a transmitter to furnish rf-irradiating energy, (3) a sample probe containing electrical coils for coupling the sample with the rf unit(s), (4) a detector to process the nmr signals, (5) a recorder to display the spectrum, and (6) a sweep generator for sweeping the magnetic field through the resonance region to produce the spectrum. These are indicated in Fig. 8-3. A nmr spectrometer differs from optical instruments in two ways. The nuclear magnetic energy levels are closely spaced so that rf energy is required to induce transitions, and rf radiation is monochromatic which eliminates gratings or prisms. Because the spacing of the energy levels is determined by the external applied magnetic field, the spectrum can be scanned by sweeping the magnetic field (*field-sweep method*) while keeping the frequency of the rf radiation constant, as well as by changing the frequency of the rf transmitter (*frequency-sweep method*) while keeping the external field constant as in conventional spectroscopy.

The majority of present nmr spectrometers operate with a field strength of 14.09 kG and rf field of 60 MHz. These are relatively inexpensive, easy-to-operate spectrometers for routine proton measurements. However, because chemical shifts and spectrometer sensitivity are field-dependent, it is often desirable to operate at the highest field strength commensurate with homogeneity and stability.[3] Available are 90, 100, 220, and 300 MHz spectrometers. For high-resolution work the magnetic field over the entire sample volume must be maintained uniform in space and time. Effective homogeneity of the field is promoted by (1) the use of large pole pieces composed of a very homogeneous alloy, (2) the polishing of pole faces to optical tolerances, and (3) the use of a narrow pole gap, that is, a smaller sample cross section and consequently a compromise with decreased sensitivity. Permanent magnets are simple and inexpensive to operate but require extensive shielding and must be thermostatted to $\pm 0.001°$. An electromagnet requires elaborate power supplies and cooling systems, but these disadvantages are offset by the opportunity to employ different field strengths to disentangle chemical shifts from multiplet structures and to study different nuclei. The upper field limit of electromagnets is about 24 kG. The superconducting solenoid requires no electrical power, but consumption of liquid helium coolant is costly. A field of 70.5 kG is the highest available in present commercial spectrometers (300 MHz).

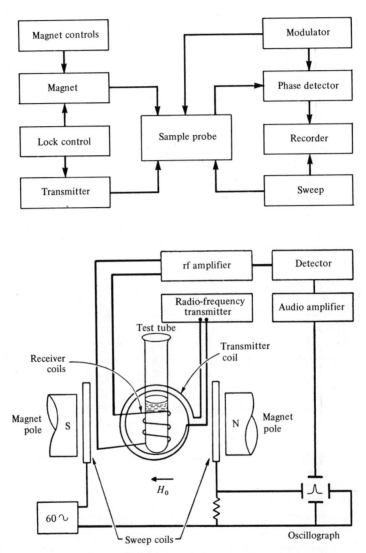

Fig. 8-3. Top, block diagram of a high-resolution nmr spectrometer. Bottom, schematic arrangement of components in vicinity of sample probe. (Courtesy of Varian Associates.)

   To flip the rotating nuclear axes with respect to the magnetic field, a linearly oscillating rf field is imposed at right angles (x-axis) to the magnetic field (z-axis). The coil that transmits the rf field is made in two halves to allow insertion of the sample. Auxiliary coils, wound around the pole pieces, allow a sweep to be made through the applied magnetic field. In the field-sweep method, the rf field is set at a fixed value, say 60 MHz, and the sweep generator periodically sweeps the main magnetic field in the immediate vicinity of 14.09 kG. The sweep range for protons is approximately from 2000 Hz downfield to 500 Hz upfield, but this must be increased to more than 10 kHz for observing the complete range of resonances of ^{13}C and ^{19}F. At the usual sweep rate of about 1 Hz per sec, a ringing, envelope-like pattern appears over the trailing edge of an absorption band. It arises from rapid sweeping through the resonance condition. The frequency of the rotating component of magnetization varies with the changing sweep field and, as a result, the induced signals are alternately in and out of phase with the applied rf field, the frequency of which is constant. The observation of ringing is a good indication of a homogeneous field.

The probe unit is inserted between the pole faces of the magnet. It houses the sample that is contained in a cylindrical, thin-walled, precision-bore, glass tube having an o.d. of 5 mm. The probe serves as the interaction point for the magnetic field, the rf field, the field sweep, and the detector. To average small magnetic field inhomogeneities throughout the sample, an air-bearing turbine rotates the sample at a rate of several hundred revolutions per minute. This spinning produces sidebands in the spectrum because the nmr peaks are modulated at the spinning frequency. A single coil probe has one coil which not only supplies the rf radiation to the sample but also serves as a part of the detector circuit for the nmr signal. Cross-coil probes have one coil for irradiating the sample and another orthogonally mounted coil for detection.

The nmr signal is an ac signal with two components 90° out of phase. The detectable signals are an absorption component and a dispersion component, shown in Fig. 8-4. By using a phase-sensitive detector which can be finely tuned to sense only one component, either the absorption or the dispersion mode can be observed. Nmr spectra are usually observed in the absorption mode whereas the dispersion mode is used in the field-frequency control.

A field modulation system is used in the detection system. For example, a 60-MHz rf signal is passed to the probe, which is fitted with small modulation coils wrapped around the pole tips of the magnet through which is passed a 6-kHz audiofrequency sig-

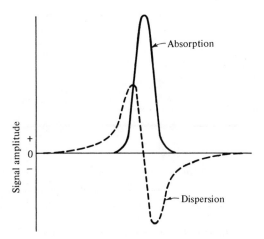

Fig. 8-4. Line shapes of the two observable nmr signals.

nal. The sample experiences the main field as well as fields positioned upfield and down-field at multiples of the modulation frequency. The sweep of the spectrum is then adjusted so that one observes the first sideband response of the nmr spectrum rather than the fundamental. By this method the signal is injected into the following rf amplifier at 60.006 MHz, and then amplified and converted to a new audio signal at 6 kHz. This system has the great advantage that a 6-kHz output can occur only when the nuclei themselves generate a signal, so that very good base-line stability is obtained, a prerequisite for accurate integrations. Because the resonance signal typically arises from only a few parts per million of nuclei, high gain amplification is required.

The nmr spectrum is recorded directly on precalibrated chart paper. This demands that the ratio of rf frequency to field strength be very stable. To obtain an nmr peak having a line width of 0.1 Hz at 60 MHz requires an overall stability of $0.1/60 \times 10^6$, or about 2 parts in 10^9. There is no problem with the rf frequency, but magnet stability is only about 10^{-7}/hr. Independent stabilization of the rf units and magnet system is difficult. The field-frequency (H/ν) relation of Eq. 8-1 is more reliably maintained by means of servo loops that lock them together (Fig. 8-5). An internal or external reference nucleus is continuously irradiated at its resonance frequency, and the nmr dispersion signal is continuously monitored while the spectrum is being swept. If the frequency of irradiation exactly equals the frequency at the center of the dispersion signal, the nmr error signal will be zero. But if the magnetic field should change, the resonance condition is no longer fulfilled and an output signal will be produced. The output error signal is amplified and fed back into correcting coils in the magnet gap to bring the error signal back to zero. In an internal lock system, a reference, such as tetramethylsilane (TMS), is added to the sample so that analytical sample and reference experience exactly the same magnetic field. By contrast, an externally located control sample, such as water for proton signals, sealed in a container in the probe, can "see" a slightly different field from that seen by the sample and is therefore less accurate than the internally locked system.

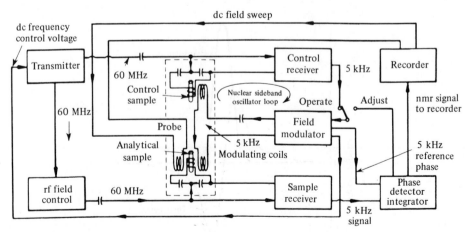

Fig. 8-5. Schematic circuit diagram of Varian A-60 nmr spectrometer. (Courtesy of Varian Associates.)

With appropriate modifications a nmr spectrometer can measure the resonance spectra of other nuclei. Usually a separate probe and rf source is required for each nucleus; retuning the spectrometer can be tedious. Nearly all spectrometers have a ^{19}F accessory, and ^{13}C capability[8] is becoming more frequent, especially with pulsed-nmr spectrometers.

With well-stabilized spectrometers, the repeatability of successive spectral scans permits the use of an on-line computer-of-average transients to improve sensitivity. The output of the detector is connected to the signal-averaging device, and n scans of the spectrum are additively accumulated in the memory of the computer. Because noise is random and therefore tends to average out while nmr signals tend to add, sensitivity can be enhanced by sweeping the spectrum many times, accumulating the signals in 1000 to 4000 separate channels (multi-channel analyzer) and averaging over all the traces. The theoretical enhancement of S/N is $n^{1/2}$; the improvement is limited only by the time stability of the spectrometer-computer system. Ten- to fifty-fold improvement in sensitivity, requiring up to a few hours of stable operation, is now routine. A Fourier transform accomplishes the same improvement in sensitivity, however, in much less time.

OTHER TYPES OF nmr SPECTROMETERS

In addition to high-resolution, continuous-wave nmr operation just discussed, there are three other modes: (a) pulsed-wave, (b) Fourier-transform, pulsed-wave, and (c) wide-line. Each will be discussed briefly.

Pulsed-Wave nmr

In pulsed-wave nmr the sample is irradiated with a short duration rf pulse having a frequency equal to the resonance frequency of the nucleus being observed. Analysis of the rate of decay of the nmr signal after termination of the pulse allows calculation of characteristic relaxation times. A modified procedure called the method of *spin echos* utilizes a carefully timed series of intense, very short duration pulses. Pulsed methods are used to determine self-diffusion coefficients in liquids and kinetic parameters.

Fourier-Transform nmr

Fourier-transform nmr consists of the application of an intense pulse of rf energy to the sample for a very short time, and the measurement of the resulting free-induction decay signal from the nuclear spins in the sample. This method irradiates the sample not with a successive series of single discrete frequencies, as in the case for the steady-state, slow passage spectrum obtained with continuous-wave nmr, but with a band of frequencies centered at the resonance frequency for the particular nuclide. The intense rf pulse excites simultaneously the entire range of precession frequencies of the chosen nuclear species. Whenever one mixes a number of signals with different frequencies simultaneously, an interference pattern is obtained. These simultaneous rf signals must be sorted. The free-induction decay signal following each repetitive pulse is digitized by a fast analog-to-digital converter, and the successive transient signals are accumulated in a laboratory computer until an adequate signal-to-noise ratio is obtained. Using Fourier algorithms, the computer then performs a fast Fourier transformation to the frequency domain to provide a normal spectral presentation of absorption vs. frequency. Because a free-induction decay signal and a continuous-wave nmr spectrum form a Fourier transform pair, given one, it is possible to calculate the other.

Simultaneous acquisition of the entire spectrum instead of successive acquisition of frequency-swept spectra can give a sensitivity enhancement or a savings in time. The time required to observe a nmr spectrum by the continuous-wave method is Δ/r (in seconds), where Δ is the total width of the spectrum and r is the resolution desired. For ^{13}C at 25 MHz, where Δ is typically about 5 kHz and the line widths are about 1 Hz, one must scan the 5 kHz region at a rate of 1 Hz/sec, or preferably slower. This requires a minimum time of 5000 sec. The only requirement on the free-induction signal is that it be recorded for a duration of $1/r$ sec. Thus a time enhancement factor of $(\Delta/r)/(1/r) = \Delta$ is achieved. The sensitivity increase is $\sqrt{5000} \simeq 70$.

The pulsed-Fourier transform technique makes possible the study of less sensitive nuclides, such as ^{13}C, ^{14}N, ^{15}N, ^{17}O, and ^{31}P, and unstable species. ^{13}C-nmr spectroscopy is proving to be a most important tool in determining the structure of organic and biological molecules. In virtually every instance where no information is available from proton nmr, ^{13}C-nmr spectra easily show differentiation between two complex molecules which may differ only in the subtle arrangement of a few carbon atoms. The reason for this is partly because the range of ^{13}C chemical shifts (200 ppm or more) is much greater than for protons, and partly because in non-^{13}C-enriched molecules only one ^{13}C nuclide is in any given molecule.

Wide-Line nmr

The wide-line mode of operation is used mainly for studies of solids. Wide-line nmr does not permit the observation of chemical shifts derived from different atomic groups, but does permit measurement of samples in their solid state for total content of a particular nucleus. Accordingly, it can yield valuable information with respect to crystal or noncrystal structures plus the internal motion of polymers.

A magnet with a field strength of 1700 to 3400 G and a homogeneity of 1 part in 10^5 is adequate. The rf range of 2 to 16 MHz suffices for studies of most isotopes. In fact, esr spectrometers may be converted to wide-line, frequency-sweep, nmr spectrometers. Because effective line widths are the same as the field inhomogeneity, all the chemical shifts are encompassed under one overlapping band. Sample sizes from 0.2 to 40 ml may be accommodated.

An early and continuing application has been the quantitative analysis of materials for total proton content from the integrated area under the resonance curve. It is a rapid, nondestructive method of analyzing for fats, oils, and moisture in many types of materials. The determination of fluorine in chemical compounds is another area of application.

SPECTRA AND MOLECULAR STRUCTURE

For most purposes, high-resolution nmr spectra can be described in terms of *chemical shifts* and *coupling constants*. Two other parameters sometimes involved are the *spin-lattice* (T_1) and *spin-spin* (T_2) *relaxation times* of the nuclei. Internal rotation, chemical exchange, and other rate processes can affect relaxation times so as to produce pronounced temperature-dependent effects on the spectra. In solids, direct magnetic dipole–dipole interactions dominate, relaxation times are long, and the nmr spectra consist of very broad lines. In liquids and gases, the direct dipole–dipole interactions usually are averaged to zero by rapid intra- and intermolecular motions, the relaxation times are much shorter, and narrow-line nmr spectra are observed.

Chemical Shifts

The resonant position of a nucleus in a nmr spectrum is dependent on its electron environment. In different chemical environments, shifts in resonant frequency arise from partial shielding of the nuclei from the applied magnetic field by the electron cloud surrounding the nuclei. The density of this cloud varies with the number and nature of the neighboring atoms. Each group of chemically equivalent nuclei is shielded to a different degree. The magnitude of the effective field felt by each group of nuclei can be expressed as follows:

$$H_{\text{eff}} = H_0(1 - \sigma) \tag{8-3}$$

where σ is a nondimensional shielding constant, and may be either a positive or negative number. Thus, the protons at various sites in a molecule are spread out into a spectrum according to the values of their shielding parameters. A field or frequency sweep will bring protons at each particular site into resonance one after another. The more the field induced by the circulating electrons shields the nucleus and opposes the applied field, the higher must be the applied field to achieve resonance if the field is varied, or the lower must be the resonance frequency if the frequency is varied. The specific locations of the shifted resonant frequencies can be used to characterize the neighbors of a given nucleus. The value of the shielding constant depends on several factors, among which are the hybridization and electronegativity of the groups attached to the atom containing the nucleus being studied. Shielding effects seldom extend beyond one bond length except with very strong electronegative groups.

Because nmr spectrometers employing different field strengths are in use, it is desirable to express the position of resonance in field-independent units and with respect to the resonance of a reference compound. For proton spectra in nonaqueous media, the reference material is tetramethyl silane, $(CH_3)_4Si$, abbreviated TMS, whose position is assigned as exactly 0.0 on the δ scale. TMS contains 12 protons but these are all chemically equivalent and therefore give rise to a single sharp signal. The magnitude of the *chemical shift* is expressed in parts per million:

$$\delta = \frac{H_{\text{sample}} - H_{\text{TMS}}}{\nu_1} \times 10^6 \tag{8-4}$$

where H_{sample} and H_{TMS} are the positions of the absorption lines for the sample and reference, respectively, expressed in frequency units (hertz); ν_1 is the operating frequency of the spectrometer. A positive δ value represents a greater degree of shielding in the sample than in the reference. Another frequently used convention, but not officially approved, is the τ scale, in which $\tau = 10.0 - \delta$.

Recommended reference materials for other nuclei include CS_2 or TMS for ^{13}C, trichlorofluoromethane (CCl_3F) for ^{19}F, and phosphoric acid for ^{31}P. The numbers on the dimensionless (shift) scale upfield from the reference are designated positive.

Proton resonances from C—H bonds are located in the range from 0.9 to 1.5 δ when only aliphatic groups are substituents. A CH_3 group usually appears at 0.9 δ when the adjacent three bonds are methylene groups; CH_2 and CH protons are slightly further downfield in that order. An adjacent unsaturated bond shifts the resonant position of CH_3 to 1.6–2.7 δ. An adjacent oxygen atom markedly shifts proton signals downfield to

TABLE 8-2 Approximate Chemical Shift of Protons

Substituent Group	Methyl Protons δ Values (τ)	Methylene Protons δ Values (τ)	Methine Protons δ Values (τ)
HC—C—CH$_2$ (or NR$_2$)	0.9 (9.1)	1.2 (8.8)	1.5 (8.5)
HC—C—CO	1.1 (8.9)	—	—
HC—C—C=C	1.1 (8.9)	1.7 (8.3)	—
HC—C—Ar	1.3 (8.7)	1.6 (8.4)	—
HC—C—S	1.3 (8.7)	1.5 (8.5)	—
HC—C—O—CO	1.4 (8.6)	—	—
HC—C—O	1.4 (8.6)	1.9 (8.1)	2.0 (8.0)
HC—C—Cl	1.5 (8.5)	1.7 (8.3)	1.5 (8.5)
HC—C—NO$_2$	1.6 (8.4)	2.1 (7.9)	—
HC—C—Br	1.7 (8.3)	1.7 (8.3)	1.8 (8.2)
HC—CH$_2$	0.9 (9.1)	1.3 (8.7)	1.5 (8.5)
cyclic CH$_2$	—	1.5 (8.5)	—
HC—C=C	1.6 (8.4)	2.3 (7.7)	—
HC—CO—O	2.0 (8.0)	2.2 (7.8)	—
HC—S	2.2 (7.8)	2.4 (7.6)	—
HC—CO—R	2.2 (7.8)	2.4 (7.6)	2.6 (7.4)
HC—NR$_2$	2.3 (7.7)	2.5 (7.5)	2.8 (7.2)
HC—Ar	2.4 (7.6)	2.7 (7.3)	2.9 (7.1)
HC—CO—OAr	2.4 (7.6)	—	—
HC—CO—Ar	2.6 (7.4)	—	—
HC—Br	2.7 (7.3)	3.3 (6.7)	3.6 (6.4)
HC—Cl	3.0 (7.0)	3.6 (6.4)	4.0 (6.0)
HC—OR	3.3 (6.7)	3.4 (6.6)	3.6 (6.4)
HC—OH	3.4 (6.6)	3.6 (6.4)	3.9 (6.1)
HC—O—CO—R	3.7 (6.3)	4.1 (5.9)	5.1 (4.9)
HC—O—Ar	3.8 (6.2)	4.1 (5.9)	—
HC—O—CO—Ar	4.0 (6.0)	4.3 (5.7)	5.0 (5.0)
HC—NO$_2$	4.3 (5.7)	4.4 (5.6)	4.6 (5.4)

3.2–3.4 δ for aliphatic entities and to 3.6–3.9 δ for aryl-O-CH situations. Many common groups produce special shielding effects because they allow circulation of electrons only in certain preferred directions within the molecule. Figure 8-6 shows shielding (+) and deshielding (−) zones in the neighborhood of triple, double, and single bonds to carbon. In C=C and C=O double bonds, the deshielding zone extends along the bond direction; even C—C bonds show some deshielding in this direction. This anisotropy of the magnetic susceptibility of chemical bonds means that the shielding or deshielding of a neighboring proton in the molecule is dependent on its distance from the bond and its orientation with respect to that bond. Aromatic rings exhibit a strong anisotropic effect. When such compounds are placed in a magnetic field, the six π-electrons circulate in two parallel doughnut-shaped orbits on each side of the ring. The resulting local magnetic field opposes H_0 in a cone-shaped zone of excess shielding extending along the hexad axis, but reinforces H_0 in a zone of deshielding extending from the edge of the ring. In aromatic

	δ Values (τ)
—C≡CH	3.1 (6.9)
cyclic $\diagup$C$\diagdown$C=CH$_2$	4.6 (5.4)
>C=C$\diagup$H $\diagdown$H	4.7 (5.3)
H$_a$ H$_c$ C=C H$_b$	(c) 5.9 (4.1) (a,b) 5.0 (5.0)
—C=CH—CO	5.9 (4.1)
—CH=C—CO	6.8 (3.2)
Ar—H	7.2 (2.8)
HCO—O	8.0 (2.0)
—CHO	9.9 (0.1)
—COOH	10–12 (0 to −2)
—SO$_3$H	11–12 (−1 to −2)
Ar—OH	4–7 (6 to 3)
R—OH	1–6 (9 to 4)
Ar—SH	2.8–3.6 (7.2 to 6.4)
R—SH	1.4 (8.6)
R—NH	0.5–5 (9.5 to 5.0)
Ar—NH	3–5 (7–5)
—NH—R	5–8 (5–2)

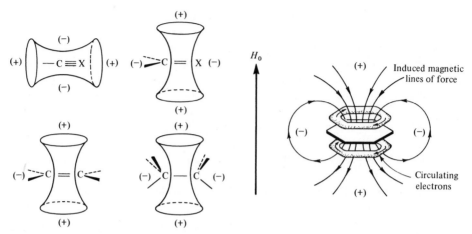

Fig. 8-6. Shielding (+) and deshielding (−) zones in the neighborhood of triple, double, and single bonds to carbon and aromatic rings.

compounds the deshielding zone is more commonly occupied; thus, protons on aromatic rings appear at much lower field (7–8 δ) than olefinic protons (5–6 δ). In acetylenes, the electron current circulates in such a way that the shielding zone extends along the bond direction, and acetylenic protons appear at high fields (1.6–3.0 δ). Table 8-2 gives some proton chemical shifts.

Processes giving rise to chemical exchange or conformational change which are complete in 1 to 0.001 sec may give rise to spectra which are time-averages in comparison with those expected in terms of instantaneous molecular conformations. If the exchange rate is high in comparison with the frequency of the chemical shifts and spin-spin couplings, the local fields seen from the nucleus of the exchanging atom will be averaged out to result in a single line, somewhat broader than normal.

Spin-Spin Coupling

Nuclei can interact with each other to cause mutual splitting of the otherwise sharp resonance lines into multiplets, called *spin-spin coupling*. These multiplets arise because magnetic moments of nuclei interact with the strongly magnetic electrons in the intervening bonds. The strength of the coupling, denoted by J, is given by the spacing of the multiplets and is expressed in hertz. Proton–proton couplings (Table 8-3) are usually transmitted only through two or three bonds, although weak couplings are often transmitted further. In certain rigid structures of favorable geometry, coupling through four bonds may be reasonably large. When unsaturated systems occur between the protons, long-range coupling may be enhanced. In allylic systems, four-bond couplings reach a maximum of about 3 Hz when the angle between the plane containing the olefinic protons and the C—H bond of the allylic carbon atom is about 90°. In H—C—C=C—C—H systems, five-bond couplings of about 3 Hz are observed. In acetylenes, allenes, and

TABLE 8-3 Spin-Spin Coupling Constants

Type of Compound	J, Hz	Type of Compound	J, Hz
$\diagup C \diagdown$ with two H (geminal)	12–15	cyclohexane H_a, H_e	$(a\text{–}a)$ 8–10 $(a\text{–}e)$ 2–3 $(e\text{–}e)$ 2–3
CH—CH (free rotation)	6–8	benzene ortho H	(o) 7–9 (m) 2–3 (p) 0–1
CH—OH (no exchange) (—NH)	5	pyridine	$(2\text{–}3)$ 5 $(3\text{–}4)$ 8 $(2\text{–}4)$ 1.5 $(3\text{–}5)$ 1.5 $(2\text{–}5)$ 1 $(2\text{–}6)$ 0
$C=C$ with two H (geminal)	0–3	pyrrole	$(1\text{–}2)$ 2–3 $(1\text{–}3)$ 2–3 $(2\text{–}3)$ 2–3 $(3\text{–}4)$ 3–4 $(2\text{–}4)$ 1–2 $(2\text{–}5)$ 1–3
$C=C$ H,H (cis/trans)	(cis) 6–14 $(trans)$ 11–18	furan	$(2\text{–}3)$ 2 $(3\text{–}4)$ 3.5 $(2\text{–}4)$ 0 $(1\text{–}2)$ 1.5
CH—CHO	1–3	fluorobenzene ortho	(o) 6–10 (m) 5–6 (p) 2
$C=C$ CH,H	4–10	$\diagup C \diagdown$ with H, F	44–81
$C=CH—CH=C$	10–13	CH—CF	3–25
CH—C≡CH	2–3	$C=C$ H,F (cis/trans)	(cis) 1–8 $(trans)$ 12–40
=CH—CHO	6		
HC,H $C=C$	1–2		
H,H $C=C$ (ring)	(5-member) 3–4 (6-member) 6–9 (7-member) 10–13		

cumulenes, observable couplings may be transmitted over many bonds, up to nine in polyacetylenes. In aromatic rings, couplings of protons in *ortho* positions (through three bonds) are 7–9 Hz, *meta* (four bonds) 2–3 Hz, and *para* (five bonds) 0.5–1.0 Hz.

Couplings depend also on geometry. The dihedral angle between planes determines the coupling of protons on adjacent carbon atoms (Fig. 8-7). Adjacent axial–axial protons, displaying a dihedral angle of 180°, are strongly coupled, whereas axial-equatorial and equatorial–equatorial protons are coupled only moderately. *Trans* and *cis* protons on olefinic double bonds show $J_{trans}/J_{cis} \simeq 2$, which can be useful in assigning structures of geometrical isomers.

The number of lines in a multiplet is given by $2nI + 1$, where n is the number of nuclei producing the splitting. For protons, this becomes $n + 1$ lines. The relative intensity of each of the multiplets, as reflected in the integral curve, is proportional to the number of nuclei in the group. Intensities of the peaks within a multiplet are given by simple statistical considerations, and are proportional, therefore, to the coefficients of the binomial expansion. Thus, one neighboring proton splits the observed resonance to a doublet (1:1), two produce a triplet (1:2:1), three a quartet (1:3:3:1), four a quintet (1:4:6:4:1), and so on.

The magnitude of J is independent of the field strength, unlike the chemical shift. Thus, as H_0 increases the multiplets move further apart but the spacing of the peaks within each multiplet remains the same. The ratio $J/\Delta\nu$, where $\Delta\nu$ is the chemical shift difference between two coupled nuclei, is the critical parameter that determines the appearance of the spectrum. When $J/\Delta\nu$ is 0.1 or less, the spectrum consists of well-separated multiplets. As $J/\Delta\nu$ approaches unity, the spectrum begins to deviate noticeably from the simple first-order appearance. New peaks may appear and intensities are no longer binomial because some spin states that were degenerate when $J/\Delta\nu$ was small split because the magnetic field mixes states. Ultimately, when the chemical shift difference $\Delta\nu$ vanishes, the multiplet will collapse to a singlet. A strongly coupled system of three or more spins is difficult to unravel by inspection alone, although certain patterns

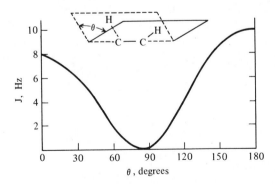

Fig. 8-7. Dependence of the coupling constant J on the dihedral angle θ in the saturated system H—C—C—H. (After M. Karplus and D. H. Anderson, *J. Chem. Phys.*, **30**, 6 (1959); M. Karplus, *J. Chem. Phys.*, **30**, 11 (1959).)

become recognizable with experience. This is one area where a 220- or 300-MHz spectrometer is valuable. As the magnetic field strength is increased, an increasing fraction of spectra become first order and can be solved by inspection (Fig. 8-8).

Other nuclei with spins of $\frac{1}{2}$ will interact with protons (and each other) and cause observable spin-spin coupling. Without deliberate isotopic substitution, significant numbers of only fluorine and phosphorus occur naturally. In fact, the presence of one of these elements may be deduced from an otherwise unexplained coupling effect. Usually J is larger than for most proton–proton couplings. The direct coupling of $^{13}C—H$ is often noticeable as sidebands on a proton nmr spectrum; these sidebands will not vary as the sample spinning rate is changed as will the spinning sidebands. To a first approximation, the magnitude of the $^{13}C—H$ coupling is proportional to percent s-character in the C—H bond.

When a proton is coupled to a nucleus which has a nonzero quadrupole moment, the latter provides efficient spin-lattice relaxation (T_1 is decreased) and this is usually sufficient to decouple, completely or partially, the spin-spin interaction with the proton. Coupling of protons with chlorine, bromine, or iodine nuclei is not observed. In the case of ^{14}N, the decoupling is usually only partially effective so that the 1:1:1 triplet result-

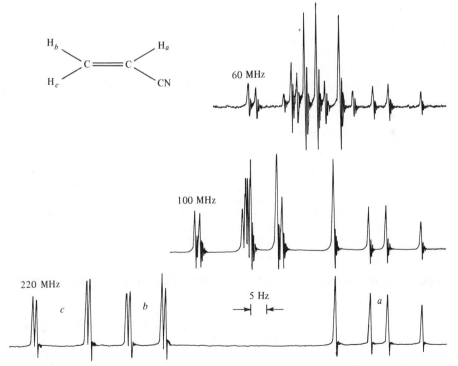

Fig. 8-8. The nmr spectra of acrylonitrile at 60, 100, and 220 MHz illustrate the primary advantage of operating at the highest attainable rf frequency and magnetic field. (Courtesy of Varian Associates.)

ing from ^{14}N—^1H coupling is normally broad and featureless, except for ammonium ion in strongly acidic media.

Integration

The area under an absorption band is proportional to the number of nuclei responsible for the absorption. A device for electronically integrating the absorption signal is a standard item on most commercial spectrometers. The integral is represented as a step function; the height of each step is proportional to the number of nuclei in that particular region of the spectrum. Accuracy is typically within ± 2%. For quantitative analysis, a known amount of a reference compound can be included with the sample. The nmr signal of the reference compound preferably should contain a strong singlet lying in a region of the nmr spectrum unoccupied by sample peaks. Exact phasing out of the dispersion signal is crucial in integration. From the two peak areas, A_{unk} and A_{std}, and the weight of the internal standard taken, W_{std}, the amount of the unknown present is calculated by

$$W_{unk} = W_{std} \times \frac{N_{std}}{N_{unk}} \times \frac{M_{unk}}{M_{std}} \times \frac{A_{unk}}{A_{std}} \tag{8-5}$$

where N's are the numbers of protons in the groups giving rise to the absorption peaks, and M's are the molecular weight of the compounds.

Whenever the empirical formula is known, the total height (in any arbitrary units) divided by the number of protons yields the increment of height per proton. Lacking this information, but deducing the assignment of a particular absorption band, one can calculate the increment per proton from the difference in elevation for the assigned group divided by the number of protons in the particular group. Unfortunately, there is no way of handling overlapping bands.

Example 8-2

A nmr spectrum shows three single peaks located at -440, -300, and -120 Hz in a field of 60 MHz with TMS as reference. The integral heights are 4.2, 1.7, and 2.5 units, respectively. Without knowledge of the empirical formula, the integral heights bear the ratio $5:2:3$ and, since no splitting is observed, this must mean a group with five protons (probably an aromatic ring from the chemical shift), a methylene group, and a methyl group, each not coupled with one another.

If one knew that the empirical formula was $C_9H_{10}O_2$, dividing the total integral height of 8.4 units by the number of protons gives 0.84 unit as the increment per proton.

ELUCIDATION OF nmr SPECTRA

Application of nmr to structure analysis is based primarily on empirical correlation of structure with observed chemical shifts and coupling constants. Extensive surveys have been published.[2,4,7,9] These tabulations can be used to predict the position of resonance lines for a postulated compound, and these predictions can be compared with the sample spectrum. Conversely, one searches compilations and published spectra to ascertain groups which might occur at the positions observed in the sample spectrum.[6,10] One of the unique advantages of nmr is that spectra can often be interpreted without reference to data from structurally related compounds.

Brief vigorous shaking with a few drops of D_2O generally results in a complete exchange of labile protons and collapse of their absorption signal. Deuterium does not absorb in the proton spectral region. Because of the smaller value of μ/I for deuterium, its coupling to hydrogen is much smaller (about one-sixth) than the corresponding H—H coupling and does not appear in the proton spectrum.

Spin-Decoupling

Double resonance often simplifies complex spectra. This technique involves imposing a second, relatively strong, rf field on the sample at the resonant frequency of the nucleus whose coupling is to be investigated. Decoupling is achieved when $v_1 v_2 / hI$ is greater than J but v_2 is still sufficiently low in power so that saturation is not approached. The spin and magnetic moment of the nucleus irradiated with v_2 is quantized in a direction perpendicular to its coupling partner, the coupling is removed, and any multiplets involved collapse to a single peak. Reversing the roles of irradiated and observed nuclei provides unequivocal identification of the spin-coupled nuclei involved.

Experimentally, in the frequency-sweep technique, the second rf field, v_2, is held at a fixed separation from the reference line (at $H_5 - H_{TMS}$ in Fig. 8-9) and the spectrum is

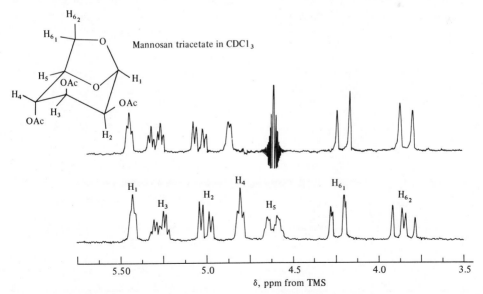

Fig. 8-9. Frequency-swept spin decoupling at 100 MHz. (Courtesy of Varian Associates.)

swept by ν_1 while the magnetic field and locking frequency are held constant. As seen in the upper spectrum of the figure, signals from protons H_3 change from a pair of quartets to a pair of triplets, indicating that H_3 and H_5 have a small long-range coupling constant. Small splittings also disappear in the patterns of H_4 and H_6. In a single experiment the effect on all other resonances of irradiating a chosen resonance may be observed.

In the field-sweep technique of decoupling, the spectrum is swept while maintaining a fixed frequency difference $\Delta\nu$ between the observing ν_1 and irradiating ν_2 frequencies. Only those coupled resonances separated by the chosen frequency difference will be observed. For each critical frequency difference a separate decoupling experiment will be required.

Dissimilar nuclei can also be decoupled. For example, by irradiation of the ^{14}N nuclei at about 4.3 MHz in a 14.09 kG field, decoupling of protons from nitrogen atoms can be made complete. Other hetero nuclei can be decoupled similarly from protons.

Spin Tickling

In *spin tickling* a weak irradiating field ν_2 is used, a field whose bandwidth is only slightly larger than the widths of individual lines. The effect of irradiating one line in the spectrum is to split other lines in the same spectrum coupled to it. Hidden lines in a complex region of the spectrum can be located by irradiating these lines one at a time while looking at the effect on other regions of the spectrum. Frequency-swept spin tickling is illustrated in Fig. 8-10. In this spectrum, one of the olefinic protons produces the set of four lines centered around 6.3 δ. The proton coupling of 14 Hz is midway between the normal *cis-* and *trans*-couplings of 10 and 17 Hz, respectively. While "sitting on" the peak around 646 Hz from TMS, an external audio oscillator ν_2 is slowly varied so

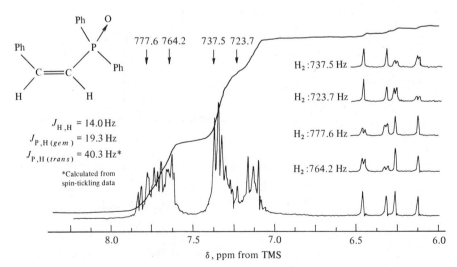

Fig. 8-10. Frequency-swept spin tickling at 100 MHz. (Courtesy of Varian Associates.)

as to sweep ν_2 through the low-field pattern of signals from 7 to 9δ. A dip in the signal at 646 Hz occurs when ν_2 is 777.6 or 764.2 Hz larger than the frequency of the oscillator used to lock to the TMS signal. Likewise, the recorder pen is positioned on the peak at about 627 Hz and the splitting effects are noted when ν_2 is 737.5 and 723.7 Hz larger than the TMS lock frequency. Thus, the four hidden lines are located and their positions in the spectrum are indicated by the arrows. From the large spacings the average value of phosphorus coupling to this proton is 40.3 Hz, as compared with the *cis*-coupling constant of 13.5 Hz, and agrees with *trans*-phosphorus-hydrogen coupling.

Solvent Influence and Shift Reagents

Proton resonance peaks will be spread across a broader range of magnetic field strength by addition of a paramagnetic compound to the solution being studied. In Fig. 8-11 the complex pattern has been resolved (see the upper trace) by the addition of 50% benzene to the CCl_4 solution in which the nmr spectrum was initially taken. Benzene solvent molecules associate with electron-deficient sites in solute molecules. Because benzene is highly anisotropic, different protons in the solute experience shielding or de-shielding depending on their orientation to the benzene ring. In this particular example, the ring protons are shielded (located above the benzene ring) and appear at higher field.

An isolated carbonyl group induces a solvent effect which changes sign near a plane drawn through the carbonyl carbon atom and perpendicular to the carbonyl group. The corresponding shift relative to pyridine changes sign near a plane drawn through the α-carbon atoms. Thus, in the case of an equatorial proton adjacent to the carbonyl group in a cyclohexanone ring, the shift will be small or zero in benzene and negative in

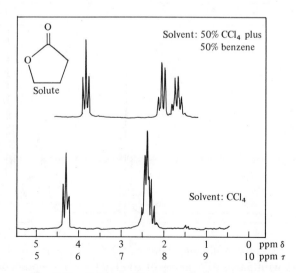

Fig. 8-11. Solvent effect on an nmr spectrum.

pyridine, whereas for axial protons or methyl groups the shift will be positive in benzene and small or zero in pyridine.[5]

The europium(III) and praseodymium(III) chelates of 1,1,1,2,2,3,3-heptafluoro-7,7-dimethyl-4,6-octanedione are Lewis acids and form complexes with a wide variety of organic Lewis bases including alcohols, esters, ketones, ethers, epoxides, azoxides, and sulfoxides. The europium chelate causes downfield shifts of proton nmr peaks, and the praseodymium chelate produces upfield shifts. These paramagnetic shift reagents can give resolution of peaks (Fig. 8-12) comparable to the resolution achievable with 100- or even 220-MHz spectrometers. The suitability of Eu(III) and Pr(III) in the paramagnetic shift reagents is traceable to their favorably short electron spin-lattice relaxation time, a property that renders them inefficient for proton relaxation and nmr line broadening.

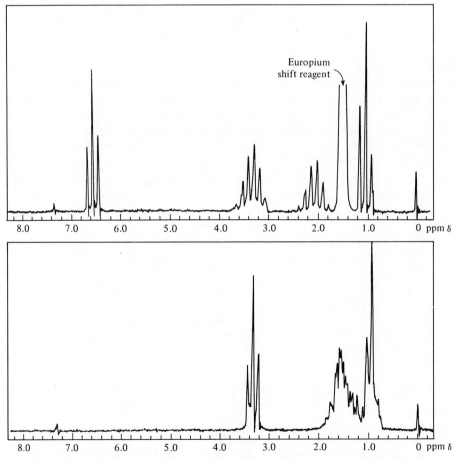

Fig. 8-12. The nmr spectra of di-*n*-butyl ether (1.0×10^{-4} moles) in 0.5 ml CCl_4 alone (*lower trace*), and with 5.0×10^{-5} moles of tris(1,1,1,2,2,3,3-heptafluoro-7,7-dimethyl-4,6-octanedione)europium(III) added (*upper trace*).

There is essentially a linear dependence of the shifts on added shift reagent up to concentrations of 7.5×10^{-5} mole in 0.5 ml at least.

Example 8-3

Returning to the information given in Example 8-2, these structures are present:

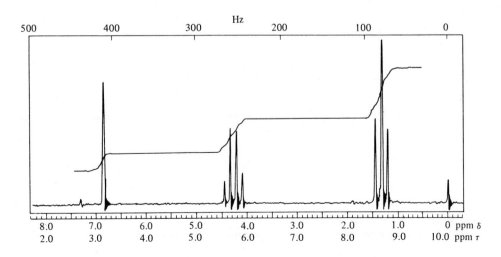

From the position of the phenyl group and the lack of multiplet structure, the benzyl group is suggested. Placing an oxygen next to the phenyl ring would shift upfield the *ortho* and *para* protons; likewise a carbonyl group would shift the *ortho* protons downfield. The position of the isolated methyl group suggests an adjacent double bond, either carbonyl or phenyl ring. The latter is impossible because a methylene group must be accommodated. Two structural candidates remain as possibilities: benzyl methyl ketone or benzyl acetate (if one can eliminate the presence of sulfur). In the former compound the methylene protons would be expected at 3.6 δ; in the latter at 5.1 δ. The observed position is 5.0 δ, and the compound is benzyl acetate.

Example 8-4

The nmr spectrum shows a single peak at δ 6.83 (τ 3.17), a quadruplet at δ 4.27 (τ 5.73), and a triplet at δ 1.32 (τ 8.68). For the multiplets the coupling constant is about 7 Hz. The integrator readings are in the ratio 1:2:3, respectively. Empirical formula is $C_8H_{12}O_4$.

The upfield methyl group is split into a triplet by an adjacent methylene group which, in turn, is split into a quadruplet—the typical ethyl pattern. This is confirmed by the integrator readings. Assignment of the low-field group is not so simple. It is not quite in the location expected for a benzene ring, nor does it contain sufficient numbers of protons. However, an olefinic proton absorption is a possibility, although further shielding is indicated. Returning to the methylene

Hz

| 500 | 400 | 300 | 200 | 100 | 0 |

| 8.0 | 7.0 | 6.0 | 5.0 | 4.0 | 3.0 | 2.0 | 1.0 | 0 ppm δ |
| 2.0 | 3.0 | 4.0 | 5.0 | 6.0 | 7.0 | 8.0 | 9.0 | 10.0 ppm τ |

group, its downfield position could be due to an adjacent $-O-(C=O)-$ structure. If the olefinic proton were alongside the carbonyl group, its absorption position would be reasonable. Summarizing our present information, we have

$$CH_3-CH_2-O-\overset{\overset{\displaystyle O}{\displaystyle \|}}{C}-CH=$$

which is exactly one-half the empirical formula. The complete structure is either diethyl fumarate or diethyl maleate. Having gotten this far we would take a nmr spectrum of each compound and would find that in diethyl maleate the low-field single peak occurs at δ 6.28 (τ 3.72). Coupling between the olefinic protons collapsed because $\delta_2 - \delta_1 = 0$.

Overlapping spectral features can be a source of difficulty in interpretation. The distribution of protons shown by the integration curve is helpful.

Example 8-5

The nmr spectrum for the compound with empirical formula $C_7H_{16}O$ shows a symmetrical heptet centered around δ 3.78 (τ 6.22) and an unsymmetrical doublet with individual peaks at δ 1.18 (τ 8.82) and δ 1.12 (τ 8.88). The integrated areas are 2 units for the heptet and 24 + 6 units for the doublet. The absence of peaks at low field indicates absence of aldehyde, unsaturation, and probably hydroxyl absorption.

The heptet spells out a probable isopropyl group whose methine proton is split by six methyl protons. Its field position indicates an adjacent oxygen. In turn, the methine proton splits the *gem*-methyl protons into a doublet. A logical presumption is an isopropyl ether group, the oxygen atom serving to isolate the methine proton from the remainder. Using the heptet proton as divisor, the integrator readings spell out nine unassigned protons in a single peak superimposed on the low-field wing of the methyl doublet in the isopropyl group. This could only mean three isolated methyl groups—a *tert*-butyl group. The compound is *t*-butyl isopropyl ether.

QUANTITATIVE ANALYSIS

The quantitative analysis by nmr deserves more attention.[11] Accurate electronic integrators with reproducibility better than 2%, available even on low-cost instruments, offer great potential as a means of quality control. Provided that at least one resonance band from each component of a mixture is free from extensive overlap by other absorptions, nmr quantitative analysis should be possible.

The determination of the ethylene oxide chain length in commercial nonionic detergents provides an example. Approximately 10% solutions of each surfactant in CCl_4 are prepared, and the spectrum and several repeat integrals of each are run (Fig. 8-13). The ratio of the ethylene oxide proton signal to the cetyl chain signal is calculated and, since the number of protons in the cetyl chain is known, the calculation of the number of $[O-CH_2-CH_2-O]$ units is a matter of simple proportion. For each detergent the total analysis time is about 10 min; the alternative is a lengthy titration procedure.

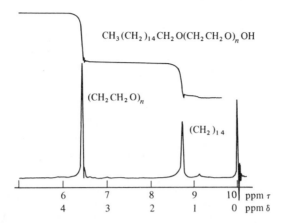

$CH_3(CH_2)_{14}CH_2O(CH_2CH_2O)_nOH$

$(CH_2CH_2O)_n$

$(CH_2)_{14}$

6	7	8	9	10
4	3	2	1	0

ppm τ
ppm δ

Fig. 8-13. Ethylene oxide chain length of nonionic surfactants by quantitative nmr.

Impure samples can be assayed using another compound as the internal standard. The nmr spectrum of

possesses a singlet peak due to the $N-CH_3$ protons at $\delta = 3.40$ and a doublet due to the $-(C=O)-CH_2-S-P-$ group at $\delta = 4.3$. Benzyl benzoate, which gives a well-resolved peak at $\delta = 5.4$, is an excellent internal standard.

Products from the air oxidation of p-cymene yield a complex mixture. The nmr spectrum, with peak assignments indicated, is given in Fig. 8-14, which also shows the position of the internal standard line from benzyl benzoate.

The determination of water in liquid N_2O_4 in the less than 0.1 weight % range, illustrates the usefulness of nmr in trace analysis and in a system for which calibration standards are not available. At $-10°C$ the exchange of protons between dissolved H_2O, HNO_2, and HNO_3 molecules is rapid, so that a single nmr peak is observed for the exchanging protons. As little as 0.03% H_2O can be detected with a relative standard deviation of 7.5% based on a single spectral scan. Accuracy is improved by using the accumulated average of 50 scans, which can be run in 90 min. With this procedure, 0.01% H_2O (100 $\mu g/ml$) can be determined. Samples of N_2O_4 are weighed into a tared nmr tube, and 30–50 μl of a benzene internal standard are added.

Residual H_2O in samples of high purity D_2O can be determined by the method of standard additions, using standard micro-volumetric techniques. The resulting nmr integrals are plotted vs. the weight percent of added H_2O and extrapolated to zero integral.

A well-known assay of a pharmaceutical formulation by nmr is that of aspirin, phenacetin, and caffeine mixtures.[1] The procedure takes about 20 min and does not

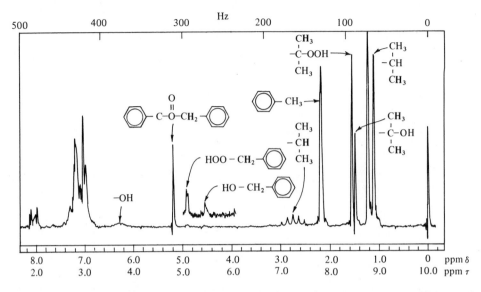

Fig. 8-14. The nmr spectrum of mixtures of products obtained from the oxidation of *p*-cymene, benzoyl benzoate added as internal standard. Integrals are not shown. (Courtesy of International Scientific Communications, Inc.)

require the separation of individual constituents. A separate analytical peak of known origin is present for each component; no calibration curves are required since the integrals for the protons giving rise to the various analytical peaks are constant and unaffected by solvent or solute interactions. For aspirin the sharp peak at about $\delta = 2.3$, which represents the ester methyl group, is used. For phenacetin (see illustration for Problem 8-21) the amide methyl group at $\delta = 2.1$ is preferable; the quartet at $\delta = 4.0$ is suitable although one of the methyl peaks of caffeine overlaps the quartet at 3.9, and a correction is necessary. For caffeine the two methyl resonances at $\delta = 3.4$ and 3.6 are used.

PROBLEMS

1. At 43°C the nmr spectrum of liquid acetylacetone shows a peak at δ 5.62 (37 units on the integrator) and a peak at δ 3.66 (19.5 units), plus additional peaks which do not concern us. Calculate the percent enol composition.

2. A hydrocarbon sample shows nmr bands over the interval δ 1.0 to 5.5. Benzophenone used as an internal standard shows nmr bands in the δ 6–7 region. The relative integrals were 228 and 184 units for 0.8023 g of benzophenone and 0.3055 g of sample. Calculate the percent hydrogen in the sample.

3. Phenol formaldehyde resins include a class prepared with excess phenol, called novolacs, which consist of phenolic nuclei linked by methylene bridges at positions

ortho or *para* to the hydroxyl group. The integration of the spectrum gives the ratio of aromatic to methylene protons as 30 to 18. Calculate the average chain length and the average molecular weight.

4. Chlorination of *o*-cyanotoluene gave *o*-cyanobenzyl chloride in about 50% yield. Attempted fractionation of the washings and mother liquor gave a liquid whose proton magnetic resonance spectrum gave three singlets upfield from the aromatic ring signals as follows (integral units in parentheses): δ 2.52 (13), δ 4.72 (20), and δ 7.01 (10). (a) Assign the nmr signals. (b) From the signal intensities, deduce the relative molar proportions and the proportions by weight of the three constituents in the liquid mixture.

5. The nmr proton spectrum of the liquid diketene, $C_4H_4O_2$, shows two signals of equal intensity. What structure is consistent with this information?

6. A sample is believed from its mass spectrum to be either of the dicyanobutenes:

$$CH_3{-}CH{=}C\diagup^{CH_2CN}_{\diagdown CN} \qquad NC{-}CH{=}C\diagup^{CH_2CN}_{\diagdown CH_3}$$

$$\text{I} \qquad\qquad\qquad \text{II}$$

What characteristic in the nmr spectrum would identify each isomer?

7. The phosphorus resonance of phosphorous acid and hypophosphorous acid is reported to be a doublet in the former and a triplet in the latter compound. Write the structures of the two acids.

8. Addition of methyldichlorosilane to vinyl acetate gives an adduct whose likely structures are

$$CH_3{-}Si(Cl)_2{-}CH_2{-}CH_2{-}O{-}CO{-}CH_3$$

or

$$CH_3{-}Si(Cl)_2{-}CH(CH_3){-}O{-}CO{-}CH_3$$

The nmr spectrum shows two bands with clearly resolved triplet splitting. Which structure is supported by the nmr evidence?

9. On the basis of the two peaks of equal strength found in the nmr spectrum of the sodium salt of Fiest's acid in D_2O, which structure is correct?

$$HOOC{-}C{\!-\!\!-\!\!-\!}CH{-}COOH \quad or \quad HOOC{-}CH{\!-\!\!-\!\!-\!}CH{-}COOH$$
$$\underset{CH_3}{\overset{C}{\diagup\diagdown}} \qquad\qquad \underset{CH_2}{\overset{C}{\diagup\diagdown}}$$

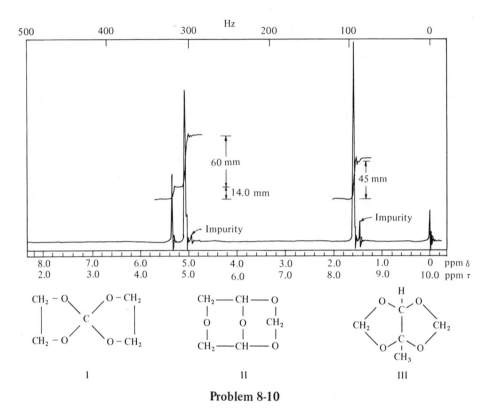

Problem 8-10

10. For an impurity isolated from an important industrial chemical by gas–liquid chromatography, the empirical formula $C_5H_8O_4$ was established by elemental analysis, and mass spectrometry provided the molecular weight of 132. An infrared spectrum revealed a strong --C--O--C-- absorption with no evidence of double bonds, carbonyl, or hydroxyl groups. From an examination of the proton nmr spectrum in the figure, which of the three structures is the correct one?

11. Isojasmone, a synthetic oil redolent of jasmine blossoms, consists of a mixture of the two isomers A and B. Isomer B has a greater odor value than isomer A. The nmr spectra and structures of the two compounds are shown in the figure. Compute the proportions of the two isomers after assigning the olefinic proton of each isomer to the proper spin-spin multiplet on the expanded trace.

12. Deduce the structure of the compound with the spectrum shown whose empirical formula is $C_{10}H_{11}NO_4$.

13. The nmr spectrum contains a single peak at δ 3.58 (τ 6.42) and another single peak at δ 7.29 (τ 2.71).. Integrated intensities are 8 and 20 units, respectively. From mass spectral information, the compound (mol. wt. 246) is known to contain two sulfur atoms. Deduce its structure.

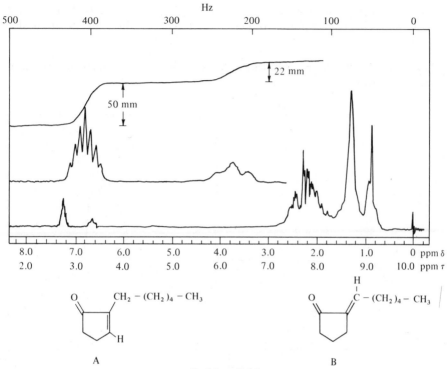

Problem 8-11

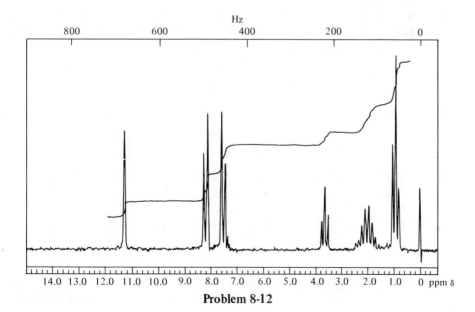

Problem 8-12

14. The nmr spectrum contains single peaks at δ 7.27 (τ 2.73), δ 3.07 (τ 6.93), and δ 1.57 (τ 8.43). The empirical formula is $C_{10}H_{13}Cl$. Deduce the structure of the compound.

15. The molecular weight is 190; deduce the structure of the compound whose nmr spectra is shown.

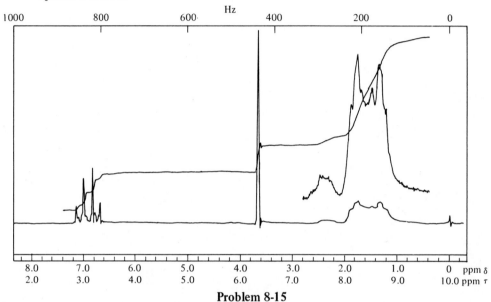

Problem 8-15

16. Overlapping multiplets always present a challenge in unraveling a nmr spectrum, such as the one for the compound $C_5H_{11}BrO_2$ shown. Deduce the structure. Ignore the small benzene peak at δ 7.32.

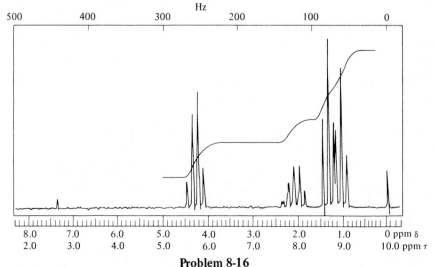

Problem 8-16

17. The nmr spectrum of compound $C_4H_7ClO_2$, shown, also has one obvious overlapping spectral feature. Deduce the structure. The small peak at δ 7.32 is a benzene marker.

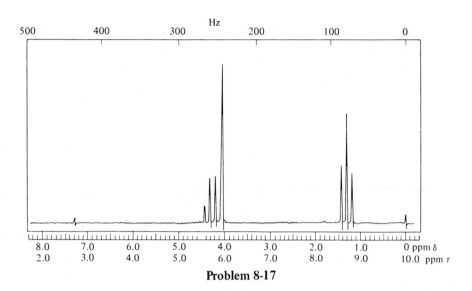

Problem 8-17

18. Deduce the structure of the compound $C_8H_{14}O_4$ from the nmr spectrum shown. Be cognizant of the requirement for spin-spin coupling.

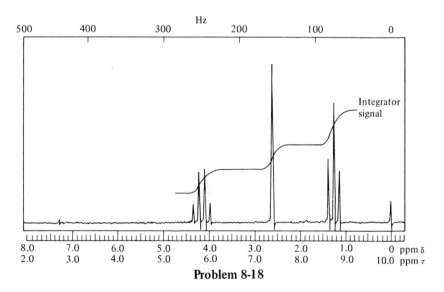

Problem 8-18

19. The compound C_3H_4 has only one peak in its nmr spectrum located at δ 1.80 (τ 8.20). Can you ascertain its structure?

20. Chart the spin-spin couplings involved in the spectrum of methyl methacrylate shown. The lower trace is an expanded scale. The resonance position of external benzene was 418.4 Hz from TMS, the internal standard, in an rf field of 60 MHz.

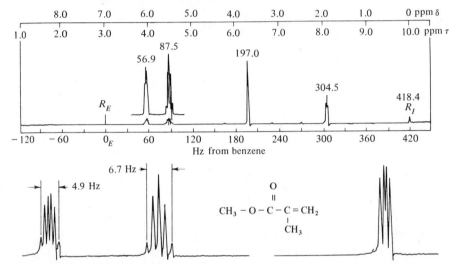

Problem 8-20. The proton nmr spectrum of methyl methacrylate (liquid) taken on Varian model A-60 spectrometer. Internal standard was TMS; external standard was benzene.

21. The nmr spectrum of compound $C_{10}H_{13}NO_2$, isolated from a headache preparation, is shown. Write its structure.

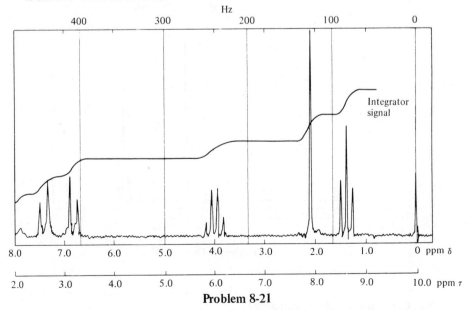

Problem 8-21

22. The nmr spectrum has single peaks at δ 7.29 (τ 2.71) and δ 2.02 (τ 7.98) plus two triplets at δ 4.30 (τ 5.70) and δ 2.93 (τ 7.07). Integrated intensities are 10, 6, 4, and 4, respectively. Empirical formula is $C_{10}H_{12}O_2$. Deduce the structural formula.

23. The nmr spectrum for the compound C_3H_6O has a triplet centered at δ 4.73 (τ 5.27) and a quintet at δ 2.72 (τ 7.28). Deduce the structure.

BIBLIOGRAPHY

Farrar, T. C., and E. D. Becker, *Pulse and Fourier Transform NMR*, Academic Press, New York, 1971.

Howery, D. G., "Nuclear Magnetic Resonance Spectrometers," *J. Chem. Educ.*, **48**, A327, A389 (1971).

Jackman, L. M., and S. Sternhell, *Applications of Nuclear Magnetic Resonance Spectroscopy in Organic Chemistry*, Pergamon, Elmsford, N.Y., 1969.

Pople, J. A., W. G. Schneider, and H. J. Bernstein, *High-Resolution Nuclear Magnetic Resonance*, McGraw-Hill, New York, 1959.

Silverstein, R. M., and G. C. Bassler, *Spectrometric Identification of Organic Compounds*, 2nd ed., Wiley, New York, 1967.

LITERATURE CITED

1. Hollis, D. P., *Anal. Chem.*, 35, 1682 (1963).
2. Jackman, L. M. and S. Sternhell, *Applications of Nuclear Magnetic Resonance Spectroscopy in Organic Chemistry*, 2nd ed., Pergamon, Elmsford, N.Y., 1969.
3. Johnson, L. F., "Nuclear Magnetic Resonance with Superconducting Magnets," *Anal. Chem.*, **43** (2), 28A (1971).
4. Mathieson, D. W., Ed., *Nuclear Magnetic Resonance for Organic Chemists*, Academic Press, New York, 1967.
5. Ronayne, J. and D. H. Williams, *J. Chem. Soc. Sect. B*, 1967, 535.
6. Sadtler Research Laboratories, *Nuclear Magnetic Resonance Spectra*, Philadelphia, Pa.
7. Silverstein, R. M. and G. C. Bassler, *Spectrometric Identification of Organic Compounds*, 2nd ed., Wiley, New York, 1967, Chapter 4.
8. Stothers, J. B., "^{13}C NMR Spectroscopy: A Brief Review," *Appl. Spectrosc.*, **26**, 1 (1972).
9. Suhr, H., *Anwendungen der Kernmagnetischen Resonanz in der Organischen Chemie*, Springer-Verlag, Berlin, 1965.
10. Varian Associates, *High Resolution NMR Spectra Catalog*, Vol. 1, 1962; Vol. 2, 1963.
11. Ward, G. A., *Am. Laboratory*, 2 (3), 12 (Mar. 1970).

Electron Spin Resonance Spectroscopy

Electron spin resonance (esr), or electron paramagnetic resonance (epr), is a branch of absorption spectroscopy in which radiation of microwave frequency induces transitions between magnetic energy levels of electrons with unpaired spins. The magnetic energy splitting is created by a static magnetic field. Unpaired electrons, relatively unusual in occurrence, are present in odd molecules, free radicals, triplet electronic states, and transition metal and rare earth ions. There is much interest in the unpaired electrons of free radicals. These electrons are generally left after homolytic fission of a covalent bond, which is often produced by ultraviolet or gamma irradiation of the sample.

ELECTRON BEHAVIOR

An electron possesses a spin S of $\frac{1}{2}$. Associated with the spin is a magnetic moment μ because the spinning electron behaves like a magnet with its poles along the axis of rotation. The magnetic moment of an electron can be written in a general form as

$$\mu = - g \, \beta_N \, M_s \tag{9-1}$$

where g, called the *spectroscopic splitting factor*, has a value which is a function of the electron's environment (a value close to two for a free electron); β_N, the *Bohr magneton*, is a factor for converting angular momentum to magnetic moment; and M_s, the angular momentum quantum number, can have values of $+\frac{1}{2}$ or $-\frac{1}{2}$.

In the absence of an external magnetic field, the free electron may exist in one of two states, $+\frac{1}{2}$ or $-\frac{1}{2}$, of equal energy (*degenerate*). Imposition of an external static magnetic field H_0 removes the degeneracy and establishes two energy levels (Fig. 9-1). The lower energy state has the spin magnetic moment aligned in the direction of the magnetic field and corresponds to the quantum number, $M_s = -\frac{1}{2}$. The difference in energy between the two levels is given by

$$\Delta E = g \, \beta_N \, H_0 = h\nu \tag{9-2}$$

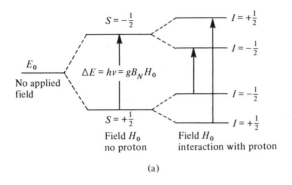

$S = -\frac{1}{2}$

$I = +\frac{1}{2}$

$I = -\frac{1}{2}$

E_0

No applied
field

$\Delta E = hv = gB_N H_0$

$I = -\frac{1}{2}$

$S = +\frac{1}{2}$

$I = +\frac{1}{2}$

Field H_0
no proton

Field H_0
interaction with proton

(a)

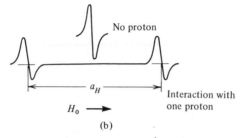

No proton

a_H

Interaction with
one proton

$H_0 \longrightarrow$

(b)

Fig. 9-1. Electron energy levels. (a) Splitting of energy levels by a magnetic field H_0 and by interaction of unpaired electron with one proton. (b) Splitting of spectral line.

Transitions from one state to the other can be induced by irradiation of the electron with electromagnetic radiation of frequency v. The interaction between the magnetic dipole of the electron and the oscillating magnetic field accompanying the electromagnetic radiation causes the transitions. When the magnetic field is expressed in kilogauss, the resonance frequency for a free electron in megahertz is given by $v = 2800\ H_0$. Most esr measurements in solution are made with an exciting frequency of 9500 MHz and a field strength of 3400 G. Typical energies involved are 6.3×10^{-17} erg molecule^{-1} or 1 cal mole^{-1}.

For a single spin system, the excess of population of the lower energy level over that in the upper state is extremely small. It is governed by the Boltzmann distribution, which is given by

$$\frac{n_{\text{upper}}}{n_{\text{lower}}} = e^{-\Delta E/kT} \tag{9-3}$$

where k is the Boltzmann constant (1.38×10^{-16} erg $^\circ$K^{-1}) and T is the absolute temperature. At 300°K for a species for which $g = 2$, and with a magnetic field of 3000 G, the relative population of the two energy levels is 0.9984. In other words, for every 1000 electrons in the low energy state, 998 are in the high energy level. Although the two levels are virtually equal in population, the absorption of electromagnetic radiation quanta by electrons in the lower level will exceed spontaneous emission from the upper level because of the Boltzmann population distribution. Consequently, there will be a net absorption of energy.

esr SPECTROMETER

The principal components of an esr spectrometer are (1) a source of microwave radiation of constant frequency and variable amplitude; (2) a means of applying the microwave power to the sample—the microwave bridge; (3) a homogeneous and steady magnetic field to provide the magnetic field (spectroscopic) splitting; (4) an ac field superimposed on the steady field so as to sweep continuously through the resonance absorption of the sample; (5) a detector to measure the microwave power absorbed from the microwave field; and (6) an oscilloscope or a graphic x-y recorder. A simplified block diagram of an esr spectrometer is shown in Fig. 9-2.

Source

Most esr spectrometers employ radiation obtained from a klystron oscillator operating in the microwave X-band (3-cm wavelength) region. In a klystron the whole oscillating circuit is within the resonant cavity of the tube. A beam of electrons flows in pulses back and forth between the cathode and a reflector filament. Power may be withdrawn from the klystron through a *wave guide* by a loop of wire which couples with the oscillating magnetic field and sets up a corresponding field in the wave guide. A wave guide consists of hollow rectangular (copper or brass) tubing, 2.2 cm by 10 cm, with silver or gold plating inside to produce a highly conducting, flat surface. Reflection of microwave power back into the klystron is prevented by an isolator—a strip of ferrite material which passes microwaves in one direction only.

Components in the microwave assembly may be coupled together by irises or slots of various sizes. Wave-guide elements are matched by using screws or stubs which can be positioned in the wave guide or across the coupling slits.

Sample Cavity

The sample, contained in a cylindrical quartz tube, is held in a cavity between the poles of the magnet. A standing wave is set up in the reflection cavity; the standing wave is composed of both magnetic and electric fields at right angles to each other. The cavity is analogous to a tuned circuit—a parallel LC combination. To minimize any influence of a high dielectric constant when such material is the sample, the sample tube is located in the cavity in a position of maximum rf magnetic field and minimum rf electric field. Tubing of 3–5 mm i.d. with a sample volume of 0.15–0.5 ml can be used with samples which do not possess a high dielectric constant. For samples with a high dielectric constant, flat cells with a thickness of about 0.25 mm, and sample volume of 0.05 ml, are often used. Rotatable cavities are used for studying anisotropic effects in single crystals and in solid samples.

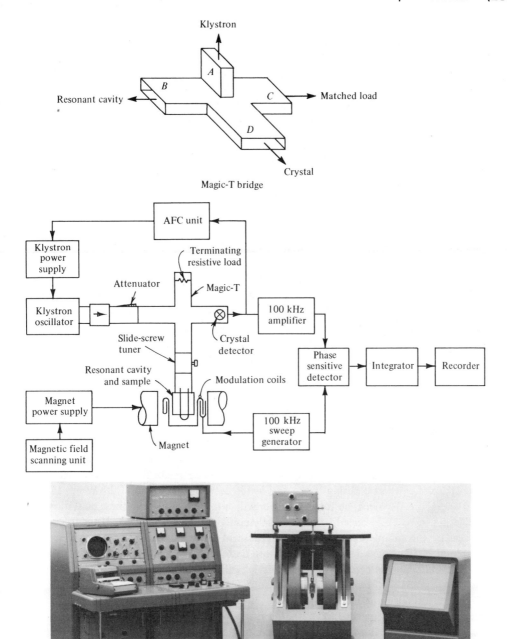

Fig. 9-2. Top, Magic-T bridge. Center, schematic diagram of an esr spectrometer. Bottom, photograph of a commercial instrument. (Courtesy of Varian Associates.)

Dual sample cavities are used for simultaneous observation of a sample and a reference material. Slots can be machined into the walls of the cavity to allow ultraviolet irradiation of the sample. The two cavity sections are separately modulated, one section by a 0.1-MHz field modulation and the other section by a 400-Hz field modulation. When the signals are fed through an appropriate 400-Hz amplifier and phase-sensitive detector, or a similar 0.1-MHz network, both resonances can be displayed on the recorder and superimposed. Both samples are literally in the same cavity. Thus, sources of error in intensity measurements are automatically compensated by comparing relative signal heights.

Magnet and Modulation Coils

An electromagnet capable of producing steady fields ranging from 500 to 5000 G is required to handle samples whose g-factor ranges from 1.5 to 6. The homogeneity of the field for solution studies should be about 1 part in 10^6. On the other hand, line widths for paramagnetic ions and for free radicals in solid matrices are rarely less than a few gauss, so that homogeneity and stability might be as low as 1 part in 10^3 before serious line broadening occurs. The sample cavity, wave guide, and magnet are arranged so that the steady field H_0 and the oscillating field of the microwave radiation are mutually perpendicular. This is attained by placing the broad face of the wave guide parallel to the pole faces of the magnet.

Microwave Bridge

The bridge, shown in Fig. 9-2, enables the microwave system to be operated as a balanced bridge with all the advantages of null methods in electrical circuits. The microwave bridge will not allow microwave power to pass in a straight line from one arm to the arm opposite. Power entering arm A will divide between arms B and C if the impedances of B and C are the same, so that no power will enter arm D. Under these conditions the bridge is said to be balanced. Arm C usually contains a balancing load. If the impedance of arm B (the sample cavity) changes because of some esr resonance absorption by a sample in it, the bridge becomes unbalanced and some microwave power enters into arm D containing the detector—a semiconducting silicon-tungsten crystal which acts as a rectifier, converting the microwave power into direct current.

A set of coils mounted on the walls of the sample cavity and fed by a 0.1-MHz sweep generator provides modulation of the dc magnetic field at the sample position. As the main magnetic field is swept slowly through resonance over a period of several minutes, a dynamically recurring imbalance in the microwave bridge is detected and amplified for presentation on a recorder as the derivative of the microwave absorption spectrum against the magnetic field (Fig. 9-3). Because of instrumental considerations associated with the signal-to-noise ratio, esr spectra are nearly always recorded as first-derivative spectra. For low-frequency modulation (400 Hz or less) the coils can be mounted outside the cavity

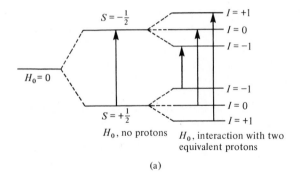

H_0, no protons H_0, interaction with two
equivalent protons

(a)

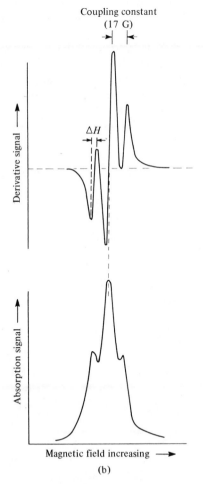

Fig. 9-3. Irradiated methanol. (a) Interaction of unpaired electron with two equivalent protons. (b) Derivative and absorption signals of the triplet spectrum.

and even on the magnet pole pieces. Higher modulation frequencies cannot penetrate metal effectively and the modulation coils must be mounted inside the sample cavity.

Sensitivity

The ultimate sensitivity at room temperature of practical X-band (9500 MHz) esr spectrometers is often expressed as

$$N_{min} = 1 \times 10^{11} \frac{\Delta H}{\sqrt{\tau}} \qquad (9\text{-}4)$$

where N_{min} is the minimum number of detectable spins per gauss, ΔH is the width between deflection points on the derivative absorption curve, and τ is the time constant of the detecting system which is inversely proportional to the bandwidth of the detection circuit. For a sample of very small dielectric loss, the minimum concentration is in the order of 10^{-9} M to barely see the line. For aqueous solutions, 10^{-7} M represents a reasonable lower limit. For structure determinations and quantitative analysis, the concentration should be about 10^{-6} M. The limiting factor is generally the noise level in the spectrometer due to crystal noise and klystron noise. Working at higher magnetic field strengths gives a higher sensitivity. For example, at 35,000 MHz (K-band spectrometer) the sensitivity is 20 times greater.

esr SPECTRA

In a homogeneous magnetic field an unpaired electron in an assemblage of other atoms, themselves possessing nuclear spins, can have a number of energy states. The possible energy states depend on the relative orientation of the magnetic moments of the unpaired electron, the closely associated nuclear spins, and the applied magnetic field. Contributions to the magnetic moment from orbital motion of the unpaired electron are not important for most organic free radicals.

Hyperfine Interaction

Since the radical electron is usually delocalized over the whole molecule or at least a large part of it, the unpaired electron comes into contact-interaction with many nuclei. Nuclei possessing a magnetic moment may interact and cause a further splitting of the electron resonance line. From the number and intensity distribution of the spectral lines, one can tell how many nuclei interact with the radical electron. The energies of a coupled level are given by

$$E = g\beta_N H_0 M_s + a_i h M_I \qquad (9\text{-}5)$$

where a_i is called the *hyperfine coupling constant* and M_I is the *spin quantum number* of the coupling nucleus. Ordinarily the hyperfine coupling constant is a small fraction of the electron splitting. On a spectrum it is the distance between associated peaks of a sub-multiplet, measured in gauss.

The selection rules for allowed electron spin resonance transitions are $\Delta M_I = 0$ and $\Delta M_s = \pm 1$. A single nucleus of spin $I = \frac{1}{2}$ will cause a splitting into two lines of equal intensity. Common nuclei with spin $\frac{1}{2}$ are ^{1}H, ^{19}F, ^{13}C, ^{15}N, and ^{31}P. Interaction with a single deuterium or nitrogen nucleus (^{2}H or ^{14}N, $I = 1$) will cause a splitting into three lines of equal intensity. Three nuclear orientations are permitted, one augmenting the external field, one diminishing it, and the other not changing it. In addition to the preceding nuclei, hyperfine splitting in polyatomic radicals by ^{6}Li, ^{7}Li, ^{10}B, ^{11}B, ^{23}Na, and ^{39}K have been reported, as have been splittings by the nuclei of transition elements, rare earths, and transuranic elements in compounds.

Hydrogen atoms that are attached to carbon atoms adjacent to the unpaired spin undergo interaction with the unpaired spin. This interaction is usually described as *hyperconjugation* and is a maximum when the carbon–hydrogen bond and orbital are coplanar. In general hyperfine splitting by hydrogen atoms is important only at the 1- or 2-position in an alkyl radical:

$$-\overset{}{\underset{4}{CH_2}}-\overset{}{\underset{3}{CH_2}}-\overset{}{\underset{2}{CH_2}}-\overset{\displaystyle\cdot}{\underset{1}{CH}}-R$$

Hyperfine splitting by β-hydrogens in acyclic radicals can be barely detected in most cases. In rigid molecules possessing a highly bridged structure, however, hydrogen atoms beta to the radical site undergo a strong interaction. If a double bond or a system of double bonds is followed by a single bond, as for example in methyl-substituted aromatic radicals, the double-bond character may be partially transferred to the single bond. In the esr spectrum of 2-methyl-1,4-benzosemiquinone radical ion, the spin density is the same at the methyl protons as at the ring protons, and the spectrum consists of seven lines with an intensity ratio of $1:6:15:20:15:6:1$, corresponding to the interaction with six equivalent protons. Sometimes the radical electron density penetrates even two C—C bonds, but with greatly diminished spin density. Other geometries can give rise to a long-range hyperfine splitting by hydrogen atoms. Any arrangement that places the back side of a carbon atom in close proximity to an orbital containing unpaired spin density will be expected to lead to interaction. A number of hyperfine splitting constants are gathered together in Table 9-1.

Example 9-1

Consider the free radical $HO_2 C—\overset{\displaystyle\cdot}{C}H—COOH$ obtained by irradiating malonic acid. The methine proton is a charged spinning particle with a nuclear spin $\frac{1}{2}$; the proton magnetic moment is $+\frac{1}{2}$ or $-\frac{1}{2}$. The unpaired electron, represented by the "dot" over the carbon atom, will be affected by the magnetic field of the proton as well as that of the applied magnetic field H_0. Consequently, each electronic sublevel is further split into two nuclear sublevels that are equally separated, for a total of four levels (Fig. 9-1). Because the nuclear moment remains fixed during electronic transitions, only two transitions are found between these electronic states. Transitions occur only between the states $M_s = +\frac{1}{2}$ or $M_s = -\frac{1}{2}$, conforming to the selection rules $\Delta M_s = \pm 1$ and $\Delta M_I = 0$. The result is a splitting of the original line into two absorption lines of equal intensity. The splittings from the carboxyl protons are too small to be detected.

If several magnetic nuclei are present, the situation is somewhat more complicated, because the electron experiences an interaction with each nucleus, and the spectrum is the result of a superposition of the hyperfine splittings for each nucleus. Several general types will be considered. When two equivalent nuclei are involved, the number of possible fields is reduced. Generally, $2nI + 1$ lines result from n equivalent nuclei; the relative intensity of these lines follow the coefficients of the binomial expansion. Two equivalent protons split each of the original electronic energy levels into three hyperfine levels in an intensity ratio $1:2:1$. An illustration is the semiquinone formed on irradiation of 2,3-dichlorobenzoquinone and the triplet pattern from irradiated methanol (Fig. 9-3).

If the unpaired electron couples with nonequivalent protons, each proton will have its own coupling constant. In general, n nonequivalent protons will produce a spectrum with 2^n hyperfine lines.

TABLE 9-1 Hyperfine Coupling Constants (in gauss)

THE *g*-FACTOR

Hyperfine splitting is independent of the microwave frequency employed. This enables one to distinguish a nuclear hyperfine interaction from the effects of differences in *g*-factors. If, for instance, the separation of two peaks is observed to vary with the microwave frequency, it follows that they correspond to two transitions with different *g*-factors, not to interaction with a nuclear spin. The *g*-factor is a dimensionless constant and equal to 2.002319 for the unbound electron. The exact value of the *g*-factor reflects the chemical environment, particularly when hetero atoms are involved, because orbital angular momentum of the electron (orbital motion about a nucleus) can have an effect on the value of the transition $\Delta M_s = \pm 1$. In many organic free radicals, the *g*-value of the odd electron is close to that of a free electron because the electron available for spin resonance is usually near or at the periphery of the species with which it is associated. However, in metal ions, *g*-values are often greatly different from the free electron value. There

TABLE 9-1 (cont.) Hyperfine Coupling Constants (in gauss)

Row 1 structures:

- CH_3 0.8 — benzene ring anion: 5.1, 4.3, 0.6
- $H-\overset{\bullet}{C}-H$ 16.4 — benzene ring: 5.2, 1.8, 6.2
- 15.2 H, OH 0.5 on $>C<$ — benzene ring: 5.2, 4.6, 1.6, 1.6, 5.9
- NO_2 9.7 — benzene ring anion: 3.4, 1.1, 4.0

Row 2 structures:

- NO_2 2.7, NO_2 0.1 (benzene anion): 1.7
- NO_2 4.0, 4.5, 2.8, 1.1, NO_2 4.0, 4.5
- NO_2 1.5, 1.1, 1.1, NO_2 (benzene anion)

Row 3 structures:

- 1.4 $HC=O$ — benzene anion: 0.4, 0.2, 2.3, 3.0, NO_2 5.1
- CH_3 0.8, $C=O$ — benzene anion: 0.5, 0.5, 2.7, 2.9, NO_2 5.9
- NO_2 12 — benzene anion: 3, 1; $\overset{N\ 1}{\underset{H\quad H\ 1}{}}$
- $\overset{\bullet}{O}$ — benzene: 1-2, 1-2, CH_3 12

$$\begin{bmatrix} CH_2\ (p)\ 9\text{--}12 \\ (o)\ 6\text{--}9 \\ (m)\ 1 \end{bmatrix}$$

[CH 6]

is a slight trend among organic radicals to higher values for free radicals containing oxygen or nitrogen, which are in turn lower than for radicals containing halogen or the peroxy group.

To measure the g-factor for free radicals, it is convenient to measure the field separation between the center of the unknown spectrum and that of a reference substance whose g-value is accurately known. A dual-sample cavity simplifies the measurement. The known may be a sample of finely powdered diphenylpicrylhydrazyl (DPPH), which is completely in the free-radical state, attached to the sample tube in a single cavity or in one chamber of a dual-cavity cell. Two signals will be observed simultaneously with a field separation of ΔH. The g-factor for the unknown is given by

$$g = g_{\text{std}} \left(1 - \frac{\Delta H}{H}\right) \tag{9-6}$$

where H is the resonance frequency. ΔH is positive if the unknown has its center at higher field.

Line Widths

Esr spectroscopy presents a time-averaged view of the geometry of a paramagnetic species. Because of the velocity of precession around the applied magnetic field of the magnetic moment connected with the unpaired electron spin, esr spectra reflect a time-averaged period about 1000 times shorter than proton magnetic resonance. As the temperature is lowered the rate of conformational interconversion decreases and certain lines in the spectrum become broader. Those lines which broaden correspond to transitions for which conformational interconversion results in a change in spin state. Further lowering of temperature causes an increase in the percentage of time during which every molecule occupies a specific conformation. Eventually the spectrum approaches that of a blocked or frozen conformation (when the conformation lifetime is more than 10^{-6} sec). For sharp esr lines, the peak-to-peak separation (ΔH) of a first-derivative spectrum should be about 0.1 gauss.

INTERPRETATION OF esr SPECTRA

It should be apparent from the theoretical treatment that the interpretation of esr spectra involves several parameters: the g-factor, the separation of the hyperfine lines and their relative intensities, the sample concentration, relaxation times, and line widths. If the absorption lines are narrow and the spectrum fully resolved, the actual measurements are straightforward, provided that the field sweep rate is known. An esr spectrum is assigned by perceiving the magnitudes of the coupling constants and correctly counting the lines. Frequently the measured spectrum will not contain all the lines expected because the g-factor and coupling values can be such that two lines come very close to one another and are not resolved. When many equivalent nuclei interact, relative peak heights become very large and it is difficult to see the smallest peaks. These smaller peaks are important because the outer portions of a spectrum are invariably the simplest, and the correct interpretation of these outer lines often provides the key to the unraveling of the more complex central parts. The best way of analyzing extremely complicated spectra is to introduce approximate coupling constants in an electronic computer and compare the computed spectra with the experimental ones.

Example 9-2

Figure 9-4, shows the spectrum of a single crystal of succinic acid, $HOOC-CH_2-CH-COOH$, after gamma irradiation at $25°C$. The coupling constants arise from the interaction of the unpaired electron with the protons on the carbon atoms. Reflecting back to the theoretical treatment in this chapter, a quartet with relative intensities $1:3:3:1$ would indicate that the three protons couple equally. Obviously this is not the case. If the two methylene protons are equivalent, the spectrum would approximate a triplet $(1:2:1)$ further split into a doublet by the CH proton. Again, this pattern is not observed. If the three protons couple unequally, an eight-line pattern would be expected as follows: $2^3 = 8$, which is the pattern observed. A pair of dividers is very useful for ascertaining coupling constants and sorting out overlapping hyperfine patterns.

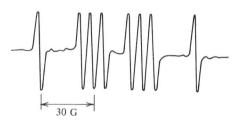

$\overleftrightarrow{30\ G}$

Fig. 9-4. Spectrum of a single crystal of succinic acid after γ-irradiation at $25°C$.

Example 9-3

Figure 9-5 shows the esr spectrum obtained at room temperature from a single crystal of irradiated ammonium perchlorate. Hyperfine interaction is expected that involves the nitrogen nucleus of spin 1 and the proton nuclei of spin $\frac{1}{2}$. The spectrum probably arises from an unpaired electron spin strongly localized on an $\overset{\cdot}{N}H_3$ molecule. The 12-line pattern could arise from contact with the nitrogen nucleus— $2nI + 1 = 3$ lines, and contact with three equivalent protons—4 lines, which splits each of the preceding three lines into a quartet of equally spaced lines with intensity ratios of $1:3:3:1$. Verification is illustrated beneath the figure. With dividers and beginning at the left, we will ascertain the coupling constant from the left-most line and the first larger (more intense) line to its right (line 3). Note that this separation is found between the following pairs of lines: 1 and 3, 3 and 6, and 6 and 9, which together constitute one quartet. The second quartet involves lines 2, 5, 8, and 11; the remaining lines constitute the third quartet. Now from the center of each quartet one locates the individual members of the 3-line pattern.

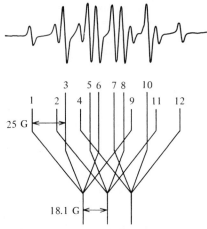

Fig. 9-5. Esr spectrum of $\overset{\cdot}{N}H_3$ formed by X-ray irradiation of ammonium perchlorate crystals.

A 3-line pattern, as in this example, and a triplet denote different environments. The former usually implies interaction with a single nucleus ($I = 1$) whereas the latter will involve interaction with two equivalent nuclei.

Example 9-4

The esr spectrum of the ethyl radical trapped in argon at $4.2°K$ consists of a quartet of relatively sharp lines, each of which is further split into a triplet. The outer lines of the triplets are relatively broad. In liquid ethane at $-170°C$ the spectrum consists of 12 very sharp lines. A discussion of the origin of the two spectra and comments on the difference between the spectra follows.

The quartet is due to CH_3 and indicates that the unpaired electron interacts equally with all three hydrogen atoms and that, therefore, the CH_3 group freely rotates about the C—C bond. The triplet is caused by the CH_2 protons which produce a dipole magnetic anisotropy which cannot be averaged out by internal rotation. End-over-end rotation does not take place at $4.2°K$. At the higher temperature, rapid tumbling of radicals in solution cancels the anisotropic effects. In crystals powerful electrostatic fields are present. These crystalline fields may be sufficient to remove the degeneracy, that is, to split the spin levels, even in the absence of an external field. This gives rise to zero-field splitting. The esr spectrum will be dependent upon the orientation of the crystalline field axis with respect to the applied laboratory magnetic field. These anisotropic interactions play important roles for radicals trapped in solids or highly viscous media. The hyperfine splitting for a radical that is not spherically symmetric depends on the orientation of the radical with respect to the magnetic field.

Example 9-5

A septet of triplets comprises the spectrum given by the semiquinone intermediate formed during the condensation of diacetyl (Fig. 9-6). The larger coupling constant is represented by the distance between the triplet centers; the smaller

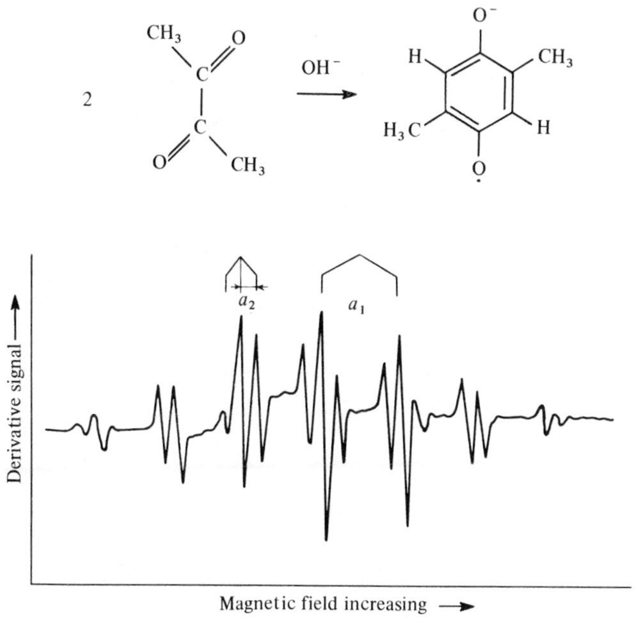

Fig. 9-6. The septet of triplets in the esr spectrum of semiquinone formed during the condensation of diacetyl.

coupling constant is given by the separation between the hyperfine pattern of each triplet. Protons in two different environments are involved. The large septet with an intensity ratio of $1:6:15:20:15:6:1$ must arise from six equivalent protons. Each of these lines is split again into a triplet by the weak coupling of two protons. The conclusion is that the condensation of diacetyl proceeds through a semiquinone intermediate whose structure must possess two methyl groups and two-ring protons. Knowledge of the magnitude of coupling constants is helpful (Table 9-1). Only relative hyperfine splittings can be obtained from a single sample. By placing a sample of known hyperfine splitting in one cavity of a dual-cavity sample holder, however, the unknown splitting can be accurately measured.

Spin Label

Biological and chemical compounds which do not contain an unpaired electron can, nevertheless, be studied when they are chemically bonded to a stable free radical.[1,2] This radical, or *spin label*, produces a sharp and simple esr spectrum that gives detailed information concerning the molecular environment of the label. Specific sites on a molecule can be tagged by choosing the appropriate spin label and reaction conditions. Five- or six-membered heterocyclic molecules incorporating a nitroxyl group whose nitrogen atom is bonded to a tertiary carbon are excellent spin-labeling compounds for biological macromolecules such as proteins and nucleic acids.

If the unpaired electron in a nitroxide radical is completely localized on the oxygen atom, there will be no isotropic coupling to the nitrogen nucleus. The observed coupling arises from the charged resonance structure

$$\underset{\displaystyle \underset{O\cdot}{|}}{\diagdown N \diagup} \quad \rightleftarrows \quad \underset{\displaystyle \underset{O}{|}}{\diagdown \overset{+}{N}\cdot \diagup}$$

where the unpaired electron is on the nitrogen atom. This charged structure is stabilized to a greater extent when the nitroxide is in a polar medium of high dielectric constant, such as water, than when it is in a nonpolar hydrocarbon environment. In a biological system, this solvent effect can be used to determine the polarity of the region surrounding a spin label.

ENDOR

Electron nuclear double resonance (*ENDOR*) is a method for improving the effective resolution of an esr spectrum.[3] The sample is irradiated simultaneously with a microwave frequency suitable for electron resonance and a radio frequency suitable for nuclear resonance. The radio frequency is swept while one point of the esr spectrum is observed under conditions of microwave saturation. The ENDOR display is esr signal height as a function of the swept nuclear radio frequency. For systems with short nuclear relaxation times, such as free radicals in solution, or with small nuclear magnetic moments, large rf fields are required. However, for systems with long relaxation times, such as solids, low power (5 W) equipment is sufficient.

The ENDOR technique is useful when a large variety of nuclear energy levels broaden the normal electron resonance line, masking the structure that contains important physical and chemical information. If the electron resonance can be held in a condition that is sensitive to nuclear transitions, the application of an rf field of the appropriate amplitude and frequency will expose the nuclear transitions responsible for the unresolved electron resonance.

ELDOR

In *electron double resonance* (*ELDOR*) the sample is irradiated simultaneously with two microwave frequencies. One of these is used to observe an esr signal at some point of the spectrum, while the other is swept through other parts of the spectrum to display the esr signal height as a function of the difference of the two microwave frequencies. The technique is useful for separating overlapping multiradical spectra and for studying various relaxation phenomena including chemical and spin exchange.

QUANTITATIVE ANALYSIS

The integrated intensity is usually related to the concentration of the paramagnetic species by comparison with a standard. The total area enclosed by either the absorption or dispersion signal is proportional to the number of unpaired electron spins in the sample. Comparison is made with a standard containing a known number of unpaired electrons and having the same line shape as the unknown (Gaussian or Lorentzian). A solid frequently used is 1,1-diphenyl-2'-picrylhydrazyl (DPPH) or solutions of peroxyl-amine disulfonate. DPPH contains 1.53×10^{21} unpaired spins per gram; substandards can be prepared by dilution with carbon black. Secondary standards include charred dextrose or synthetic ruby attached to the cavity. A dual-sample cavity is used to minimize difficulties with actual physical interchange of standard and sample.

Direct analytical applications of esr are gradually being developed. The analysis of vanadium in petroleum products has proven to be a rapid and convenient method that covers the range 0.1 to 50 μg/ml. Vanadyl(IV) etioporphyrin dissolved in heavy oil distillate serves as a standard. A continuous process analyzer has been described in which the sample circulates continuously through the sample compartment. Manganese(II) ion can be determined in aqueous solutions over the range from 10^{-6} M to 0.1 M. Other ions that can be handled quantitatively by esr include copper(II), chromium(III), gadolinium(III), iron(III), and titanium(III).

Polynuclear hydrocarbons, such as anthracene, perylene, dimethylanthracene and naphthacene, have been determined after conversion to radical cations and adsorption onto the surface of an activated silica–alumina catalyst.

One problem inherent in the esr method is that a relatively large fraction of the total area under the absorption (or first-derivative) curve tails off very slowly. The result is that overlap in the wings of a spectrum with peaks of another spectrum is more serious than immediately expected. Randolph[4] presents a critical evaluation of quantitative esr methods.

PROBLEMS

1. For an esr spectrometer that operates at a frequency of 35,000 MHz (K-band), calculate the static magnetic field required and the value of ΔE.

2. From the Heisenberg uncertainty principle, $\Delta E \Delta t \geqslant h/2$, where ΔE is the energy of the transition and Δt the time available for the measurement, estimate the range of Δt necessary to achieve reasonably sharp lines ($\Delta H \simeq 0.1$ G) for a radical with $g = 2$.

3. Which valency states of copper and silver will show a strong esr signal?

4. The esr spectra obtained during the enzymic oxidation by peroxidase–H_2O_2 of three substrates is shown. Match the spectrum with the free radical diagrammed below.

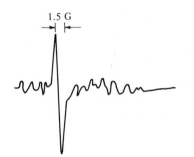

A. From reductic acid

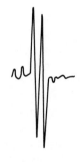

B. From dihydroxyfumaric acid

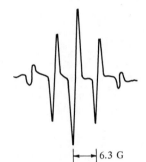

C. From ascorbic acid

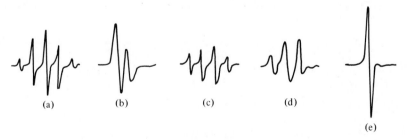

Spectrum 1   Spectrum 2   Spectrum 3

Problem 9-4

5. For each spectrum shown, identify the semiquinone from which it arises. The five compounds are *p*-semiquinone itself and the following derivatives: monochloro-, 2,3-dichloro-, trichloro-, and tetrachloro-. How does the splitting of each spectrum arise?

(a)   (b)   (c)   (d)

(e)

Problem 9-5

6. For each spectrum in the accompaning diagram, deduce the structure of the free radical and diagram the energy levels. (a) γ-Irradiation of dimethyl sulfone at 77°K, (b) alkali metal reduction of benzene in ethereal solution at −70°C, (c) γ-irradiation of t-butyl iodide at 77°K, (d) γ-irradiation of ethyl chloride at 77°K, (e) reaction of OH radical with glycollic acid, and (f) the general nature of the group in the 4-position of the semiquinone from 2,6-di-t-butyl phenol.

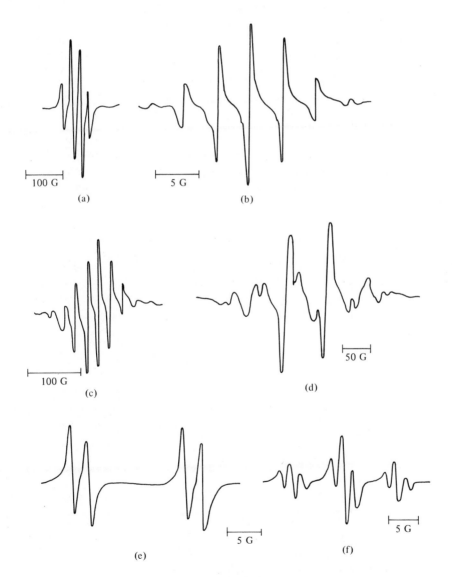

Problem 9-6

7. Analyze the spectrum shown for the hypothetical radical $H \cdot X^+$ ($I = 1$ for X).

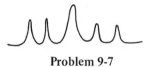

Problem 9-7

8. When malonic acid, $CH_2(COOH)_2$, is X-irradiated at room temperature, the spectrum appears to consist of a dominant doublet and a less intense, overlapping triplet. On standing for a few days, only a doublet remains. Determine the two products that are formed.

9. A single resonance line less than 1 gauss in width is observed at the position of the free electron resonance in an aqueous solution of sodium hydrosulfite, $Na_2S_2O_4$. (a) What species is responsible? (b) Why is no hyperfine structure observed in the spectrum? (c) How would a sample enriched in ^{33}S verify the assignment?

10. The ion $(SO_3)_2NO^{2-}$ in an aqueous solution of the potassium salt gives the esr spectrum shown in the figure: three sharp equally spaced lines (13-G separation) of equal intensity. After 50-fold amplification two additional lines, denoted by an asterisk, appear in the spectrum. Comment on the origin of the spectrum. [*Note:* ^{15}N ($I = \frac{1}{2}$) has a magnetic moment 0.6943 as large as that of ^{14}N.]

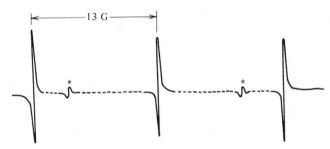

Problem 9-10

11. When 2,6-di-*t*-butyl-4-methyl-4-*t*-butyl peroxycyclohexa-2,5-diene-1-one is decomposed at 130°C in xylene within the sample cavity of an esr spectrometer, the spectrum shown is obtained. Deduce the principal radical intermediate.

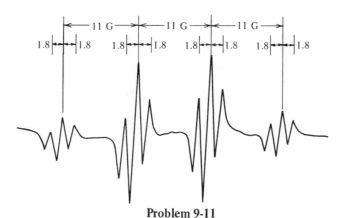

Problem 9-11

12. When polyethylene is γ-irradiated in vacuum, a six-line spectrum is obtained with approximately a binominal intensity distribution and a splitting of 26 G. However, when stretched polyethylene samples are arranged with the direction of stretch perpendicular to the magnetic field, a spectrum of five doublets is obtained, with a doublet splitting of 13 G. (a) What is the most likely radical? (b) Comment on the coupling of the α- and β-protons.

13. Diagram the splitting of the energy levels giving rise to the esr spectrum of the semi-naphthoquinone shown.

Problem 9-13

14. Assign the spectrum lines in the accompanying figure to the proper ring protons of naphthalene; the negative ion spectrum results from contact with alkali metal in an inert solvent.

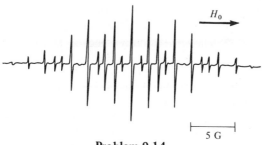

Problem 9-14

15. Photolysis of a solution of hydrogen peroxide in isopropyl alcohol at $110°K$ leads to free radical formation and an esr spectrum consisting of seven lines with a hyperfine splitting of 20 gauss, a line width of 10 gauss, and an approximate intensity distribution of $1:6:15:20:15:6:1$. Which radical is responsible?

16. Match the individual structure of the three isomeric methylcyclohexanones with the esr spectra shown. The radical anions were detected by exposing a dimethyl sulfoxide solution to air for 15–25 sec. The solution contained 0.1 M potassium *tert*-butoxide and a particular methylcyclohexanone. (*Note:* An alkyl group confers conformation stability on a cyclohexene ring in terms of esr frequencies.)

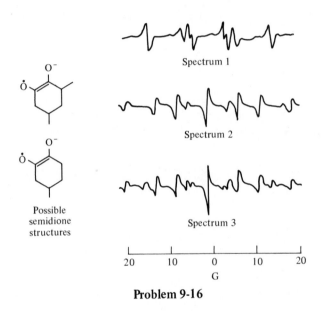

Spectrum 1

Spectrum 2

Possible
semidione
structures

Spectrum 3

Problem 9-16

17. What difference would be anticipated between the esr spectra of the following aryloxy radicals: 2,6-di-*t*-butyl-4-methyl derivative (I) and 2,6-di-*t*-butyl-4-ethyl derivative (II).

18. Predict the esr spectrum obtained from the aryloxy radicals formed from each of these compounds: (a) 2,6-di-*t*-butyl-4-hydroxymethylene phenol, (b) 2,4,6-tri-*t*-butyl phenol, (c) 2,6-dimethyl-4-*t*-butyl phenol, and (d) 2-methyl-4,6-di-*t*-butyl phenol.

19. For each of the following compounds, determine the structure of the free radical that is formed upon treatment with PbO_2: (a) 2,4-dimethyl-6-*t*-butyl phenol, spectrum shows a quartet of triplets; (b) 2-methyl-4-ethyl-6-*t*-butyl phenol, spectrum shows a quartet of triplets; (c) 2-ethyl-4-methyl-6-*t*-butyl phenol, spectrum shows a triplet of triplets; and (d) 2,4,6-trimethyl phenol, spectrum shows a septet of triplets.

BIBLIOGRAPHY

Alger, R. S., *Electron Paramagnetic Resonance, Techniques and Applications*, Wiley-Interscience, New York, 1968.

Ayscough, P. B., *Electron Spin Resonance in Chemistry*, Methuen, London, 1967.

Bersohn, R. and J. C. Baird, *An Introduction to Electron Paramagnetic Resonance*, Benjamin, Menlo Park, Calif., 1966.

Ingram, D. J. E., *Free Radicals as Studied by Electron Spin Resonance*, Academic Press, New York, 1958.

LITERATURE CITED

1. Griffith, O. H. and A. S. Waggonner, *Accounts Chem. Res.*, **2**, 17 (1969).
2. Hamilton, C. L. and H. M. McConnell in *Structural Chemistry and Molecular Biology*, A. Rich and N. Davidson, Eds., W. H. Freeman, San Francisco, 1968.
3. Kedzie, R. W., *Am. Laboratory*, **1** (12), 19 (1969).
4. Randolph, M. L. in *Biological Applications of Electron Spin Resonance*, H. M. Swartz, J. R. Bolton, and D. C. Borg, Eds., Wiley, New York, 1972, Chapter 3.

10

X-Ray Methods

When an atom is excited by removal of an electron from an inner shell, it usually returns to its normal state by transferring an electron from some outer shell to the inner shell with consequent emission of energy as X rays, that is, photons of high energy and short wavelengths in the order of tenths of angstroms to several angstroms. Eventually a free electron will be captured by the ion.

X Rays can be used in chemical analysis in several ways. One method uses the fact that the X rays emitted by an excited element have a wavelength characteristic of that element and an intensity proportional to the number of excited atoms. Thus emission methods can be used for both qualitative and quantitative work. The excitation can be carried out in several ways: by direct bombardment of the material with electrons (direct emission analysis and electron probe microanalysis) or by irradiation of the material with X rays of shorter wavelength (fluorescent analysis).

A second method of X-ray analysis utilizes the differing absorption of X rays by different materials (absorption analysis). Major discontinuities in the absorption of X rays by an element occur when the energy of the X rays become sufficient to knock an electron out of the inner levels of an atom.

A third method of using X rays in analytical work is the diffraction of X rays from the planes of a crystal (diffraction analysis). This method depends upon the wave character of X rays and the regular spacing of planes in a crystal. Although diffraction methods can be used for quantitative analysis, they are most widely used for qualitative identification of crystalline phases.

In 1913 Moseley first showed the extremely simple relation between atomic number, Z, and the reciprocal of the wavelength, $1/\lambda$, for each spectral line belonging to a particular series of emission lines for each element in the periodic table. This relationship is expressed as

$$\frac{c}{\lambda} = a(Z - o)^2 \tag{10-1}$$

where a is a proportionality constant and σ is a constant whose value depends on the particular series.

X-Ray emission and absorption spectra are quite simple because they consist of very few lines compared to the emission (Chapter 13) or absorption spectra observed in the visible or ultraviolet regions (Chapters 3 and 4). This relative simplicity arises because the X-ray spectra result from transitions between energy levels of the innermost electrons in the atom. There are only a few electrons in these inner shells and the resulting energy levels are limited, thus giving rise to only a few permitted transitions. There is only one K shell. The L electrons are grouped according to their binding energy into three sublevels: L_I, L_II, and L_III; the complete M shell consists of five sublevels. X-Ray emission or absorption spectra are dependent only on atomic number and not on the physical state of the sample nor on its chemical composition, except for the lightest atoms, because the innermost electrons are not involved in chemical binding and are not significantly affected by the behavior of the valence electrons.

PRODUCTION OF X RAYS AND X-RAY SPECTRA

An X-ray tube is basically a large vacuum tube containing a heated cathode (electron emitter) and an anode, or target (Fig. 10-1). Electrons emitted by the cathode are

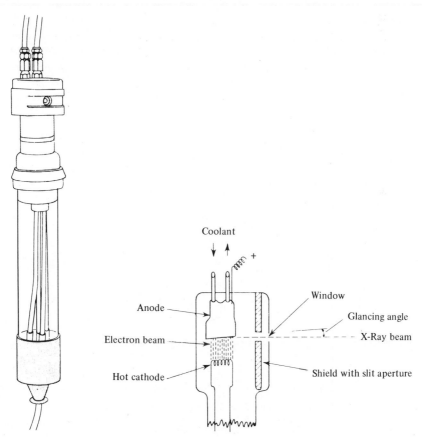

Fig. 10-1. An X-ray tube with schematic of filament and target. (Courtesy of Machlett Laboratories, Inc.)

accelerated through a high-voltage field between the target and cathode. Upon impact with the target, the stream of electrons is quickly brought to rest. The electrons transfer their kinetic energy to the atoms of the material making up the target. Part of the kinetic energy is emitted in a continuous spectrum of X rays covering a broad wavelength range (Fig. 10-2) with a broad maximum in intensity and falling off to a definite short-wavelength limit. The broad range of X-ray photon energies is due to deceleration of the impinging electrons by successive collisions with the atoms of the target material. Consequently, the emitted quanta are of longer wavelength than the short-wavelength cutoff, λ_0, which is independent of the target element and depends only upon the voltage across the X-ray tube. At the cutoff wavelength all of the energy of the electron is converted, at one impact, to a photon. The relationship between the voltage and λ_0 is given by the Duane-Hunt equation[5]

$$\lambda_0 \text{ (in Å)} = \frac{hc}{eV} = \frac{12{,}400}{V} \tag{10-2}$$

where V is the X-ray tube voltage in volts, e is the charge on the electron, h is Planck's constant, and c is the velocity of light. An increase in tube voltage results in an increase in the total energy emitted and a movement of the spectral distribution toward shorter wavelengths. The wavelength of maximum intensity is about 1.5 times the short-

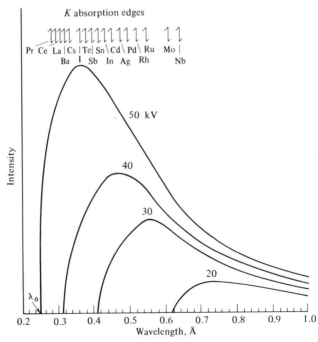

Fig. 10-2. X-Ray continuum from a target operated at voltages specified. Along the top edge are indicated the wavelengths of K absorption edges for elements 41 to 59.

wavelength limit. The intensity of the spectrum increases with atomic number of the target element.

If sufficient energy is available, the transfer of energy from the impinging electron beam may eject an electron from one of the inner shells of the atoms constituting the target material. The place of the ejected electron will then promptly be filled by an electron from an outer shell whose place, in turn, will be taken by an electron coming from still farther out. Thus the ionized atom returns to its normal state in a series of steps in each of which an X-ray photon of definite energy is emitted, that is, fluorescent radiation. These transitions give rise to the characteristic line spectrum of the material in the anode or of a specimen pasted on the target. When originating in an X-ray tube, these lines will be superimposed on the continuum. The K series of lines is observed when an electron in the innermost K shell is dislodged; it arises from electrons dropping down from L or M orbitals into the vacancy in the K shell. Corresponding vacancies in the L shells are filled by electron transitions from outer shells and give rise to the L series.

The characteristic X-ray spectrum of an element can also be excited by irradiation of a sample with a beam of X rays, provided that the primary X radiation is sufficiently energetic to remove an electron from an inner shell of the element. Because the inner electron must be completely removed from the element, the energy required is greater than that of any emission lines in the element's spectra, for emission lines result when that series of electrons falls into a vacant inner level from higher energy levels within the atom. When the energy of the exciting radiation just equals the energy required to remove an electron from the element, the exciting radiation is strongly absorbed, that is, there is a sharp rise in the absorption of the exciting radiation (Fig. 10-3). This is known as an *absorption edge*. If an X-ray tube is used to produce the exciting radiation, λ_0, the short-

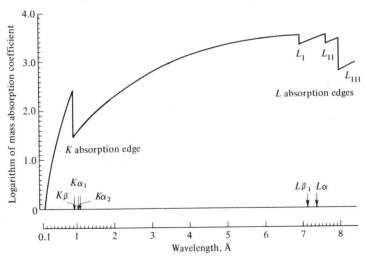

Fig. 10-3. X-Ray absorption spectrum of bromine. The characteristic emission lines of the K and L series are shown with arrows.

wavelength cutoff, must equal or be shorter than the wavelength of the absorption edge and thus there is a critical potential which must be applied to the tube.

Besides bombardment of a target with electrons, as in an X-ray tube, and irradiation of a target with energetic photons, as in the production of fluorescent radiation from a sample, X rays are also produced during the decay of certain radioactive isotopes. Many isotopes emit gamma rays which are the same as short-wavelength X rays. Other isotopes decay by K capture. In this process the nucleus captures a K electron, thus becoming an element with one less atomic number. The vacant K shell is filled by electrons falling in from outer shells, thus emitting characteristic X rays. An example of a radioactive isotope that decays by K capture and may be useful as a source of essentially monoenergetic X rays is ^{55}Fe. The K lines of manganese are emitted. (See Chapter 11 for a more complete description of radioactive isotopes.)

Example 10-1

To calculate the short-wavelength limit for an X-ray tube operated at 50 kV,

$$\lambda_0 = \frac{12,400}{50,000} = 0.248 \text{ Å}$$

Upon irradiation with this energy europium ($Z = 63$), whose K absorption edge lies at 0.255 Å, would emit its characteristic K series of lines, although with low intensity. However, the energy is insufficient for excitation of the K lines of gadolinium ($Z = 64$), whose K absorption edge lies at 0.247 Å.

As the wavelength of the incident radiation is decreased or as the potential across an X-ray tube is increased, there is successive ionization: first of electrons in the M shells of the sample or target, then of electrons in the L shells as the L_{III}, L_{II}, and L_I absorption edges are progressively exceeded, and finally culminating in the K shell's absorption edge. The K spectra are generally used for the detection and analysis of elements up to about neodymium ($Z = 60$); the L spectra are used from lanthanum to *trans*-uranium elements when an X-ray tube that has a maximum rating of 50 kV produces the exciting radiation. The wavelengths of selected spectral lines and absorption edges of a number of elements are shown in Table 10-1.

Example 10-2

Let us consider a vacant orbital in a bromine atom produced by the ejection of an electron from the innermost K shell of electrons. The energy required just to lift a K electron out of the environment of the atom must exceed the energy of the K absorption edge at 0.918 Å, or

$$V = \frac{12,400}{0.918} = 13,475 \text{ V } (13.475 \text{ kV})$$

The wavelength of the K absorption edge is always shorter than that of the K emission lines. The $K\beta_1$ line at 0.934 Å arises when an electron drops from the M shell; the $K\alpha_1$ and $K\alpha_2$ lines, a closely spaced doublet at 1.048 and 1.053 Å, arise from sublevels of slightly different energies within the L shell. In energy units, the

TABLE 10-1   Characteristic Wavelengths of Absorption Edges and Emission Lines for Selected Elements

Element	Minimum Potential for Excitation of K Lines, kV	K Absorption Edge, Å	$K\beta$, Å	$K\alpha_1$, Å	L_{III} Absorption Edge, Å	$L\alpha_1$, Å
Magnesium	1.30	9.54	9.558	9.889	247.9	251.0
Titanium	4.966	2.50	2.514	2.748	27.37	27.39
Chromium	5.988	2.070	2.085	2.290	20.7	21.67
Manganese	6.542	1.895	1.910	2.102	19.40	19.45
Cobalt	7.713	1.607	1.621	1.789	15.93	15.97
Nickel	8.337	1.487	1.500	1.658	14.58	14.57
Copper	8.982	1.380	1.392	1.541	13.29	13.33
Zinc	9.662	1.283	1.295	1.435	12.13	12.26
Molybdenum	20.003	0.620	0.632	0.709	4.912	5.406
Silver	25.535	0.484	0.497	0.559	3.698	4.154
Tungsten	69.51	0.178	0.184	0.209	1.215	1.476
Platinum	78.35	0.158	0.164	0.186	1.072	1.313

SOURCE: J. A. Dean, Ed., *Lange's Handbook of Chemistry*, 11th ed., McGraw-Hill, New York, 1973.

$K\alpha_1$ line represents the difference: K edge minus L_{III} edge. Thus, for bromine

$$K\alpha_1 \text{ (in keV)} = 13.475 - 1.522 = 11.953$$

The absorption and emission spectra for bromine are shown in Fig. 10-3, and the energy-level diagram is shown in Fig. 10-4.

The bond character in molecules and solids affects the X-ray spectra of the light elements whose emission lines originate from the valence electron shell, and even those lines and absorption edges from the next innermost shell. In general, relative to the lines

TABLE 10-2   Mean Wavelengths and Shifts of K Lines of Sulfur for the Different Oxidation States of Sulfur

Oxidation State	λ, XU[a]		Mean Shift	
	$K\alpha_1$	$K\alpha_2$	$\Delta\lambda$, XU	ΔE, eV
S^{6+}	5358.08	5360.89	−2.76	+1.19
S^{4+}	5358.63	5361.47	−2.20	+0.95
S^{2+}	5360.13	5362.93	−0.72	+0.31
S^{0}	5360.83	5363.66	0	0
S^{2-}	5361.15	6363.99	+0.33	−0.14

[a]One angstrom $\equiv 1002.02$ XU, where 1 XU = 1/3029.45 the spacing of the cleavage planes of a calcite crystal, a former standard wavelength unit.

SOURCE: A. Faessler, "X-ray Emission Spectra and the Chemical Bond," in *Proceedings of the Xth Colloquium Spectroscopicum Internationale,* Spartan Books, Washington, 1963, pp. 307–319.

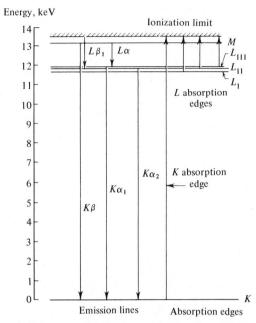

Fig. 10-4. Energy level diagram of bromine ($Z = 35$) showing the transitions that give rise to the absorption discontinuities and the emission lines.

of the free element, the lines of the atom in a compound are shifted toward shorter wavelengths if the atom has a positive charge, and toward longer wavelengths if it has a negative charge in the compound. Similar fine structure may be observed at an absorption edge if a high-resolution spectrometer is employed. Mean wavelengths and shifts in the emission lines for the different oxidation states of sulfur are given in Table 10-2.

INSTRUMENTAL UNITS

Instrumentation associated with X-ray methods in general is outlined schematically in Fig. 10-5. Many of the components will be discussed more fully in subsequent sections.

X-Ray Generating Equipment

The modern X-ray tube is a high-vacuum, sealed-off unit, as shown in Fig. 10-1, usually with a copper or molybdenum target, although targets of chromium, iron, nickel, silver, and tungsten are used for special purposes. The target is viewed from a very small angle above the surface. If the focal spot is a narrow ribbon, the source appears to be very small when viewed from the end, which leads to the sharper definition demanded in diffraction studies. For fluorescence work the focus is much larger, about 5×10 mm, and is viewed at a larger angle (about $20°$). Because it becomes very hot, the target is

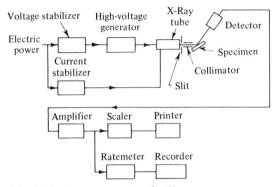

Fig. 10-5. Instrumentation for X-ray spectroscopy.

cooled by water and is sometimes rotated when a very intense X-ray beam is generated. The X-ray beam passes out of the tube through a thin window of beryllium or a special glass. For wavelengths from 6 to 70 Å, ultrathin films (1-μm aluminum or cast Parlodion films), separate the X-ray tube from the remainder of the equipment, which must be evacuated or flushed with helium.

Associated equipment includes high-voltage generators and stabilizers. Voltage regulation is accomplished by regulating the main ac supply. Current regulation is achieved by monitoring the dc X-ray tube current and controlling the filament voltage. Either full-wave rectification or constant high potential may be used to operate the X-ray tube. In full-wave rectification the voltage reaches its peak value 120 times a second but only persists at that value for a small fraction of the time. Constant high potential, obtained through electronic filtering, increases the output of characteristic X rays from a specimen, particularly with elements emitting at short wavelengths. With a tube operated at 50 kV the gain is twofold for elements up to about atomic number 35 (Br), increasing to fourfold for atomic number 56 (Ba). Commonly, X-ray tubes are operated at 50 or 60 kV. Tubes of 100 kV rating are available and extend the range of elements whose K series can be excited and the sensitivity, because on increasing the voltage the intensities of all lines increase.

Collimators

Radiation from an X-ray tube is collimated either by a series of closely spaced, parallel metal plates or by a bundle of tubes, 0.5 mm or less in diameter. In a fluorescence spectrometer, one collimator is placed between the specimen and the analyzer crystal to limit the divergence of the rays that reach the crystal. The second collimator, usually coarser, is placed between the analyzer crystal and the detector, where it is particularly useful at very low goniometer angles for preventing radiation that has not been reflected by the crystal from reaching the detector. Increased resolution can be obtained by decreasing the separation between the metal plates of the collimator or by increasing the length of the unit (usually a few centimeters), but this is achieved at the expense of intensity.

Filters

When the wavelengths of two spectral lines are nearly the same and there is an element with an absorption edge at a wavelength between the lines, that element may be used as a filter to reduce the intensity of the line of shorter wavelength. In X-ray diffractometry it is common practice to insert a thin foil in the primary X-ray beam to remove the $K\beta$ line from the spectrum while transmitting the $K\alpha$ lines with a relatively small loss of intensity. Filters for the common targets of X-ray tubes are listed in Table 10-3. Background radiation (the continuum) is reduced by the same method. Usually it makes no difference whether the filter is placed before or after the specimen unless the specimen fluoresces; if so, the filter is placed at the entrance slit of the goniometer.

TABLE 10-3   Filters for Common Targets of X-Ray Tubes

Target Element	$K\alpha_1$, Å	$K\beta$, Å	Filter	K Absorption Edge Filter, Å	Thickness[a] mm	Percent Loss of $K\alpha_1$
Mo	0.709	0.632	Zr	0.689	0.081	57
Cu	1.541	1.392	Ni	1.487	0.013	45
Cr	2.290	2.085	V	2.269	0.0153	51
	$L\alpha_1$, Å	$L\beta_1$, Å				$L\alpha_1$
Pt	1.313	1.120	Zn	1.283	–	–
W	1.476	1.282	Cu	1.380	0.035	77

[a]To reduce the intensity of $K\beta$ line to 0.01 that of the $K\alpha_1$ line.

Analyzing Crystal

Virtually monochromatic radiation is obtained by reflecting X rays from crystal planes. The relationship between the wavelength of the X-ray beam, the angle of diffraction θ, and the distance between each set of atomic planes of the crystal lattice d, is given by the Bragg condition[2]:

$$m\lambda = 2d \sin \theta \qquad (10\text{-}3)$$

where m represents the order of the diffraction. The geometric relations are shown in Fig. 10-6. For the ray diffracted by the second plane of the crystal, the distance $\overline{CBD}$ represents the additional distance of travel in comparison to a ray reflected from the surface. Angles CAB and BAD are both equal to θ. Therefore,

$$\overline{CB} = \overline{BD} = \overline{AB} \sin \theta \qquad (10\text{-}4)$$

and

$$\overline{CBD} = 2\,\overline{AB} \sin \theta \qquad (10\text{-}5)$$

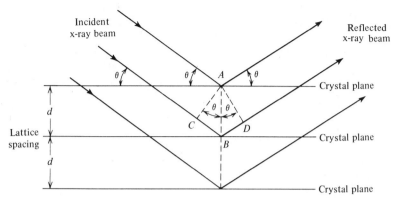

Fig. 10-6. Diffraction of X-rays from a set of crystal planes.

where $\overline{AB}$ is the interplanar spacing, d. In order to observe a beam in the direction of the diffracted rays, $\overline{CBD}$ must be some multiple of the wavelength of the X rays so that the diffracted waves will be in phase. Note that the angle between the direction of the incident beam and that of the diffracted beam is 2θ. In order to scan the emission spectrum of a specimen, the analyzing crystal is mounted on a goniometer, an instrument for measuring angles, and rotated through the desired angular region, as shown in the schematic diagram of a fluorescent spectrometer (Fig. 10-15).

The range of wavelengths usable with various analyzing crystals is governed by the d-spacings of the crystal planes and by the geometric limits to which the goniometer can be rotated. The d-value should be small enough to make the angle 2θ greater than approximately $8°$ even at the shortest wavelength used, otherwise excessively long analyzing crystals would be needed in order to prevent the incident beam from entering the detector. A small d-spacing is also favorable for producing a larger dispersion, $\partial\theta/\partial\lambda$, of the spectrum, as can be seen by differentiating the Bragg equation:

$$\frac{\partial\theta}{\partial\lambda} = \frac{m}{2d\cos\theta} \tag{10-6}$$

On the other hand, a small d-value imposes an upper limit to the range of wavelengths that can be analyzed, because at $\lambda = 2d$ the angle 2θ becomes $180°$. Actually, the upper limit to which goniometers can be rotated is mechanically limited to a 2θ value of around $150°$. For longer wavelengths a crystal with a larger d-spacing must be selected. Crystals commonly used are listed in Table 10-4. These crystals are all composed of light atoms; only sodium chloride, quartz, and the heavy metal fatty acids* have elements heavier than $Z = 9$, so that their own fluorescent X rays will not interfere with measurements. Higher order reflections, m greater than 1, from the analyzing crystal may result in the overlap of lines originating from different elements.

*The lead palmitate and strontium behenate analyzers are prepared by repeatedly dipping an optical flat into the film of the metal fatty acid, that is, the Langmuir-Blodgett technique.

TABLE 10-4 Typical Analyzer Crystals

Crystal	Reflecting Plane	Lattice Spacing d in Å	Useful Range in Å Maximum[a]	Minimum[b]
Topaz	303	1.356	2.62	0.189
Lithium fluoride	200	2.014	3.89	0.281
Aluminum	111	2.338	4.52	0.326
Sodium chloride	200	2.821	5.45	0.393
Calcium fluoride	11$\bar{1}$	3.16	6.11	0.440
Quartz	10$\bar{1}$1	3.343	6.46	0.466
Ethylenediamine d-tartrate (EDDT)	020	4.404	8.51	0.614
Ammonium dihydrogen orthophosphate (ADP)	200	5.325	10.29	0.742
Gypsum	020	7.60	14.70	1.06
Mica	002	9.963	19.25	1.39
Lead palmitate		45.6	78.3	6.39
Strontium behenate		61.3	121.7	8.59

[a]Maximum $2\theta = 150°$ $m\lambda = 2d \sin 75°$
[b]Minimum $2\theta = 8°$ $m\lambda = 2d \sin 4°$

Analyzing crystals have presented problems in the extension of X-ray analysis beyond a few angstroms. Potassium acid phthalate has made the determination of magnesium and sodium more practical. To extend the analytical capabilities beyond 26 Å, multiple monolayer soap film "crystals" are used. So far, a lead stearate decanoate "crystal" has proved to be superior for elements $Z = 9$ to $Z = 5$.

Detectors

In addition to photographic film, which is used in diffraction studies, the Geiger counter, the proportional counter, the scintillation counter, and certain semiconductors are used to measure X radiation. Details of their construction are given in Chapter 11. The spectral sensitivity of each detector varies with the wavelength of the X radiation, as shown in Fig. 10-7. The choice is governed by the nature of the X radiation to be detected.

The argon-filled Geiger counter, using halogen as a quenching gas, has a sensitive volume wide enough to detect nearly the entire large-area beam used in some X-ray optics. The tube is relatively insensitive to scattered hard radiation and, thus, its background intensity is low. Its quantum efficiency is about 60–65% in the range from 1.5 to 2.1 Å, and decreases to 40% at 1.4 to 2.9 Å. Its principal limitation lies in its long dead time—about 270 μsec, which produces counting losses at high intensities, and in its low sensitivity for the shorter wavelengths. The Geiger counter is rarely used for measuring intensities in excess of about 500 counts/sec; no practical correction is possible for coincidence losses when scanning across a peak to obtain integrated line intensities. Pulse height discrimination is impossible because pulses are all the same strength.

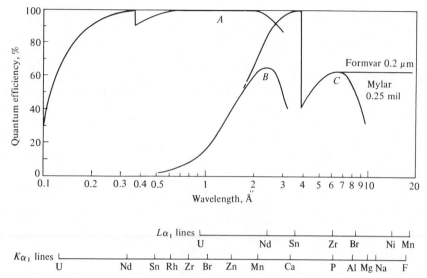

Fig. 10-7. Quantum efficiencies of detectors commonly used in X-ray spectrometry. *A.* Scintillation counter with Tl activated, NaI scintillator. *B.* Argon-filled Geiger counter with Be window. *C.* Gas-flow proportional counter; 90% Ar, 10% CH_4; 0.25 mil Mylar window. Lower scales indicate the wavelengths of representative emission lines.

The proportional counter has about the same spectral sensitivity characteristics as the Geiger counter but a very short dead time—about 0.5 μsec, and its response is linear up to extremely high count rates. The flow proportional counter can be equipped with an extremely thin window, usually 0.25-mil Mylar film, which naturally decreases the losses due to window absorption of the very soft X rays. Its range extends to 12 Å and it is the counter of choice for long-wavelength X radiation.*

Detector windows present challenging problems in work at very long wavelengths (to 70 Å), because these windows must be transparent to very low energy photons and, in addition, must be capable of supporting atmospheric pressure. Typical windows include 1-μm (sign painter's) aluminum dipped in Formvar (usable for sodium and magnesium), 1-μm hydrocarbon (cast Formvar, Parlodion, or collodion) films, and 0.1-μm hydrocarbon films. The 0.1-μm films must be supported on a 70% optical transmission grid or on the 0.5 mm spacing blade on a flow detector collimator. Their use is required for oxygen, nitrogen, and boron. The lifetime of unsupported, 1-μm films never exceeds 8 hr.

The scintillation counter has the shortest dead time of the three counters—0.25 μsec. It has a nearly uniform and high quantum efficiency throughout the important wavelength region—0.3 to 2.5 Å—and is usable to possibly 4 Å. Longer wavelengths are absorbed in the coating covering the crystal, necessitated by the hygroscopic properties of sodium iodide.

---

*Windows of 0.1-μm Formvar or "thin" nitrocellulose (often supported by screens) extend the transmission to approximately 120 and 160 Å, respectively.

The signal produced when an X-ray quantum is absorbed by the proportional and scintillation counters is extremely small and requires both a preamplifier and a second stage of amplification before the signal can be fed to a scaler or recorder. To diminish noise pickup, the preamplifier is located immediately after the detector in the latter's housing. However, the pulse strength produced in each counter is proportional to the quantum energy (and inversely proportional to the wavelength of the X-ray quantum). With circuits for pulse height discrimination, it is possible to discriminate electronically against unwanted wavelengths of different elements. Discrimination between elements eight to ten atomic numbers apart is possible with a scintillation counter. A proportional counter, because of its narrower pulse amplitude distribution, can discriminate between elements four to six atomic numbers apart. Unlike a filter, a pulse height analyzer can be used to pass either line of superposed spectral lines, serving, in effect, as a secondary monochromator. It is particularly useful for rejecting higher-order scattered radiation from elements of higher atomic number when determining the elements of lower atomic number. Modern pulse height analyzers with ever-improving resolution make finer and finer nondispersive analyzers.

Example 10-3

The use of a pulse height analyzer for Si $K\alpha_1$ radiation will illustrate the step-by-step operations. A relatively pure sample of silicon is inserted into the sample

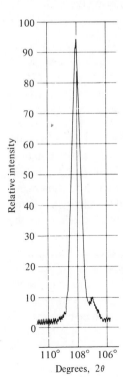

Fig. 10-8. Relative intensity of the SI $K\alpha_1$ line as a function of goniometer setting (2θ). Analyzing crystal: ethylenediamine *d*-tartrate.

holder (see Fig. 10-15). Scanning with the goniometer from 106° to 110° provides the graph shown in Fig. 10-8. From an ethylenediamine *d*-tartrate crystal, Si $K\alpha_1$ radiation is reflected at $2\theta = 108°$. Next, with the goniometer set manually at the peak of the silicon radiation, the distribution of pulses due to the silicon X-ray quanta is obtained by scanning the pulse amplitude base line and using a 1-V window. The integral curve of intensity versus pulse amplitude is shown in Fig. 10-9. From this information, the base of the pulse height discriminator would be set at 8.5 V and the window width at 13.0 V, since it is noticed that no pulses are detected until the upper-line setting approaches 21 V. In this example, the silicon radiation was peaked at 15 V. Naturally, if the silicon pulses were peaked at a lower or higher voltage, then the window and base-line settings would be different. The peak distribution (in volts) is a function of the dc voltage on the counter and the amplifier gain.

Semiconductors can be used for the detection of ionizing radiation. When an X-ray photon, for example, is absorbed by a semiconductor it leaves behind a trail of electron-hole pairs, each one of which requires about 3.5 eV for formation. Because this energy is about 1/10 of that required to form a free electron by ionization of a gaseous molecule, radiation absorbed in semiconductors will form about 10 times as many measurable

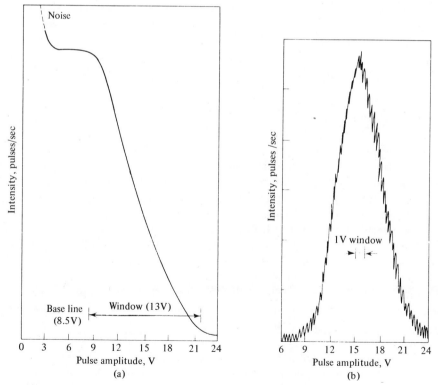

Fig. 10-9. Pulse amplitude distribution of SI $K\alpha_1$ radiation. (a) Integral curve and (b) differential curve. Goniometer set at 108°.

charges and thus the statistics of counting are improved and, furthermore, higher resolution on the basis of incident photon energy can be achieved.

Most semiconductor detectors consist of a thin layer of n-type material on the surface of a large piece of p-type material. A reverse bias potential is applied with the + voltage applied to the n-type layer and the − voltage to the p-type. Thus there is created a layer near the surface, at the junction, which is essentially depleted of carriers. Very little current flows unless ionizing radiation goes into the depletion zone and creates electrons and holes which are immediately collected by the bias voltage applied and result in a pulse of current. Because a very thin n-type layer is used, most of the action occurs in the p-type depletion zone just beneath the detector surface.

DIRECT X-RAY METHODS

The process of exciting characteristic spectra by electron bombardment was applied many years ago in the investigation of characteristic spectra of the elements by Siegbahn and others. In this manner the element hafnium was discovered by Von Hevesy and Coster in 1923. The specimen must be plated or smeared on the target of the X-ray tube. This has disadvantages: the X-ray tube must be reevacuated each time the specimen is changed; a demountable target is required; and the heating effect of the electron beam may cause chemical reaction, selective volatilization, or melting. These difficulties virtually prohibit the large-scale application of the direct method to routine analysis, except for electron probe microanalysis.

Electron Beam Probe

Electron probe microanalysis, developed by Castaing[3] (1951), is a method for the nondestructive elemental analysis from an area only 1 μm in diameter at the surface of a solid specimen. A beam of electrons is collimated into a fine pencil of 1-μm cross section and directed at the specimen surface exactly on the spot to be analyzed. This electron bombardment excites characteristic X rays essentially from a point source and at intensities considerably higher than with fluorescent excitation. The limit of detectability (in a 1-μm size region) is about 10^{-14} g. The relative accuracy is 1–2% if the concentration is greater than a few percent and if adequate standards are available.

Three types of optics are employed in the microprobe spectrometer: electron optics, light optics, and X-ray optics (Fig. 10-10). Of these, the most complex is the electron optical system, a modified electron microscope, which consists of an electron gun followed by two electromagnetic focusing lenses to form the electron beam probe. The specimen is mounted as the target inside the vacuum column of the instrument and under the beam. A focusing, curved-crystal X-ray spectrometer is attached to the evacuated system with the focal spot of the electron beam serving as the source of X radiation. A viewing microscope and mirror system allow continuous visual observation of the exact area of the specimen where the electron beam is striking. Point-by-point microanalysis is accomplished by translating the specimen across the beam.

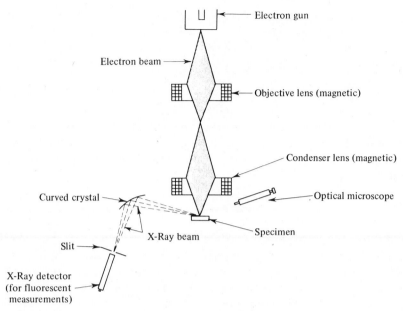

Fig. 10-10. Schematic of an electron probe microanalyzer. X-Ray beam can be passed directly into detector or reflected from analyzer crystal.

The method is used in the study of variations in concentration occurring near grain boundaries, the analysis of small inclusions in alloys or precipitates in a multitude of products, and corrosion studies where excitation is restricted to thin surface layers, because the beam penetrates to a depth of only 1 or 2 μm into the specimen.

X-RAY ABSORPTION METHODS

Because each element has its own characteristic set of K, L, M, and other, absorption edges, the wavelength at which a sudden change in absorption occurs can be used to identify an element present in a sample, and the magnitude of the change can be used to determine the amount of the particular element present. The fundamental equation for the transmittance of a monochromatic, collimated X-ray beam is

$$P = P_0 e^{-(\mu/\rho)\rho x} \tag{10-7}$$

where P is the radiant power of P_0 after passage through x cm of homogeneous matter of density ρ and whose linear absorption coefficient is μ. The parenthetical term μ/ρ is the mass absorption coefficient, often expressed simply as μ_m. It depends upon the wavelength of the X rays and the absorbing atom; that is,

$$\mu_m = CZ^4 \lambda^3 \frac{N_A}{A} \tag{10-8}$$

where N_A is Avogadro's number, A is the atomic weight, and C is a constant over a range between characteristic absorption edges. It is significant that the mass absorption coefficient is independent of the physical or chemical state of the specimen. In a compound or mixture it is an additive function of the mass absorption coefficients of the constituent elements, namely,

$$\mu_{m_T} = \mu_{m_1} W_1 + \mu_{m_2} W_2 + \cdots \tag{10-9}$$

where μ_{m_1} is the mass absorption coefficient of element 1 and W_1 is its weight fraction, and so on for all the elements present. Because only one element has a change in mass absorption coefficient at the edge, the following relationship is obtained for the ith element,

$$2.3 \log \frac{P}{P_0} = (\mu''_{m_1} - \mu'_{m_1}) W_i \rho x \tag{10-10}$$

where the term in the parentheses represents the difference in mass absorption coefficient at the edge discontinuity. Thus, the logarithm of the ratio of beam intensities on the two sides of an absorption edge depends only upon the change in mass absorption coefficients of the element characterized by this edge and on the amount of the particular element in the beam; ρx is the mass thickness of the sample in grams per square centimeter. There is no matrix effect, which gives the absorption method an advantage over X-ray fluorescence analysis in some cases.

In analogy with absorption measurements in other portions of the electromagnetic spectrum, one would expect to obtain a representative set of transmittance measurements on each side of an absorption edge with an X-ray spectrometer and extrapolate to the edge. However, X-ray absorption spectrophotometers that provide a continuously variable wavelength of X radiation are not commercially available. Instead, only a single attenuation measurement is made on each side of the edge. A multichannel instrument is required.

The general procedure will be illustrated by the determination of lead tetraethyl and ethylene dibromide in gasoline. Four channels are required. One channel is used as a reference standard; the other three channels provide the analyses for lead, bromine, and a correction for variations of the C/H ratio and the presence of any sulfur and chlorine. Primary excitation is provided by an X-ray tube operated at 21 kV. The secondary targets for each channel are as follows, with the fluorescent X-ray lines employed:

Channel 1: RbCl, Rb $K\alpha_1$
Channel 2: RbCl, Rb $K\alpha_1$
Channel 3: SrCO$_3$, Sr $K\alpha_1$
Channel 4: NaBr, Br $K\alpha_1$

The relationship between the pertinent absorption edges and the target fluorescent emission lines is shown in Fig. 10-11. In operation, a nominal sample is sealed in the sample cell in Channel 1; the sample to be analyzed is placed in the cells in the remaining channels. The exposure is started and automatically terminated when the integrated intensity in Channel 1 reaches a predetermined value (perhaps 100,000 counts in a time

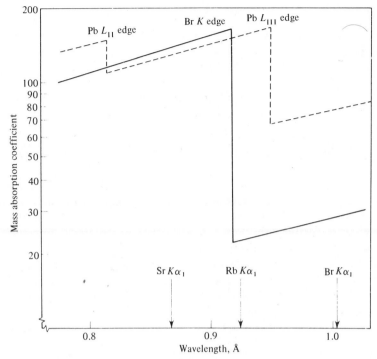

Fig. 10-11. Absorption edges and emission lines pertinent to the X-ray absorption analysis of lead tetraethyl and ethylene dibromide in gasoline.

interval of 100 sec). The integrated intensities accumulated in the other channels are then recorded. Initially the four channels are adjusted to reach 100% transmittance with nominally pure gasoline. Results for bromine are computed from the difference in counts between Channels 3 and 2; for lead from Channels 2 and 4.

Microradiography

Another application employing the different absorbing powers of different elements toward an X-ray beam permits the gross structure of various types of small specimens to be examined under high magnification. Positions where there are elements that strongly absorb the X rays will appear light, and positions where there are elements which do not absorb the X rays will appear dark on a film placed behind the sample.

Clark and Gross[4] developed a method employing ordinary X-ray diffraction equipment. Any of the commonly employed targets operated at 30–50 kV can be used. No vacuum camera is necessary. The microradiographic camera, shown in Fig. 10-12, is designed to fit as an inset in the collimating system of any commercial X-ray equipment. Special photographic film, which possesses an extremely fine grain, makes magnifications up to 200 times possible without loss of detail from graininess. Sample thicknesses vary

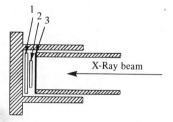

Fig. 10-12. Schematic of microradiographic camera: (1) film, (2) sample, (3) black paper.

from 0.075 mm for steels up to 0.25 mm for magnesium alloys. Only a few seconds of exposure is necessary.

Various techniques are possible depending upon the specimen. Biological specimens may be impregnated with a material of high molecular weight to characterize particular structures. Occasionally the necessary density variations are initially present within the sample. More often, the use of various selective monochromatic wavelengths from different target elements must be employed.

Another technique uses a tube with a very small focal spot as an X-ray source. If the focal spot approaches a point source, magnification in the microradiograph is obtained by simple geometry with considerable sharpness. Actually, focal spots as small as 5 μm in diameter can be obtained and magnifications up to 50 times or so are possible. The magnification is the ratio of the distance of film to target to the distance of object from the target.

Nondispersive X-Ray Absorptiometer

The general arrangement of a nondispersive X-ray absorptiometer is shown in Fig. 10-13. A tungsten target X-ray tube is operated at 15–45 kV. In the X-ray beam is a

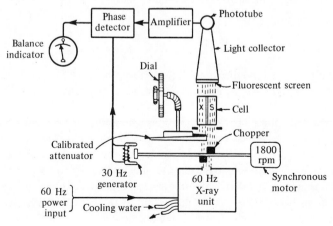

Fig. 10-13. Nondispersive X-ray absorptiometer. (Courtesy of General Electric Co.)

synchronous motor-driven chopper which alternately interrupts one-half of the X-ray beam. A variable-thickness aluminum attenuator (in the shape of a wedge) is placed between the chopper and reference sample compartment. Duplicate reference and sample cells up to 65 cm in length can be accommodated; those for liquids and gases can be arranged for continuous flow of process streams. Both halves of the X-ray beam fall on a common phosphor-coated photomultiplier tube which is protected from visible light by a thin metallic filter.

In operation, a reference sample is placed in the appropriate cell and the specimen to be analyzed in the sample tube. The attenuator is adjusted until the absorption in the two X-ray beams is brought into balance. The change in thickness of aluminum required for different samples is a function of the difference in composition. Prior calibration enables a determination in terms of the solute in an unknown. Liquids are simplest to handle. The thickness of solid specimens, and the density of powder samples, must be uniform to a precision greater than that expected in the result.

Polychromatic absorptiometry can be used to determine chlorine in hydrogen. Sulfur in crude oil can be distinguished from the carbon–hydrogen residuum. Other examples include barium fluoride in carbon brushes, barium or lead in special glass, and chlorine in plastics and hydrocarbons. In fact, the method is applicable to any sample that contains one element markedly heavier than the others and when the matrix is essentially invariant in concentration.

X-RAY FLUORESCENCE METHOD

Characteristic X-ray spectra are excited when a specimen is irradiated with a beam of sufficiently short-wavelength X radiation. Intensities of the resulting fluorescent X rays are smaller by a factor of roughly 1000 than an X-ray beam obtained by direct excitation with a beam of electrons. Only availability of high-intensity X-ray tubes, very sensitive detectors, and suitable X-ray optics renders the fluorescent method feasible. The intensity is important because it influences the time that will be necessary to measure a spectrum. A certain number of quanta has to be accumulated at the detector in order to reduce sufficiently the statistical error of the measurement. The sensitivity of the analysis, that is, the lowest detectable concentration of a particular element in a specimen, will depend on the peak-to-background ratio of the spectral lines. Relatively few cases of spectral interference occur because of the relative simplicity of X-ray spectra.

X-Ray Fluorescence Spectrometer

The general arrangement for exciting, dispersing, and detecting fluorescent radiation with a plane-crystal spectrometer is shown pictorially in Fig. 10-14 and diagrammatically in Fig. 10-15. The specimen in the sample holder (often rotated to improve uniformity of exposure) is irradiated with an unfiltered beam of primary X rays, which causes the elements present to emit their characteristic fluorescence lines. A portion of the scattered fluorescence is collimated by the entrance slit of the goniometer and directed onto the

Fig. 10-14. Typical X-ray fluorescence spectrometer. (Courtesy of Philips Electronic Instruments.)

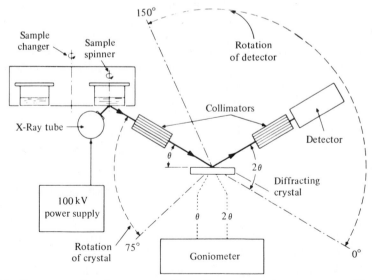

Fig. 10-15. Geometry of a plane–crystal X-ray fluorescence spectrometer. (Courtesy of Philips Electronic Instruments.)

plane surface of the analyzing crystal. The line radiations, reflected according to the Bragg condition, pass through an auxiliary collimator (exit slit) to the detector, where the energy of the X-ray quanta is converted into electrical impulses, or counts.

The primary slit, the analyzer crystal, and secondary slit are placed on the focal circle so that Bragg's law will always be satisfied as the goniometer is rotated, the detector being rotated at twice the angular rate of the crystal. The analyzer crystal is a flat single-crystal plate, 2.5 cm in width and 7.5 cm in length. The specimen holder is often an aluminum cylinder, although plastic material is used to examine acid or alkaline solutions. A thin film of Mylar supports the specimen, and an aluminum mask restricts the area irradiated (often a rectangle 18 mm by 27 mm). Intensity losses caused by the absorption of long-wavelength X rays by air and window materials can be reduced by evacuating the goniometer chamber. Another method for reducing losses is to enclose the radiation path in a special boot which extends from the sample surface to the detector window and then displace the air by helium, which has a low absorption coefficient. Vacuum spectrometers are used where helium is scarce and for elements boron ($Z = 5$) to sodium ($Z = 11$).

Focusing spectrometers that involve reflection from or transmission through a 10-cm or 28-cm curved crystal have been described.[1]   Collimators are not required, and the increase in intensity obtained by focusing the fluorescence lines makes the technique suitable for the analysis of small specimens. In the curved-crystal arrangement (Fig. 10-16) the analyzing crystal is bent to a radius of curvature twice that of the focal circle, and then the inner surface is ground to the radius of curvature of the focal circle. A slit

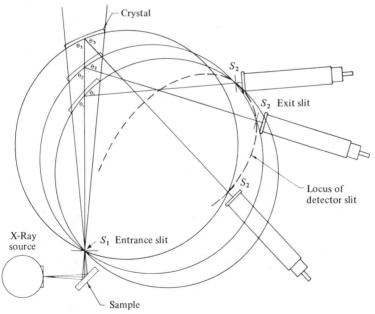

Fig. 10-16. Focusing X-ray optics, employing a curved analyzing crystal. (Courtesy of Applied Research Laboratories, Inc.)

on the focusing circle acts as a divergent source of polychromatic radiation from the specimen. All of the radiation of one wavelength diverging from the slit will be diffracted at a particular setting of the crystal, and the diffracted radiation will converge to a line image at a symmetric point on the focusing circle. The angular velocity of the detector is twice that of the crystal and, as the two of them move along the periphery of the circle, the X-ray spectral lines are dispersed and detected just as in the flat-crystal arrangement.

In order to excite fluorescence the primary radiation must obviously have a wavelength shorter than the absorption edge of the spectral lines desired. Continuous as well as characteristic radiation of the primary target can serve the purpose. To get a continuous spectrum of short enough wavelength and of sufficient intensity, one may calculate the required voltage of the X-ray tube from Eq. 10-2, remembering that the wavelength of maximum intensity is approximately $1.5\lambda_0$. In qualitative analyses it is usually desirable to operate the X-ray tube at the highest permissible voltage in order to ensure that the largest possible number of elements in the specimen will be excited to fluoresce. It will also ensure the greatest possible intensity of fluorescence for each element in quantitative analyses. In two cases, however, the X-ray tube voltage should be made lower than the available maximum: (1) when it is desirable not to excite fluorescence of all elements in the specimen, but rather employ selective excitation conditions, and (2) when very long-wavelength spectral lines are to be excited—in order to minimize scattering of primary radiation through the system by holding down the intensity of the short-wavelength continuum.

Analytical Applications

For qualitative analysis, the angle θ between the surface of the crystal and the incident fluorescence beam is gradually increased; at certain well-defined angles the appropriate fluorescence lines are reflected. In automatic operation the intensity is recorded on a moving chart as a series of peaks, corresponding to fluorescence lines, above a background that arises principally from general scattering. The angular position of the detector, in degrees of 2θ, is also recorded on the chart. Additional evidence for identification may be obtained from relative peak heights, the critical excitation potential, and pulse height analysis.

For quantitative analysis, the intensity of a characteristic line of the element to be analyzed is measured. The goniometer is set at the 2θ angle of the peak, and counts are collected for a fixed period of time, or the time is measured for the period required to collect a specified number of counts. The goniometer is then set at a nearby portion of the spectrum where a scan has shown that only the background contributes. For major elements, 200,000 counts can be accumulated in 1 or 2 min. Background counts will require much longer time—a very low background may require 10 min to acquire 10,000 counts. The net line intensity, that is, peak minus background, in counts per seconds, is then related to the concentration of the element via a calibration curve.

Particle size and shape are important and determine the degree to which the incident beam is absorbed or scattered. Standards and samples should be ground to the same mesh size, preferably finer than 200 mesh. Errors from differences in packing density can be handled by addition of an internal standard to the sample. Powders are pressed into a wafer in a metallurgical specimen press or converted into a solid solution by fusion with borax. Samples are best handled as liquids. If they can be conveniently dissolved, their analysis is greatly simplified and precision is greatly improved. Liquid samples should exceed a depth that will appear infinitely thick to the primary X-ray beam—about 5 mm for aqueous samples. The solvent should not contain heavy atoms; in this respect HNO_3 and water are superior to H_2SO_4 or HCl.

Before relating the intensity of fluorescent emission to concentration of emitting element, it is usually necessary to correct for matrix effects. Matrix dilution will avoid serious absorption effects. The samples are heavily diluted with a material having a low absorption, such as powdered starch, lithium carbonate, lampblack, gum arabic, or borax (used in fusions). The concentration, and therefore the effect, of the disturbing matrix elements is reduced, along with a reduction of the measured fluorescence. However, the most practical way to apply a systematic correction is by an internal standard. Even so, the internal standard technique is valid only if the matrix elements affect the reference line and analytical line in exactly the same way. The choice of a reference element depends on the relative positions of the characteristic lines and the absorption edges of the element to be determined, the reference element, and the disturbing elements responsible for the matrix effects. If either the reference line or the analytical line is selectively absorbed or enhanced by a matrix element, the internal standard line to analytical line ratio is not a true measure of the concentration of the element being determined. Preferential absorption of a line would occur if a disturbing element had an absorption edge between the comparison lines. The intensity of a line can be enhanced if a matrix element absorbs primary radiation and then, by fluorescence, emits radiation which, in turn, is absorbed by a sample element and causes the sample to fluoresce more strongly. Thus, if the matrix fluorescence lies between the absorption edges of the analytical and internal standard elements, selective enhancement might result.

The X-ray fluorescence method, inherently very precise, rivals the accuracy of wet chemical techniques in the analysis of major constituents. On the other hand, it is difficult to detect an element present in less than one part in 10,000. The method is attractive for elements that lack reliable wet chemical methods, for example, elements such as niobium, tantalum, sodium, and the rare earths. It often serves as a complementary procedure to optical emission spectrography, particularly for major constituents, and also for the analysis of nonmetallic specimens, because the sample need not be an electrical conductor. To overcome air absorption for elements of atomic number below 21, operating pressure must be 0.1 torr. Even so, below magnesium the transmission becomes seriously attenuated although the method has been extended to boron.

Simultaneous analysis of several elements is possible with automatic equipment, such as the Applied Research Laboratory Quantometer. Instruments of this type have semi-fixed monochromators with optics mounted around a centrally located X-ray tube and sample position. Each crystal is adjusted to reflect one fluorescence line to its associated detector. A compatible recording unit permits both optical and X-ray units to be recorded with the same console.

Nondispersive Spectrometers

For some samples where only very few elements are present and their X-ray lines are widely separated in wavelength, the crystal analyzer may be eliminated and pulse height discrimination employed in its place.   For compositions greater than about 1%, and elements separated by a few atomic numbers, nondispersive analysis is very useful because the intensities are increased about a thousandfold.

X-RAY DIFFRACTION

Every atom in a crystal scatters an X-ray beam incident upon it in all directions. Because even the smallest crystal contains a very large number of atoms, the chance that these scattered waves would constructively interfere would be almost zero except for the fact that the atoms in crystals are arranged in a regular, repetitive manner.   The condition for diffraction of a beam of X rays from a crystal is given by the Bragg equation, Eq. 10-3. Atoms located exactly on the crystal planes contribute maximally to the intensity of the diffracted beam; atoms exactly halfway between the planes exert maximum destructive interference and those at some intermediate location interfere constructively or destructively depending on their exact location but with less than their maximum effect. Furthermore, the scattering power of an atom for X rays depends upon the number of electrons it possesses.   Thus the position of the diffraction beams from a crystal depends only upon the size and shape of the repetitive unit of a crystal and the wavelength of the incident X-ray beam whereas the intensities of the diffracted beams depend also upon the type of atoms in the crystal and the location of the atoms in the fundamental repetitive unit, the unit cell.   No two substances, therefore, have absolutely identical diffraction patterns when one considers both the direction and intensity of all diffracted beams; however, some similar, complex organic compounds may have almost identical patterns. The diffraction pattern is thus a "fingerprint" of a crystalline compound and the crystalline components of a mixture can be identified individually.

Reciprocal Lattice Concept

Diffraction phenomena can be interpreted most conveniently with the aid of the reciprocal lattice concept.   A plane can be represented by a line drawn normal to the plane; the spatial orientation of this line describes the orientation of the plane. Furthermore, the length of the line can be fixed in an inverse proportion to the interplanar spacing of the plane that it represents.

When a normal is drawn to each plane in a crystal and the normals are drawn from a common origin, the terminal points of these normals constitute a lattice array.   This is called the *reciprocal lattice* because the distance of each point from the origin is reciprocal to the interplanar spacing of the planes that it represents.   Figure 10-17 shows, near the origin, the traces of several planes in a unit cell of a crystal, namely the (100), (001), (101), and (102) planes.   The normals to these planes, also indicated, are called the

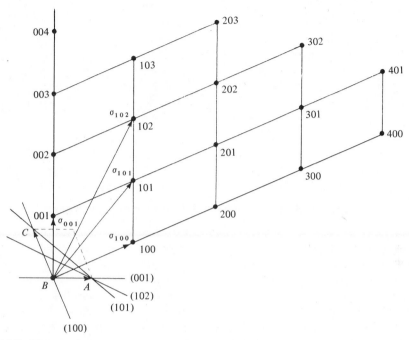

Fig. 10-17. The side view of several planes in the unit cell of a crystal with the normals to these planes indicated.

reciprocal lattice vectors σ_{hkl} and are defined by

$$\sigma_{hkl} = \frac{\lambda}{d_{hkl}}$$

In three dimensions, the lattice array is described by three reciprocal lattice vectors whose magnitudes are given by

$$a^\star = \sigma_{100} = \frac{\lambda}{d_{100}}$$

$$b^\star = \sigma_{010} = \frac{\lambda}{d_{010}}$$

$$c^\star = \sigma_{001} = \frac{\lambda}{d_{001}}$$

and whose directions are defined by three interaxial angles $\alpha^\star, \beta^\star, \gamma^\star$.

Writing the Bragg equation in a form that relates the glancing angle θ most clearly to the other parameters, we have

$$\sin \theta_{hkl} = \frac{\lambda/d_{hkl}}{2} \tag{10-11}$$

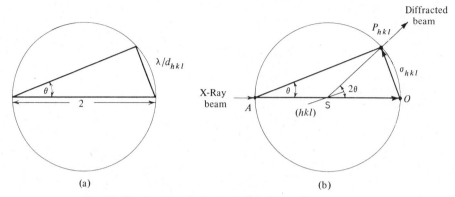

Fig. 10-18. Representation of the diffraction condition.

The numerator can be taken as one side of a right triangle with θ as another angle and the denominator as its hypotenuse, Fig. 10-18(a). Because of the physical meaning of the quantities in Eq. 10-11, the construction can be interpreted as shown in Fig. 10-19(b). The diameter of the circle $(\overline{ASO})$ represents the direction of the incident X-ray beam. A line through the origin of the circle, parallel to $\overline{AP}$ and forming the angle θ with the incident beam, represents a crystallographic plane that satisfies the Bragg diffraction condition. The line $\overline{SP}$, also forming the angle θ with the crystal plane and 2θ with the

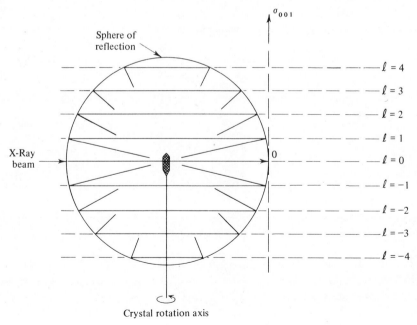

Fig. 10-19. The reciprocal lattice construction for a rotating crystal.

incident beam, represents the diffracted beam's direction. Then the line $\overline{OP}$ is the reciprocal lattice vector to the reciprocal lattice point P_{hkl} lying on the circumference of the circle. The vector σ_{hkl} originates at the point on the circle where the direct beam leaves the circle. The Bragg equation is satisfied when and only when a reciprocal lattice point lies on the "sphere of reflection," a sphere formed by rotating the circle upon its diameter $\overline{ASO}$.

Thus, the crystal in a diffraction experiment can be pictured at the center of a sphere of unit radius, and the reciprocal lattice of this crystal is centered at the point where the direct beam leaves the sphere, as shown in Fig. 10-19. Because the orientation of the reciprocal lattice bears a fixed relation to that of the crystal, if the crystal is rotated, the reciprocal lattice can be pictured as rotating also. Whenever a reciprocal lattice point intersects the sphere, a reflection emanates from the crystal at the sphere's center and passes through the intersecting reciprocal lattice point.

Diffraction Patterns

If the X-ray beam is monochromatic, there will be only a limited number of angles at which diffraction of the beam can occur. The actual angles are determined by the wavelength of the X rays and the spacing between the various planes of the crystal. In the *rotating crystal method*, monochromatic X radiation is incident on a single crystal which is rotated about one of its axes. The reflected beams lie as spots on the surface of cones which are coaxial with the rotation axis. If, for example, a single cubic crystal is rotated about the (001) axis, which is the equivalent to rotation about the $c^{\star}$ axis, the sphere of reflection and the reciprocal lattice are as shown in Fig. 10-19. The diffracted beam directions are determined by intersection of the reciprocal lattice points with the sphere of reflection. All the reciprocal lattice points lying in any one layer of the reciprocal lattice layer perpendicular to the axis of rotation will intersect the sphere of reflection in a circle. The height of the circle above the equatorial plane is proportional to the vertical reciprocal lattice spacing $c^{\star}$. By remounting the crystal successively about different axes, one can determine the complete distribution of reciprocal lattice points. Of course, one mounting is sufficient if the crystal is cubic, but two or more may be needed if the crystal has lower symmetry.

In a modification of the single-crystal method, known as the *Weissenberg method*, the photographic film is moved continuously during the exposure parallel to the axis of rotation of the crystal. All reflections are blocked out except those which occur in a single layer line. This results in a film that is somewhat easier to decipher than a simple rotation photograph. Still other techniques are used; one, the precession method, results in a photograph which gives an undistorted view of a plane in the reciprocal lattice of the crystal.

In the *powder method*, the crystal is replaced by a large collection of very small crystals, randomly oriented, and a continuous cone of diffracted rays is produced. There are some important differences, however, with respect to the rotating-crystal method. The cones obtained with a single crystal are not continuous because the diffracted beams occur only at certain points along the cone, whereas the cones with the powder method

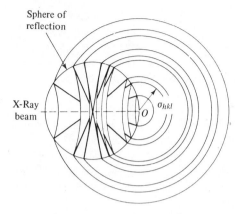

Sphere of
reflection

X-Ray
beam

σ_{hkl}

O

Fig. 10-20. Origin of powder diffraction diagrams in terms of the series of concentric spheres generated from the reciprocal lattice points about the origin, O, of the reciprocal lattice and their intersection with the sphere of reflection.

are continuous. Furthermore, although the cones obtained with rotating single crystals are uniformly spaced about the zero level, the cones produced in the powder method are determined by the spacings of prominent planes and are not uniformly spaced. The origin of a powder diagram is shown in Fig. 10-20. Because of the random orientation of the crystallites, the reciprocal lattice points generate a sphere of radius σ_{hkl} about the origin of the reciprocal lattice. A number of these spheres intersect the sphere of reflection.

Camera Design

Typical cameras for X-ray powder diffraction work are shown in Fig. 10-21. Cameras are usually constructed so that the film diameter has one of the three values, 57.3 mm, 114.6 mm, or 143.2 mm. The reason for these values can be understood by considering the calculations involved. If the distance between corresponding arcs of the same cone of diffracted rays—for example, the distance between points A and B of Fig. 10-22—is measured and called S, then

$$4\theta_{rad} = \frac{S}{R} \tag{10-12}$$

where θ_{rad} is the Bragg angle measured in radians, and R is the radius of the film in the camera. The angle, θ_{deg}, measured in degrees, is then

$$\theta_{deg} = \frac{57.295S}{4R} \tag{10-13}$$

where 57.295 equals the value of a radian in degrees. Therefore, when the camera diameter ($2R$) is equal to 57.3 mm, $2\theta_{deg}$ may be found by measuring S in millimeters. When the diameter is 114.59 mm, $2\theta_{deg} = S/2$, and when the diameter is 143.2 mm, $\theta_{deg} = 2(S/10)$.

Once angle θ has been calculated, Eq. 10-3 can be used to find the interplanar spacing, using values of wavelength λ from Table 10-1. Sets of tables[11] are available that

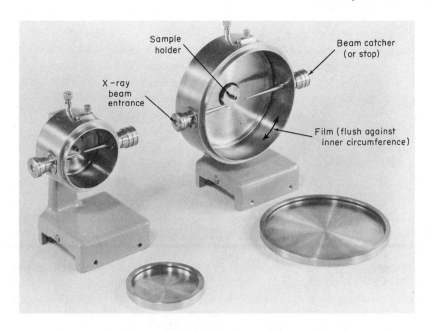

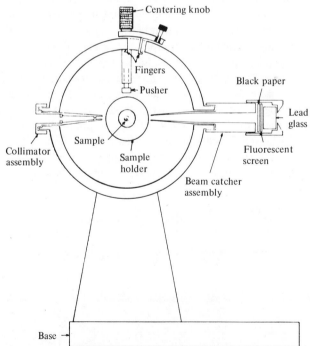

Fig. 10-21. X-Ray powder diffraction cameras, 57.3-mm and 114.6-mm diameter, and schematic of internal arrangement. (Courtesy of Philips Electronic Instruments.)

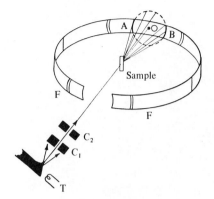

Fig. 10-22. Schematic of powder diffraction patterns. T, X-ray tube; C_1 and C_2 collimating slits; S, powdered crystalline sample; A, a line on film, left portion, and B, right portion; O, intersection of undiffracted beam with film. Angle $ASO = 2\theta$.

give the interplanar spacing for the angle 2θ for the types of radiation most commonly used.

Cameras of larger diameter make it easier to measure the separation of lines provided that the lines are sharp. The sharpness of the lines depends to a large extent upon the quality of the collimating slits and the size of the sample. The slits should produce a fine beam of X rays with as small a divergence as possible. The sample size should be small so that it will act as a small source of the diffracted beam. On the other hand, smaller samples, finer pencils of incident X rays, and the cameras of larger diameter all tend to require longer exposure times, so that in practice a compromise must be made.

For very precise measurements of interplanar spacings, the diameter of the film and the separation of the lines must be very accurately known. Several methods of measurement have been proposed. The effective camera diameter can be determined by calibration with a material such as sodium chloride whose interplanar spacings are accurately known. In another method, the Straumanis method,[7] the film is inserted in the camera so that the ends of the film are at about $90°$ from the point of emergence of the beam from the camera. The developed film appears as in Fig. 10-23. If a_1, a_2 and b_1, b_2 represent the two sides of arcs on the left and right sides of the film, then the two averages determine the positions of the entering and emerging beams.

Therefore,

$$\frac{b_1 + b_2}{2} - \frac{a_1 + a_2}{2} = \text{distance corresponding to } 180° \tag{10-14}$$

or

$$360° = b_1 + b_2 - a_1 - a_2 \tag{10-15}$$

Fig. 10-23. X-Ray powder diffraction photograph of sodium chloride. Film mounted in Straumanis method.

and the angle 4θ associated with any pair of lines can then be calculated:

$$\frac{a_2 - a_1}{b_1 + b_2 - a_1 - a_2} = \frac{4\theta}{360} \qquad (10\text{-}16)$$

Choice of X Radiation

Two factors control the choice of X radiation, as can be seen by rearranging the terms of the Bragg equation

$$\theta = \sin^{-1}\left(\frac{\lambda}{2d}\right) \qquad (10\text{-}17)$$

Because the ratio in the parentheses cannot exceed unity, the use of long-wavelength radiation limits the number of reflections that can be observed. Conversely, when the unit cell is very large, short-wavelength radiation tends to crowd individual reflections very closely together.

The choice of radiation is also affected by the absorption characteristics of a sample. Radiation having a wavelength just shorter than the absorption edge of an element contained in the sample should be avoided, because then the element absorbs the radiation strongly. The absorbed energy is emitted as fluorescent radiation in all directions and increases the background (which would result in darkening on a film, making it more difficult to see the diffraction maxima sought). It is obvious, then, why one commercial source provides radiation sources from a multiwindow tube with anodes of silver, molybdenum, tungsten, or copper.

Specimen Preparation

Single crystals are used for structure determinations whenever possible because of the relatively larger number of reflections obtained from single crystals and the greater ease of their interpretation. A crystal should be of such size that it is completely bathed by the incident beam. Generally, a crystal is affixed to a thin glass capillary which, in turn, is fastened to a brass pin, as shown in Fig. 10-24(a).

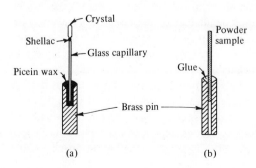

Fig. 10-24. Specimen mounts for X-ray diffraction. (a) Single crystal and (b) powdered sample.

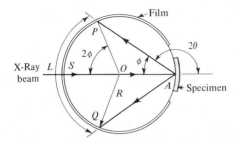

Fig. 10-25. Geometry of the back-reflection symmetrical focusing camera.

When single crystals of sufficient size are not available or when the problem is merely the identification of a material, a polycrystalline aggregate is formed into a cylinder whose diameter is smaller than the diameter of the incident X-ray beam. Metal samples are machined to a desirable shape, plastic materials can often be extruded through suitable dies, and all other samples are best ground to a fine powder (200–300 mesh) and shaped into thin rods after mixture with a binder (usually collodion). The mount is shown in Fig. 10-24(b).

Although liquids cannot be identified directly, it is frequently possible to convert them to crystalline derivatives that have characteristic patterns. Many of the classical derivatives can be used: identification of aldehydes and ketones as 2,4-dinitrophenyl-hydrazones, fatty acids as p-bromoanilides, and amines as picrate derivatives.

In order to obtain diffraction patterns from large, dense samples, a back-reflection camera can be used. The geometry of the camera is shown in Fig. 10-25. The X rays pass through an opening in the center of the film and impinge on the sample. Beams diffracted over a range of Bragg angles extending from 59° to 88° are registered on the circular film.

Automatic Diffractometers

Results are achieved rapidly and with much better precision when automatic diffractometers are used to record diffraction data. A diffractometer to record data from powdered samples is built much like the one shown in Fig. 10-15. The X-ray tube furnishes the radiation directly (filters are generally used to get more nearly mono-chromatic radiation). The diffracting crystal is replaced by the powdered or metallic sample. To increase the randomness of orientation of the crystallites, the sample may be rotated in its own plane, that is, the plane perpendicular to the bisector of the angle between the source and detector beams. Note also that as the sample is rotated in the other plane to sweep through various θ angles, the detector must be rotated twice as rapidly to maintain the angle 2θ with the irradiating beam.

When used as detectors, scintillation and proportional counters, with their associated circuitry, are far superior to photographic film in regard to the number of reflections per day that can be recorded. They can achieve a precision of 1% or better. Even with the best darkroom and photometric procedures the relative degree of blackness of each spot or cone on the film cannot be estimated with an accuracy of much more than 10%, and often the error in estimation is greater than this. The principal advantage of photographic film over counters is that it provides a means of recording many reflections at one time.

Automatic single-crystal diffractometers are quite complex. A cradle assembly provides a wide angular range for orienting and aligning the crystal under study. A precision-diffractometer assembly allows the detector to transverse a spherical surface from longitude $-5°$ to $150°$ and from latitude $-6°$ to $60°$ on the Norelco instrument. The complete unit provides four rotational degrees of freedom for the crystal and two for the detector. Various crystal and counter angles are set on the basis of programmed information and the resulting diffraction intensity is measured as a function of angle. Single-crystal diffractometers are becoming widely used for gathering the necessary data for crystal structure determinations.

X-Ray Powder Data File

If only the identification of a powder sample is desired, its diffraction pattern is compared with diagrams of known substances until a match is obtained. This method requires that a library of standard films be available. Alternatively, d values calculated from the diffraction diagram of the unknown substance are compared with the d values of over 5000 entries, which are listed on plain cards, Keysort cards, and IBM cards in the X-ray powder data file.[12] An index volume is available with the file. The cataloging scheme[6] used to classify different cards lists the three most intense reflections in the upper left corner of each card. The cards are then arranged in sequence of decreasing d values of the most intense reflections, based on 100 for the most intense reflection observed. A typical card is shown in Fig. 10-26.

5-0628

d	2.82	1.99	1.63	3.258	NaCl			★
I/I_1	100	55	15	13	SODIUM CHLORIDE		HALITE	

Rad. Cu λ 1.5405 Filter Dia.	dÅ	I/I_1	hkl	dÅ	I/I_1	hkl
Cut off I/I_1	3.258	13	111			
Ref. Swanson and Fuyat, NBS Circular 539, Vol. II, 41 (1953)	2.821	100	200			
	1.994	55	220			
Sys. Cubic S.G. O_H^5 – Fm3m	1.701	2	311			
a_0 5.6402 b_0 c_0 A C	1.628	15	222			
α β γ Z 4 Dx 2.164	1.410	6	400			
Ref. Ibid.	1.294	1	331			
	1.261	11	420			
	1.1515	7	422			
$\varepsilon\alpha$ n $\omega\beta$ 1.542 $\varepsilon\gamma$ Sign	1.0855	1	511			
2V D mp Color	0.9969	2	440			
Ref. Ibid.	.9533	1	531			
	.9401	3	600			
An ACS reagent grade sample recrystallized twice from	.8917	4	620			
hydrochloric acid.	.8601	1	533			
X-ray pattern at 26°C.	.8503	3	622			
	.8141	2	444			
Replaces 1-0993, 1-0994, 2-0818						

Fig. 10-26. X-Ray data card for sodium chloride. (Courtesy of American Society for Testing Materials.)

To use the file to identify a sample containing one component, the d value for the darkest line of the unknown is looked up first in the index. Since more than one listing containing the first d value probably exists, the d values of the next two darkest lines are then matched against the values listed. Finally, the various cards involved are compared. A correct match requires that all the lines on the card and film agree. It is also good practice to derive the unit cell from the observed interplanar spacings and to compare it with that listed in the card.

If the unknown contains a mixture, each component must be identified individually. This is done by treating the list of d values as if they belonged to a single component. After a suitable match for one component is obtained, all the lines of the identified component are omitted from further consideration. The intensities of the remaining lines are rescaled by setting the strongest intensity equal to 100 and repeating the entire procedure.

Reexamination of the cards in the file is a continuing process in order to eliminate errors and remove deficiencies. Replacement cards for substances bear a star in the upper right corner.

X-Ray diffraction furnishes a rapid, accurate method for the identification of the crystalline phases present in a material. Sometimes it is the only method available for determining which of the possible polymorphic forms of a substance are present—for example, carbon in graphite or in diamond. Differentiation among various oxides—such as FeO, Fe_2O_3, and Fe_3O_4, or between materials present in such mixtures as $KBr + NaCl$, $KCl + NaBr$, or all four—is easily accomplished with X-ray diffraction, whereas chemical analysis would show only the ions present and not the actual state of combination. The presence of various hydrates is another possibility.

Quantitative Analysis

X-Ray diffraction is adaptable to quantitative applications because the intensities of the diffraction peaks of a given compound in a mixture are proportional to the fraction of the material in the mixture. However, direct comparison of the intensity of a diffraction peak in the pattern obtained from a mixture is fraught with difficulties. Corrections are frequently necessary for the differences in absorption coefficients between the compound being determined and the matrix. Preferred orientations must be avoided. Internal standards help but do not overcome the difficulties entirely.

Structural Applications

A discussion of the complete structural determination for a crystalline substance is beyond the scope of this text. It will suffice to point out that, with careful work, atoms can be located to a precision of hundredths of an angstrom unit or better.

In polymer chemistry a great deal of information can be obtained from an X-ray diffraction diagram. Fibers and partially oriented samples will show spotty diffraction patterns rather than uniform cones; the more oriented the specimen, the spottier the

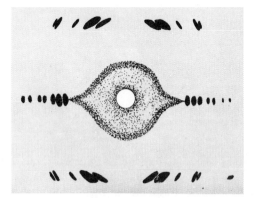

Fig. 10-27. Fiber diagram of polyethylene. (After A. Ryland, *J. Chem. Educ.*, **35**, 76 (1958). Courtesy of the *Journal of Chemical Education.*)

pattern. Figure 10-27 shows the fiber diagram of polyethylene. The center row of spots in the pattern is called the equator, and the horizontal rows parallel to the equator are called the layer lines. The equatorial spots arise by diffraction from lattice planes that are parallel to the fiber axis. The layer line spots arise by diffraction from planes that intersect the fiber axis. The repeat distance along the polymer chain can be calculated from the distances of the layer lines from the equator and their separation from one another. In the simplest cases the repeat distance will correspond to that of a fully extended chain of the known chemical composition.

Crystal Topography

There are a number of experimental diffraction techniques, developed in recent years, by which the microscopical defects in a crystal can be shown. Most crystals are far from perfect crystals and exhibit regions (grains) with somewhat differing orientations, or they may contain individual defects such as dislocations or faults distributed throughout the crystal. Studies of these defects are important in understanding the nature of stress in metals, the nature and behavior of "doped" crystals used in transistors, the production of "perfect" crystals, and other phenomena.

Microradiographic methods are based on absorption and the contrast in the images is due to differences in absorption coefficients from point to point. X-Ray diffraction topography depends for image contrast upon point to point changes in the direction or the intensity of beams diffracted by planes in the crystal.

One much used method of X-ray diffraction topography is known as the Berg-Barrett method. The experimental arrangement is shown in Fig. 10-28. The crystal is set so as

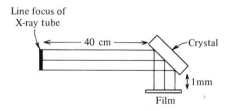

Line focus of
X-ray tube

40 cm

Crystal

1 mm

Film

Fig. 10-28. Experimental arrangement for Berg-Barrett method of X-ray diffraction topography.

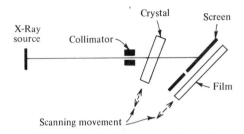

Fig. 10-29. Experimental arrangement for Lang method of X-ray diffraction topography.

to reflect the X-rays at the Bragg angle for some plane. Geometric resolutions of about 1 μm can be achieved and single dislocations can be resolved. The contrast on the film is due to variations of the reflecting power due to imperfections in the crystal.

Another method for X-ray diffraction topography is known as the Lang method. The experimental setup is shown in Fig. 10-29. A ribbon X-ray beam is collimated to such a small angular divergence that only one characteristic wavelength is diffracted by the crystal. Simultaneous movement of the crystal and film allow a large area of the crystal to be investigated.

X-RAY PHOTOELECTRON SPECTROSCOPY OR ESCA (ELECTRON SCATTERING FOR CHEMICAL ANALYSIS)

When X rays impinge upon matter, electrons may be ejected as has been described previously. If monoenergetic X rays are used, the kinetic energy of the ejected electrons, E_k, is determined by the difference between the binding energy of the electron, E_b, and the energy of the X-ray photon, $h\nu$. If, then, the kinetic energy of the ejected electrons is determined, the binding energy of the electrons can be evaluated by the relationship

$$E_b = h\nu - E_k \tag{10-18}$$

Since the binding energies of electrons in the outer region of the atomic core depend upon the chemical environment, this method, described by Siegbahn,[10] can be used to analyze several types of mixtures that would be difficult by other means. For example, Swartz and Hercules[9] have developed a quantitative measurement for MoO_2 and MoO_3 mixtures. There is a 1.7 eV increase in the binding energy of the $3d_{3/2}$ and $3d_{5/2}$ electrons ejected from the Mo(VI) oxide as compared to the Mo(IV) oxide. The $3d_{3/2}$ electrons have a binding energy about 3.1 eV higher than do $3d_{5/2}$ electrons.

In practice, the samples are irradiated by monoenergetic X rays and the emitted electrons are analyzed by passage through a double-focusing electron spectrometer similar to those used to study beta-ray energies. Swartz and Hercules[9] used the aluminum $K\alpha_1$ X-ray line at 1487 eV for their work on molybdenum oxides. A Bendix Channeltron Electron Multiplier detector was used to count the electrons. For the molybdenum oxide analyses, synthetic samples were used to obtain the calibration curve relating concentration of MoO_2 to the ratio of counts of $Mo(3d_{3/2})$ electrons in MoO_3 to $Mo(3d_{5/2})$ electrons in MoO_2.

The ESCA method of analysis is primarily suitable for surface analyses because the ejected electrons are easily stopped by even a minute thickness of solid. Thus the analysis is characteristic of the top few monolayers of a surface. All elements except hydrogen may be identified, and the oxidation state and bonding of the element is usually determinable. ESCA has been used to study the changes responsible for the poisoning of catalysts, the surface contaminants on semiconductors which may cause poor performance, surface reactions on metals, characterization of textile fiber surfaces, and many other surface phenomena including analysis of human skin composition.

LABORATORY WORK

General Instructions for Operation of X-Ray Diffraction Units

1. Read the instruction manual furnished by the manufacturer, if available. The general procedure for most X-ray units is outlined below.
2. Make sure that all unused ports are covered with lead shields and that the proper filter is in place on all ports to be used.
3. Mount and align the cameras on all ports to be used.
4. Turn on the main line switch (a key switch on some instruments).
5. Set the desired exposure on the timer.
6. Make sure that the overload relay is in the ON position.
7. Press the LINE ON button. Water should now flow through the X-ray unit.
8. Press the X-ray ON button. After a time-delay in the instrument operates, current should flow through the high-voltage circuit and be indicated on the milliammeter.
9. Adjust the voltage to the desired setting with the voltage control. For copper tubes, 30 kV is recommended, and for molybdenum tubes, 45 kV is recommended.
10. Adjust the tube current with the current control. For copper targets, 15 mA is usually suitable, and for molybdenum target tubes, 19 mA is often employed.
11. When shutting the apparatus off, follow the reverse procedure of turning the apparatus on. Turn voltage and current controls to their lowest limit and be sure these controls are in this position before starting the apparatus again.

Experiment 10-1   Identification of Substances by the Powder Diffraction Method

1. A very small sample of a crystalline material will be furnished. The sample must be very finely powdered (finer than 200 mesh) if it is not already so.
2. Prepare a small-diameter, thin-walled melting-point tube (diameter 0.7 mm or less) and fill it with the powdered sample. Place the tube in the chuck at the center of the powder camera and fix it in place with a drop of wax. Line the sample tube up so

that it rotates without wobbling.   Alternatively, the sample may be coated on the outside of a very fine glass rod, using collodion or vaseline or other noncrystalline material to stick it on.

3. Take the camera into the darkroom and load it with film of the proper size. A punch is available for perforating the film so that the collimator and beam trap can be inserted into the camera. Insert the collimator and beam trap and place the cover on the camera.

4. Place the camera on the X-ray unit. An exposure of 20–30 min is recommended for a first trial. If this results in too light or too dark an exposure, estimate the required time for another exposure and repeat the whole procedure.

5. Remove the film in the darkroom. Develop for 4 min in X-ray developer and fix for twice the time required for the film to clear. Wash thoroughly and dry.

6. Measure the distance between lines of the film, using the film measuring device. From the radius of the camera calculate the interplanar spacings creating the observed lines.

   If the distance between corresponding arcs of the same cone of diffracted rays—for example, the distance between A and B of Fig. 10-21—is measured and called S, then:

$$\frac{S}{R} = 4\theta_{rad}$$

where R = radius of camera, and θ_{rad} is the angle of incidence, measured in radians. The angle, θ_{deg}, measured in degrees, is then:

$$\theta_{deg} = \frac{S}{4R} \times 57.295$$

Eq. 10-3 can then be used to calculate the spacing d, using the values of λ for $K\alpha_1$ from Table 10-1. The Geological Survey has prepared sets of tables[8] which give the d spacing directly if copper or iron target X-ray tubes are used and if the angle 2θ is calculated. These tables are very convenient to use.

7. For precise measurements of the camera radius and for compensation for film shrinkage during development, a pattern of sodium chloride should be taken. The main spacings in the sodium chloride pattern are 2.821, 1.99, 1.63, and 1.260 Å.

8. Estimate the relative intensity of the lines produced on the film. Refer to the A.S.T.M. X-ray tables of compounds to identify the unknown. Turn in all calculations, films, and the identification of the unknown to the instructor.

PROBLEMS

1. What is the short-wavelength limit for a 100-kV X-ray tube? What is the atomic number of the element for which just insufficient energy is available for excitation?

2. Write the transition relations for each of these lines: (a) $K\alpha_2$, (b) $K\beta_1$, and (c) $L\alpha_1$.

3. Calculate the critical excitation potentials for the K and L series of these elements:

Element	K Absorption Edge, Å	L_{III} Edge, Å
Al	7.951	170
Cr	2.070	20.7
Zr	0.688	5.58
Nd	0.285	1.995
W	0.178	1.215
U	0.107	0.722

4. Calculate the wavelengths of the $K\alpha_1$ lines for the elements in Problem 3.

5. For what elements will Mo $K\alpha_1$ prove sufficiently energetic to excite their L_{III} spectra? What is the limit for K spectra?

6. What causes the discontinuity in the efficiency curve of the NaI scintillation counter and of the argon-filled flow proportional counter (Fig. 10-7)?

7. Compute the goniometer setting (2θ) for each of the following emission lines when the analyzing crystal is (a) LiF, (b) CaF_2, (c) EDDT: The $K\alpha_1$ lines of Al, S, Ca, Cr, Mn, Co, Br, Sr Ag, Mo, W.

8. Discuss four ways that might be employed for the separation of interfering spectral lines.

9. For the determination of uranium in aluminum by measurement of U $L\alpha_1$ fluorescence, these counting relationships were obtained:

U (wt %)	Count Rate, cps
2	436
5	835
10	1262
15	1533
20	1720

Determine the slope of the calibration curve (cps/% U) over each interval of uranium concentration and the counting time (in minutes) required to achieve a 95% confidence level for 1% precision.

10. Sulfur (0.4–6.0%) has been determined in carbon materials by X-ray fluorescence using the S $K\alpha$ line (5.36Å) which under a particular set of operating conditions gave 1 cps equivalent to 0.014% S. Background radiation is equivalent to 0.05% S. Select a proper analyzing crystal, the goniometer setting, the excitation conditions (X-ray tube voltage), and counting times to achieve results whose deviation does not exceed 5% at 95% confidence level.

11. What is the reflection angle (2θ) for Cu $K\alpha_1$ radiation from each analyzing crystal in Table 10-4?

12. The pulse amplitude distributions for these elements: Mg, Al, Si, P, S, and Ca, show a peak of pulse distribution at 11.0, 13.0, 15.0, 17.4, 19.0, and 31.8 V, respectively. For each the width at one-half peak height is 2.5 V, and the base width is 9.5 V. What base line and window settings (in volts) would be employed in these situations?
   (a) The determination of Mg in the presence of P.
   (b) The determination of S in the presence of Mg.
   (c) The determination of P in the presence of Al.
   (d) The separation of calcium from all the others.
   (e) The total Al plus Si in a sample, all other elements absent.

13. Suggest methods for handling each pair of overlapping X-ray spectral lines whose wavelength in angstroms are enclosed in brackets:
   (a) Mn $K\alpha_1$ [2.103] – Cr $K\beta$ [2.085]
   (b) Zn $K\alpha_1$ [1.435] – Re $L\alpha_1$ [1.433]
   (c) Nb $K\alpha_1$ [0.746] – W $L\alpha_1$ [1.476]

14. Graphically represent the following disturbing effects in the use of an internal standard element (S) for the determination of element (E), the disturbing element being (D). Plot all absorption edges and emission lines. (a) Selective absorption of "S", (b) selective absorption of "E", (c) enhancement of "S", and (d) enhancement of "E".

15. Strontium has been determined in sediments of oil-bearing formations using yttrium as an internal standard. The spectral characteristics of these elements are:

$$Sr\ K\alpha_1, 0.877;\quad K\beta, 0.783;\quad K\ \text{edge}, 0.770$$
$$Y\ K\alpha_1, 0.831;\quad K\beta, 0.740;\quad K\ \text{edge}, 0.727$$

(a) Compute the critical voltage for each element. (b) Using a LiF crystal, compute the Bragg angle (2θ) for the emission lines. (c) Calibration data obtained is tabulated:

Sr, wt. %	Measurement of Y $K\alpha_1$; Time (sec) for 6400 counts	Measurement of Sr $K\alpha_1$; Time (sec) for 6400 counts
0.0000	41.1	80.1
0.1000	40.1	60.4
0.2000	40.2	49.5
0.3000	40.0	41.6
0.4000	42.4	38.3

Plot intensity ratio vs. concentration of strontium. Unknown samples gave these intensity ratios—Sr : Y—A, 0.8860; B, 0.7802; C, 0.6011.

16. What is the difference in wavelength (and Bragg angle) between Hf $L\alpha_1$ and the second order of Zr $K\alpha_1$, 1.566 Å and 0.784 Å (in first order, respectively) with a LiF analyzing crystal? Suggest a method for eliminating the second-order Zr line.

17. Oil paintings have been tested for authenticity by examining the pigment composition with the electron probe and a hypodermic needle core. With an ammonium dihydrogen phosphate crystal (101 plane), $d = 10.62$, the following characteristic X-ray spectra (and relative intensity) were obtained:

Top White Layer, 2θ	Bottom White Layer, 2θ
27.6° (8)	34.0° (12)
30.0° (60)	36.0° (100)
58.0° (4)	71.0° (2)
63.0° (18)	

Was the painting produced before or after the time when titanium-white pigments were available?

18. From the data given in Table 10-3, estimate the mass absorption coefficients (a) of Zr for the Mo $K\alpha_1$ radiation, (b) of Ni for the Cu $K\alpha_1$ radiation, and (c) of Cu for the W $L\alpha_1$ radiation.

19. Calculate the reduction in intensity of an X-ray beam from Mo $K\alpha_1$ line (0.709 Å) resulting from 1 ml of 1% TEL liquid in n-octane. The aviation mixture consists of 61.5% $Pb(C_2H_5)_4$ and 38.5% $C_2H_4Br_2$ per milliliter of TEL. The mass absorption coefficients are C, 0.64 cm^2/g; H, 0.38 cm^2/g; Pb, 140 cm^2/g; and Br, 79 cm^2/g. Density = 0.72 g cm^{-3}.

20. Repeat Problem 19 using the Cu $K\alpha_1$ line (1.542 Å). The mass absorption coefficients are C, 4.6 cm^2/g; H, 0.43 cm^2/g; Pb, 241 cm^2/g; and Br, 88 cm^2/g. Note the increase in absorption by lead, but also by the carbon atoms.

21. Identify the emission lines and, from these, the base wire and plate metal in each sample. Spectrometer employed a LiF crystal and tungsten target operated at 50 kV.

Sample	Plate Metal, 2θ	Base Wire, 2θ
1	48.64°	41.30° (2nd)
2	99.87° (2nd)	63.88° (3rd)
3	31.19°	110.92° (2nd)
4	16.75°	110.92° (2nd)

22. A sample ground and pelleted with lithium carbonate and starch gave these emission lines (2θ) using a LiF crystal. Identify each line.

$$111.0°, \ 100.2°, \ 57.8°, \ 48.8°, \ 45.1°, \ 44.0°, \ 38.0°$$

23. Suggest an X-ray method for each of these determinations: (a) the thickness of electroplated metal films, such as successive layers of Cu, Ni, and Cr on steel (chrome plate), (b) the thickness of SrO and BaO on evaporated electrode coatings, and (c) the concentration of fillers and impregnants, such as BaF_2 in carbon brushes.

24. Iron-55 decays by K-electron capture to stable ^{55}Mn with the attendant emission of K-line X rays of Mn. The half-life of ^{55}Fe is 2.93 years. For which elements would this isotope be a convenient source of X radiation for absorption analysis?

25. An unknown material was placed in the sample holder of an X-ray fluorescence unit which used a tungsten target tube operated at 60 kV to furnish the exciting radiation. A mica crystal was used in the analyzer. The lattice spacings of mica is 9.948 Å. Reflections were observed at angles (2θ) of 9°34′, 12°8′, 19°12′, 24°24′, and 38°58′. Calculate the wavelength of the fluorescent lines and identify the elements present.

26. An unknown powder was placed in a sample tube, and the X-ray diffraction pattern was observed in a camera of radius 57.3 mm. The X-ray unit was fitted with a nickel target tube with a cobalt filter. The distances between corresponding arcs of the three strongest lines observed on the developed film were measured as 77.5, 89.9, and 130.4 mm. These lines seemed to the eye to be of about equal intensity. Calculate the spacings d of the crystal in angstrom units and identify the substance by reference to the A.S.T.M. card file.

27. In the case of polyethylene, the repeat distance obtained from the X-ray diffraction pattern of a fiber diagram was 2.54 Å. What type of chemical structure is implied?

BIBLIOGRAPHY

Bertin, E. P., *Principles and Practice of X-ray Spectrometric Analysis*, Plenum, New York, 1970.

Birks, L. S., *X-ray Spectrochemical Analysis*, Wiley-Interscience, New York, 1959.

Birks, L. S., *Electron Probe Microanalysis*, Wiley-Interscience, New York, 1963.

Buerger, M. J., *X-ray Crystallography*, Wiley, New York, 1942.

Bunn, C. W., *Chemical Crystallography*, 2nd ed., Oxford University Press, New York, 1961.

Clark, G. L., *Applied X-rays*, 4th ed., McGraw-Hill, New York, 1955.

Henke, B. L., J. B. Newkirk, and G. R. Mallett, Eds., *Advances in X-ray Analysis*, Vol. 13, Plenum, New York, 1970, and other volumes in this series.

Liebhafsky, H. A., H. G. Pfeiffer, E. H. Winslow, and P. D. Zemany, *X-ray Absorption and Emission in Analytical Chemistry*, Wiley, New York, 1960.

Liebhafsky, H. A., H. G. Pfeiffer, and E. H. Winslow, *X-Ray Methods: Absorption, Diffraction and Emission*, in *Treatise on Analytical Chemistry*, Vol. 5, Part I, I. M. Kolthoff, and P. J. Elving, Eds., Wiley-Interscience, New York, 1964, Chapter 60.

Robertson, J. M., *Organic Crystals and Molecules*, Cornell University Press, Ithaca, New York, 1953.

Sproull, W. T., *X-rays in Practice*, McGraw-Hill, New York, 1946.

Wittry, D. B., X-Ray Microanalysis by Means of Electron Probes, in *Treatise on Analytical Chemistry*, Vol. 5, Part I, I. M. Kolthoff, and P. J. Elving, Eds., Wiley-Interscience, New York , 1964, Chapter 61.

LITERATURE CITED

1. Birks, L. S., E. J. Brooks, and H. Friedman, *Anal. Chem.*, **25**, 692 (1953).
2. Bragg, W. L., *The Crystalline State*, Macmillan, New York, 1933.
3. Castaing, R., Thesis, University of Paris, 1951.
4. Clark, G. L. and S. T. Gross, *Ind. Eng. Chem.*, *Anal. Ed.*, **14**, 676 (1942).
5. Duane, W. and F. L. Hunt, *Phys. Rev.*, **66**, 166 (1915).
6. Hanawalt, J. D., H. W. Rinn, and L. K. Frevel, *Ind. Eng. Chem.*, *Anal. Ed.*, **10**, 457 (1938).
7. Ievins, A. and M. E. Straumanis, *Z. Krist.*, **94A**, 40, 48 (1936); *see also* M. E. Straumanis, *Anal. Chem.*, **25**, 700 (1953).
8. Jack, J. J. and D. M. Hercules, *Anal. Chem.*, **43**, 729 (1971).
9. Swartz, W. E., Jr., and D. M. Hercules, *Anal. Chem.*, **43**, 1774 (1971).
10. Siegbahn, K. et al., *ESCA: Atomic, Molecular and Solid State Structure Studied by Means of Electron Spectroscopy*, Almquist and Wiksells, Uppsala, 1967.
11. Switzer, G., J. M. Axelrod, M. L. Lindberg, and E. S. Larsen, "Tables of Spacings for Angle 2θ, Cu $K\alpha_1$, Cu $K\alpha_2$, Fe $K\alpha$, Fe $K\alpha_1$, Fe $K\alpha_2$," Circular 29, Geological Survey, U. S. Dept. of Interior, Washington, D. C., 1948; "Tables for Conversion of X-ray Diffraction Angles to Interplanar Spacings," Publication AMS 10, Government Printing Office, Washington, D. C.
12. "Alphabetical and Grouped Numerical Index of X-ray Diffraction Data," *Am. Soc. Testing Materials, Spec. Tech. Publ.* 48E (1955).

11

Radiochemical Methods

Until the advent of the cyclotron, and more recently the chain-reacting pile, most of the work with radionuclides was with the heavy, naturally occurring radioactive elements. It is now possible to obtain artificially produced radionuclides of most of the elements and to obtain many of these in large quantities and with extremely high activity. Quite naturally the availability of radionuclides has provided great impetus to their use.

NUCLEAR REACTIONS AND RADIATIONS

A radionuclide is characterized by its half-life, the type of transition involved when it decays, and the type and energy of the radiation emitted. Such information is essential for the recognition and understanding of the problems associated with the measurement of radionuclides.

Particles Emitted in Radioactive Decay

The heavy, naturally occurring radioactive elements, such as thorium, uranium, and the like, emit, among other products, doubly ionized helium particles known as *alpha particles.* Alpha particles have only a slight penetrating power, being stopped by thin sheets of solid materials and penetrating only 5–7 cm of air. Their energies, however, are generally very high, and may exceed 10 MeV (million electron volts). As a result, the ionizing power of an alpha particle is high; that is, on passing through material, a large number of ion pairs are produced along the linear path traversed by an alpha particle. For example, a 5-MeV alpha particle, which is stopped in 3.5 cm of air, produces about 25,000 ion pairs per centimeter of travel. Due to the greater ionizing power of alpha particles, they can generally be distinguished from beta or gamma radiation on the basis of pulse amplitude. The ionization chamber is the preferred detector. Alpha-particle activity can be measured in the presence of considerable beta activity by first measuring the total activity, then interposing a filter to absorb the alpha particles, and measuring the beta activity.

A radioactive element formed by neutron capture usually has a higher neutron/proton ratio than its stable isobars, and therefore it frequently decays by beta particle emission. A *beta particle* is a very energetic electron or positron. Beta particle spectra are continuous. Few have the energy, E_{max}, which is the upper limit of the spectrum corresponding to the transition energy, often expressed as penetrating power or range. A 0.5-MeV beta particle has a range of 1 m in air and produces about 60 ion pairs per centimeter of its path. Above 0.4 MeV beta particles have sufficient energy to penetrate the windows of most counting devices, and measurement is not difficult. Below this energy value, however, special techniques are required. Very thin-window counters may be employed; the sample may be introduced directly into the active volume of the counter; or it may be dissolved in a liquid scintillator. Because corrections for self-absorption, self-scattering, and back-scattering are necessary but difficult to obtain, beta particle counting should be avoided when possible.

Gamma rays, actually high-energy photons, are monoenergetic, and gamma-ray spectra consist of discrete lines. The penetrating power of gamma radiation is much greater than that of either alpha or beta particles, but the ionizing power is less. The sensitivity of the detector must be increased by using longer chambers, by gas fillings under pressure that possess higher atomic number, or by thicker scintillator material. A filter of sufficient thickness to absorb all beta particles, and inserted between the sample and detector, permits the measurement of gamma radiation exclusively from a mixture of activities.

A long-lived positron-emitting nucleus may decay by capturing one of its own orbital K electrons; this is called *internal conversion*. The excess energy is emitted as a gamma ray, and the resulting ion with a vacant K orbital then emits X radiation characteristic of the new element.

Interaction of Nuclear Radiation with Matter

Since the radiations from radionuclides are detected by means of their interactions with matter, a brief summary of these modes of interaction is presented here.

Beta particles interact primarily with the electrons in the material traversed by the particle. The molecules may be dissociated, excited, or ionized. It is the ionization, however, which is of primary interest in the detection of beta particles. As a beta particle slows down while moving through matter, the specific ionization, that is, the number of ion pairs produced per unit track length, increases and reaches a maximum near the end of the track. On the average, each ion pair produced represents a loss of 35 eV. Actually, the beta particle may lose a large part of its energy in a single interaction, but if it does, the ions produced have so much excess energy that they in turn produce additional ion pairs. The absorption of beta particles in matter follows approximately the exponential relation given by Eq. 11-2 for gamma radiation, up to a certain thickness where the absorption finally exceeds that predicted by the exponential law and soon becomes infinite. This maximum thickness is known as the range of the beta particle (Fig. 11-1). For energies 0.5 to 3 MeV, the following range-energy relationship gives the maximum energy of the beta particle:

$$\text{range (in mg cm}^{-2} \text{ of absorber)} = 520 \, E_{max} - 90 \qquad (11\text{-}1)$$

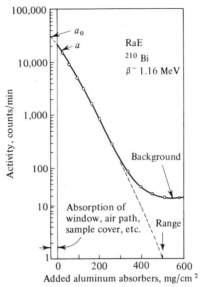

Fig. 11-1. Absorption of beta particles.

or, approximately $\frac{1}{2} E_{max}$ in MeV. In lower energy regions it is best to use a range-energy curve.

Whenever a beta particle comes close to a nucleus, it may have its direction of travel greatly changed, so that the path of a single beta particle in traversing matter may be a very devious one. Scattering due to the material supporting the radionuclide, called *back-scattering*, increases the lower the energy of the beta particle and the higher the atomic number of the scatterer. Increasing the thickness of the support up to 0.2 range of a beta particle increases the back-scattering factor to a maximum value. Lead walls and doors of counting and shielding equipment should be lined with Lucite. Self-absorption and self-scattering also occur with beta particles. For very thin sources, 0.1-0.2 mg cm^{-2}, both factors are negligible—this situation prevails with carrier-free sources mounted on thin plastic film. For thicker samples, the same amount of inert carrier must be used or a correction curve prepared for different quantities of carrier.

The continuous X radiations produced when electrons are decelerated in the coulomb fields of atomic nuclei are called *bremsstrahlung*. This type of radiation is produced whenever fast electrons pass through matter; the efficiency of the conversion of kinetic energy into bremsstrahlung goes up with increasing electron energy and with increasing atomic number of the material.

Alpha particles lose energy by the same mechanisms as beta particles. However, owing to their large relative mass and higher charge, the specific ionization is much larger than for beta particles. Because the alpha particles are emitted from the nucleus with discrete energies and because the alpha particles lose only a small fraction of energy in collision with electrons, nearly all the alpha particles from a given nucleus will traverse the same distance in an absorber. Thus a determination of the range of alpha particles is a simple matter, and the energy of the alpha particles can be found by reference to plots.

Although the range-energy curve is not linear, a useful rule of thumb is that for alpha emitters between 5 and 8 MeV in energy, the range in air is approximately equal numerically to the energy in MeV.

Gamma rays lose energy on passage through matter in three ways: by the *photoelectric effect*, by the *Compton effect*, and by *pair production*. The photoelectric effect is important for heavy absorbing elements and for low gamma-ray energies. In this process gamma rays are absorbed by inner electrons bound in an atom, and the energy carried by the gamma ray is transferred completely to the electron with the resultant ejection of the electron from the atom (compare X-ray absorption). The Compton effect consists of a gamma ray interacting with an electron and transferring part of its energy to the electron. The electron is ejected from the atom and a new photon of lower energy proceeds from the collision in an altered direction. The Compton effect is important with light target elements and with gamma rays possessing energies less than 3 MeV. Pair production of a positron and an electron results when a high-energy gamma ray is annihilated following interaction with the nucleus of a heavy atom. Such a process is important with heavy elements and gamma rays of energies greater than 1.02 MeV, the energy corresponding to twice the rest mass of an electron. Conversely, when a positron and electron meet, the two are annihilated, and two gamma rays with energies of 0.51 MeV each are produced. Some of these spectral features are illustrated later in Fig. 11-12.

Each ion pair which results from the passage of radiation through air represents an average energy loss of about 35 eV. The number of ion pairs per centimeter of travel is known as the *specific ionization*. In the case of gamma rays, the ion pairs formed are almost entirely produced by secondary processes—the photoelectrons, the Compton electrons, and the positrons and electrons. The attenuation of gamma radiation is given by

$$P = P_0 e^{-\mu x} \tag{11-2}$$

where P is the radiant power of P_0 which is transmitted through an absorber of thickness x, and μ is the linear absorption coefficient. Often the energy of a gamma ray is expressed as the thickness of absorber required to diminish P to $\frac{1}{2} P_0$; this half-thickness is given by

$$x_{1/2} = \frac{0.693}{\mu} \tag{11-3}$$

Absorber thicknesses are frequently given in units of surface density, g/cm^2; that is, ρx where ρ is the density of the absorber. Equation 11-2 becomes

$$P = P_0 e^{-(\mu/\rho)\rho x} \tag{11-4}$$

where μ/ρ is known as the mass absorption coefficient.

Neutrons do not carry a charge and therefore they do not interact with electrons. Because of their lack of charge, neutrons can more readily enter the nucleus of an absorber element, and it is by secondary reactions that the neutron is detected and measured; for example,

$$^{10}B + n \rightarrow {}^7Li + \alpha \tag{11-5}$$

It is the alpha particle which is actually detected.

Radioactive Decay

The decay of a radionuclide follows the well-known first-order rate law, which may be written in differential form as follows:

$$\frac{dN}{dt} = -\lambda N \tag{11-6}$$

where N is the number of radionuclides remaining at time t, and λ is the characteristic decay constant (in time^{-1}). The activity A is related to N by the equation

$$A = \lambda N \tag{11-7}$$

and is usually the quantity observed or computed. The rate equation may be integrated to yield

$$A = A_0 e^{-\lambda t} \tag{11-8}$$

or

$$\ln A = \ln A_0 - \lambda t \tag{11-9}$$

where A_0 is the activity at zero time.

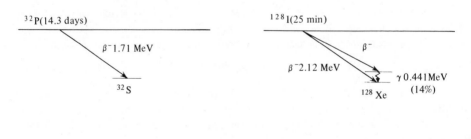

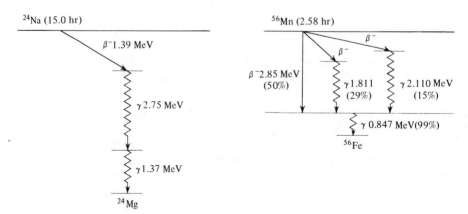

Fig. 11-2. Some decay schemes of radioisotopes.

The time required for one-half of the radioactive material to decay—the *half-life* of the radionuclide, is generally used in describing radioactive emitters, namely,

$$t_{1/2} = \frac{1}{\lambda} \ln \frac{A}{A/2} = \frac{0.693}{\lambda}$$

(11-10)

The half-life of the radioactive element is that time required for half of the radioactive nuclei of the element to disintegrate and release its energy. For example, strontium-90 has a half-life of approximately 30 years. In 30 years half of the nuclei will disintegrate and release their energy, in another 30 years half of the remainder of the nuclei will decay. This process continues such that every 30 years half of the remainder of the element decays and the other half remains stable for future decay. After 10 half-lives only 0.1% of the radioactive nuclei remain. An accurate knowledge of the characteristic decay constant (λ) is essential when working with short-lived radionuclides in order to correct for the decay while the experiment is in progress. Selected decay schemes are shown in Fig. 11-2.

Units of Radioactivity

Activity is expressed in terms of the curie, where 1 Ci is 3.700×10^{10} disintegrations per second (dps). Specific activity, the activity per unit quantity of radioactive sample, is expressed in a variety of ways by dps per unit weight or volume, or in units such as microcurie or millicurie per milliliter, per gram, or per millimole. The last is preferable for labeled compounds.

MEASUREMENT OF RADIOACTIVITY

Radiation from radionuclides can be detected and measured in many ways. The best method to employ in any particular situation depends upon the nature of the radiation and the energy of the radiation or particles involved.

Photographic Emulsion

Any ionizing particle or radiation will cause activation and, on development, darkening of a photographic plate. The intensity of the blackening can be used to measure the activity, but better methods are available. The photographic plate is useful, however, in studies of the distribution of radioactive material in a thin section of a substance—autoradiography. A slice of tissue, for example, when placed on a plate, would cause blackening at the places where uptake of a radionuclide had occurred and thus would indicate the distribution of the tracer in the biological material. Film badges are used to measure the total exposure of workers to radiation.

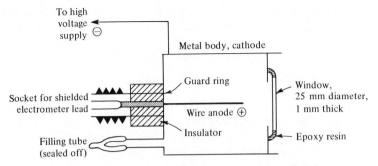

Fig. 11-3. Schematic diagram of an ionization chamber.

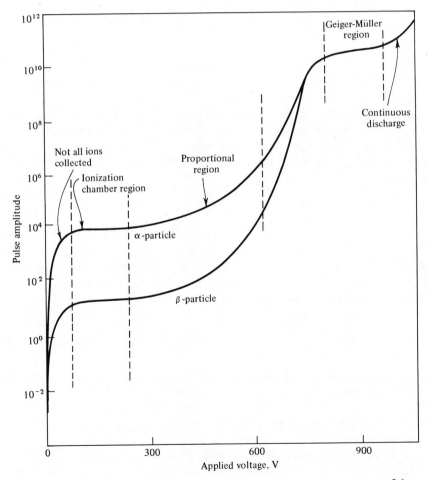

Fig. 11-4. Pulse amplitude as a function of applied voltage for ionization type of detectors.

The Ionization Chamber

In the *ionization chamber* (Fig. 11-3) an electric field is applied between two electrodes across a volume of gas—air for alpha particles, krypton or xenon under pressure for gamma radiation. The potential across the electrodes is adjusted to minimize recombination of the ion pairs without causing gas amplification (Fig. 11-4). An ionization chamber is an accurate, quick-acting detector even for very weak radiation. The sample is placed outside the window or inserted in a well extending into the chamber volume. For each ionizing event *pulse ion chambers* produce an electronic pulse proportional to the number of electrons released by the ionizing radiation. *Current ion chambers* integrate the events and provide a dc current.

The Geiger Counter

The Geiger counter, also called a Geiger-Müller or G-M tube, is shown schematically in Fig. 11-5. A potential of 800-2500 V is applied to a central wire anode surrounded by a cylindrical cathode—a glass wall which has been silvered or a brass cylinder. The two electrodes are enclosed in a gas-tight envelope typically filled to a pressure of 80 mm of argon gas plus 20 mm of methane or ethanol or 0.1% of chlorine. A thin end-window of mica, about 2.5 cm in diameter and 2–3 mg/cm^2 in thickness, or a glass wall in dipping counters, is the point of entry of the radiation.

When an ionizing particle enters the active volume of the Geiger counter, collision with the filling gas produces an ion pair. This is followed by migration of these particles

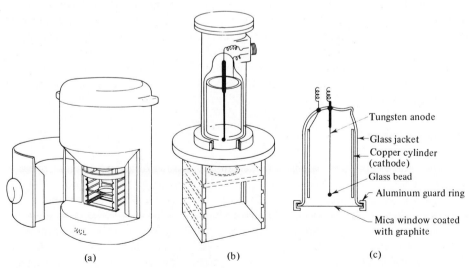

Tungsten anode
Glass jacket
Copper cylinder (cathode)
Glass bead
Aluminum guard ring
Mica window coated with graphite

(a)   (b)   (c)

Fig. 11-5. (a) End-window type of Geiger counter. (b) Counter and sample holder with shielding removed. (c) Schematic of counter.

toward the appropriate electrodes under the voltage gradient. The mobility of the electron is quite high, and under the influence of the potential gradient, it soon acquires sufficient velocity to produce a new pair of ions upon collision with another atom of argon. Under these conditions, which are repeated many times, each original ionizing particle entering the active volume of the counter gives rise to an avalanche of electrons traveling toward the central anode. Photons, emitted when the electrons strike the anode, spread the ionization throughout the tube. These processes produce a continuous discharge which fills the whole active volume of the counter in less than a microsecond. Each discharge builds up to a constant pulse of maximum amplitude 10 V and 50–100 μsec duration. These pulses can be counted precisely with the aid of scaling circuits or a ratemeter with no intermediate amplification; this is the principal advantage of the Geiger counter.

During the time the electron avalanche is collected on the anode, the positive ions, being much heavier, have progressed only a short distance on their way to the cathode. Their travel time is about 200 μsec and during most of this time their presence as a virtual sheath around the anode effectively lowers the potential gradient to a point where the counter is insensitive to the entry of further ionizing particles—the *dead time* of the counter. Because the halogen or organic gas molecules have a lower ionization potential than argon, after a few collisions the ions moving toward the cathode consist only of these entities. In contrast to argon ions, these positive ions do not produce photons when neutralized at the cathode. Consequently, photons which could initiate a fresh discharge are prevented from forming and the counter is self-quenching. Upon being neutralized, the organic filling gas dissociates to various molecular fragments and, eventually, the quenching agent is exhausted. Counter life is limited to about 10^{10} counts. Because chlorine atoms merely recombine, the quencher is available for further use. A halogen-quenched counter has a life in excess of 10^{13} counts.

Counting rates are limited to about 15,000 counts/min because of the long dead time of 200–270 μsec and the large reduction in true count rate as the maximum count rate is approached. Sensitivity for beta particles is excellent, but for gamma radiation the sensitivity is less than that of the scintillation counter. There is no possibility of using pulse height discrimination with a Geiger counter because the pulses are all the same amplitude, nor is there any practical correction for coincidence losses when scanning across a spectrum.

Proportional Counters

When the electric field strength at the center electrode of an ionization chamber is increased above the saturation level, but under that of the Geiger region (Fig. 11-4), the size of the output pulse from the chamber starts to increase but is still proportional to the initial ionization. A device operated in such a fashion is called a *proportional counter*.

Pulse formation is identical with that described for Geiger counters, but gas amplification is approximately 1000 × less. Consequently, a preamplifier (× 10) is needed and is mounted together with the detector to avoid reduction in pulse size through capacitance in connecting cables. In the proportional region few, if any, photons are released. Consequently, the total number of secondary electrons is proportional to the number of

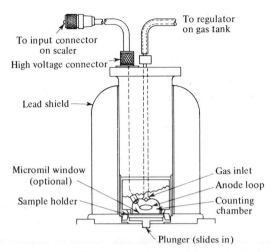

To regulator
on gas tank

To input connector
on scaler

High voltage connector

Lead shield

Micromil window
(optional)

Gas inlet

Anode loop

Sample holder

Counting
chamber

Plunger (slides in)

Fig. 11-6. Schematic diagram of a flow counter mounted in a shield; the very thin window is optional. The sample is inserted into the active volume by means of the lateral slide holder. (Courtesy of Nuclear-Chicago Corporation.)

primary ion pairs produced by the original ionizing particle. Furthermore, the discharge is limited to the immediate environment of the entering ionizing particle and the path traversed by the ion pair plus their secondary electrons and positive ions. The dead time is thus very short, about 0.25 μsec. Multiplication factors from 10 to 10^5 are possible; they are dependent on applied voltage, gas pressures, and counter dimensions.

The sample may be placed outside a sealed counter or inside a windowless counter, thus avoiding losses due to window absorption (Fig. 11-6). With the latter type of counter the chamber is purged with a rapid flow of counter gas, often 10% methane in argon, and a steady flow of gas is maintained during counting. Counter life is virtually unlimited since the filling gas is constantly undergoing replenishment. Such a counter is particularly suited for distinguishing and counting low-energy alpha and beta particles.

Proportional counters are useful for counting at extremely high counting rates—50,000–200,000 counts/sec; the upper limit is imposed by the associated electronic circuitry. The signal produced is extremely small and requires both a preamplifier and a second stage of amplification before the signal can be fed to a scaler. Excellent plateaus of about 100 V can be obtained whose slopes are as low as 0.1% counting-rate change per 100 V (whereas values of less than 1% variation are uncommon with Geiger counters). Proportional counters are useful for counting alpha particles, since their ionizing power is so much greater than that of beta particles that one can distinguish the two.

Scintillation Counting

Scintillators are chemicals used to convert radioactive energy into light. When an ionizing particle is absorbed in any one of several transparent scintillators, some of the energy acquired by the scintillator is emitted as a pulse of visible or near-ultraviolet light. The light is observed by a photomultiplier tube, either directly or through an internally reflecting light pipe. The combination of a scintillator and photomultiplier tube is called a *scintillation counter* (Fig. 11-7). A good match should exist between the emission spectrum of the scintillator and the response curve of the photocathode. The decay time for scintillators is very short; 250 nsec for a sodium iodide crystal, 20 nsec for anthracene, and 10 nsec for liquid organic systems. The signal from a scintillation counter is proportional to the energy dissipated by the radiation in the scintillator so that this counter may be used with pulse height discrimination.

For counting alpha particles the best scintillator is a thin layer of silver-activated zinc sulfide, which may be coated on the envelope of the photomultiplier tube.

Scintillation crystals of anthracene or stilbene (wavelength of emission: 4450 and 4100 Å, respectively) affixed by a good optical liquid to an end-window photomultiplier tube are suitable for beta particles of moderate and high energy. However, organic liquid scintillators are often preferred because of their shorter decay times. Low-energy beta emitters, such as ^3H, ^{14}C, and ^{35}S, are commonly counted by dissolving the compound containing the radionuclide in the liquid scintillator. The bulk solvent may be either an alkylbenzene, such as toluene or xylene, or an aliphatic ether, such as 1,4-dioxane, to which is added 4–10 g/liter of a primary scintillator plus 0.1 g/liter of a secondary scintillator, the latter often acting as a wavelength shifter. Typical primary scintillators are 2,5-diphenyloxazole (PPO) and 2-(4'-t-butylphenyl)-5-(4"-biphenyl)-1,3,4-oxadiazole (butyl PBD). Two common secondary scintillators are 1,4-bis-(5-phenyloxazol-2-yl)-

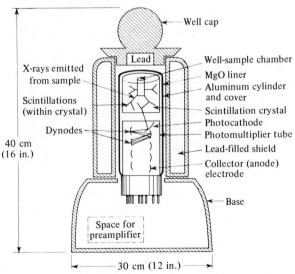

Fig. 11-7. Well-crystal scintillation counter and shield.

benzene (POPOP) and 1,4-bis-(4-methyl-5-phenyloxazol-2-yl)-benzene (DM-POPOP). The fluorescence quantum yield of PPO is 1.00, that of POPOP is 0.93, and that of the singlet toluene monomer is 0.17. Therefore, if the singlet–singlet energy transfer were complete, the luminescence from the sensitized system (toluene plus PPO) would be 6 times brighter than the luminescence from pure toluene. The role of POPOP in the liquid scintillation solution is to undergo a trivial energy transfer by an efficient absorption–reemission process.   Since the POPOP absorption spectrum overlaps well with the PPO emission spectrum (Fig. 11-8), and since POPOP has a very large molar absorptivity in the overlap region, only a very small amount of POPOP is necessary for nearly complete absorption of the PPO fluorescence.   The POPOP fluorescence at 4100 Å closely matches the response of the photomultiplier tube.

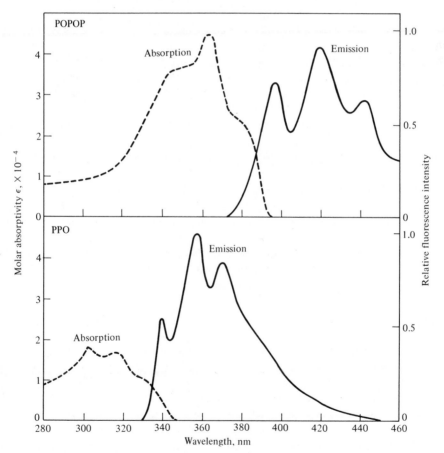

Fig. 11-8. *Upper*, absorption and fluorescence emission spectra of POPOP. *Lower*, absorption and fluorescence emission spectra of PPO.

Chemical impurities may interfere with the transfer of energy from solvent to solute to produce chemical quenching, or they may absorb the light emitted from the solute molecules to produce color quenching. Chemiluminescence will give yet a third change in the spectrum. Color quenching can be reduced or eliminated by digestion with hydrogen peroxide and perchloric acid. For ^3H and ^{14}C, complete combustion of the sample to water and a soluble carbonate produces a simple counting method. To minimize the oxygen quenching effect, which will vary with varying amounts of the dissolved air, a large excess of PPO is normally used. This shortens the mean lifetime of the toluene excimer by a factor of 37–50.

Sample containers for liquid scintillation counting are constructed of fabricated glass, nylon, quartz, or polyethylene; the latter vials, made from petrochemicals, are used when the minimum background must be achieved.

To measure gamma radiation and bremsstrahlung from high-energy beta emitters, an inorganic scintillator, such as sodium iodide crystal doped with 1% thallium(I) iodide, is best. This scintillator has a large photoelectric cross section, a high density which provides a high probability of absorption, and a high transparency to its own radiation (the optical emission lines of thallium) which enables large thicknesses to be used for absorption of gamma radiation. When gamma radiation interacts with a NaI(Tl) crystal, the transmitted energy excites the iodine atom and raises it to a higher energy state. When the iodine atom returns to its ground electronic state, this energy is reemitted in the form of a light pulse in the ultraviolet which is promptly absorbed by the thallium atom and reemitted as fluorescent light at 4100 Å. The crystal is sealed from atmospheric moisture and protected from extraneous light by an enclosure of aluminum foil, which also serves as an internal reflector.

The "well-type" scintillation crystal increases the counting efficiency by surrounding the sample with the detector crystal. The sample is placed in a well drilled into a crystal 5–10 cm in diameter; the size is chosen so that it contains the entire path of the ionizing particle or radiation and so measures the total energy. The resolution of a NaI(Tl) counter spectrometer is relatively poor (peak width of 6% at 1 MeV and 18% of 100 keV), but the efficiency approaches 100%, and the instrument is a multichannel device because the entire spectrum can be recorded at one time.

Semiconductor Detectors

Semiconductor detectors have revolutionized gamma-ray spectroscopy by providing energy resolution unattainable with previous methods. In these detectors the charge carriers produced by ionizing radiation are electron-hole pairs rather than ion pairs. The ionizing radiation lifts electrons into the conduction band and these electrons travel toward the positive electrode with high mobilities. The positive charge travels in the opposite direction by successive exchanges of electrons between neighboring lattice sites. Two general types of semiconductor detectors will be discussed: the *surface barrier silicon detector* and the *lithium-drifted germanium detector*.

A surface barrier detector consists of a *p-n* junction formed at the surface of a slice of silicon. At the junction there is a planar region, in which no charge carriers are present, where an electric field exists. This region is called the *depletion region*. A thin depletion depth is present when no bias voltage is applied across the *p-n* junction. If a reverse bias is applied, the depletion depth is increased, and is given by $d \simeq 0.5 \sqrt{\rho V}$ (in micrometers) where ρ is the silicon resistivity in ohm-centimeters and V is the bias in volts. Thus, with higher bias and higher resistivity, deeper depletion depths are formed. If charges are injected into the depletion region, they will be swept out of it by the electric field, and a voltage pulse will appear across the *p-n* junction. If energy measurements are to be made, selection of a detector for a particular purpose requires selecting a depletion depth greater or equal to the range of the particle of interest.

The lithium-drifted germanium spectrometer consists of a virtually windowless Ge(Li) crystal, a vacuum cryostat maintained by cryosorption pumping, a liquid-nitrogen Dewar, and a preamplifier. The Ge(Li) crystal is fabricated by drifting lithium ions (a donor) into and through *p*-type germanium. This is performed under the influence of a high electric field at 400°C. This process results in compensation of all acceptors within the bulk material, yielding a very high resistivity (or intrinsic) region which acts like ultrapure germanium within the bulk material. The drifting process is discontinued while a layer of *p*-type germanium still remains (Fig. 11-9). This intrinsic or compensated volume becomes the radiation-sensitive region. When ionizing radiation enters the intrinsic layer, electron-hole pairs are created, and the charge produced is rapidly collected under the influence of the bias voltage. The completed detector must be maintained at 77°K at all times to prevent precipitation of the lithium, since the lithium drift process is not stable at normal room temperature. At this low temperature thermal noise is greatly reduced and the resolution capabilities are vastly increased.

The energy resolution of semiconductor detectors is intrinsically good because of the large number of electron-hole pairs formed in comparison with the number of ion-pairs formed in a gas ionization chamber. The average energy for electron-hole pair production is 2.95 eV for germanium and 3.65 eV for silicon; this is far less than the 500 eV required per photoelectron in a NaI(Tl) scintillation detector. Thus, for a given amount of energy absorbed about 170 times as many electron-hole pairs are formed as ion pairs. Since the relative resolution is proportional to the square root of the signal, the resolution of the

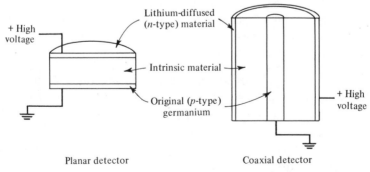

Planar detector   Coaxial detector

Fig. 11-9. Schematic diagrams of two common types of lithium-drifted germanium detectors.

Ge(Li) detector is about a factor of 13 better than the NaI(Tl) detector (see illustrations accompanying Problems 34 and 35). Typical quoted figures for energy resolution of a semiconductor detector 3 mm thick are 3.8 keV for electrons, 0.6 keV for X rays, and 20 keV for photons. This is the width at half-maximum of a peak in the energy spectrum. The rise time is about 10 nsec. However, the semiconductor detectors fall short of the efficiency of NaI(Tl) detectors, and is approximately 1% per millimeter thickness.

Auxiliary Instrumentation

Counters require auxiliary electronic equipment including a high-voltage supply, an amplifier (often plus a preamplifier), a scaler, and a count-registering unit. The required stability of the high-voltage supply and the required sensitivity and linearity of the amplifier are dictated by both the detector and the application.

From the counter the output pulse, after amplification, is fed into a scaling circuit which, in reality, is an electronic divider. The circuit is arranged so that the output is a single pulse for each $2, 4, 8, \ldots, 2^n$ incident particles. By a system of glow lamps the events withheld can be numbered. In the scale of two (binary) type, the overall scaling factor is 2^n, where n is the number of binary stages incorporated. More convenient and rapid reading is achieved on decade scalers. Each stage passes on every tenth pulse so that the instrument reads decimally. With either type the output from the scaler operates a mechanical register. Timing is done with built-in electric clocks which start and stop the count for a preset time interval, or, after a predetermined number of counts have been accumulated, the elapsed time is noted.

A significant amount of radiation is always present in the vicinity of a detector from natural radioactive elements and cosmic rays. Insertion of the counter into a shield of lead 2-3 in. in thickness reduces the background counting rate appreciably (Fig. 11-5). Further improvement can be achieved with anticoincidence circuits.

Pulse Height Discrimination

Whenever the amplitude of the pulse is proportional to the energy dissipation in the detector, the measurement of pulse height is a useful tool for energy discrimination. Current pulses are fed into a linear amplifier of sufficient gain to produce voltage output pulses in the amplitude ranges of 0-100 V.

One method of analysis of the pulse spectra is by use of a single-channel analyzer. The base-line discriminator passes only those pulses above a certain amplitude and eliminates pulses below this amplitude. It is useful for excluding scattered radiation and amplifier noise. Pulses associated with a particular radioactive emission must be amplified sufficiently so that their amplitudes exceed the discriminator setting. In practice this is accomplished by a combination of adjustment of the gain of the amplifier and the dc voltage applied to the detector.

A pulse height analyzer also contains a second discriminator called variously the window width, the channel width, or the acceptance slit. Now all pulses above the sum of the base-line and window setting are also rejected.   Only pulses with an amplitude within the confines of these settings will be passed on to the counting stages.   These operations are outlined schematically in Fig. 11-10.

The analyzer slowly scans the pulse distribution with a window a few volts in width. For example, when the nominal slit setting is at 75 V (Fig. 11-11), there are few pulses of this amplitude in the pulse spectrum.   As the slit sweeps on to the 65-V region, there are a large number of pulses which are of proper amplitude to pass the acceptance slit and be recorded.   As the acceptance slit completes its scan down to zero energy, the complete pulse height distribution is plotted.   The curve shown in Fig. 11-11, known as the *differential scan*, contains several features.   The sharp symmetrical peak is the result of total absorption of the gamma energy by the NaI(Tl) crystal, and is normally referred to as the full-energy peak or photopeak.   The continuous curve below the full-energy peak is due to Compton interaction, wherein the gamma photon loses only part of its energy to the crystal.   The location of the full-energy peak on the pulse amplitude or energy axis is proportional to the energy of the incident radiation, and is the basis for the qualitative application of the scintillation spectrometer or proportional counter.   The area under the full-energy peak is related to the number of photons or particles interacting with the detector, and is the basis for the quantitative application of the detector.   In the analysis of gamma photons in the energy range from 0 to 100 keV, there appears to be an escape peak 28 keV below the full-energy peak.   This is due to photoelectric interactions wherein the resulting iodine K X ray escapes from the crystal in a NaI(Tl) scintillation counter. Above 1.5 MeV the effect of pair production becomes apparent in a gamma scan.   In addition to the full-energy peak, there are peaks in the scan at full-energy minus 0.51 MeV and full-energy minus 1.02 MeV.   Also a small peak occurs at 0.51 MeV due to detection of an annihilation photon resulting from pair production in the shielding material.   The manganese-56 spectrum (Fig. 11-12) shows these features.

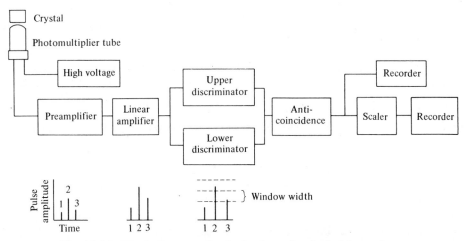

Fig. 11-10. Block diagram of a single-channel pulse height analyzer.

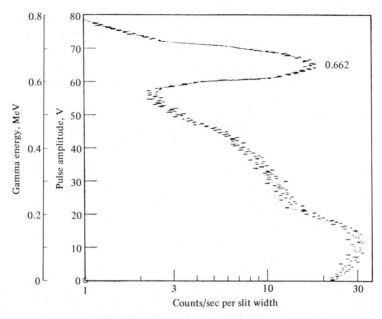

Fig. 11-11. Pulse height spectrum (differential curve) for cesium-barium-137 (γe^- 0.662 MeV).

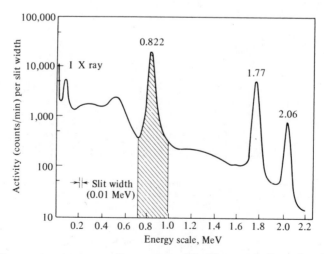

Fig. 11-12. The gamma spectrum of manganese-56. The area indicated under the photopeak at 0.822 MeV would be used in quantitative work.

With a single-channel analyzer the resulting curve is generally plotted automatically over a period of from 15 min to 2.5 hr, depending on sample counting rate and required precision. The pulse height scale is usually calibrated by the use of gamma (or beta) emitters of known energy; it is conveniently set to be a multiple of the energy peaks in the gamma spectrum. Instruments with 20 to over 1000 channels are commercially available. The multichannel analyzers of the 100- and 256-channel types all have provision for automatic analog readout.

The net area under each photopeak is directly proportional to the absolute gamma emission rate of the corresponding isotope. This is usually determined by taking the product of peak height, width at half-maximum, and the normalizing factor, 1.07—based on the assumption that the full-energy peak is Gaussian. The peak area can also be integrated with a planimeter. For the analysis of mixed gamma-emitting radioisotopes, the characteristic lower energy curve of the most energetic full-energy peak must be drawn in from previously recorded standard curves detailing the Compton continuum region vs. the principal full-energy peak area, as shown in Fig. 11-3. By subtraction, the net height of the second most energetic gamma photon can be established and its width measured. The Compton continuum of the second photopeak is then drawn in and the area of the third peak can be evaluated, and so on. These operations can be performed automatically on multichannel analyzers with standard curves stored on magnetic tapes.

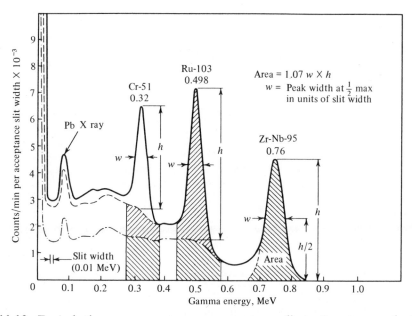

Fig. 11-13. Typical three-component gamma spectrum illustrating two methods of quantitative analysis.

Statistics in Measurement of Radioactivity

The random nature of nuclear events requires that a large number of individual events be observed to obtain a precise value of the count rate. Several factors must be considered when attempting to measure any activity. The ionizing particle may never reach the active volume of the detector; instead it may be absorbed in the walls or air path. The detector may not be perfectly efficient because of several conditions—the detector may not have recovered from a previous event (the dead time), the particle from the source may not produce an ion in the sensitive volume of the detector, and some regions of the detector are more sensitive than others.

Statistical laws will predict the magnitude of the deviations about a mean value to be expected, as well as the probability of occurrence of deviations of a given magnitude. One standard deviation, or 1σ (sigma), is the maximum deviation from the "true" value that may be expected in 68% of a large series of measurements. It serves as a measure of the precision of a single observation or a series of observations. In activity measurements the standard deviation in the total count is equal to the square root of the number of counts N taken:

$$\sigma = \sqrt{N} \tag{11-11}$$

The standard deviation in the counting rate r is

$$\sigma_r = \frac{1}{t}\sqrt{N} = \sqrt{\frac{r}{t}} \tag{11-12}$$

Expressed in this manner, the standard deviation is a measure of the scatter of a set of observations around their mean value. The relative standard deviation (fractional 0.68 error), often given in percent, is

$$100\sigma = \frac{100}{\sqrt{N}} \tag{11-13}$$

It is a measure of the precision of an observation. It can be expected that one standard deviation (or smaller) will arise in about 7 out of 10 determinations. The following confidence limits apply to representative multiples of the standard deviation:

Deviation		$\pm0.68\sigma$	$\pm\sigma$ $\pm2\sigma$	$\pm3\sigma$
Population mean (limits)		$\pm0.68\sigma$	$\pm\sigma$ $\pm1.96\sigma$	$\pm2.58\sigma$
Probability that observation (or confidence level) lies within this deviation, %		50	68 95	99

Example 11-1

If 8100 counts are timed, the standard deviation is

$$\sigma = \sqrt{8100} = 90 \text{ counts}$$

Expressed as relative standard deviation,

$$100\sigma = \frac{90}{8100}(100) = \frac{100}{\sqrt{8100}} = \pm 1.11\%$$

The probable error will be 0.68σ or 0.74%.

Example 11-2

To ascertain the number of counts which must be taken so that the deviation in 95% of the determinations (2σ) will not exceed 2.0%, proceed in this manner.

$$\frac{2\sqrt{N}}{N} = 0.02$$

$$\sqrt{N} = 100$$

$$N = 10,000 \text{ counts}$$

If a source has an activity of 200 counts/sec, it will take 50 sec to accumulate 10,000 counts.

A counter always shows some background activity. The statistical fluctuation of the background B must be included in any estimate of the standard deviation of the source whose activity A will include any background activity recorded simultaneously. The standard deviation of the net source activity, that is, $A - B$, is given by

$$\sigma = \sqrt{\sigma_A^2 + \sigma_B^2} \tag{11-14}$$

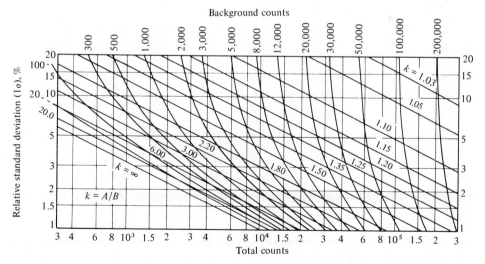

Fig. 11-14. Optimum number of counts for minimizing combined counting time when relative standard deviation is assigned and total counter-to-background ratio (A/B) is known. (After R. Loevinger and M. Berman, *Nucleonics*, **9** (No. 1) 29 (1951).)

Background radiation becomes significant when the peak-to-background ratio, $A:B$, is less than 20, and it becomes very difficult to measure a source accurately when the counting rate is just a little greater than the background rate. The optimum division of available time between background and source counting is given by

$$\frac{t_B}{t_A} = \sqrt{\frac{r_B}{r_A}} \tag{11-15}$$

Here counting rates have to be obtained from a preliminary run. The minimum combined time necessary to achieve a predetermined precision can be estimated from curves shown in Fig. 11-14.

Example 11-3

Suppose it is desired to determine with a precision of ±3% the activity of a sample which has a net count of 15 counts/min when the background is 30 counts/min. Thus, the ratio, $A:B = 45:30 = 1.50$, and from Fig. 11-14 the 3% line intersects the oblique line for $A/B = 1.50$ at a total count of about 18,000 to 19,000. The point of intersection lies about three-fourths of the distance between the background-count curves for 8,000 and 12,000. Thus, the background count should be about 11,000. The counting periods are about 400 min for the total count and 367 min for the background count. Figure 11-14 is very valuable whenever a preset number of counts is being accumulated.

Coincidence Correction

In order to correct for counting losses at high counting rates caused by the finite resolution time of counters, two courses are open. One method is to construct a calibration curve from a series of dilutions of a strong sample or from measurements of a series of standards of known strengths. A second procedure is to measure two samples separately and then measure the two samples together. The resolving time of the counter τ is given by

$$\tau = \frac{r_1 + r_2 - r_{1,2} - r_B}{2r_1 r_2} \tag{11-16}$$

where r_1 and r_2 represent the counting rate of source 1 plus background and source 2 plus background, respectively; $r_{1,2}$ is the measured counting rate of source 1 plus source 2 and the background r_B.

If the resolving time of the counter is greater than that of any other part of the measuring circuit, then the true counting rate r_0 is related to the observed counting rate r and the resolving time τ, namely,

$$r_0 = \frac{r}{1 - r\tau} \tag{11-17}$$

Geometry

The arrangement of counter and source should always be reproducible so that the solid angle subtended by the counter with respect to the source remains unchanged. The "geometry" of any arrangement may be obtained by measurement of standard sources. Insertion of the source into the active volume of a counter approximates 2π-geometry; 4π-geometry is approximated in well-type detectors.

APPLICATIONS OF RADIONUCLIDES

The introduction of a radioactive-labeled material into a sample system or a measurement of the natural or induced radioactivity of a system become very useful techniques for rapid and economical methods of analysis for elements or materials. Isotope dilution with radioactive tracers, labeled reagents, activation analysis, or the use of radioactive tracers for procedure development have much use in analytical chemistry.

The chain-reacting pile is without a peer as a tool for the general quantitative production of radioactivity because of the magnitude and spatial extent of the thermal neutron flux it is able to sustain. The pile excels the cyclotron in respect to both intensity and the magnitude of the effective flux produced.

Preformed generator sources, such as radionuclide cobalt-60 source or a beryllium neutron source, can be useful in meeting many analytical requirements when a pile is not conveniently near at hand.

A selection of radionuclides and their characteristics is given in Table 11-1. Nuclides with very short half-lives will decay too rapidly to be generally useful; on the other hand, a long-lived nuclide will be difficult to measure because disintegrations are too infrequent.

Preparation and Mounting of Samples

The active source must be free of interfering substances, in a suitable chemical and physical form, and disposed in a definite and fixed position relative to the detector. In virtually all applied radiochemistry, relative intensities of two or more samples are all that need be determined, thus making absolute measurements unnecessary. The principal requirement is reproducibility and, if this is not possible, the effects of the variable factors must be determined and a correction applied.

A chemical separation from inactive, and occasionally active, contaminants generally precedes the activity measurement. Ordinary analytical techniques and reactions form the basis for most radiochemical separations. Carrying a microconstituent by coprecipitation with a macroconstituent (a carrier), solvent extraction, volatilization, adsorption, ion exchange, electrodeposition, and chromatography are useful for handling very low concentrations of material. If the radiations are intense and penetrating, shielding and remote-control equipment are required. All work with appreciable quantities of radioactivity should be done with rubber gloves, protective clothing, and adequate ventilation

TABLE 11-1 Nuclear Properties of Selected Radioisotopes

Radio-isotope	Half-life	Target Isotope		Thermal Neutron Cross Section, barns	Major Radiations, Energies in MeV (γ Intensities, %)
			Natural Abundance, %		
^{3}H	12.26 y				β^- 0.0186; no γ
^{14}C	5730 y	^{13}C	1.108	0.0009	β^- 0.156; no γ
^{15}O	123 s				β^+ 1.74; γ 0.511
^{22}Na	2.62 y				β^+ 1.820, 0.545; γ 0.511, 1.275(100)
^{24}Na	14.96 h	^{23}Na	100	0.53	β^- 1.389; γ 1.369(100), 2.754(100)
^{28}Al	2.31 m	^{27}Al	100	0.235	β^- 2.85; γ 1.780(100)
^{32}P	14.28 d	^{31}P	100	0.19	β^- 1.710; no γ
^{35}S	87.9 d	^{34}S	4.22	0.27	β^- 0.167; no γ
^{36}Cl	3.08×10^5 y	^{35}Cl	75.53	44	β^- 0.714; γ 0.511
^{38}Cl	37.29 m	^{37}Cl	24.47	0.4	β^- 4.91; γ 1.60(38), 2.17(47)
^{40}K	1.26×10^9 y		0.118	70	β^- 1.314; β^+ 0.483; γ 1.460(11)
^{42}K	12.36 h	^{41}K	6.77	1.2	β^- 3.52; γ 0.31, 1.524(18)
^{45}Ca	165 d	^{44}Ca	2.06	0.7	β^- 0.252
^{51}Cr	27.8 d	^{50}Cr	4.31	17	γ 0.320(9); e^- 0.315
^{56}Mn	2.576 h	^{55}Mn	100	13.3	β^- 2.85; γ 0.847(99), 1.811(29), 2.110(15)
^{55}Fe	2.60 y	^{54}Fe	5.84	2.9	Mn X-rays
^{59}Fe	45.6 d	^{58}Fe	0.31	1.1	β^- 1.57, 0.475; γ 0.143(1), 0.192(3), 1.095(56), 1.292(44)
^{60}Co	5.263 y	^{59}Co	100	19	β^- 1.48, 0.314; γ 1.173(100), 1.332(100)
^{65}Ni	2.564 h	^{64}Ni	1.16	1.5	β^- 2.13; γ 0.368(5), 1.115(16), 1.481(25)
^{64}Cu	12.80 h	^{63}Cu	69.1	4.5	β^- 0.573; β^+ 0.656; e^- 1.33; γ 0.511, 13.4(1)
^{65}Zn	245 d	^{64}Zn	48.89	0.46	β^+ 0.327; e^- 1.106; γ 0.511, 1.115(49)
^{69m}Zn	13.8 h	^{68}Zn	18.6	0.10	γ 0.439(95); e^- 0.429
^{76}As	26.4 h	^{75}As	100	4.5	β^- 2.97; γ 0.559(43), 0.657(6), 1.22(5), 1.44(1), 1.789, 2.10(1)
^{80}Br	17.6 m	^{79}Br	50.52	8.5	β^- 2.00; β^+ 0.87; γ 0.511, 0.61(7), 0.666(1)
^{80m}Br	4.38 h				γ 0.037(36); e^- 0.024, 0.036, 0.047

TABLE 11-1   (*Continued*)

| Radio-isotope | Half-life | Target Isotope | | | Major Radiations, Energies in MeV (γ Intensities, %) |
			Natural Abundance, %	Thermal Neutron Cross Section, barns	
^{82}Br	35.34	^{81}Br	49.48	3	β^- 0.444; γ 0.554(66), 0.619(41), 0.698(27), 0.777(83), 0.828(25), 1.044(29), 1.317(26), 1.475(17)
^{90}Sr	27.7				β^- 0.546; no γ
^{90}Y	64.0 h				β^- 2.27; no γ
^{110mi}Ag	255 d	^{109}Ag	48.65	89	β^- 1.5; γ 0.658(96), 0.68(16), 0.706(19), 0.764(23), 0.818(8), 0.885(71), 0.937(32), 1.384(21), 1.505(11)
^{122}Sb	2.80 d	^{121}Sb	57.25	6	β^- 1.97; β^+ 0.56; γ 0.564(66), 1.14(1), 1.26(1)
^{124}Sb	60.4 d	^{123}Sb	42.75	3.3	β^- 2.31; γ 0.603(97), 0.644(7), 0.72(14), 0.967(2), 1.048(2), 1.31(3), 1.37(5), 1.45(2), 1.692(50), 2.088(7)
^{128}I	24.99 m	^{127}I	100	6.4	β^- 2.12; γ 0.441(14), 0.528(1), 0.743, 0.969
^{137}Cs	30.0 y				β^- 0.511, 1.176; γ 0.662 (85)
^{137m}Ba	2.554 m				γ 0.662(89); e^- 0.624, 0.656
^{198}Au	2.697 d	^{197}Au	100	98.8	β^- 0.962; γ 0.412(95), 0.676(1), 1.088
^{204}Tl	3.81 y	^{203}Tl	29.5	11	β^- 0.766

SOURCE: J. A. Dean, Ed., *Lange's Handbook of Chemistry*, 11th ed., McGraw-Hill, Inc., New York, 1973.

NOTE: The following abbreviations have been used: s, seconds; m, minutes; h, hours; d, days; and y, years.

to remove active vapors, dusts, and sprays. Of scarcely less importance is the danger of contamination by active materials of laboratories, equipment, and detecting instruments. Establishment and enforcement of suitable regulations and careful attention to cleanliness of operation are integral aspects of the proper technique of working with radioactivity.

The active material must be spread in a uniform layer over a definite area unless 4π-geometry is employed. Thin uniform layers of solids may be spread as slurries in water or other solvent which is later evaporated. If the deposit is not coherent but tends to fall apart and flake off, it may be stabilized by the use of a suitable binder (collodion), or it may be held in place with a covering layer of Scotch tape, aluminum foil, or similar material. The sample is frequently obtained in a solid form by precipitation from solution and separation by centrifugation. Crystalline precipitates are best, whereas flocculent precipitates, which are highly hydrated, tend to give unsatisfactory deposits because on drying they contract into a number of isolated, dense particles.

Evaporation of solutions in cup-shaped containers, or for small amounts, on flat surfaces, is convenient. The method is limited to nonvolatile activities. Electrodeposition onto a flat surface gives excellent deposits for many metals; a thin film of plastic sprayed with gold can serve as the active electrode.

Samples can be mounted on flat foils or discs; several tenths of a milliliter of a liquid can be held in place by surface tension or within a ring of silicone grease. When backscattering by the mount is objectionable, a thin film of low atomic number may be used—Mylar, polystyrene, and similar plastics. An arrangement is needed for holding samples in a definite position relative to the detector during measurement. Counters and ion chambers are generally equipped with an arrangement having one or more shelves or sets of slots in which sample holders of a standard size can be placed. Radionuclides emitting sufficiently penetrating beta and gamma radiations are conveniently assayed in solution with dipping counter tubes or counters surrounded by hollow jackets, or by insertion in the well of an ionization chamber or scintillation counter. Active gases can be introduced into proportional counters or ionization chambers equipped with stopcocks and pressure manometers.

Tagging Compounds

Since the radioactive isotopes are chemically identical with their stable isotope counterpart, they may be used to "tag" a compound. The tagged compound may then be followed through any analytical scheme, industrial system, or biological process. It is essential that a compound be tagged with an atom, however, which is not readily exchangeable with similar atoms in other compounds under normal conditions. For example, tritium could not be used to trace an acid if it were inserted on the carboxyl group where it is readily exchanged by ionization with the solvent.

It is not always true that the radionuclide will resemble exactly the normal isotope. Differences in weight do cause slight changes in the reactivity of molecules. For ordinary purposes, however, these isotope effects are slight except for the lightest elements (such as tritium). Many multiple decays are found in radionuclides produced in chain-reacting piles. If the parent (^{140}Ba, $t_{1/2}$ = 12.8 days) is longer-lived than the daughter (^{140}La, $t_{1/2}$ = 40 hr), a state of radioactive equilibrium is reached after about eight half-lives (13 days). Thus, if ^{140}Ba is to be used as a tracer for barium, the isolated samples either should be freed chemically of the daughter ^{140}La and counted without delay, or the samples should be kept for two weeks before counting.

In biological investigations the question of purity of the isolated material is important. One must be certain that the radioactivity of the compound isolated is due to the compound itself and not to some minor contaminant which may have a high specific activity.

In ordinary analytical work, radionuclides have been used to study errors resulting from adsorption and occlusion in gravimetric methods, and to devise methods of preventing coprecipitation, adsorption, and occlusion.

Analyses with Labeled Reagents

Radiometric methods employing reagent solutions or solids tagged with a radionuclide have been used to determine the solubility of numerous organic and inorganic precipitates, or as a radioreagent for titrations involving the formation of a precipitate. In this type of application it is necessary to establish the ratio between radioactivity and weight of radionuclide plus carrier present. This may be established by evaporating an aliquot to dryness, weighing the residue, and measuring the radioactivity.

In solubility studies, the compound of interest is synthesized, using the radionuclide, and a saturated solution of the compound is prepared. A measured volume of the saturated solution is evaporated to dryness, and the radioactivity of the residue determined. From the previously established relationship between weight and radioactivity, the amount of the compound present can be calculated.

Procedures have been described for the use of radioreagents as titrants. For example, phosphorus-32 was converted into a soluble phosphate and incorporated into a standard solution of disodium hydrogen ortho phosphate. This solution was used to titrate several inorganic ions. After each addition of phosphate a sample of the clear, filtered solution was withdrawn (by means of a filter-stick) and the activity was determined. After the equivalence point was passed, the activity rose rapidly with additions of radioreagent. From the intersection of the activity curves, the end point was determined.

The efficiency of an analytical procedure can be determined by adding a known amount of a radioisotope before analysis is begun. After the final determination of the element in question, the activity of the precipitate is determined and compared with the activity at the start.

Chemical yields need not be quantitative in an analytical procedure when the results are corrected by the recovery of the radioisotope. To the mixture of ions, a known amount of radioisotope is added, or to a mixture of activities a known amount of carrier element is added, then separated in the necessary state of chemical and radiochemical purity but without attention to yield. The isolated sample is determined by any suitable method, and the activity is measured.

Isotope Dilution Analyses

This technique measures the yield of a nonquantitative process, or it enables an analysis to be performed where no quantitative isolation procedure is known. To the unknown mixture, containing a compound with inactive element P, is added a known weight W_1 of the same compound tagged with the radioactive element $P^\star$. The specific activity A_1 of the tagged compound of weight W_1 is known. A small amount of the pure compound is isolated from the mixture and the specific activity measured (A). The amount isolated need be only a very small fraction of the total amount present, merely a sufficient quantity for weighing or determining accurately. The extent of dilution of the radiotracer shows the amount W of inactive element (or compound) present, as given by the expression

$$W = W_1 \left(\frac{A_1}{A} - 1 \right) \tag{11-18}$$

The method has proven valuable in the analysis of complex biochemical mixtures and in the radiocarbon dating of archeological and anthropological specimens.

Neutron Activation Analysis

Neutron activation analysis involves the production of a radioactive isotope by the capture of neutrons by the nuclei of the substance to be analyzed. Most elements when irradiated by thermal neutrons give rise to a radioactive species of the same atomic number but one mass unit greater than its progenitor through an (n, γ) reaction. Immediately after neutron capture a gamma ray is emitted whose energy is equal to the neutron binding energy plus the kinetic energy of the neutron. The reaction with sodium is

$$^{23}\text{Na} + n \rightarrow {}^{24}\text{Na} + \gamma \tag{11-19}$$

The resulting sodium isotope ^{24}Na is radioactive, decaying into ^{24}Mg with a half-life of 14.96 hr and emitting a beta particle of maximum energy equal to 1.389 MeV and two gamma rays of 1.369 and 2.754 MeV, respectively (see Fig. 11-2). Therefore the presence of sodium in a sample, irrespective of its chemical combination, can be detected by analyzing the radioactivity induced in the sample after being exposed to neutrons for a certain period of time. The identification can be done by measurement of the half-life or the energy of the beta particles, or by analyzing the gamma-ray spectrum using scintillation methods or a semiconductor detector and a multichannel analyzer.

Neutron activation analysis has several advantages: (1) it is a very sensitive method that detects impurities at the level of 10^{-3} to 10^{-7} μg/g; (2) it is quite specific because the radioactivity induced is characteristic of the substance analyzed although unresolved overlapping photopeaks do occur; and (3) interference from other elements during the analysis is minimized. After the sample has been irradiated, its integrity can be threatened only by contamination with radioactive material. Neutron activation analysis has some limitations because it may happen that the cross section for neutron absorption by the substance analyzed is very small, or the isotope produced is not radioactive, or its half-life is very long or very short, or its radiation is not amenable to precise measurement because of very weak beta particles or electron capture. One rather intractable problem is presented by the positron emitters, of which there are over 50, all producing 0.511-MeV annihilation radiation.

Irradiation is accomplished by placing the sample to be analyzed in an intense flux of either thermal or fast neutrons for a length of time sufficient to produce a measurable amount of the desired radioisotope. The capture rate will be proportional to the neutron flux Φ, in neutrons/cm^2-sec, and to the number of target nuclei N available; the proportionality constant is known as the *capture cross section*, σ, expressed in barns (10^{-24} cm^2/nucleus). A correction must be made for the fractional abundance f of the target nuclide. The amount of product activity A_0, expressed in disintegrations per second, existing at the end of the irradiation is given by

$$A_0 = N\Phi\sigma fS \tag{11-20}$$

where S, the *saturation factor*, is

$$S = 1 - 10^{-0.693\, t_i/(2.3\, t_{1/2})} \tag{11-21}$$

where t_i is the duration of the irradiation. During neutron activation of finite duration some of the radioisotope produced will decay according to Eq. 11-8, and the saturation factor represents the ratio of the amount of activity produced during the irradiation period to that produced in infinite time. At $t_i/t_{1/2}$ values of 1, 2, 3, 4, 5, 6, ..., ∞, S has corresponding values of 0.5, 0.75, 0.87, 0.94, 0.97, 0.98, ..., 1.00. The radioactivity will reach 98% of the saturation value for irradiation times equal to six half-lives.

In terms of the weight, W, of the element, Eq. 11-20 may be written as

$$W = \frac{A_0 M}{N_A \Phi\sigma fS} \tag{11-22}$$

where N_A is Avogadro's number and M is the atomic weight of the element.

In quantitative determinations the comparative method is used. The comparators are weighed and packed in duplicate, together with samples of appropriate size, and both are irradiated in the same physical location and under the same flux conditions. Sometimes a determination can be made by direct measurement, following a suitable decay period for short-lived isotopes with interfering photopeaks. When overlapping photopeaks are suspected, it is necessary to have supplemental information. For example, several successive spectra will give half-life information that will enable one to calculate the individual contributions. Qualitative analysis by resolution of the decay curve is illustrated in Fig. 11-15. Oftentimes some radiochemical processing may be required. Chemical separations are usually devised to yield the elements to be determined, either in a radiochemically pure form that can be counted by simple counting equipment, or into several groups which can then be further examined by gamma-ray spectrometry. Inactive carriers are usually present during solution of the sample to enable a correction to be made for losses occurring during chemical processing. The general procedure is well illustrated in the neutron activation of aluminum-base alloys.[1]

Thermal neutron fluxes obtainable in chain-reacting piles range from 10^{11} to 10^{14} neutrons cm^{-2}-sec^{-1}. Portable sources are available for laboratories not having access to atomic reactors. A 1-g radium–beryllium source provides 10^7 neutrons cm^{-2}-sec^{-1}; an antimony-124–beryllium source gives 3×10^6 neutrons Ci^{-1}-sec^{-1}; and 1-g of californium-252 emits 2.34×10^{12} neutrons sec^{-1}.

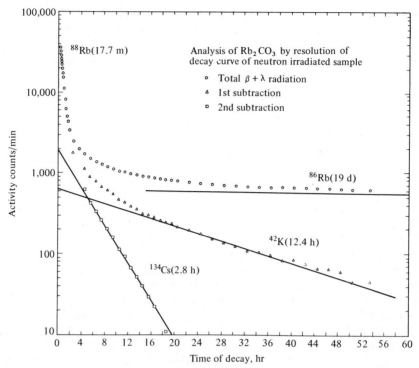

Fig. 11-15. Qualitative analysis by resolution of decay curve following radioactivation of a sample of rubidium carbonate. Impurities are cesium and potassium. (After G. E. Boyd, *Anal. Chem.*, **21**, 344 (1949).)

Example 11-4

Calculate the activity of a 0.00100-g sample of an aluminum alloy containing 0.041% of manganese after a 30-min irradiation in a flux of 5×10^{11} neutrons cm^{-2} sec^{-1}. Other necessary information is obtained from Table 11-1 for insertion in Eq. 11-20;

$$A = \frac{(0.00041)\,(0.00100)\,(1.00)\,(6.02 \times 10^{23})\,(5 \times 10^{11})\,(13.3 \times 10^{-24})\,(S)}{54.94}$$

where

$$S = 1 - e^{-(0.693)(30)/(2.58)(60)}$$

$$= 1 - 0.878 = 0.122$$

and therefore,

$$A = 3680 \text{ disintegrations/sec}$$

About 2 hr after discharge from the reactor, all the aluminum-28 (2.3 min) and most of the other short-lived isotopes will have decayed to negligible activity. The activity of manganese-56 has dropped to

$$A = 3680 e^{-\lambda t} = \text{disintegrations/sec}$$

Manganese could be determined by direct measurement of the 0.822-MeV gamma radiation. If it is, the energy region from 0.70 to 1.00 MeV is scanned: the background is taken as the base of the photopeak (see Fig. 11-12).

Example 11-5

In the determination of iron in aluminum alloys, the 1.29-MeV gamma of iron-59 was measured. A decay period of one week before chemical processing allowed for the decay of the 15.0-hr sodium-24 formed from the reaction $^{27}Al(n, \alpha)^{24}Na$.

$$A = A_0 e^{-(0.693)(168)/(15.0)} = 0.00235 \, A_0$$

After an interval of seven days, the sodium-24 activity has decayed to 0.00235 of its original activity. The gamma radiation from sodium-24 at 1.369 and 2.754 MeV would no longer constitute an interference.

Some elements can be detected with more sensitivity or more conveniently by activation with fast neutrons than with thermal neutrons. The list of applicable elements includes N, O, F, Al, Si, P, Cr, Mn, Fe, Cu, Y, Mo, Nb, and Pb. In particular, the nondestructive determination of oxygen is now frequently done by activation with 14-MeV neutrons to form 7.35-sec ^{16}N via the $^{16}O(n, p)^{16}N$ reaction. The gamma radiation emitted by ^{16}N is mostly 6.13 MeV, with a little 7.13 MeV, which is exceptionally high so that there are essentially no interferences (except for F which also undergoes the $^{19}F(n, \alpha)^{16}N$ reaction). Total analysis time for oxygen is less than 1 min; the limit of detection is 1 $\mu g/g$ in a 10-g sample. For six elements, O, Si, P, Fe, Y, and Pb, the sensitivity is greater using fast rather than thermal neutrons.

As a neutron source, the fast-neutron (fission-spectrum) flux that is present in the core of a nuclear reactor can be used. Samples are wrapped with cadmium foil to prevent thermal neutrons from reaching the sample. A small accelerator source is a 200-keV Cockcroft-Walton deuteron accelerator with a metallic tritium target and a D^+ beam which forms neutrons by the $^3H(d, n)^4He$ reaction with energies of about 14 MeV. Pulsed generators provide very high, fast neutron fluxes, well over 10^6 neutrons/cm^2-sec, of about 30-msec duration. Pulsed operation favors the formation of short-lived (less than 50-sec) radioisotopes. The amount of a particular activity A_p produced in a pulse, relative to the amount of activity produced in the usual steady-state operation of the same reactor to saturation activity A_s, is approximately $A_p/A_s = 70/t_{1/2}$.

MÖSSBAUER SPECTROSCOPY

Nuclear gamma-ray resonance spectroscopy, better known as *Mössbauer spectroscopy*, is based on the fact that certain nuclides, when solidly built into a crystal lattice and at temperatures significantly below the Debye temperature,* can emit recoil-free gamma

---

*The Debye temperature is given by $T_D = hf_D/k$, where f_D is the elastic frequency of atoms in a lattice; hf_D is a phonon (analogous to a photon). Changes in the state of the lattice must correspond to the emission or absorption of one or more phonons.

rays. The crystal absorbs the recoil momentum, and there is virtually no loss of energy, unless vibrations are started in the lattice during emission. These conditions only prevail when the energy of the emitted gamma ray does not exceed about 150 keV. The mono-energetic gamma ray is uniquely capable of undergoing a resonance absorption, again without recoil, by the proper stable nuclide bound in a solid matrix. Many of the chemical applications involve ^{57}Fe and ^{119}Sn, although the Mössbauer effect has been observed with about 50 nuclides.

Iron-57 is the most popular Mössbauer nuclide. Its radioactive parent ^{57}Co has a moderately long half-life of 270 days and a high specific activity. The decay (Fig. 11-16) proceeds by electron capture through a highly excited state which decays to the first excited state with a spin of $\frac{3}{2}$. This level decays in 91% of the cases to the ground state with a spin of $\frac{1}{2}$. Our concern is with this transition. When the lattice is not excited, so that the entire energy of the nuclear transition appears in the gamma ray, the recoilless Mössbauer radiation has a spread of energies limited only by the Heisenberg uncertainty principle. A nuclear lifetime of the $\frac{3}{2}$ energy level of 10^{-7} sec corresponds for the 14.4-keV state to a line width of

$$\Delta E = \frac{h}{2\pi\tau} = \frac{6.626 \times 10^{-34} \text{ J} \cdot \text{sec}}{2(3.14)(1.0 \times 10^{-7} \text{ sec})(1.6 \times 10^{-19} \text{ J/eV})} \tag{11-23}$$

$$= 6.6 \times 10^{-9} \text{ eV}$$

If the absorbing nuclide is in a different electronic (chemical) environment from that of the source, the transition energy is changed, destroying resonance. To restore resonance, additional energy must be added to or subtracted from the gamma radiation of the source. Relative motion between the absorber and the source adds Doppler motion energy

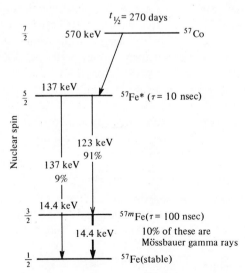

Fig. 11-16. Decay scheme of ^{57}Co to stable nuclide ^{57}Fe.

to the system, permitting resonance conditions to be reestablished. Devices to provide this relative motion include constant-acceleration cams, rotating wheels, and electro-mechanical transducers such as high-fidelity loudspeakers. For a first-order Doppler shift,

$$\frac{\Delta E}{E} = \frac{v}{c} \tag{11-24}$$

and the required Doppler velocity for ^{57}Fe is

$$v = \frac{6.6 \times 10^{-9} \text{ eV}}{14.4 \times 10^3 \text{ eV}} (3 \times 10^{10} \text{ cm/sec}) = 1.4 \times 10^{-2} \text{ cm/sec} \tag{11-25}$$

When the intensity of the resonance gamma-ray signal is plotted against a series of veloci-ties, Mössbauer spectra are obtained.

The major components of a Mössbauer spectrometer are a radioactive source, a Doppler scanning device, the sample, an energy-selective gamma-ray detector, and pulse-handling electronics. In the point-by-point procedure, the source or the absorber is moved at a constant velocity, relative to the other, and the transmitted resonance radiation is counted for a convenient time. The velocity is then changed. Commercial equipment usually operate in a constant-acceleration mode where a whole range of veloci-ties is scanned from zero to a preset maximum. This scanning is accomplished in synchro-nization with a multichannel analyzer. Each channel accumulates the number of trans-mitted counts for the same given velocity increment during each cycle. Cryogenic acces-sories permit low-temperature operations. For bulky samples, such as plates, surface formations, or samples too thick to use in the transmission mode, the back-scattering technique may be used. In this arrangement, the 6.3-keV X rays that are scattered from the sample of iron are counted. These X rays result from the internal conversion of the absorbed 14.4-keV gamma ray by the ^{57}Fe in the sample.

The energy of a nuclear level may be influenced by the s-electron density, and by the electric and magnetic fields either internal or external to the sample. These influences give rise to three basic patterns of Mössbauer spectra. For a chemist the *isomer shift* (Fig. 11-17), which is the displacement of the resonance dip from zero Doppler velocity, is the most basic measurement derived from a Mössbauer spectrum. The isomer shift

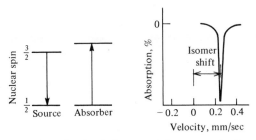

Fig. 11-17. Isomer shift of $K_4[Fe(CN)_6] \cdot 3H_2O$ (0.21 mm/sec) at 25°C; reference ab-sorber, $Na_2[Fe(CN)_5(NO)] \cdot 2H_2O$).

arises because the nucleus is not a point source but interacts as a region of charge space with the electrostatic environment. The magnitude of the isomer shift is a linear function of the *s*-electron density about the nucleus; and it is indirectly sensitive to the other electrons which may be involved in chemical bonds (for example, the *p*-electrons in SnX_4 halides through sp^3 hybridization). The *d*- and *f*-shell electrons may provide a shielding effect, altering the *s*-electron density at the nucleus. In the case of ^{57}Fe, the magnitude of the isomeric shift is largely determined by the occupation of the $3d$ and $4s$ orbitals and any external influences upon these from covalent bond character or fields from neighboring ions. It is customary to report isomer shifts observed for a particular source and a reference absorber (see Table 11-2).

The excited state spin of $\frac{3}{2}$ can be split in an electrostatic field into a $\pm\frac{3}{2}$ and $\pm\frac{1}{2}$ level (Fig. 11-18). This *quadrupole splitting* of the resonance line arises from interaction of the electric field gradient around the nucleus with the electric quadrupole moment of the excited ^{57}Fe nucleus (present due to a nonspherical *s*-electron distribution). The difference between the velocities of the two peaks in the Mössbauer spectrum is the quadrupole splitting value, and the isomer shift is taken as the displacement of the centroid of the spectrum from zero velocity. Because the electron field gradient around the nucleus depends on the electronic configuration of the nucleus and on its environment, conclusions can be drawn from the quadrupole splitting and isomer shift about the nature of the chemical bonding, the crystal lattice, the oxidation state, the site symmetries, the distortion of bonds, or anything that may change the *s*-electron distribution about the nucleus. Even if the electric charge distribution in the absorber is spherically symmetrical, and therefore has no quadrupole moment, the emitting nuclei which are in the excited state are likely to have one. More than two peaks may result if the Mössbauer transition is other than a $\pm\frac{3}{2}$ to $\pm\frac{1}{2}$ transition. If both emitter and absorber have quadrupole moments, the experimental spectra are complex.

An interaction of the magnetic dipole moment of the nuclide with an internal or

TABLE 11-2   Relationship Between Mössbauer Spectral Parameters and Oxidation States of Iron[a]

Oxidation State	Isomer Shift	Quadrupole Splitting	Q_T[b]
0	~2.3		
Organo complexes	0.0 to 0.7	0.0 to 2.5	0.0 to 0.1
2+, ionic compounds,	1.45 to 1.65	1.7 to 3.2	0.2 to 0.5
covalent compounds	0.00 to 0.40	0.0 to 2.0	~0
3+, ionic compounds,	0.60 to 1.15	0.0 to 0.9	~0
covalent compounds	0.00 to 0.40	0.0 to 2.0	0.03 to 0.40
Standard reference absorber, $Na_2[Fe(CN)_5(NO)] \cdot 2H_2O$	0.00		

[a] All values in mm/sec at $25°C$.
[b] Variation of quadrupole splitting with temperature, mm sec^{-1} $100°K^{-1}$.

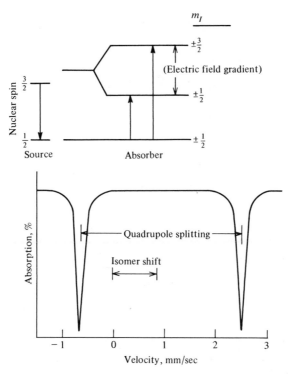

Fig. 11-18. Mössbauer spectrum of $FeSO_4 \cdot 7H_2O$; isomer shift = 3.19 mm/sec. Reference absorber, $Na_2[Fe(CN)_5NO] \cdot 2H_2O$, showed an isomer shift = -0.61 mm/sec. All data for 25°C.

external magnetic field (crystal field and electron shell around the nucleus) will result in a *magnetic hyperfine structure*. Also called the nuclear Zeeman effect, the spontaneous magnetic field present in metallic iron (about 330 kG) removes the ground-state and excited-state degeneracy. This splits the excited state of ^{57}Fe into four energy levels and the ground state into two energy levels (Fig. 11-19). The six possible transitions between the levels of the excited and ground state, arising from consideration of the selection rule $\Delta m = 1, 0$, or -1, result in the six resonance lines in the spectrum of metallic iron. This interaction can be used to measure dipole moments or the strengths of hyperfine magnetic fields as well as determining Curie, Morin, and Nèel temperatures.

Applications of the Mössbauer effect are found in structural chemistry, surface chemistry and catalysis, metallurgy, biological systems, and mineralogy.[2,3,4] The magnitude of the isomer shift is related to the oxidation state of the nuclide (Table 11-2). In the case of iron, it is relatively easy to determine the 2+ and 3+ oxidation states for ionic compounds. For the covalent compounds, however, an evaluation of the variation of the quadrupole splitting with temperature is required because the isomer shift and the quadrupole splitting fall within the same range of values. Iron and its many alloys can be

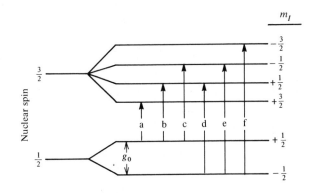

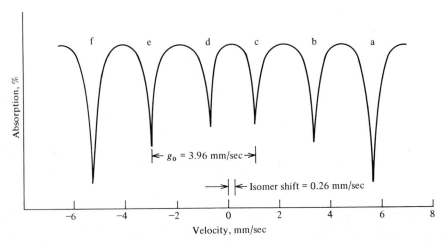

Fig. 11-19. *Upper,* Energy-level diagram for magnetic splitting of the ground and the first excited state of ^{57}Fe in metallic iron. g_0 is ground-state splitting. *Lower,* Mössbauer spectrum of metallic iron relative to reference absorber $Na_2 [Fe(CN)_5 NO] \cdot 2H_2O$.

studied and the type and amount of each phase can be determined since, for example, martinsite and austinite give completely different Mössbauer patterns. Through the use of the back-scattering technique, films of material on surfaces can be examined. In this manner rust on steel surfaces has been identified as β-FeOOH. The method yields information concerning the structure of the metal or metal ion with regard to the surface of a catalyst. It is possible to estimate relative concentrations of constituents in a particular type of sample from the peak areas. An exponential relationship with concentration similar to Beer's law is followed. Examples include the relative concentrations of iron(II) and iron(III) in minerals and iron–phosphate glasses, the ratio of corroded to uncorroded iron in thin foils, and the ratio of olivine iron to pyroxene iron for meteorites.

LABORATORY WORK

Radioactive Materials

Standards for calibration of counters can be obtained from the National Bureau of Standards (U.S.). These include RaD + E beta particle standards and ^{60}Co and radium gamma radiation standards. Directions for the use of these are furnished with the standards.

Radionuclide sets of radioactive reagents are commercially available and provide a variety of radiation types and energies at AEC license-exempt quantities. One set contains the following isotopes: ^{14}C as uniformly labeled fructose, ^{32}P as phosphoric acid, ^{131}I as the potassium salt, ^{35}S as sulfuric acid, and chlorides of ^{65}Zn, ^{22}Na, ^{60}Co, and $^{55/59}$Fe.

Two standards, easily prepared and safe to handle, involve potassium and uranium salts. Potassium-40 occurs naturally to the extent of 0.118% in all potassium salts; the specific activity amounts to 0.919 disintegrations per mg per min. Known amounts can be weighed into planchets and covered with a layer of Scotch tape. The counter efficiency for potassium-40 is often about 25%.

Pure U_3O_8 is prepared by precipitating uranyl hydroxide with aqueous ammonia, filtering, and igniting the hydroxide to the oxide. The specific activity of U_3O_8 is 636 disintegrations per mg per min. Known amounts of the oxide are weighed, then spread out in a very thin uniform layer about 1 cm in diameter. Cover with two layers of Scotch tape (approximately 10 mg/cm^2 per layer). If the uranyl salt has not recently been separated from its daughters, ^{234}Th and ^{234}Pa (UX$_1$ and UX$_2$), these will be in radioactive equilibrium with the parent uranium. Uranium emits alpha particles which can be absorbed by aluminum foil 0.025 mm in thickness. Thorium-234 emits beta particles which are absorbed by the double layer of Scotch tape. Only the ^{234}Pa radiation will affect the counter.

Induced Radioactivity

Certain radioactive isotopes can be produced by irradiation with neutrons generated in a radiometric-beryllium source according to the reaction

$$^9_4Be + ^4_2He \rightarrow ^{12}_6C + ^1_0n$$

A source containing 5 mg of radium mixed with 50 mg of beryllium provides a flux of about 10^3 neutrons cm^{-2} sec^{-1}. Activation is limited to elements with a high capture cross section ($>50 \times 10^{-24}$ cm^2) and suitable isotopic abundance; essentially only gold-197 and indium-115. The concurrence of an appreciable number of high-energy gamma rays together with the alpha emission from radium sources introduces a considerable inconvenience in handling this source.

An antimony-124 (60 days half-life)-beryllium source provides a convenient source whose flux is comparable to that from a radium-beryllium source. Its disadvantage is the relatively short half-life of antimony-124.

A 1 Ci plutonium–beryllium source provides a somewhat larger flux of 10^5 neutrons $cm^{-2} sec^{-1}$. A hexagonal gridwork holds uranium slugs.

Neutrons from any of these sources must be slowed down, or moderated, to thermal velocities by surrounding the source with purified water or paraffin. The sample to be activated may be mixed with, or dissolved in, the moderator or simply inserted at a favorable position in the moderator.

Radioisotope generators for introductory laboratory use have been described by Crater et al. [H. L. Crater, J. B. Macchione, W. J. Gemmill, and H. H. Kramer, *J. Chem. Educ.*, **46**, 287 (1969).]

The Szilard-Chalmers chemical enrichment process extends the usefulness of a low neutron flux to elements with cross sections down to 1×10^{-24} cm^2. To be successful the element must be capable of existence in two or more mutually stable and separable forms which do not undergo rapid isotopic exchange, and the target molecule must be difficult or impossible to reform from the fragments produced by neutron capture.

Irradiation of aqueous permanganate solutions of pH less than 10 yields active $^{56}MnO_2$ with high efficiency. Even the small amount of MnO_2 formed by passing the irradiated solution through filter paper is sufficient to act as carrier for the active precipitate and to retain it on the filter. [W. F. Libby, *J. Am. Chem. Soc.*, **62**, 1930 (1940).]

Successful enrichment of the antimony-122 activity is achieved through a short irradiation of triphenyl stibine dissolved in ether. A large fraction of the active antimony is ejected into an aqueous solution in contact with the ether phase. [R. R. Williams, Jr., *J. Phys. & Colloid Chem.*, **52**, 603 (1948).]

Halogen activities which yield about 10^3 disintegrations per min per mg of halogen can be prepared by irradiating ethyl iodide for the iodine activity or ethyl bromide or ethylene dibromide for the bromine activities. [R. H. Schuler, R. R. Williams, Jr., and W. H. Hamill, *J. Chem. Educ.*, **26**, 667 (1949).] The radioactive halogen is prepared daily (or as needed) by placing the neutron source in a spherical 1-liter flask containing the ethyl halide and a small amount of the corresponding free halogen. After 3 or 4 half-lives, or overnight, the free halogen is extracted with 5 ml of a 10% solution of sodium hydroxide and is converted to silver halide. The active silver halide is collected in several portions on filter paper and is given immediately to the students. [R. R. Williams, Jr., W. H. Hamill, and R. H. Schuler, *J. Chem. Educ.*, **26**, 210 (1949).]

One-inch circles of indium metal foil, weighing approximately 275 mg, can be irradiated to saturation in about 150 min with a radionuclide-beryllium source placed in a paraffin block. [W. H. Hamill, R. R. Williams, Jr., and R. H. Schuler, *J. Chem. Educ.*, **26**, 310 (1949).]

Experiment 11 - 1   Geiger Counter Characteristics

Connect the Geiger counter to the high-voltage supply of the scaling circuit, attaching the positive side to the central wire and the negative side (grounded) to the outer cylinder. With the voltage control set to its minimum value, turn on the electronic circuit and set the count switch to ON.

Place a radioactive sample under the counter and increase the applied voltage until impulses begin to register.   Record this voltage (the starting potential).   If excessive counting rates are observed ($> 25{,}000$ counts/min), lower the voltage or remove the sample immediately to avoid damage to the counter.

Increase the applied voltage by 24 V and measure the counting rate over a period long enough to collect about 3600 counts.   For reasons of convenience the activity of the source should be at least 1000 counts/min.   Immediately beyond the starting potential a rapid rise in counting rate will be noticed until the "knee" of the Geiger plateau is attained.   Thereafter determine the counting rate at 50-V intervals until the counting rate increases as much as 10% between voltage settings. The usual plateau extends over a 100- to 300-V range.   Do not continue to increase the voltage beyond the plateau or the counter will be damaged.

Express the data graphically as activity (counts per minute) versus applied voltage. All subsequent counting with this particular counter should be performed at one voltage near the middle of the plateau (the operating voltage of the tube).   The slope of the Geiger plateau is expressed as the difference of two mean counting rates taken 100 V apart, and divided by the average counting rate over the particular interval.

Experiment 11 - 2   Statistical Fluctuations in Counting Rate

Choose a sample of low activity (500–1000 counts/min), or place one of higher activity on a lower shelf. Obtain 20 1-min counts.

Compute the mean counting rate, the deviations of each observation, the standard deviation of an individual observation, and the standard deviation of the mean. Evaluate the probable error, and the 2-σ and 3-σ variations of the mean.   Tabulate your results. What percentage of the 1-min counts lies within each mode of expressing the deviation?

Determine the background counting rate by removing all active samples from the vicinity of the counter, and collect enough counts to give a 10% standard deviation about the mean. From Fig. 11-14, estimate the number of counts which must be collected from the source and the background to achieve a standard deviation equal to 2%.

Experiment 11 - 3   Dead Time of Counter

Use two samples of small size and high activity (approximately 10,000 counts/min, if using a Geiger counter), mounted on thin Mylar.

Count sample A for a period of time sufficient to achieve 1% standard deviation. Without disturbing sample A, insert sample B and count again. Remove sample A, without changing the relative position of sample B, and count sample B alone. If desired, this procedure may be repeated in some other geometric arrangement, but without exceeding a counting rate of 25,000 counts/min for the combined samples.

Correct the counting rates for the background counting rate. Compute the dead time from Eq. 11-16.

Experiment 11 - 4   Counter Geometry

The counting rate observed with any detector varies with the distance and the subtended angle from the detector window to the source. Because Geiger and proportional counters are usually mounted on a shelf arrangement, it is necessary to know the percentage of the activity recorded by the counter for each of the various shelf positions.

Place a radioactive standard with a known rate of disintegration on the uppermost shelf and record the activity. Repeat with the standard placed on each of the remaining shelf positions. Correct the observed activity for background and for absorption losses in the air path and window (see Experiment 11-5). Calculate the percentage of disintegrations registered by the counter for each shelf position.

Experiment 11 - 5   Absorption Curve for Beta or Gamma Emitters

Set the operating voltage of the counter at the predetermined value, and measure the background counting rate.

Place a sample of rather high activity in a fixed position; shelf 2, numbering downward, is convenient. Measure the counting rate. Add absorbers of known value in a position approximately midway between the source and the counter (uppermost shelf is convenient). Measure the counting rate for successive thicknesses of absorbers until the background counting rate or a constant counting rate is attained. With each absorber count for 5 min or until 5000 counts have been accumulated. The absorption curve for beta emitters is obtained with aluminum foil and massive metal of various thicknesses. The absorption curve for gamma radiation and high-energy beta emitters is studied by positioning lead absorbers of varying thicknesses between the counter and source. It is preferable to use radioisotopes that are single beta or gamma emitters.

Plot the results on semilog graph paper—the net activity as counts per minute, corrected for background and dead time loss, on the ordinate logarithmic scale vs. the absorber thickness in milligrams per square centimeter on the abscissa. Extrapolate the plot to zero absorber thickness. Consider the absorption of the sample itself, any mounting cover (Scotch tape, 10 mg cm^{-2}), and the air path (1 mg cm^{-2}/cm) between the sample and counter plus the window of the counter (marked on counter). Often these total approximately 50 mg cm^{-2}. From the activity at zero absorber thickness and that at 50 mg cm^{-2}, estimate the correction to be added to each measured counting rate.

Estimate by visual extrapolation the range of the beta particles having the maximum energy. From Eq. 11-1 calculate the maximum energy of the beta particles, or use the curves given in Friedlander and Kennedy's book for the energy of the gamma radiation and the maximum energy of the beta particles. From the absorption curve, determine the half-thickness in aluminum (or in lead) for the particular emitter.

The self-absorption of the beta-radiation of ^{63}Ni in metallic nickel sources as a demonstration experiment has been described. [W. J. Gelsema, L. Donk, J. H. T. F. P. v. Enckevort, and H. A. Blijleven, *J. Chem. Educ.*, **46**, 528 (1969).]

Experiment 11-6   Back-Scattering of Beta Particles

Use a source that is mounted on a piece of cellophane (not over 1.5 mg cm^{-2}) that is cut to fit over a 2-cm hole in a sample holder. Place the source on the bottom of the holder with Scotch tape.

Measure the counting rate of the source with no added backing. Repeat with successive pieces of plastic (5.0 mg cm^{-2}) taped carefully to the back of the sample holder. Reposition the holder carefully each time to ensure reproducibility with respect to the counter.

Measure the counting rate of the source with various thicknesses of aluminum taped underneath the sample holder. Use thicknesses up to at least 0.4 range for E_{max}. Repeat with thick pieces of various metals taped to the back of the holder—aluminum, copper, silver, platinum, and lead of sufficient thickness to ensure saturation.

Plot the counting rate vs. the thickness (in mg cm^{-2}) of added backing material. Separate curves will be obtained for the plastic and aluminum. Show where the saturation thickness is effectively achieved with aluminum.

Plot the percentage increase in net counting rate against the atomic number of the backing material for saturation backing. At the instructor's discretion, determine the atomic number of an unknown material by using it in a back-scattering experiment.

Experiment 11 - 7   Identification of a Radionuclide from Its Decay Curve

The decay of a short-lived radionuclide is followed by measuring the activity at appropriate intervals of time, always placing the sample in the same position relative to the counter each time a counting rate is determined. Suitable elements include manganese-56, antimony-122, bromine-80, iodine-128, and indium-116, methods for which have been described in the section on induced radioactivity. In addition, barium-137m ($t_{1/2}$ = 2.63 min) can be separated from its cesium-137 parent by elution from Dowex 50W-x4 (20–50 mesh) in the potassium form at pH 11 with 0.05 to 1.0% (w/v) EDTA. [R. L. Hayes and W. R. Butler, Jr., *J. Chem. Educ.*, **37**, 590 (1960).]

Longer-lived radionuclides can also be used, but it will necessitate measurements over a period of days or weeks.

Plot the net activity (corrected for background) against time on semilog paper. Determine the half-life over several time intervals and report the mean value. An unknown radioisotope (from among a restricted list) can also be identified from its absorption curve.

When a mixture of activities is present, a resolution of the decay curve is performed by the method of successive differences. First, extrapolate the linear portion (the "tail" of the decay curve) to zero time. Subtract the activity associated with the longer-lived component from the gross decay curve to obtain the activity of the shorter-lived component at different time intervals and consequently its decay curve. If the gross residual decay curve is now linear, the half-life of the longer-lived isotope can be computed; if not linear, the extrapolation process is repeated once more.

Experiment 11 - 8   Isotope Dilution Analysis

Experiments involving isotope dilution necessarily depend on the radioisotope and sample available. Directions for the use of iron-55/59 will illustrate the possibilities.

Prepare a solution of ferric chloride with a known concentration of iron and containing a small amount of iron-55/59. Add 1 ml of this solution, containing approximately 15 mg of iron, to a centrifuge tube which contains about 5 ml of water. Precipitate the iron with aqueous ammonia, centrifuge, wash with water, and recentrifuge. Transfer a portion of the precipitate to a weighed metal planchet and dry under an infrared lamp. Weigh the sample and then count with an aluminum absorber (at least 200 mg cm^{-2}) above it. Record the weight (W_1) and the activity (A_1).

Next, take 1 ml of an iron solution of unknown concentration and add to it 200 μl of the known solution. Mix the solution thoroughly and treat the sample in the same manner as the standard. Record the final weight (W) and counting rate (A). Compute the amount of iron in the unknown solution by means of Eq. 11-18.

Relative isotope dilution analysis using samples of infinite thickness and involving calcium-45 has been described. [C. B. Johnston, G. W. Drake, and W. E. Wentworth, *J. Chem. Educ.*, **46**, 284 (1969).]

Experiment 11 - 9   Solubility of Precipitates

A 0.250-g sample of potassium iodide is spiked with 5 μCi of iodine-131 and then precipitated with an excess of lead ion. The precipitate is filtered, washed, and suspended in 100 ml of water (or an aqueous solution with a slight excess of either common ion). After filtration the activity of the filtrate is measured and the solubility of the lead iodide calculated.

An analogous experiment could involve measurement of the solubility of strontium sulfate, using about 0.200 g of potassium sulfate that has been spiked with 50 μCi of sulfur-35. Other experiments should suggest themselves.

PROBLEMS

1. The following measurements were obtained with a proportional flow counter:

Applied Voltage, V	Observed Count Rate, counts/min	Applied Voltage, V	Observed Count Rate, counts/min
1200	225,000	1600	231,000
1300	231,700	1700	231,500
1400	232,400	1800	233,000
1500	231,400	1900	271,000

Plot the results on graph paper and select an operating voltage.

2. Using a gamma emitter, these measurements were obtained with an ionization chamber:

Applied Voltage, V	Observed Count Rate, counts/min	Applied Voltage, V	Observed Count Rate, counts/min
1200	19,400	1470	41,000
1245	29,100	1500	40,900
1290	35,700	1545	41,100
1335	38,900	1590	41,500
1380	40,500	1635	42,200
1425	40,600	1680	45,100
		1725	48,700

Plot the results on graph paper and select an operating voltage.

3. A 50.0-mg sample of U_3O_8, compressed into virtually a point source, was placed in each of five shelf positions and the net activity recorded:

Shelf	1 (uppermost)	2	3	4	5
Activity, counts/min	11,130	2,540	1,130	478	286

What is the geometry of each position?

4. Calculate the weight of (a) 1 Ci of ^{32}P, (b) 1 mCi of ^{36}Cl, (c) 5 mCi of ^{131}I, (d) 10 μCi of ^{193}Au, and (e) 10 mCi of ^{60}Co.

5. How many microcuries are in a 50-mg sample of U_3O_8?

6. A sample of ^{35}S contains 10 mCi. After 174 days, how many disintegrations per minute occur in the sample?

7. How much activity of the ^{32}P, the ^{131}I, and the ^{198}Au (in Problem 4) remains (a) after 14 days, (b) after 30 days, and (c) after 60 days?

8. Calculate the probable error, the 1-σ, and the 2-σ variations (in percent) for each of these total number of counts: (a) 3200, (b) 6400, (c) 8000, (d) 25,600, and (e) 102,400.

9. Compute the dead time of the Geiger counter and the corresponding counting losses from this information: sample A gave a count rate of 9.728 counts/min, sample B gave a rate of 11,008 counts/min, and together samples A plus B gave a rate of 20,032 counts/min.

10. Assuming that the dead time of a Geiger counter is 200 μsec, and that there are no other counting losses, what is the efficiency of the counter for (a) 2500 ionizing particles per second; (b) 1000, (c) 200, and (d) 5?

11. What is the useful range of counting rates if the dead time of the detector is (a) 0.25 μsec, (b) 1.0 μsec, (c) 5 μsec, and (d) 270 μsec?

12. A sample gave a counting rate of 200 counts/min in a 10-min count. The background gave a counting rate of 40 counts/min in a 5-min count. What is the fractional 0.95 error (2-σ) of the sample corrected for background?

13. For the sample in Problem 12, how much time should be devoted to counting the sample and the background if the standard deviation of each measurement, corrected for background, is to be (a) 5%, (b) 2%?

14. Using a RaD-E sample mounted on a thin film of plastic, the count rate in a particular shelf position was 6531 counts/min (air). When different materials were placed immediately in back of the sample, the count rate was 6755 with aluminum, 6819 with copper, and 7609 with lead. Determine the scattering effect and report it as the percent change in count rate relative to air.

15. The following experimental data were obtained when the activity of a beta-active sample was measured at the intervals shown.

Time, min	Activity, counts/min	Time, min	Activity, counts/min
10	542	80	56
20	315	90	50
30	200	100	44
40	135	120	37
40	100	140	31
60	77	160	27
70	64	180	23

Plot the decay curve on semilog paper and analyze it into its components. What are the half-lives and the initial activities of the component activities?

16. The decay of a particular halogen, subjected to several hours of irradiation, provided the following data:

Time, min	Activity, counts/min	Time, min	Activity, counts/min
10	1800	50	650
18	1400	60	550
24	1215	80	430
32	970	120	330
36	880	180	270
40	800	240	230

Plot the decay curve on semilog paper and analyze it into its components. What are the half-lives and the initial activities of the component activities? Can you identify the particular halogen?

17. The following data were obtained when the activity of a beta emitter was measured at the intervals shown.

Time, min	Activity, counts/min	Time, min	Activity, counts/min
0	10,000	120	560
10	6,425	140	468
20	4,695	160	425
40	2,340	180	370
60	1,320	200	332
80	894	240	265
100	680		

Plot the results on semilog paper and analyze it into its components. What are the half-lives and the initial activities of the component activities?

18. In a certain measuring arrangement, the beta particles of ^{136}Cs are absorbed as follows (correction made for gamma radiation):

Thickness of Aluminum, mg/cm^2	Activity, counts/min	Thickness of Aluminum, mg/cm^2	Activity, counts/min
0	10,000	53	270
12	4,700	72	45
27	1,700	85	10
41	730	100	10

Find the maximum energy of the beta radiation. What is the aluminum half-thickness?

19. The absorption of RaE (^{210}Bi) beta radiations produced the data below, uncorrected for counting losses or background. Dead time of counter is 200 μsec. Assume the absorption of the sample, air path, and counter window to be 34 mg/cm^2. Background counting rate is 30 counts/min.

Thickness of Aluminum, mg/cm^2	Activity, counts/min	Thickness of Aluminum, mg/cm^2	Activity, counts/min
0	19,100	163	1,620
25	13,680	200	850
57	8,720	265	270
90	4,820	335	40
123	3,080	1000	30

Determine the maximum energy of the beta radiation.

20. The visual range of absorber thickness for ^{32}P is 780 mg Al/cm^2. Determine the maximum energy of the beta particle.

21. Absorption data taken for ^{36}Cl indicated an aluminum half-thickness of 28 mg cm^{-2}. What is its maximum beta energy?

22. To a crude mixture of organic compounds containing some benzoic acid and benzoate was added 40.0 mg of benzoic acid-7-^{14}C (activity = 2,000 counts/min). After equilibration, the mixture was acidified and extracted with an immiscible solvent. The extracted solid, following removal of solvent, was purified by recrystallization of the benzoic acid to a constant melting point. The purified material weighed 60.0 mg and gave a count rate of 500 counts/min. Compute the weight of benzoic acid in the crude mixture.

23. A fermentation broth was known to contain some Aureomycin. To a 1000-g portion of the broth was added 1.00 mg of Aureomycin containing carbon-14 (specific activity = 150 counts/min/mg). From the mixture, 0.20 mg of crystalline Aureomycin was isolated which had a net activity of 400 counts in 100 min. Calculate the weight of Aureomycin per 1000 g of broth.

24. If a 10.0-mg precipitate of $BaSO_4$ contained 0.1 μCi of ^{35}S, what fraction of the precipitate contains radiosulfur?

25. Argon ionization detectors, employed in gas–liquid chromatography, utilize as radioisotope source ^{90}Sr and its daughter ^{90}Y. Estimate the range of the beta particles emitted in air and in iron (the material of construction of cell walls).

26. A 10.0-ml volume of a chloride-ion solution was added to a 50-ml volumetric flask and precipitated with 10.0 ml of 0.0440 N silver nitrate solution which contained silver-110. After the precipitate coagulated, the flask was filled to the mark and mixed thoroughly. A 20-ml aliquot of the clear supernatant liquid, after filtration or centrifugation, was counted and it gave a count rate of 924 counts/min. A 5.0-ml volume of the standard silver solution, diluted to 20 ml and counted, gave a count rate of 7555 counts/min. The background amounted to 100 counts/min. What is the chloride-ion concentration in the unknown solution?

27. In the analysis of mixtures of sodium and potassium carbonates, the half-lives are too nearly the same to permit a resolution of the composite gross decay curve if the total beta radiations were counted. Suggest a method for the analysis of this binary mixture. (*Hint:* the beta particle emitted in the decay of the sodium activity possesses a maximum energy of 1.39 MeV compared with the potassium activity, where beta particles of a maximum energy of 3.52 MeV are radiated.)

28. If a 10.0-mg sample of aluminum foil were irradiated for 30 min in a neutron flux of 5×10^{11} neutrons cm^{-2} sec^{-1}, how long should the sample be allowed to "cool" before chemical processing or counting in order that the strong aluminum activity will have decayed to less than 1 count/min?

29. For the irradiation time and flux stated in Problem 28, what is the limit of detection (40 counts/sec) for traces of sodium as sodium-24 in "pure" aluminum foil after the aluminum activity has decayed to less than 1 count/min. Counting geometry is 100%. Assume no other activities are present and ignore corrections for absorption of sodium beta particles by the aluminum foil.

30. What weight of sample should be taken for the activation analysis of an aluminum alloy which contains 0.019% zinc if the irradiation time is 62 hr with a flux of 5×10^{11} neutrons cm^{-2} sec^{-1}, followed by a cooling period of 24 hr? A counting rate of 1000 counts/min is desirable.

31. In a particular aluminum alloy, these elements are present in the following percentages: Cu, 0.30; Mn, 0.30; Ni, 0.59; Co, 0.0053. If all samples were 10.0 mg in weight, how long should the irradiations be continued for the determination of each element? Assume a counting rate of 10,000 counts/min in a 5-min counting period is desirable after a cooling period of 0.7 day. Flux is 5×10^{11} neutrons/cm²/sec.

32. Following the irradiation conditions for copper in Problem 31, how many hours should elapse before a direct determination of copper (without intermediate chemical processing) is attempted? What thickness of aluminum absorber will attenuate completely all beta radiation from other elements when measuring the gamma radiation from radiocopper?

33. With a neutron flux of 10^5 neutrons cm^{-2} sec^{-1}, what activities (disintegrations per minute per milligram) can be anticipated for iodine-128, bromine-80, antimony-122, and manganese-56?

34. From the gamma-ray spectrum of neutron-activated sea water taken with a Ge(Li) detector, identify the elements present. Photopeak energies are expressed in keV.

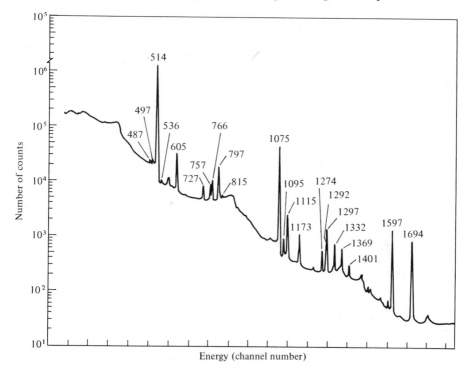

Problem 11-34

35. From the gamma-ray spectrum taken with a Ge(Li) detector, decide whether a bullet produced the circular opening from which the sample was taken.

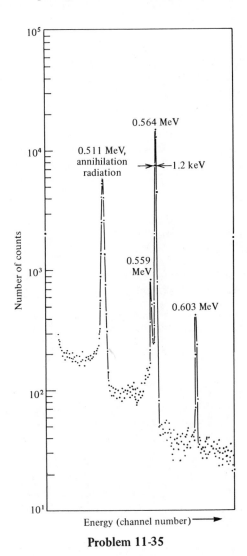

Problem 11-35

36. For tin-119, calculate the line width and Doppler velocity required to achieve resonance absorption. The transition energy of the $\frac{3}{2}$ level is 23.8 keV; the half-life is 19 nsec.

37. Iron(II) salts show very marked quadrupole splitting, as does $K_3[Fe(CN)_6]$, whereas 3+ iron salts show only minor quadrupole splittings, and $K_4[Fe(CN)_6]$ does not show any quadrupole splitting (Fig. 11-17). Suggest reasons for these observations.

BIBLIOGRAPHY

Choppin, G. R., *Experimental Nuclear Chemistry*, Prentice-Hall, Englewood Cliffs, N.J., 1961.

Crouthamel, C. E. and R. R. Heinrich, "Radiochemical Separations," in *Treatise on Analytical Chemistry*, Vol. 9, Part 1, I. M. Kolthoff and P. J. Elving, Eds., Wiley-Interscience, New York, 1971, Chapter 96.

DeSoete, D., R. Gybels, and J. Hoste, *Neutron Activation Analysis*, Wiley-Interscience, New York, 1972.

Finston, H. L., "Radioactive and Isotopic Methods of Analysis: Nature, Scope, Limitations and Interrelations," in *Treatise on Analytical Chemistry*, Vol. 9, Part 1, I. M. Kolthoff and P. J. Elving, Eds., Wiley-Interscience, New York, 1971, Chapter 94.

Finston, H. L., "Nuclear Radiations: Characteristics and Detection" in *Treatise on Analytical Chemistry*, Vol. 9, Part 1, I. M. Kolthoff and P. J. Elving, Eds., Wiley-Interscience, New York, 1971, Chapter 95.

Friedlander, G., J. W. Kennedy, and J. M. Miller, *Nuclear and Radiochemistry*, 2nd ed., Wiley, New York, 1964.

Goldanskii, V. I. and R. H. Herber, Eds., *Chemical Applications of Mössbauer Spectroscopy*, Academic Press, New York, 1968.

Guinn, V. P., "Activation Analysis" in *Treatise on Analytical Chemistry*, Vol. 9, Part 1, I. M. Kolthoff and P. J. Elving, Eds., Wiley-Interscience, New York, 1971, Chapter 98.

Lyon, W. S., Ed., *Guide to Activation Analysis*, Van Nostrand Reinhold, New York, 1964.

May, L., Ed., *An Introduction to Mössbauer Spectroscopy*, Plenum Press, New York, 1971.

Seaman, W., "Tracer Techniques" in *Treatise on Analytical Chemistry*, Vol. 9, Part 1, I. M. Kolthoff and P. J. Elving, Eds., Wiley-Interscience, New York, 1971, Chapter 97.

LITERATURE CITED

1. Brooksbank, W. A., G. W. Leddicotte, and J. A. Dean, *Anal. Chem.*, **30**, 1785 (1958).
2. Fluck, E., W. Kerler, and W. Neuwirth, *Angew. Chem. Internat. Ed. Engl.*, **2**, 277 (1963).
3. May, L., *Appl. Spectrosc.*, **23**, 204 (1969).
4. Seidel, C. W., *Am. Laboratory*, **1**,(2), 22 (Feb. 1969).

12

Flame Emission and Atomic Absorption Spectrometry

Three flame spectrometric methods: flame emission, atomic absorption, and atomic fluorescence—will be discussed in this chapter. In each method a fine spray of the sample solution is introduced into a flame where it is desolvated, vaporized, and atomized. The use of a solution spray permits a uniform distribution of sample throughout the body of the flame and the introduction of a representative portion of each sample into the flame. Figure 12-1 shows schematically the operational techniques for each of the methods. Non-flame absorption devices are useful when the sample is limited or in the solid state.

In *emission flame spectrometry* atoms and molecules are raised to an excited electronic state through thermal collisions with the constituents of the burned flame gases. Upon their return to a lower or ground electronic state, the excited atoms and molecules emit radiations that are characteristic for each element. The emerging radiation passes through a monochromator that isolates the desired spectral feature. The spectrum is then registered by a photodetector, the output of which is amplified and measured on a meter or recorder. Correlation of the emission intensity with the concentration of the test substance in the solution spray forms the basis of quantitative evaluation.

In *atomic absorption spectrometry* radiation from an external light source, emitting the spectral line(s) that correspond to the energy required for an electronic transition from the ground state to an excited state, is passed through the flame. The flame gases are treated as a medium containing free, unexcited atoms capable of absorbing radiation from an external source when the radiation corresponds exactly to the energy required for a transition of the test element from the ground electronic state to an upper excited level. Unabsorbed radiation then passes through a monochromator that isolates the exciting spectral line of the light source and into a detector. The absorption of radiation from the light source depends on the population of the ground state, which is proportional to the solution concentration sprayed into the flame. Absorption is measured by the difference in transmitted signal in the presence and absence of the test element.

Atomic fluorescence involves irradiation of the atomic vapor in the flame with a suitable illuminating source placed at right angles to the flame and optical axis of the spectrometer. Some of the incident radiation is absorbed at wavelengths corresponding to atomic absorption lines. Immediately afterwards the energy is released as fluorescence,

FLAME EMISSION

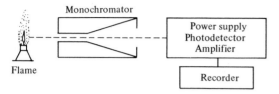

ATOMIC ABSORPTION

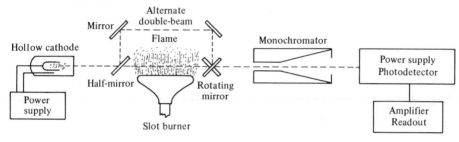

ATOMIC FLUORESCENCE

Fig. 12-1. Schematic diagrams showing the measuring techniques for flame emission, atomic absorption, and atomic fluorescence spectrometry.

which is emitted in all directions, of characteristic wavelength. The intensity of the fluorescence is directly proportional to the concentration of test element in the sample. This method is still in its development stage.

INSTRUMENTATION

For all three flame spectrometric methods the following components are required: pressure regulators and flow meters for the fuel and oxidant gases; nebulizer and burner assembly; optical system and photosensitive detector(s); and amplifier and readout system with attendant power supplies. In addition, for atomic absorption and atomic fluorescence, a suitable light source is needed for each element being determined; frequently, a light interrupter is also placed between the light source and the flame.

Pressure Regulators and Flow Meters

To maintain a constant thermal environment in the flame, it is imperative that the gas pressures and gas flows be held constant while the flame spectrometer is in use. With double-beam instruments this requirement can be relaxed somewhat. Double-diaphragm pressure regulators—10 psi for the fuel and 30 psi for the oxidant supply—followed by a rotameter and needle valve (or orifice interface) should be installed in the lines from the gas cylinders to the nebulizer–burner assembly. Usual flows range from 1 to 4 liters/min. A knowledge of the individual flows of fuel and oxidant enables an operator to choose various mixtures ranging from lean to fuel-rich types of flames.

Nebulizer-Burner System

The most important component in any flame emission or atomic absorption spectrometer is the nebulizer–burner system. This system converts the test substance in the sample solution to atomic vapor and, also in emission, excites the neutral atoms (or molecules) to emit their characteristic radiation. No memory effect should exist; that is, the content of one sample should not affect the result from the next. Other requirements of the system are ease of cleaning, freedom from corrosion, and ease of adjustment.

Since the optimum flame region to be used varies for different elements, it is essential that the burner height be adjustable to secure the maximum emission or absorption signal. It is advisable to remove the products of combustion and heat by an exhaust hood over the burner.

Premixed or Laminar-Flow Burner   In the premixed gas or laminar-flow burner, the aerosol is produced within a mixing chamber where the coarse and fine droplets are separated. The fine droplets, virtually a fog, are mixed with the flame gases; then they are vaporized. Baffles (or spoilers) inside the expansion chamber ensure adequate mixing and smooth delivery to the flame. The burner head consists either of an array of holes or a slot at which the flame burns; they are circular in flame emission to concentrate the emitted light, but elongated in atomic absorption to get the maximum amount of flame gases in the light path. This type of burner serves well for flames with low burning velocities, such as air-propane, air-acetylene, and nitrous oxide-acetylene. A typical design is shown in Fig. 12-2. The burner is usually a 10-cm slot, single or triple, 0.38-0.6 mm in width, in a head fabricated from aluminum or stainless steel; the slot is shortened to 5 cm when employing a nitrous oxide-acetylene flame. To prevent flashback down the burner port and an explosion in the mixing chamber, the streaming velocity of the fuel-oxidant mixture through the burner port must be at least equal to the burning velocity; however,

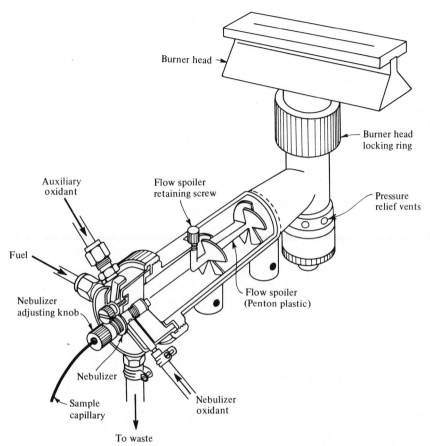

Fig. 12-2. A slot burner and expansion chamber. (Courtesy of Perkin-Elmer Corporation.)

to ensure a margin of safety and achieve a reasonably stiff flame, the streaming velocity should be several times the burning velocity. The burning velocity of a fuel-oxidant mixture is a function of its composition, and usually reaches its maximum value near the stoichiometric ratio (Table 12-1).

Turbulent or Sprayer-Burner The turbulent or sprayer-burner injects the sample directly into the flame. The fuel, the combustion supporting gas, and the aerosol are all passed through separate channels to an opening on which the flame rests (Fig. 12-3). The flame is tall and narrow, with the hottest portion probably no more than 1 cm across. It is well suited for emission work but provides a relatively short light path for absorption studies. The path length can be increased in several ways: by using three burners in series, by passing the light beam 3–5 times back and forth through the flame, and by inclining and pointing the burner into the end of a long ceramic tube. All the sample reaches the flame although the larger droplets may travel through the flame without being desolvated.

TABLE 12-1 Adiabatic Flame Temperatures and Burning Velocities of Common Premixed Flames

Fuel	Oxidant	Temperature, °C	Burning Velocity, cm/sec
Natural gas	Air	1700-1900	55
Propane	Air	1925	—
Propane	Oxygen	2800	—
Hydrogen	Air	2000-2050	320-440
Acetylene	Air	2120-2400	160
Hydrogen	Oxygen	2550-2700	915
Acetylene	Nitrous oxide	2600-2800	460
Acetylene	Oxygen	3050-3130	1100

Audible noise is a problem and arises from the turbulence created as the gases are mixed and burned in the same local area. The turbulence of the sprayer-burner and the presence of unvaporized droplets in the flame must count as disadvantages for atomic absorption studies. Because flashback is impossible, the direct-injection burner is particularly well suited to safe operation with gas combinations having high burning velocities, such as oxygen-hydrogen or oxygen-acetylene.

Nonflame Atom Cells Nonflame atom cells fill an important ancillary role in atomic absorption spectrometry. The conventional nebulizing system wastes sample, and the residence time of metal atoms in the light path in a conventional flame is very short, on the order of milliseconds. The heated-graphite (Massmann) furnace consists of a hollow graphite cylinder 50 mm long and 10 mm in diameter placed so that the sample beam passes through it (Fig. 12-4). In the middle of the cylinder top there is a hole about 2 mm in diameter through which the sample (1 to 100 μl) is pipetted. The cylinder is purged with an inert gas, either nitrogen or argon. An electric current, in three stages, is passed through the cylinder walls to evaporate the solvent, to ash the sample, and finally to raise

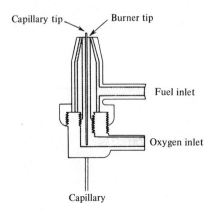

Capillary tip

Burner tip

Fuel inlet

Oxygen inlet

Capillary

Fig. 12-3. Integral-aspirator burner. (Courtesy of Beckman Instruments, Inc.)

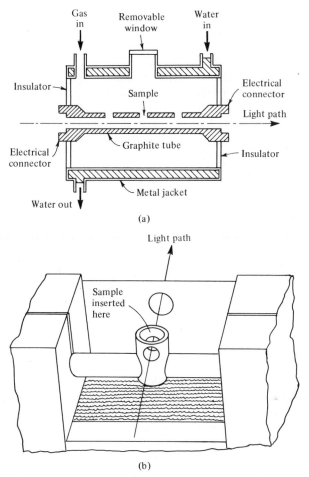

Fig. 12-4. Nonflame atom cells. (a) Cross section of heated graphite furnace. (Courtesy of Perkin-Elmer Corporation.) (b) Carbon rod and cup (mini-Massmann) furnace. (Courtesy of Varian Techtron.)

the unit to incandescence to atomize the sample. A transient absorption signal, produced as the metal is atomized, is measured on a recorder. The time spent by atoms in the light path as they diffuse down the graphite tube is on the order of 2-3 sec.

A miniature version, often referred to as the "mini-Massmann furnace," consists of a carbon or graphite rod with a sample compartment in the upper center (Fig. 12-4). Sample volumes up to 5 μl can be accommodated. Variations on this design use a vertical cup or a tantalum strip (as a V-shaped trough). A chimney placed beneath the rod or strip permits the introduction of a laminar flow of argon, hydrogen, or nitrogen. The sequence of evaporation, ashing, and atomization operations is similar to that with the larger furnace, but the time required (and power consumption) is less—a few seconds for each step. Residence time of atoms in the light beam is 0.5 sec or less, but because atomization occurs so rapidly the peak absorption value is high. Surrounding the rod with a hydrogen diffusion flame markedly reduces interelement interferences and nonatomic absorption, and gives improved atomization for less volatile elements.

The most attractive features of nonflame cells are high sensitivity (10^{-8} to 10^{-11} g absolute), the capability of handling very small sample volumes or amounts, direct analysis without pretreatment (particularly for biological samples), and low noise signal. Matrix interference effects are often greater than in flame systems, and the precision, typically 5-10%, compares unfavorably with that of flames, typically 1%.

Light Sources for Atomic Absorption and Atomic Fluorescence

Source requirements for atomic absorption are somewhat different than those for atomic fluorescence. For atomic absorption the resonance spectral lines should be narrow as compared with the width of the absorption line to be measured, bright against a very low background, and stable. The light of a specific element must not suffer interference from other spectral lines that are not resolved by the spectrometer, such as lines originating from impurity elements, electrode materials, or carrier gases. For atomic fluorescence the main requirement is high intensity of the resonance line at the absorption wavelength peak. All sources will have a finite warm-up time before their light output becomes essentially constant, approximately 15-30 min.

Hollow-Cathode Lamp   The hollow-cathode discharge lamp consists of an anode and cylindrical cathode enclosed in a gas-tight chamber, as shown in Fig. 12-5. For elements difficult to fabricate as a solid cathode, a liner or pool of molten metal suffices. Alloys can be used provided they do not require a higher current for adequate excitation, nor suffer from selective volatilization which would cover the surface of the alloy. The tube is evacuated and filled with an ultra-pure monatomic carrier (or filler) gas, usually argon or neon, at about 1-5 torr. Applying a high potential across the electrodes creates a low-pressure discharge. The inside of the cathode is filled with a negative glow. Positive ions are formed through ionization of the carrier gas atoms by electron impact. These newly formed ions are accelerated through the high electrical gradient toward the cathode, and upon collision with the metal in the cathode cavity sputter the metal into the discharge region. Atoms of the metal are then excited to emission by collisions with gas ions and electrons that have been slowed down enough by previous collisions to excite atomic spectra. Surfaces where the discharge is undesirable are insulated (or shielded). Argon is preferred as carrier gas. The lower excitation potential of this gas compared with that of neon favors the production of resonance lines (as contrasted to ionic lines), and the greater mass of the argon atom facilitates the sputtering action. However, atomic lines are significantly brighter in neon, although the ionic spectra are also more intense, especially as the pressure increases (above 5 torr). Lamps fail as a result of the loss of the carrier gas by absorption on the interior surfaces of the lamp (called *gas cleanup*).

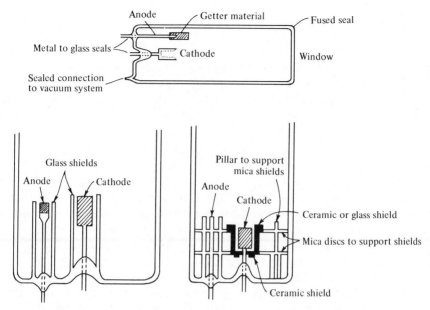

Fig. 12-5. Schematic diagram of shielded-type hollow-cathode lamp.

Vapor Discharge Lamp   For low-melting elements, a vapor discharge (Osram, Philips, or Wotan) lamp produces an arclike spectra which makes it a convenient light source for elements such as the alkali metals, zinc, cadmium, and mercury. The requisite vapor pressure of the element is produced by electrical heating, and this vapor is then excited by a low-voltage electrical discharge. Primary electrons released from an oxide-coated cathode are accelerated as in any other type of glow discharge. Secondary electrons released as a result of collisions between the primary electrons and the carrier gas (often neon) do not have sufficient energy to ionize carrier gas atoms further, but can excite metal atoms with low excitation potentials. The current density through the tube and the density of metal atoms present are the main contributing factors to the intensity of radiation.

Other Sources   An intense continuum source, such as the xenon arc, finds use in atomic fluorescence. The microwave-excited electrodeless discharge lamp* is well suited for atomic absorption and atomic fluorescence work. A few milligrams of metal or its halide is placed in a thin quartz tube under a few torr pressure of inert gas, and the tube is placed in a cavity within a microwave field—often 2450 MHz, as supplied by a medical diathermy unit.

*For preparative details, see R. M. Dagnall and T. S. West, *Applied Optics* 7, 1289 (1968).

The Optical System

The function of the optical system is to select a given line in the emission spectrum and to isolate it from all other lines. Inexpensive flame (emission) photometers employ interference filters to isolate the radiation characteristic of a given element. These instruments are limited in application to those samples and to those elements that, upon excitation by a flame, provide a simple spectrum in the visible portion of the spectrum, such as the alkali metals.

Better isolation of spectral energy can be achieved with a prism or a grating monochromator, which also provides a continuous selection of wavelengths and an opportunity to measure background radiation adjacent to an analytical line. Both the Czerny-Turner and Littrow mounts are widely employed. In a general-purpose instrument a resolution of the order of 0.5 Å is desirable. Large dispersion and high resolving power in monochromators are advantageous for resolving spectra with discrete emission features, whereas emission band spectra will show up more clearly with instruments of low dispersion.

In atomic absorption the spectral slit width of the monochromator is determined by the source emission line width. The monochromator is used simply to isolate the desired line from the other source emission lines, and to lower drastically the total light flux reaching the detector. When permissible, the slits are widened to admit more light. This translates into an improvement in precision and detection limit.

Photodetector, Amplifier, and Readout System

Use of a barrier-layer cell is restricted to systems emitting a large amount of radiant energy and to instruments with an optical system that permits a wide band of radiant energy to strike the detector. These instruments are generally filter flame photometers for measuring sodium and potassium through their emission signals.

For high sensitivity and precision the multiplier phototube with its associated power supply and measuring system is essential. The spectral response of several commercial multiplier phototubes is shown in Fig. 3-8. Stabilized power supplies are required. Short-term fluctuations can sometimes be handled with an integrating system in which the detector output is fed either to a condenser for a fixed period of time (perhaps 30 sec) and the resulting output measured, or passed to a recorder for 1-2 min and the tracings averaged.

The troublesome effect of radiation from within the flame in atomic absorption and atomic fluorescence is eliminated by inserting a mechanical chopper between the light source and the flame, or operating the lamp from a pulsed dc supply and using an ac measuring circuit sharply tuned to the modulating frequency. Care must be taken never to exceed the saturation limit of the photodetector by flooding it with light emanating from within the flame and arising from sample components present in high concentration even though their signals are eliminated by modulation.

When the electronic noise in the detector-amplifier system can be reduced to a negligible level, scale expansion can profitably be employed. The zero of the readout meter or recorder is displaced off scale. In this way the readout device can be used as the upper end of a much longer scale and the reading for small decreases in transmitted light may be increased manyfold. Analogously, a zero suppressor circuit is useful in emission work for bucking out a reading resulting from the flame background.

Commercial Instruments

A single-beam instrument contains only one set of optics. Light emitted from the flame, or transmitted through the flame from the atomic absorption light source, is either (1) focused by a lens of heat-resistant glass and passed through interchangeable interference filters onto a single photodetector, or (2) passed into a monochromator and radiation leaving the exit slit is focused onto the detector. One or more stages of amplification is necessary before the signal is presented to the readout device. For recording purposes in the emission mode, the wavelength dial is driven by a drive mechanism that is synchronized with the chart drive of the recorder. In ac measuring systems, radiation leaving the exit slit of the monochromator is chopped with a rotating disc and amplified by an ac amplifier.

To adapt a single-beam instrument into an atomic absorption spectrometer (see Fig. 12-1) requires only a light source placed ahead of the flame, with a mechanical chopper placed between the light source and the flame (unless the lamp is operated from an ac or or pulsed dc power supply). The output of the detector-amplifier unit is arranged to give full-scale deflection with no sample in the flame. It is important that the power supply to the lamp and photodetector be highly regulated and that the gain of the amplifier be independent of power line variations to ensure that the conditions do not change between the time that standards and samples are sprayed into the flame.

In double-beam instruments the emission signal is split into two beams, the mode differing between flame emission and atomic absorption. For emission instruments a second light path is provided for the light emitted from the flame by the internal standard element that is added in a fixed amount to samples and standards alike. Appropriate interference filters (or a dual monochromator) isolate the emission lines of the test substance and internal standard, and each isolated line is focused onto a separate detector. The signal of one detector opposes that of the other as is done with double-beam filter photometers. (Also see Fig. 12-7 and the discussion of a dual double-beam optical system.)

In atomic absorption spectrometers, one beam is directed through the flame while the other bypasses it. After recombination, the two beams pass through a monochromator to a detector and readout system. In the double-beam, time-shared grating spectrometer, shown in Fig. 12-6, the rotating sector (or chopper) has alternate reflecting and transparent sectors separated by opaque portions. The signal from the detector is amplified and then separated into sample and reference channels by a vibrating-reed chopper. The reference voltage is then attenuated by a slidewire (the null potentiometer) and recombined with the sample voltage in such a way that only the difference between them remains. This difference voltage is amplified, rectified, and fed to a microammeter. The

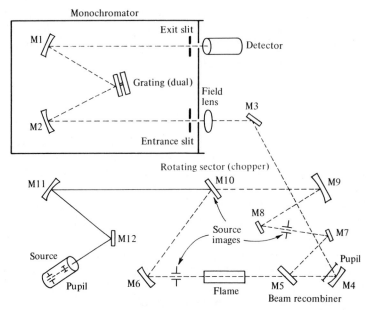

Fig. 12-6. Schematic diagram of the double-beam optical system of Perkin-Elmer model 303. (Courtesy of Perkin-Elmer Corp.)

operator turns the slidewire until the meter reads a null; the amount of attenuation required is equivalent to the percent absorption $(100 - \%T)$ in the sample beam. Alternatively, the difference signal is fed to a recorder and appears as a series of plateaus—the "zero" trace with blank spraying and the sample or standard signal with the appropriate solution spraying. The double-beam mode overcomes the effect of lamp drift and change in detector sensitivity with time, but it cannot compensate for instability and noise from the flame because the flame is in only one of the two beams. Variations in the power supply to the lamp must not be of such magnitude that they cause an appreciable change in the width of the line from the source.

A dual double-beam optical system (Fig. 12-7), with each monochromator independent of the other, provides the analyst with a number of choices: (1) two independent channels with each analyzing a separate element simultaneously; (2) one channel operates in the emission mode while the other operates in the atomic absorption mode; (3) one channel operates in the conventional absorption mode while the other channel determines the degree of background absorption at a nearby nonabsorbing line (or deuterium lamp); (4) a two-line method in which each channel is tuned to two separate absorbing wavelengths of the same element; and (5) one channel measures the signal from an added internal standard while the other channel measures the unknown signal. In a less sophisticated dual-wavelength system, two interference filters are alternately interposed in the optical path between the flame and the detector by means of a rotating wheel.

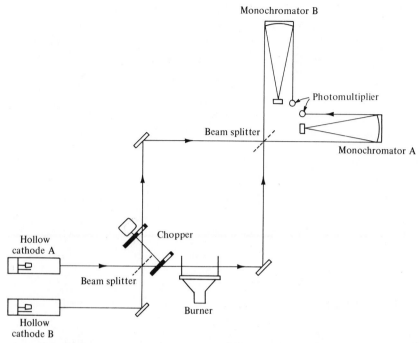

Fig. 12-7. Schematic optical diagram of a dual double-beam flame spectrometer. (Courtesy of Jarrell-Ash Division, Fisher Scientific Co.)

THEORETICAL PRINCIPLES

Flames and Flame Temperatures

The flame is used (1) for transforming the test substance from the liquid or solid state into the gaseous state—vaporization, (2) for conversion of the molecular entities into an atomic vapor—atomization, and (3) in emission studies for excitation of the atomic (and molecular) vapor to emission. Many different fuel-oxidant mixtures have been utilized. These have been ably reviewed by Gaydon,[3] Kniseley,[4] and Mavrodineanu and Boiteux.[5] The two main requirements of a satisfactory flame are that it has the proper temperature or fuel-oxidant ratio to carry out the enumerated functions of the flame, and that the spectrum of the flame itself does not interfere with observation of the emission or absorption feature being measured. Components of the flame gases limit the usable range to wavelength to those that lie outside the atmospheric-absorption region, that is, to wavelengths longer than about 2100 Å.

The structure of a premixed flame is shown in Fig. 12-8. Emerging from region A, the unburned hydrocarbon gas mixture passes into a region of free heating about 1 mm in

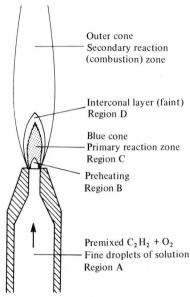

Outer cone
Secondary reaction
(combustion) zone

Interconal layer (faint)
Region D

Blue cone
Primary reaction zone
Region C

Preheating
Region B

Premixed $C_2H_2 + O_2$
Fine droplets of solution
Region A

Fig. 12-8. Schematic diagram of a stoichiometric oxygen–acetylene flame.

thickness (region B), in which it is heated by conduction and radiation from reaction region C and by diffusion of radicals into it, which initiate the combustion. Flame gases travel upwards from the reaction zone with speeds of the order of 1–10 mm/msec. Gases emerging from region C consist mainly of CO, CO_2, and H_2O (and N_2 if air is one of the original gases), with lesser amounts of the species H_2, H, O, and NO. The actual composition of the gases varies with the initial mixture composition. In the inner cone, the gases are not in thermal equilibrium, the amounts of the radicals being too high (the precursors of chemi-excitation reactions). Thermal equilibrium is achieved almost completely in region D. Combustion is completed in the outer mantle, assisted by entrainment of the surrounding air. Hydrogen flames show no obvious signs of an inner cone.

Turbulent flow is used in sprayer–burners, where the fuel and oxidant are mixed at the surface of the burner. The turbulence is necessary for thorough mixing of the gases within the combustion zone. No well-defined zones are observed other than a luminous inner cone with hydrocarbons and an outer flame mantle.

In order to attain equilibrium in flame gases the particles of the system must suffer a sufficient number of collisions with each other. At atmospheric pressure, a molecule in the flame makes about 10^6 collisions with the surrounding molecules in the time interval of 1 msec. The establishment of equilibrium for various energy conditions of a molecule or atom require the following number of collisions; vibration, 10^5; excitation and disso-ciation, 10^7; and ionization, 10^9. Thus, time intervals of interest in atomic excitation involve more than 1 msec and several mm of travel in the flame gases. Concentrations of atomic and molecular species vary with height above the burner. The flame front propagation rate for an oxygen-acetylene flame is very high, approximately 1200 cm/sec; therefore, the residence time of the atoms is very short. However, the burning velocity for the air-acetylene or nitrous oxide-acetylene flame is 160 or 460 cm/sec, respectively. The decrease in burning velocity permits a longer residence time for the atoms in the observed area and gives an increase in sensitivity. Burning velocities for various gas mixtures are given in Table 12-1.

When thermal and chemical equilibria are fully achieved, a definite adiabatic temperature exists in the hot flame gases (Table 12-1). The exact value depends on the fuel-oxidant ratio and is generally highest for a stoichiometric mixture. An air-fuel flame, because of its relatively low temperature, is only energetic enough to excite about a dozen elements, chiefly the alkali and alkaline earth metals. However, many additional elements are largely atomized, rendering this type of flame suitable for atomic absorption studies of the metals of groups IA, IB, and IIB, together with Ga, In, Tl, Pb, Te, Mn, Ni, and Pd. The hotter nitrous oxide-acetylene and oxygen-acetylene flames are advantageous when handling elements that form refractory compounds or whose volatilization is inhibited by concomitants in the sprayed solution.

The chemical environment within the flame influences the production of free atoms. Although approximately 150°K cooler than the normal stoichiometric flame, the fuel-rich acetylene flame is capable of producing intense atomic line spectra from elements that form relatively stable monoxides at ordinary flame temperatures. These same flames also increase spectrochemical sensitivity in atomic absorption. These increases may occur through reduction of metallic oxide molecules by carbon-containing species that are present in high concentrations in the fuel-rich flame, particularly in the interconal gases, and through a decrease in the rate of formation of MO molecules due to the deficiency of atomic oxygen within the flame gases.

Flame temperature is affected by the volume of aerosol injected into the flame gases. When using direct-injection burners, aqueous solutions may lower the flame temperature of the oxygen-hydrogen flame by 200-300°K, and that of the hotter oxygen-acetylene flame by as much as 500°K. The cooling effect of an organic solvent is less because heat evolved in the combustion of the solvent partially compensates for the energy consumed in the dissociation of the solvent. Little temperature change is observed when using a mixing chamber-nebulizer combination; evaporation within the chamber reduces the size of the droplets and less solvent reaches the flame.

Self-Emission of Flames

The emission spectrum of the flame gases is important because background from the flame will often contribute noise in both atomic emission and atomic absorption spectrometry. This noise, if excessive, will lead to poorer detection limits and lower precision of analysis.

The hydrogen flame gives the best signal-to-background ratio. Little background emission is present except for prominent OH band structures that occur in the 2800-2950, 3060-3200, and 3400-3480 Å regions. Between 8,000 and 12,500 Å there is a broad emission from H_2O molecules.

The acetylene flame shows, like the outer cones of most organic flames, principally the OH band spectrum and the continuum produced by dissociating CO molecules. Superimposed on this background are the fainter bands of the CH radical from 3870 to 4385 Å, several bands of the C_2 radical at 5636, 5165, 4600, and 4780 Å, and bands attributable to the CN molecule with heads at 3600, 3900, 4200, and 6100 Å. The CH, C_2, and CN bands are very strong in the spectrum of the reaction zone which, in a fuel-rich flame, may become quite elongated and luminous. In the far ultraviolet the reaction zone exhibits the atomic emission line of carbon at 2478 Å and a continuous background due to the incandescent carbon particles, upon which is superimposed the molecular band system of CO. The CN band systems are very prominent in the nitrous oxide-acetylene flame, whereas the C_2 and CH bands are considerably lower in intensity.

Nebulization

Although solids have been volatilized directly using induction heating or electron bombardment, virtually all commercial flame spectrometers rely on pneumatic nebulization of a liquid sample to deliver a steady flow of aerosol to a flame that effects atomization. The sample orifice is positioned either concentric with, or at right angles to, the annulus from which the aspirating gas exits. Liquid, drawn through the same capillary by the pressure differential generated by the high-velocity gas stream as it passes over the sample orifice, is set into oscillation. Filaments of liquid are drawn out from the bulk of the solution. The filaments collapse to form droplets, and the larger droplets may break up further under the continued action of the high-velocity gas stream. The mode of breakdown is suggested in Fig. 12-9. Sometimes an impact bead or a counter gas-jet is placed in the path of the droplets to enhance the disruptive action. The final aerosol will consist of droplets of various sizes.

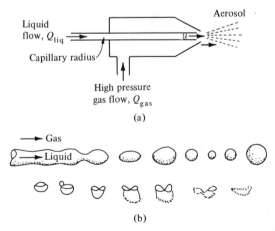

Fig. 12-9. (a) Construction of pneumatic nebulizer, and (b) breakdown of a liquid filament into droplets.

The most satisfactory equation relating to droplet production is an empirical one given by Nukiyama and Tanasawa[6];

$$d_0 = \frac{585}{v}\left(\frac{\gamma}{\rho}\right)^{0.5} + 597\left[\frac{\eta}{(\gamma\rho)^{0.5}}\right]^{0.45}\left(1000\,\frac{Q_{liq}}{Q_{gas}}\right)^{1.5} \tag{12-1}$$

where d_0 = the volume-surface or Sauter mean diameter of droplets, in μm;
   γ = the surface tension of the solution, in dynes/cm;
   η = the liquid viscosity, in dynes/cm^2;
   v = the velocity of the aspirating gas, in m/sec;
   ρ = the density of the liquid, in g/cm^3;
   Q_{liq} = the rate of liquid flow, in ml/min; and
   Q_{gas} = the rate of aspirating gas flow, in ml/min.

Thus viscosity, surface tension, and density, all physical properties of a sample solution, may be determinative factors in the drop-size distribution. When these factors are altered by concomitants, some variation in the rate of effective sample introduction into the flame and a consequent change in the signal intensity may be expected. Viscosity, the primary force opposing aspiration, appears in Eq. 12-1 both directly and indirectly through its effect on aspiration rate. Its net effect on an emission or absorption signal is difficult to predict.

When water is aspirated, a mean droplet diameter of about 20 μm, with a range from 5 to 65 μm, is found. Droplets larger than about 20 μm are lost in the spray chamber or fail to desolvate completely before leaving the reaction zone of a turbulent flame. This loss accounts for about 90% of the aspirated volume. The fraction of solution volume in

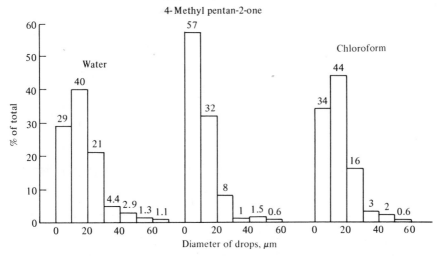

Fig. 12-10. Representation of drop-size distribution from an integral-aspirator burner. (After J. A. Dean and W. J. Carnes, *Anal. Chem.*, **34**, 192 (1962). Courtesy of American Chemical Society.)

droplets smaller than 20 μm in diameter is nearly doubled when solutions of methyl isobutyl ketone are nebulized (Fig. 12-10).[1] Surface tension is drastically reduced by low concentrations of surface-active agents, but this is effective with premix burners only. There is no effect in turbulent burners because a surface is not maintained. Errors due to unanticipated changes in viscosity and surface tension can be avoided by preparing samples and standards similarly, and avoiding total acid or salt concentrations greater than about 0.5%. Delivery of sample solution, independent of the variables associated with aspiration, can be obtained by forced delivery to the nebulizer by means of a motor-driven syringe or a peristalsis pump.

Intensity Relations in Emission and Absorption

Some of the atomic and molecular entities formed at the atomization step will undergo electronic excitation from thermal collisions with radicals and molecules in the flame gases. Thermal transfer involves conversion of internal molecular vibrational energy of a flame molecule to electronic energy. This energy transfer is facilitated by the fact that there is nearly always an energy level of the excited flame molecule that closely matches

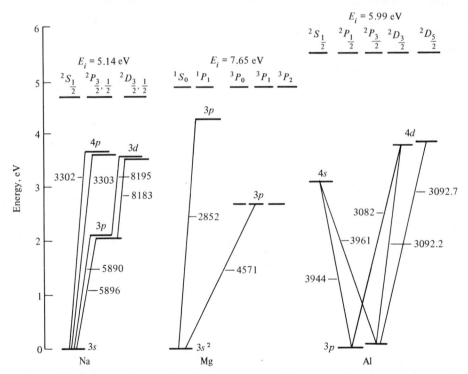

Fig. 12-11. Energy level diagram of low-lying electronic states for sodium, magnesium, and aluminum.

the excited level of the metal atom or molecule, owing to the large number of possible energy levels in a molecule. Thus there is a sort of resonance in which good efficiency may be expected. For atoms, the energy-level diagrams are of the type shown in Fig. 12-11. Lines arising from transitions involving an excited state and still higher excited states are usually very weak in an emission and absent in an absorption spectrum. Nevertheless, spectral interferences frequently arise from these transitions in emission flame spectrometry. Emission lines arising from singly ionized atoms are observed for several easily ionized elements.

Band spectra arise from electronic transitions involving molecules. For each electronic transition there will be a whole suite of vibrational levels involved. This causes the emitted radiation to be spread over a portion of the spectrum rather than be concentrated in a discrete line. Band emissions attributed to triatomic hydroxides (CaOH) and monoxides (AlO) are frequently observed and occasionally employed in flame emission spectrometry (Fig. 12-12).

When an excited atom returns to lower energy levels and luminous radiation is emitted, definite quantum jumps or energy changes are observed. Corresponding to each quantum jump there is an emission of appropriate frequency in accordance with the relationship: $\Delta E = h\nu$, where ΔE is the difference in the energies of the two levels. Under conditions of thermal equilibrium, the occupational number of atoms in an excited level, N^*, attained by a transition direct from the ground state, is related to the number of atoms present in the ground state per unit volume, N_0, which, in turn, is proportional to

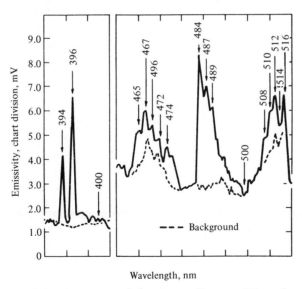

Fig. 12-12. Flame emission spectrum of aluminum. Present, 25 μg aluminum per ml in methyl isobutyl ketone. Bandpass is 0.3 nm at atomic lines and 0.5 nm at band systems. (After H. C. Eshelman, J. A. Dean, O. Menis, and T. C. Rains, *Anal. Chem.*, **31**, 183 (1959). Courtesy of American Chemical Society.)

the concentration of test substance in the sample solution. The Boltzmann equation expresses this relationship:

$$\frac{N^*}{N_0} = \frac{g^*}{g_0} \exp\left(\frac{-\Delta E}{kT}\right) \tag{12-2}$$

where g^*, g_0 are the statistical weights of the states, k is the Boltzmann constant, and T is the absolute temperature. The population of the excited state is defined by the energy of the particular state and the temperature; the actual mechanism of excitation and de-excitation are completely irrelevant. Values of the ratio N^*/N_0 are given in Table 12-2. The fraction of atoms in still higher excited states is much less, partly because of the greater energy required for excitation, but mainly because of a low transition probability.

The intensity of a spectral emission line, I_ν, is determined by the number of atoms in which identical transitions take place simultaneously. It is given by the expression

$$I_\nu = V A_{nm}\, h\nu\, N_0 \left(\frac{g_m}{g_n}\right) \exp\left(\frac{-\Delta E}{kT}\right) \tag{12-3}$$

where V is the flame volume seen by the detector and A_{nm} is the Einstein transition probability that the excited atom will spontaneously make the transition $(E_m \rightarrow E_n)$ in unit time (1 sec) with emission of a photon.

In addition to the preceding mechanism, transitions from the ground electronic state to the first excited state take place when radiation of frequency exactly equal to the res-

TABLE 12-2 Values of N^*/N_0 for Various Resonance Lines

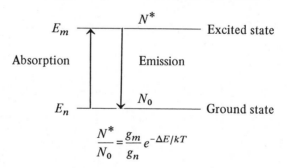

Resonance Line		g_m/g_n	ΔE, eV	N^*/N_0 2000°K	3000°K
Cs	8521 Å	2	1.45	4.44×10^{-4}	7.24×10^{-3}
Na	5890 Å	2	2.10	9.86×10^{-6}	5.88×10^{-4}
Ca	4227 Å	3	2.93	1.21×10^{-7}	3.69×10^{-5}
Fe	3720 Å		3.33	2.29×10^{-9}	1.31×10^{-6}
Cu	3248 Å	2	3.82	4.82×10^{-10}	6.65×10^{-7}
Mg	2852 Å	3	4.35	3.35×10^{-11}	1.50×10^{-7}
Zn	2139 Å	3	5.80	7.45×10^{-15}	5.50×10^{-10}

onance frequency, $(E_m - E_n)/h = \nu$, passes through the flame gases into which the sample has been sprayed. A part of the radiant energy of the incident light beam, P_0, will be absorbed. The transmitted intensity, P, may be written:

$$P = P_0 \exp(-k_\nu d) \tag{12-4}$$

where k_ν is the absorption coefficient and d is the average thickness of the absorbing medium. Figure 12-13 shows a typical curve of k_ν vs. ν. Around the central frequency ν_0 the curve presents a width. The distance between the two points where k has fallen to one-half of its maximum value, k_0, is called the *half-width* of the absorption line and is denoted by $\Delta\nu$. The principal causes of absorption line broadening are *Doppler* broadening and *Lorentz* broadening. Doppler broadening is due to the relative motion of the radiating atoms, as a result of thermal activity, and the receiving device. For a given atomic line,

$$\Delta\nu_D = 7.16 \times 10^{-7} \nu_0 \sqrt{T/M} \tag{12-5}$$

where M is the molecular or atomic weight in grams. Lorentz (or pressure) broadening is due to collisions of the absorbing molecules or atoms with other molecules or atoms. It is given by

$$\Delta\nu_L = (2\sqrt{2/\pi})n' \sqrt{RT/\mu} \left(\frac{\sigma + \sigma'}{2}\right)^2 \tag{12-6}$$

where n' is the number of foreign gas molecules or atoms per unit volume, μ is the reduced mass of the foreign gas atom and the absorbing species, and σ, σ' are the diameters of the two colliding species. Both half-widths are of the same order of magnitude and range from 6 to 48 mÅ for combustion flames.

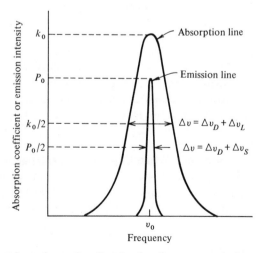

Fig. 12-13. Profiles of an absorption line in the flame gases and an emission line from a hollow-cathode lamp.

Relative values of the number of absorbing species per unit volume, N, can be obtained by measuring P_0 and P for the central frequency of the absorbing line. The integral of k_ν over the spectral range of the absorption line, that is, the integrated absorption, becomes

$$K = \int k_\nu \, d\nu = \frac{h\nu}{c}(B_{nm}N_n - B_{mn}N_m) \tag{12-7}$$

where B_{nm} is the transition probability of absorption, B_{mn} is the transition probability of induced emission, and N_m, N_n are the number of species in the excited and ground energy states, respectively. In flames the ratio N_m/N_n (or N^*/N_0) is always very small (Table 12-2). Thus, the integrated absorption is essentially proportional to the number of atoms in the ground energy state. Equation 12-7 then becomes

$$\int k_\nu \, d\nu = \frac{\lambda_0^2}{8\pi} A_{nm} \frac{g_n}{g_m} N_0 = \left(\frac{\pi e^2}{mc}\right) f_{nm} N_0 \tag{12-8}$$

where f_{nm} is the *oscillator strength* in absorption.* From Eqs. 12-4 and 12-8, a plot of log (P_0/P) against N_0 will be a straight line. The number of neutral atoms, N_0, will depend on the kinetics and energetics of the desolvation, vaporization, and dissociation processes, and also on the ease of ionization. Atomic absorption is independent of the excitation energy, and compared to emission intensity, the effect of flame temperature is less because broadening of line widths varies only as the square root of the temperature.

To obtain the integrated absorption it is necessary to send through the absorbing medium a radiation of frequency strictly centered on the absorption frequency and whose half-width is small compared with that of the absorption line (Fig. 12-13). In the hollow cathode lamps used in atomic absorption, the Doppler and self-absorption broadening effects reduce sensitivity most. For a given atomic line the Doppler broadening is proportional to the square root of the temperature. Therefore, the temperature of the radiating plasma is kept as low as possible by keeping the lamp current low or providing cooling for the electrodes. Self-absorption broadening is due to absorption of radiation by nonradiating atoms *in the source*—the exact phenomenon utilized for atomic absorption in the flame gases. It can be reduced by shortening the path length and the concentration of vapor through which the light must pass within the confines of the lamp.

Fluorescent radiation of metal atoms may be generated when the flame that contains the metal vapor is irradiated by a strong light source emitting the metal spectrum. The metal atoms in the flame will partly absorb the resonance lines of the incident radiation, and a fraction of the absorbed photons will be reemitted. The fluorescent light is specific for the test element. Spectral interference from background radiation or neighboring metal lines can be eliminated by modulating the incident light beam. For a given flame and metal line, the flux, W_f, of the fluorescent radiation received by the detector at low

---

*Oscillator strength is defined for the transition n to m as the number of harmonically vibrating electrons that would together have the same absorption as one atom in the state n.

metal concentration, is given by

$$W_f = \text{constant} \cdot N_0 B_{mn} P_0 \qquad (12\text{-}9)$$

A characteristic difference between atomic fluorescence and atomic absorption methods is that with the fluorescence method the detector receives only nonmodulated, weak background radiation from the flame when no metal is present. Also the intensity of fluorescence is linearly proportional to the exciting radiant flux. These differences are significant near the detection limit and thus for trace analysis.

Distribution Patterns of Atomic Concentration in Flames

The concentration of normal and excited atoms varies in different parts of a flame and, also, with the fuel–oxidant ratio employed. Atomic and molecular distribution studies may be made by measuring absorption, emission, or fluorescence in a flame containing dispersed atoms when the flame is moved vertically or horizontally relative to the light path of the optical system. Figure 12-14 shows the distribution of atoms in a 10-cm air–acetylene flame by absorption measurements. Contours are drawn at intervals of 0.1 absorbance unit with maximum absorbance in the center.[2,8] Neither the area of observation nor the fuel–oxidant ratio is critical for silver, whereas for molybdenum the region of maximum absorption is sharply localized. The height of maximum absorption marks the level where the increased atomization with height is just balanced by the rate of decrease in the concentration of free atoms through dilution by the flame gases and formation of oxides and hydroxides.

The distribution pattern obtained in emission very often differs from that obtained from absorption measurements. For a number of elements, emission lines that are absent or very weak in the outer mantle of a stoichiometric flame appear in unusual strength in the reaction zone of a fuel–rich acetylene flame. Generally, the test substance should be dissolved in an organic solvent to enhance the effect. A one-dimensional distribution pattern, taken along the center line of the flame, is shown in Fig. 12-15. In the illustration, mixture strength of a flame is defined as the actual volume of oxygen introduced into the flame divided by that required for the complete oxidation of the combustibles.

Spectral Interferences

Spectral interferences are closely associated with the resolving power of the optical system employed. In emission flame spectrometry, any radiation which coincides with or overlaps that of the test element, whether it stems from the sample or from the excitation source, constitutes a spectral interference. Interferences of this type are especially prevalent when filters are used to isolate the atomic spectra from molecular band spectra. With a medium-dispersion (16 Å/mm) monochromator, interference is much less. Emission intensity of the orange band system of CaOH (5430–6220 Å) interferes with the sodium lines at 5890 and 5896 Å and with the barium line at 5536 Å. With broad bandpass filters, interference is so serious that various chemical expedients have been proposed to elimi-

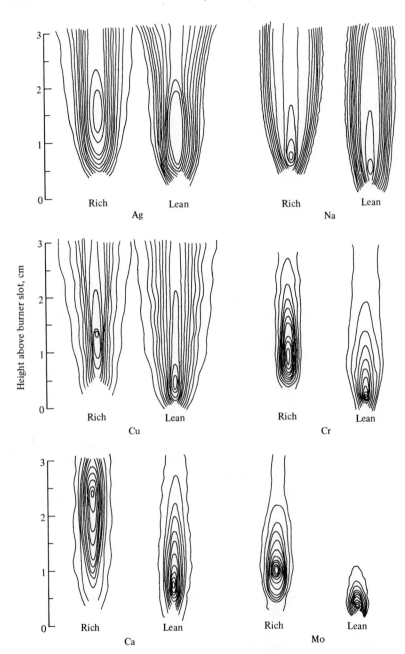

Fig. 12-14. Distribution of atoms in a 10-cm air–acetylene flame. Fuel-rich and fuel-lean results are shown. Contours are drawn at intervals of 0.1 absorbance unit with maximum absorbance in center. (After C. S. Rann and A. N. Hambly, *Anal. Chem.*, **37**, 879 (1965). Courtesy of American Chemical Society.)

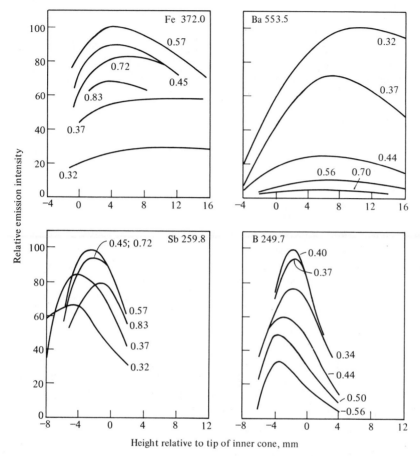

Fig. 12-15. Flame emission intensity as a function of height of observation with mixture strength as parameter. (After J. A. Dean and J. E. Adkins, *Analyst*, **91**, 709 (1966). Courtesy of Society for Analytical Chemistry.)

nate the calcium emission. Even with monochromators the problem remains, although to a lesser degree, because the calcium band system underlies the atomic line emissions. Difficulties in the determination of copper arise from the remnants of the strong OH band system from the flame gases; the copper line at 3248 Å lies on the side of an OH band.

Spectral interference may be caused by adjacent line emissions when the test substance and interferent have nearly the same wavelength and fail to be resolved by the monochromator. In this case, the lines will be read together in proportion to the degree of overlap. A serious overlap involves the manganese triplet (4031, 4033, 4035 Å), the gallium line (4033 Å), the potassium doublet (4044, 4047 Å), and a lead line at 4058 Å. Often the difficulty must be overcome by selecting other spectral lines or by prior chemical separation. In atomic absorption and atomic fluorescence, spectral interferences originating from sample matrices or flame components can be largely eliminated by working with ac amplifiers tuned to the frequency at which the source is chopped or modulated. Signals from the flame, not being modulated, are rejected. However, emission from strong band systems of the flame gases, such as the OH and CN systems, should not coincide with an absorption line.

Physical Interferences

Particulate matter in the flame is responsible for light scattering and molecular absorption. This may arise from unevaporated droplets, but more likely from unevaporated or refractory salt particles left following desolvation. Scattering need not be considered in emission measurements, although the presence of hot particles in the frame increases the background continuum. In atomic fluorescence, light scattered from particles appears as a signal which cannot be distinguished from that due to the test substance if a monochromatic source is used. With continuous sources a wavelength scan provides the necessary background correction. Correction for scattered and absorption losses in atomic absorption are easily made by measuring the absorbance at wavelengths adjacent to the resonance line using a nonabsorbed line from the lamp; alternatively, the continuous emission from an auxiliary deuterium lamp that is passed alternately through the flame may be measured.

Sample solutions very high in total solids suffer a reduction in signal. Concentrated solutions tend to be viscous and dense. Salt encrustation around the orifices impairs aspirator performance, which may become quite erratic. Large salt particles from concentrated aerosols hinder solvent vaporization. Frequent aspiration of pure solvent helps to remove encrusted salts from burner orifices. In real samples, the detection limit is often set by the matrix concentration that can be tolerated.

Dissociation of Metal Compounds

In the flame, metal atoms can be bound to other "partners"; these may be a constituent of the flame gases, such as O atoms, or a component of the solution sprayed, such as chlorine in the form of HCl. In equilibrium at a given temperature, the fraction of metal atoms tied up as molecules depends on the bond strength as well as on the concentration of the partner in the flame. A rather stable bond can be expected under flame conditions if the bond strength exceeds a few electron volts.

Metal compounds in the flame are usually simple diatomic molecules, such as CaO, or triatomic molecules, such as CaOH. Elements such as Na, Cu, Tl, Ag, and Zn are practically completely atomized in the flame; that is, they do not form molecular compounds with flame partners in noticeable proportions. Metal monoxides are the most common compounds found in flames burning with air, oxygen, or nitrous oxide. Whereas the alkali metals form practically no oxides, a major fraction of the alkaline earth elements is present as monoxides unless flames very rich in fuel are used. Certain other metals, such as La, Al, and Ti, form refractory oxides which are extremely stable. As a consequence, the free atomic concentrations of these elements are virtually negligible in flames of stoichiometric composition and moderate temperature. However, in fuel-rich, hot, nitrous oxide–acetylene flames, these oxides may be sufficiently dissociated to enable analysis of these elements by atomic emission or absorption spectrometry.

Ionization of Metal Species

A loss in spectrochemical sensitivity results when a free metal atom is split into a positive ion and an electron:

$$M \rightleftharpoons M^+ + e^- \tag{12-10}$$

Alkali metals, which require a comparatively low energy for ionization, about 4–5 eV, often occur in noticeable proportions as ions in flames with $T \geqslant 2500°K$. In the hot nitrous oxide–acetylene flames, elements with ionization energies of 6 eV, or higher, may be markedly ionized also.

The degree of ionization, α_i, is defined as

$$\alpha_i = \frac{[M^+]}{[M^+] + [M]} \tag{12-11}$$

At equilibrium, when the ionization and recombination rates are balanced, K_i (in atm) is given by

$$K_i = \frac{[M^+][e^-]}{[M]} = \left(\frac{\alpha_i^2}{1 - \alpha_i^2}\right) p_{\Sigma M} \tag{12-12}$$

where $p_{\Sigma M}$ (in atm) is the total atom concentration of metal in all forms in the burned gases.* The ionization constant, K_i, can be calculated from the Saha equation:

$$\log K_i = -5040 \frac{E_i}{T} + \frac{5}{2} \log T - 6.49 + \log \frac{g_{M^+} g_{e^-}}{g_M} \tag{12-13}$$

where E_i is the ionization energy of the metal in eV, T is the absolute temperature of the flame gases, and the g terms are the statistical weights of the ionized atom, the electron, and the neutral atom. For the alkali metals the final term is zero; for the alkaline earth metals, it is 0.6. The percent ionization for a selected group of elements at several assumed flame temperatures is given in Table 12-3.

The ionization of a metal can be suppressed by the addition of another easily ionized metal—a *deionizer* or radiation buffer. The partial pressure of free electrons is increased by the addition of a second ionizable element and this shifts the equilibrium in Eq. 12-10 in the direction of an increased partial pressure of neutral atoms. The data in Table 12-4 illustrate the effect of alkali metals in brightening the cesium signal. As the ionization potential of the deionizer decreases, a smaller quantity is required to repress the ionization of the test substance. To ensure that ionization interference is eliminated, the product

---

*For a derivation, see M. N. Saha, *Phil. Mag.*, **40**, 472 (1920) or M. W. Zemansky, *Heat and Thermodynamics*, Addison-Wesley, Reading, Mass., 1957.

TABLE 12-3   Percent Ionization of Elements in Flames[a]

Element	Ionization Potential, eV	Air–Propane, 2200°K	Oxygen–Hydrogen, 2450°K	Oxygen–Acetylene, 2800°K	Nitrous Oxide–Acetylene, 3300°K
Lithium	5.391	<0.01	2.8	16.1	
Sodium	5.139	1.1	5.1	26.4	
Potassium	4.340	9.3	33.4	82.1	
Rubidium	4.177	13.8	45.1	88.8	
Cesium	3.894	28.6	71.0	96.4	
Beryllium	9.32	0	0	0	0
Magnesium	7.646	—	—	<0.01	6
Calcium	6.113	<0.01	1.0	7.3	43
Strontium	5.694	<0.01	2.8	17.2	84
Barium	5.211	1.9	8.8	42.3	88

[a] Partial pressure of metal atoms in the flame assumed to be 1×10^{-6} atm for all but acetylene–nitrous oxide, which was approximately 10^{-8} atm.

$(K_i)_M p_M$ of the deionizer must be such that variation of the product for the test element does not influence the value of the square root of the sum in the expression:

$$p_{e^-} = \sqrt{(K_i)_{M,1} p_{M,1} + (K_i)_{M,2} p_{M,2}} \qquad (12\text{-}14)$$

Chemical Interference in the Condensed Phase

This is the most troublesome variety of interference in both emission and absorption. Interference of this type arises when a concomitant combines with the test element to form a compound which, after crystallizing out of the evaporating droplets of aerosol, vaporizes only with difficulty or, perhaps, is converted to the gaseous state more easily than would be the test element in the absence of the concomitant. For example, salts of high-melting compounds, such as refractory oxides, will not be converted rapidly into the

TABLE 12-4   Relative Enhancement of Cesium Signal by Addition of Different Alkali Metals

Molar Concentration of the Added Element	Cs, 0.0001 M Solution			Cs, 0.001 M Solution		
	Li	Na	K	Li	Na	K
0.0001	103	105	150	100	100	100
0.001	119	145	444	100	105	125
0.01	170	352	1200	100	107	195
0.1	340	765	1980	105	140	290
1.0	780	1270	2260	150	205	325

SOURCE: N. S. Poluektov and R. A. Vitkun, *Zhur. Anal. Khim.*, 16, 260 (1961). All values pertain to a premixed air–acetylene flame.

gaseous state. A depression of the analysis signal will result which sometimes is so severe as to obliterate completely an emission or absorption signal. The strong depression of the calcium (and other alkaline earth elements) signal in the presence of phosphate, aluminate, silicate, and similar anions is well known. In fact, over limited intervals of concentration, the depression is linear and has formed the basis for an indirect determination of the depressant in the presence of a standard amount of calcium. Interference is less in hotter flames and becomes progressively less the farther the observation area recedes from the burner.

The use of releasing agents or protective chelating agents often circumvents or lessens this type of interference (Fig. 12-16). A releasing agent preferentially combines with the interferent or simply by mass action denies the test element the interferent, thereby leaving the test element free to vaporize in the flame. Lanthanum is frequently employed for the restoration of the calcium signal in the presence of phosphate and sulfate. In protective chelation or masking, the element sought is masked to prevent it from combining with the interferent in the solution phase although, of course, subsequently the masked species must be promptly decomposed in the flame. Various polyhydroxy alcohols or EDTA have been used to mask calcium in the presence of phosphate; control of pH is also important. When applicable, solvent extraction is effective. Rains[7] has prepared a list

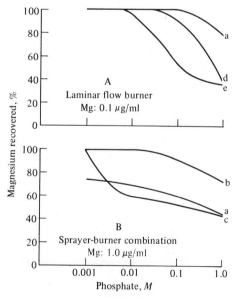

Fig. 12-16. Effect of phosphate on magnesium in absorption with two different burners. A, air–acetylene flame; B, oxygen–hydrogen flame. (a) H_3PO_4; (b) 10% (v/v) glycerol and 0.1 M $HClO_4$; (c) lanthanum, 1 mg/ml; (d) $(NH_4)_2HPO_4$ with 10% (v/v) glycerol, and 0.1 M $HClO_4$; (e) $(NH_4)_2HPO_4$. (After T. C. Rains, "Chemical Interferences in Condensed Phase," in *Flame Emission and Atomic Absorption Spectrometry*, Vol. 1, J. A. Dean and T. C. Rains Eds., Marcel Dekker, New York, 1969.)

of releasing and protective chelating agents. When selecting the releasing or masking agent, one must consider its emission spectrum, purity, and cost, as well as efficiency. Combinations of reagents are beneficial and often preclude excessively high salt concentrations that are necessary when each is employed alone.

Detection Limit and Sensitivity

The *detection limit* is defined as the concentration (usually in μg/ml) of an element which will shift the absorbance (or emission) signal an amount equal to the peak-to-peak noise of the base line (or background signal). This is the concentration that gives a signal-to-noise (S/N) ratio of 2, as illustrated in Fig. 12-17. Unless the type of data processing and the degree or type of filtering is stated, a comparison of detection limits cannot be made between laboratories. However, the detection limit is a useful specification.

The *sensitivity* in atomic absorption is defined as the concentration of test element in solution (or the amount in weight units per sample weight) that produces an absorbance of 0.0044 (1% absorption). It is usually stated in terms of μg/ml/1% Abs (or μg/g/1% Abs).

Sensitivity and detection limit are compared in Fig. 12-17. For element A the sensitivity and detection limit of the system are the same. For element B there is very little noise present in the system and the detection limit is markedly improved although the sensitivity remains unchanged. In the latter case, if reserve amplification is available in the instrument, the signal could be increased (with equivalent increase in the noise probably) until the noise level equal to that observed for element A is attained. If this is done, the sensitivity for element B would be much greater than for element A.

EVALUATION METHODS

Generally, a set of concentrated stock solutions, each containing, for example, 1000 μg/ml of a single element, is prepared. These standards are prepared best in $HClO_4$ or

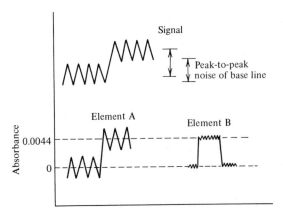

Fig. 12-17. *Upper*, detection limit for $S/N = 2$. *Lower*, sensitivity of elements A and B.

HNO_3 to minimize anion effects, and from high-purity metals or reagent grade chemicals. From them, dilutions to any desired concentration are made as needed. For storage it is advisable to use bottles of polyethylene or resistant-glass composition. Care must be exercised in the handling of diluted samples and standards to avoid contact with contaminated pipets, syringes, glassware, and dust particles in the air.

Calibration Curve

In flame emission the test substance concentration is deduced from the intensity of the line-above-background signal. Preferably one scans over the wavelength of interest and records directly both the line intensity and the background reading on either side of the line. A blank solution should also be scanned over the same wavelength region. Net emission intensities—line intensity minus background reading—are plotted versus the concentrations of the standard solutions.

In atomic absorption a curve is prepared to cover the appropriate concentration range. Usually, this means the preparation of standards which produce an absorption of 0–80%. The wavelength setting is peaked on the line, and the 100% signal level is defined with a blank—by widening the slits to the optimum value and adjusting the source intensity. Then the source is blocked off and the zero adjustment (0% T) is made with the zero suppression control. The amount of test element is inferred from the reduction in signal caused by its presence. For an atomic absorption spectrometer equipped with a linear readout system, the correct method for plotting data is to convert "percent absorption" to absorbance; that is, $A = \log (P_0/P) = \log 100 - \log \%T$, where P_0 is the source intensity with the blank aspirated and P the source intensity with sample or standard aspirated. On linear coordinates, absorbance is plotted against concentration. The curve may become nonlinear at high absorbance values, and if so, the number of standards in this region should be increased.

Standard-Addition Method

Where it is not possible to suppress interferences from matrix elements completely, the *standard-addition method* may be used provided the absorbance–concentration or emission–concentration curve is known to be a straight line passing through the origin. The net absorbance (or emission reading) of the sample solution of concentration x is A_x (or R_x), and that of a similar solution to which a known concentration a of the metal has been added is A_1 (or R_1), then x can be calculated from the relation

$$\frac{x}{x+a} = \frac{A_x}{A_1} \tag{12-15}$$

Appropriate correction must be made for any background signal. It is advisable to check the result with a second standard addition. A graphic representation of the standard-addition method is shown in Fig. 12-18. Normally additions which are equal to twice and

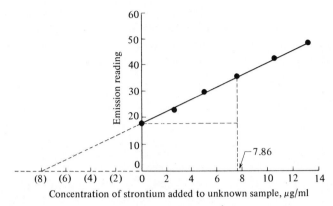

Fig. 12-18. Graphic representation of the standard-addition method of evaluation (see Table 12-5).

one-half the original amount are optimum statistically. All solutions should be diluted to the same final volume, for then any interferent will be present at the same concentration in all solutions and will affect equally the neutral atom population either originally present or added as the standard increment. Table 12-5 contains illustrative data.

Internal-Standard Method

In the *internal-standard method* a fixed quantity of internal-standard element is added to samples and standards alike. Upon excitation, radiant energy emitted by the test element and internal-standard element is measured simultaneously by dual detectors or by scanning successively the two emission lines. The ratio of the emission intensity of the analysis line to that of the internal-standard line is plotted against the concentration of the analysis element on double logarithmic coordinates. The intensity of each line is cor-

TABLE 12-5   Evaluation by Method of Standard Addition

Meter Readings, Strontium Standards	Strontium, μg/ml Present or Added	Meter Readings, Sea Water Sample
0	0	18
16.0	2.62	24
32.5	5.24	30
47.1	7.86	36
63.7	10.50	42
79.2	13.10	48

NOTE: The net emission readings are given for a series of pure strontium standards and for a series of solutions containing equal volumes of sea water and added standard solution of strontium. Sample contained 7.86 μg/ml of strontium.

rected for the background radiation in which it lies. The plot of log (emission ratio) versus log (concentration of test element in standard solutions) should give a straight line over limited concentration intervals. On most double-beam instruments used for emission flame photometry, the ratio is given directly by the reading of the balancing potentiometer; calibration curves on linear coordinate paper suffice. Use of this evaluation method is limited essentially to instruments designed for the determination of sodium and potassium that use lithium as the internal standard element. This method overcomes variations in flow rates of fuel and oxidant, and alterations in nebulization.

LABORATORY WORK

General Instructions

Read the instruction manual accompanying the particular instrument to be used before attempting any laboratory operations.

Lighting the flame Turn on the air (or oxygen) pressure and then the fuel (city gas, propane, hydrogen, or acetylene) pressure. Immediately bring a lighted match alongside the burner and slightly above the tip of the burner. The flame makes a loud noise when lighted and while burning. Allow the flame to burn several minutes before proceeding. **When extinguishing the flame, always turn off the fuel before turning off the oxygen or air**.

Optimum fuel and oxygen (or air) pressures Adjust the oxygen (or air) and fuel pressure to the value recommended in the instruction manual furnished by the manufacturer. Spray a solution of the test element into the flame. Vary the fuel pressure in steps of 0.5 psi and measure the emission of the test element and of the solvent alone when each is sprayed into the flame. Plot the net emission of the test element against the fuel pressure.

If a maximum value is attained, set the fuel pressure at this value and vary the oxygen (or air) pressure in steps of 1 psi. If no maximum is attained for fuel pressure, choose an intermediate value of fuel pressure and use it throughout the remainder of your work. Plot the emission as a function of the oxygen (or air) pressure.

Repeat these steps for each test element and at each wavelength of the test element, particularly if the emission originates from an atomic line in one case, but from a molecular species in another.

Slit width With flame spectrometers the slit width may be varied. For emissions emanating from atomic lines, the slit width is usually set at 0.02–0.05 mm. For band spectra, the slit width can sometimes profitably be set at values as large as 0.2 mm. For prism spectrometers, consult the dispersion curve supplied by the manufacturer.

Wavelength setting With filter photometers, insert the proper filter or filter combination in the holder and rotate the units into the light beam.

With monochromators, it is necessary to find the exact wavelength-drum setting at which the instrument gives maximum response. Spray a solution of the test element into the flame and scan the wavelength region where the spectral lines or bands are located (see Appendix E). Slowly vary the wavelength until the maximum signal is obtained. If possible, use a different slit width, and rescan.

Summary Select the optimum working conditions after considering the emission intensity, the resolution of the spectral lines (or bands) from each other and from adjacent spectral features of other sample constituents, the magnitude of the background radiation, and the steadiness (noise) of readings.

Experiment 12-1   Determination of Sodium and Potassium

Standard solutions Dissolve 2.5420 g of sodium chloride in deionized water and dilute to 1 liter. Dissolve 1.9070 g of potassium chloride or 2.586 g of potassium nitrate in deionized water and dilute to 1 liter. Each solution contains 1000 μg/ml of the respective cation.

Weigh out 2.473 g of lithium carbonate and transfer to a 1-liter volumetric flask. Add approximately 300 ml of deionized water and then add slowly 15 ml of concentrated hydrochloric acid. After the CO_2 has been released, dilute the solution to 1 liter. This solution contains 1000 μg/ml of Li_2O. Most samples of lithium salts are contaminated with considerable sodium and some potassium, and therefore the same batch of stock lithium solution should be used in preparing the working standards used in the internal-standard method.

Working standards Prepare a set of six standards that contain 5, 10, 25, 50, 75, and 100 μg/ml of potassium (or sodium). These solutions will be used in the direct-intensity method.

Prepare a set of seven standards that contain 0, 5, 10, 25, 50, 75, and 100 μg/ml of potassium (or sodium), with each solution also containing 100 μg/ml of Li_2O. This series will be used in the internal-standard method.

Procedure Determine the calibration curve for potassium using the direct-intensity method. Use deionized water and the strongest standard solution to adjust the reading scale to zero and 100 divisions (or full scale). Without changing the instrument controls, determine the emission reading for the other concentrations. These directions presuppose that a single-beam flame photometer is available. If not, use the internal-standard method.

When using the internal-standard method, set the instrument zero with the standard containing 100 μg/ml of Li_2O but with potassium absent. Be sure any reading scale knob is positioned at the zero mark. Next, set the 100-division (full-scale) reading with the strongest standard aspirating; again be sure the reading dial control is positioned at the 100 mark. Recheck each reading once again. Without changing the instrument controls, determine the emission reading for the other concentrations.

At the instructor's discretion, unknown samples may be compared by either or both procedures.

Plot emission intensity vs. concentration of potassium from the direct-intensity data on rectilinear graph paper and also on log-log graph paper. Mark the regions where the calibration curve is linear and the log-log plot has a slope of unity.

Experiment 12-2   Influence of Fuel and Oxygen (or Air) Pressure

Use one of the solutions prepared for Experiment 12-1.

Adjust the instrument to read zero when deionized water is aspirated, and to read near midscale when the standard solution is aspirated (direct-intensity method); or use the appropriate zero solution and standard solution, each containing lithium, when using the internal-standard method.   Insert the standard solution whose concentration lies immediately below and one whose concentration lies above the original standard solution; measure the emission intensity of each.

Change the fuel pressure by 0.5 psi but do not change any other instrument controls. Measure the emission intensity of the three standard solutions. Continue to vary the pressure in 0.5-psi increments over a range of 2–3 psi and measure the emission of these three solutions.

Plot the emission readings vs. the fuel pressure. Explain the difference between the results obtained for the direct-emission and the internal-standard methods.

If provisions are available for varying the air, or oxygen, pressure, this might be done while keeping the fuel pressure at a fixed value. Since this is usually also the aspirating gas, the range of pressures over which the aspirator and burner will operate properly will be limited.

Experiment 12-3   Influence of Solvent

Standard solutions Prepare a set of standards, each containing 25 $\mu g/ml$ of potassium (or sodium), and 0, 10, 30, 50, and 70 volume % of isopropyl alcohol (or ethanol). Also prepare a solution that contains 30 volume % glycerol. Include lithium if using an internal-standard method.

Procedure Adjust the instrument to read zero with the appropriate solution (water or 100 $\mu g/ml$ Li_2O) and to a scale reading of 20 divisions with the aqueous potassium standard. Now measure the emission intensity of each of the alcoholic standards and the glycerol standard solution.

Plot the emission intensity (direct-reading method) vs. the volume % of alcohol. Explain why little change in emission is noted in the internal-standard method. Does the viscosity of a solution exert an influence on the emission intensity?

With a single-beam photometer a number of organic solvents may be investigated.

Experiment 12-4   Operating Parameters in Atomic Absorption

1. Investigate designated lamp lines for absorption sensitivity by constructing calibration curves and noting slopes. To convert percent absorption ($\% A$) to absorbance, consult Appendix F.

2. For the most sensitive absorption line, investigate the effect of different slit openings on the nature of the calibration curve.

3. For the most sensitive line, investigate the effect of flame richness: lean, normal, and fuel-rich. Do this by varying the fuel flow while keeping the air flow constant.

4. Study the flame height at which absorption is observed at two heights 1 cm apart by raising the burner relative to the optical path.

5. For an absorption line, perhaps a less sensitive one in order to avoid excessive dilutions of standards, push the limit of detection as far as possible by use of the sensitivity scale (1X, 2X, 5X, and 10X). Note the noise level for each reading. The signal/noise ratio will be needed to calculate the detection limit.

6. For designated elements, such as iron, add some H_3PO_4 to a portion of one standard and note the new absorption reading.

7. For designated elements, such as potassium, investigate the effect of organic solvents on the absorption.

8. For an element such as potassium, investigate the effect of an ionization suppressor by addition of 2000 and 5000 $\mu g/ml$ of sodium to 5 $\mu g/ml$ of potassium.

PROBLEMS

1. For the analysis of cement samples, a series of standards were prepared and the emission intensity for sodium and potassium was measured at 590 nm and 768 nm, respectively. Each standard solution contained 6300 $\mu g/ml$ of calcium as CaO to compensate for the influence of calcium upon the alkali readings. The results are shown below:

Concentration, $\mu g/ml$	Emission Reading Na_2O	K_2O
100	100	100
75	87	80
50	69	58
25	46	33
10	22	15
0	3	0
Cement A	28	69
Cement B	58	51
Cement C	42	63

For each cement sample 1.000 g was dissolved in acid and diluted to exactly 100 ml. Calculate the percent of Na_2O and K_2O.

2. In Problem 1, what contributed to the emission reading of the blank at the analytical wavelength for sodium, but did not for potassium? A Beckman model DU spectrometer was employed to obtain the results. Would the blank reading be larger, smaller, or the same if a filter photometer equipped with glass absorption filters had been employed? [See *Anal. Chem.*, **21**, 1296 (1949).]

3. Indium has a strong emission line at 451.1 nm. Using the Beckman model DU with a slit aperture of 0.02 mm, the background read 5.5 units and the line read 350 units, whereas for a slit aperture of 0.05 mm, the background read 35 units and the line read 1400 units. (a) What is the line-to-background ratio for each slit setting? (b) Estimate the dependence of line emission and background reading (due to molecular band systems) upon the slit aperture?

4. Boron gives a series of fluctuation bands due to the radical BO_2 which lie in the green portion of the spectrum. Although the overlapping band systems present a problem in the measurement of the flame background, the minimum between adjacent band heads can be used. These results were obtained:

Boron Present, $\mu g/ml$	Emission Reading	
	518-nm Peak	505-nm Minimum
0	36	33
50	44	36
100	52	39
150	60.5	42.5
200	68.5	45.5

What are the concentrations of boron in these unknowns?

A	45	36.5
B	85	65
C	66	50

[See *Anal. Chem.*, 27, 42 (1955).]

5. A calibration curve for strontium, taken at 460.7 nm, was obtained in the presence of 1000 $\mu g/ml$ of calcium as CaO and also in the absence of added calcium. These results were

Strontium Present, $\mu g/ml$	Emission Reading	
	No Calcium	Calcium Added
0	0	13
0.25	2	18.5
0.5	6	24
1.0	16	36
2.5	44	70
5.0	94	125
7.5	150	181
10.0	200	238

(a) Graph the calibration curve on rectilinear graph paper and also on log-log paper. (b) What might be the cause of the upward curvature in the region of low concentrations on the rectilinear graph when calcium is absent? (c) Why does the addition of calcium straighten the calibration curve and increase the net emission reading for strontium?

6. Typical emission readings for magnesium at its atomic line are as follows, each reading corrected for background:

Magnesium, μg/ml	Emission in Scale Units	Magnesium, μg/ml	Emission in Scale Units
1.25	3.5	50	41
2.5	7.0	75	50
5.0	12.4	100	57
10	19.6	150	69.5
15	24.2	200	80
25	30.8	250	87

Plot the data on log-log graph paper. Note the slope of the graph at very low concentrations and also at concentrations above approximately 10 μg/ml of magnesium. What is the explanation for the shape of the log-log plot?

7. In the determination of manganese at 403.3 nm, solution A, containing an aliquot of the unknown solution, gives a meter reading of 45. Solution B, containing the same quantity of unknown solution plus 100 μg/ml of added manganese, gives a meter reading of 83.5. Each reading has been corrected for background. Calculate the quantity of manganese in solution A.

8. Under the same conditions as Problem 7, solution A gave a net reading of 31. Solution B, with 75 μg/ml of added manganese, gave a net reading of 68. Calculate the quantity of manganese in solution A.

9. A sample of mineral ash gave a meter reading of 37. Solutions B and C, containing the same quantity of unknown solution plus 40 and 80 μg/ml of added potassium, respectively, gave net meter readings of 65 and 93. Calculate the quantity of unknown potassium in the original solution.

10. A metal naphthenate sample, ashed and diluted to a fixed volume, gave a reading of 29. Solutions B and C, containing the same quantity of unknown solution plus 25 and 50 μg/ml of barium gave readings of 53 and 78, respectively. Calculate the quantity of barium in the original solution.

11. To illustrate the effect of aqueous–organic solvents upon droplet size, calculate the mean droplet diameter for (a) water, (b) 50% methanol–water, and (c) 40% glycerol–water. Pertinent data follow:

System	Surface Tension, dynes/cm	Viscosity, dynes/cm^2	Density, g/cm^3	Velocity of Aspirating Gas, m/sec	Q_{air}/Q_{liq}
Ethanol, 50%	28	0.029	0.934		
Glycerol, 40%	68.6	0.039	1.102	279	2540
Methanol, 50%	30.6	0.027	0.946	198	9540
Methyl isobutyl ketone	24.6	0.0051	0.801		
Water	73	0.010	1.00	198	6400

12. For (a) water, (b) 50% (v/v) ethanol–water, and (c) methyl isobutyl ketone as solvents, plot the droplet diameters for solution flow rates ranging from 0.1 to 5 ml/min. As values for a typical nebulizer, assume the velocity of the aspirating gas to be 333 m/sec and Q_{gas} to be 8.5 liters/min. Other data are given in Problem 11.

13. Calculate the fraction of cesium atoms ionized in a flame at 2000°K when the total cesium concentration in the flame gases is (a) 10^{-4} atm, (b) 10^{-6} atm, (c) 10^{-7} atm.

14. Calculate the fraction of lithium atoms ionized in a premixed laminar flame at 3000°K when the concentration sprayed into the flame is (a) 10^{-2} M, (b) 10^{-3} M, and (c) 10^{-4} M.

15. What individual amounts of (a) cesium, (b) rubidium, (c) potassium, or (d) lithium should be added individually to a flame at 2500°K and at 2800°K to suppress the ionization of a solution containing 0.23 μg/ml of sodium?

	K_i, atm	
Element	2500°K	2800°K
Li	1.48×10^{-9}	2.63×10^{-8}
Na	4.8×10^{-9}	7.40×10^{-8}
K	1.8×10^{-7}	2.08×10^{-6}
Rb	3.9×10^{-7}	3.98×10^{-6}
Cs	1.45×10^{-6}	1.32×10^{-5}

16. Calculate the iron content in a diethyldithiocarbamate extract using the following data:

Absorbance Units		Iron Added,
Blank	Sample	μg/200 ml
0.0020	0.0090	None
0.0214	0.0284	2.00
0.0414	0.0484	4.00
0.0607	0.0677	6.00

17. If lithium (6708 Å) is used as an internal standard for the determination of sodium (5890 Å) and potassium (7665 Å), what is the maximum permissible temperature variation of the flame if one desires to maintain deviations in intensity ratios less than 1%? Assume a flame temperature of 2000°K.

18. From the absorbance traces shown for arsenic by atomic absorption, estimate (a) the signal-to-noise ratio, (b) the sensitivity of the method, and (c) the detection limit.

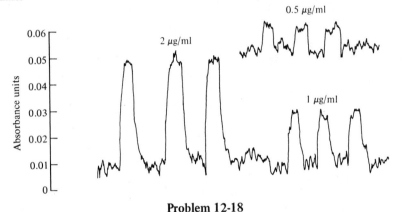

Problem 12-18

BIBLIOGRAPHY

Christian, G. D. and F. J. Feldman, *Atomic Absorption Spectroscopy: Applications in Agriculture, Biology, and Medicine*, Wiley-Interscience, New York, 1970.

Dean, J. A., *Flame Photometry*, McGraw-Hill, New York, 1960.

Dean, J. A. and T. C. Rains, Eds., *Flame Emission and Atomic Absorption Spectrometry*, Vol. 1: *Theory*, 1969; Vol. 2: *Components and Techniques*, 1971; Vol. 3: *Elements and Matrices*, 1974, Marcel Dekker, New York.

Elwell, W. T., and J. A. F. Gidley, *Atomic Absorption Spectrophotometry*, 2nd ed., Pergamon, Elmsford, N.Y., 1966.

Mavrodineanu, R., Ed., *Analytical Flame Spectroscopy: Selected Topics*, Macmillan, New York, 1970.

Mavrodineanu, R. and H. Boiteux, *Flame Spectroscopy*, Wiley, New York, 1965.

LITERATURE CITED

1. Dean, J. A. and W. J. Carnes, *Anal. Chem.*, **34**, 192 (1962).
2. Dean, J. A. and T. C. Rains, Eds., *Flame Emission and Atomic Absorption Spectrometry*, Vol. 1, Marcel Dekker, New York, 1969, Chapter 8.
3. Gaydon, A. G., *Flames: Their Structure, Radiation, and Temperature*, 2nd ed., Chapman and Hall, London, 1960.

4. Kniseley, R. N., "Flames for Atomic Absorption and Emission Spectrometry" in *Flame Emission and Atomic Absorption Spectrometry*, Vol. 1, J. A. Dean and T. C. Rains, Eds., Marcel Dekker, New York, 1969.
5. Mavrodineanu, R. and H. Boiteux, *Flame Spectroscopy*, Wiley, New York, 1965.
6. Nukiyama, S. and Y. Tanasawa, *Trans. Soc. Mech. Eng. Japan*, Reports 4, 5, 6 (1938–1940). Translations available through the Defence Research Board, Dept. of National Defense, Ottawa, Canada.
7. Rains, T. C., "Chemical Interferences in Condensed Phase" in *Flame Emission and Atomic Absorption Spectrometry*, Vol. 1, J. A. Dean and T. C. Rains, Eds., Marcel Dekker, New York, 1969.
8. Rann, C. S. and A. N. Hambly, *Anal. Chem.* **37**, 879 (1965).

13

Emission Spectroscopy

The spectrograph as an analytical tool has had a long period of development since the discovery of Bunsen and Kirchhoff that the spectra of flames colored from metallic salts were characteristic of the metals. Modern automatic recording spectrographs are capable of giving the percentage of a number of elements directly on dials and in only a few minutes. Qualitatively, the spectrograph is capable of detecting 0.001% or less of most of the metallic ions and of certain nonmetals, that is, P, Si, As, C, and B, in a sample of only a few milligrams. Quantitative determination of these elements is also readily accomplished. One spectroscopist can usually analyze as many different samples of the same type of material as five or more men by routine wet procedures. Consequently, the spectrograph has largely replaced the older, wet analytical procedures for the routine determinations of lesser components of steels, metallic alloys, and other substances.

The constituent parts of the spectrograph, including the energy sources and the registering devices, are considered in the following pages. A few typical commercially available instruments are then described. Finally the various procedures used in qualitative and quantitative work are presented.

Origin of Spectra

A complete discussion of the origin of emission spectra is beyond the scope of this book. For the present purpose, it is sufficient to understand that there are three kinds of emission spectra: *continuous spectra*, *band spectra*, and *line spectra*. The continuous spectra are emitted by incandescent solids and are characterized by the absence of any sharply defined lines. The band spectra consist of groups of lines that come closer and closer together as they approach a limit, the head of the band. Band spectra are caused by excited molecules. Line spectra consist of definite, usually widely and apparently irregularly spaced lines. This type of spectrum is characteristic of atoms or atomic ions that have been excited and are emitting their extra energy in the form of light of definite wavelengths.

The quantum theory predicts that each atom or ion has definite energy states in which the various electrons can exist. In the normal or ground state, the electrons are in the lowest energy state. On addition of sufficient energy by thermal, electrical, or other means, one or more electrons may be removed to a higher energy state farther from the nucleus. These excited electrons tend to return to the ground state and in so doing emit the extra energy as a photon of radiant energy. Since there are definite energy states and since only certain transitions are possible, there are a limited number of wavelengths possible in the emission spectrum. The greater the energy in the excited source, the higher the energy of the excited electron and therefore the more numerous the lines that may appear. However, the wavelengths of the existing lines will not be changed. An abbreviated pattern is shown in Chapter 12 (Fig. 12-11).

The intensity of a spectral line depends mainly on the probability of the required energy "jump" or transition taking place. Self-absorption occasionally decreases the intensity of some of the stronger lines. Self-absorption is caused by the reabsorption of energy by the cool, gaseous ions in the outer regions of the source. When high-energy sources are used, the atoms may be ionized by the loss of one or more electrons. The spectrum of an ionized ion is quite different from that of the neutral atom; in fact, the spectrum of a singly ionized ion will bear a strong resemblance to that of the neutral atom of atomic number one less.

EXCITATION METHODS

The flame, an ac arc, a dc arc, and the ac spark are the common methods of excitation. Each has special advantages and special applications. However, the function of each excitation unit is to introduce the sample into the source in a vaporized form and to excite electrons in the vaporized atoms to higher energy levels.

Flame

The flame furnishes a rather low-energy source and excites only a few lines, but this may be an advantage. For a complete discussion of flame excitation, refer to Chapter 12.

dc Arc

The dc arc, produced by a voltage of from 50 to about 300 V, is a common method for introducing the sample into the discharge. Vaporization occurs from the heating caused by the passage of current. Arc temperatures range from $4000°$ to $8000°$ K. The emission lines produced are primarily those due to neutral atoms and are indicated in tables by the symbol of the atom. (Those due to singly ionized ions are indicated by II following the atomic symbol.) The necessary components for a dc arc are a direct current power supply, a variable resistor, and a discharge gap (Fig. 13-1). Current across the discharge gap ranges from 1 to 30 A. One difficulty with an arc is its tendency to wander and flicker, especially when struck between carbon or graphite electrodes. This unsteadiness can be reduced somewhat by including a reactor in the circuit. A second difficulty involves selected volatility wherein the more volatile components may be selectively vaporized during the early portion of the arcing period. Consequently, line intensities should be monitored during the entire burning period in order to establish the optimum time for observation. To obviate this difficulty the sample may be burned to completion.

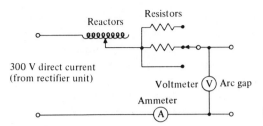

Fig. 13-1. Circuit for dc arc.

The dc arc is a very sensitive source with a good line-to-background ratio. It is generally used for the determination and identification of substances present in very small concentrations because it can detect elements below the limit of detection of a spark. A comparatively large amount of the substance being analyzed passes through the arc, and therefore an average or more representative value of the concentration is shown, provided that the complete sample is burned. The light from the center portion of an arc is the portion usually employed, because there may be a local concentration of certain ions near the electrodes. However, since the cathode region gives higher excitation energy, especially for lines arising from un-ionized atoms, it is more sensitive for these atoms and is sometimes employed for illuminating the slit of the spectrograph. When an arc is operated between carbon electrodes in air, some cyanogen $[(CN)_2]$ molecules are formed and, being excited by the arc, emit typical molecular band spectra in the region from 3600 to 4200 Å.

Stallwood Jet

An attachment frequently used with the dc arc to enhance sensitivity is the Stallwood Jet, shown in Fig. 13-2. In the Stallwood Jet of Spex Industries, gas is forced into

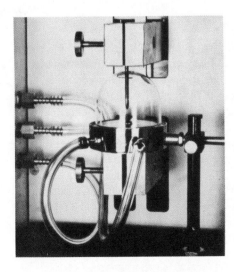

Fig. 13-2. The Spex Stallwood jet. (Courtesy of Spex Industries, Inc.)

a swirl chamber and then upward through an orifice surrounding the sample electrode. As the electrode burns away, it may be advanced with respect to the curtain of gas. Instead of the gas entering the open atmosphere as it leaves the orifice, it enters a quartz enclosure where, by building up a slight positive pressure, it excludes the ambient air. Illumination from the arc is through a small window in the side of the enclosure. Gas compositions generally used are argon–oxygen mixtures ranging from an 80–20 to a 60–40 mixture. Enhanced sensitivity arises from two causes. One is that cyanogen bands are eliminated and elements hidden in the cyanogen bands are more easily detected; these elements include molybdenum, the rare earths, iron, and gallium. The other reason for improved sensitivity relates to volatility. The excitation efficiency of a sample can be markedly increased if it is forced to burn slowly and is cooled with a gas stream. This is particularly helpful with volatile materials when they are placed in a deep cratered electrode.

ac Arc

The high-voltage ac arc employs a potential difference of 1000 V or more (Fig. 13-3). In the illustration, provision is made for a variable resistance in the circuit. The arc is drawn out to a distance of only 0.5–3 mm. For reproducible results the separation of the electrodes, the potential, and the current must be carefully controlled. The whole assembly must be carefully shielded so as to protect the operator from the dangerously high voltages. In comparison with the dc arc, the ac arc is steadier and more reproducible because reignition each half cycle on a random part of the sample provides good sampling technique.

ac Spark

The ac spark gives much higher excitation energies than the ac arc with less heating effect. The spark is produced by connecting a high-voltage transformer (10–50 kV) across two electrodes. A condenser is usually connected in parallel with the spark gap in order to increase the current. An inductor is also desirable in the circuit, because this has been found to decrease the excitation of lines and bands of the air molecules. Large values of inductance decrease the excitation energy and make the spark more arc-like in

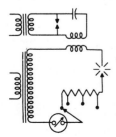

Fig. 13-3. Schematic diagram of an ac arc source.

its characteristics. The relationship between the current, i, potential, V, capacitance, C, and the inductance, L, when a spark first jumps, is given by the equation

$$i = V \sqrt{\frac{C}{L}} \qquad (13\text{-}1)$$

Thus the characteristics of the spark depend upon the capacitance and inductance. Rather elaborate devices are available for allowing variations in these values. The circuit proposed by Feussner[2] (Fig. 13-4) employs an auxiliary rotating spark gap which is driven by a synchronous motor. The gap is closed for only a brief instant at the peak of each half-cycle, and thus the number of decay cycles in the spark is controlled, which leads to more uniform and reproducible excitation conditions.

The spark is the preferred source whenever high precision rather than extreme sensitivity is required. Several types of spark sources have been proposed, each with its special uses. They differ mainly in the mechanism used to trigger the breakdown of the analytical gap so as to provide a series of discrete and identical breakdowns. In general, the spark source excites predominately ionic spectra. It is more reproducible and stable than the arc. Less material is consumed, and consequently the spark source is employed for higher concentrations of material than is the arc. Because the heating effect is less than that of the arc, it is well adapted for the analysis of low-melting materials. The spark source is also free from the troublesome cyanogen bands. However, the spark may strike to a particular spot on an electrode and thus give a nonrepresentative indication of the concentration of substance being determined.

The spark source is readily adapted to the analysis of solutions. The solution, rendered conducting by the addition of hydrochloric acid, is allowed to flow over a lower electrode while a spark is struck to the electrode, or rather to the thin layer of solution above the lower electrode. A bundle of fine wires may also serve as the lower electrode and draw the solution up to the top of the bundle by capillary attraction.

For the excitation of gases such as He, Ar, Kr, Ne, Xe, H_2, O_2, N_2, S_{vapor}, and Na_{vapor}, a gaseous discharge tube, that is, the common Geissler tube, is employed.

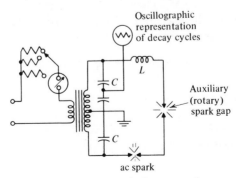

Fig. 13-4. Feussner circuit for ac spark source.

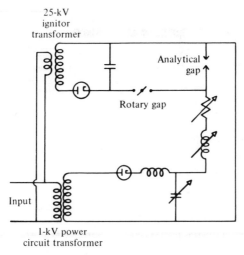

25-kV
ignitor
transformer

Analytical
gap

Rotary gap

Input

1-kV power
circuit transformer

Fig. 13-5. Basic circuit for a multisource source unit.

Multisource Unit

The modern commercial spectroscopic unit usually combines several sources into one unit, the *multisource* unit. This unit provides a variety of discharges. As diagrammed in Fig. 13-5, the unit consists of a 1000-V transformer, which charges a variable capacitance placed in parallel with the analytical gap. The power circuit also contains a variable inductance and resistance in series with the analytical gap. The discharge is initiated by a low power 25,000-V ignitor circuit connected across the analytical gap and controlled by the synchronous interrupter. By proper synchronization of the ignitor, the unit will charge the condenser during one half-cycle and the condenser will discharge during the next half-cycle. Because a relatively low voltage is used, larger capacitances are feasible and very high instantaneous currents can be obtained. Likewise, variation in the resistance can produce short duration "spark-like" discharges as well as long duration "arc-like" discharges.

Laser Method

An optical ruby laser can be used to excite spectral emission from samples, even from nonconducting materials that formerly required analysis in a cup electrode or by a graphitic pelletizing procedure. The operation of a laser microprobe is simple, requiring no sample preparation. The fundamental principle involves the absorption by the sample of the laser beam, which is concentrated onto the sample by a microscope objective lens (Fig. 13-6). Intense heat vaporizes all samples exposed to the beam and, for some of them, also achieves excitation energies in the vapor. However, the essence of the Jarrell-Ash microprobe (illustrated) is the raising of the vaporized sample to a useful spectral

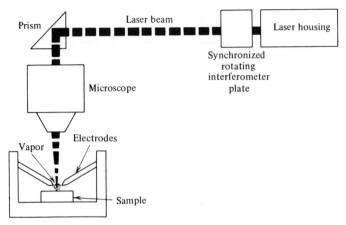

Fig. **13-6**. Schematic diagram of a laser microprobe. (Courtesy of Jarrell-Ash Co.)

emission by causing the vapor to short circuit an auxiliary electrical gap positioned be-tween the sample and microscopic objective. A sample spot as small as 50 μm may be selected.

Electrodes

The electrodes may be composed of the material being investigated if the material is a conductor and will withstand high temperatures. If not, the sample is usually placed in a small core in a carbon or graphite electrode ground to a blunt point in a pencil sharpener or a special grinding tool. A variety of preformed electrode shapes have been devised in order to steady the arc or increase its temperature. Figure 13-7(a) illustrates the usual form of the lower electrode. Figure 13-7(b) shows a form with a center post to help prevent wandering of the arc. Figure 13-7(c) shows a form used to increase the arc temperature by decreasing the conduction of heat away from the crater containing the sample.   Steadier arcs can be produced if the sample is mixed with pure, powdered graphite.   Metallic electrodes have some advantages over graphite electrodes, but of course the lines of the electrode metal will appear in the spectrum. These may serve as a reference spectrum.

Several techniques have been employed for handling solutions. In the porous cup technique a hole about 3.2 mm in diameter is drilled in the center of a 6.5 mm diameter

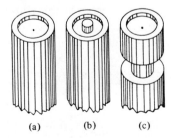

(a)   (b)   (c)

Fig. **13-7.** Several forms of carbon electrodes.

graphite electrode, the hole extending to within about 1 mm from the bottom of the electrode. The solution is placed in the hole, and the electrode is then employed as the upper electrode. A solid graphite rod forms the lower electrode. Either an arc or spark may be used. The solution slowly diffuses through the millimeter of graphite into the arc or spark. Still another device is to use a rotating graphite disc as the lower electrode. The disc rotates through the solution placed in a small boat and carries the liquid into the arc or spark which is struck to the upper part of the rotating disc. Solutions can also be evaporated in the craters of the electrodes. If carbon or graphite is employed, the crater is frequently waterproofed by dipping it into redistilled kerosene or collodion. Methods mentioned above for spark sources are also used.

The lower electrode is usually made the positive electrode. The arc is started by touching the two electrodes together and drawing them apart. If this does not start the arc, a graphite rod may be drawn across the gap.

Graphite is a better conductor than carbon, can be obtained in purer form, and is to be preferred in most cases for the electrode material, although carbon gives a somewhat steadier arc. The electrodes can be purified in several ways which cannot be discussed here.[4] The impurities present in Acheson spectrographic graphite electrodes generally will not cause more than a few faint lines to appear in the blanks.

SPECTROGRAPHS

Every spectrograph will have a dispersing medium, that is, either a grating or a prism, a slit, and a camera or other recording device.

The spectral lines recorded are replica images of the slit. The slit should, therefore, be straight and have parallel and sharp edges to avoid reflection from the edges. It should be kept clean and free from nicks, because dust particles or imperfections will be reproduced in the images. The slit should be adjustable, preferably continuously so, and should also be bilateral; that is, both sides should open or close rather than just one side. The center of a unilateral slit changes with width.

When using a prism and with some grating mountings, lenses are required to render the light parallel and to focus the light on the camera or detector. Since the focal length of a lens varies with the wavelength, light of different wavelengths will be brought to focus at different distances from the lens. If the lens can be constructed from two or more different materials it can be made to bring all wavelengths to focus at about the same distance; but in the ultraviolet region, where practically only quartz is available; that is not possible. The plate or film must then be tilted and curved somewhat to compensate for this characteristic of the lens. Lenses also show other errors such as spherical aberration which may require grinding of aspherical surfaces for correction or the use of only small lens apertures.

Prism Instruments

The various spectrographs may be distinguished chiefly by whether they employ a prism or grating as the dispersing medium and by the particular type of mounting of the prism or grating.

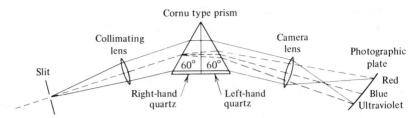

Fig. 13-8. Optical diagram of a Cornu-type spectrograph.

The Cornu mounting (Fig. 13-8) requires two pieces of quartz, one of right-handed and one of left-handed circularly polarizing quartz. This is necessary because quartz possesses the property of rotating the plane of polarized light and will also separate an unpolarized beam into two beams, circularly polarized in opposite directions. Since the index of refraction for the two beams is different, two images will result unless the two different types of quartz crystals are used or unless the beam is returned in an opposite direction through the crystal. Glass, because it is isotropic, does not show this effect, and only one piece is required. It is used frequently in instruments designed for the visible and near infrared or near ultraviolet regions. Glass exhibits greater dispersion than quartz in the visible region and is often used for this reason.

The Littrow mounting employs only one piece of quartz with the back surface of the prism metallized. Since the light passes back and forth through the same prism and lens, polarization effects are eliminated. Littrow mounting results in a compact instrument (Fig. 13-9).

A small prism spectrograph usually covers the range from 2000 to 8000 Å in one 7.5-cm photograph. By contrast, a large Littrow spectrograph covers the same spectral range in three photographs on 25-cm plates. The dispersion of the large type is about

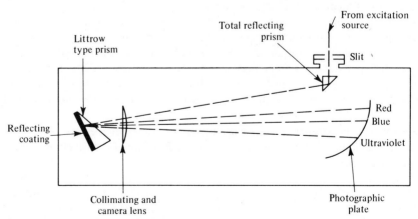

Fig. 13-9. Schematic optical diagram of a Littrow-type spectrograph (Courtesy of Bausch & Lomb Optical Co.)

0.4 mm/Å at 2500 Å and decreases to 0.10 mm/Å at 4000 Å. Sometimes an interchangeable glass prism is available for increased dispersion in the visible region.

Measures Used for the Comparison of Prism Instruments

The index of refraction of substances varies with the wavelength; over short ranges,
somewhat removed from regions of anomalous dispersion, it can be expressed by
Hartmann's formula[6]:

$$n = n_0 + \frac{c}{\lambda - \lambda_0} \tag{13-2}$$

where n = index of refraction
λ = wavelength
n_0, λ_0, and c = constants

The *angular dispersion*, the change in angle of the dispersed beam with a change in
wavelength, is equal to $d\theta/d\lambda$. Substances generally show greater dispersion at wavelengths near their "cut-offs" or regions of strong absorption. Glass and quartz show increasing dispersion as one moves from the visible toward the ultraviolet region. The
dispersion may also be conveniently written as the *linear dispersion*, $\Delta x/\Delta \lambda$, where
Δx represents the distance in millimeters on the plate between two lines $\Delta \lambda$ apart. A
typical curve relating wavelength to scale reading (in millimeters) for a small Littrow
prism spectrograph is shown in Fig. 13-10. This nonlinear dispersion of prismatic instruments causes some extra work in calculating the wavelengths of unknown lines. The
determination of wavelengths is made either from a graph constructed for the instrument
or by measuring the position of three known lines and the unknown line and employing
the formula of Hartmann in the form

$$\lambda = \lambda_0 + \frac{c}{d_0 - d} \tag{13-3}$$

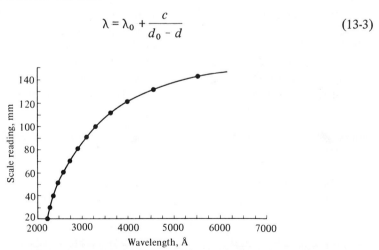

Fig. 13-10. Typical calibration curve for the small Littrow spectrograph.

in which λ_0, c, and d_0 are constants. A reference point on the spectrum plate or film is chosen near the unknown line and three lines of known wavelength. The distances from the reference point to the lines of known wavelength and the distance from the reference point to the unknown line are measured. The constants are evaluated by substitution of the known wavelengths for λ and the measured distances for d. The unknown wavelength is then easily computed from Eq. 13-3.

The *resolving power*, Rs, indicates the ability of a spectrograph to resolve two spectral lines separated by wavelength

$$Rs = \frac{\bar{\lambda}}{\partial\lambda} = t\,\frac{\partial n}{\partial\lambda} \qquad (13\text{-}4)$$

where $\bar{\lambda}$ = the average wavelength of two lines just distinguishable from each other
   $\partial\lambda$ = the difference in wavelength of these two lines
   t = thickness of the base of the prism
   n = index of refraction of the prism

Since $\partial n/\partial\lambda$ remains nearly constant for most prism materials, the thicker the prism, the greater will be the resolving power. Resolving power is also affected somewhat by the width and shape of the slit and by the other characteristics of the optical system.

Grating Instruments

A grating consists of a large number of parallel, equally spaced lines ruled upon a glass surface or a metal coating (often aluminum) on glass. Replica gratings made from original masters by coating the original with a plastic and then stripping off the plastic (Fig. 3-16) are in common use. A discussion of gratings and their optical properties will be found in Chapter 3.

Commercial grating spectrographs are available with a variety of mountings. The Rowland mounting is so arranged that the film and grating are at right angles to the slit, and the length of the arms is such that the three components all lie on the Rowland circle (Fig. 13-11). The Rowland circle is a circle with radius of curvature half that of the grating itself. If the slit and grating lie on the Rowland circle, the images of the slit are brought to a focus somewhere on this circle. To scan the spectrum, both the plate and

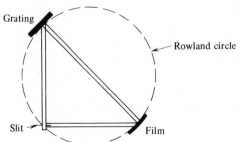

Fig. 13-11. Rowland mounting for a concave grating.

grating must be moved mechanically. This arrangement finds use mainly in X-ray spectrographs.

The Wadsworth mounting (Fig. 13-12) requires a mirror so that the grating may be illuminated by parallel light. The light-gathering power of the arrangement is high. Furthermore, the arrangement is stigmatic; that is, light arising from horizontal lines and from vertical lines is brought to a focus at the same distance from the grating. Most other grating mountings are astigmatic, and it is then necessary to find some position beyond the slit in which to place any device which would limit the length of the lines produced by the slit. There are two positions in an astigmatic mounting in which lines will be in focus at the camera: one position for vertical lines (the slit edges) and one position for horizontal lines (slit limiting devices). An example of a Wadsworth mount is found in the Jarrell-Ash 1.5-m spectrograph. The grating has 600 lines/mm, covers the range from 2200 to 7800 Å in first order with a dispersion of 10.9 Å/mm. The camera photographs 50 cm of spectrum.

The Eagle mounting, or modifications thereof (Fig. 13-13), is very popular in spite of the fact that rather complicated adjustments are needed for the film and grating. Astigmatism is slight. This mounting is the most compact of the concave grating mountings. The Bausch & Lomb 1.5-m spectrograph employs an Eagle mounting, has fixed slits of 10, 20, and 50 μm, and provides a dispersion of 16 Å/mm in the first order. It covers the range from 2250 to 6250 Å.

The Seya-Namioka mounting (Fig. 13-14) has an angle of $70°15'$ between the entrance and exit slits. Only slight defocusing occurs when the grating is rotated. This arrangement provides an excellent scanning monochromator.

The Ebert mounting uses a plane grating rather than a concave one, thus making the ruling of the grating easier. Also, the mounting is compact. This arrangement requires a large concave mirror to render the radiation striking the grating parallel and to intercept the dispersed beam and focus it on the plate at the camera. A modification, the Czerny-Turner mounting (Fig. 13-15), has two smaller concave mirrors in place of a single large mirror. The Ebert mounting is stigmatic and also achromatic, so that the rays of all wavelengths are brought to focus at the camera without changing the camera-to-mirror

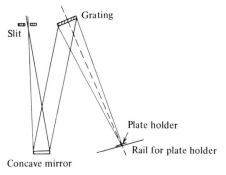

Fig. 13-12. Wadsworth mounting for a concave grating.

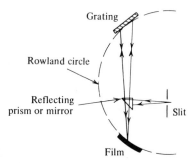

Fig. 13-13. Eagle mounting for a concave grating.

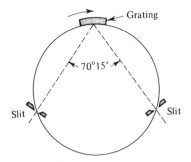

Fig. **13-14.** Seya-Namioka mounting.

distance. This makes it easy to change wavelengths merely by rotating the grating. Higher orders are readily accessible. Today almost all large spectrographs (3-m instruments) have plane gratings with modified Ebert mountings. Standard gratings have 600 to 1200 lines/mm with resolution ranging from 5.1 Å/mm in first order to 0.7 Å/mm in third order and a range of 1800 to 30,000 Å. Some instruments (Jarrell-Ash) have an "order sorter"—a fore-prism arrangement which stands between the source and the main slit of the instrument and which serves to place the various orders of spectra one above another on the photographic plate.

A combination of an echelle grating with a Littrow prism provides a high-resolution spectrograph. The principle of operation is illustrated in Fig. 13-16, and a schematic diagram of the optical system is shown in Fig. 13-17. Figure 13-16 shows what an echelle grating does. The iron triplet at 3100 Å is shown as it would appear when resolved by

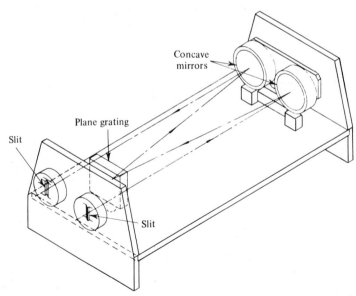

Fig. **13-15.** Czerny-Turner modification of the Ebert mounting, which has a single large mirror.

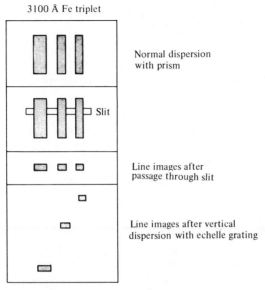

3100 Å Fe triplet

Normal dispersion
with prism

Slit

Line images after
passage through slit

Line images after vertical
dispersion with echelle grating

Fig. 13-16. How an echellogram is produced. (Courtesy of Bausch & Lomb Optical Co.)

the Littrow prism. The slit is then rotated 90° to the horizontal position and placed across the spectrum. When all sections of the line images that will not pass through the horizontal slit are removed, the image looks like the third section of the illustration. Now if this image is passed to an echelle grating (78 lines/cm), vertical dispersion is intro-

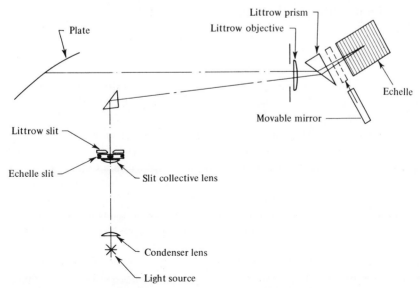

Plate

Littrow prism

Littrow objective

Littrow slit

Echelle slit

Slit collective lens

Echelle

Movable mirror

Condenser lens

Light source

Fig. 13-17. Littrow-Echelle system. (Courtesy of Bausch & Lomb Optical Co.)

duced with a result about 10 times superior to that of the prism alone. With this arrangement it is possible to obtain resolution better than 0.5 Å/mm in a compact, eight-foot instrument.

Measures Used for the Comparison of Grating Instruments

The dispersion, both angular and linear, is defined in the same manner as that for a prism instrument. For a grating, however, the dispersion is constant or very nearly so and does not vary with wavelength as it does for a prism. The dispersion is said to be "normal." This is a decided advantage in the identification of lines.

The resolving power Rs of a grating can be shown to be

$$Rs = nN \tag{13-5}$$

where n is the order of the spectrum, and N is the number of lines in the illuminated portion of the grating. Resolving power also depends upon the quality of a grating. A grating may show faint displaced images of lines—"ghost lines"—due to imperfections in the ruling. Usually it will also show some faint higher-order spectra overlapping the desired spectrum.

The particular spectrograph to be used in any situation depends on the type of material and the nature of the work to be performed. For materials with fair amounts of iron, cobalt, nickel, manganese, uranium, chromium, and the like, an instrument with high dispersion is required, because the spectra of these elements consist of a very great number of closely spaced lines. For alloys of aluminum, lead, tin, copper, silver, magnesium, and other metals, an instrument with reciprocal dispersions of only 10–20 Å/mm will suffice, because only a few lines appear.

The Photographic Process

In all spectrographs except the direct readers, the intensity of the spectral lines is registered on a photographic emulsion. The nature of the photographic process is therefore of considerable importance.

Photographic materials consist of a light-sensitive emulsion coated on a support (glass plate or plastic film). This emulsion contains light-sensitive crystals (grains) of silver halides suspended in gelatin. When the material is exposed in a camera of the spectrograph, it shows no visible effect, but an invisible change occurs—a latent image is produced. Treatment of the material in a developer solution converts the exposed silver halide crystals into metallic silver, which forms a visible and usable image. After development, the emulsion still contains the sensitive silver halides that were not utilized in producing the image, and on exposure to light these undeveloped crystals would eventually darken and obscure the image. Therefore, in order to make the image permanent, the material is "fixed" in a solution that dissolves the undeveloped silver halides but does not appreciably affect the silver image. After fixing, the material must be washed thoroughly to remove the chemicals used in developing and fixing. The entire series of operations must follow rigidly controlled conditions in respect to time, temperature, and chemicals. Recommended developing times and temperatures are given in data sheets supplied by the manufacturer of the film. Many types of automatic processing machines, most of which produce a quality superior to random manual agitation, are in use in spectrographic work.

If one plots the density of a film, that is, the logarithm of the ratio of the intensity of a beam of light passing through a clear portion of the film to the intensity of the same beam after passing through an exposed portion of the film, as a function of the logarithm of the exposure, curves such as the one shown in Fig. 13-18 result. It will be noted that there is a region B–C over which the density is directly proportional to the logarithm of the intensity of exposure. This is the useful range of the film. The slope of this straight portion of the curve is known as the γ of the emulsion.

$$\gamma = \tan \theta \tag{13-6}$$

High values of γ indicate a high degree of contrast, and low values of γ indicate low contrast.

The density of a film can be measured by passing a beam of light through a clear portion of the film and measuring the intensity of the transmitted beam by a phototube. The beam is then passed through the blackened portion of the film and the intensity is recorded. The logarithm of the ratio of the intensity of light passing through the clear film and through the blackened film is computed. An example of a commercial densitometer is shown in Fig. 13-19, and a schematic of the optical system is shown in Fig. 13-20. Essential parts include a source of light (tungsten lamp), a slit to detect the desired portion of the emulsion, a holder for the photographic plate (or film), a phototube, and a circuit and meter (or recorder) for reading the photocurrent. A projection system provides visual inspection of a portion of the spectrum. A racking mechanism provides a means of moving the plate horizontally (scanning) and vertically (from one spectrum to another).

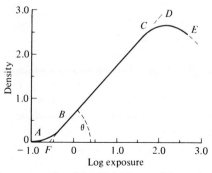

Fig. 13-18. Characteristic curve of a photographic emulsion: A, threshold exposure; B–C, linear portion of curve; D–E, reversal region; F, inertia of emulsion.

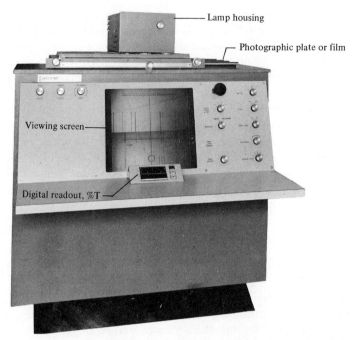

Fig. 13-19. Densitometer-comparator. (Courtesy of Baird-Atomic, Inc.)

The shape of the characteristic curve of a photographic emulsion varies from emulsion to emulsion, with wavelength, and with the conditions of development. In order to determine the curve for any given emulsion, it is first necessary to standardize the conditions of development, that is, the time, the type of developer, the temperature and other factors. A step-wedge transmitting known relative intensities of light through the various steps may be placed directly on the film and the film exposed through the wedge. After development the densities of the steps are measured and plotted against the known intensities. In order to calibrate at several wavelengths it is better to place the wedge in front of the slit of the spectrograph and expose with a metallic arc, such as a copper arc. Each line will then show a reproduction of the wedge.

Instead of a wedge or a series of filters with known transmittancies, a rotating step-sector disc or a log-sector disc (Fig. 13-21) may be placed just in front of the slit of a stigmatic spectrograph or at the second (horizontal) focus position of an astigmatic spectrograph. The different parts of the lines will show definite graduated intensities depending upon the construction of the sector. Photographic emulsions, however, show what is known as an *intermittancy effect*; that is, several short exposures may not produce the same effect as one long exposure. In other words, the relationship

$$\text{exposure} = Pt \qquad (13\text{-}7)$$

where P is the intensity and t is the time of exposure, breaks down if intermittent exposures are employed. If the frequency is high, this effect may be negligible. Therefore, sectors are run by high-speed motors. The particular emulsion employed should always be studied to discover its behavior in this respect.

A photographic emulsion is sensitive basically to radiation that is absorbed by the silver salts, that is, basically to blue, violet, and shorter-wavelength radiation. However, an emulsion can be sensitized optically by the addition of suitably chosen dyes that absorb radiation of longer wavelengths. By this means it is possible to extend the sensitivity

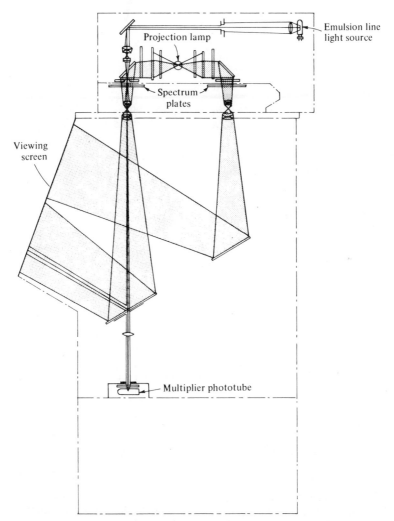

Fig. 13-20. Typical optical system of a microdensitometer. (Courtesy of Baird-Atomic, Inc.)

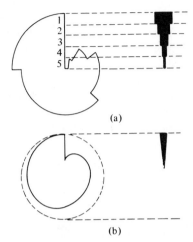

(a)

(b)

Fig. 13-21. (a) Step-sector disc, (b) log-sector disc, and schematic illustration of a line spectrum obtained with each. Blackness is indicated by the width of the line.

through the green (orthochromatic film), through the green and red (panchromatic film), and into the infrared regions of the spectrum.

QUALITATIVE IDENTIFICATION

The elements present in a sample can be determined by comparing the spectrum of the unknown with that of pure samples of the elements or by measuring the wavelengths

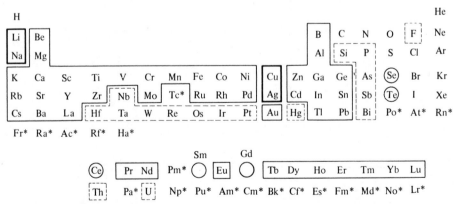

Fig. 13-22. Elements detectable using a dc arc source. In the range of 2000–9000 Å elements detectable in the < 1 µg/g limit are enclosed in heavy solid lines; those in the 1–10 µg/g limit are enclosed in solid lines; those in 10–100 µg/g range enclosed in dashed lines; and those in the 100–1000 µg/g range are circled. Elements identified with an asterisk occur only as radioactive isotopes.

of the lines and looking up the corresponding elements in tables. If only certain elements are being sought, the spectra of these elements may be taken on one film along with the spectrum of the unknown. It is then easy to compare lines in the known samples with lines in the unknown. It is usually considered sufficient proof that an element is present in an unknown if three or more sensitive lines of the element in question can be definitely identified in the spectrum of the unknown sample. Elements shown in Fig. 13-22 are readily detectable using a dc arc as source of excitation. To detect the sensitive arc lines of carbon, phosphorus, and sulfur, air must be excluded from the optical path of the instrument either through use of a vacuum or displacement of air by helium.

R.U. Lines and R.U. Powder

Those lines of each element that are the last to disappear as the concentration of the element is gradually decreased are known as R.U. (*raies ultimes*) or persistent lines. These are the lines most useful in detecting small concentrations of impurities. The Johnson-Mathey Company, London, England, prepares a powder containing more than 50 elements in such concentrations that only the R.U. lines of most of the elements appear in the arc spectrum. A spectrogram of this powder taken on each spectrograph and with all lines identified is a useful aid in identification of elements.

Wavelength Measurement and Tables

The wavelength of unknown lines may be determined by linear interpolation between lines of known wavelengths if the dispersion of the instrument is "normal." For prismatic dispersion, the formula of Hartmann (Eq. 13-1) must be employed. Many instruments have wavelength scales that may be impressed on the film. Such scales are useful for rough identification of the wavelengths of lines.

For exact measurements of the distances of lines on a film, a magnifying glass with a built-in scale is useful. For distances greater than a few millimeters, a measuring microscope may be employed.

There are many tables available that list the wavelengths of the spectral lines and the corresponding elements. Many of these tables are listed in the selected references at the end of this chapter. Once an element is definitely located by identification of three or more lines, the tables that list the lines under each element are useful in eliminating the remainder of the lines due to that element before proceeding with the identification of the second, third, and all other constituents.

QUANTITATIVE METHODS

Early workers in spectrography attempted rough estimations of the concentration of elements in various ways. Hartley[5] correlated the concentration of solutions with the number of lines appearing in the spark spectrum.

De Gramont[1] and later Meggers, Kiess, and Stimson[7] employed a series of standard electrodes with known concentrations of the substance to be determined. Spectra of the various standards and of the unknown are photographed alternately on the same plate and under the same conditions. The concentration of the desired constituent can then be estimated by comparing the blackening of the lines of this constituent with the same lines in the standards. Photometric or simple visual comparison of the blackening of the lines is possible. The accuracy depends on the number of standard samples available and on the maintenance of constant excitation and exposure conditions.

Internal Standards

In the above-mentioned methods and in any procedure that depends on the measurement of the intensity of only the lines of the unknown element the excitation conditions, the time and nature of the exposure, and the conditions of development must all be carefully controlled. In order to eliminate the effects of variations in these factors, the modern methods of spectrographic analysis measure the intensity of an unknown line relative to that of an internal standard line. The internal standard line may be a weak line of the main constituent, or it may be a strong line of some material added in a definite concentration to the sample. The ratio of the intensities of these lines, that is, the analysis line and the internal standard line, will be unaffected by exposure conditions and development conditions. Gerlach and Schweitzer[3] were the first to propose this method of "internal standards."

Homologous Pairs

In order that variations in excitation conditions may not affect the relative intensities of the two lines, it is necessary that the two lines constitute what is known as a "homologous pair." Such pairs are lines that respond in the same way to changes in excitation conditions. Both lines should arise from the same type of excitation, that is, atoms or ions. Homologous pairs may be selected by experiment or be chosen on the basis of recommendations of others as recorded in the literature. Two lines that change intensities quite differently with variations in the conditions of excitation are called a "fixation pair" and such a pair of lines is sometimes observed as a check that the excitation conditions remained constant during each exposure.

Gerlach and Schweitzer prepared tables listing homologous pairs of lines that had equal blackening at given concentrations of the desired element. Such a table for the determination of cadmium in tin is shown in Table 13-1. This method suffers from the defect that only definite limited steps are available.

The most obvious and the best method of comparing the intensities of the unknown and the internal standard lines is to measure the density of the two lines on the film or plate by means of a densitometer. The intensity of the light striking the plate and creating the two lines is then calculated by means of the characteristic curve for the emulsion under the chosen conditions. Either the ratio of the intensities of the homologous

TABLE 13-1 Homologous Pairs for Determination of Cadmium in Tin

Cadmium Line, Å	Tin Line, Å	Equal Intensity at Cd Concentration, %
3404	3331	10
3404	3656	2
3404	3142	1.5
3404	3219	0.5
3466} 3468	3656	0.3
3611} 3615	3656	0.2
3404	3224	0.1
3466} 3468	3224	0.05
2288	2282	0.01

pair of lines is plotted against concentration, or the log of the ratio is plotted against the log of the concentration. Either plot should result in a nearly straight line, because intensity of light is proportional to the concentration of the responsible atom or ion in the source. A less precise method would be to plot the ratio of the densities of the lines directly against the log of the concentration—assuming, of course, that one is working on the strictly straight-line portion of the characteristic curve of the emulsion.

Log- and Step-Sector Methods

There are other methods of comparing the relative intensities of two lines when a densitometer is not available. If a step-sector or a log-sector is run in front of the slit during the exposure, the resulting lines will have different lengths (Fig. 13-23). The strong lines will be long, because even at small exposure times sufficient light will reach the film to cause a visible blackening. On the other hand, the weak lines will be short, because a visible blackening of the film will result only where the sector allows radiation to pass onto the slit for a long time. A measurement of the length of the lines indicates the intensity. Thus, if C is the concentration, D is the density, P is the intensity of char-

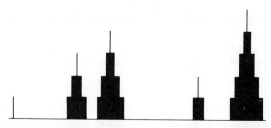

Fig. 13-23. Schematic diagram of line spectra obtained from a log-sector in front of the slit. Width indicates blackness.

acteristic line in the source, and h is the height of line image on the photographic plate, then

$$\log C \text{ is proportional to } \log P \text{ or } D \tag{13-8}$$

and if the sector is logarithmic, that is,

$$h \text{ is proportional to } \log P \tag{13-9}$$

then

$$h \text{ is proportional to } \log C \tag{13-10}$$

Actually, the difference in height of the internal standard and unknown lines is plotted against the logarithm of the concentration. A nearly straight line should result (Fig. 13-24). The main difficulty with this procedure is the determination of the exact length of the lines. It is very difficult to judge just when a line disappears.

The step-sector method and the photometric measurement of density method can be combined for very precise work. If a step sector is employed, the densitometer can be used to measure the densities of the two lines at steps where the two lines are of nearly equal blackening. The densitometer measurements are more precise when the densities are nearly equal. Since the ratios of intensities of the various steps are accurately known, one can easily calculate the original intensities of the lines.

In any case a series of several standard samples must be run to establish points on the "*working curve*." Once this working curve has been established, similar unknown samples can be quickly analyzed. A new working curve must be established for each type of material. The presence of elements not originally present in the standards from which the working curve was established will usually affect the determination.

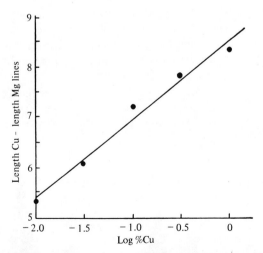

Fig. 13-24. Example of a working curve, log-sector method: determination of copper in an alumina–silica mixture; magnesium, 1%, added as internal standard. Lines compared: Cu 3275 and Mg 2852.

Direct Readers

Optical direct readers can be used whenever spectrographic methods are applicable. The essential difference between a spectrograph and a direct reader is that with the latter the photosensitive emulsion is eliminated (and thus the need for developing equipment, comparator densitometers, and calculating equipment) and replaced by multiplier phototubes and electronic circuitry. Basically, the receiver system is an arrangement of semifixed slits on the focal circle of the spectrometer with cylindrical mirrors to focus the light coming through the slits onto multiplier phototubes, as illustrated in Fig. 13-25. Although the photographic technique is inherently more versatile, the direct reading technique is faster and more precise.

Direct readers are usually custom built for a specific analytical requirement by carefully placing exit slits and multiplier phototubes to measure specific lines of elements of interest. Of course, physical size of the receiver system limits the number of individual lines that can be monitored. Instruments generally measure from 8 to 24 elements in a single matrix. The receivers are often grouped into several bridges, one for each type of matrix, so that the instrument can be used to analyze several types of samples successively.

Photomultiplier readout is accomplished in several ways. In one method the photomultiplier outputs are integrated by charging individual capacitors during a fixed exposure period. Although the resulting voltages are a function of the concentrations of the elements in the sample, simple voltage measurement is not enough to obtain good precision of analysis. Rather the internal standard method is used, in which the voltage ratio of each of the unknown lines is measured relative to an internal standard line. The voltage ratio measurement is performed during the period immediately following the exposure period; the mechanism of turning off the spark at the end of the exposure and of beginning the measure period is entirely automatic. At the beginning of the measure period, the capacitors, which have been charged by the photomultipliers, are rearranged in the circuit and are connected through sensitive amplifier-trigger circuits to individual dial indicators (or digital readout tapes). In the measure period, the capacitors for unknown elements maintain their voltages and do not discharge; only the internal standard capacitor discharges through a fixed resistance. At the beginning of the measure period all the indicators start to run, but whenever the reference capacitor reaches the voltage of an element capacitor, the indicator associated with that capacitor stops. The time that any dial indicator will run can be expressed as

$$t = RC \ln \frac{E_s}{E_x} \tag{13-11}$$

where R is the resistance through which the capacitor C is discharged, and E_s, E_x are the voltages to which the reference and unknown capacitors are charged. Thus the time scale can be replaced by a scale reading directly in percent concentration for a particular element.

To standardize the direct reader the operator places a standard sample in the electrode holder and pushes the start button. After the dial-indicator hands are stopped, the operator notes whether or not the dials indicate the known analysis of the standard. If

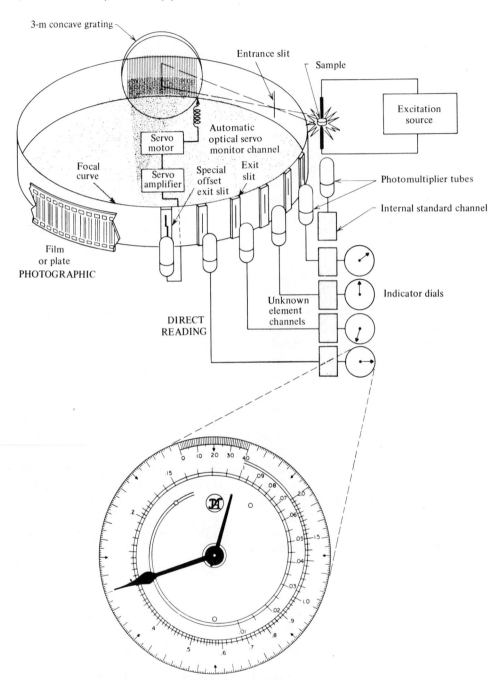

Fig. 13-25. Schematic optical diagram of a direct reader with an enlarged view of a dial-indicator. (Courtesy of Baird-Atomic, Inc.)

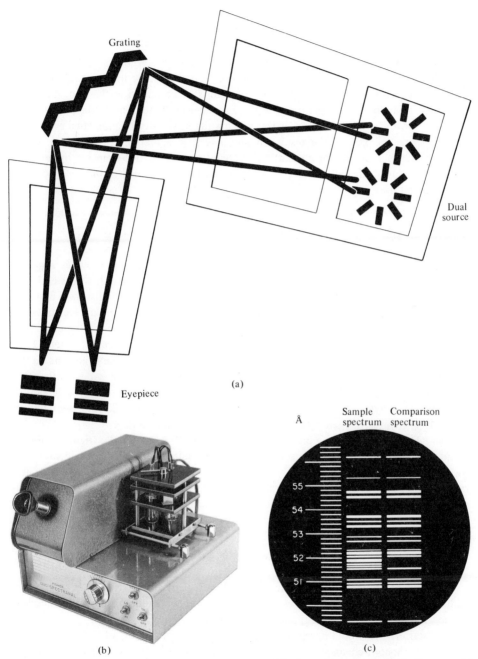

Grating

Dual
source

Eyepiece

(a)

Å   Sample   Comparison
       spectrum   spectrum

(b)   (c)

Fig. 13-26. The Fisher Duo-Spectranal, showing (a) the schematic optical diagram, (b) the instrument and electrode jig, and (c) a view through the eyepiece. (Courtesy of Fisher Scientific Co.)

any hand does not point to the correct scale reading, the operator rotates the scale so that it does so. Thus, any change in sensitivity causes a parallel displacement of the calibration curve but not a change in its slope (the constant RC in Eq. 13-11).

Other methods devised for presentation of results involve recording a fixed number of counts by the multiplier tube monitoring the internal standard line. At this point the excitation period is terminated. The ratio of counts accumulated in the phototubes receiving lines of individual elements to the fixed number in the reference tube is correlated with the known analysis of standards (by varying the sensitivity of individual counters).

Direct readers present the results of an analysis in a matter of 1–2 min. About 30 min is usually required for a quantitative analysis of about five elements by spectrographic procedures. Direct readers with the optical system sealed in pipe are available for production floor installation.

Visual Spectroscope

The Fisher Duo-Spectranal is an inexpensive spectroscope that provides a quick, easy way to make qualitative and semiquantitative analyses for elements whose emission lines lie in the visible portion of the spectrum. The sample is usually dissolved in dilute nitric acid with a small amount of potassium nitrate added. Two platinum electrodes are immersed in about 2 ml of the acid solution: the excitation electrode is submerged to a depth of no more than 2 mm; the other electrode is submerged more deeply. When 115-V alternating current is passed between the electrodes, a spark-like discharge appears on the surface of the shorter electrode. This discharge consists of a rapid series (60/sec) of microscopic hydrogen–oxygen explosions, taking place in the film of mixed gases produced by electrolysis at the surface of the excitation electrode. Thus, essentially an electrically generated oxygen–hydrogen flame is the excitation source.

The instrument employs a plane reflection grating (55 Å/mm) and a stigmatic optical system brings the spectra of two sources to a focus, side by side, in the field of a five-power Ramsden-type eyepiece. The field of view covers about 700 Å; by tilting the grating, the operator can sweep spectra and eyepiece scale through the field of view. An optical schematic and typical view through the eyepiece is shown in Fig. 13-26.

LABORATORY WORK

It would be foolhardy to attempt to teach a student all the intricacies of spectrographic methods in a few laboratory periods. However, individual instructors, endowed with specific pieces of equipment, may wish to devise specific experiments.

As a starter, one should purchase and study the latest edition of A.S.T.M.'s "Methods for Emission Spectrochemical Analysis," American Society for Testing Materials, Philadelphia, Pa., 1964. A useful spectrochemical primer has been published in *The Spex Speaker*, **9** (2) (June, 1964), a house organ of Spex Industries, authored by R. F. O'Connell and A. J. Mitteldorf.

PROBLEMS

1. Calculate the wavelength of the unknown line from the following data taken from a spectrogram recorded on a prism spectrograph: distances from reference point to lines of wavelength 3247.54, 3262.33, and 3273.96 Å, were respectively 0.50, 6.42, and 11.00 mm. The distance from the reference point to the unknown line was 8.51 mm.

2. What is the theoretical resolving power of a spectrograph equipped with a 2.5-in. grating of 15,000 lines/in. when used in the first order? Theoretically, what order would have to be employed to resolve the iron doublet at 3099.90 and 3099.97 Å.

3. A sample of an unknown light-metal alloy was placed on a spark stand and a spectrum was recorded. Observation of the spectrogram revealed lines at the following wavelengths: 6438, 5184, 5173, 4810, 4722, 4680, 3838, 3832, 3829, 3611, 3466, and 3403 Å, plus many lines of aluminum. What elements, besides aluminum, are present?

4. In the spectrographic determination of lead in an alloy, using a magnesium line as internal standard, these results were obtained:

Solution	Densitometer Reading		Concentration of Lead, mg/ml
	Mg	Pb	
1	7.3	17.5	0.151
2	8.7	18.5	0.201
3	7.3	11.0	0.301
4	10.3	12.0	0.402
5	11.6	10.4	0.502
A	8.8	15.5	
B	9.2	12.5	
C	10.7	12.2	

(a) Prepare a calibration curve on log-log paper. (b) Evaluate the concentrations for solutions A, B, and C.

5. A step-sector with arc lengths (or angle subtended by each step at the center) in the ratio of $1:2:4:8:16:32$ was rotated in front of the slit of a spectrograph while a sample of a tin alloy containing lead was being arced in the source unit. After the plate was developed, fixed, and dried, the density of a suitable tin line was measured at each step with a microphotometer.

   The values of I_0/I obtained for each step were 1.05, 1.66, 4.68, 13.18, 37.15, and 52.5. Plot the characteristic curve for the film and determine the γ and the inertia of the emulsion.

6. Several standard samples of the tin alloy mentioned in Problem 5 were prepared by chemical analysis for the lead content. These alloys were then employed as electrodes as in Problem 5. The ratio of the density of the tin line at 2761 Å and the density of the lead line at 2833 Å were measured on the microphotometer. The results are

listed below:

Sample No.	% Lead	$D_{\text{tin line}}$	$D_{\text{lead line}}$
1	0.126	1.567	0.259
2	0.316	1.571	1.013
3	0.708	1.443	1.546
4	1.334	0.825	1.427
5	2.512	0.447	1.580

Using the results of Problem 5, plot a "working curve" of log percent lead as abscissa and log $(I_{\text{Pb}}/I_{\text{Sn}})$ as ordinate. An unknown tin alloy sample was treated in the same way as the standards. The 2761-Å tin line had a density of 0.920 on the photographic plate, while the 2833-Å lead line had a density of 0.669. What was the percentage of lead in the alloy?

7. Approximately what is the concentration of cadmium in a cadmium–tin alloy if the 3224 Å line of tin appears brighter than the 3404 Å line of cadmium but weaker than the 3466 + 3468 Å unresolved lines of cadmium?

BIBLIOGRAPHY

Ahrens, L. H. and S. R. Taylor, *Spectrochemical Analysis*, 2nd ed., Addison-Wesley, Reading, Mass., 1961.

ASTM, "Methods for Emission Spectrochemical Analysis," American Society for Testing Materials, Philadelphia, Pa., 1964.

Brode, W. R., *Chemical Spectroscopy*, 2nd ed., Wiley, New York, 1943.

Gerlach, W. and E. Schweitzer, *Foundations and Method of Chemical Analysis by the Emission Spectrum* (authorized translation *of Die chemische Emissionspektralanalyse*, Vol. I, L. Voss, Leipzig, 1929), Adam Hilger, Ltd., London, 1929.

Gibb, T. R. P., *Optical Methods of Chemical Analysis*, McGraw-Hill, New York, 1942, pp. 1-69.

Grove, E., Ed., *Analytical Emission Spectroscopy*, Marcel Dekker, New York, Part I, 1971, Part II, 1972.

Harrison, G. R., R. C. Lord, and J. R. Loofbourow, *Practical Spectroscopy*, Prentice-Hall, Englewood Cliffs, N. J., 1948.

Meggers, W. F., C. H. Corliss, and B. F. Scribner, "Tables of Spectral-line Intensities," Parts I and II, *Nat. Bur. Std. (U.S.), Monograph* **32** (1961–1962).

Nachtrieb, N. H., *Principles and Practice of Spectrochemical Analysis*, McGraw-Hill, New York, 1950.

Sawyer, R. A., *Experimental Spectroscopy*, 2nd ed., Prentice-Hall, Englewood Cliffs, N. J., 1951.

Scribner, B. F. and Margoshes, M., "Emission Spectroscopy" in *Treatise on Analytical Chemistry*, Vol. 6, Part I, Kolthoff, I. M. and P. J. Elving, Eds., Wiley-Interscience, New York, 1965, Chapter 64.

Slavin, Morris, "Emission Spectrochemical Analysis" in *Chemical Analysis*, Vol. 36, Elving, P. J. and I. M. Kolthoff, Eds., Wiley-Interscience, New York, 1971.
Twyman, F., *Metal Spectroscopy*, C. Griffin, London, 1951.

LITERATURE CITED

1. De Gramont, A., *Compt. Rend.*, **159**, 6 (1917); **171**, 1106 (1920).
2. Feussner, O., *Arch. Eisenhuttenw.*, **6**, 551 (1921).
3. Gerlach, W. and K. Schweitzer, *Foundations and Methods of Chemical Analysis by the Emission Spectrum*, Adam Hilger, Ltd., London, 1929.
4. Gibb, T. R. P., *Optical Methods of Chemical Analysis*, McGraw-Hill, New York, 1942, p. 10.
5. Hartley, W. N., *Phil. Trans. London*, **175**, 326 (1884).
6. Hartmann, J., *Astrophys. J.*, **3**, 218 (1898).
7. Meggers, W. F., C. C. Kiess, and F. S. Stimson, *Nat. Bur. Std. (U.S.)*, *Sci. Papers* 444 (1922).

14

Refractometry and Interferometry

THEORY

When a ray of light passes obliquely from one medium into another of different density, its direction is changed on passing through the surface. This is called *refraction*. If the second medium is optically denser than the first, the ray will become more nearly perpendicular to the dividing surface. The angle between the ray in the first medium and the perpendicular to the dividing surface is called the angle of incidence, i, whereas the corresponding angle in the second medium is called the angle of refraction, r. Sin i and sin r are directly proportional to the velocities of the light in the two media. The ratio sin i/sin r is called the *index of refraction*, n. If the incident ray is in the denser medium, n will be less than 1; if in the rarer, greater than 1. Commonly n is taken as greater than 1, the ray passing from the optically rarer medium (usually air) to the denser.

The index of refraction for two given media varies with the temperature and the wavelength of light and also with the pressure, if we are dealing with gases. If these factors are kept constant, the index of refraction is a characteristic constant for the particular medium and is used in identifying or determining the purity of substances and for determining the composition of homogeneous binary mixtures of known constituents.

The refractive index is theoretically referred to vacuum as the first medium, but the index referred to air differs from this by only 0.03% and, for convenience, is more commonly used. The refractive index of a transparent substance gradually decreases with increasing wavelength except at regions of absorption where the refractive index changes abruptly. The change of refraction with wavelength is known as *dispersion*. Because of dispersion, the wavelength must be specified when refractive indices are stated. The symbol n_D^{20} means the index of refraction for the D lines of sodium* measured at 20°C.

When the beam of light passes from a denser to a rarer medium, the angle r will be greater than the angle i. As angle i increases, the ratio sin i/sin r remaining constant, the angle r must also increase and remain greater than i. If angle i is increased to the value where r becomes 90°, the beam of light will no longer pass from the first medium to the second, but will travel through the first medium to the dividing surface and then pass along this surface, thus making an angle of 90° with the perpendicular to the surface (Fig.

*The yellow doublet at 5890/5896 Å.

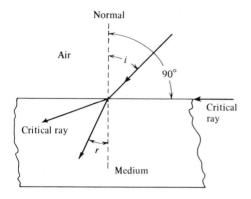

Fig. 14-1. Angles of incidence (i) and refraction (r).

14-1). This is called the *critical ray*. In Fig. 14-1, i and r would then be interchanged, and the direction of the arrows would be reversed. If i is smaller than this particular value, light will pass through the second medium; if greater, all light will be reflected from the surface back into the first medium. This furnishes the basis for the reference line used in several refractometers. Total reflection can occur only when light passes from the denser to the rarer medium.

The refractive index of a liquid varies with temperature and pressure, but the *specific refraction*, r_D,

$$r_D = \frac{n^2 - 1}{n^2 + 2} \cdot \frac{1}{\rho} \tag{14-1}$$

where ρ is the density, is independent of these variables. This relationship is known as the Lorentz and Lorenz equation. The molar refraction is equal to the specific refraction multiplied by the molecular weight. It is a more or less additive property of the groups or elements comprising the compound. Tables of atomic refractions are available in the literature; an abridged set of values is given in Table 14-1. Thus, specific refraction is

TABLE 14-1 Atomic Refractions

Group	Mr_D	Group	Mr_D
H	1.100	Br	8.865
C	2.418	I	13.900
Double bond (C=C)	1.733	N (primary aliphatic amine)	2.322
Triple bond (C≡C)	2.398	N (*sec* aliphatic amine)	2.499
O (carbonyl) (C=O)	2.211	N (*tert* aliphatic amine)	2.840
O (hydroxyl)(O—H)	1.525	N (primary aromatic amine)	3.21
O (ether, ester)(C—O—)	1.643	N (*sec* aromatic amine)	3.59
S (thiocarbonyl)(C=S)	7.97	N (*tert* aromatic amine)	4.36
S (mercapto)(S—H)	7.69	N (amide)	2.65
F	1.0	—NO$_2$ group (aromatic)	7.30
Cl	5.967	—C≡N group	5.459

valuable as a means for identification of a substance and as a criterion of its purity. In a homologous series of compounds the specific refraction of higher members generally increases fairly regularly with increasing length of the carbon chain.

The values of refractive index for organic liquids range from about 1.2 to 1.8, those for organic solids from about 1.3 to 2.5.

Dispersion is also sometimes useful in the identification of compounds. Dispersion is often taken as the Abbé number, v, defined as

$$v = \frac{n_D - 1}{n_F - n_C} \tag{14-2}$$

where n_F and n_C are the refractive indices for the F and C lines of hydrogen ($\lambda = 4861$ Å and $\lambda = 6563$ Å, respectively).

Example 14-1

The refractive index of acetic acid at $20°C$ is 1.3698, the density at $20°C$ is 1.049 g cm^{-3}, and the molecular weight is 60.0. From Eq. 14-1 the specific refraction is found:

$$r_D = \frac{[(1.3698)^2 - 1]}{[(1.3698)^2 + 2]} \cdot \frac{1}{1.049} = 0.2155 \text{ cm}^3\text{g}^{-1}$$

The molar refraction is

$$Mr_D = (60.0)(0.2155) = 12.93 \text{ cm}^3\text{mole}^{-1}$$

The molar refraction is a constitutive property depending upon the structural arrangement of the atoms within the molecule. From the atomic and group refractions in Table 14-1, the molar refraction of acetic acid can be computed and compared with the experimental value as follows:

Acetic Acid		Methyl Formate	
2 carbons	= 4.836	2 carbons	= 4.836
4 hydrogens	= 4.400	4 hydrogens	= 4.400
1 carbonyl oxygen	= 2.211	1 carbonyl oxygen	= 2.211
1 hydroxyl oxygen	= 1.525	1 ester oxygen	= 1.643
	12.972		13.090

Methyl formate possesses the same empirical formula as acetic acid and differs only slightly in structure. This difference is apparent in the molar refraction values, although the difference only amounts to about 0.90%. To distinguish between the two compounds would require a precision of ±0.006 in refractive index and ±0.005 in density. For methyl formate, $n_D^{20} = 1.344$ and $d^{20} = 0.974$.

REFRACTOMETERS

There are three types of refractometers — the Abbé, the immersion or dipping, and the Pulfrich instruments. The last named is used infrequently and will not be discussed.

The Abbé Refractometer

The instrument and its essential parts are shown in Fig. 14-2. Light reflected from a mirror passes into the illuminating prism P_1, the upper surface of which is rough ground. This rough surface acts as the source of an infinite number of rays which pass through the 0.1-mm layer of liquid in all directions. These rays then strike the surface of the polished prism P_2 and are refracted. The critical ray forms the border between the light and dark portions of the field when viewed with the telescope which moves with the scale. The scale is provided with a reading telescope.

The range of available models is from $n = 1.30$ to 1.71 and from 1.45 to 1.84. Except on some of the newer models, the range cannot be changed. The reproducibility of the individual readings is ±0.0002 in refractive index. The instrument reads the refractive index directly, is durable, requires only a drop of sample, and gives a good approximation of the value of $(n_F - n_C)$, the difference in refractive index for the blue and red lines of hydrogen, a measure of the dispersion. Except for precision models, however, it is not well suited for accurate measurement of solutions with a volatile component or of powders. Solid samples can be attached to the lower surface of the Abbé prism. The clear plane face of the specimen is brought into optical contact with the prism face by placing

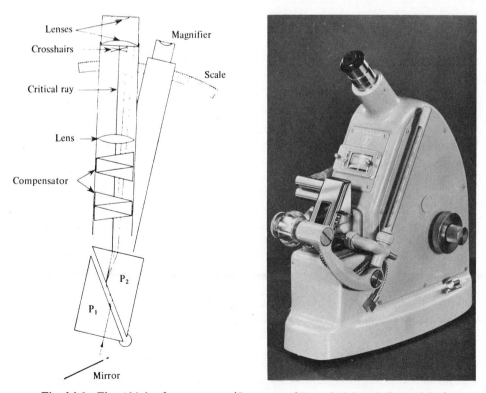

Fig. 14-2. The Abbé refractometer. (Courtesy of Bausch & Lomb Optical Co.)

a drop of liquid on the prism surface and carefully pressing the solid into place. As contacting liquid, 1-bromonaphthalene (n_D = 1.68) is commonly used.

White light is used, and to prevent a colored, indistinct boundary between the light and dark fields due to the differences in refractive indices for light of different wavelengths, two direct-vision prisms, called *Amici prisms*, are placed one above the other in front of the objective of the telescope. These are constructed of different varieties of glass and are so designed as not to deviate a ray of light corresponding to the sodium *D* line. Rays of other wavelengths are, however, deviated, and by rotating these Amici prisms it is possible to counteract the dispersion of light at the liquid interface.

The temperature should be controlled within ±0.2°C. The instrument is fitted with hollow prism casings through which water may be passed. A short thermometer is inserted into the water jacket. The most satisfactory temperature control is obtained by using a small circulating pump to pass water from a thermostat through the prism casing.

Improved reproducibility is obtained with a precision Abbé refractometer. Three ranges are available: 1.30-1.50, 1.40-1.70, and 1.33-1.64. Refractive index readings are reproducible within $\pm 2 \times 10^{-5}$ to $\pm 6 \times 10^{-5}$ when the temperature is maintained within ±0.02°C. Improved precision is obtained by dispensing with compensating prisms and by use of unusually large and precise Abbé prisms mounted on a long, vertical, taper bearing. The instrument is usually calibrated with known standards when one is working a few degrees above or below the calibration temperature, as corrections for changes in prism index and in the state of the reference medium become very important.

The Immersion Refractometer

This type is the simplest to use but requires 10-15 ml of sample. It uses white or artificial light and contains an Amici compensator as already described. The single prism is mounted rigidly in the telescope containing the compensator and eyepiece as shown in Fig. 14-3. The scale is mounted below the eyepiece inside the tube. The lower surface of the prism is immersed in a small beaker containing the sample with a mirror below to reflect light up through the liquid. The complete instrument in position with the water bath for maintaining a constant temperature is also shown.

The scale, situated at the focal plane of the eyepiece, is graduated from −5 to +105. The field will be partly dark and partly light, separated by a sharp line as already explained (Fig. 14-1). The position of this line is read on the scale, and the tenths of a division are found by turning a micrometer screw at the top of the instrument, which slides the scale toward the border line until it covers the lower numerical scale division previously noted. The figure on the micrometer drum then shows the decimal to be added. A change of 0.01 division corresponds to ±0.000037 in n_D. The immersion refractometer therefore gives greater precision in its readings than any other type except the interference refractometer.

Since the refractive index changes with the temperature, a standard temperature must be chosen. This, unfortunately, is 17.5°C, which is rather difficult to maintain. The solution to be tested is placed in a very small, specially designed beaker placed in a rack in a water bath illuminated through the bottom. A current of water at the proper temperature is passed through the bath. This may be done by running tap water from a constant-level tank into the bath at the proper rate; or various types of constant-temperature baths may be used.

The correctly adjusted refractometer should show for distilled water at various temperatures the following readings:

15°	15.5	18°	14.9	22°	14.0
16°	15.3	19°	14.7	23°	13.75
17°	15.1	20°	14.5	24°	13.5
17.5°	15.0	21°	14.25	25°	13.25

The temperature should not vary more than 0.1°C, because readings are reported to an estimated 0.01 scale division. In order to be of any value the reading must be converted into concentration by means of published tables, the most comprehensive of which are those by Wagner, obtainable from suppliers of the instrument. These tables apply only to

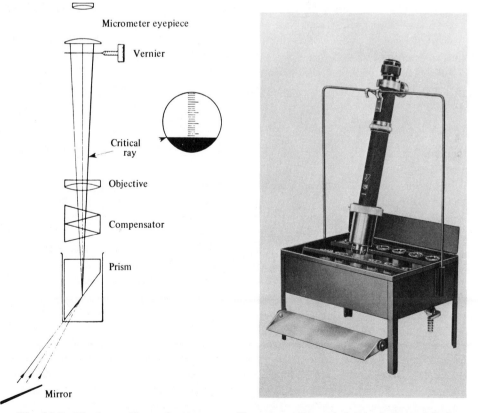

Micrometer eyepiece

Vernier

Critical ray

Objective

Compensator

Prism

Mirror

Fig. 14-3. The immersion refractometer. (Courtesy of Bausch & Lomb Optical Co.)

17.5°C, and there is no formula for converting them to other temperatures. There are, however, tables for methyl and ethyl alcohol at other temperatures. Leach and Lythgoe[3] have published complete data for 20°C. The table by Andrews[1] covers only the range 70-100% ethyl alcohol at 25°C. In all tables, readings are given only in scale divisions. Readings may be converted into index of refraction by reference to tables furnished with the instrument or to tables in *Lange's Handbook of Chemistry*, 11th ed.

The range of the instrument with prism 1 is 1.325-1.367; This covers all ordinary salt solutions and alcohols. For higher values, special auxiliary prisms are furnished extending the range to 1.492. Thus the range of this refractometer is much narrower than that of the Abbé, but this gives it the advantage of greater sensitivity.

A disadvantage of refractometric analysis is the necessity for carefully regulating the temperature. An attempt has been made by Clemens[2] to avoid this, but considerable precision is necessarily sacrificed.

The refractometer measures concentration more accurately and readily than can be done by ordinary density measurements with a hydrometer. For example, assuming a sufficiently accurate temperature control, 0.02 scale division (which is about the best one can do in reading the instrument, estimating the nearest 0.01 division) corresponds to the following weight of substances per 100 ml: methyl alcohol, 24 mg; ethyl alcohol, 12 mg; ammonium chloride, 4 mg; perchloric acid, 10 mg.

If both density and refractometer readings are determined, it is possible to determine each of two components, such as methyl and ethyl alcohol, with a fair degree of accuracy if nothing else is present. It should be noted that both density and refractive index are measures of the total amount of substance in solution, no matter how many different ones there may be.

Applications

The immersion refractometer is especially useful in determining the concentration of aqueous and alcoholic solutions. Wagner[5] describes precautions to be used, such as constancy of temperature, rinsing the prism with water of the same temperature, wiping lightly, and allowing 2 min before reading.

Shippy and Barrows[4] showed how the index of refraction could be used to determine the composition of solutions of sodium chloride and potassium chloride. A curve was constructed by plotting percentage of sodium chloride against index of refraction. A fair degree of accuracy was attained.

In physiological chemistry the refractometer is very important. In only 2 ml of serum it can be used to determine nonalbuminous constituents, total globulins, insoluble globulins, albumens, and total albumen, with great accuracy. The action of ferments can be followed with the refractometer. The refractometer is also useful in controlling the analysis of commercial products, in identifying unknown substances, and in distinguishing substances of the same boiling point and compounds of the same nature, such as halogenated hydrocarbons.

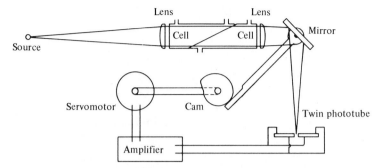

Fig. 14-4. Schematic optical system of a recording refractometer.

Recording Refractometers

Instruments have been designed for continuous and automatic recording of refractive indices (or differences between a reference and specimen). These instruments utilize servomechanisms which track the position of a slit image or critical boundary, or operate a compensating mechanism to maintain a constant position of the image or boundary. Figure 14-4 is a schematic representation of the latter type of instrument. Light from the source is defined by a slit and chopped by a rotating sector (not shown), then passes through the double-prism cell. The refracted beam then strikes a mirror (or beam-splitting arrangement) which focuses the light on twin detectors. Imbalance is removed by the servomotor which is geared to the mirror. The amount of rotation, which is proportional to the refractive index, is correlated with known standards. The instrument may be used to compare the refractive index of two process streams.

THE INTERFEROMETER

Refractometers are dependent upon the refraction of light when passing from one substance to another. The finest measurement of refractive index, however, is based upon the interference of light. In this method there is no diffraction. The light enters and leaves the solution at right angles. The interferometer, reduced to its simplest terms, may be represented by Fig. 14-5.

Parallel light passes through two small openings, R_1 and R_2 in Fig. 14-5. Since $R_1 O$ and $R_2 O$ are of equal length, the two beams arrive at O in phase and a bright spot results. At other points on the screen the lengths of the two beams are not the same. Thus at some point X_1 the two beams differ by half a wavelength. Interference of the two beams produces a dark spot here. At a point a little farther along, Y_1, the difference is one wavelength and a bright spot is formed. With monochromatic light, this succession of light and dark spots (maximum and minimum) continues indefinitely. Now, if a substance of slightly greater refractive index is placed at C, the optical length of the beam $R_2 O$ is increased by an amount Δb because the velocity of light through C is decreased. The

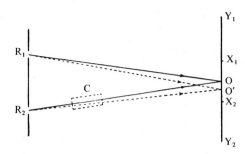

Fig. 14-5. Optical principle of an interferometer.

magnitude of this increase depends on the thickness of the sample and upon its refractive
index, where

$$\Delta b = b(n - n_0) \tag{14-3}$$

b = thickness of sample
n = refractive index of sample
n_0 = refractive index of medium (air)

The velocities of light in two media are proportional to their indices of refraction.

The two beams will no longer arrive in phase at O, but at some other point O', which
is now optically equally distant from R_1 and R_2. The entire band system will be shifted
by this amount. For light of wavelength λ, the distance between O and O', measured in
numbers of fringes, N (each made up of a dark and a light band), is

$$N = \frac{\Delta b}{\lambda} \tag{14-4}$$

$$N = \frac{b(n - n_0)}{\lambda} \tag{14-5}$$

If N is greater than 1, it is impossible to tell how many whole numbers of bands
greater it is, because all bands are alike. This difficulty is avoided by using white light
instead of monochromatic. Now the central band is the only one which is pure white.
The bands on either side of this maximum of the first order are fringed with blue toward
the center of the system and with red along the outer edge. This is due to the different
wavelengths that make up white light. The next adjacent bands are even more highly
fringed. After six or seven bands, the diffusion is so great that the rest of the field is again
uniformly white. Thus, with substance at C in the path of one of the beams, by counting
the number of bands which the central band has been shifted, we may determine the
value of $b(n - n_0)$. Any one of these terms can then be calculated if the others are known.
If two plates of equal thickness were placed in the two beams, the number of bands that
the central band shifts would be a means of calculating the refractive index of one of the
plates, provided that the value of the other one was known. The interferometer is, however, not used primarily for measuring index of refraction but for comparing and measuring concentrations of solutions and gases.

In one type of instrument, shown in Fig. 14-6(a), the optical length of the two beams is equalized by means of a glass plate in the path of each beam P_1 and P_2 at an angle of about $45°$ to the beam, one plate being fixed and the other attached to a lever by which it can be rotated, thus increasing or decreasing its effective thickness. The movement is measured by a micrometer screw. This is turned until the central achromatic bands of the two systems correspond, that is, the optical path of the two beams is the same length. It is possible to match them to $\frac{1}{20}$ of a band, corresponding to a reading of about one scale division on the micrometer screw. This instrument was originally used for measuring changes in the composition of gases, and when the gas chambers are 1 m long, one scale division corresponds to a change in n_D of 0.000000015. It is capable, therefore, of measuring quantities of such substances as CO_2 and CH_4 present in air in amounts as low as 0.02%.

In a later type of portable gas-and-water interferometer, made by Zeiss and shown in Fig. 14-6(b), the light is reflected by the mirror M so that it passes twice through the chambers, and the bands are observed at the same end as the light source. In this way it is possible to obtain the same precision with half the length. The light is furnished by a small electric lamp and is focused on a slit. The interference bands are not as brilliant as those of the other instrument but are just as plain and as easily set. The chambers can be jacketed with an air thermostat or with water, and gases as well as liquids can be used. With the latter, the cells vary from 1 to 40 mm in length. The scale reading is proportional to the thickness of the liquid, so that the range and precision of the instrument may be varied by changing the length of the cell.

A diagram in plane and elevation of the Zeiss water interferometer is shown in Fig. 14-7. White light is furnished by the small 4-V tungsten lamp F. By means of the lens A, a mirror, and the totally reflecting prism K, the image of the filament F is focused on the narrow slit S (see Fig. 14-7, lower half); this slit acts as a (very narrow) secondary source, the light from which is rendered parallel [just as in Fig. 14-6(b)] by the lens L. The light then passes through the two compartments C_1 and C_2 of the water chamber and the rectangular apertures R_1 and R_2; from there it goes to a mirror M, where the two beams of light are reflected back upon themselves, pass through the water chamber again, and finally, by means of the lens L, are reunited at O, forming a series of interference fringes. These fringes are viewed by the cylindrical ocular E, which gives a magnification of 50 diameters, but in the horizontal direction only.

Besides the two interfering beams of light already considered, another pair proceed

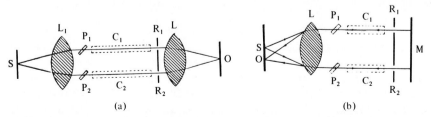

(a)   (b)

Fig. 14-6. Optical schematic of interferometers: (a) extended version and (b) folded type.

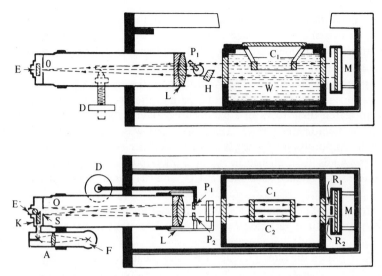

Fig. 14-7. Diagrams in plane and elevation of the Zeiss water interferometer. (Courtesy of Zeiss Instrument Co.)

from the slit S in a precisely similar way, except that they pass *below* and not through the chamber C, likewise forming at O a second system of interference fringes. This latter fringe system is (practically) fixed in position; its sole purpose is to furnish a set of fiducial lines which take the place of the crosshairs ordinarily used as reference marks in optical instruments. Accordingly, if the eye is placed at E, one sees two sets of alternate bright and dark bands, the two sets being separated by a narrow line. In each set of fringes only one of the bright bands is pure white; the bands adjacent to it are bordered with blue toward the center and with red on the outside. It is this central achromatic band which constitutes the reference point of each system.

One set of bands can be displaced relative to the other set by tilting the movable inclined plate P_1 (P_2 is fixed); this is effected by turning the micrometer screw with attached drum D. Thus the two achromatic bands can be brought to coincidence, and the corresponding reading on the drum can be observed.

If water is put in one-half of the cell and a dilute solution of salt in the other, the number of scale divisions of displacement will be determined by the difference in refractive indices of the solution and the pure solvent. A calibration curve can be obtained by making up solutions of known concentration and comparing them with water. Plotting scale readings against concentration, a line is obtained which is almost straight. The deviation from a straight line is due, not to an inconstant variation of refractive index, but because a variation of 10 scale units at one end of the scale may increase the optical length of the beam more than it would at the other end of the scale. Stated in other words, the thickness of band varies with the scale division because of properties inherent in the lever arm action. The band thickness may vary as much as 10%. But since this variation is quite regular, a few points will be sufficient for calibration. The range of the 5-mm chamber is from n_D 1.33320 to 1.34010 for 3000 divisions using water as the comparison liquid. This corresponds to a range of 15.0-33.0 on the immersion refractometer. Assuming that the latter can be read to 0.05 division, the precision with this particular chamber is about 10 times that of the refractometer. With the 40-mm chamber it would be about 80 times as great, but the range would be $\frac{1}{8}$ as great. With this chamber one division corresponds to 1.5-3.0 mg of solute per liter for most aqueous solutions. The greatest differences of concentration which can be directly compared are therefore from 0.45 to 0.90%. The range of the measurement with the interferometer can be increased by comparing the solution against solutions having a known amount of solute present. This does not decrease the precision of the measurement. Thus any concentration of solute can be determined if a series of known solutions has been prepared so that each solution of the series differs by no more than 3000 scale divisions from the preceding one.

There are two procedures which one may follow when using the interferometer for analysis. First, it may be used as a direct-reading instrument, as just outlined. A calibration curve is constructed by making up a number of solutions and comparing them with water, preferably the same water as that used in making up the solutions. The readings are plotted against concentration and connected to make a smooth curve. When the unknown is compared with water, its concentration can be read from the curve.

In the other method, the interferometer is used as a zero-reading instrument. In this method, no previous calibration is necessary but an approximate knowledge of the concentration of the unknown solution is required. It is then compared with two solutions, one slightly more and one slightly less concentrated. This method is slightly more laborious for a single determination, but one gains in precision what one loses in convenience. This is so because only a limited portion of the scale is used (the solutions should not differ by more than 200 scale divisions). In addition to not requiring a previous calibration, this method is not subject to the error caused by the apparent shifting of the achromatic band of the interference system. When comparing a solution of a salt with water, it must be remembered that the central band is brought back to its zero position by turning a glass plate. The increase in the refractive index of the liquid is counterbalanced by decreasing the effective thickness of the compensator plate. Since the dispersion power of the solution differs from that of glass, the band system changes its appearance, so that after the concentration of salt has increased sufficiently (usually about 300 scale divisions), there is an apparent shift in the position of the achromatic band which, if not considered, would cause an error of one band width (18 scale divisions). This shows the advantage of working over a very limited portion of the scale.

To secure a precision of ±0.000001 with the refractometer, the temperature must be regulated to 0.01°C. Since the interferometric method is a differential one, no special regulation of temperature is required in order to determine the difference in n_D of two solutions to 0.0000001. The sensitivity of the instrument, in terms of average parts of solute per million of solvent, is as follows for one scale division:

Refractometer (temperature to 0.01°C) 200-300
Interferometer (simple temperature control) 40-mm chamber 1.5-3.0

The interferometer is not entirely independent of temperature because there is a slight difference between the temperature coefficients of solution and solvent. Thus, a solution of potassium chloride giving a reading of 200 when compared with water at 25°C will give a reading of 202 when the two are compared at 20°C. With water solutions a variation of ±0.5°C is permissible even in very accurate work. It is *absolutely necessary*, however, that the *two chambers* should be at exactly the same temperature. With organic liquids accurate control of temperature is required.

Applications

The use of the interferometer in analyzing gases has already been mentioned. It has been used to determine the permeability of balloon fabrics to hydrogen and helium. Many applications are possible in the analysis of gases. Using a 1-m chamber, 0.02% of carbon dioxide or methane in air can be determined.

The interferometer has been used in the investigation of sea water to chart ocean currents and in the analysis of dilute solutions used for freezing-point determinations. It can be used to determine potassium and sodium in a mixture of their sulfates or chlorides with a precision of ±0.1 mg on a 50-mg sample. The mixture is dissolved in exactly 200 times its weight of water and compared with a standard solution of pure potassium sulfate dissolved in 200 times its weight of water. The reading will range from 430 to 0 as the composition of the mixture ranges from pure sodium sulfate to pure potassium sulfate, and a calibration curve is constructed.

The instrument has been used in water investigations, in measuring absorption, in investigating colloidal solutions, sewage, fermented liquids, and milk and in biological problems such as measurement of serums, CO_2 in blood, ethyl alcohol in blood, ferment activity, and concentration of heavy water. It has been used to determine the end point in titrations and to follow the velocity of reactions. In acidimetric and precipitation reactions it gives as accurate results as good visual methods. It is necessary to plot the straight lines showing the change in reading with solution added; at the end point a sharp angle occurs. The interferometer is particularly valuable in measuring small changes in the composition of mixtures of two organic liquids as a result of preferential adsorption.

LABORATORY WORK

General Directions for the Abbé Refractometer

1. Be sure that the instrument is clean and in working condition. Make certain that the prisms are clean and dry.
2. Start the temperature-controlling device and adjust to 20°C ± 1°, or better if possible.
3. Turn the milled head to separate the prisms. Introduce a drop of distilled water at the funnel-shaped opening between the prisms, or place it on the lower prism and lock the prisms together. *Special care* must be taken not to touch or scratch the prism.
4. Set the scale near 1.33. Adjust the light source and mirror or tilt the instrument on its bearing until the illumination is as bright as possible.
5. Adjust the dispersion screw on the telescope so that the dividing line between the light and dark halves of the field is as sharp as possible. If the dividing line is not sharp and the two fields are not readily distinguished from each other, place a sheet of white paper over the mirror. This gives a better source of diffused white light.

6. Move the arm carrying the reading telescope until the dividing line cuts the inter-section of the crosshairs. Focus the eyepiece so that the crosshairs are clearly seen.

7. Read the index of refraction on the scale, estimating the fourth place. Focus the reading telescope so that the divisions are clearly discernible. The dispersion may be read on the rotating drum.

8. Turn off the *heating element*, if one is being used, and then turn off the water.

9. Open the prisms and clean them with soft tissue moistened with a *little* alcohol. Do not pour or spill alcohol all over the prisms.

10. Close the prisms and replace the cover on the instrument.

11. Do not change any of the adjustments. The instrument has been adjusted to give correct readings.

   Be careful not to scratch the prisms.

Following the foregoing directions determine the refractive index of a liquid supplied by the instructor. Make five settings and take the average. Suitable liquids are sugar solution, alcohol, or an oil. From the refractive index and density of each pure liquid, calculate the specific refractivity and the molar refractivity. Compare your results with theoretical values.

General Directions for the Immersion Refractometer

This experiment consists in determining the percentage by volume or weight of an unknown ethyl alcohol solution and the grams per 100 ml of solute in some salt solution. The only data available for a salt at 25°C are those shown in Fig. 14-8, which is a curve constructed for potassium nitrate from a large number of student results. For other salts Wagner's data at 17.5°C must be used.

For volatile liquids the instrument is provided with a metal container constructed with a glass bottom which fits over the prism and locks with a partial turn. It is advisable to use this instead of a beaker when alcohol solutions are used. It is important that the dot on the attachment correspond to the dot above the prism; otherwise the pins will not fit into the slots properly.

The best way to control the temperature in the water bath (see Fig. 14-3) is to circu-late through it, by means of a small centrifugal pump, a stream of water from a large

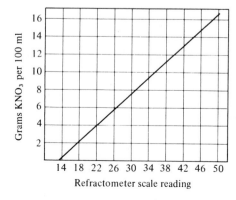

Fig. 14-8. Determination of potassium ni-trate by immersion refractometer at 25°C.

thermostat maintained at the proper temperature. The water bath may be insulated on the side with sheet asbestos or felt to decrease heat exchange.

1. When a constant temperature has been attained, as shown by the protected thermometer in the water bath, place in the holes in the rack provided for this purpose the small beakers of special design containing the solutions to be analyzed, and cover them with crucible covers. Place beside them a beaker of distilled water. The temperature of the latter may be tested with a very small thermometer.
2. When the temperature is constant, place the prism in the beaker of distilled water to attain the proper temperature.
3. Adjust the position of the instrument and of the mirror until the maximum contrast is obtained between the light and dark parts of the field.
4. Adjust the compensator until the dividing line is free from color.
5. Focus by turning the eyepiece until a sharp line is obtained.
6. The reading should be the same as that given in the table on p. 421. Fractions of a division are obtained by turning the round vernier until the line just coincides with the lower numerical scale division. Make a series of five settings and take the average. If there is a small correction, it is best to note this and apply it in all measurements. If it is considerable, ask the instructor to reset the instrument.
7. It is desirable to keep the prism in distilled water at the proper temperature when not testing a solution, because then it requires less time to reach the temperature of the solution.
8. Remove the prism from the water, wipe it dry with a soft cloth, and place it in the solution to be analyzed. Make a series of readings and take the average.
9. Refer to the proper table to find the concentration corresponding to the reading.
10. Report the concentration and pertinent data to the instructor.
11. Rinse the prism with distilled water of the same temperature and wipe it dry with a soft cloth. Be very careful not to scratch the prism.

Notes

If a wavy border line is obtained, the temperature of the prism is not uniform.

It is well to take the temperature in the beaker before and after the readings in a given solution. It is recommended to use the volatile liquid container instead of an open beaker for alcohol solutions.

Use either artificial light or daylight, whichever is more convenient.

A change in reading shows that the temperature is changing.

PROBLEMS

1. A substance having the analysis C_3H_6O might be either acetone or allyl alcohol. Determine which of these two substances it is from the fact that the molar refraction is 16.97.

2. The refractive index of carbon tetrachloride at $20°C$ is 1.4573, and the density at $20°C$ is 1.595 g cm^{-3}. Calculate the molar refraction.

3. From the atomic refractions, calculate the refractive index to 20°C of nitrobenzene. The density is 1.210 g cm^{-3}.

4. Calculate the specific refraction and the molar refraction for each of these liquids:

Compound	n_D^{20}	d^{20}
Benzene	1.4979	0.879
Ethanol	1.3590	0.788
Ethyl acetate	1.3701	0.901
Toluene	1.4929	0.866
Nitrobenzene	1.5524	1.21
Water	1.3328	0.998

5. For D_2O, $n_D^{20} = 1.32830$, and for water, $n_D^{20} = 1.33280$. If the refractive index for a sample is 1.32980, calculate the percent D_2O present.

6. What is the refractive index of a mixture of 10 ml of benzene and 40 ml of nitrobenzene? See Problem 4 for necessary data.

7. A 120-ml sample of wine was distilled to remove all the alcohol; the distillate was diluted to a volume of 100 ml. The reading on the immersion refractometer at 25°C was 26.8. What was the percentage by volume of alcohol in the wine? Necessary tables are in J. A. Dean, Ed., *Lange's Handbook of Chemistry,* 11th ed., pp. 10–251, McGraw-Hill, New York, 1973.

BIBLIOGRAPHY

Bauer, N., K. Fajans, and S. Z. Lewin, in *Physical Methods of Organic Chemistry*, 3rd. ed, A Weissberger, Ed. Vol. 1, Part II, Wiley-Interscience, New York, 1960, Chapter 28.

Lewin, S. Z. and N. Bauer in *Treatise on Analytical Chemistry*, Part I, Vol. 6, Kolthoff, I. M. and P. J. Elving, Eds., Wiley-Interscience, New York, 1965, Chapter 70.

Maley, L. E., "Refractometers," *J. Chem. Educ.*, **45**, A467 (1968).

Tilton, L. W. and J. K. Taylor, in *Physical Methods in Chemical Analysis*, W. C. Berl, Ed., Vol. I, Academic Press, New York, 1950.

LITERATURE CITED

1. Andrews, L. W., *J. Am. Chem. Soc.*, **30**, 353 (1908).
2. Clemens, C. A., *J. Ind. Eng. Chem.*, **13**, 813 (1921)
3. Leach, A. E. and H. C. Lythgoe, *J. Am. Chem. Soc.*, **27**, 964 (1905).
4. Shippy, B. A. and G. H. Barrows, *J. Am. Chem. Soc.*, **40**, 185 (1918).
5. Wagner, B., *Z. Angew. Chemie*, **33**, 262 (1920).

15

Polarimetry, Circular Dichroism, and Optical Rotatory Dispersion

Polarimetry, the measurement of the change of the direction of vibration of polarized light when it interacts with optically active materials, is one of the oldest of the instrumental procedures. Much of the work on the development of prisms and other devices for the production of polarized light was done in the early part of the nineteenth century.[*] A small, rough polarimeter seems to be a simple piece of apparatus, but a precision polarimeter is an example of complicated optical equipment.

POLARIMETRY THEORY

Ordinary, natural, unreflected light behaves as though it consisted of a large number of electromagnetic waves vibrating in all possible orientations around the direction of propagation. If, by some means, we sort out from the natural conglomeration only those rays vibrating in one particular plane, we say that we have *plane-polarized light*. Of course, since a light wave consists of an electric and a magnetic component vibrating at right angles to each other, the term "plane" may not be quite descriptive, but the ray can be considered planar if we restrict ourselves to noting the direction of the electrical component. *Circularly polarized light* represents a wave in which the electrical component (and therefore the magnetic component also) spirals around the direction of propagation of the ray, either clockwise ("right-handed" or dextrorotatory) or counterclockwise ("left-handed" or levorotatory).

If one combines a polarimeter with a monochromator so that measurements of optical rotation can be made at various, known wavelengths, then the optical rotatory dispersion, ORD, can be determined. On the other hand, if a polarimeter or some other device is used to produce d and l circularly polarized light, a spectrophotometer can be used to measure the differing absorption by certain compounds at certain wavelengths and thus determine the circular dichroism, CD. Both the ORD and CD characteristics are useful in structural determinations of optically active compounds.

[*]See, for example, J. B. Biot, *Mem. prem. classe Inst. France* **13**, 218 (1812); W. Nicol, *Edinburgh New Phil. J.* **6**, 83 (1828).

It is possible, and for many explanations of the interaction of light and matter it is quite enlightening, to represent a plane-polarized ray as the vector sum of two circularly polarized rays, one moving clockwise and one counterclockwise and with the same amplitude of vibration. It is obvious from Fig. 15-1 that at zero time the sum of C_{l_0} and C_{d_0}, the left and right circularly polarized rays, equals P_0, the plane-polarized ray. At the time when C_l is at C_{l_1}, C_d is at C_{d_1}, and the vector sum is P_1, and so on around the circle.

If, following the passage of the plane-polarized ray through some material, one of the circularly polarized components—say, the left circularly polarized ray—has been slowed down, then the resultant would be a plane-polarized ray rotated somewhat to the right from its original position. Figure 15-2 illustrates the case where the right-handed ray is $90°$ ahead of the left-handed ray. The plane-polarized resultant ray is rotated $45°$, that is, $90°/2$, from the original position. The rotation α is just one-half of the phase difference φ of the two circular components.

The index of refraction, n, represents the ratio of the velocity of a ray of light in a vacuum, c, to its velocity in a medium, v. That is,

$$n = \frac{c}{v} \tag{15-1}$$

If a substance showed different indices of refraction for the l and d components of a plane-polarized ray, then one beam would be slowed down on passage through the medium, and the plane of polarization of the ray would be rotated. Thus:

$$n_l = \frac{c}{v_l} \quad \text{and} \quad n_d = \frac{c}{v_d} \tag{15-2}$$

$$\frac{v_l}{v_d} = \frac{n_d}{n_l} \tag{15-3}$$

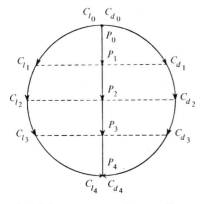

Fig. 15-1. Representation of a plane-polarized ray as the sum of two circularly polarized rays.

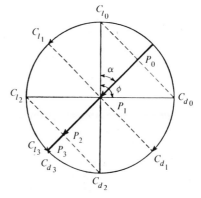

Fig. 15-2. Rotation of the plane of polarized light due to slowing down of one of the circular components.

If we let b represent the length of the column of material traversed by the ray; λ_0, the wavelength of the light; ν, the frequency of rotation (or vibration) of the light; and c, the velocity of light in a vacuum; then the difference in degrees, φ, between the two rays is given by Eq. 15-4.

$$\varphi = \frac{2\pi b\nu}{\nu_d} - \frac{2\pi b\nu}{\nu_l} \tag{15-4}$$

$$= \frac{2\pi b\nu c}{\nu_d c} - \frac{2\pi b\nu c}{\nu_l c} \tag{15-5}$$

$$= \frac{2\pi b n_d}{\lambda_0} - \frac{2\pi b n_l}{\lambda_0} \tag{15-6}$$

$$= \frac{2\pi b}{\lambda_0} (n_d - n_l) \tag{15-7}$$

or

$$\alpha = \frac{\pi b}{\lambda_0} (n_d - n_l) \tag{15-8}$$

and

$$n_d - n_l = \frac{\alpha \lambda_0}{\pi b} \tag{15-9}$$

A solution of 3.45 g of sucrose per 100 g of aqueous solution at 18°C shows a rotation α of 99.8° for light of 5000 Å. The tube length b is 10 cm. Using Eq. 15-9, one calculates

$$n_d - n_l = \frac{99.8° \times 5000 \times 10^{-8} \text{ cm}}{180° \times 10 \text{ cm}}$$

$$n_d - n_l = 2.77 \times 10^{-6}$$

Thus relatively small differences in the index of refraction for right and left circularly polarized light cause appreciable rotation of the plane of polarized light. The difference in indices of refraction for right and left circularly polarized light is known as *circular birefringence*.

Pasteur,[9] van't Hoff,[10] and Le Bel[8] worked out the principles which chemists now recognize as requirements in order that a given molecule possess "optical activity," that is, rotate the plane of polarized light. A compound is optically active in solution if its structure cannot be brought to coincide with that of its mirror image, that is, the compound does not possess a plane or a center of symmetry. If a tetrahedral carbon atom is substituted with four different groups, it is said to be asymmetric and would lead to optical activity unless a second similar asymmetrically substituted carbon atom is contained in the molecule. For example, mesotartaric acid and other meso compounds are not optically active. In the case of many nonplanar compounds, such as spiro compounds, allylenic compounds, and certain substituted biphenyl compounds, dissymmetrical structures with optical activity may result without an asymmetric carbon atom in the molecule. Likewise, optical activity is not limited to carbon compounds but may occur in any dissymmetrical three-dimensional compound.

Some substances show optical activity only in the crystalline, solid state. In noncubic crystals, there are at least two primary directions in the crystal that show different spatial arrangements of the atoms and thus different force fields. Radiation is transmitted at different velocities in different directions. Such crystals are known as anisotropic crystals and rotate the plane of polarized light.

Optical Rotatory Dispersion and Circular Dichroism Theory

Cotton[3] discovered an interesting connection between rotatory power and light absorption in optically active compounds. As one approaches certain optically active absorption bands in a compound from the long-wavelength side, the rotatory power at first increases strongly, then falls off and changes sign. This effect is known as the *Cotton effect*. Within the absorption band, the molar absorptivity for right- and left-hand circularly polarized light are different; that is, $(\epsilon_d - \epsilon_l) \neq 0$. This effect changes linearly polarized light into elliptically polarized light and is known as *circular dichroism*.

If one assumes that a substance near an absorption band absorbs left circularly polarized light, the l component, more strongly than the right circularly polarized light, the d component, that is, $\epsilon_l > \epsilon_d$, then the amplitude of the d component will be greater than the l. Furthermore, if one assumes that $n_d > n_l$, then the d component will be retarded more than the l component. The resulting elliptically polarized light is represented in Fig. 15-3.

The angle between the major axis of the ellipse and the plane of the original radiation is the angle of rotation, α. The ellipticity, that is, the angle whose tangent is the ratio of the minor axis of the ellipse, OB, to the major axis, OA, is known as θ. The molecular ellipticity, $[\theta]$, can be shown to be given by the relationship

$$[\theta] = 3305 \, (\epsilon_l - \epsilon_d) \tag{15-10}$$

Circular dichroism graphs are plots of $[\theta]$ against wavelength and might resemble the curve in Fig. 15-4.

The specific rotation $[\alpha]$ defined below changes with wavelength, and the rate of change of specific rotation with wavelength is known as *optical rotatory dispersion*.[4] Drude[6] has shown that the specific rotation may be expressed as a function of wavelength by an equation with several terms:

$$[\alpha] = \frac{k_1}{\lambda^2 - \lambda_1^2} + \frac{k_2}{\lambda^2 - \lambda_2^2} + \frac{k_3}{\lambda^2 - \lambda_3^2} + \cdots \tag{15-11}$$

where λ is the wavelength of measurement and $k_1, k_2, k_3, \ldots$ are the constants that can be identified with the wavelengths of maximum absorption of the optically active absorption bands.

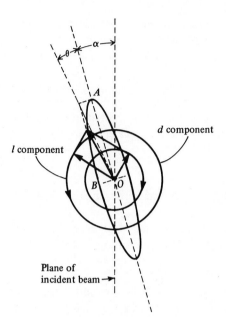

Fig. 15-3. Elliptically polarized light produced when $n_d > n_l$ and $\epsilon_l > \epsilon_d$.

In a region far removed from an optically active absorption band, the dispersion is normal, and the equation of Drude can be simplified to Eq. 15-12,

$$[\alpha] = \frac{k}{\lambda^2 - \lambda_0^2} \tag{15-12}$$

where λ_0 is a constant representing the wavelength of the nearest optically active absorption band. When $\lambda \gg \lambda_0$, Eq. 15-12 may be reduced further to

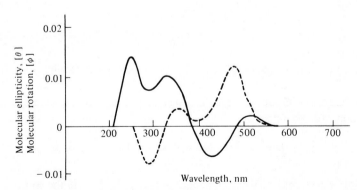

Fig. 15-4. Circular dichroism, dotted line, and optical rotatory dispersion, solid line, for a hypothetical substance with absorption bands in the region 200–600 nm.

$$[\alpha] = \frac{k}{\lambda^2} \tag{15-13}$$

A plot of molecular rotation, $[\phi]$, (see Eq. 15-20) versus wavelength for a hypothetical compound is shown in Fig. 15-4. Note that both molecular ellipticity and molecular rotation may be either positive or negative depending on the relationships between ϵ_d and ϵ_l or n_d and n_l.

Measurement of Optical Rotation

The rotation exhibited by an optically active substance depends on the thickness of the layer traversed by the light, the wavelength of the light used for the measurement, and the temperature. If the substance measured is a solution, then the concentration of the optically active material is also involved, and the nature of the solvent may also be important. There are certain substances that change their rotation with time. Some are substances that change from one structure to another with a different rotatory power and are said to show *mutarotation*. Mutarotation is common among the sugars. Other substances, owing to enolization within the molecules, may rotate so as to become symmetrical and thus lose their rotatory power. These substances are said to show *racemization*. Mutarotation and racemization are influenced not only by time but by pH, temperature, and other factors. In expressing the results of any polarimetric measurement, it is therefore very important to include all experimental conditions.

The results of polarimetric measurements are reduced to a set of standard conditions. The length employed as standard is 10 cm for liquids and 1 mm for solids. The standard wavelength is that of the green mercury line (5461 Å), although the sodium doublet (5890 Å + 5896 Å) has been widely employed, especially in the older measurements. The standard temperature is 20°C.

Thus, if b is the layer thickness in decimeters, C is the concentration of solute in grams per 100 ml of solution, α is the observed rotation in degrees, and $[\alpha]$ is the specific rotation or rotation under standard conditions, then

$$[\alpha] = \frac{100\alpha}{bC} \tag{15-14}$$

The temperature of the measurement is indicated by a superscript and the wavelength by a subscript written after the brackets.

For a pure liquid, the concentration is unimportant, but temperature changes cause expansion and contraction of the liquid and a consequent change in the number of active molecules in the path of the light. For pure liquids unit density is assumed as standard, and the definition of specific rotation becomes

$$[\alpha] = \frac{\alpha}{b\rho} \tag{15-15}$$

where ρ is density.

Temperature changes have several effects upon the rotation of a solution or liquid. An increase in temperature increases the length of the tube; it also decreases the density, thus reducing the number of molecules involved. It causes changes in the rotatory power of the molecules themselves due to association or dissociation, increased mobility of the atoms, and affects other properties. In general, the effect of temperature may be expressed by Eq. 15-16:

$$[\alpha]^t = [\alpha]^{20} + z(t - 20) \qquad (15\text{-}16)$$

where z = temperature coefficient of rotation
t = temperature in degrees Celsius

Substances vary widely in their values of z.

Any liquid or solution, when placed in a magnetic field, rotates the plane of polarized light because of the effect of the magnetic field upon the motion of the electrons in the molecules. This effect was discovered by Faraday[7] and is known as the *Faraday effect*. The rotation χ is positive if in the same direction as the magnetizing current. For most substances χ is positive and is appreciable for many organic substances.

For analytical purposes, the chief interest in polarimetry is to determine the concentration of substances, although abundant correlation between rotation and chemical structure has been found.[5] The relationship between rotation and concentration of a solution is, unfortunately, not strictly linear, so that the specific rotation of a solution is not a constant. The concentration of the measurement should always be stated when $[\alpha]$ is given. The values of the specific rotation extrapolated to infinite dilution may be employed.

The relationship between $[\alpha]$ and concentration may usually be expressed by one of the three equations proposed by Biot.

$$[\alpha] = A + Bq \qquad (15\text{-}17)$$

$$[\alpha] = A + Bq + Cq^2 \qquad (15\text{-}18)$$

$$[\alpha] = A + \frac{Bq}{C + q} \qquad (15\text{-}19)$$

where q = percentage of solvent in the solution
A, B, C = constants

Equation 15-17 represents a straight line, Eq. 15-18 a parabola, and Eq. 15-19 a hyperbola. The constants A, B, and C are determined from several measurements at different concentrations.

Calculations of Polarimetry and Saccharimetry

The equations for the calculation of specific rotation, $[\alpha]$, from measurement of the angle of rotation have already been given (Eqs. 15-14 and 15-15). Often it is desired to calculate the molecular rotation $[\phi]$, which is given by the equation

$$[\phi] = \frac{[\alpha] \times \text{mol wt}}{100} \tag{15-20}$$

The polarimeter is widely used as a saccharimeter in sugar analysis. In the determination of the concentration of sucrose in a substance containing no other optically active material except sucrose, it is convenient to take 75.2 g of unknown as sample and dissolve it in enough water to make 100 ml of solution. Then, with a 2-dm tube, the rotation in degrees is numerically equal to the concentration of sucrose in percent by weight. This follows from Eq. 15-14 using $[\alpha]_D^{20} = 66.5°$ for sucrose.

This sample, 75.2 g, is rather large, and consequently most modern saccharimeters are not graduated in degrees but in smaller divisions. Several types of graduation have been used. Most modern saccharimeters are graduated in the "International" scheme in which one division equals 0.3462°. It is obvious that one should check the normal weight of any saccharimeter, preferably by calibration with pure sucrose, before any unknowns are run.

Most raw sugar samples contain other optically active substances besides sucrose. When sucrose is heated with acid or with the enzyme invertase, it is "inverted" to form one molecule of fructose and one of glucose. Sucrose has a specific rotation of +66.5°, fructose has a specific rotation of $-93°$, and glucose +52.5°. Thus the specific rotation changes from +66.5° to $(-93° + 52.5°)/2 = -20.2°$ upon inversion. By measuring the change in rotation upon inversion it is possible to determine sucrose in the presence of other optically active substances. The formula used for the determination of sucrose is known as the *Clerget formula*, Eq. 15-21. See Browne and Zerban[1] for a more complete discussion of this formula and of sugar analysis in general.

$$\text{percent sucrose} = \frac{100\,(a - h)}{144 - \dfrac{t}{2}} \times \frac{W}{w} \tag{15-21}$$

where a = rotation in Ventzke degrees of sucrose solution before inversion
h = rotation in Ventzke degrees after hydrolysis
t = temperature in degrees centigrade
W = normal weight for saccharimeter employed
w = weight of sample taken per 100 ml of solution

The Clerget factor, 144 in Eq. 15-21, varies slightly with concentration and with the details of the method of hydrolysis employed. The factor is derived by considering the experimentally determined Ventzke readings for inverted sugar, the dilution introduced by the addition of hydrochloric acid (10 ml per 100 ml of solution) necessary for the hydrolysis, and the fact that one molecule of water is consumed for each molecule of sucrose hydrolyzed.

APPLICATIONS OF OPTICAL ROTATORY DISPERSION AND CIRCULAR DICHROISM

The main applications of both optical rotatory dispersion and of circular dichroism lie in the area of structure determination of such optically active substances as amino acids, polypeptides and proteins, steroids, antibiotics, terpenes, and metal–ligand complexes. Many applications to date are empirical in nature and depend upon knowledge of the behavior of compounds similar to those under investigation. Some general rules are evident, however.

Aliphatic amino acids show a unique Cotton effect, the sign of which reflects the stereochemistry at the asymmetric center. α-Amino acids of levo configuration show a positive Cotton effect around 215 nm, whereas their dextro enantiomers display a negative Cotton effect. In polypeptides it is possible to estimate the percent of α-helix structure by measurements of optical rotatory dispersion.

In steroids, *cis* and *trans* ring junctions lead to different types of optical rotatory dispersion curves. In one form the specific rotation increases with decreasing wavelength until it reaches a peak and reverses (a positive curve), and in the other the specific rotation decreases with decreasing wavelength until it reaches a trough and reverses (a negative curve). The location of carbonyl groups in steroids also can often be narrowed down to a limited number of possibilities by noting the sign of the Cotton-effect curve and the wavelength and specific rotation of the peak or trough.

Theoretical studies of radiant energy absorption by chromophores asymmetrically surrounded in a molecule have led to the so-called "octant" rule. The octant rule states that the sign of the contribution that a given atom at point $P(x, y, z)$ makes to anomalous rotatory dispersion will vary as the simple product, $x \cdot y \cdot z$, of its coordinates. As an il-

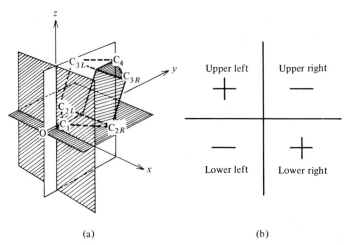

(a)   (b)

Fig. 15-5. Illustration of the octant rule for cyclohexanone: (a) the octants and (b) contributions of groups in the far octants.

lustration, let's look at the carbonyl chromophore group in the cyclohexanone molecule. Represent the cyclohexanone molecule in the chair form with coordinates defined by the x–y, x–z, and y–z planes as shown in Fig. 15-5. The carbonyl group lies in the x–y plane and is bisected by the y–z plane. Atoms or substituent atoms can now be located in one of the eight octants defined by $\pm x$, $\pm y$, and $\pm z$. Atoms or substituents in the $+x, +y, +z$; $+x, -y, -z$; $-x, -y, -z$ or $-x, -y, +z$ octants have positive contributions to the Cotton effect. Atoms or substituents in the other four octants have negative effects and atoms or groups which lie on or near a plane have little or no effect.

THE POLARIMETER

The polarimeter consists of the following basic parts:

1. A light source
2. A polarizer
3. An analyzer
4. A graduated circle to measure the amount of rotation
5. Sample tubes

Except in the simplest instruments a half-shade device is also included. Some polarimeters may be equipped with photocells or other devices for measuring the intensity of light emerging from the instrument, although most polarimeters are designed for visual observation.

The most common light sources for polarimetry are sodium vapor lamps and mercury vapor lamps. The sodium lamp emits light of wavelengths 5890 Å and 5896 Å plus a little continuous background which can be largely eliminated with a filter of 7% potassium dichromate used in a 6-cm-thick layer. The mercury lamp emits light of several wavelengths, the prominent visual lines being at 4358, 4916, 5461, 5770, and 5791 Å. The proper choice of filters will permit the isolation of each line. If a continuous light source can be employed, then ordinary sunlight or light from a tungsten filament lamp can be used.

The polarizer (and the analyzer) may be of several different types. One type consists of a crystal, usually calcite or quartz, cut diagonally at such an angle that one component of the light is totally reflected. The second component passes through the second half of the crystal and thus emerges, going in the same direction as the original beam (Fig. 15-6). The two halves of the prism are cemented together with a cement having an index of refraction as near as possible to 1.4865, which is the value of n_e, the extraordinary index of

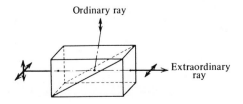

Fig. 15-6. The Glan-Thompson prism.

refraction for calcite. The index of refraction for calcite at right angles to the above-mentioned ray is $n_0 = 1.6584$.

Several different varieties of polarizing (and analyzing) prisms are known. They vary in the angles of the faces of the prism and of the cut diagonally through the prism. The Glan-Thompson prism (Fig. 15-6) and the Nicol prism (Fig. 15-7) are the most common. The Nicol prism requires smaller pieces of calcite and is cheaper but is not as good as the Glan-Thompson prism. The light emerging from a true Nicol prism is displaced from the original beam and will revolve in a circle as the prism is rotated. Again, two Nicols used together will not produce total extinction at any point for the whole field and thus will introduce uncertainty in the balance point. With either a Nicol or Glan-Thompson prism the entering light must be essentially parallel; otherwise some unpolarized light will be transmitted. Light must not, therefore, be concentrated on the prism by using a converging beam.

Light can also be polarized by reflection from a mirror at the proper angle—Brewster's angle. If light strikes a mirror at such an angle i that

$$\tan i = n \tag{15-22}$$

where n = refractive index of the mirror material, then only the component vibrating perpendicular to the plane of incidence (parallel to the mirror surface) will be reflected. Reflection is not used in modern polarimeters to produce polarized light. It is interesting to note, however, that light emerging from a monochromator is partially polarized, with the greatest intensity perpendicular to the exit slit. Thus if a monochromator precedes a polarimeter, the slit should be perpendicular to the direction of transmission of the polarizing prism.

A third method of producing polarized light is by Polaroid filters. Polaroid filters are composed of strongly dichroitic crystals oriented in a plastic material. These crystals strongly absorb light vibrating in one direction and only weakly absorb light vibrating in the perpendicular direction. Polaroids can never give 100% polarization; also, the light must lie in the region from about 5000 Å to 6800 Å. Polaroids are used, therefore, only on less expensive instruments.

When two prisms, a polarizer and an analyzer, are used together, the intensity of light transmitted through the combination is given by the law of Malus:

$$I = KI_0 \cos^2 \theta \tag{15-23}$$

where I = emerging intensity from analyzer
I_0 = incident intensity on analyzer
θ = angle between the directions of transmission of the two prisms

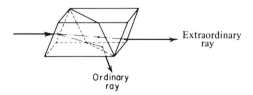

Extraordinary
ray

Ordinary
ray

Fig. 15-7. The Nicol prism.

K = factor taking into account reflection and absorption losses in the analyzing prism; K is approximately equal to 1.

The graduated circle is fitted with a vernier for more precise measurement of the angle through which the analyzing prism has been rotated. Special reading devices employing a pair of parallel index lines are used on the most precise instruments. A tangent screw with graduated drum allows the borders of an etched line on the main scale to be made to coincide with the two hairlines. With such a device readings can be made to $0.002°$.

The polarimeter tubes must have plane and parallel glass discs at the ends. The glass must be free from strain; otherwise the discs will produce a partial circular polarization of the light, and complete extinction of light will be impossible. Each tube should be tested by filling it with water, placing it between crossed prisms, and noting whether the dark field remains dark. The length of polarimeter tubes may be determined by measuring the rotation of a known, strongly rotating liquid or solution—for example, nicotine in ethyl alcohol—at a definite temperature.

In cheaper instruments one measures the position of the analyzer required to give a minimum intensity without the sample in the tubes, and again with the sample in the tubes. The difference in the two readings is the rotation caused by the introduction of the sample. The human eye, however, is much better at *matching* light intensities than is the human mind at *remembering* intensities, as it must if the minimum intensity is to be determined. Consequently, the more precise polarimeters make use of so-called half-shade devices which result in matching two half-fields for a balance point.

Many different half-shade devices are available. Each has its own advantages and disadvantages. The Jellett-Cornu prism (Fig. 15-8) is constructed by sawing a Glan-Thompson prism in two lengthwise, grinding one face down a little, and cementing the parts back together. When light passes through this polarizing prism, the two halves will produce polarized light beams tilted slightly with respect to each other. Rotation of the analyzer prism in front of such a polarizing prism would result in complete extinction, first of one-half of the field and then of the other half. At some intermediate position the two halves of the field would appear of equal brightness. This point is taken as the balance point. The Jellett-Cornu device does not allow variation of the half-shade angle, the angle between the two prisms. Variation of this angle is desirable, because large angles are necessary for precise balancing with weak light sources and small angles give more precise balancing with strong light sources.

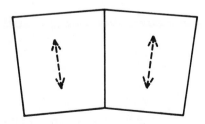

Fig. 15-8. The Jellett-Cornu prism, end view.

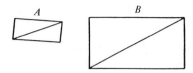

Fig. 15-9. The Lippich prism, side view.

The Lippich prism (Fig. 15-9) is a popular half-shade device. A small polarizing prism A precedes the large polarizing prism B. With such an arrangement one-half of the field can be rotated slightly with respect to the other half. The analyzer must be at some intermediate position in order to achieve equal illumination of both halves of the field. Sometimes two Lippich prisms are used, dividing the field into three parts that match at the balance point. The half-shade angle of any Lippich arrangement can be varied by rotating the Lippich prism.

A popular, inexpensive half-shade device is the Laurent half-wave plate. A thin plate of quartz cut parallel to its optic axis is placed over one-half of the field of the polarizer. The quartz plate is cut just thick enough that for a given wavelength of light (usually sodium D lines) the slow ray lags exactly one-half wavelength behind the fast ray. This results in a slight rotation of the light passing through the part of the polarizer covered by the plate. The amount of rotation can be varied by rotating the quartz plate with respect to the polarizing prism, and thus the half-shade angle is variable. The quartz plate is suitable, however, only for the wavelength for which it was constructed. At other wavelengths it becomes more difficult to find the balance point due to lack of contrast in the fields as the analyzer is rotated.

The optical arrangement of a precision polarimeter is shown in Fig. 15-10 and an example of a commercial instrument is shown in Fig. 15-11. The Rudolph instrument has a Lippich double field polarizer.

In the type of instruments described above, the rotation is measured by rotating the analyzer with respect to the polarizer. It is also possible to measure rotatory power by leaving the analyzer permanently crossed with respect to the polarizer and compensating any rotation caused by the sample with a piece of quartz which rotates light in the opposite direction to that of the sample. The design of such an instrument is shown in Fig. 15-12. Wedges I and II are made of levorotatory quartz and are ground to the same

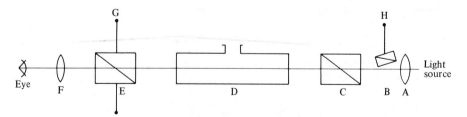

Fig. 15-10. Optical arrangement of a polarimeter with a Lippich half-shade device; A, collimating lens; B, Lippich half-shade prism; C, polarizing prism; D, tube; E, analyzing prism; F, eyepiece; G, scale; H, lever to adjust half-shade angle.

Fig. 15-11. Rudolph polarimeter. (Courtesy of O. C. Rudolph & Sons.)

angle. Wedge I is stationary, but II is movable. Moving wedge II thus varies the thickness of the block of levorotatory quartz. Block III is made of dextrorotatory quartz and is of thickness equal to that of wedges I and II when II is in an intermediate position. Thus both positive and negative rotations of the sample can be compensated by moving the wedge in or out from its intermediate position. Compensating polarimeters of the type described above are used largely for sugar analyses and are known as saccharimeters. Fortunately, the rotatory dispersion of quartz, sucrose, and a few other sugar solutions is very nearly the same. Thus if white light is used as a source, the quartz can compensate the rotation of the sugar solution at all wavelengths. If the dispersions of quartz and sugar were not the same, only light of one wavelength would be compensated completely, and the field would appear colored rather than dark.

Automatic Recording Spectropolarimeters

Automatic recording spectropolarimeters, necessary for the determination of optical rotatory dispersion, may be placed broadly in two classes: instruments that use a null-

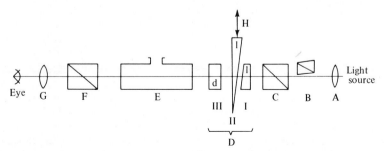

Fig. 15-12. Quartz-wedge compensating polarimeter: A, collimating lens; B, Lippich half-shade prism; C, polarizing prism; D, quartz-wedge compensator; E, tube; F, analyzer prism (position fixed); G, eyepiece; H, scale and movement device for compensator.

point method and instruments that use a ratio method. At present all commercially available instruments fall into the first class. They differ among themselves in how the null point is achieved. Two will be described in some detail.

The Rudolph spectropolarimeter works on the null-point principle, the null point being ascertained by an imposed mechanical oscillation ($\pm\epsilon$) of the analyzer, whose mean angular position is orthogonal with respect to the plane of polarization of the entering light beam (Fig. 15-13). At this point, angular changes of $+\epsilon$ and $-\epsilon$ produce the same current in the phototransducer system. The optical rotation α produced by the introduction of a sample between the polarizer and the analyzer is measured by the angular rotation ($-\alpha$) of the polarizer that is required to restore the balance of the signal output. The analyzer prism is mechanically oscillated to produce a 20-Hz modulation of the light beam striking the photomultiplier tube. A portion of its signal is separated, by means of a chopper, into two signals, corresponding to the right and left oscillations of the analyzer. The difference between these two signals is fed into a null-point-seeking servo system, which drives the polarizer to a position that compensates for the rotation of the sample. The recording of the optical activity is obtained by means of a linkage between the angular position of the polarizer and an X-Y recording system.

The Cary spectropolarimeter is similar in principle to the Rudolph instrument, except that in the Cary instrument the mechanical oscillation of the analyzer is replaced by

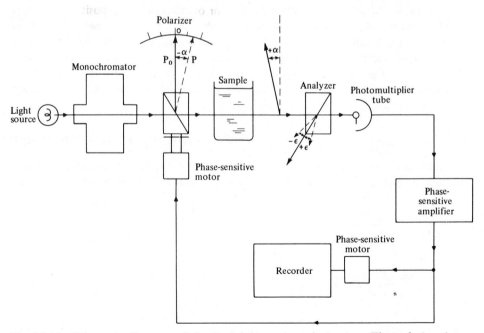

Fig. 15-13. Schematic diagram of the Rudolph spectropolarimeter. The polarizer is rotated through an angle ($-\alpha$) that is equal and opposite to the angle of rotation of the sample. The angle of scan, $\pm\epsilon$, is induced by a mechanical oscillator. P_0, initial position of polarizer axis; ϵ, angle of scan. (Courtesy of O. C. Rudolph & Sons.)

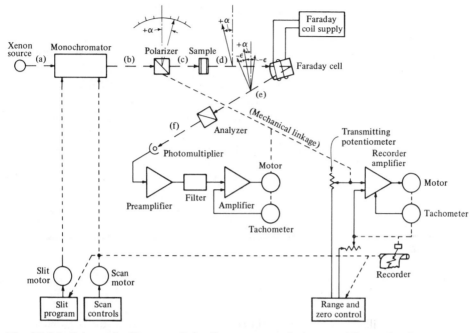

Fig. 15-14. Schematic diagram of the Cary spectropolarimeter. The angle of scan, $\pm\epsilon$, is induced by a Faraday cell. (a) Undispersed, nonpolarized beam; (b) monochromatic non-polarized beam; (c) monochromatic polarized beam; (d) beam "c" rotated by sample; (e) beam "d" cyclically displaced by Faraday cell; (f) component transmitted by analyzer. (Courtesy of Applied Physics Corp.)

an oscillation brought on by a magneto-optical effect. To achieve this, a Faraday cell is placed ahead of the analyzer (Fig. 15-14). The Faraday cell consists of a silica cylinder surrounded by a coil. An alternating current (60 Hz) passes through the coil, thus cyclically displacing the plane of polarization of the beam. A motor energized by the amplified current from the photomultiplier moves the polarizer by means of a mechanical linkage.

INSTRUMENTS FOR CIRCULAR DICHROISM MEASUREMENT

An ordinary spectrophotometer can be adapted to measure circular dichroism. It is only necessary to provide some means of producing d and l circularly polarized radiation. For this purpose a plane-polarized beam can be passed through a quarter-wave plate. If the plate is rotated from $+45°$ to $-45°$, first d and then l circularly polarized light is produced. In order to cover a large wavelength region, several quarter-wavelength plates are necessary. Other devices can also be used to produce circularly polarized beams so that the relative absorption can be measured by the spectrophotometer.

LABORATORY WORK

Experiment 15-1   Determination of the Specific Rotation of a Substance

Weigh out from 10 to 20 g of the optically active substance assigned by the instructor. Dissolve in water and dilute to the mark in a 50-ml volumetric flask.

Determine the zero reading of the polarimeter with a tube filled with distilled water. Average several readings for precision.

Fill a 200-mm tube with the solution under investigation. Note the temperature of the solution and determine the rotation. Average several readings for precision. Calculate the specific rotation from the data, using Eq. 15-15.

Dilute 10.00 ml of solution to 25 ml in a volumetric flask. Repeat the determination of $[\alpha]$ on this solution. Repeat the dilution and determination of $[\alpha]$ once again. Plot specific rotation $[\alpha]$ against concentration.

Experiment 15-2   Determination of Concentration of Sucrose Solutions

This is a simplified experiment designed to illustrate the use of the polarimeter for sucrose analysis, employing inversion of the sucrose. For procedures employed in actual determinations of sucrose in commercial products, see the books by Browne and Zerban, the A.O.A.C., and Bates.[2]

Procedure   A solution of sucrose in water will be furnished as the unknown. Measure the rotation, α, at room temperature, t, in a 200-mm tube.

To 50 ml of solution in a flask, add exactly 5 ml of concentrated hydrochloric acid. Insert a thermometer and place the flask in a water bath. Heat the bath slowly at such a rate that the temperature of the sugar solution reaches 68°C in 15 min. Cool the flask quickly. Read the rotation in a 200-mm tube when the temperature has returned to its original value, t. Multiply the polarimeter reading on the invert sugar by $\frac{11}{10}$ to allow for the dilution by the acid. This rotation value equals α' in the formula below. Calculate grams of sucrose per 100 ml of solution, C, by the formula

$$C = \frac{\alpha - \alpha'}{1.9175 - 0.0066\,t}$$

PROBLEMS

1. One gram of an organic substance was dissolved in 50.00 ml of water. This solution, in a 20 cm tube, read +2.676° in a polarimeter, while distilled water in the same tube read +0.016°. Calculate the specific rotation of the substance.

2. Exactly 10 g of raw sugar was dissolved in water and made up to a volume of 100 ml. In a 20 cm tube at 25°C this solution read 12.648° in a polarimeter. After inversion, according to the directions of Experiment 15-2, the solution read −3.922°. Calculate the percentage of sucrose in the raw sugar.

3. Calculate Brewster's angle for borosilicate glass of index of refraction $n_D = 1.47$. Devise an experiment to check this result.

4. Approximately what fraction of light would be removed from a beam of light passing through two polaroids set at 45° with respect to each other?

5. Substitute Beer's law into Eq. 15-10. Assume that initial beam intensity, P_0, for d radiation equals P_0 for l radiation and thus develop the equation for determining molecular ellipticity using a spectrophotometer.

6. Calculate the difference in indices of refraction for right and left circularly polarized light for a substance giving a rotation of −10.5° at 6000 Å in a 10 g per 100 ml solution in a 20-cm polarimeter tube.

7. Draw curves like Fig. 15-3 for the situation where $n_l > n_d$ and $\epsilon_d > \epsilon_l$, and for the situation where $n_l > n_d$ and $\epsilon_l > \epsilon_d$.

8. Ten grams of a compound was dissolved in water and made up to a volume of 100 ml. What is the molecular ellipticity if the solution placed in a 10-cm tube showed spectrophotometer readings of 40% transmittance for d circularly polarized light and 42% for l circularly polarized light. The incident beams are equal in intensity.

BIBLIOGRAPHY

Carroll, B. and I. Blei, *Science*, **142**, 200 (1963).

Djerassi, C., *Optical Rotatory Dispersion*, McGraw-Hill, New York, 1960.

Heller, W. and D. D. Fitto in *Physical Methods of Organic Chemistry*, Vol. I, Part II, 3rd ed., A. Weissberger, Ed., Wiley-Interscience, New York, 1960, Chapter 33.

Velluz, L., M. Legrand, and M. Grosjean, *Optical Circular Dichroism*, Academic Press, New York, 1965.

LITERATURE CITED

1. Browne, C. A. and F. W. Zerban, *Physical and Chemical Methods of Sugar Analysis*, Wiley, New York, 1941.

2. Browne, C. A. and F. W. Zerban, *Official and Tentative Methods of Analysis*, Association of Official Agricultural Chemists, Washington, D.C., 1945; F. J. Bates, *Polarimetry, Saccharimetry, and the Sugars*, Government Printing Office, Washington, D.C., 1942.

3. Cotton, A., *Compt. Rend.*, **120**, 989, 1044 (1895); *Ann. Chim. Phys.*, **8**, 347 (1896).

4. Djerassi, C., *Optical Rotatory Dispersion*, McGraw-Hill, New York, 1960.
5. Djerassi, C., *Science*, **134**, 649 (1961).
6. Drude, P., *The Theory of Optics*, Longmans, New York, 1929.
7. Faraday, M., *Phil. Mag.*, **28**, 294 (1846).
8. Le Bel, J. A., *Bull. Soc. Chim.*, **22**, 337 (1874).
9. Pasteur, L., *Ann. Chim. Phys.*, **24**, 442 (1848).
10. van't Hoff, J. H., *La chimie dans l'espace*, Rotterdam, 1874.

16

Mass Spectrometry

The first mass spectrometer dates back to the work in England of J. J. Thompson in 1912 and of F. W. Aston in 1919, but the instrument that served as a model for more recent ones was constructed in 1932. The mass spectrometer produces charged particles consisting of the parent ion and ionic fragments of the original molecule, and sorts these ions according to their mass/charge ratio. The mass spectrum is a record of the numbers of different kinds of ions—the relative numbers of each are characteristic for every compound, including isomers. Sample size requirements for solids and liquids range from a few milligrams to submicrogram quantities as long as the material can exist in the gaseous state at the temperature and pressure existing in the ion source. The average sample size for routine gas analysis is about 0.1 ml at standard conditions, but with special instrumentation, samples of 10^{-8} ml can be analyzed.

Mass spectrometers are a powerful tool for extracting a wealth of information concerning the structure of organic compounds and the elemental analysis of solid state samples, and for analyzing complex organic mixtures. A detailed interpretation of the mass spectrum frequently makes it possible to place functional groups into certain areas of the molecule and to see how they are connected to one another. The molecular weight can be determined directly, even to ten thousandths of a mass unit on more sophisticated spectrometers. In addition, a mass spectrometer is an essential adjunct to the use of stable isotopes in investigating reaction mechanisms and in tracer work. Moreover, mass spectrometry has contributed greatly to a more detailed understanding of kinetics and mechanisms of unimolecular decomposition of molecules.

COMPONENTS OF MASS SPECTROMETERS

There is no universal mass spectrometer. Certain designs and configurations lend themselves to the solution of specific problems better than do others. However, common to most mass spectrometers are four units: (1) the inlet system; (2) the ion source; (3) the electrostatic accelerating system; and (4) the detector and readout system. In addition, there must be provision for maintenance of a high vacuum throughout the spectrometer from the inlet to the detector. Interfacing of a gas chromatograph with a mass spectrometer is often desirable.

Inlet Sample System

To handle all types of material, different sample systems are required (Fig. 16-1). Introduction of gases involves merely transfer of the sample from a gas bulb into the metering volume. The latter is a small glass manifold of known volume (about 3 ml), coupled to a mercury manometer and attached by a port to the inlet manifold. A sample is metered in the standard volume and then expanded into a reservoir volume (perhaps 3 liters) immediately ahead of the sample "leak." The meter pressure ranges from 30 to 50 torr; after expansion the pressure ranges from 10^{-3} to 10^{-1} torr.

Liquids are introduced in various ways—by break-off devices (see Fig. 16-6), by touching a micropipet to a sintered glass disk under mercury or gallium, or by hypodermic needle injection through a silicone rubber dam. The low pressure in the reservoir draws in the liquid and vaporizes it instantly.

Heated inlet systems extend the usefulness of mass spectrometry to polar materials, which are prone to be adsorbed on the walls of the chambers at room temperature, and to less volatile compounds insofar as they possess a vapor pressure of the order of 0.02 torr at the temperature of the sample reservoir, usually 200°C. The temperature is limited by the materials of construction and by thermal degradation. Above 200°C most com-

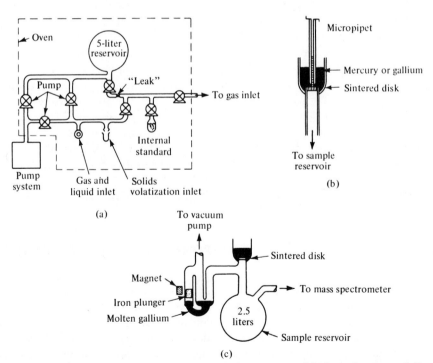

Fig. 1. (a) Inlet sample system for mass spectrometer. (b) Introduction of liquids through a sintered disk. (c) Magnetically actuated, gallium cutoff valve.

pounds containing oxygen or nitrogen are thermally decomposed. Solids melting below the reservoir temperature can be introduced directly. With a direct sample introduction probe, the sample is loaded into a short length of melting point capillary, placed in the well at the end of the probe, and inserted to within a few millimeters of the ion source through a vacuum lock that maintains the vacuum-tight arrangement. Then the sample temperature is raised until sufficient vapor pressure is indicated by the total ion current indicator or by appearance of a spectrum. Oftentimes a small amount of chemistry suffices to convert a compound, itself not volatile, into a derivative which still retains all the important structural features but has sufficient vapor pressure. Magnetically actuated gallium cutoffs are employed as valves (Fig. 16-1).

From the sample reservoir the gases diffuse through a molecular leak into the ion source. The leak is a pinhole restriction (about 0.013–0.050 mm in diameter) in a gold foil. The preferred type of flow into the ion source depends on the purpose for which the instrument is intended. For analytical work, conditions for molecular flow are usually employed in which collisions between molecules and the walls are much more frequent than collisions between molecules. However, this type of leak is less desirable in instruments designed primarily for isotope work, since repeated measurements are made of the relative concentrations of two members of a mixture. In isotope studies viscous flow is preferred in which a gas molecule is more likely to collide with other gas molecules than with the surfaces of the container—thus there is no tendency for various components to flow differently from the others.

In continuous-monitoring inlet systems the sample must be admitted to the instrument at or near atmospheric pressure. Consequently, it is necessary to drop the pressure by a viscous flow system to a range in which molecular flow can be achieved by the use of leak perforations of reasonable size. In one system a pair of viscous leaks is arranged through which gas is drawn by an auxiliary mechanical pump; the leaks are so proportioned that the pressure intermediate between them is about 1 torr. The sample is admitted to the mass spectrometer through a perforated foil from the region of intermediate pressure between the two leaks.

Ionization Sources[4]

The ion source is of primary importance and must be considered as the heart of the mass spectrometer. In fact the ion source might be regarded as a chemical reaction vessel, and each source must be chosen as appropriate to the sample. Ion sources have the dual function of producing ions without mass discrimination from the sample and accelerating them into the mass analyzer with only a small spread of kinetic energies prior to acceleration. Sources that produce a large spread of energies in the ion beam must be used with double-focusing mass analyzers to obtain sufficient resolution.

The ionization efficiency of a source must be high so that a large portion of the neutral sample particles presented will become ions to be analyzed and detected as a mass spectrum. High efficiency is particularly important for the analysis of nanogram quantities of sample material and trace impurities in solids. An ion beam current of 10^{-10} A is a desirable source output.

The ability to obtain useful mass spectral data at electron energies low enough to avoid fragmentation is important for analyses of mixtures and isotopically labeled compounds. Interpretation of the mass spectrum of a mixture of compounds is greatly facilitated when there are no fragment ions present. It is particularly important to eliminate the $M - 1$ and $M - 2$ peaks (i.e., the parent peak minus one and two atomic mass units) in the analysis of isotopically labeled molecules.

Electron-Impact Ionization The electron-impact ion source is the most commonly used and highly developed ionization method. As shown in Fig. 16-2(a), once past the molecular leak, the neutral molecules find themselves in a chamber that is maintained at a pressure of 0.005 torr and at a temperature of $200 \pm 0.25°C$. Located perpendicular to the incoming gas stream is an electron gun. Electrons emitted from a glowing filament (rhenium, thoriated iridium, or carbonized tungsten) are drawn off by a pair of positively charged slits through which the electrons pass into the body of the chamber. An electric field maintained between these slits accelerates the electrons. The number of electrons is controlled by the filament temperature, whereas the energy of the electrons is controlled by filament potential. Molecules in the gas phase are ionized by these energetic electrons that strike or pass close by. This results in a Franck-Condon transition producing a molecular ion, which has an odd number of electrons and is usually in a high state of electronic and vibrational excitation.

The electric field can be varied from 6 to 100 V. A range from 6 to 14 V is employed in molecular weight determinations, whereas the field is kept in the range from 50 to 70 V to obtain a mass spectrum. In the latter voltage range sufficient energy is available for rupture of any bond in a molecule. At the same time, the voltage lies on the plateau of the ionization efficiency curves of compounds and, therefore, small variations in the electric field have a minimum effect on peak intensity.

After passage through the ion source, the electron beam is collected on an anode. The current to this trap is used to control the electron-beam intensity, so that the rate of formation of ions will be as nearly constant as possible. It is customary to collimate the

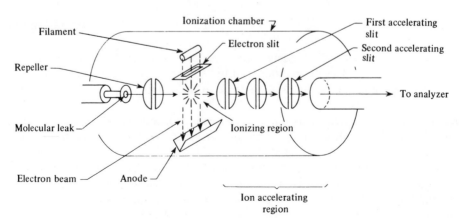

Fig. 16-2(a). Ionization sources. Electron-impact ion source and ion accelerating system.

beam of ionizing electrons by a small magnetic field (of the order of 100 G) which is confined to the ionization region.

A serious limitation of the electron-impact source is the drastic method used to create molecular ions. Molecules that are extremely unstable with respect to electron impact show no molecular ion peaks in their mass spectra. Reducing the energy of the impinging electrons to a value near the ionization potential only increases the relative molecular ion intensity; the absolute ion intensity is greatly reduced. In addition the electron-impact process is inefficient and requires that the sample be in the vapor state. Also all background gases are ionized with the same efficiency as the sample.

Knudsen Cell   The Knudsen cell [Fig. 16-2(b)] combines thermal and electron-bombardment excitation. Upon heating, the sample, contained in a crucible, is vaporized through a small effusion orifice. The vapor is collimated by slits in both the heat sink and the entrance slit of the mass spectrometer ion source. Molecules of the sample stream out through the slit directly into the space traversed by the electron beam. The molecular beam emerging from the crucible, the ion beam, and the ionizing electron beam are mutually orthogonal. The cell is heated by either radiation or electron bombardment to

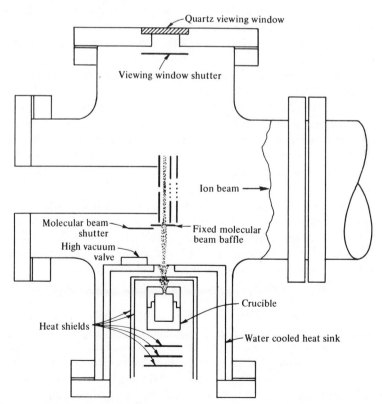

Fig. 16-2(b). Ionization sources. Knudsen cell. (Courtesy of Bendix Corp.)

attain temperatures continuously adjustable from ambient to 2500°C. The Knudsen cell inlet system is designed for thermodynamic studies, and special analyses of solids and low vapor pressure liquids.

Laser Microprobe The laser microprobe offers vaporization with minimum thermal degradation. The laser supplies a controlled heating flux of about 10^8 W cm^{-2} which is focused onto the target area that may be as small as $100 \, \mu m^2$. Such a microprobe can be focused on very small inclusions and impurities in heterogeneous solid materials. The vapor plume is ionized subsequently by the electron-impact technique. The few nanograms of solid that are vaporized are sufficient for analysis by any mass spectrometer; but since the heating time is so short (about 0.4 msec), only a time-of-flight spectrometer can produce a complete spectrum from a single laser pulse.

Field Ionization[2] Field ionization occurs when a molecule is brought near a metal surface in the presence of a high electric field (10^8 V/cm). A thin wire of a few micrometers diameter (or a sharp metal blade) constitutes the anode and is located 1 mm away and immediately behind the exit slit of the ion chamber which serves as the cathode and opposite field-forming member. The remainder of the source consists of the focusing slits common to all ion sources; in fact, the field ionization unit can be combined with the electron-impact source in a single assembly.

Electrons are drawn to the field anode from molecules that pass through the high electric field due to the quantum mechanical tunnel effect. Only a very small amount of internal energy is excited in the molecule by this process, leading to few or no fragment ions. This makes it possible to determine the molecular weight and, with a high-resolution spectrometer, the empirical formula for labile compounds; also the quantitative measure of compounds in mixtures is simplified. Field ionization is a valuable technique for studying surface phenomena, such as adsorbed species and the results of chemical reactions on surfaces. Limitations of the technique include fragile source elements and sensitivities an order of magnitude below those of electron-impact ionization.

Spark Source Ionization Analysis of high-melting solids is frequently accomplished by use of a radio-frequency (800 kHz) spark of high intensity. Samples are formed into two electrodes held in small movable vises [Fig. 16-2(c)]. A potential, usually in excess of 100 kV, is applied and a discharge initiated. Pulse length and repetition rate are controlled for selective volatilization and to avoid overheating the sample. Ion optics are used to collimate the ion beam because the energy spread is broad. To maintain adequate resolution, double-focusing analyzers must be used.

Spark source mass spectrometry as a technique for the analysis of solids has grown in use enormously in recent years because of two inherent advantages. First, its detection sensitivity is high whether defined in terms of the low concentration of an impurity that can be detected in a matrix or in terms of the total amount of sample needed to detect impurities. The second outstanding advantage of the spark source as an analytical tool is its nonselectivity. Its mass spectrum reveals all elements in the sample with approximately the same detection sensitivity.

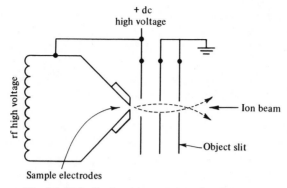

Fig. 16-2(c). Ionization sources. Spark source.

It is an erratic ion source and the ion current fluctuates widely with time. The erratic nature of the spark source requires an integrating detector, usually a photographic plate. The need for the entire periodic table to be encompassed in the recorded mass spectrum has resulted in the widespread choice of the Mattauch-Herzog geometry of double focusing in commercial instruments.

The rf spark can be used to analyze many different kinds of materials. With electrical conductors and semiconductors, the spark can be formed directly between electrodes of the sample. Techniques have been developed for sintering powders or for mixing insulators in powder form with suitable conductors, which after compacting into rod form are suitable for sampling with the spark source. The preparation, cleaning, and handling of samples is critical because a monatomic layer of surface contamination is readily detected.

Surface Ionization Another type of ion source used for inorganic solid materials is the surface ionization method. Samples are coated on a ribbon (tungsten) filament and then heated until they evaporate. When an atom or molecule is evaporated from a surface (2000°C) it has a certain probability of being evaporated as a positive ion; this probability is predictable and is a function of the ionization potential of the sample and the work function of the filament material. Ionization potentials of the sample material, I, must be below the work function of the filament material, W. The relationship for ratio of ions, m^+, to neutral particles, $m^°$, is given by

$$\frac{m^°}{m^+} \propto \exp\left(\frac{I - W}{kT}\right)$$

This technique is very appropriate for inorganic compounds that generally have low ionization potentials (3-6 eV). Surface ionization is especially useful in determining isotope ratios in inorganic compounds for geochemical applications or in studies of elements involved in nuclear chemistry. No ionization of the background gases in the mass spectrometer occurs. On the other hand, surface ionization is difficult for organic compounds with ionization potentials of 7–16 eV.

Chemical Ionization[11]   In chemical ionization a reaction gas, such as methane, is introduced into the ionization chamber at a pressure of 1 torr. Electron impact produces ionization of methane, and these ions react further with neutral methane molecules to form products which are chemically reactive species and have such chemical properties as Lewis-acid or Lewis-base behavior. These products of ion molecule reactions are the reagents which can react with the sample molecule. Hydride transfers are common. A quasimolecular ion, often the only peak, is observed in which the molecular weight is increased by the mass of one hydrogen atom. Many times this is observed for compounds which do not normally show a molecular ion in electron-impact sources. The nature of the chemical ionization process is such that the product ions formed contain even numbers of electrons. The amount of energy involved in a reaction tends to be low; this accounts for the simpler cracking patterns and generally high intensity of the quasimolecular ion. It is possible to use many different gases as reactants and thus produce spectra from reactant ions of different energies.

Accelerating System

The positive ions formed in the ionization chamber are drawn out by a small electrostatic field between the large repeller plate (charged positive) behind them—the original entrance to the ion source which did not affect the molecules while they were yet unionized—and the first accelerating slit (charged negative) [see Fig. 16-2(a)]. A strong electrostatic field between the first and second accelerating slit of 400–4000 V accelerates the ions of masses m_1, m_2, m_3, etc., to their final velocities. The ions emerge from the final accelerating slit as a collimated ribbon of ions with velocities and kinetic energies given by

$$eV = \tfrac{1}{2} m v_1^2 = \tfrac{1}{2} m v_2^2 = \tfrac{1}{2} m v_3^2 = \cdots \tag{16-1}$$

At the start of each spectrum scan, the second of the two accelerating slits is charged to an initial potential of perhaps 4000 V. This charge is then allowed to leak off to ground at a controlled rate over a period of perhaps 20 min. The electrostatic voltages must be stabilized better than 0.01%, with a field uniformity of better than 0.1%.

Ion-Collection Systems

Resolved ion beams, after traveling through an analyzer, sequentially strike a collector and are detected. In one system the ion beam impinges on an insulated cup (Faraday cage). As each positive ion strikes the collector, it picks up an electron so that a minute electron current flows to the collector. Ion currents, ranging from 10^{-15} to 10^{-11} A, are amplified by an electrometer amplifier or a field-effect transistor.

For ion currents below about 10^{-15} A, an electron multiplier is necessary (Fig. 16-3). The ion beam strikes the conversion dynode of a 16-stage, Be–Cu electron multiplier. Secondary electrons are constrained by a magnetic field to follow circular paths, causing them to hit the same electrode from which they were emitted, but at a different point.

Successively emitted electrons then move along by a series of semicircular jumps. Most of the path lies between a pair of glass plates coated with a high-resistance metallic film. An electric gradient is impressed along the lengths of these plates to produce the needed acceleration. The magnetic field is produced by a number of small permanent magnets. Use of electron multiplier detectors permits very rapid scanning of peaks across the collector slit because the time constant of the system can be made small. The limiting factor is either the system noise level (on the order of 10^{-18} or 10^{-19} A) or the system background.

The readout display usually consists of a direct writing-recording oscillograph with 3–5 galvanometers with relative sensitivities of 1, 3, 10, 30, and 100. These are easy to operate, reliable and of low initial cost. The resultant spectra often need considerable manual manipulation because each mass number must be counted, and the amplitudes measured and normalized before identification can be made.

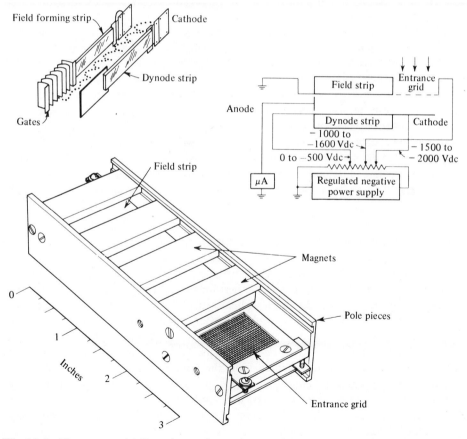

Fig. 16-3. Electron multiplier phototube and typical electrical circuit for operating the tube. (Courtesy of Bendix Corporation.)

A photographic plate can give greater resolution than an electrical detector on the same instrument. It is more cumbersome though. Because the photographic plate is a time-integrating device, it can provide the highest sensitivity of any detector. A 36-to-1 mass range can be recorded simultaneously. Spectra from extremely small samples, samples with low vapor pressure, and ions of short life can be detected. All of these might be missed with an electrical detector. Photographic plate detection is used with most radio-frequency spark instruments. Because the number of ions reaching the collector per unit time varies drastically, some means of integrating the ion beam is required.

Vacuum System

A good vacuum system is required to create a suitable low-pressure environment for ionization to take place so that ions are not neutralized by collisions. Usually oil diffusion pumps (and a molecular sieve trap for hydrocarbons) are used in the source and analyzing regions of the spectrometer, backed by a single roughing pump, to provide a vacuum of 10^{-5} to 10^{-7} torr. For spark source work, much higher vacua are needed, and this usually requires the use of ion pumps or cryogenic pumping methods. Automatic bakeouts must be provided for regeneration of the molecular sieve material and for bakeout of the mass analyzer. The latter is often done routinely at the end of the day.

A very high-speed pumping system, on the order of 1000 liters/sec, is necessary to maintain the requisite low pressure surrounding the source and ensure that there is no cross contamination between successive samples. This is particularly important in gas chromatography–mass spectrometry combinations, where new samples are continually being introduced to the ion source on a second-to-second basis. Differential pumping that allows pressure differentials of 10 to 100 to be maintained between the higher pressure inside the ionization chamber relative to the surrounding source housing is used.

Interfacing Gas Chromatograph–Mass Spectrometer

The aim of an interfacing arrangement is to operate both a gas chromatograph and a mass spectrometer without degrading the performance of either instrument.[10,12] Gas chromatography is an ideal separator, whereas mass spectrometry is unexcelled for identifications. Compatibility is a problem. Gas chromatographs normally operate at pressures greater than 760 torr, whereas the mass spectrometer functions best at about 10^{-5}-10^{-6} torr. Each has one important facet in common; the amount of material that each can usefully handle is in the same range. The interface between the two instruments must therefore attenuate the carrier gas to an acceptable pressure; however, it is often desirable to transfer directly to the mass spectrometer all the material being chromatographed. Furthermore, the interface should not introduce any broadening of the gas chromatograph peak.

Complete mass spectra can be scanned and recorded in less than a second with the writing speed of the light beam oscillograph. This permits repeated scanning during a single gas chromatographic peak and often results in detection of changing composition which is not apparent in total ion recordings from flame ionization or thermal conductivity detectors, or from the total ion current monitor on the mass spectrometer. Moreover, it permits taking the mass spectrum at the exact maximum of the gas chromatographic peak.

Effluent Splitter Based on the concept of enrichment by diffusion, the effluent splitter relies on the fact that carrier gas molecules usually are much lighter than those of the sample and therefore can be removed preferentially by vacuum pumps through an effusion chamber. The carrier gas with organic vapors present flows across, or is in contact with, a porous barrier. The lighter helium permeates the effusion barrier in preference to the heavier organic molecules and is removed by the vacuum system. About 10–20% of the sample passes into the ion source. The separation procedure increases the sample-to-carrier gas ratio 50–100 times and also provides the necessary pressure reduction. Porous barriers include porous etched glass tubing, porous stainless steel tubing or diaphragm, Teflon tubing, and porous silver diaphragm (Fig. 16-4).

Jet/Orifice A precisely aligned, supersonic jet/orifice system is also effective in removing the carrier gas by the effusion principle. As the gas chromatograph effluent passes through a small jet, the stream is directed toward an orifice. During the passage between the jet and the orifice, the carrier gas preferentially moves in a direction off the axis of the main stream. When the jet of gas reaches the aperture (Fig. 16-4), the concentration of carrier gas increases away from the center line while the concentration of sample tends to increase toward the center. The orifice intercepts only the center portion of the stream. Two such jet/orifice assemblies may be used in series, or one may be used alone. The mechanical design and gas flow rate are the most critical parameters. At high flow rates (20 ml/min), about 20% of the sample is transmitted.

Molecular Separator A second type of molecular separation takes advantage of large differences in the permeability between most organic molecules and the carrier gas (usually helium) through silicone rubber. The column effluent from the gas chromatograph flows past a thin rubber membrane. Helium, or any permanent gas, has a low permeability and is not adsorbed by the membrane, whereas the organic molecules are adsorbed and pass through the membrane and directly into the high vacuum of the mass spectrometer. The permeability of the membrane is determined by solubility and diffusion rate. For helium the product of these two quantities is very small even though diffusion for such a small molecule as helium is very rapid. The major disadvantages of this technique are the time lag of about 0.1 sec while the sample molecules pass through the membrane, the dead volume, and sample discrimination. Sample recovery is 50–90% and the enrichment factor is 500–1000.

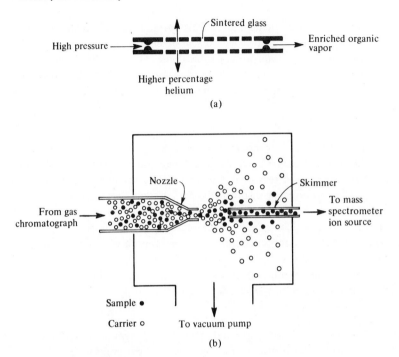

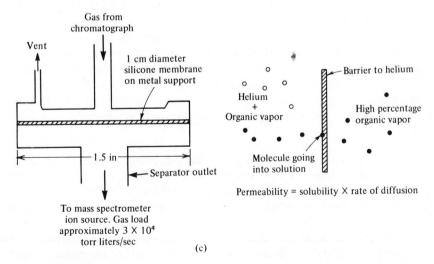

Fig. 16-4. Gas chromatograph/mass spectrometer interfaces. (a) Porous barrier separator or effluent splitter. (b) Jet/orifice separator. (c) Molecular separator using a permeable membrane.

MASS SPECTROMETERS

Resolution

The ability to separate ions of different masses reaching the detector ranges from 1 part in 20 to better than 1 part in 30,000 on high-resolution instruments. Generally, re- solving powers below 1000 are used for routine mixture analyses. The resolving power of a mass spectrometer is defined as the ratio $M/\Delta M$, where M and $M + \Delta M$ are the mass numbers of two neighboring peaks of equal intensity in the mass spectrum. For example, to distinguish oxygen of mass 31.9988 from sulfur of mass 32.06, a resolution of 533 is necessary. Similarly, to resolve the $CH_4—O$ doublet at mass 100 fragment, $M/\Delta M \geqslant$ 3165. On the other hand, a resolution of 1 part in 200 adequately distinguishes between mass 200 and mass 201.

The "valley definition" of the resolving power is based on the relative height of the valley formed between two overlapping peaks. A figure of 10% is commonly used, with

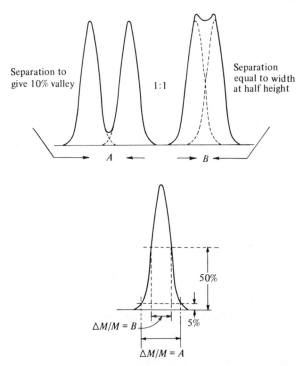

Fig. 16-5. *Upper left*, resolution equal to 10,000 on 10% *valley definition*; that is peaks with separation $\Delta M/M = A$ = equals peak width at 5% height points (*lower curve*). *Upper right*, resolution equal to 19,500 on *width at half height* definition; that is, peaks drawn with separation $\Delta M/M = B$ = equals peak width at 50% height points (*lower curve*).

each peak contributing 5% to the valley; that is, peaks with separation $\Delta M/M$ equal the peak width at 5% height points, as shown in Fig. 16-5. This definition is unduly pessimistic for distinguishing doublets. The doublet will generally be distinguishable if the two peaks are separated by their "width at half-height," provided their intensity ratio is not greater than 10 to 1.

Resolution is strongly influenced by the pressure in the spectrometer and is a function of slit widths, deflection radius, and homogeneity of the ion source. Variation of the source and collector slit widths is the usual method of changing the resolving power. Of course, a decrease in the slit widths to improve the resolving power results in a decrease in sensitivity.

Magnetic-Deflection Analyzer Systems

The most widely used type of instrument depends on deflection of the ion beam in a static magnetic field. In the Dempster (180°) design, shown in Fig. 16-6, the rapidly moving ions, after being accelerated by the electrostatic slits, are diverted into circular paths by a magnetic field parallel to the slits and perpendicular to the ion beam. Ions of mass m (in grams) and charge e (in abcoulombs), on passage through an accelerating electric field,

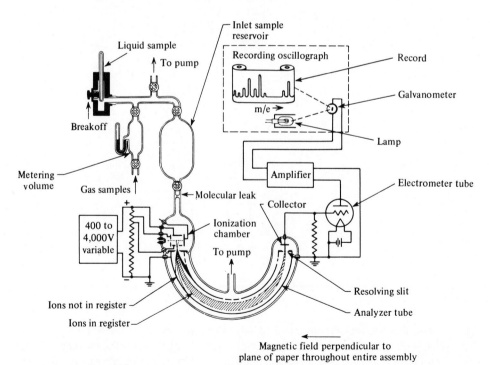

Fig. 16-6. The Dempster (180°) magnetic-deflection analyzer system. (Courtesy of Consolidated Electrodynamics Corp.)

attain a velocity v, which can be expressed in terms of the accelerating potential V (in abvolts) and the kinetic energy of the individual ion as it leaves the electric field,

$$\tfrac{1}{2}mv^2 = eV \tag{16-2}$$

or, solving for the velocity term,

$$v = \sqrt{\frac{2eV}{m}} \tag{16-3}$$

Equating the centripetal and the centrifugal forces to which the ion beam is subjected on entering the uniform magnetic field H (in gauss),

$$mv^2 r = Hev \tag{16-4}$$

Each ion follows its own circular trajectory of radius r, which is proportional to momentum and given by

$$r = \frac{mv}{eH} \tag{16-5}$$

Substituting Eq. 16-3, the radius is also expressed by

$$r = \frac{1}{H}\sqrt{2V\frac{m}{e}} \tag{16-6}$$

Only those ions which follow the path which coincides with the arc of the analyzer tube in the magnetic field are brought to a focus on the exit slit where the detector is located. Ions of other mass/charge (m/e) ratio strike the analyzer tube (which is grounded) at some point, are neutralized, and are pumped out of the system along with all other unionized molecules and uncharged fragments. Thus, the magnetic field classifies and segregates the ions into beams, each of a different m/e, where

$$\frac{m}{e} = \frac{H^2 r^2}{2V} \tag{16-7}$$

To obtain the mass spectrum, the accelerating voltage or the magnetic field strength is varied at a constant rate. Usually the accelerating voltage is varied and each m/e ion from light to heavy is successively brought to a focus on the exit slit. A magnetic analyzer will also focus ions that are of the same mass and velocity but of different initial directions. Thus, this type of instrument resolves ions of different masses, and maximizes the resolved ion beam intensity by focusing.

The rather bulky magnet required in the $180°$ design prompted the development of sector instruments in which the ion source, the collector slit, and the apex of the sector-shaped magnetic field are collinear, as shown in Fig. 16-7 for a Nier $60°$ sector instrument. Sector instruments are unique because the ion source and collector are completely removed from the magnet region. This isolation permits the use of unusual and diverse ion source constructions, and the use of conventional electron multipliers for ion detection. The ion transit time from the accelerator slits to the magnetic field is significantly longer in the sector-type instrument and, consequently, any peaks resulting from metastable

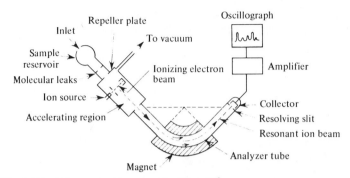

Fig. 16-7. Schematic diagram of a Nier 60° sector mass spectrometer.

transition products are several times larger than the corresponding peaks in a 180°-type spectrometer.

With magnetic analyzer systems a resolution of 1 in 200 mass units can be obtained. Narrow slits, strict alignment of components, and additional signal amplification enables mass peaks in the range from 200 to 600 to be resolved—the ultimate limit for this class of spectrometer.

Instruments designed for process control use a permanent Alnico magnet with a field of 4000–6000 G. The overall range is restricted. For example, one commercial instrument has a range from 2 to 80 mass units and adequate resolution for separation of adjacent peaks up to about 35. Small, portable spectrometers with a radius of curvature of 5 cm can withstand temperatures of 450°C and operate in the ultrahigh vacuum range (10^{-15} torr).

Cycloidal Focusing Spectrometer

The cycloidal analyzer utilizes a homogeneous electric field at a right angle to and superimposed on a magnetic field (Fig. 16-8). Ions created in the source are injected into the mass analyzer by electric fields within the source. These ions are accelerated and decelerated by the homogeneous electric field of the analyzer while being turned by the magnetic field. Because of the angular and energy-focusing characteristics of the cycloidal method, ions of a given m/e value emerging from the source with different energies and/or angles will describe different cycloidal trajectories, but all will arrive at the collector slit. The electric potential of the mass analyzer is increasingly negative from front to back with the electric field plate which contains the source and collector slits at ground potential. Uniformity of the electric field is increased by means of a stack of guard rings which are held at successive potentials by means of a voltage divider. As either of the two crossed fields is varied, ions of identical m/e ratio focus on the fixed collector slit even though not equivalent in energy. For ions focused in the cycloidal analyzer,

$$\frac{m}{e} = \frac{k H^2}{2\pi V} \tag{16-8}$$

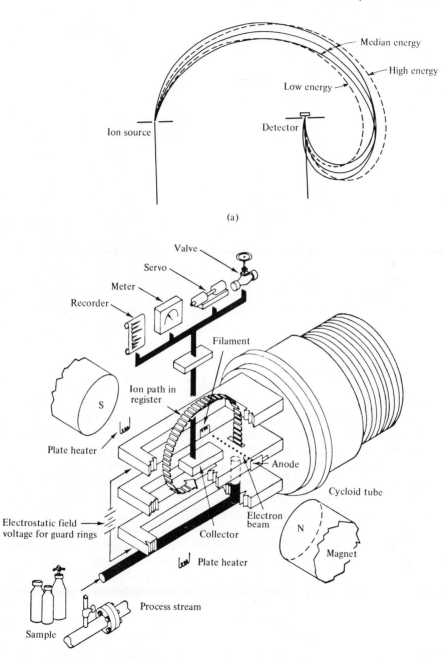

(a)

(b)

Fig. 16-8. (a) Trajectories of ion beam in focus. (b) Schematic diagram of a cycloidal-focusing mass spectrometer. (Courtesy of Consolidated Electrodynamics Corp.)

where k is a constant. The mass range extends from 10 to 2000; the resolution is at least 1000 (10% valley definition) throughout the range.

Isotope-Ratio Spectrometer

A less expensive adaptation of the usual mass spectrometer, the isotope-ratio mass spectrometer, has been made available for work in these fields. In the modified instrument the ion currents from two ion beams—for example, the ion beams from $^{12}CO_2$ and $^{13}CO_2$—are collected simultaneously by means of a double exit slit and are amplified simultaneously by two separate amplifiers. The larger of the two amplified currents is then attenuated by the operator, with a set of decade resistors, until it will exactly balance the smaller current from the other amplifier. The ratio of the two currents is determined from the resistance required. This is a null method and practically eliminates the effect of other variables in the system. A tracer material can be detected even after great dilution. Medical researchers study body functions using isotopically labeled tracers. Precise age dating is based on the rate of decay of radioactive nuclides. If the decay rate of ^{238}U to ^{206}Pb, ^{40}K to ^{40}Ar, and ^{87}Rb to ^{87}Sr is known, and the ratio of the isotopes of one of these pairs is measured, then the age of minerals and rocks can be determined.

Double-Focusing Spectrometers

In a double-focusing mass spectrometer the electrostatic analyzer provides energy selection; that is, it will separate ions according to energy, while the magnetic portion provides momentum (thus mass) selection. Ions are focused both for velocity and direction. Such an analyzer is required when the ion source produces a beam with a wide energy spread, for instance in a rf spark source, and for high-resolution work where otherwise the small energy spread in a fairly monoenergetic beam limits resolution to about 2000 or less with a single focusing analyzer. Two different designs are in vogue, the Nier-Johnson and the Mattauch-Herzog geometries.

In the Mattauch-Herzog geometry (Fig. 16-9), the ions are passed through a radial electric field followed by a 90° magnetic sector. The angle subtended by the electric field is 31°50′. Ions of a given m/e ratio are collimated into a parallel beam so that ions leaving the electric sector are dispersed according to energy and focused at infinity. If the strength of the radial electrostatic sector is expressed as E (volts per centimeter), then

$$mv^2 = eEr \quad \text{or} \quad r = \frac{mv^2}{eE} \qquad (16\text{-}9)$$

which shows that ions of a given kinetic energy will follow the same path, regardless of their initial kinetic energy spread. These ions are now brought to a focus by the magnetic analyzer along a plane surface where a photographic plate can be located. Alternatively, the spectrum can be scanned across an exit slit and detected with an electron multiplier. The improved focus enables entrance and exit slits to be narrowed with a corresponding

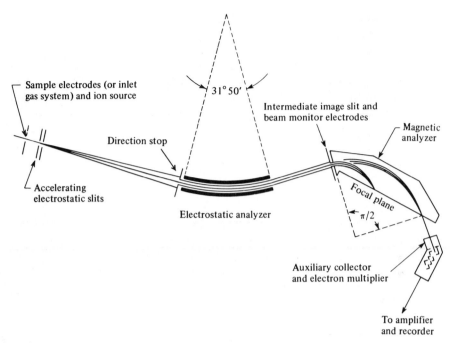

Fig. 16-9. Schematic diagram of a double-focusing mass spectrometer showing basic Mattauch-Herzog geometry. The mass spectrum may be photographed in one exposure. Alternatively, the mass spectrum may be scanned with an electron multiplier.

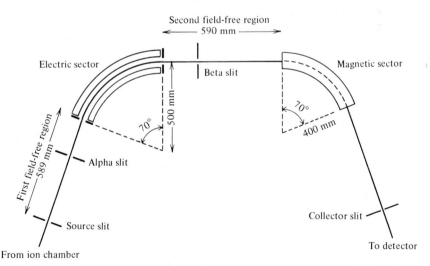

Fig. 16-10. Schematic diagram of a double-focusing mass spectrometer of modified Nier-Johnson geometry. (Courtesy of Perkin-Elmer Corp.)

increase in resolving power up to at least 20,000. The ion energy widths are restricted by means of an energy-defining mask.

The Nier-Johnson geometry (Fig. 16-10) is another version of a double-focusing instrument. The ions are led through a radial electric sector, followed by a magnetic sector. Since the ion beams converge as they leave the electric sector, an intermediate slit can be mounted between the electric and magnetic sectors. Ions of only one value of m/e are sharply in focus at any given combination of field strengths; hence this geometry is not suitable for photographic detection. Resolution is limited essentially by the width of the slits.

Radio-Frequency Spectrometers

If the ion beam emerging from the source and accelerator slits contains ions of uniform energy, then an assembly of grids arranged as in Fig. 16-11 permits selection of ions according to their velocities. Alternate grids are connected together to a steady potential; the other set of alternate grids are connected to a radio-frequency source. Ions in phase with the rf field will attain a velocity v equal to the product of the frequency of the field f (in the order of 10 to 100 MHz) and the spacing between adjacent grids s (in centimeters), namely

$$v = sf \tag{16-10}$$

The field-free spaces permit further discrimination between the ions in phase and those of nearby velocities which may have succeeded in getting past the first set of grids. Subsequent modular stages can exert a greater selection between these resulting bunches of ions, and at the same time reject ions with harmonically related velocities. An energy selector is located in front of the detector. The retarder grid is given a positive voltage that is sufficient to repel (or deflect) all those ions that have received less than a specified

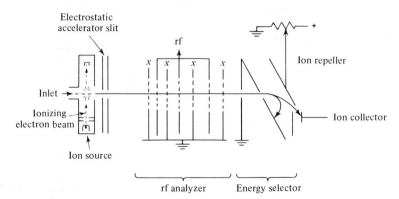

Fig. 16-11. Schematic diagram of an rf mass spectrometer. All grids marked x are at the same dc potential. (Courtesy of Beckman Instruments, Inc.)

fraction of the total available energy. The mass spectrum is scanned by varying the frequency of the rf section of the analyzer. From Eqs. 16-1 and 16-10, the m/e ratio reaching the detector, when reduced to practical cgs units, is given by

$$\frac{m}{e} = \frac{0.266\ V}{s^2 f^2} \qquad (16\text{-}11)$$

For negative ion analyses the potentials are reversed.

The rf spectrometer is only about 25 cm in length, lightweight, simple in construction, and does not require a magnet. Its mass range is from 2 to 100.

Time-of-Flight Spectrometer

The essential principle of time-of-flight (TOF) mass spectrometry is that if ions of different mass are given the same kinetic energy, they will acquire different velocities and will therefore have a time of flight which is mass dependent. Sample molecules are ionized by electron impact. An electron beam is pulsed through the ionization region for 1 μsec at some preselected energy, typically 70 eV. Immediately following this ionization pulse, the first accelerating grid is given a pulse of negative charge which starts the positive ions through the stages of acceleration. The ion beam reaches drift energy, typically 2700 eV, in a distance of less than 2 cm. Because all ions have essentially the same energy at this point, their velocities will be inversely proportional to the square roots of their masses. The ions are now allowed to enter and move down a field-free region (Fig. 16-12), 1 m in length, with whatever velocity they may have acquired. The lighter ions speed on ahead while the heavier ions travel at lower velocities. Hence the original beam

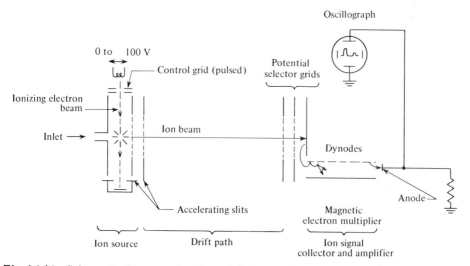

Fig. 16-12. Schematic diagram of a time-of-flight mass spectrometer. (Courtesy of Bendix Corp.)

becomes separated into "wafers" of ions according to their masses. The wafers of ions impact sequentially on the flat cathode of the ion detector. A cathode-ray oscilloscope is synchronized with the pulse repetition rate of the spectrometer. The transit time t (in microseconds) of ions through a distance L (in centimeters) is given by

$$t = L \sqrt{\left(\frac{m}{e}\right)\left(\frac{1}{2V}\right)} \quad \text{or} \quad \frac{m}{e} = \frac{2Vt^2}{L^2} \tag{16-12}$$

A complete mass spectrum of a sample can be repeated 20,000 times in 1 sec; thus a complete mass spectrum is generated each 50 μsec. The instrument is excellent for kinetic studies of fast reactions, for direct analysis of effluent peaks from a gas chromatograph, and for observation of samples volatilized by a single pulse from a laser. Resolution is about 1 part in 400. A disadvantage is the possible overlap of pulse masses. Insertion of an energy selector grid before the detector limits ions that are allowed to reach the detector to only selected masses.

Quadrupole and Monopole Analyzers

A quadrupole field is formed by four electrically conducting, parallel rods of radius r_p that are oriented symmetrically around and tangent to a circle of radius r_0, such that $r_p = 1.16 r_0$. Opposite pairs of surfaces are connected together (Fig. 16-13). To the two pairs of rods are applied equal but opposite potentials, each potential having dc and rf voltage components. The equipotential surfaces in the region between the four poles appear as oscillating hyperbolic potentials.

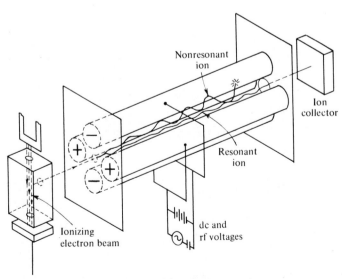

Fig. 16-13. Quadrupole mass analyzer.

An ion injected down the longitudinal axis will undergo transverse motion in the plane perpendicular to the longitudinal axis in addition to its injection velocity down this axis. There are no field gradients along the device, so the ions travel in the axial direction. The dc electric fields tend to focus positive ions in the positive plane, and defocus them in the negative plane. When an alternating rf field is superimposed, an ion of light mass responds to the changes in the electric field without striking an electrode. As the resultant field becomes negative during part of the negative half-cycle of the alternating field, the positive ion will be accelerated toward the electrodes and will achieve a substantial velocity. The following positive half-cycle will have an even greater influence on the motion of the ion, causing it to reverse its direction and accelerate even more. The ion will exhibit oscillations with increasing amplitudes until it finally strikes on the electrodes. The lighter the ion in mass, the smaller the number of cycles before it is collected by the electrode. On the other hand, heavy positive ions will gradually drift toward the electrodes because they will not respond to any significant extent to the small repulsive force existing during part of the positive half-cycle of the alternating field. Only particles with one particular m/e ratio from a collection of injected ions will undergo oscillatory motion and be transmitted through the analyzer. The device can be transformed into a scanning mass spectrometer by varying the electrical parameters so that successively different masses are transmitted. The most common mode of sweeping is a simultaneous variation of the rf and dc voltages, with the frequency held constant. With fixed electrode geometry, the field is determined by electrical parameters only; these can be adjusted and changed very rapidly so that the sweep can be as rapid as 1000 amu/sec.

The quadrupole analyzer is not restricted to detection of monoenergetic sources. Ions are accepted within a 60° cone around the axis. The quadrupole analyzer therefore does not require focusing slits, which results in higher sensitivity. The resolution is a function of the number of cycles an ion spends in the field. Increasing the pole length (usually 5–20 cm) increases the resolution and the capability to handle ions of higher energies. If the rf frequency is increased, the length of the analyzer can be reduced. Pole diameters are also a factor: increasing the pole diameter increases the sensitivity by a large factor, whereas decreasing the pole diameter increases the mass range. The rods are constructed of specially heat-treated stainless steel that are ground and polished to precise tolerances better than 0.00025 cm, and are mounted by means of boron nitride insulators.

The equation of motion of ions in a quadrupole field, the Mathieu equation, has been discussed.[6]

The monopole uses a single rod and a grounded v-block electrode, as shown in Fig. 16-14. Virtual images of the rod behind the grounded plane of the v-block form the quadrupole field. The symmetry of the quadrupole indicates that there exist two mutually perpendicular planes between the rods, which are at zero potential, both dc and ac. Both rf and dc potentials are applied to the single rod. In operation the monopole is quite different from the quadrupole. To transverse the analyzer successfully, the ion must not strike the v-block. This means that the y coordinate of the ion is always positive, and the x coordinate is always less than y. The tube length must be no longer than half of the beat length; ions of lower mass have a shorter beat length and strike the v-block,

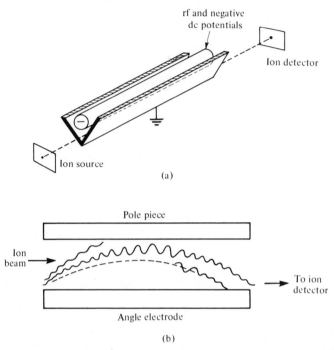

Fig. 16-14. (a) Schematic diagram of monopole mass analyzer. (b) Ions of higher mass, correct mass, and lower mass traversing the analyzer region.

whereas ions of higher mass are unstable in the y direction. By contrast with the quadrupole mass analyzer, the monopole can operate as a focusing spectrometer, with an image of the entrance slit formed near the exit aperture. Because of the focusing property, the monopole does not share the advantage of the quadrupole in being insensitive to a large spread in ion entrance energies. The monopole uses less rf power and exhibits less mass discrimination than the quadrupole.

Ion Cyclotron Resonance Spectrometer

In a magnetic field the centrifugal force acting on an ion is balanced by the force from the field; that is,

$$mv^2 r = \frac{eHv}{c} \quad \text{or} \quad \frac{v}{r} = \frac{eH}{mc} = \omega_c \tag{16-13}$$

which causes the ion to travel in a circle. Angular velocity ω_c is the number of radians transversed per unit time, and the frequency of revolution in cycles per unit time is $\omega_c/2\pi$. Subjecting the ion simultaneously to an electric field applied at right angles to the magnetic field causes the ion to proceed in a cycloidal path at right angles to both fields. Frequency is now the number of complete cycloids per unit time. If a rapidly alternating electric field is applied, as the frequency of alternation approaches the cyclotron frequency of an ion which is present, the ion will absorb energy and be accelerated, causing the radius of its cycloidal path to increase. For ions of between 1 and 200 amu, with a magnetic field of 15,000 G, the frequencies range from 50 kHz to about 25 MHz.

Instead of collecting the ions, a marginal oscillator is used to detect the energy that is absorbed by the ion. A mass spectrum can be obtained by plotting the output of the detector circuit vs. a time domain quantity, either the rate of scan of the magnetic field at constant frequency or, alternatively, the rate of change of the electric field frequency at constant magnetic field. To constrain the ions to a motion down the center of the analyzer, trapping voltages are applied to side plates. A positive potential serves to trap positive ions while pulling negative ions and scattered electrons out of the analyzer chamber. When negative ions are studied, the trapping potential is simply reversed.[8]

Figure 16-15 shows an exploded view of the electrodes that comprise the mass analyzer. This arrangement is mounted in a vacuum chamber shaped to fit between the poles of an electromagnet.

If the applied oscillating electric field is not at cyclotron frequency, a fixed phase relationship cannot be maintained and the average acceleration is small. Physically, this

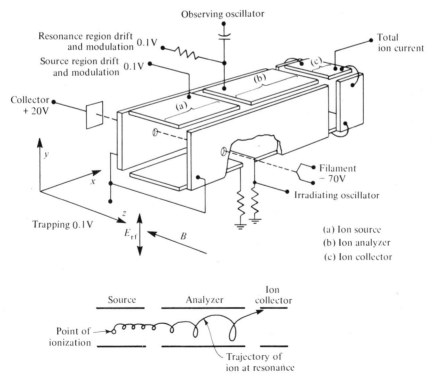

Fig. 16-15. Ion cyclotron resonance spectrometer. (Courtesy of Varian Associates.)

results in the radius of a nonresonant ion increasing, collapsing, and repeating this cycle. A compromise must be made between mass range and resolution. For example, at ω_c = 765 Hz a resolution of approximately 5000 can be achieved in the absence of collisions; however, the highest mass that can be observed is 30 amu (H = 15,000 G). At 153 Hz the mass range is increased to 150 amu, but the resolution is decreased to 1000.

Because drift velocities on the order of 500 cm/sec are common, typical ion residence times are 0.5 to 25 msec, and typical path lengths for thermal velocity ions are 1–50 m. This is 1000 times longer than in a sector mass spectrometer, and an ion will suffer 1000 times as many collisions at any given pressure in an ion cyclotron spectrometer. It is this feature that accounts for the usefulness of an ion cyclotron spectrometer for the study of ion molecule reactions at low pressures. Ion molecule reactions[1,9] are found to be much more sensitive to small differences in molecular structure, as with isomeric and tautomeric compounds, than are the corresponding high-energy electron-impact fragmentation spectra.

Omegatron

The omegatron is a small unit which operates on the principle of the ion cyclotron resonance spectrometer. A cylindrical beam of electrons is ejected through a small opening parallel to the magnetic field and perpendicular to the rf electric field. The ions are formed by electron impact near the center of the square electrode structure (Fig. 16-16), which is totally immersed in the magnetic field, and subject also to a rf electric field. Only ions of a m/e ratio in resonance with the oscillating field are successively accelerated, and they travel in a spiral trajectory of ever increasing radius until they strike the ion collector. Ions with other m/e ratios are forced into smaller circular paths near the axis of the electron-impact beam.

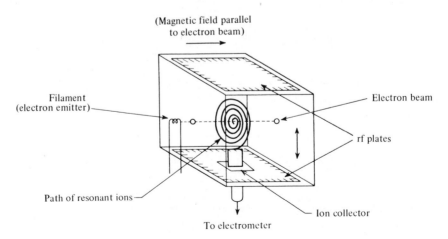

Fig. 16-16. Simplified diagram of an omegatron.

Commercial omegatrons are available as small units intended to be connected to vacuum systems for residual gas analysis. An instrument about 2 cm on a side, operating at 5000 G with a permanent magnet, and with a frequency from 150 kHz to 4 MHz, will cover 50 amu. Resolution, about 1 part in 40, is proportional to the number of revolutions the ions can make. For higher resolution a larger number of revolutions are required, but then space charge effects among the ions become very large. Because of this, most omegatrons have useful resolution only to about mass 50.

CORRELATION OF MASS SPECTRA WITH MOLECULAR STRUCTURE

When bombarded by electrons, every substance ionizes and fragments uniquely. A molecule may simply lose an electron or it may fragment into two smaller units, an ionized fragment and a neutral particle, the sum of whose masses equals their precursor. A molecular or parent ion is generally observed in considerable intensity when the gaseous molecules are bombarded with electrons of energy just sufficient to cause ionization, but not bond breakage, which equals about 8–14 eV for most organic molecules. As the electron energy is increased further, the molecular ion can be formed with excess energy in its electronic and vibrational degrees of freedom. Because redistribution of energy between the bonds takes place rapidly, all of the bonds are affected simultaneously. As soon as the excess energy over the ground-state energy possessed by the molecular ion becomes equal to the dissociation energy of some particular bond, the appropriate fragment ion can be formed. Although the peak intensities are extremely sensitive to ionizing voltage at low values of the ionizing voltage, the relative peak intensities become fairly constant once the ionization voltage exceeds 50 eV. At higher ionizing energies the total production of ions is higher, but the net effect of higher overall intensity and the resultant severe fragmentation is an increase in relative intensity of the fragment peaks at the expense of the parent peak.

Molecular Identification

In identification of a compound the most important single item of information is the molecular weight. The mass spectrometer is unique among analytical methods in being able to provide this information very accurately. At ionizing voltages ranging from 9 to 14 V it can be assumed that no ions heavier than the molecular ion will be formed and, therefore, the mass of the heaviest ion, exclusive of isotopic contributions, gives the nominal molecular weight.

Restriction on the number of possible molecular formulas can be achieved by study of the relative abundance of natural isotopes for different elements [Table 16-1 (a) and (b)] at masses 1 and 2 or more units larger than the parent ion. Observed values are compared with those calculated for all possible combinations of the naturally occurring heavy isotopes of the elements. For a compound $C_w H_x O_z N_y$, a simple formula allows one to calculate the percent of the heavy isotopic contributions from a monoisotopic peak, P_M,

TABLE 16-1(a) Abundances of Some Polyisotopic Elements (%)

Element	% Abundance	Element	% Abundance	Element	% Abundance
^{1}H	99.985	^{16}O	99.76	^{33}S	0.76
^{2}H	0.015	^{17}O	0.037	^{34}S	4.22
^{12}C	98.892	^{18}O	0.204	^{35}Cl	75.53
^{13}C	1.108	^{28}Si	92.18	^{37}Cl	24.47
^{14}N	99.63	^{29}Si	4.71	^{79}Br	50.52
^{15}N	0.37	^{30}Si	3.12	^{81}Br	49.48

TABLE 16-1(b) Selected Isotope Masses

Element	Mass	Element	Mass
^{1}H	1.0078	^{31}P	30.9738
^{12}C	12.0000	^{32}S	31.9721
^{14}N	14.0031	^{35}Cl	34.9689
^{16}O	15.9949	^{56}Fe	55.9349
^{19}F	18.9984	^{79}Br	78.9184
^{28}Si	27.9769	^{127}I	126.9047

SOURCE: *Lange's Handbook of Chemistry*, J. A. Dean, Ed., 11th ed., McGraw-Hill Book Co., New York, 1973.

to the P_{M+1} peak:

$$100 \frac{P_{M+1}}{P_M} = 0.015x + 1.11w + 0.37y + 0.037z$$

and to the P_{M+2} peak:

$$100 \frac{P_{M+2}}{P_M} = 0.20z + 0.006w(w-1) + 0.004wy + 0.0002wx$$

Tables of abundance factors have been calculated by Beynon[3] for all combinations of C, H, N, and O up to mass 500. Table 16-2 illustrates the spectral peak contributions at nominal masses 135 and 136 for a few of the compounds having a parent peak (or possibly a fragment peak) of mass 134. Once the empirical formula is established with reasonable assurance, hypothetical molecular formulas are written. One can utilize the entries in the formula indices of Beilstein and Chemical Abstracts.

Compounds containing chlorine, bromine, sulfur, or silicon are usually apparent from the prominent peaks at masses P_{M+2}, P_{M+4}, etc. The abundance of heavy isotopes is treated in terms of the binomial expansion $(a + b)^m$, where a is the relative abundance of light isotopes, b is the relative abundance of heavy isotopes, and m is the number of atoms of the particular element present in the molecule. When two elements are present, the binomial expansion $(a + b)^m (c + d)^n$ is used.

If the mass of the parent ion is measured with a high-resolution mass spectrometer,

TABLE 16-2   Heavy-Isotope Contributions to Parent Peak of Mass 134 (%)

Empirical Formula	P_{M+1} Peak	P_{M+2} Peak
$C_5H_{10}O_4$	5.72	0.94
$C_5H_{14}N_2O_2$	6.47	0.58
$C_6H_{14}O_3$	6.83	0.80
$C_8H_6O_2$	8.82	0.74
$C_9H_{10}O$	9.93	0.64
$C_9H_{12}N$	10.30	0.48
$C_{10}H_{14}$	11.03	0.55

the number of possible empirical formulas can be still further restricted. Because the masses of the elements are not exactly integral multiples of a unit mass, a sufficiently accurate mass measurement alone enables the elemental composition of the ion to be determined. For combinations of C, H, N, and O, the relationship is

$$\frac{\text{exact mass difference from nearest integral mass} + 0.0051x - 0.0031y}{0.0078} = \text{number of H's}$$

For example, a crystalline solid containing only C, H, and O, gave the mass 134.0368 for the molecular ion. Thus,

$$\frac{0.0368 + 0.0051x}{0.0078} = 6 \text{ H's when } x = 2 \text{ oxygen atoms}$$

and the empirical formula is $C_8H_6O_2$. One substitutes integral numbers for x (oxygen) and y (nitrogen) until the divisor becomes an integral multiple of the numerator within 0.0002 mass units.

Two general rules aid in writing formulas. If the molecular weight of a C, H, N, and O compound is even, so is the number of hydrogen atoms it contains; if the molecular weight is divisible by four, the number of hydrogen atoms is also divisible by this number. When nitrogen is known to be present in any compound of C, H, O, As, P, S, Si, and the halogens that have an odd molecular weight, the number of nitrogen atoms must be odd. Once the exact molecular formula has been decided, the sum total of the number of rings and double bonds can be determined by the formula

$$R = \tfrac{1}{2}(2w - x + y + 2)$$

when covalent bonds make up the molecular structure. From the formula $C_8H_6O_2$, $R = 6$; one strong possibility is a phenyl ring (4) plus two additional double bonds.

Metastable Peaks

A one-step decomposition process may be indicated by an appropriate *metastable* peak in the mass spectrum. Metastable peaks arise from ions which decompose in the field-free region after they are accelerated out of the ion source but before entering the analyzer. Their lifetime is about 10^{-6} sec. A metastable ion transition takes the general form: Original ion → daughter ion + neutral fragment. The metastable peak $m*$ will appear, as a weak, diffuse peak, usually at a nonintegral mass, given by:

$$m* = \frac{(\text{mass of daughter ion})^2}{\text{mass of original ion}}$$

In a spectrum which is linear with respect to m/e values, the distance of $m*$ below the daughter ion will be of similar magnitude to the distance of the daughter ion below the original ion. The foregoing relationship holds only for ions decomposing in a small portion of the accelerating region and will be more frequently observed with 60° and 90° sector instruments where a field-free region exists after the accelerating slits and before the magnetic analyzer. Of course, the absence of a metastable peak from the spectrum does not preclude a particular decomposition.

Example 16-1

A compound of molecular formula $C_9H_{12}S$ gave a mass spectrum with the parent peak at mass 152 (45%) and fragment ion peaks at masses 137 (7%), 110 (100%), 77 (7%), 66 (11%), 65 (8%), and 43 (12%). Metastable peaks were located at 123, 79.6, and 54.1. These correspond to the transitions:

$$m*$$
$$123 \qquad 152^+ \rightarrow 137^+ + 15$$
$$79.6 \qquad 152^+ \rightarrow 110^+ + 42$$
$$54.1 \qquad 110^+ \rightarrow 77^+ + 33$$

which suggests the elimination of CH_3, $CH_3—CH{=}CH_2$, and HS, respectively. The loss of HS from the base peak also suggests that —SH was connected to a phenyl ring (mass 77). Furthermore, the base peak at 110 is the result of a rearrangement because it is an even-mass ion originating from an even-mass molecular ion. This establishes the presence of a mass 43 group attached to a mass 109 entity. The initial loss of a methyl group strongly suggests an isopropyl group, although this should be confirmed from reference spectra or from a nmr spectrum. Finally, the strength of the molecular peak hints at a stable molecule. The compound is isopropylphenylthioether.

Mass Spectra

The mass spectrum of a compound contains the masses of the ion fragments and the relative abundance of these ions plus often the parent ion. The uniqueness of the molecular fragmentation aids in identification. Because sufficient molecules are present and dissociated for the probability law to hold, the dissociation fragments will always occur in the same relative abundance for a particular compound. The mass spectrum becomes a "fingerprint" for each compound, as no two molecules will be fragmented and ionized in exactly the same manner on electron bombardment. There are sufficient differences in these molecular fingerprints to permit identification of different molecules in complex mixtures. To a considerable extent the breakdown pattern can be predicted. Conversely, the size and structure of the molecule can often be reconstructed from the fragment ions in the spectrum of a pure compound. For example, Table 16-3 indicates the relative abundance of the significant fragments produced from three isomeric octanes. In the structural formulas the asterisk indicates the bond which is broken in the most probable process of scission, and the plus sign indicates the next most probable process. Favored sites for bond rupture in the molecule parallel chemical bond lability. The mass 114 corresponds to the parent ion formed by the loss of a single electron from the parent compound; mass 99 corresponds to the loss of a methyl group plus an electron; mass 71, to the loss of a propyl group plus an electron; mass 57, to the loss of a butyl group; and mass 43, to the loss of an amyl group.

It is usual practice in reporting mass spectra to normalize the data by assigning the most intense peak (the so-called *base peak*) a value of 100; other peaks are reported as percentages of the base peak.

When working from a spectrum it is advisable to tabulate the prominent ion peaks, starting with the highest mass, and also to record the group lost to give these ion peaks. All possible molecular structures are listed, employing a file of common fragment ions encounted in mass spectra. Finally, one attempts to predict the mass spectral features from available correlation data and to check these features against the actual spectrum. Usually only one bond is cleaved; in succeeding fragmentations a new bond is formed for each additional bond that is broken. When fragmentation is accompanied by formation of a new bond as well as by breaking an existing bond, a *rearrangement* process is said to have occurred. The migrating atom will almost exclusively be hydrogen; six-membered cyclic transition states are most important although alternative ring sizes also operate.

Some general features of the mass spectra of compounds can be predicted from general rules for fragmentation patterns:

1. Cleavage is favored at branched carbon atoms: tertiary > secondary > primary, with the positive charge tending to stay with the branched carbon (carbonium ion).
2. Double bonds favor cleavage beta to the bond (but see Note 8).

TABLE 16-3 Mass Spectral Pattern of Trimethylpentanes

	Relative Abundances, %		
Mass/Charge Ratio	2,3,3-Trimethylpentane $C \quad C$ $+ \quad \mid$ $C-C*C+C-C$ $\mid$ C	2,2,4-Trimethylpentane $C \qquad C$ $+ \qquad \mid$ $C-C*C-C-C$ $\mid$ C	2,3,4-Trimethylpentane $C \qquad C$ $\mid \qquad \mid$ $C-C*C*C-C$ $\mid$ C
114	0.1	0.02	0.3
99	3	5	0.1
71	1	1	40
57	70	80	9
43	15	20	50

SOURCE: H. W. Washburn, H. F. Wiley, S. M. Rock, and C. E. Berry, *Ind. Eng. Chem. Anal. Ed.*, **17**, 75 (1945).

3. A substance having a strong parent peak often contains a ring, and the more stable the ring the stronger the parent peak.
4. Ring compounds usually contain peaks at mass numbers characteristic of the ring.
5. Saturated rings lose side chains at the alpha carbon. The peak corresponding to the loss of two ring atoms is much larger than for the loss of one ring atom.
6. In alkyl-substituted ring compounds, cleavage is most probable at the bond beta to the ring if the ring has a double bond next to the side chain.
7. A hetero-atom will induce cleavage at the bond beta to it:
8. Compounds containing a $\diagup$C=O group tend to break at this group, with the positive charge remaining with the carbonyl portion.

The presence of Cl, Br, S, and Si is easy to deduce from the unusual isotopic abundance patterns of these elements. These and other elements, such as P, F, and I, are also detectable from the unusual mass differences that they produce between some fragment ions in the spectrum.

OTHER APPLICATIONS

Quantitative Analysis of Mixtures

The system employed in quantitative analysis by mass spectrometry is basically the same as that employed in infrared or ultraviolet absorption spectrometry. Spectra are recorded for each component. Consequently, samples of each compound must be available in a fairly pure state. From inspection of the individual mass spectra known or suspected to be present in a mixture, analysis peaks are selected on the basis of both intensity and freedom from interference by the presence of another component. If possible, monocomponent peaks (perhaps parent-ion peaks) are selected. The sensitivity is usually given in terms of the height of the analysis peak per unit pressure—obtained by dividing the peak height for the analysis peak by the pressure of the pure compound in the sample reservoir.

Calculation of sample compositions is simplified if the components of the mixture give spectra with at least one peak whose height is due entirely to the presence of one component. The height of the monocomponent peak is measured and divided by the appropriate sensitivity factor to give its partial pressure. Division then by the total pressure in the sample reservoir at the time of analysis yields the mole percent of the particular component.

If the mixture has no monocomponent peaks, simultaneous linear equations are then set up from the coefficients (percent of base peak) at each analysis peak, one equation for each compound in the mixture with n terms (unless one or more terms are zero) when n components are in the mixture. For the analysis of a butanol mixture,[7] the equations at four masses are set up from coefficients listed in Table 16-4 for the particular instrument:

$$90.58x_1 + 1.47x_2 + 1.02x_3 + 2.46x_4 = M_{56} = 126.7$$

TABLE 16-4 Mass Spectral Data (Relative Intensities) for the Butyl Alcohols

	Percent of Base Peak (italic)			
m/e	n-Butyl	sec-Butyl	t-Butyl	Isobutyl
15	8.39	6.80	13.30	7.47
18	2.18	0.23	0.49	2.05
27	50.89	15.87	9.87	42.20
28	16.19	2.98	1.67	5.94
29	29.90	13.94	12.65	21.17
31	*100.00*	20.31	35.53	63.10
33	8.50	—	—	53.40
39	15.63	3.36	7.70	19.03
41	61.57	10.13	20.82	55.68
42	32.36	1.64	3.32	60.46
43	61.36	9.83	14.45	*100.00*
45	6.59	*100.00*	0.59	5.03
55	12.29	2.06	1.55	4.35
56	90.58	1.02	1.47	2.46
57	6.68	2.74	9.02	3.89
59	0.26	17.78	*100.00*	4.98
60	—	0.64	3.26	0.57
74	0.79	0.29	—	9.06

SOURCE: A. P. Gifford, S. M. Rock, and D. J. Comaford, *Anal. Chem.*, **21**, 1026 (1949).

$$0.26x_1 + 100.00x_2 + 17.78x_3 + 4.98x_4 = M_{59} = 301.5$$
$$6.59x_1 + 0.59x_2 + 100.00x_3 + 5.03x_4 = M_{45} = 322.6$$
$$0.79x_1 + 0 + 0.29x_3 + 9.06x_4 = M_{74} = 14.8$$

To achieve greater speed in computation, the matrix of coefficients is inverted, yielding a set of equations in terms of each unknown and the analytic masses:

$$x_1 = 110.70M_{56} - 1.625M_{59} - 0.7442M_{45} - 28.77M_{74} \quad (n\text{-butyl})$$
$$x_2 = 1.39M_{56} + 100.08\ M_{59} - 17.67\ M_{45} - 45.53M_{74} \quad (t\text{-butyl})$$
$$x_3 = -6.83M_{56} - 0.489M_{59} + 100.31\ M_{45} - 53.56M_{74} \quad (sec\text{-butyl})$$
$$x_4 = -9.39M_{56} + 0.157M_{59} - 3.17\ M_{45} + 1108.0\ M_{74} \quad (\text{isobutyl})$$

Peaks from the mixture spectrum are substituted into the inverse matrix equations, yielding the number of divisions of base peak due to each component. Division by the appropriate sensitivity factor (Table 16-5) yields the partial pressure of each component. Each partial pressure is divided by the total computed pressure, yielding mole percent. The sum of the partial pressures determined in this way should equal the total sample pressure. A discrepancy would indicate an unsuspected component or a change in operating sensitivity.

An outstanding feature of mass spectrometric analysis is the large number of components that can be handled without need for fractionation or concentration. Mixtures containing up to as many as 30 components can be analyzed, and quantities of material

TABLE 16-5   Analysis of a Mixture of Butyl Alcohols[a]

Component	Value of x	Sensitivity, divisions/10^{-3} torr	Partial Pressures, 10^{-3} torr	Mole %
n-Butyl	$x_1 = 12{,}871$	1151	11.18	24.4
t-Butyl	$x_2 = 23{,}976$	2093	11.46	25.0
sec-Butyl	$x_3 = 30{,}555$	2698	11.33	24.8
Isobutyl	$x_4 = 14{,}234$	1205	11.81	25.8

SOURCE: A. P. Gifford, S. M. Rock, and D. J. Comaford, *Anal. Chem.*, **21**, 1026 (1949).

[a]Mass peaks used: 45, 56, 59, and 74.

as low as 0.001 mole percent can be detected in hydrocarbon mixtures. Calculations are usually carried out on high-speed automatic computers. More complex mixtures, covering a wide boiling range, may require a rough or simple distillation before analysis. Precision normally falls within the range of ±0.05 to 1.0 mole percent.

Solid State Analyses

Solid state mass spectroscopy, in contrast with optical emission, has a very simple spectrum. The mass spectrum of each element consists of a principal line repeated at fractional mass values, but at reduced intensities, due to ions with multiple charges. The intensity of lines of any element at the same atomic concentration is roughly equal (within a factor of two) for most compounds, greatly facilitating semiquantitative analysis. Signal/noise ratios of 10^6 to 1 are available with double-focusing spectrometers, so that not only can trace impurities be detected, but sample sizes in the submicrogram range are feasible. Surface contaminants can be distinguished from bulk impurities, since the vacuum spark initially samples the surface of a solid. Techniques are available for handling insulators, and even the solid residue remaining from an evaporated liquid drop can be analyzed. Sensitivity is about 10 parts per billion, and is limited chiefly by the scattering of the ions enroute through the instrument by residual gas atoms.

Use of Stable Isotopes

Stable isotopes can be used to "tag" compounds and thus serve as tracers to determine the ultimate fate of the compound in chemical or biological reactions. The mass spectrum displays amounts of the added isotope in the fragment ions as well as in the parent ion. Thus, the position of the tracer isotope in the molecule can often be determined without laborious chemical degradation techniques.

A number of stable isotopes in sufficiently concentrated form are available, including practically all of the isotopes of the lighter elements: H, B, C, N, O, S, and Cl. These isotopes complement the relatively larger number of radioactive isotopes.

The isotope-dilution method, described in Chapter 11 for radioactive isotopes, can be employed equally well with stable isotopes to determine the amount of substance present in a complex mixture. It is only necessary to know the ratio of isotopes present in the added sample of the substance, the ratio present in the final sample isolated from the mixture, and the weight of the added sample. It is not necessary that the test substance be separated quantitatively from the mixture; only a few milligrams of pure substance are necessary.

Leak Detection

The helium mass spectrometer, widely used to detect leaks, consists of a magnetic deflection type mass spectrometer that is tuned to maximum sensitivity in the presence of helium, and is coupled to a vacuum system. This is an inexpensive, compact, portable instrument. For determination of exact leak locations (within 1.0 mm), the exterior of the evacuated test piece is carefully sprayed with a hand-held helium jet. When the jet passes over a leak, a small amount of helium enters the object, passes into the spectrometer, and is detected. Sensitivity is as low as 2×10^{-11} atm ml/sec of helium. To determine total leakage, an evacuated test piece is placed within a helium-filled hood connected to the leak detector. If leakage is present, the sum total of helium admitted into the object travels through the system and is measured. When objects contain helium under pressure or can be pressurized with helium, a sampling probe "sniffer" is passed over the test piece. Escaping helium is drawn into the instrument and the exact location of the leak determined.

Ion Microprobe Mass Analyzer

A schematic diagram of an ion microprobe mass analyzer is shown in Fig. 16-17. The impinging ion beam is generated in a hollow-cathode duoplasmatron ion source which is capable of producing ions of a wide variety of gases. These ions, which can be either positively or negatively charged, are accelerated to energies ranging from 5 to 22.5 keV and passed through the primary mass spectrometer. This spectrometer selects and purifies a specific ionic species for sample bombardment. The purified ion beam is focused to a small probe in an electrostatic lens column consisting of a condenser lens and an objective lens, and allowed to impinge on the surface of the sample. Probe diameter can vary from 2 to 500 μm. An auxiliary optical microscope permits the sample area being analyzed to be viewed. Bombardment of a solid sample with the high-energy beam of ions causes the atoms at the sample surface to be sputtered away. A fraction of the sputtered particles is electrostatically charged. These ions (which have an energy dispersion) are collected and analyzed in a double-focusing mass spectrometer. The ion beam leaving the mass spectrometer is detected with a high-gain device of the Daly type[5]; ions eject secondary electrons at the conversion electrode, and these are accelerated toward the scintillator of a photomultiplier tube. The resolved ion signals can be read as count rates in scalers or a direct current on chart recorders.

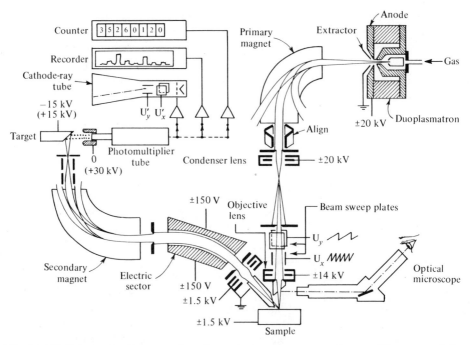

Fig. 16-17. Schematic diagram of an ion microprobe mass analyzer. (Courtesy of Applied Research Laboratories.)

In addition to fixed-point bombardment, the impinging ion beam can be swept over a selected area of the sample using the sweeping plates in the primary lens column. The detected signals, when viewed on a cathode-ray tube, give a two-dimensional distribution of the sputtered element in the selected sampling area. Depth analysis, with a resolution of about two atomic layers, can be accomplished as successive atomic layers of the surface are eroded away. The method complements other microanalytic techniques, such as the electron microprobe X-ray analyzer, by providing information on the concentration and distribution of the isotopes of the elements in the surface of a solid. Electron microprobes can handle detection of elements with ease only down to sodium in the periodic table; the ion probe handles lighter elements as well, covering the entire range from hydrogen to uranium. With the ion microprobe analyses can be made, for example, of interfaces in geological specimens, of grain boundaries in metallurgical samples, and of particles in air pollution particulate analysis. The instrument provides a point-to-point analysis with spatial resolution of less than 2 μm to 500 μm.

PROBLEMS

1. For a field strength of 2400 G in 180° magnetic-deflection spectrometer, what electrostatic voltage range suffices for scanning from mass 18 to mass 200? The radius of curvature of the 180° analyzer tube is 12.7 cm.

2. For a drift length of 100 cm in a time-of-flight mass spectrometer, what is the difference in arrival time between ions of $m/e = 44$ and $m/e = 43$ when the accelerating voltage is 2800 V?

3. What rf field is necessary to handle a range of masses from 4 to 40 units in a rf mass spectrometer if the accelerating voltage varies from 400 to 4000 V and the spacing between grids is 1 cm?

4. The parent peak spectrum of tridecylbenzene (260.2504), phenyl undecylketone (260.2140), 1,2-dimethyl-4-benzoyl naphthalene (260.1201), and 2,2-naphthyl benzothiophene (260.0922) would require what resolution for quantitative analysis based on the parent peak?

5. What resolving power is needed to separate (a) the CH_2N—C_2H_4 doublet at mass 200; (b) the N_2—CO doublet at mass 150; and (c) the CH_2—N doublet at mass 200?

6. A peptide was admitted to a high-resolution mass spectrometer and the parent peak mass was measured relative to the parent peak in the spectrum of dibromobenzene (236.8638). The measured ratio of unknown mass/reference mass was 1.001197 ±0.000002. Compute the exact weight of the peptide and deduce the molecular formula.

7. From the following exact molecular weights, estimate the empirical formulas assuming only C, H, O, or N is present unless otherwise indicated: (a) 164.0473, (b) 120.0575, (c) 180.0939, (d) 94.0531, (e) 109.0528, (f) 190.9540 (contains Cl), (g) 181.0891, (h) 334.0873 (contains S), and (i) 177.0426.

8. (a) In the high-resolution spectrum of methionine, a quartet of peaks appear at nominal mass 88: 88.0220, 88.0345, 88.0335, and 88.0267. Deduce the fragment ion responsible for each peak. (b) Methionine also gives a doublet at nominal mass 75. One line corresponds to $C_2H_4NO_2$; the other has m/e 75.0267. Outline the process leading to the fragment $C_2H_4NO_2$. Deduce the fragment ion of m/e 75.0267.

9. What is the probable composition of a molecule of mass 142 whose $P + 1$ peak is 1.1% of the parent peak?

10. Deduce the number and type of halogen atoms present in a molecule from the abundance of heavy isotopes and the intensity ratios, given in the following table:

	P	$P + 2$	$P + 4$	$P + 6$	$P + 8$
Compound A	30	29	10	1	—
Compound B	13	30	19	6	1
Compound C	5	20	30	19	5
Compound D	23	30	7	—	—
Compound E	18	30	14	2	—

11. From the isotopic abundance information, what can be deduced concerning the empirical formula of each of the following compounds?

m/e	% of Base Peak		m/e	% of Base Peak		m/e	% of Base Peak
90(P)	100		89(P)	17.12		206(P)	25.90
91	5.61		90	0.58		207	3.24
92	4.69		91	5.36		208	2.48
			92	0.17			

m/e	% of Base Peak		m/e	% of Base Peak		m/e	% of Base Peak
230(P)	1.10		140(P)	14.8		151(P)	100
232	2.12		141	1.40		152	10.4
234	1.06		142	0.85		153	32.1
						154	2.9

12. The significant portion of the mass spectral data is given for the individual alcohols. Select appropriate analytical masses and write a series of four equations in terms of divisions of base peak due to each of the four alcohols.

	% of Base Peak				Unknown Mixtures		
m/e	Methyl	Ethyl	n-Propyl	Isopropyl	A	B	C
15	35.48	9.44	3.77	10.70			
19	0.29	3.13	0.90	6.51			
27	–	21.62	15.20	15.50			
29	58.80	21.24	14.14	9.49			
31	100	100	100	5.75			
32	68.03(P)	1.14	2.25	–	600	600	2350
33	0.98	–	–	–			
39		–	4.00	5.52	4800	3000	3000
43		7.45	3.18	16.76			
45		37.33	4.39	100			
46		16.23(P)	–	–	1000	1100	698
59			9.61	3.58	4000	2300	5000
60			6.36(P)	0.44(P)			
Sensitivity, divisions/10^{-3} torr	8.76	17.98	26.51	23.47			

13. The mixture peaks for three unknown mixtures of alcohols are shown in Problem 12. For each mixture compute the mole percent of each alcohol.

14. A material containing only C, H, and O, and in the form of leaflets melting at 40°C, possesses a rather simple mass spectrum with the parent peak at m/e 184

(10%), the base peak at m/e 91, and small peaks at m/e 77 and 65. Metastable peaks appear at m/e 45.0 and 46.5. Deduce the structure of the compound.

15. The mass spectrum possesses a strong parent peak at m/e 122 (35%) plus peaks at m/e 92 (65%), m/e 91 (100%), and m/e 65 (15%). In addition there are metastable peaks at 46.5 and 69.4 mass units. Deduce the compound's structure.

16. A solid, melting at 33°C, has the following mass spectrum: Parent ion at m/e 200 with $P+1$ exhibiting a strength equal to 10.55% of the parent peak, and $P+2$, 5.77% of the parent peak. The major ion peaks occur at these mass units: 172 (18%), 155 (61%), 108 (10%), 107 (11%), 92 (30%), 91 (100%), and 65 (30%). Metastable peaks appear at 46.5, 53.5, 67.9, 106.3, 121.7, and 147.9 mass units. Deduce the structure of the compound insofar as possible.

17. Deduce the structural formula for each of these compounds from the mass spectral data:

$C_4H_3O_2$		$C_4H_6O_2$	
m/e	% of Base Peak	m/e	% of Base Peak
27	39.3	15	27.7
29	19.8	26	22.4
39	14.8	27	68.1
41	23.7	29	13.0
42	24.7	42	11.8
43	22.3	55	*100*
45	19.1	58	8.4
60	*100*	59	5.2
73	27.1	85	12.3
88(*P*)	1.6	86(*P*)	2.1

C		D	
m/e	% of Base Peak	m/e	% of Base Peak
29	18	63	22
39	23	64	20
51	29	65	33
65	18	92	82
78	50	93	18
91	*100*	120	*100*
105	41	121	34
134(*P*)	57.4	152(*P*)	45.0
135	5.80	153	4.1
136	0.41	154	0.4

18. Deduce the complete structural formula of the compound from the mass spectrum in the figure.

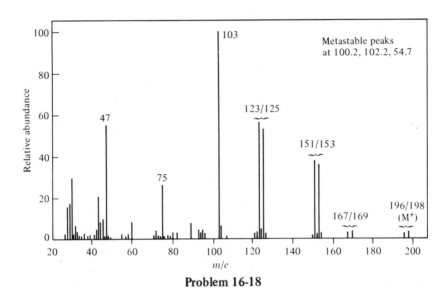

Problem 16-18

19. Deduce the structural formula of the compound (b.p. 74°C) whose mass spectrum is shown.

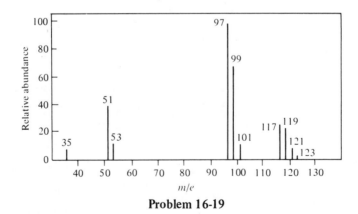

Problem 16-19

20. A low melting solid with molecular formula $C_5H_8O_4$ gave the mass spectrum shown. Deduce the structural formula of the compound.

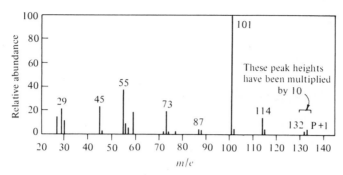

Problem 16-20

BIBLIOGRAPHY

Ahearn, A. J., "Spark Source Mass Spectrometric Analysis of Solids," in *Trace Characterization*, W. W. Meinke and B. F. Scribner, Eds., National Bureau of Standards Monograph 100, Washington, D.C., 1967.

Beynon, J. H., *Mass Spectrometry and Its Application to Organic Chemistry*, Elsevier, Amsterdam, 1960.

Biemann, K., *Mass Spectrometry: Applications to Organic Chemistry*, McGraw-Hill, New York, 1962.

Budzikiewicz, H., C. Djerassi, and D. H. Williams, *Mass Spectrometry of Organic Compounds*, Holden-Day, San Francisco, 1967.

McLafferty, F. W., *Interpretation of Mass Spectra*, Benjamin, 2nd ed., Menlo Park, Calif., 1973.

LITERATURE CITED

1. Baldeschwieler, J. D., Science, **159**, 263 (1968).
2. Beckey, H. D., *Research/Development*, p. 26 (November 1969).
3. Beynon, J. H. and A. E. Williams, *Mass and Abundance Tables for Use in Mass Spectrometry*, Elsevier, Amsterdam, 1963.
4. Chait, E. M., *Anal. Chem.*, **44** (3), 77A (1972).
5. Daly, N. R., *Rev. Sci. Instr.*, **31**, 264 (1960).
6. Dawson, P. H. and N. R. Whetten, *Research/Development* p. 46 (Feb. 1968).
7. Gifford, A. P., S. M. Rock, and D. J. Comaford, *Anal. Chem.*, **21**, 1062 (1949).
8. Gross, M. L. and C. L. Wilkins, *Anal. Chem.*, **43** (14) 65A (1971).
9. Henis, J. M. S., *Anal. Chem.*, **41** (10) 22A (1969).
10. Karasek, F., *Anal. Chem.*, **44** (4), 32A (1972).
11. Munson, B., *Anal. Chem.*, **43** (13) 28A (1971).
12. Updegrove, W. S. and P. Haug, *Am. Laboratory*, **2** (2) 9 (Feb. 1970).

17

Thermal Analysis

Heat may be used as a reagent. Taken through a wide range of temperatures, a substance may undergo physical and chemical changes, react with the ambient atmosphere, or jettison water of crystallization and other fragments. All changes are accompanied by the absorption or release of energy in the form of heat. Some changes involve a weight gain or loss, and there may be thermomechanical or electrical conductivity changes. The rate and temperature at which materials undergo physical and chemical transitions as they are heated and cooled, and the energy and weight changes involved is the subject of thermal analysis. Thermometric titrimetry, also discussed in this chapter, involves changes in solution temperature that are plotted as a function of time or volume of titrant.

DIFFERENTIAL THERMAL ANALYSIS AND DIFFERENTIAL SCANNING CALORIMETRY

In *differential thermal analysis* (*DTA*) the temperature of a sample and a thermally inert reference material are measured as a function of temperature (usually sample temperature). Any transition which the sample undergoes will result in liberation or absorption of energy by the sample with a corresponding deviation of its temperature from that of the reference. This differential temperature (ΔT) versus the programmed temperature (T) at which the whole system is being changed tells the analyst the temperature at which the transition occurs and whether the transition is exothermic or endothermic.

Closely related to DTA is *differential scanning calorimetry* (*DSC*). In this method the sample and reference material are also subjected to a closely controlled programmed temperature. In the event that a transition occurs in the sample, however, thermal energy is added to or subtracted from the sample or reference containers in order to maintain both sample and reference at the same temperature. Because this energy input is precisely equivalent in magnitude to the energy absorbed or evolved in the particular transition, a recording of this balancing energy yields a direct calorimetric measurement of the transition energy.

The information obtained from DTA and DSC techniques, coupled with thermomechanical analysis, X-ray diffraction patterns, and chemical analysis of residues and any evolved gases, provides a quantitative and qualitative estimation of solid-state reactions. Comparison of data can be made from successive runs utilizing various environmental conditions and pressures.

496

Instrumentation

Typical DTA equipment is illustrated in Fig. 17-1. The furnace contains a sample block with identical and symmetrically located chambers. Each chamber contains a centered thermocouple. The sample is placed in one chamber and a reference material, such as α-Al_2O_3, is placed in the other chamber. The furnace and sample block temperature are then increased at a linear rate, most often 5° to 12°C per minute, either by increasing the voltage through the heater element by a motor-driven variable transformer or by a thermocouple-actuated feedback type of controller. The difference in temperature between sample and reference (S, R) thermocouples, connected in series-opposition, is continuously measured. After amplification (about 1000 times) by a high-gain, low-noise, dc amplifier for the microvolt-level signals, the difference signal is recorded on the y-axis of a millivolt recorder. The temperature of the furnace is measured by a separate thermocouple which is connected to the x-axis of the recorder, frequently through a reference ice junction or room-temperature compensator. Because the thermocouple is placed directly in the sample, or attached to the sample container, the DTA technique provides the highest thermometric accuracy of all of the thermal methods. The area under the output curve, however, is not necessarily proportional to the amount of energy transferred in or out of the sample. If maximum calorimetric accuracy is desired, as in DSC,

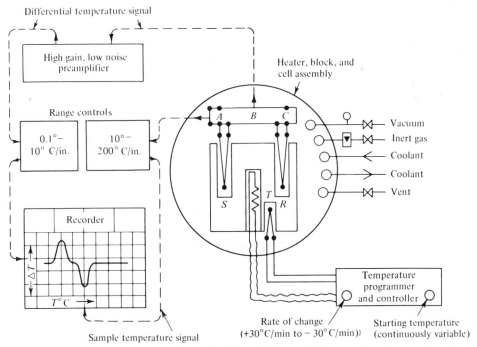

Fig. 17-1. Schematic diagram of the Du Pont differential thermal analyzer. (Courtesy of E. I. Du Pont de Nemours, Inc.)

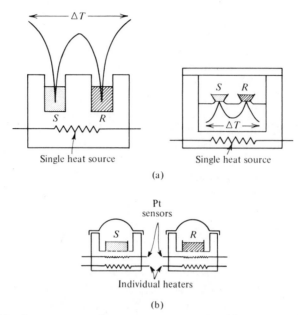

Fig. 17-2. Arrangement of temperature sensors in (a) DTA and (b) DSC.

the sample and reference thermocouples are removed from direct contact with the sample (see Fig. 17-2).

The temperature range is between $-190°$ and $1600°C$. Sample sizes range from 0.1 to 100 mg. Sensitivities down to $0.002°C$ in DTA, or 2 μcal/sec in DSC, are common. By using a suitable pressure container, these techniques may be extended to pressures up to 1000 psi and down to the 10^{-5} torr range.

THERMOGRAVIMETRIC ANALYSIS

Thermogravimetric analysis (TGA) provides the analyst with a quantitative measurement of any weight change associated with a transition. For example, TGA can directly record the loss in weight with time or temperature due to dehydration or decomposition. Thermogravimetric curves are characteristic for a given compound or system because of the unique sequence of physico-chemical reactions which occur over definite temperature ranges and at rates that are a function of the molecular structure. Changes in weight are a result of the rupture and/or formation of various physical and chemical bonds at elevated temperatures that lead to the evolution of volatile products or the formation of heavier reaction products. From such curves data are obtained concerning the thermodynamics and kinetics of the various chemical reactions, reaction mechanisms, and the intermediate and final reaction products.

In *differential thermogravimetry (DTGA)* the actual measurement signal appears as a derivative plot of the weight loss or gain which aids in the accurate assignment of the end and beginning points of overlapping reactions which often appear as combined peaks in TGA.

Instrumentation

For TGA the sample is continuously weighed as it is heated to elevated temperatures. Samples are placed in a crucible or shallow dish that is attached to an automatic-recording balance. The automatic null-type balance incorporates a sensing element which detects a deviation of the balance beam from its null position. One transducer is a pair of photocells, a slotted flag connected to the balance arm, and a lamp (Fig. 17-3). Once an initial balance has been established, any changes in sample weight cause the balance to rotate. This moves the flag so that the light falling on each photocell is no longer equal. The resulting nonzero signal is amplified and fed back as a current to a taut-band torque motor (the pivot-point of the balance) to restore the balance to equilibrium. This current is proportional to the weight change and is recorded on the y-axis of the recorder. Changes in mass can also be detected by contraction or elongation of a precision helical spring whose movement is detected by the movement of an attached core in a linear variable differential transformer (Fig. 17-4). With either type balance the sample container is mounted inside a quartz or pyrex housing which is located inside the furnace. Furnace temperature is continuously monitored by a thermocouple whose signal is applied to the x-axis of the recorder. Linear heating rates from $5°$ to $10°C$ per minute are generally

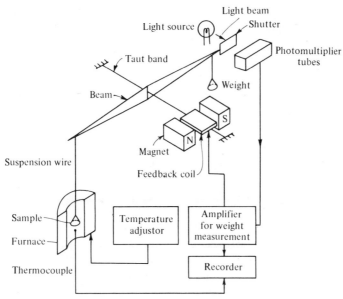

Fig. 17-3. Schematic diagram of TGA equipment with optical sensor. (Courtesy of Shimadzu Seisakusho, Ltd.)

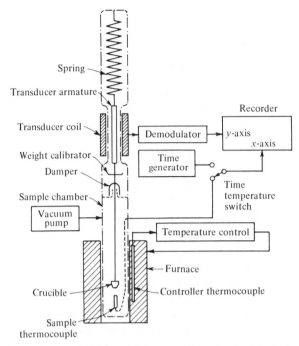

Fig. 17-4. Modular diagram of TGA equipment with spring and transducer coil as sensor. (Courtesy of American Instrument Co.)

employed. In differential thermogravimetry the actual measurement signal is derived from a solid-state resistance-capacitance circuit which uses the direct output of the electrical weight-change signal from the thermobalance for the primary signal input. The resulting output is the derivative, $\Delta w/\Delta t$, which is used in kinetic interpretations. The usual temperature range for TGA is from ambient to 1200°C. Sample sizes range from 1 to 300 mg, and sensitivities down to a few micrograms of weight change are common.

METHODOLOGY OF DSC (OR DTA) AND TGA

The weight-change curve for calcium oxalate monohydrate is shown in Fig. 17-5. Water is evolved beginning slightly above 100°C. At about 250°C a break is obtained in the curve at the stoichiometry corresponding to that of the anhydrous salt. Further heating gives definite weight plateaus for the carbonate (from 500° to 600°C) and finally the oxide (above about 870°C). Exact locations of the weight plateaus are dependent on the heating rate (a slower heating rate will shift values to lower temperatures) and the ambient atmosphere around the sample particles. The curve is quantitative in that calculations can be made to determine the stoichiometry of the compound at any given temperature.

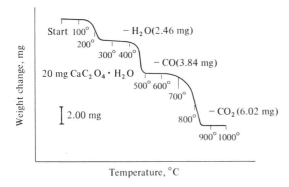

Start 100°   – H₂O(2.46 mg)
    200°
        300° 400°
                    – CO(3.84 mg)
20 mg CaC₂O₄ · H₂O
            500° 600°
                700°
    ⎡ 2.00 mg   – CO₂ (6.02 mg)
    ⎣   800°
                900° 1000°

Temperature, °C

Weight change, mg

Fig. 17-5. Thermogravimetric evaluation of calcium oxalate monohydrate; heating rate 6°C/min.

Thermal analysis will be affected by the experimental conditions. Deviations caused by instrumental factors include furnace atmosphere, size and shape of the furnace and sample holder, sample holder material and its resistance to corrosive attack, wire and bead size of the thermocouple junction, heating rate, speed and response of the recording equipment, and location of the thermocouples in the sample and reference chambers. Another set of factors influencing results depend on the sample characteristics; these include layer thickness, particle size, packing density, amount of sample, thermal conductivity of sample material, heat capacity, ease with which gaseous effluents can escape, and the atmosphere surrounding the sample. Details concerning these factors should be sought in the treatises cited in the Bibliography.

Thermogravimetric analysis, a valuable tool in its own right, is perhaps most useful when it complements differential thermal analysis studies. Virtually all weight-change processes absorb or release energy and are thus measurable by DTA or DSC, but not all energy-change processes are accompanied by changes in weight. This difference in the two techniques enables a clear distinction to be made between physical and chemical changes when the samples are subjected to both DSC (or DTA) and TGA tests.

In general, each substance will give a DSC or DTA thermogram whose number, shape, and position of the various endothermic and exothermic features serve as a means of qualitative identification of the substance. When an endothermic change occurs, the sample temperature lags behind the reference temperature because of the heat in the sample. On the thermogram, the initiation point for a phase change or chemical reaction is the point at which the curve first deviates from the baseline. When the transition is complete, thermal diffusion brings the sample back to equilibrium quickly. The peak (or minimum) temperature indicates the temperature at which the reaction is completed. When the break is not sharp, a reproducible point may be obtained by drawing one line tangent to the baseline and another tangent to the initial slope of the curve.

Various behaviors that may be deduced from a DSC thermogram are shown in Fig. 17-6. The heat capacity at any point is proportional to the displacement from the blank baseline. A broad endotherm indicates a slow change in heat capacity. A "second order or glass" transition, observed as a baseline shift (T_1), denotes a decrease in order within the system. Molecular segments begin to rotate and, in so doing, engender the flexibility and elastomer qualities of plastics, textiles, and greases. Endotherms generally represent

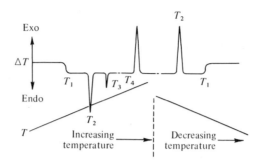

Fig. 17-6. DTA thermogram of a hypothetical substance contrived to illustrate the discussion.

physical rather than chemical changes. Sharp endotherms (T_3) are indicative of crystalline rearrangements, fusions, or solid-state transitions for relatively pure materials. Broader endotherms (T_2) cover behavior ranging from dehydration, temperature-dependent phase behaviors, to melting of polymers. Exothermic behavior (without decomposition) is associated with the decrease in enthalpy of a phase or chemical system. Narrow exotherms usually indicate crystallization (ordering) of a metastable system, whether it be supercooled organic, inorganic, amorphous polymer, or liquid, or annealing of stored energy resulting from mechanical stress. Broad exotherms denote chemical reactions, polymerization, or curing of thermosetting resins. Exotherms with decomposition can be either narrow or broad depending on kinetics of the behavior. Explosives and propellants are sharpest, and "unzipping" of polyvinylchloride is rapid, while oxidative combustion and decomposition are generally broad.

On cooling one would expect the reverse of features observed on the heating cycle (see Fig. 17-6). Since T_4 is not found to recur on cooling, the reaction is obviously non-reversible (perhaps a pyrolytic decomposition). Instead of taking the system up to T_4, let us begin the cooling cycle before that temperature. As it cools, the substance is seen to lose its transition peak at T_3. Judging from the area under the T_2 peak, the transition energy of T_3 has been added to T_2. This indicates a metastable condition at T_3, the retained energy being released in one large step at a lower temperature. Further along the cooling curve, the glass transition at T_1 falls properly into place to complete the cycle.

Unravelling the significance of thermograms is not always straightforward. A reference library of thermograms of interest to a particular laboratory is vital. Temperature data on commercial products or melting points for pure substances reported in the literature are of little value when comparison with a dynamically scanned thermal profile is desired. Complementary techniques are valuable. Establishing whether a gaseous product is evolved at a corresponding DTA or DSC transition, and its identification, often assists in elucidating the decomposition route. Gas chromatographs and mass spectrometers[3] can be coupled to thermal analysis equipment for repetitive analysis of gaseous decomposition products. Analysis of the evolved gases by chemical means is also possible. Thermal decompositions in inert, oxidative, or special atmospheres provide clues through changes in the thermograms or displacement of thermic features.

Example 17-1

The TGA and DTA thermograms of manganese hypophosphite monohydrate are shown in Fig. 17-7. The weight-loss data (TGA curve) from a 200-mg sample run under vacuum and with the analysis of effluent gases showed the loss of one mole of water at $150°C$, one mole of phosphine at $360°C$, and the slow loss of another mole of water starting around $800°C$. In comparison with the DTA curve, two major peaks remain unidentified: the large exotherm at $590°C$ and the endotherm at $1180°C$, plus several smaller thermic features. Thermogravimetric data obtained from runs performed under vacuum and in a nitrogen atmosphere failed to show any loss associated with these peaks. Each sample was measured for its real density; the resulting data are shown on the DTA curve. Undoubtedly the sharp DTA exotherm at $590°C$ represents a phase change. The relatively small endotherm starting above $900°C$ must represent a recrystallization exotherm following the elimination of water which is superimposed on the latter endotherm. The peak at $1180°C$ is due to melting. With this information the thermal decomposition reactions and phase changes are:

$$Mn(H_2PO_2)_2 \cdot H_2O(s) \rightarrow Mn(H_2PO_2)_2(s) + H_2O(g)$$

$$Mn(H_2PO_2)_2(s) \rightarrow MnHPO_4(s) + PH_3(g)$$

$$\alpha\text{-}MnHPO_4(s) \rightarrow \beta\text{-}MnHPO_4(s)$$

$$2MnHPO_4(s) \rightarrow Mn_2P_2O_7(s) + H_2O(g) \text{ (and recrystallization)}$$

$$Mn_2P_2O_7(s) \rightarrow Mn_2P_2O_7(l)$$

Example 17-2

The thermograms for calcium acetate monohydrate (Fig. 17-8) illustrate the influence of different atmospheres. The first endothermal peak is unaffected by change in atmosphere. The weight-loss data indicate loss of water and thus conversion of the monohydrate to the anhydrous salt. The next feature on the DTA curve is an endotherm in CO_2 and Ar, but an exotherm in an O_2 atmosphere; the weight loss corresponds to one mole each of CO_2 and CO. In an oxygen atmosphere the highly exothermic reaction must be the oxidation of carbon monoxide. The final stage is the decomposition of calcium carbonate to calcium oxide, which is a function of the partial pressure of CO_2 in contact with the sample and consequently is shifted to higher temperatures in the CO_2 atmosphere.

Thermal studies with polymers can predict a product's performance in use; that is, its stiffness, toughness, or stability. Melting-point, phase-transition, pyrolysis, and curing temperatures can be accurately measured. Once a polymer has been broadly classified by other methods, thermograms often can be used to establish, by comparison with known reference materials, the degree of polymerization, the thermal history of the sample, crystal perfection and orientation, the effect of different coreactants and catalysts, the percentage of crystalline polymer, and the extent of chain branching. For example, thermograms for a low molecular weight, nonlinear, branched-chain polymer will show a continuous series of rather broad and low-melting endotherms whereas a high molecular weight, stereo-regular, linear polymer will reveal a single narrow and higher-melting endotherm. If a polymer has been incompletely cured, the heating cycle may reveal an exotherm at a temperature close to the one employed for the polymerization reaction. An

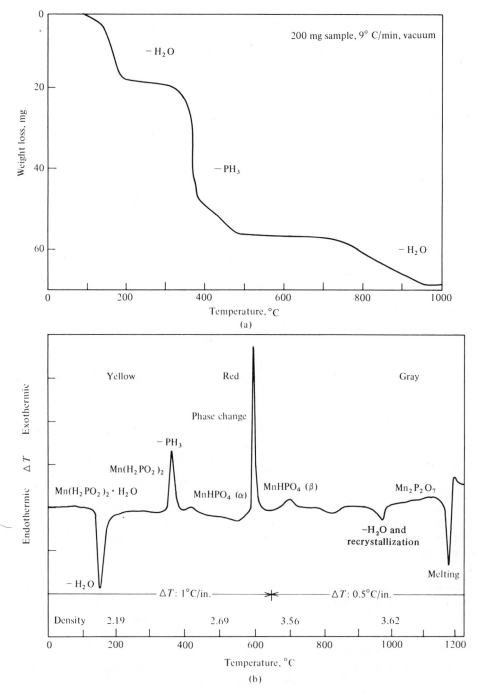

Fig. 17-7. (a) TGA and (b) DTA curves for $Mn(H_2PO_2)_2 \cdot H_2O$. (Courtesy of American Instrument Co.)

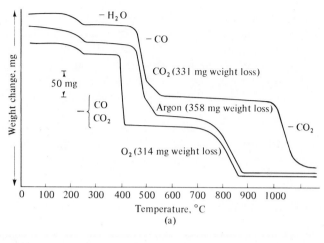

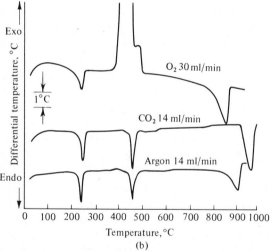

Fig. 17-8. Thermograms for calcium acetate monohydrate. (a) TGA, 6°C/min. (b) DTA, 12°C/min. Particular atmosphere above the solid phase is indicated for each thermogram. (Courtesy of American Instrument Co.)

exotherm occurring just below the melting temperature can indicate "cold crystallization," which results if a sample is quenched quickly after being melted. On reheating, crystallites form rapidly and exothermically just prior to remelting of the polymer. Annealing temperatures are similarly revealed as exotherms.

If the molecular weight or density of a polymer has been established by appropriate (often lengthy) methods, subsequent determination of its melt temperature (a 15-min process) can be related to molecular weight or density. Product quality can be maintained subsequently by simply examining thermograms of polymer materials to obtain molecular weights or densities from an appropriate calibrated graph.

Instead of using the traditional method of preparing a derivative from the organic sample and a reagent, the sample can be heated with a specific reagent at a programmed heating rate in a selected atmosphere. The DTA or DSC curve will show the derivative-forming reaction, the physical transitions of the sample or reagent (whichever is in excess), and the physical transitions of the intermediates and final products. When one reactant is volatile and in excess, a rerun of the thermogram will usually show only the derivative characteristics.[1]

The area of exotherms or endotherms can be used to calculate the heat of the reaction or the heat of a phase transition. Suitable calibration is necessary with DTA equipment, but the values are given directly with DSC instruments.

THERMOMECHANICAL ANALYSIS

Thermomechanical analysis (*TMA*) provides measurements of penetration, expansion, contraction, and extension of materials as a function of temperature. Typical apparatus, diagrammed in Fig. 17-9, consists of a probe connected mechanically to the core of a linear variable differential transformer (LVDT). The core is coupled to the sample by means of a quartz probe containing a thermocouple for measurement of sample temperature. Any movement of the sample is translated into a movement of the transformer core and results in an output that is proportional to the displacement of the probe, and whose sign is indicative of the direction of movement. The temperature range is from that of liquid nitrogen to 850°C.

In the penetration and expansion modes, the sample rests on a quartz stage surrounded by the furnace. Under no load, expansion with temperature is observed. Calculation of the thermal coefficient of linear expansion may be made directly from the slope of the resulting curve. A weight tray attached to the upper end of the probe allows a predetermined force to be applied to the sample to study variations under load. Probes of small tip diameter and a loaded weight tray are used when the sensitive detection of softening temperatures, heat distortion temperatures, and glass transitions are of interest. Larger tip diameters and zero loading are used in the expansion mode when coefficients of expansion and dimensional changes due to stress relief are the objects of investigation. Sample sizes may range from a 0.1-ml coating to a 0.5-in. thick solid. Sensitivities down to a few microinches are observable.

For measurement of samples in tension, sample stage and probe are replaced by a sample holder system consisting of a stationary and movable hook constructed of fused silica. This permits extension measurements on films and fibers. Holes about 0.6 cm apart are punched into injection-molded pieces or solution-cast or extruded films; also a fiber fused into a loop can be used for this test. The double-hook probe is designed to grasp a pair of aluminum spheres that are crimped onto either end of a fiber sample. Measurements made with these probes can be related to the tensile modulus of a sample.

Volume expansion characteristics of samples are measured by placing the sample in a quartz cylinder fitted with a flat-tipped quartz probe in a cylinder-piston arrangement. Sample volume changes are translated into linear motion of the piston.

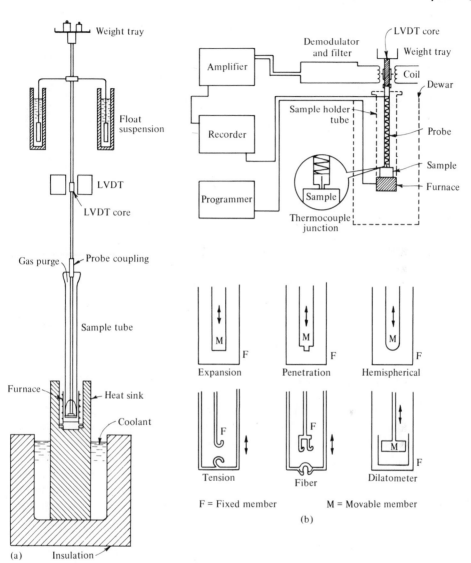

Fig. 17-9. (a) Thermomechanical analyzer (Courtesy of Perkin-Elmer Corp.). (b) Probe configurations. (Courtesy of E. I. Du Pont de Nemours, Inc.)

Example 17-3

The expansion and heat capacity behavior of herring oil is shown in Fig. 17-10. The expansion characteristics show changes at $-108°$, $-80°$, $-47°$, $-27°$, and $-15°C$, at which point the material is apparently fluid. These volume changes confirm the changes in heat capacity measured by DSC at $-106°$, $-53°$, $-26°$, and $-13°C$, and emphasize the need for using more than one mode of thermal analysis to illustrate the thermal response of a system.

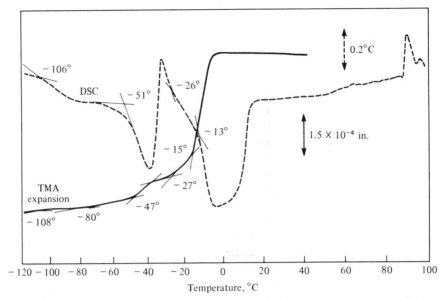

Fig. 17-10. Thermomechanical (expansion) behavior and differential scanning calorimetry of herring oil.

ELECTRICAL THERMAL ANALYSIS

Electrical thermal analysis (ETA) measures the electrical conductivity of a material as a function of temperature. In the method a selected voltage is applied across two electrodes immersed in a sample. The resulting current is amplified, converted into a linear signal, and plotted on the y-axis of a recorder while the temperature is plotted on the x-axis. The method finds broad application in the thermal analysis of electrical insulating materials and solid-state electronics. In some applications, ETA is more sensitive than DTA, as for example, detecting desorption of gases. Electrical thermal analysis is most useful in polymer studies, gas absorption on surfaces, moisture and plasticizer content, and trace amounts of impurities in polymer systems.

TOTAL ORGANIC CONTENT ANALYZER

In this system a high-temperature furnace is coupled closely to a high-temperature flame ionization detector and associated electronic and pneumatic systems needed for the control and measurement of the effluent gases from the sample boat (Fig. 17-11). The sample boat, attached to a thermocouple probe, is initially held in a cool portion of the furnace to allow the furnace to be swept by a gas purge prior to the start of the temperature program. After the initial purge, the sample boat is moved into the heated portion of the furnace. Gases evolved from the sample are swept immediately into the flame

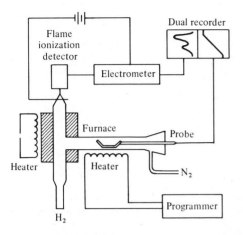

Fig. 17-11. Total organic content analyzer. (Courtesy of Carle Instruments, Inc.)

ionization detector, which is maintained at a temperature above the final oven temperature. Temperature may be programmed through any desired range from ambient to 530°C, or maintained at any selected isothermal value. Carbon evolution rates as low as 10^{-9} g/min can be detected. Approximately 1 mg of sample is sufficient for determining decomposition rates of polymers down to rates as low as 0.01%/min. Reweighing the sample boat after a pyrolysis run gives the residue, which in turn can be exposed to an oxidizable atmosphere and again reweighed to show the oxidizable residue. Thus this technique can be used for determining the amount of oxidizable and nonoxidizable inorganic materials used as fillers in polymeric materials. The differential properties of the flame ionization detector permit the determination of trace amounts of organic material in natural surface waters, cooling tower water, and condensate water (for leakage of organic materials). Another area of application is the determination of vapor pressure and vapor pressure curves of organic compounds. The analyzer has particular application for materials that cannot be analyzed by gas-liquid chromatography because the fractions are too high boiling or too unstable to be completely eluted from the column.

THERMOMETRIC TITRIMETRY

Thermometric titrimetry and *titration calorimetry* are techniques in which the temperature of a system is measured as a function of titrant added. The resultant temperature–volume curve is similar to other linear titration curves.

Instrumentation

The equipment (Fig. 17-12) consists basically of a motor-driven automatic buret, an adiabatic titration chamber, a thermistor and Wheatstone bridge circuit, and a strip-chart recorder. To minimize heat transfer between the solution and its surroundings, the titrations are performed under as near adiabatic conditions as possible in an insulated beaker

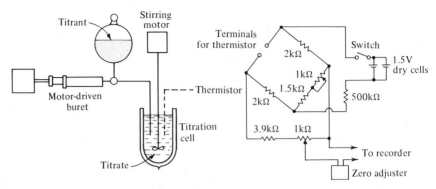

Fig. 17-12. Schematic titration assembly and bridge circuit for conducting thermometric titrations. (After H. W. Linde, L. B. Rogers, and D. N. Hume, *Anal. Chem.*, **25**, 494 (1953). Courtesy of American Chemical Society.)

or Dewar flask of 100- to 250-ml capacity that is closed with a stopper provided with holes for the buret tip, a glass stirrer, and the thermistor. The titrant is delivered at flow rates of 0.1 to 1.0 ml/min. To obviate volume corrections and to minimize temperature variations between the titrant and sample, the titrant concentration is usually 100 times larger than that of the reactant. Amounts of sample are selected so that a volume of titrant not exceeding 1–3 ml is required.

Because temperature changes in the course of a titration range between 0.01° and 0.2°C, the accuracy of the temperature measurement must be about 10^{-4}°C. For a thermistor having a resistance of 2 kΩ and a sensitivity of -0.04 deg^{-1} Celsius in the 25°C temperature range, a change of 0.01°C corresponds to an imbalance potential of 0.157 mV. Temperatures of titrant and sample should be within 0.2°C before a titration is begun. A small heating element, located inside the titration vessel, can be used to warm the sample to the temperature of the titrant or as a calibrating device when estimating heats of reaction or mixing.

In a differential thermometric apparatus, temperature-sensing elements are placed in both the sample and blank (pure solvent plus titrant) solutions. Sensitivity is improved and extraneous heat effects, such as stirring and heats of dilution, are minimized.[11]

Methodology

Contrary to potentiometric titrations of various types that depend solely on equilibrium constants and, hence, the free energy of the reaction, $\Delta G°$, or

$$-\Delta G° = RT \ln K \tag{17-1}$$

thermometric titrations depend only on the heat of the reaction, ΔH, or

$$\Delta H = \Delta G + T \Delta S \tag{17-2}$$

Thus, a thermometric titration may be feasible when all "free energy" methods fail. This point is clearly shown in Fig. 17-13, where a comparison is given of the potentiometric

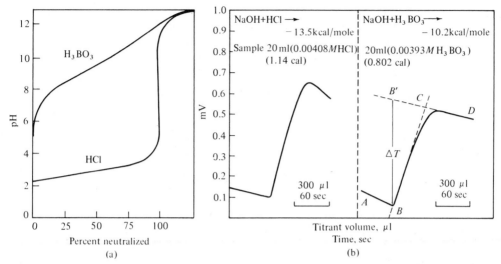

Fig. 17-13. (a) Potentiometric and (b) thermometric titration curves for hydrochloric and boric acids with $0.2610\,M$ sodium hydroxide.

and thermometric titration curves for HCl and H_3BO_3. In contrast to the potentiometric curve, the thermometric titration curve has a well-defined end point for the weak acid. The change in temperature of the titration is dependent on the heat of reaction of the system, according to the equation

$$\Delta T = \frac{N \, \Delta H}{Q} \qquad (17\text{-}3)$$

where N represents the number of moles of water formed by neutralization, ΔH is the molar heat of neutralization, and Q is the heat capacity of the system. In practice, ΔH and Q are constant throughout the titration so that ΔT is proportional to N.[4]

On the thermometric titration curve shown in Fig. 17-13, point A is where the temperature readings were begun, and line AB is a trace of the temperature of the solution before the addition of titrant. If the line AB shows a marked slope, it is an indication of excessive heat transfer between the solution and its surroundings. At point B the addition of titrant was begun; line BC shows the gradual evolution of the heat of reaction. Point C is the end point. Line CD may slope downward or upward. The linear portions of the curve are extrapolated to give the initial and equivalence points, and the distance between them is measured along the volume (or time) axis of the chart to ascertain the volume of titrant consumed in the reaction. The vertical line BB' is the temperature difference (ΔT) used to evaluate enthalpies (Eq. 17-3).

Applications

Applications of thermometric titrimetry include the determination of the concentration of an unknown substance, determination of reaction stoichiometry, and the determination of the thermodynamic quantities: ΔG, ΔH, and ΔS. The first application is perhaps the most useful to the analytical chemist. Precision and accuracy of measurements depend largely on enthalpy of the reaction involved, and range from 0.2 to 2%. About 0.0001 M is the lower limit of concentration that can be successfully titrated in the more favorable cases.

All acids with $K_a \geqslant 10^{-10}$ can be titrated thermometrically in 0.01 M solution with a precision of 1% if the heat of neutralization is 13 ± 3 kcal/mole. The extension to acids too weak to titrate potentiometrically is clearly demonstrated by the curves in Fig. 17-13. Good end points are obtained for other weak acids and bases, even in emulsions and thick slurries.

Nonaqueous systems are well suited for thermometric titrations, although attention must be paid to the heat of mixing of solvents and dilution. The lower specific heat of many organic solvents introduces a favorable sensitivity factor. Under strictly anhydrous conditions, even diphenylamine, urea, acetamide, and acetanilide are readily titratable with perchloric acid in glacial acetic acid.[8] Lewis bases, such as dioxane, morpholine, pyridine, and tetrahydrofuran, have been titrated with the Lewis acid $SnCl_4$ in the solvents CCl_4, benzene, and nitrobenzene.[2]

Thermometric titrations are very useful in titrating acetic anhydride in acetic acid–sulfuric acid acetylating baths, water in concentrated acids by titration with fuming acids, and free anhydrides in fuming acids. In fact, methods based on heats of reaction offer one of the few approaches to the analysis of concentrated solutions of these materials.[10]

Good results can be obtained in precipitation and ion-combination reactions such as the halides with silver or mercury(II), and cations with EDTA and oxalate. Silver titration of halides has been done at elevated temperatures in molten salts.[7]

When the titration reaction is appreciably incomplete in the vicinity of the equivalence point, actual titration curves exhibit curvature from which equilibrium constants and corresponding free energies can be calculated. The temperature rise that occurs during an exothermic reaction can be used to determine constituents. For example, benzene has been determined rapidly and with good precision in the presence of cyclohexane by measuring the heat of nitration when a standard nitrating acid mixture is added to the sample; the temperature rise is a direct function of the benzene present. In a similar manner, heats of reaction have been used to estimate the heats of successive steps in the formation of metal-ammine complexes,[9] the heats of chelation,[5,6] and heats of reaction in fused salts under virtually isothermal conditions.[7]

LABORATORY WORK

General Procedure for Thermometric Titrations

Motor-driven burets, 5 ml in capacity, with delivery rate of about 0.01 ml/sec, are suitable. However, other size burets and different delivery rates can be employed. Efficient stirring permits a titrant addition rate up to 10 ml/min.

The solution of the sample (and blank or solvent in differential work) is placed in suitable-size Dewar flasks or polyethylene cups mounted in Styrofoam plastic and suspended in a rigid metal framework.

With the sample container in position, each container should enclose a temperature-sensing element, a capillary buret tip, and the glass rod of a motor-driven stirrer, all of which are installed in holes drilled through the cover of the container. The cover should fit snuggly into the container.

Suitable thermistors will have characteristics similar to these specifications: cold resistance at 25°C, 2000–4000 Ω; temperature coefficient at 25°C, -4.0% °C; thermal time constant in still water, about 1 sec. Several thermistors connected in parallel enable one to utilize a lower input impedance recorder and still have the advantage of the high temperature coefficient of high-resistance thermistors.

Experiment 17-1   Determination of Quantity of Reactant Present

Place a 50- to 75-ml portion of a sample solution in the container and place the container in its insulated compartment. Insert the titration assembly, buret, thermistor, and stirring rod. Adjust the bucking voltage (zero adjust) until a suitable base line is obtained on the recorder. Start the chart drive and the buret about 10 sec later. About 10 sec after the titration is complete, as indicated by a change in slope of the chart recording, stop the buret. Suggested systems include 0.01 to 0.1 N solutions of any acid or base titrated with an appropriate titrant whose concentration is about 10- to 100-fold greater.

Experiment 17-2 ΔH Determination

The procedure, up to this point, is similar to that in the preceding experiment. With the bucking circuit, readjust the base line to a suitable level on the recorder. Turn on a heating coil immersed in the solution. After a temperature change similar to that observed in the titration has occurred, turn off the heating coil. Adjust the base line again and make another recording of the heating rate. During the recording of each heating curve, the voltages across the heating coil and a standard resistor are measured with a potentiometer.

The product of the two measured voltages, when divided by the resistance of the standard resistor, gives the heating rate of the coil in joules per second. From the slopes of all heating curves and of the titration curve, the heat of the reaction can be calculated from the equation,

$$\Delta H = \frac{E_1 E_2 S_1}{(4.185) R \, NFS_2} = \text{calories per milliequivalent}$$

where E_1 and E_2 are the voltages across the heating coil and standard resistor, respectively, R is the resistance of the standard resistor, N is the normality of titrant whose flow rate F is in milliliters per second, and S_1, S_2 are slopes of titration heating curves.

Experiment 17-3   Benzene in Cyclohexane by Heat of Reaction

Assemble the titration apparatus and electrical circuit described in Experiment 17-1. Omit the buret. Secure a stopwatch or timer.

Prepare the nitrating acid by mixing 2 volumes of 70 percent nitric acid ($d = 1.41$) with 1 volume of 95% sulfuric acid ($d = 1.82$).

1. Weigh 50 g of sample into a 4-oz Bakelite screw-cap bottle. Place the bottle containing the sample and a bottle containing the nitrating acid in a thermostat at about $20°C$ until the contents have attained an equilibrium temperature.
2. Transfer 50 ml of nitrating acid to the Dewar flask and insert the motor stirrer. Wait 3 min and then start the motor stirrer. After 1 min, record the initial temperature.
3. Stop the motor. Pour the sample into the flask and start the stirrer. Take readings of the temperature after an elapsed time of 1, 2, 3, and 5 min.
4. Construct a calibration curve of temperature rise in a 3-min interval vs percent benzene in cyclohexane. Run pure cyclohexane and standards containing 0.5–5.0 percent benzene by weight.

Note A thermometer graduated to $0.1°C$ is adequate for temperature measurements of the range of benzene contents suggested. Lower limits could be achieved through the use of the thermistor circuit.

PROBLEMS

1. Formulate the solid state reaction of sodium bicarbonate when heated. It decomposes between $100°$ and $225°C$ with the evolution of water and carbon dioxide. The combined loss of water and carbon dioxide totaled 36.6% by weight, whereas the weight loss due to carbon dioxide alone was found to be 25.4%.

2. A definite relation exists between decomposition temperature of $CaCO_3$ and equilibrium partial pressure of CO_2. A series of thermograms were obtained with a dynamic flow of CO_2 in the pressure range from 40 to 600 torr. Since pure CO_2 was used, the partial pressure is equivalent to the system pressure. Estimate the heat of dissociation from the following data:

Initial decomposition temperature, $°C$	926	895	840	802	759	749
Pressure CO_2, torr	600	400	200	100	50	40

3. Ascertain the glass transition of a polycarbonate resin from the following heat capacity measurements:

Temperature Range, $°K$	Specific Heat	Temperature Range, $°K$	Specific Heat
400.0–402.5	0.345	412.5–415.0	0.373
402.5–405.0	0.349	415.0–417.5	0.385
405.0–407.5	0.355	417.5–420.0	0.417
407.5–410.0	0.361	420.0–422.5	0.449
410.0–412.5	0.367		

4. The heat of fusion of a mixture of AgCl–AgBr can be used for the analysis of Cl–Br mixtures because these ions form ideal solid solutions in all proportions. ΔH_{fusion} (in cal/g) was found to be 12.1 for pure AgBr and 22.0 for pure AgCl. What weight percent of AgCl is present in a mixture which has the following values of heat of fusion: (a) 14.4, (b) 20.0, (c) 16.9, (d) 16.05, and (e) 19.6?

5. On a DSC thermogram obtained with 10.2 mg of dotriacontane and an external 12.1-mg standard of indium whose ΔH_{fusion} = 6.8 cal/g, the following areas, as measured by a planimeter, were obtained: chain rotation (65°C), 158 units; fusion of dotriacontane (72°C), 439 units; and fusion of indium, 93 units. Calculate the transition energies of dotriacontane.

6. A mixture of 95% Ar and 5% O_2 was passed through a DTA oven which was heated at 10°C/min. A sample of 1.000 g of UO_2 registered an exotherm at 360°C whose peak area was 25.6 cm². When a current of 2.1 A at 3.6 V was passed for 30 sec, a calibration peak of 15.6 cm² was obtained. Determine the energy liberated in the reaction: $3UO_2 + O_2 \rightarrow U_3O_8$.

7. In the accompanying figure curve A is the weight-loss thermogram from pure $CaCO_3$, curve C shows a similar trace from $MgCO_3$, and curve B is the thermogram of a limestone sample. (a) Derive an expression for the direct quantitative analysis of CaO and MgO. (b) Write equations for the solid-state decomposition of $MgCO_3$. (c) Calculate the percent CaO and MgO in the limestone sample.

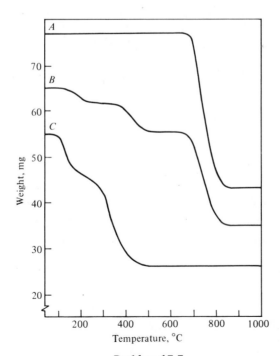

Problem 17-7

8. The decomposition reactions of a 100-mg sample of nickel oxalate dihydrate vary in different atmospheres. In both flowing and stationary air, successive weight losses of 19 and 39 mg were observed. However, in flowing CO_2 and in flowing N_2 the successive weight losses were 19 and 49 mg. The same temperature program was employed in the four runs. Write the decomposition reactions.

9. In fused lithium nitrate–potassium nitrate at 158°C, the shape of the titration curve obtained in 8.6×10^{-4} molal solution of potassium chloride with 1.40 molal silver nitrate showed that precipitation at the equivalence point was about 20% incomplete. In contrast, precipitation of 1.17×10^{-2} molal solution of KCl was 98.5% complete. Estimate the molal solubility product of AgCl in the eutectic salt melt.

10. Demonstrate, by discrete ionization and neutralization steps and suitable calculations, why an acid as weak as boric acid gives an adequate thermometric titration curve. For boric acid neutralization, ΔS is -31.1 e.u.

11. Estimate the values of ΔH and sketch the hypothetical titration curve for a mixture of calcium and magnesium ions titrated with EDTA. Thermodynamic characteristics at 25°C of chelation equilibria with EDTA are:

Cation	$pK_{stability}$	$\Delta S°$, e.u.
Ca^{2+}	-11.0	$+31$
Mg^{2+}	-9.1	$+60$

12. From the TG, DTG, and DTA curves shown in the figure, deduce the decomposition route for NH_4VO_3.

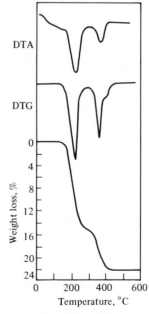

Problem 17-12

13. A simultaneous DTA–TGA curve for manganese(II) carbonate in a porous crucible is shown (solid lines). (a) What are the transitions involved at each peak on the DTA trace, and what are the products at each TGA plateau? (b) Another laboratory, using a controlled atmosphere with 13 atm CO_2, obtained the curves shown (dashed line). Why is the initial oxide different?

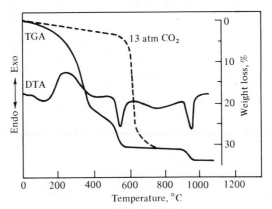

Problem 17-13

14. Contrast the two sets of DTA–TGA curves in the figure for the decomposition of $CaSiF_6 \cdot 2H_2O$ and $ZnSiF_6 \cdot 6H_2O$. Express the solid-state decomposition for both fluorosilicates.

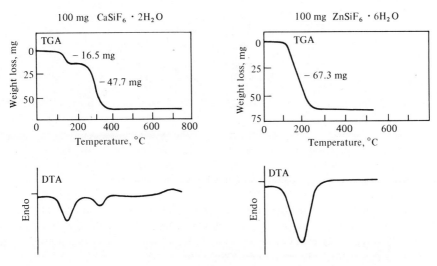

Problem 17-14

15. Using the figure write the solid-state reactions involved in the thermal decomposition of sodium hydrogen sulfate monohydrate and its intermediate decomposition products. Deduce the phase transitions.

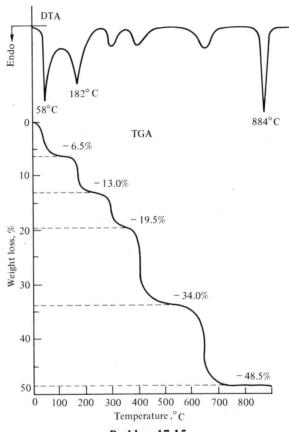

Problem 17-15

16. From the thermomechanical penetration curve shown, deduce the nature of the two transitions.

17. From the penetration and expansion measurements shown in the figure, (a) deduce the nature of the two transitions and (b) estimate the thermal coefficient of linear expansion for the linear segments of the TMA expansion curve.

18. Three successive runs on the same sample of a fiber-glass mat impregnated with an uncured epoxy resin are shown. The scan in run A was stopped at 90°C, and the sample cooled and rerun (run B). Run C is the sample from run B after cooling. Discuss the features observed in the DSC scans.

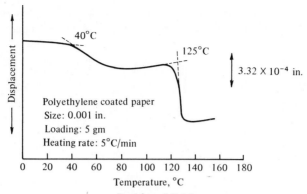

Problem 17-16

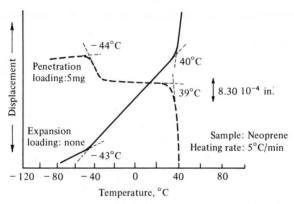

Problem 17-17

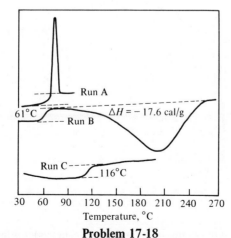

Problem 17-18

19. The figure shows the micro DTA thermograms of the two-phase system involving picrylchloride and hexamethylbenzene; construct the phase diagram.

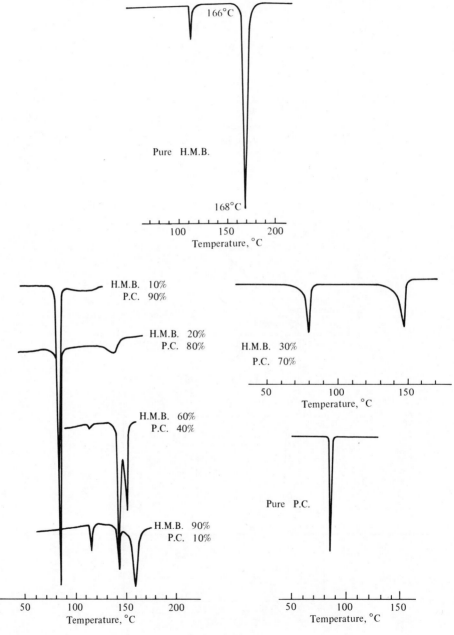

Problem 17-19. DTA thermograms of picrylchloride–hexamethylbenzene systems. Experimental data: sample, 3.2 gm; atmosphere, air; heating rate, 5°C/min.

20. Calculate the specific heats of the thoriated nickel, Zytel 61 (nylon), and gold from their heat capacity curves and that of the sapphire standard reference material, a disc-shaped sample whose specific heat at 360°K is 0.212.

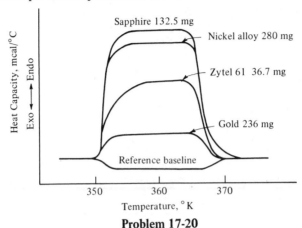

Problem 17-20

BIBLIOGRAPHY

Duval, C., *Inorganic Thermogravimetric Analysis*, 2nd ed., American Elsevier, New York, 1963.

Garn, P. D., *Thermoanalytical Methods of Investigation*, Academic Press, New York, 1965.

Mackenzie, R. C., Ed., *Differential Thermal Analysis*, Academic Press, London, 1970.

Schwenker, R. F., Jr. and P. D. Garn, Eds., *Thermal Analysis*, Vol. 1: *Instrumentation, Organic Materials and Polymers*, 1969; Vol. 2: *Inorganic Materials and Physical Chemistry*, 1968, Academic Press, New York.

Wendlandt, W. W., *Thermal Methods of Analysis*, Wiley-Interscience, New York, 1964.

LITERATURE CITED

1. Chiu, J., *Anal. Chem.*, **34**, 1841 (1962).
2. Cioffi, F. J. and S. T. Zenchelsky, *J. Phys. Chem.*, **67**, 357 (1963).
3. Gohlke, R. S. and H. G. Langer, *Anal. Chem.*, **37** (10) 25A (1965); ibid., **37**, 433 (1965).
4. Jordan, J., *Record of Chem. Prog.*, **19**, 193 (1958).
5. Jordan, J. and T. G. Alleman, *Anal. Chem.*, **29**, 9 (1957).
6. Jordan, J., J. Meir, E. J. Billingham, and J. Pendergrast, *Anal. Chem.*, **31**, 1439 (1959).
7. Jordan, J., J. Meir, E. J. Billingham, and J. Pendergrast, *Anal. Chem.*, **32**, 651 (1960).
8. Keily, H. J. and D. N. Hume, *Anal. Chem.*, **36**, 543 (1964).
9. Poulsen, I. and J. Bjerrum, *Acta Chem. Scand.*, **9**, 1407 (1955).
10. Somiya, T., *J. Soc. Chem. Ind.* (Japan), **32**, 306, 490 (1929).
11. Tyson, B. C., W. H. McCurdy, and C. E. Bricker, *Anal. Chem.*, **32**, 1640 (1961).

18

Gas Chromatography

Gas–liquid chromatography accomplishes a separation by partitioning a sample between a mobile gas phase and a thin layer of nonvolatile liquid held on a solid support. *Gas–solid chromatography* employs a solid adsorbent as the stationary phase.[*]

The sequence of a gas chromatographic separation is as follows. A sample containing the solutes is injected into a heating block where it is immediately vaporized and swept as a plug of vapor by the carrier gas stream into the column inlet. The solutes are adsorbed at the head of the column by the stationary phase and then desorbed by fresh carrier gas. This partitioning process occurs repeatedly as the sample is moved toward the outlet by the carrier gas. Each solute will travel at its own rate through the column, and consequently a band corresponding to each solute will form. The bands will separate to a degree that is determined by the partition ratios of the solutes and the extent of band spreading. The solutes are eluted, one after another, in the increasing order of their partition ratios and enter a detector attached to the column exit. If a recorder is used, the signals appear on the chart as a plot of time versus the composition of the carrier gas stream. The time of emergence of a peak identifies the component, and the peak area reveals the concentration of the component in the mixture. Although the gas chromatographic method is limited to volatile materials (about 15% of all organic compounds), the availability of gas chromatographs working at temperatures up to 450°C, pyrolytic techniques, and the possibility of converting many materials into a volatile derivative extend the applicability of the method.

GAS CHROMATOGRAPHS

Basically a gas chromatograph consists of six parts: (1) a supply of carrier gas in a high-pressure cylinder with attendant pressure regulators and flow meters, (2) a sample injection system, (3) the separation column, (4) the detector, (5) an electrometer and strip-chart recorder (and integrator, perhaps), and (6) separate thermostated compartments for housing the column and the detector so as to regulate their temperature. The components are shown schematically in Fig. 18-1.

Helium is the preferred carrier gas, but it is not readily obtainable in some countries. Nitrogen is an alternative.

[*]Liquid chromatography is discussed in J. A. Dean, *Chemical Separation Methods*, D. Van Nostrand, New York, 1969.

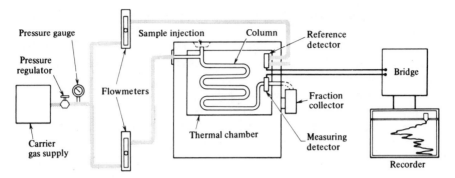

Fig. 18-1. Schematic of a gas chromatograph.

Sample Injection System

The most exacting problem in gas chromatography is presented by the sample injection system. The sample must be introduced as a vapor in the smallest possible volume and in a minimum amount of time without either decomposing or fractionating the sample, or upsetting the equilibrium conditions of the column. Both quantity of sample introduced and the manner of introduction must be reproducible with a high degree of precision.

Liquid samples are usually injected by a microsyringe through a self-sealing rubber septum into a metal block that is heated by a controlled-resistance heater. Here the sample is vaporized as a "plug" and carried into the column by the carrier gas stream (Fig. 18-2). Because the entire sample must undergo instantaneous vaporization to attain plug flow, the injection-zone temperature must exceed the boiling points of all components.

Although gas samples can be injected by a gas-tight syringe, the most accurate and precise method for gas samples uses calibrated sample loops and switching valves. Loops are available in volumes ranging from several microliters to several milliliters. A T-shaped injection port permits syringe injection of a liquid sample into a sealed hot zone. After a short period of time to allow for complete vaporization, the valves are switched to sweep the sample from the hot zone into the column.

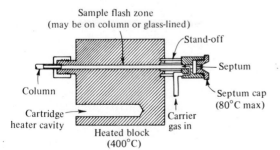

Fig. 18-2. Cross section of sample inlet system.

Wall-coated open tubular columns can accomodate only thousandths of a microliter of sample. To obtain the desired sample size, a sample splitting device is placed between the injection block and column. A typical system uses a 0.5 μl injection split 100 to 1, where 1 part enters the column and 99 parts are exhausted to the atmosphere.

Pyrolysis has become a widely accepted method for handling solids. In the tube-type pyrolyzer, the sample is placed in a sample boat that is positioned in the center of the tube. The tube is then sealed and pulse-heated electrically at a controlled rate of temperature rise to the desired final temperature. Then the tube is opened and the vapors swept into the column with the carrier gas stream. Operation, based on the Curie principle, utilizes filaments of certain alloys, which when subjected to intense rf energy, rapidly heat to a specific temperature unique to the alloy. Rise times to temperatures as high as 800°C are on the order of nanoseconds. A constant flow of carrier gas is maintained over the sample. Because the filaments must be coated with the sample, this technique is restricted to material that will coat out of solution.

Chromatographic Column

Two basic types of chromatographic columns are in general use: packed and open tubular columns. Packed columns are tubes that have been filled with an inert support coated with a nonvolatile liquid phase for use in gas–liquid chromatography, or the filling may be an adsorbent or molecular sieve in gas–solid chromatography. Open tubular columns differ from packed columns in that the gas path through the column is an unrestricted hole through the tubing and the separating medium is coated on the wall of the tubing. The pressure drop is orders of magnitude smaller than that of packed columns of the same length; this permits the use of very long columns. Packed columns normally are used in lengths of 0.7 to 2 m; open tubular columns run anywhere from 30 to 300 m in length. Columns are made of tubing coiled into an open spiral or a flat pancake shape.

The separating ability per meter of an open tubular column does not differ greatly from that of packed columns; however, the use of very long columns coupled with relatively rapid analysis times provides the chemist with a means of separating compounds that have small differences in their physical characteristics. The major drawback to wall-coated open tubular columns (WCOT) is the small amount of liquid phase that the wall is capable of holding. This objection is overcome by increasing the surface area of the column by coating the wall with a finely divided support upon which a much larger amount of liquid phase can be coated, yet without increasing the film thickness. This is the support-coated open tubular (SCOT) column. The prime advantage of an open tube, namely low pressure drop that makes long lengths feasible, is preserved but now the beta value ($\beta = V_{gas}/V_{liq}$) approaches that of packed columns, and column capacity is greatly increased. The tubing is wider (0.5 mm), the flow rates faster (4–10 ml/min), and the dead volume of connections is less critical. Sample splitting is useful but not required; sample size is 0.5 μl or less. If a separation demands more than 10,000 plates, a SCOT column is probably the best solution.

Supports The support in general use for packed columns is diatomaceous earth which has been crushed and calcined above 900°C (pink support) or which has been mixed with a small amount of flux, calcium carbonate, and calcined above 900°C (white support). The best particle size is 100/120 mesh for 2-mm ($\frac{1}{8}$-in.) columns and 80/100 mesh for a 4-mm column; for effective packing the i.d. of the tubing should be at least 8 times the diameter of the support particles. The white support has pore sizes of about 9 μm, whereas the pink support has a pore size of about 2 μm. A column of the pink support will hold more of the liquid phase and hold it in smaller and shallower pools, which require shorter transit times for the solute. Both types of supports perform quite well for the analysis of relatively nonpolar samples but tend to be too active for polar samples. The surface of diatomaceous supports are covered with silanol (SiOH) groups that tend to adsorb sample molecules, particularly when lightly loaded with liquid phase or when nonpolar liquid phases are used. This effect can be reduced greatly if the silanol groups are converted to silyl ethers by reaction with dimethyldichlorosilane. As a result, surface activity and peak tailing are reduced considerably. At the same time, however, silanization reduces the surface area of the support and limits stationary-phase loading to a maximum of 10%. Thus there is a constant compromise in the choice of solid support. Solid supports having surface energy low enough to prevent tailing are wetted poorly by the liquid phase, resulting in low efficiency. Those that are easily wetted usually cause tailing of polar samples.

Special supports are popular for certain applications. Very lightly loaded glass beads are used for very rapid analyses well below the boiling point of the sample components. The surface of the beads is roughened by several techniques to obtain better wetting and increased liquid phase capacity. Most phase loadings are 0.05 to 0.2% of the liquid phase. Teflon can be used for analysis of corrosive samples. Porous polymer beads find use in gas–solid chromatography. These beads have very low affinity for water and very polar sample molecules, which therefore elute very rapidly from the columns as symmetrical peaks. These porous polymer beads are sold under the trade names Poropak and Chromosorbs.

Liquid Phases Literally thousands of partition liquids have been described in the literature. While it is true that many offer unique separations, it is also true that most separations can be performed on a relatively few liquid phases. Several of the more common liquid partitioning liquids, ranging from nonpolar to very polar, and their properties are given in Table 18-1.

The general rule "like dissolves like" is a good one to follow when selecting liquid phases; thus, polar liquid phases are best for polar samples and, conversely, nonpolar liquid phases are best for nonpolar samples. Examples are squalane for hydrocarbons, Carbowax for alcohols, and polyesters for fatty acid methylesters. Liquid phases which are similar to the component retard these components compared to liquid phases which are dissimilar. Thus aromatic compounds (induced polarity) are retarded on Carbowax whereas paraffins (nonpolar) are eluted more rapidly. Conversely, alcohols are eluted rapidly on squalane relative to paraffins. This property can be used to make group sepa-

TABLE 18-1 Typical Liquid Phases

Liquid Phase	Type	Minimum/Maximum Temperature, °C	Rohrschneider Constants				
			x	y	z	u	s
Squalane[a]	2,6,10,15,19,23-Hexamethyl-tetracosane	20/200	0	0	0	0	0
OV-1, OV-101, SE-30	Methyl silicone	100/350	0.16	0.20	0.50	0.85	0.48
Apiezon L	Hydrocarbon	50/250	0.32	0.39	0.25	0.48	0.55
OV-3	Phenyl silicone	/350	0.42	0.81	0.85	1.52	0.89
OV-7	Phenyl silicone	/350	0.70	1.12	1.19	1.97	1.34
DEHS	Di-(2-ethylhexyl) sebacate	/125	0.73	1.65	1.15	2.20	1.24
DNP	Dinonyl phthalate	20/150	0.84	1.76	1.48	2.70	1.53
QF-1	Fluorosilicone	0/250	1.09	1.86	3.00	3.94	2.41
OV-11, Silicone DC 710	Phenyl silicone	/350	1.13	1.57	1.69	2.57	1.95
UCON 550X	Polypropyleneglycol	/200	1.14	2.76	1.68	3.12	2.08
OV-17	Phenyl silicone	0/375	1.30	1.66	1.79	2.83	2.47
TCP	Tricresyl phosphate	20/125	1.74	3.22	2.58	4.14	2.95
Silicone GE XE-60	Cyanosilicone	/250	2.08	3.85	3.62	5.43	3.45
Carbowax 4000	Polyglycol	45/200	3.22	5.46	3.86	7.15	5.17
DEGS, LAC-2-R-446	Diethyleneglycol succinate	20/200	4.93	7.58	6.14	9.50	8.37
B, B'-OXY	β,β'-Oxypropionitrile	0/75	5.88	8.48	8.14	12.58	9.19
			649	384	531	457	695

[a] Kovats indices for squalane:

rations or to accelerate or retard components with respect to others in the sample (also see Table 18-3).

Unless the sample dissolves well in the liquid phase, little or no separation occurs. The gas phase is inert, and separation occurs only in the liquid phase. A reasonable guideline is that a retention of five times the air peak is required to show any reasonable selectivity. Lower temperatures will almost always increase the solubility and the selectivity. The minimum temperature limit is important and is determined by the melting point or the viscosity of the liquid phase. As viscosity increases the absolute solubility decreases and the mass transfer between the gas and liquid phases becomes so slow that little separation occurs.

The upper temperature limit of liquid-phase stability and volatility has been a barrier. Dexsil-300 GC, a polycarboranesiloxane that can be used from 50° to 500°C, is a methyl silicone chain that is interrupted periodically by the inclusion of a carborane structure which acts as an energy sink to prevent destruction of the molecule. Bonding a liquid phase directly to a solid support provides polar liquid phases with no appreciable vapor pressure. For example, Carbowax 400 and β, β'-oxypropionitrile have been bonded to the surface of a porous silica bead to produce Si—O—C linking (Durapak series). Bonding directly to a controlled surface-porosity glass bead, using the Si—O—Si bond, produces the Zipax series of liquid phases.

The sum of intermolecular forces determine solute–solvent interaction and therefore retention time. The Kovats retention index (R.I.)[2,4] indicates where a compound will appear on a chromatogram with respect to n-paraffins. By definition, the R.I. for pentane is 500, for hexane 600, for heptane 700, and so on, regardless of the column used or the conditions. For example, suppose on a squalane column the retention times for n-hexane, benzene, and n-octane were 15, 17, and 25 min, respectively. On a plot of log retention time of the n-paraffins versus their retention indices, a retention index of 649 for benzene is read off the graph. Thus it is only necessary to run a standard hydrocarbon sample on a column to determine retention times of compounds for which Kovats retention index information has been collected.

If the retention indices for a compound are determined on both a polar and a nonpolar phase the difference can be calculated: $\Delta I = I_{\text{polar}} - I_{\text{nonpolar}}$. This difference gives a measure of solute–solvent interaction of a liquid phase due to all the intermolecular forces other than the dispersion or London forces, which are the principal solute–solvent effects in squalane. By proper choice of standard test compounds, the effects due to hydrogen bonding, dipole moment, and molecular configuration can be measured and used as a means for the orderly classification of the many liquid phases in use. In the Rohrschneider system,[5] benzene (x), ethanol (y), methylethylketone (z), nitromethane (u), and pyridine (s) were chosen. Now

$$\Delta I = ax + by + cz + du + es \qquad (18\text{-}1)$$

where x is the polarity of a column when benzene is analyzed and is equal to $\Delta I/100$ for benzene, and similarly for the other index standards. All must be determined under identical conditions of temperature and column loading. Rohrschneider constants are given for the liquid phases included in Table 18-1. For example, tricresyl phosphate is more polar than QF-1 for benzene and ethanol, but the reverse is true for methylethyl-ketone. The high z-value for QF-1 illustrates numerically that this phase selectively retards ketones.

Detectors

Located at the exit of the separation column, the detector senses the arrival of the separated components as they leave the column and provides a corresponding electrical signal. The temperature of the detector compartment must be sufficiently high to prevent condensation of sample vapors, yet not cause sample decomposition.

Thermal Conductivity Cell The thermal conductivity detector (TCD) is made of four filaments arranged in an electrical bridge network. Each helical filament is situated in a separate cavity in a brass block which serves as a heat sink, as shown in Fig. 18-3. The electrical circuitry is shown in Fig. 18-4. Two filaments in opposite arms of the bridge are surrounded by the carrier gas; the other two filaments are surrounded by the effluent from the column. The temperature of the filaments is determined by the rate of heat loss by conduction through the carrier gas; the gas flows around the filaments and through the cavities. As components elute from the column, the composition of the gas changes. The resultant change in thermal conductivity produces a change in filament temperature, and in turn a change in the resistance of the filament, causing an electrical output from the Wheatstone bridge circuit. Aside from hydrogen, helium has the highest thermal conductivity of all gases (see Fig. 27-4 and Table 27-1) and, as a carrier gas, provides the highest sensitivity of detection for all other gases.

Because of its simplicity, the thermal conductivity detector often is preferred for survey work and for moderate sensitivity work in all areas. It responds to all types of inorganic and organic compounds. Because it is nondestructive, it is particularly suitable for preparative or fraction-collecting work. The cavity cell volume is 2.5 ml for detectors associated with 4-mm columns; this is decreased to 0.25 ml in the microcell designed for use with SCOT and small-diameter packed columns. One annoying limitation of the detector is its low resistance to oxidation and sometimes to chemical attack of the gold-sheathed tungsten or Teflon-coated tungsten filaments. The detection limit is about $5 \, \mu g/ml$ of sample gas concentration or 10 ng of sample weight and the range is about 10^5.

Flame Ionization Detector The flame ionization detector (FID) currently is the most popular detector because of its high sensitivity, wide range, and great reliability. It

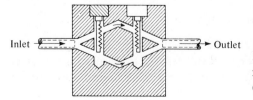

Inlet ——→   ——→ Outlet

Fig. 18-3. Bypass design of hot-wire thermal conductivity cell for rapid response. (Courtesy of Gow-Mac Instrument Co.)

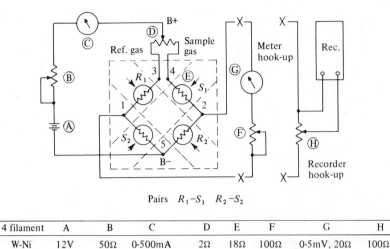

Pairs R_1-S_1 R_2-S_2

4 filament	A	B	C	D	E	F	G	H
W-Ni	12V	50Ω	0-500mA	2Ω	18Ω	100Ω	0-5mV, 20Ω	100Ω
W-2	18-36V	100Ω	0-500mA	4Ω	40Ω	200Ω	0-5mV, 100Ω	200Ω

Fig. 18-4. Circuitry for four-filament, hot-wire, thermal conductivity cell. (Courtesy of Gow-Mac Instrument Co.)

consists of a small hydrogen flame burning in an excess of air and surrounded by an electrostatic field (Fig. 18-5). Column effluent enters the burner base through a millipore filter and is mixed with the hydrogen entering the burner. Organic compounds eluted from the column are burned. During the combustion, ionic fragments and free electrons are formed; these are collected, producing an electric current proportional to the rate at which the sample enters the flame. Many variations in design are available. Some instruments add makeup gas to the column effluent in order to speed up the gas velocity and sweep the column effluent through the "dead" volume between the column exit and detector entrance, thus minimizing extra-column band spreading. The FID responds only to oxidizable carbon atoms, and response is proportional to the number of carbon atoms in the sample component. There is no response from fully oxidized carbons such as carbonyl or carboxyl groups (and thio analogs), and response diminishes with increasing substitution of halogens, amines, and hydroxyl groups. The FID does not respond to inorganic compounds apart from those easily ionized (see Table 12-3). Insensitivity to water, the permanent gases, CO, and CO_2 is advantageous in analysis of aqueous extracts and in air-pollution studies. Of course, CO and CO_2 can be converted easily to methane by reduction with hydrogen over a nickel catalyst.

Detection is about 20 pg of sample weight or about 5 ng/ml of sample gas concentration. Linear range is 10^7. Precise temperature control is not a requirement for this detector, an obvious advantage in programmed-temperature applications.

Electron-Capture Detector   The principle of the electron-capture detector (ECD) is based on electron absorption by compounds having an affinity for free electrons. These are compounds that have an electronegative element or group. Upon exposure to a

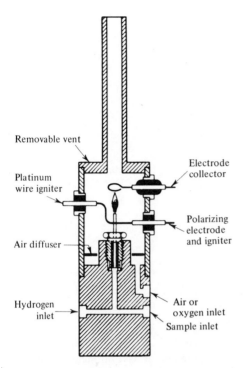

Fig. 18-5. Schematic diagram of a flame ionization detector. (Courtesy of Beckman Instruments, Inc.)

source of low-energy electrons, these compounds tend to attach or capture electrons to form negative ions. A schematic diagram and pin-cup design is shown in Fig. 18-6. As nitrogen carrier gas flows through the detector, beta particles from a tritium source (or nickel-63) ionize the nitrogen molecules and form "slow" electrons which migrate to the anode under a fixed potential which can be varied from 2 to 100 V. These collected electrons produce a steady baseline current flowing across the cell from A to B. The production of secondary electrons occurs only in area E; methane is mixed with the carrier

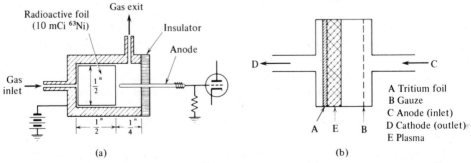

Fig. 18-6. Electron capture detector. (a) Pin-cup design. (b) Schematic diagram. (Courtesy of Varian Aerograph.)

gas and serves as a quench gas to reduce the energy of the electrons for more efficient capturing. When an electron-capturing component emerges from the column and moves from C to E, it reacts with an electron to form either a negative molecular ion, or a neutral radical and a negative ion. These are swept out by the gas flow. The net result is the removal of an electron from the system and a decrease in the standing current. The decrease is recorded as a negative peak on the recorder. A pulsed power supply, with a pulse duration only long enough to collect electrons (but ignore the negative ions), is commonly used.

Response is essentially nonlinear, following an exponential relation similar to Beer's law. By using feedback and varying the pulse rate, operation is linearized, giving a linear range of 500. Detection is at the 0.1 pg level. More important is the selectivity toward traces of halogen-containing compounds, anhydrides, peroxides, conjugated carbonyls, nitriles, nitrates, ozone, oxygen, organometallics, and sulfur-containing compounds whereas the detector is virtually insensitive to hydrocarbons, amines, and ketones. Selectivity ratios of 100,000 to 1 are quite common. The most common application of the ECD is the determination of chlorinated pesticides; others are halogenated anesthetics, polynuclear carcinogens, and sulfur hexafluoride tracers in meteorology. Hydroxy or amino compounds can be converted to heptafluorobutyryl or trifluoroacetyl derivatives, which are helpful in biomedical work.

Other Detectors   The flame photometric detector employs a photomultiplier tube to view the emission from a reversed, fuel-rich hydrogen flame through an interference filter. Single detectors are used either for the HPO band system of phosphorus at 5260 Å or the S_2 band system of sulfur at 3940 Å; alternatively, a dual detector monitors both signals simultaneously. The burner assembly is similar to that used in the flame ionization detector except that air and carrier gas flow through the jet and hydrogen is supplied by a separate annulus around the jet. Response to phosphorus is linear over a range of about 10,000 to 1 with a detection limit of 10 pg. Sulfur response is proportional to the square of the quantity and is nearly linear on a log–log scale over the range from 200 pg to 100 ng. This detector finds use in air-pollution studies for sulfur entities and in pesticide analysis for phosphorus.

A regular sprayer-burner or slot burner, plus appropriate interference filters or grating monochromator and photomultiplier detector, can be used to detect many other inorganic elements, including silicon from silanized derivatives. Other detection devices are also available, but are not used as widely in gas chromatography. The gas chromatograph has been combined with other instruments, particularly the mass spectrometer.

Dual-Detectors   In dual-detector operation the effluent from a single column is split and sent to two different detectors simultaneously [Fig. 18-7(a)]. A two-pen recorder allows simultaneous presentation of the signals. Two detectors, each with their individual selectivity, make possible a more detailed characterization of the sample than can any single detector. It aids in qualitative analysis by distinguishing components comprising overlapping peaks.

In the differential mode, matched columns and identical detectors are used. This ar-

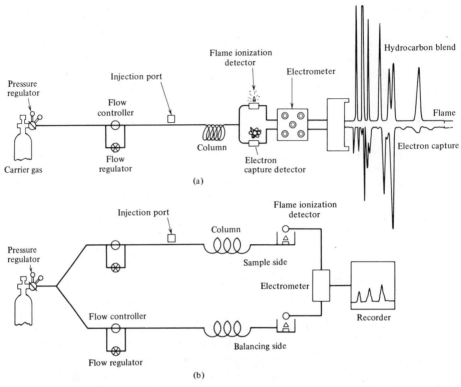

Fig. 18-7. Dual-detector operation. (a) Two different detectors operated simultaneously. (b) Differential operation with identical detectors.

rangement [Fig. 18-7(b)] eliminates baseline drift caused by bleeding of the liquid phase, particularly during temperature programming.

Quantitative Evaluation

Ordinarily the chromatogram is traced by a strip-chart recorder connected to the output of the detector-amplifier unit. Three conditions must prevail: (1) The output of the detector-recorder system must be linear with concentration. This range is expressed by the usable range of the detector and, coupled with sensitivity, gives the concentration limits. (2) The carrier-gas flow rate must be constant so that the time abscissa may be converted to volume of carrier gas. (3) The pen response of the recorder must match the speed of the detector's response or one must couple automatic integration devices directly to the electronics of the detector. In other words, if the peak width is 0.2 sec, one cannot use a recorder with a 1-sec response time. Under these conditions, peak area can be used as a quantitative measure of component present.

Triangulation Methods Condal-Bosch[1] has discussed triangulation methods which are very simple and precise if certain precautions are taken. These methods consist of extending the approximately straight segments about the inflection points of the elution peak. These two lines, together with the baseline, constitute a triangle whose area is obtained as one-half the product of the base times either the actual peak height or the vertical height of the apex of the triangle formed by the tangents. In this way one obtains an area that is equal to about 97% of the actual area of the chromatographic peak. Standard deviation is about 4.0%. The area can also be obtained as one-half the product of the peak height times the width at half the peak height. The only geometric construction consists of drawing the baseline, measuring the peak height, noting its midpoint, drawing a horizontal through this point, and measuring the peak width on this horizontal. The area calculated in this way corresponds to 91% of the actual area for a Gaussian curve. The standard deviation is about 2.5%.

For asymmetric peaks, a method which gives the area without requiring any correction factor at all, and applicable to curves of every degree of symmetry, resorts to a trapezoid construction. One width is taken at 0.15 and the other at 0.85 of the peak height. The formula for the area is half of the sum of the widths multiplied by the peak height.

Cut and Weigh The peak area is determined by cutting out the chromatographic peak and weighing the paper on an analytical balance. The accuracy of the method depends on the care used in cutting and on the constancy of the thickness and moisture content of the chart paper (or preferably a Xerox copy of the chromatogram to preserve the original).

Automatic Integrators Two types of automatic integrators will be discussed. The digital integrator is an all-electronic system. It consists of an electronic decimal counter preceded by an impulse generator, the output rate of which is proportional to the input voltage signal for the detector-amplifier combination. The incoming signal is received from beyond the zero adjustment in the case of a thermal conductivity detector, or from the amplifier in the case of an ionization-type detector. The signal passes through an input-signal attenuator, then to the preamplifier and finally is supplied to a peak-and-valley sensor and voltage-frequency converter. The latter is the key component that develops an output pulse having a frequency that is directly proportional to the magnitude of the incoming dc signal. Thus, the total number of counts over a period of time is proportional to the peak area. Elapsed time for each peak from the start of the upward curvature to peak crest is measured, displayed, and printed out on paper tape alongside the associated peak-area value. A finite time is required to recognize peak start and peak end; therefore it is necessary to keep the signal to the integrator large. To secure high accuracy on small peaks the count rate must be high, at least 1000 counts/sec/mV. At the same time the counter capacity needs to be high to accommodate the wide linear range.

A digital differentiator-type slope detector must sense valleys and shoulders of unresolved peaks, also peaks varying in width from less than 1 sec to several minutes. Peak-

and-valley sensors, sensitive to signal rates of change as small as 0.1 μV/sec, accomplish this feat. Adjustment of slope sensitivity enables the operator to discriminate between noise and actual gas chromatographic peaks. Standard deviation is 0.44%.

The ball-and-disc integrator (illustrated in Fig. 18-8) is an automatic mechanical type of integrator. A ball positioned on a rotating flat disc will rotate at a speed proportional to its distance from the center of rotation. The ball is positioned on the disc at a distance from the center in the same relationship as the position of the recorder pen to the base line. If the disc is rotated at a constant speed (time), then the ball will rotate at a speed proportional to the position of the recorder pen from zero, provided the recorder zero and the center of the rotating disc have been exactly aligned. This speed is then transmitted to a roller through a second ball which, by means of "spiral in" and "spiral out" cam, actuates the integrator pen at a speed directly proportional to the position of the recorder pen. The drive between the disc and the ball is by traction through an oil film. Although this hydrostatic phenomenon is not clearly understood, the oil film acts similar to an induction motor, where slip is proportional to the driven load.

To read the integrator pen trace (refer to Fig. 18-9), first establish the desired chart time interval from the recorder pen trace and then project directly down to the integrator pen trace (*arrows*). The value of an interval is obtained by counting the chart graduations (not time lines) crossed by the integrator pen trace. A full stroke of the saw-tooth pattern in either direction represents 100 counts. Every division has a value of 10. Values less than 10 are estimated. On some models the space between "blips" is equivalent to 600 counts, making it possible to record up to 24,000 counts per inch of chart. For example, the interval for the main peak (left) on Fig. 18-9 is 1083 counts. With care the pattern can be read within 2 counts. The pattern (on Fig. 18-9, right) illustrates the means for estimating the baseline correction when the peak baseline does not coincide with the recorder baseline. Standard deviation is 1.3%.

Correlation of Area and Quantity Areas under a gas chromatographic peak can be correlated with similar areas obtained with known concentrations of the particular solute run under conditions that are as similar as possible. This is usually done by the construction of a graph of quantity versus peak area. Errors in the injection technique normally limit the accuracy to 0.5% at best.

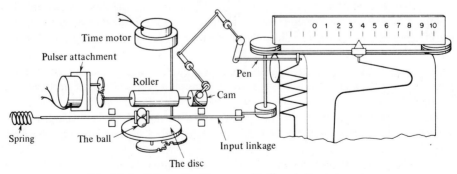

Fig. 18.8. Schematic diagram of ball-and-disc integrator.

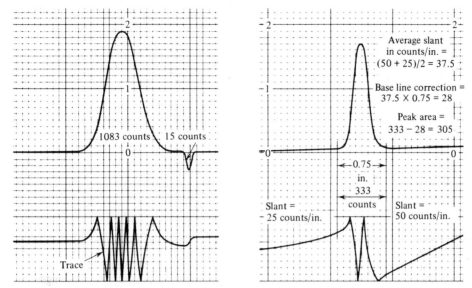

Fig. 18-9. Estimation of peak areas with ball-and-disc integrator. *Right*, method for handling baseline correction.

To express the result of the analysis in percentages, one can assume that all of the components of the sample are present in the chromatogram. The sample size can then be obtained by summing up the individual components. Individual quantities are referred to this sum. This procedure is termed *normalization*. It fails for minor components when the major peaks do not register adequately, and when detectors insensitive to certain components are used.

GAS CHROMATOGRAPHY THEORY

Gas chromatography theory cannot be discussed here in detail, nor can the complex interactions of all the variables be considered. For details the literature should be consulted. On the other hand, a brief treatment of basic parameters should help in understanding the technique.

Retention Behavior

The volume of carrier gas necessary to convey a solute band the full length of a column is the retention volume—the fundamental quantity measured in gas chromatography. It reflects the distribution of the solute between the eluent gas and the stationary liquid phase. For a given column operated at temperature T_c and carrier gas flow rate F_c, the length of time that each component spends in the column, called the retention time, is a constant. The retention time will be the same whether the solute is pure or in a mixture. On a chromatogram the distance on the time axis from the point of sample injection to the peak of an eluted component is called the *uncorrected retention time, t_R*. The corresponding *retention volume* is the product of retention time and flow rate (expressed as volume of gas per unit time:

$$V_R = t_R F_c \qquad (18\text{-}2)$$

Gas flow rate must be corrected to column temperature and outlet pressure. With wet flow meters allowance must be made for the vapor pressure of water, and with capillary meters for the pressure drop across the capillary.

The air spike, shown in Fig. 18-10, measures the transit time for a nonretained substance. Converted to volume (V_M), it represents the interstitial volume of gaseous phase in the column plus the effective volume contributions of the injection port and detector. Retention volume measured from the air peak provides an *adjusted retention volume*, V_R', which is corrected for the dead space:

$$V_R' = t_R F_c - t_{air} F_c = V_R - V_M \qquad (18\text{-}3)$$

The adjusted retention volume changes slightly with amount of sample; consequently, the significant adjusted retention volume is one obtained by extrapolation to zero sample size.

In gas chromatography the mobile phase is compressible. Since gas moves more slowly near the inlet than at the exit of the column, a *pressure-gradient correction* or *compressibility factor, j*, must be applied to the adjusted retention volume to give the *net retention volume, V_N*, namely,

$$V_N = j V_R' \qquad (18\text{-}4)$$

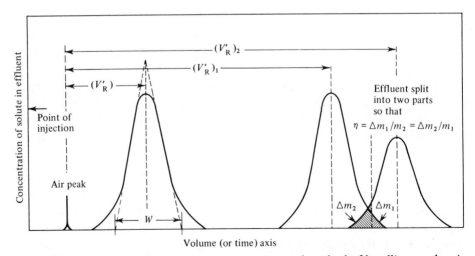

Fig. 18-10. Idealized elution peaks showing notation and method of handling overlapping bands.

The compressibility factor is given by the expression[*]

$$j = \frac{3}{2} \frac{[(P_i/P_o)^2 - 1]}{[(P_i/P_o)^3 - 1]} \qquad (18\text{-}5)$$

where P_i is the carrier gas pressure at the inlet to the column, and P_o that at the outlet. This factor stresses the importance of a small pressure drop across the column.

When the solute enters the column it immediately equilibrates between the stationary solvent phase and the mobile gaseous phase. The concentration (or weight) in each phase is given by the *partition coefficient*

$$K_d = \frac{C_S}{C_M} \qquad (18\text{-}6)$$

where C_S, C_M are the concentrations of solute in the stationary solvent phase and mobile gas phase, respectively. For example, when $K_d = 1$, the solute will distribute itself evenly between the two phases and thus will spend half the time in the gas phase and half the time in the liquid phase, emerging at a retention time equal to twice the retention time of the air peak. If the partition isotherm is linear, the partition coefficient will be a constant independent of the solute concentration. Generally this is true at the low concentration prevailing in gas-liquid chromatography.

At the appearance of a peak maximum at the column exit, one-half of the solute has eluted in the retention volume V_R, and half remains in the volume of the gaseous phase V_M, plus the volume of the stationary liquid phase V_S at the column temperature. Thus,

$$V_R C_M = V_M C_M + V_S C_S \qquad (18\text{-}7)$$

Rearranging and inserting the partition coefficient, we obtain the fundamental equation for gas chromatographic separations

$$V_R' = V_M + K_d V_s \quad \text{or} \quad V_R - V_M = K_d V_S \qquad (18\text{-}8)$$

provided the compressibility factor, j, is applied. Retention volumes will be primarily determined by the term $K_d V_S = K_d(w_L/\rho_L)$, where w_L is the weight and ρ_L the density of the liquid phase at the temperature of the column. Since the coefficient of cubical expansion of most organic liquids lies between 0.5×10^{-3} and 1.5×10^{-3}, it will usually be adequate to obtain the density of the liquid phase by measurement of the density at room temperature and estimation with a coefficient of $10^{-3}/°C$.

When the amount of liquid is increased either by using a thicker layer or a longer column, the retention volumes are increased. In order to take into account the weight of liquid phase in a column, the specific retention volume, V_g, is defined as

$$V_g = \frac{273}{T_c} \frac{V_N}{w_L} \qquad (18\text{-}9)$$

[*]For a derivation of the compressibility factor, see W. E. Harris and H. W. Habgood, *Programmed Temperature Gas Chromatography*, Wiley, New York, 1966, p. 49.

It corresponds to the volume of carrier gas required to remove half of the solute from a hypothetical column at a specified temperature ($0°C$ unless otherwise specifically stated) which contains 1 g of a liquid phase and which has no pressure drop or apparatus dead space. Now from Eqs. (18-6) and (18-9),

$$K_d = V_g \rho_L \left(\frac{T_c}{273}\right) \quad \text{or} \quad V_g = \frac{273}{\rho_L} \frac{K_d}{T_c} \tag{18-10}$$

The *capacity factor* (or partition ratio), k, is a measure of sample retention by the column, in terms of column volumes

$$k = \frac{V_R - V_M}{V_M} = \frac{C_S V_S}{C_M V_M} \tag{18-11}$$

or the right-hand expression in terms of the amount of solute in the two phases. Since the volumetric phase ratio $\beta = V_M/V_S$, $k = K_d V_S/V_M$.

Column Efficiency

When the chromatogram is obtained under linear elution conditions, ideally each solute will produce a Gaussian distribution, as shown in Fig. 18-10. For good resolution narrow baseline widths, W, are desirable. Here the peak width at the base is the segment of baseline cut out by the tangents drawn to the inflection points of the Gaussian peak; it is equivalent to four standard deviations (4σ).

Plate Number   An empirical measure of column efficiency is defined as the theoretical plate number, N, in which

$$N = 16 \left(\frac{V_R'}{W}\right)^2 \tag{18-12}$$

The narrower the peak, the larger is N and the more efficient the column system. Typical values of N are 1500–3000 plates/meter for packed columns and 3000 plates/meter for open tubular columns.

An alternative expression for N is

$$N = 5.54 \left(\frac{V_R'}{W_{1/2}}\right)^2 \tag{18-13}$$

where $W_{1/2}$ is the width at one-half the peak height as measured from the baseline. Using this method to evaluate N, the need to construct tangents to the peak slopes is eliminated and accuracy is considerably improved.

Band spreading arises from the partition processes in the column (the terms in the plate-height equation) and from extra column causes. Extra column band spreading arises from five causes: (1) sample volume is too large, resulting in an excessive original bandwidth; (2) the sample injection is carried out too slowly, thus increasing the width of the "plug," (3) too much dead volume exists between the injection port and column entrance and (4) between the column exit and detector, which results in band spreading due to diffusion of the sample vapor into the carrier gas; and (5) the detector's gas volume is large. To evaluate the magnitude of the last cause, actual gas volumes corresponding to the peak width at half-height should be calculated from the chromatogram and compared with the detector's gas volume. For example, standard thermal conductivity detectors can be applied with no difficulty to packed columns 4 mm in diameter, but cannot be tolerated in fast analyses with a SCOT column where the detector volume may not exceed about 0.2 ml. The development of the more sensitive ionization class of detector was necessary before open tubular columns could be exploited.

Plate Height   Plate height is given by

$$H = \frac{L}{N} \tag{18-14}$$

where L is the column length. Plate height commonly is expressed at constant temperature as a function of average linear carrier-gas velocity, u:

$$H = A + \frac{B}{u} + Cu \tag{18-15}$$

The gas velocity may be estimated by dividing the column length by the gas holdup time (corrected for pressure gradient), that is, $u = L/jt_M$.

The A term, attributed to eddy diffusion, represents the distance that a streamline persists before its velocity is drastically changed by the packing. It is proportional to the average particle diameter of the support. In open tubular columns the A term is zero.

The B term contains the gaseous diffusion coefficient of the solute. It describes the band spreading due to molecular diffusion in the flowing gas stream along the direction of flow.

The C term describes the resistance to mass transfer in the liquid and gas phases. In reality the C coefficient involves two groups of terms: $C_{liq} + C_{gas}$. The C_{liq} portion is given by $(8/\pi^2) [k/(1 + k)^2] (2d_f^2/D_{liq})$, where d_f is the effective thickness of the film of liquid phase and D_{liq} is the diffusion coefficient of the solute in the stationary liquid film. The C_{gas} term is proportional to the square of the diffusion distance in the gas and inversely proportional to the diffusion coefficient of the solute in the gas phase. The C terms show that the plate height depends on the capacity factor of the species and will be different for different species.

A plot of plate height versus carrier gas velocity, often referred to as the Van Deemter plot, is shown in Fig. 18-11. The resistance to mass transfer determines the limiting slope at high carrier-gas velocities; the A term is the intercept of the slope. At low velocities, the plate height is controlled by the molecular diffusion in the mobile phase. At some point the curve achieves a minimum, which represents the optimum velocity for the minimum plate height. In actuality the minimum is quite broad for the usual 10% liquid-phase loading in packed columns. In such cases the gas velocity can be set to the optimum value for the most difficult solute pair in the sample, or column efficiency can be traded for speed of analysis, discussed in a later section.

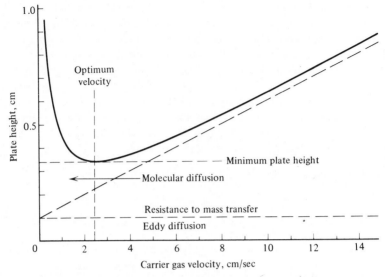

Fig. 18-11. Plot of plate height vs. carrier gas velocity.

Resolution

The relative capacity factors of two components are a measure of the column's ability to separate them. This is expressed as the *relative retention*, α, defined as

$$\alpha = \frac{k_2}{k_1} = \frac{V'_{R,2}}{V'_{R,1}} \tag{18-16}$$

If the relative retention is unity, the peaks coincide and no separation has occurred. A meaningful separation in gas–liquid chromatography can be achieved with α values as low as 1.02.

Two factors influence resolution: the peak-to-peak separation and the average peak width. Increasing the distance between peak maxima and/or keeping peaks narrow improves separation. These are taken into account in the expression for resolution

$$R = \frac{V_{R,2} - V_{R,1}}{0.5\,(W_2 + W_1)} \tag{18-17}$$

A value of $R = 1.5$ represents baseline resolution, whereas in the case of $R = 1.0$, the resolution is about 94% complete, which corresponds to a 3% overlap of peak areas. If equal quantities of two solutes in adjacent bands are assumed, resolution can also be expressed in terms of k, α, and N

$$R = \frac{1}{4}\left(\frac{\alpha - 1}{\alpha}\right)\left(\frac{k_2}{1 + k_2}\right)\sqrt{N} \tag{18-18}$$

Thus, as two adjacent solutes move through the column, their elution curves separate to a degree that is determined by their relative retention, the capacity factor, and the number of theoretical plates. Upon rearrangement, the number of theoretical plates required to achieve a resolution of two adjacent curves can be calculated, and if the plate height of the column is known, the necessary column length can be directly established

$$\frac{L_{req}}{H} = N_{req} = 16R^2 \left(\frac{\alpha}{\alpha - 1}\right)^2 \left(\frac{k_2 + 1}{k_2}\right)^2 \tag{18-19}$$

In a given system, we have to quadruple column length to double resolution. However, the analysis time is not necessarily proportionately longer, as will be discussed later.

Degree of Cross Contamination In practical chromatography only enough theoretical plates are required to reduce the degree of cross contamination in the adjacent peaks to a desired experimental level η. Glueckauf[3] has published curves (Fig. 18-12) relating the

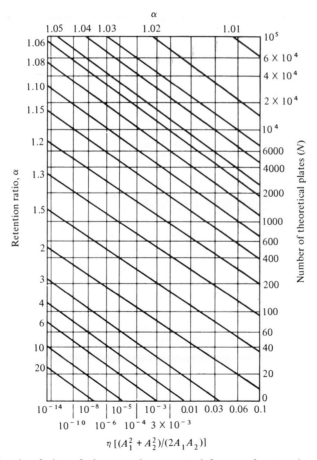

Fig. 18-12. Glueckauf plot of plate number required for any fractional impurity, η, and relative amounts of adjacent solutes, A_1, A_2, with the relative retention ratio, α, as parameter. (After E. Glueckauf, *Trans. Faraday Soc.*, **51**, 34 (1955).)

cross contamination of two adjacent peaks as a function of the number of theoretical plates for various values of the relative retention ratio when using packed columns.

The extent of cross contamination can be estimated from the experimental elution curves. The effluent from the column (peak areas) is divided into two portions to give products of equal purity, that is, $\eta_1 = \Delta m_2/m_1$, $\eta_2 = \Delta m_1/m_2$, and $\eta_1 = \eta_2$. When the solute concentrations are unequal, the fractional impurity η must be multiplied by the factor: $(A_1^2 + A_2^2)/2A_1A_2$, where A_1, A_2 are the areas of elution peaks. In general, the division of effluent should be made not half-way between the peaks, but nearer the peak involving the least solute (see Fig. 18-10).

Example 18-1

In Fig. 18-10, assume that the normalized areas are $A_1 = 0.6$ and $A_2 = 0.4$, and that the relative retention ratio is 1.8. To resolve these peaks with fractional impurity 0.01 (1.0%) in each band, first calculate the product:

$$\eta \frac{(A_1^2 + A_2^2)}{2A_1A_2} = 0.01 \frac{(0.36 + 0.16)}{0.48} = 0.0108$$

The required number of theoretical plates is obtained from the Glueckauf plot (Fig. 18-12) by extending a vertical line from the value 0.0108 on the abscissa to the corresponding diagonal line for $\alpha = 1.8$. The ordinate corresponding to this intersection is approximately 60 plates.

From the peak dimensions for the second component, the actual number of theoretical plates in the column is obtained by means of Eq. 18-13 or 18-14. If the actual number is less than the required number of plates, a longer column is indicated, or some other adjustment in the parameters affecting the resolution must be made.

Relative Retention In the expression for resolution (Eq. 18-17), the first term is very sensitive to changes in the value of the relative retention, as shown in Table 18-2, which also lists N_{req} and the necessary column length. From this data, a relative retention

TABLE 18-2 Values Related to the Relative Retention

α	$\left(\dfrac{\alpha}{\alpha - 1}\right)^2$	N_{req} for $R = 1.5$ and $k = 2$	L_{req}, m for $H = 0.6$ mm
1.01	10,201	826,281	495
1.02	2,601	210,681	126
1.03	1,177	95,377	52
1.04	676	54,756	33
1.05	441	35,721	21
1.10	121	9,801	5.8
1.15	58	4,418	2.6
1.20	36	2,916	1.7
1.25	25	2,025	1.2
1.30	19	1,514	1.0

TABLE 18-3   Relative Retentions (*n*-Pentane = 1) of Various Compound Types
'at Three Boiling-Point Levels for Various Liquid Substrates

Substrate	Convoil-20			Tricresyl Phosphate			Poly(diethylene glycol succinate)		
Operating Temp., °C	100			100			100		
Boiling-Point Level, °C	60	100	140	60	100	140	60	100	140
Hydrocarbon types									
n-Paraffins	1.80	5.6	17.0	1.65	4.6	13.5	1.5	2.9	5.6
2-Methyl paraffins	1.75	5.4	16.5	1.60	4.3	11.5	1.0	1.9	—
Type I olefins	1.90	5.6	17.5	2.25	5.8	16.5	1.7	3.4	—
Type III olefins	1.90	5.6	17.5	2.50	5.7	—			
Cyclopentanes	2.33	6.7	19.8	2.25	5.8	16.0			
Cyclohexanes	—	7.0	20.0	—	6.7	18.3			
Cyclo-olefins	2.5	7.2	20.5	3.9	9.7	23			
Diolefins	—	—	—	2.6	7.1	18.3			
Acetylenes	1.50	4.5	15.0	3.6	9.7	23			
Alkylbenzenes	—	6.7	19.8	—	16.4	43.5		18.4	50
Oxygenated types									
Alcohols									
Primary normal	0.29	1.4	10	3.3	11.5	—	33	54	140
Secondary	0.2	1.9	—	2.2	11.5	—	30	46.8	145
Tertiary	0.28	2.6	—	2.2	11.5	—	24	45.6	160
Ketones	0.72	3.1	12.0	4.8	13.5	37.5	20.5	37	80
Ethers	1.35	5.4	22	—	—	—	9.2	56	—
Esters									
Formates	1.0	3.3	13	4.6	12.5	43			
Acetates	1.0	3.3	13	4.6	12.5	43	16	30	61
Aldehydes	1.0	3.3	13	5.2	16.3	—	14.3	30	61
Acetals	—	—	—	—	—	—			
Retention volume for *n*-pentane (at 25°C, 1 atm), ml		77			34			13	

as large as possible is seen to be desirable. If the relative retention is below 1.05–1.10, the use of a packed column is impractical. The relative retention depends on two conditions: the nature of the liquid phase and the column temperature. The liquid phase is the most important selection, and one should always try to choose as selective a phase as possible.

In Table 18-3 retentions for various solute classes are tabulated for three types of phases: nonpolar, one of intermediate polarity, and a polar phase. With this type of information one can determine in advance whether a mixture can be separated with a particular stationary liquid phase. This information also shows what classes of compounds can readily be separated from other classes.

Example 18-2

On which liquid substrate would the lower acetate esters be separated best? For the three stationary liquid phases given below, these relative retentions

(n-pentane = 1) are available at a column temperature of $100°C$:

	B.P., °C	Silicone Oil DC 200	Di-2-ethyl-hexyl sebacate	Poly (diethylene glycol succinate)
n-Propyl acetate	102°	4.88	6.42	30.4
		⟩ 1.95	⟩ 2.10	⟩ 1.50
Ethyl acetate	77°	2.50	3.05	20.2
		⟩ 1.62	⟩ 1.83	⟩ 1.26
Methyl acetate	57°	1.54	1.67	16.0

On the basis of the relative retentions: propyl/ethyl and ethyl/methyl, the di-2-ethylhexyl sebacate column is slightly superior to the silicone oil substrate; however, the silicone oil column is somewhat faster. Because of extremely long retention times, the glycol column is not considered.

   Generally a graph of log retention volume (or time) against the number of carbon atoms for members of a homologous series yields a smooth curve. Often the regular variation of boiling points parallels the retention volume. Several homologous series are graphed in Fig. 18-13. By utilizing the selectivity of particular liquid phases, one can gain a great deal of information about an unknown or a mixture of unknowns. On each of two columns—one containing a polar and the other a nonpolar liquid phase—a series of compound classes (n-alkanes, alkylbenzenes, alcohols, ketones, etc.) are run to determine the retentions of each compound. By plotting the retention values for the two stationary phases against each other, lines radiating from the origin are obtained (one for each homologous series). The slopes of the lines are dependent on the relative interactions of

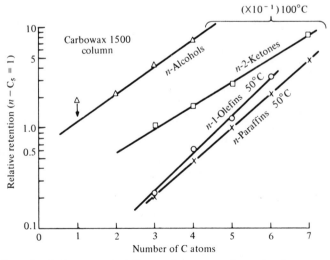

Fig. 18-13. Plot of relative retention (log scale) vs. number of carbon atoms for several homologous series.

each series with the two stationary phases. Radial areas on this diagram (Fig. 18-14) may then be allocated to the various types of molecular structure. A disadvantage of this plot is that points are spaced along the lines in a distribution logarithmic to molecular weight and thus tend to be crowded into the corner near the origin. By plotting the logarithms of the retentions against each other, a corresponding series of approximately parallel lines is obtained, with points spaced linearly with molecular weight.

The Rohrschneider system is valuable for predicting separations. Referring back to Eq. 18-1, the terms a, b, c, d, and e are constants for the substance being analyzed. These constants can be determined for a variety of substances on a column for which Rohrschneider constants are available. The steps are: (1) determine the retention index (Kovats) for the substance on each column, and (2) calculate the difference between the retention index on the column and the retention index on a squalane column. There must be a minimum of five ΔI determinations, each obtained on a different column in order to calculate the a, b, c, d, and e constants. Once obtained, these constants are valid for the substance on *any column*.

Temperature Dependence The relationship between relative retention and column temperature is expressed by the equation:

$$\log \alpha = -\frac{(\Delta H_2 - \Delta H_1)}{RT} \qquad (18\text{-}20)$$

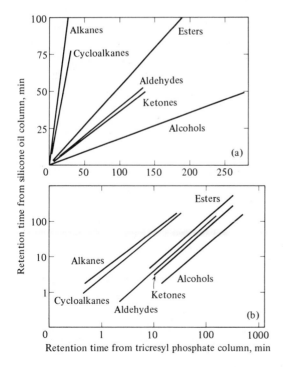

Fig. 18-14. Two-column plots: (a) linear and (b) logarithmic. (After J. S. Lewis, H. W. Patton, and W. I. Kaye, *Anal. Chem.*, 28, 1370 (1956). Courtesy of American Chemical Society.)

Because the transfer of solute molecules from the moving to the stationary phase is an exothermic process in many cases, ΔH_1 and ΔH_2 are often negative. Moreover, the parenthetical term will be positive because the final component of an adjacent pair will have the higher partial molal heat of solution in the liquid state. Thus the relative retention will decrease with increasing temperature; however, there are cases where the opposite is true. Increasing the temperature lowers the retention time on the order of 5 percent per degree increase; roughly a decrease of 20°C will approximately double the retention time. A knowledge of ΔH and V_g for a given solute immediately makes it possible to draw the graph of inverse temperature versus log V_g because the former quantity enables one to calculate the slope of the line, whereas the latter provides an ordinate value at one temperature. Among members of a homologous series, the solutes have the same specific retention volume at the column temperature corresponding to their boiling points.

The results for the separation of several alkylbenzenes on a column of tricresyl phosphate are shown in Fig. 18-15. In particular, the degree of separation for the pairs toluene/p-xylene and p-xylene/isopropylbenzene varies considerably with temperature. At about 190°C ($10^4/T = 21.6$) p-xylene is eluted before isopropylbenzene and the relative retention is about 1.3 (requiring a column with 200 plates to achieve a resolution of 1.0, or 3% overlap of peak areas); whereas at 120°C the relative retention has decreased to about 1.2 (requiring about 450 plates). Extrapolation of the plot for isopropylbenzene shows that it intersects that for p-xylene at 70°C. If further extrapolation is justified, the order of elution below 70°C would be reversed, whereas at 70°C separation could never be affected in a column of this liquid phase. Note also that at higher temperatures than 70°C, although the separation of p-xylene/isopropylbenzene improves, that of toluene/p-xylene worsens.

When selecting the column temperature, one should not forget the limitations imposed by the maximum/minimum temperature range for the liquid phase (Table 18-1). Selection of the temperature of an isothermal column is a complex problem, and usually one has to make a compromise, often dictated by the pair of solutes most difficult to separate.

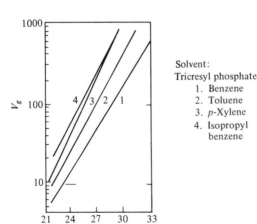

Solvent:
Tricresyl phosphate
1. Benzene
2. Toluene
3. p-Xylene
4. Isopropyl benzene

Fig. 18-15. Temperature dependence of aromatic compounds. Solvent: tricresyl phosphate; 1. benzene, 2. toluene, 3. p-xylene, 4. isopropyl benzene. (After D. Ambrose, A. I. M. Keulmans, and J. H. Purnell, *Anal. Chem.*, **30**, 1582 (1958). Courtesy of American Chemical Society.)

Optimization of Operating Conditions

After the number of plates required for a desired degree of resolution has been determined, one can proceed to design optimum operating conditions. The time required for separation is t_p, the time needed to get the solute band through one plate, multiplied by the number of plates required. Thus,

$$t = Nt_p \qquad (18\text{-}21)$$

and t_p is given by the plate height divided by the band (or zone) velocity, $u/(1 + k)$, or

$$t_R = N_{\text{req}}(1 + k)\frac{H}{u} \qquad (18\text{-}22)$$

where $1/(1 + k)$ is the retardation factor, that is, the fraction of time spent by a solute molecule in the mobile phase. The H/u ratio can be obtained directly from the experimental plate height/velocity (Van Deemter) graph; it is the slope of the line drawn from the origin to a point on the graph, as shown in Fig. 18-16. This slope decreases as flow velocity increases, but a point of diminishing return is reached at a velocity much greater than twice the optimum velocity. Once the velocity has been selected, the column length necessary for the separation is NH. It is always better to use higher gas velocities, with corresponding increase in column length. Although a longer column is needed, the analysis time is less. The value of the H/u ratio will generally be larger for a packed column than for a wall-coated tubular column, and the value of the latter column will be larger than for a surface-coated open tubular column. Thus, in respect to the gas velocity, except for a few cases, a separation can always be performed faster on an open tubular column than on a packed column, and a SCOT column generally permits faster analysis than the WCOT column.

Combining Eq. 18-22 with 18-19, the following relationship is obtained:

$$t_{R,2} = 16R^2 \left(\frac{\alpha}{\alpha - 1}\right)^2 \left[\frac{(1 + k_2)^3}{k_2^2}\right] \frac{H}{u} \qquad (18\text{-}23)$$

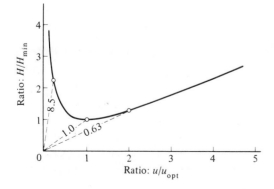

Fig. 18-16. Plate height–velocity curve redrawn as $H/H_{\min}$ vs. v/v_{opt}. Shortest possible separation time for any point on the curve is proportional to the slope of the line joining that point to the origin.

Let us examine the capacity factor that is contained in the third term, namely $[(1 + k_2)^3 / k_2^2]$. This term has a minimum at $k = 2$. Conditions should therefore be selected where the capacity factor is between 1 and 4 (because in this range the value of this quantity in brackets varies between 6.75 and 8). Of course this is not an easy task because the capacity factor depends on the type of column and the temperature. If the liquid phase is fixed, k is inversely proportional to the phase ratio, β. A packed column has a phase ratio of about 15–20, a WCOT column a value around 100, and a SCOT column a value of about 60–70. Thus, the capacity factor (partition ratio) will always be smaller on an open tubular column than on a packed column prepared with the same liquid phase and operated at the same temperature. For example, if $k = 2$ on a packed column, the value at the same temperature would be 0.3–0.4 for a WCOT column and about 0.5–0.6 for a SCOT column. In column of high β, solutes with short retention times would be difficult to separate since from Eq. 18-19 the term would be near zero. The whole advantage of the high plate numbers given by open tubular columns may be lost if they are operated under conditions in which k is fractional. For this reason, open tubular columns are operated at appreciably lower temperatures than are packed columns. Also open tubular columns are particularly adaptable to the analysis of samples having components whose volatilities vary widely and whose late peaks become very wide and flat when using packed columns at isothermal temperatures.

Programmed-Temperature Gas Chromatography

The separation of constituents in samples composed of compounds with a wide range of boiling points can be improved and accelerated by raising the temperature of the entire column at a uniform rate during the analysis. The particular advantage in raising the temperature is that temperature has a greater effect on the chromatographic process than any other single variable. The sample is injected into a relatively cool column in the normal manner. Earlier peaks, representing low-boiling constituents, emerge essentially as they would from an isothermal column operated at a relatively low temperature. However, the volatilities of the higher-boiling materials in the stationary phase are so low that these substances are almost completely immobilized at the inlet of the column. As the temperature is raised, volatilities will increase, and the remaining compounds will successively reach temperatures at which they have significant vapor pressures and will begin to migrate and ultimately emerge from the column. On a low-temperature iso-thermal column, these higher-boiling materials would emerge as flat peaks, often undetected, whereas by temperature programming they will be kept bunched by the rapidly increasing temperature. As a consequence, an extremely wide boiling range of compounds may be separated in less time, and the peaks on the chromatogram are sharper and more uniform in shape. In fact peak heights may be used as the basis for accurate quantitative analysis with programmed-temperature gas chromatography.

Programming of column temperature is achieved by electrically or mechanically raising the set point of a temperature controller. Using a linear temperature program, the temperature rises at a rate of $r°/min$. Heating from ambient to 400°C within 8 min is possible, with cooling from $400°$ to $100°C$ in 3 min. A low thermal mass of the oven is essential and is achieved by constructing the inner compartment from thin-gauge stainless steel. It is separated from the outer insulated compartment by an air space through which air is drawn rapidly on the cooling cycle. In one arrangement the heater is suspended around the blower. Temperature gradients may be kept within $1°C$ with commercial units. Subambient programming extends the technique to extremely volatile components.

A step-function approximation can be used to follow the solute migration as a function of temperature. Let us assume that the temperature increases in steps of $30°C$ to halve the partition ratio k each unit time. Further, let us assume the air peak time t_M to be 0.5 unit and $k = 40$. The isothermal retention time is given by

$$t_R = (k + 1)t_{air}$$

$$= (40 + 1)0.5 = 20.5 \text{ units}$$

In unit time the fraction traveled by the band is $L/20.5 = 0.049L$ of the column. During the next step, when the effective k is 20,

$$t_R = (20 + 1)0.5 = 10.5$$

and the fraction of column traveled by the band is $1/10.5 = 0.095$; the total distance being $0.049 + 0.095 = 0.144$ column length. During the succeeding steps,

$$t_R = (10 + 1)0.5 = 5.5$$

and the fractional distance is 0.18 for a total distance of 0.32 column length;

$$t_R = (5 + 1)0.5 = 3.0$$

and the fractional distance is 0.33 for a total distance of 0.65 column length; and finally

$$t_R = (2.5 + 1)0.5 = 1.75$$

and the fractional distance traveled is 0.57. However, only 0.35 column length remains to be traversed, or $0.35/0.57 = 0.61$ of a time unit. Thus, total retention time is 4.61 units.

The distance migrated by the peak in successive $30°C$ temperature increments is shown in Fig. 18-17. In the last $30°C$ interval the distance migrated is about half the

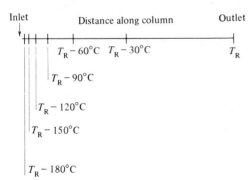

Fig. 18-17. Schematic diagram of distance migrated by a peak in successive $30°C$-increments in programmed-temperature chromatography from $80°$ to $260°C$.

column length. Generally, the heating rate r can be estimated from the relation: $rt_M \leqslant 12°C$.

LABORATORY WORK

Experiment 18-1   Efficiency of Chromatographic Column

Inject 1 μl of some pure compound (and include some air) into the column. Operate the column at some convenient temperature. Adjust the flow of carrier gas to about 140 ml/min for the run (assuming 0.25-in. i.d. column diameter).

When the peak has eluted, lower the flow rate to 120 ml/min, inject a fresh sample, and record the elution curve. Repeat the experiment at flow rates of 100, 80, 60, and 40 ml/min.

Calculate the plate number for your compound at each of the flow rates. From the column length, calculate the plate height at each flow rate. Plot the values of plate height against the linear gas velocity (which is calculated by dividing the length of the column by the retention time for the air peak). Estimate the coefficients of the terms in the plate height—velocity equation. Determine the optimum linear gas velocity and the minimum plate height.

Experiment 18-2   Effect of Temperature on Retention Behavior

Stabilize the column at 150°C (if compatible with the liquid-phase packing). Adjust the carrier-gas flow to the optimum value, and inject a sample containing several adjacent members of a homologous family, or members separated by two carbon atoms for higher molecular weights.

Repeat the experiment at different column temperatures: 120°, 100°, 80°C.

Tabulate the retention time for the components at each column temperature. Graph the logarithm of the retention time (corrected for the air peak) vs. the carbon number.

If sufficient information is available, calculate the partition coefficient of each component. Graph the logarithm of the partition coefficient vs. the reciprocal of the absolute temperature. From the slope of the graph, calculate the heat of vaporization for each component.

Experiment 18-3   Qualitative Analysis

Stabilize the column temperature and adjust the carrier-gas flow. Inject a 2-μl sample of some commercial fluid (such as lighter fluid) or a synthetic mixture of organic liquids. Also run chromatograms of appropriate known compounds. For lighter fluid, these compounds should be run: benzene, toluene, and at least two normal paraffins.

Establish the identity of each peak in the chromatogram of the unknown sample by reference to runs of known materials and by comparison with tabulated relative retentions. To use tabulated retention values, it may be advisable to adjust the column temperature and carrier-flow rate to conform with the tabulated conditions. Use of two columns packed with a polar and a nonpolar substrate will materially aid in the identification.

Experiment 18-4   Resolution of Mixtures

Stabilize the column temperature and adjust the carrier-gas flow to the optimum value. Inject a mixture known to give overlapping peaks under the particular operating conditions.

Repeat the experiment using unequal amounts of the overlapping components in the mixture.

Determine the plate number for each component and calculate the resolution between the two peaks. Estimate the fractional impurity for each peak. Revise operating conditions to improve the resolution.

PROBLEMS

1. The following data were obtained on a 25% dinonyl phthalate, 80/100 mesh Chromosorb P column, 91.5 cm in length, operated at 53°C, which contained 2.50 g of dinonyl phthalate (density $= 0.9712$ g/cm^3 at 20°C) and 7.50 g of oven-dried Chromosorb. Recorder speed was 2.54 cm/min.

Compound	F_c (ml/min)	t'_R (cm)	P_0 (cm)	P_i (cm)
Benzene	48.0	52.0	75.8	122.2
Cyclohexene	51.7	42.1	75.2	123.9
Cyclohexane	52.6	31.0	75.2	124.4

For each compound, calculate (a) the adjusted retention volume, (b) the net retention volume, (c) the specific retention volume (at 0°C), and (d) the partition coefficient.

2. In Problem 1, the base width of the peaks (in cm) was: benzene, 9.1; cyclohexene, 7.4; and cyclohexane, 5.4. Calculate (a) the average number of plates and (b) the plate height for the column.

3. From the information in Problems 1 and 2, what plate number would be needed to resolve adjacent compounds within (a) 1.0% and (b) 0.1%? What length column, packed as in Problem 1, would provide the separation?

4. The data in Table 18-4 were obtained from a 360-cm column packed with 30/50 mesh Chromosorb P and various amounts of hexadecane as the liquid substrate.

Carrier gas was hydrogen. Column operating temperature was 30°C. Solute was propane. Chart speed was 61.0 cm/hr.

TABLE 18-4

Column Description	Flow Rate, liters/hr	P_i, mm Hg	P_0, mm Hg	Adjusted Retention Volume, mm	Peak Width (Base), mm
31% (w/v) substrate	1	799	766	204.2	25.2
V_S = 22.7 ml	2	785	722	108.1	12.7
V_M = 59.5 ml	4	788	667	57.3	7.7
Column volume, 100 ml	6	787	613	40.9	6.1
Liquid cross section,	10	799	532	26.7	4.7
0.0629 cm²					
Free gas cross section,					
0.165 cm²					
23% (w/v) substrate	1	794	768	157.4	19.2
V_S = 15.3 ml	2	783	731	78.0	8.1
V_M = 68.6 ml	3	785	711	54.1	5.3
Column volume, 102 ml	4	785	686	40.6	4.0
Liquid cross section,	5	779	655	32.8	3.3
0.0426 cm²	10	785	563	18.1	2.1
Free gas cross section,					
0.191 cm²					
13% (w/v) substrate	2	769	715	60.1	6.3
V_S = 9.37 ml	4	769	661	31.2	2.7
V_M = 77.4 ml	6	788	626	21.4	1.8
Column volume,	10	790	542	14.5	1.25
105.5 ml					
Liquid cross section,					
0.0233 cm²					
Free gas cross section,					
0.215 cm²					

(a) For each column, calculate the plate number for each flow rate. (b) From column length and plate number, compute the plate height for each value of linear carrier-gas velocity at column temperature and average column pressure. (c) Graph the values of plate height against the linear velocity. (d) For each column, estimate the terms in the equation of plate height vs. linear gas velocity. (e) Estimate the optimum linear gas velocity for each column. Comment on the change in shape of the plate-height plot as the amount of liquid substrate is decreased. (f) Calculate the values of the partition coefficient and the partition ratio for propane on each of the columns.

5. The performance of methylnaphthalenes on columns of 80/100 mesh Chromosorb P and on 200/230 mesh microbeads is given in Table 18-5. All columns were 152 cm in length and 0.6 cm in diameter. Liquid substrate was silicone oil 710 [density = $1.0 - (0.0008t_c)$].

TABLE 18-5

Column Number	1	2	3	4
Solid support	Glass Beads		Chromosorb P	
Liquid loading, %(w/w)	0.16	3	10	30
Weight of liquid, g	0.060	0.40	1.44	5.55
Temperature, °C	90°	100°	142°	182°
Pressure ratio, P_i/P_0	5.33	4.33	3.26	3.67
Flow rate, ml/min	155	415	192	208
Linear velocity, cm/sec	41.7	49.3	27.2	36.1
V_N, ml 1-MeN	365	1350	870	1020
V_N, ml 2-MeN	325	1200	780	910
Base width, ml 1-MeN	32	127	72	92
Base width, ml 2-MeN	34	115	64	77
V_M, ml	10	22	18	15

For each methylnaphthalene on each column, calculate (a) the partition coefficient at the column temperature, (b) the partition ratio, (c) the plate number, (d) the phase ratio, and (e) the resolution obtained.

6. What number of plates would be required to effect a separation with 0.1% impurity in adjacent bands ($\alpha = 1.1$) in these 3 cases: Case 1, using ordinary packed columns with relatively large values of partition ratio ($k = 5$–100). Case 2, using open tubular columns with $k = 0.5$. Case 3, using open tubular columns with $k = 3$.

7. Tabulated are elution data obtained on a liquid paraffin column:

Solute	ΔH_S, kcal/mole	V_g (at 100°C)
n-Hexane	6.80	17.8
n-Heptane	7.32	40.0
1-Hexene	7.45	18.2
1-Heptene	8.78	39.5
2-Hexene	7.60	26.0
2-Heptene	8.52	55.0

(a) Graph the inverse temperature plot for each solute. (b) Select an isothermal operating temperature that provides an optimum separation of all solutes from each other. To achieve a fractional impurity of 0.01 (1%), how many plates would be needed? (c) On the n-alkane curves, note the specific retention value corresponding to their boiling points. (d) Given that V_g (at 100°C) is 8.2 ml/g for n-pentane (whose boiling point is 36°C), calculate the ΔH_S value.

8. The relative retentions (n-pentane = 1.00) of n-alkanes and alkylbenzenes at three boiling-point levels are given for Convoil-20 (a saturated hydrocarbon oil) and tricresyl phosphate substrates on columns operated at 100°C:

Boiling-Point Level, °C	Convoil-20			Tricresyl Phosphate		
	60	100	140	60	100	140
n-Alkanes	1.80	5.6	17.0	1.65	4.6	13.5
Alkylbenzenes	—	6.7	19.8	—	16.4	43.5

(a) Construct a plot of boiling point (°C) vs. log relative retention for each hydrocarbon type and for each substrate (4 graphs in all). (b) Construct a graph of log relative retention on Convoil-20 vs. log relative retention on tricresyl phosphate for each class of compounds. (c) Utilize these graphs to identify the specific n-alkane or alkylbenzene from these relative retentions:

Relative Retentions

Solute	Convoil-20	Tricresyl phosphate
A	2.3	2.10
B	66.0	130
C	8.7	21
D	11.2	9.0
E	1.0	0.95
F	17.5	39
G	5.2	4.4
H	34	70

9. Adjusted retention volumes are given for a series of n-alcohols and acetates. Column temperature was 77°C; helium flow was 89 ml/min.

V'_R, ml

	15% Carbowax 400	15% Nujol
Acetates		
Methyl	43.2	24.0
Ethyl	60.5	48.4
Propyl	106	120
Alcohols		
Methyl	72.3	8.0

Ethyl	110	14.1
n-Propyl	207	32.7
n-Butyl	408	85.5

(a) Prepare a graph of log retention volume against carbon number for each homologous series on each column substrate. (b) Plot the log retention volume on one column against the log retention volume on the other for each family. (c) Estimate the adjusted retention volume on each column for *n*-butyl acetate and for *n*-amyl alcohol.

10. A series of methyl esters of the fatty acids were chromatographed. Time (in min) at peak maximum for known saturated esters were: $C_{12} = 2.65$, $C_{14} = 4.6$, and $C_{20} = 27.0$. On a sample, run under identical conditions, peaks were observed at 2.55, 8.3, 15.2, 26.9, and 48 min. Which esters were present in the sample?

11. Specific retention volumes (in ml) for some chlorinated hydrocarbons and benzene at several column temperatures on three column substrates are listed:

Temp., °C	CH_2Cl_2	$CHCl_2-$ CH_2	$CHCl_3$	CCl_3CH_3	CCl_4	$CCl_2=$ $CHCl$	$CCl_2=$ CCl_2	C_6H_5Cl	C_6H_6
Paraffin									
74	29.5	51.3	74.3	108	141	189	568	677	136
97	17.0	29.5	41.5	57.8	76.5	98.2	273	323	74.3
125	8.08	14.2	19.6	28.0	34.5	42.6	105	124	—
Tricresyl Phosphate									
74	56.1	71.8	137	96.3	99.8	181	359	826	133
97	30.3	38.1	68.5	51.6	52.7	92	170	372	69.4
125	13.9	17.9	29.6	24.2	24.7	38.8	69.5	143	—
Carbowax 4000									
74	69.6	54.8	138	57.1	55.8	122	165	548	89
97	31.9	26.3	59.3	28.5	27.4	57.1	78.7	235	43.3
125	13.5	12.1	24.2	13.5	13.5	23.1	31.1	83.6	—

All columns were made of coated Chromosorb packed into 200 cm × 60 mm copper tubing. The paraffin column contained 4.58 g on 13.93 g of support; the tricresyl phosphate column, 4.46 g on 13.71 g; and the Carbowax 4000 column, 4.38 g on 13.1 g. The densities of the liquid phases at 74°C/4°C in g/cm^3 are: paraffin, 0.768; tricresyl phosphate, 1.128; and Carbowax 4000, 1.081. (a) Graph the data and compute the heat of vaporization for each solute on each of the stationary liquid phases. (b) Calculate the partition coefficient at each temperature for each solute on each column substrate.

12. Adjusted retention times at various column temperatures were obtained for N-methylaniline on a 20% by weight of Ucon 50-HB-5100 as stationary liquid phase. These were: 17.2 min at 117°C, 7.1 min at 138°C, and 2.3 min at 175°C. (a) Calculate the heat of vaporization of N-methylaniline. (b) Estimate the retention times at 150°, 200°, and 225°C.

13. Given, at various column temperatures, are retention times relative to N-methylaniline. Other pertinent data are contained in Problem 12.

	Retention Times, min		
Compound	117°C	138°C	175°C
Ethylenediamine	0.057	0.078	
1,3-Propanediamine	0.114	0.141	
1,4-Butanediamine	0.229	0.251	
1,5-Pentanediamine	0.406	0.433	
1,6-Hexanediamine	0.74	0.74	
1,7-Heptanediamine		1.26	1.12
1,8-Octanediamine		2.11	1.76
1,9-Nonanediamine		3.51	2.72
Triethylenediamine	0.328	0.388	
Diethylenetriamine		0.81	0.84

(a) Graph the log retention times against carbon number for terminal diamines. (b) Graph log t_R' vs. $1/T$ for each compound. (c) How many plates are required for separation of each adjacent peak at 138° and 175°C? The difficulty centers around the 1,5-, 1,6-, and 1,7-terminal diamines and diethylenetriamine and triethylenediamine. (d) If the 1-m column had an efficiency of 1000 plates, would one expect the peaks for triethylenediamine and diethylenetriamine to be adequately resolved from the adjacent terminal diamines?

14. From the information tabulated, graph the relation between log relative retention volume on poly(ethylene glycol adipate) and log relative retention volume on Apiezon M for the methyl esters of saturated and unsaturated fatty acids.

		Relative Retention Values (16:0 = 1.00)	
Common Name of Parent Acid	No. of C Atoms: No. of Double Bonds	Poly(ethylene glycol adipate) at 197°C	Apiezon M at 197°C
Lauric	12:0	0.30	0.18
Palmitic	16:0	1.00	1.00
Stearic	18:0	1.82	2.39
Oleic	18:1	2.04	2.0
Linoleic	18:2	2.44	1.9

Linolenic	18:3	3.13	1.9
	20:0	3.31	5.6
	20:1	3.67	4.7
	20:2	4.46	4.5
	20:3	5.02	3.9

15. Using the double logarithmic graph prepared in Problem 14, identify the specific compounds from the following pairs of relative retention values obtained on poly-(ethylene glycol adipate) and on Apiezon M at 197°C: compound A (0.549, 0.42); compound B (1.15, 0.89); compound C (0.300, 0.18); compound D (0.62, 0.38); compound E (1.43, 0.85); compound F (5.95, 13.0); compound G (1.64, 0.77); compound H (1.33, 1.55); compound I (1.97, 0.74); compound J (11.2, 31.0); compound K (0.155, 0.069).

16. Predict relative retention values for these methyl esters of fatty acids from the graph prepared in Problem 14, plus a graph of log relative retention volume on poly-(ethylene glycol adipate) vs. carbon number for the particular homologous series: compound A (15:0); compound B (18:4); compound C (20:4); compound D (22:0); compound E (22:1).

17. The following data were obtained on an o-nitrophenetole column, 4.25 m in length of 60 mm diameter coiled copper tubing containing 4.0 g of substrate on 10 g of Celite 545. Carrier gas was helium; flow rate, 50 ml/min. t_R is the retention time from injection point to peak maximum. Δt_b is baseline intercept cut by tangents to peak inflections.

Component	Temp., °C	t_R, min	Δt_b, min
Air	25	1.32	—
Butene-1	25	5.95	0.46
trans-Butene-2	25	7.56	0.57
cis-Butene-2	25	8.55	0.65
1,3-Butadiene	25	10.42	0.76
Air	16	1.34	—
Butene-1	16	7.14	0.56
trans-Butene-2	16	9.24	0.70
cis-Butene-2	16	10.55	0.80
1,3-Butadiene	16	12.99	0.95

For each compound, calculate (a) the adjusted retention volume, (b) the plate number and plate height, and (c) the resolution of adjacent pairs of peaks.

18. A series of saturated aldehydes (total amount = 3.0 mg) were chromatographed on a column consisting of 3.3% silicone oil DC 156, operated at 180°C. Adjusted retention times, base widths, and peak heights were:

Component	C_5	C_6	C_7	C_8	C_9	C_{10}
t_R', min	10.1	13.25	18.75	24.3	32.75	45.0
Δt_b, min	3.0	4.5	4.5	5.7	8.5	13
Peak height, mm	16	17	16.5	17	14.5	12.5

(a) Calculate the amount of each aldehyde in the sample. (b) For each peak estimate the theoretical number of plates. (c) For the peaks incompletely separated, how many plates would be required to reduce the impurity in each peak to 3.0%? To 0.1%?

19. Determine the time of analysis for each of these three columns and the assumed conditions: (1) a packed column with a phase ratio of 20; (2) a wall-coated open tubular column with a phase ratio of 120; and (3) a support-coated open tubular column with a phase ratio of 60. Assume that $H = 0.6$ mm on each of these columns, and that the column temperature is 75°C. The partition ratio k of the second peak on the packed column is to be taken as 1, 6, and 30, respectively, in the three comparative cases. Also assume a relative retention of 1.10 for the pair of compounds and that the desired resolution is 1.5. The three columns are operated at the following average linear gas velocities with helium as carrier gas: packed column, 6 cm/sec; WCOT column, 12 cm/sec; and SCOT column, 18 cm/sec.

20. Adjusted retention times are given for the following compounds on a particular column. What is the Kovats retention index of each of these compounds on this column?

n-Butane	0.95	2-Methylbutane	1.20
n-Pentane	1.80	Butene-1	0.80
n-Hexane	3.50	Hexene-1	2.95
n-Heptane	6.95	Benzene	3.75
n-Octane	13.7	n-Butanol	8.4

21. Predict the ΔI values on (a) an OV-1 column, (b) a tricresyl phosphate column, and (c) a Carbowax 4000 column for these substances:

Rohrschneider Coefficients

Substance	a	b	c	d	e
Cyclohexane	32.06	-22.47	-21.64	4.07	29.72
Toluene	108.33	3.77	8.75	-7.01	-7.61
Acetone	-5.30	-4.61	94.94	7.90	5.64
Chloroform	69.71	28.91	-72.62	53.05	-6.29
Dioxane	45.86	-2.89	40.20	-7.49	40.24

22. The separation of C_1-C_4 formates and acetates on Porapak Q is made according to total carbon number. Except for peak 1 (methanol), identify the numbered peaks in the chromatogram shown. Among the formate esters, only n-alkyl members are present.

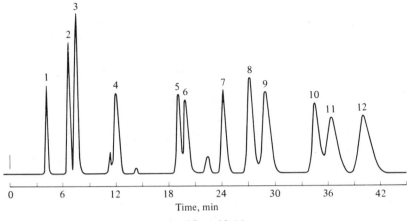

Problem 18-22

23. Castor oil contains principally glycerides of ricinoleic acid. After conversion to the methyl esters by transmethylation, the accompanying chromatogram was obtained as shown in the figure. From the lower trace made by the disk integrator, determine the percentage composition for this mixture.

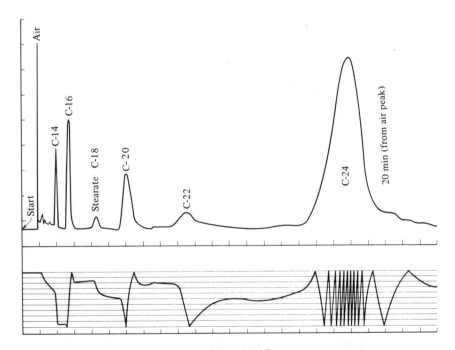

Problem 18-23

BIBLIOGRAPHY

Dal Nogare, S. and R. S. Juvet, Jr., *Gas-Liquid Chromatography*, Wiley-Interscience, New York, 1962.

Ettre, L. S. and W. H. McFadden, *Ancillary Techniques of Gas Chromatography*, Wiley-Interscience, New York, 1969.

Harris, W. E. and H. W. Habgood, *Programmed Temperature Gas Chromatography*, Wiley, New York, 1966.

Helfferich, F. and G. Klen, *Multicomponent Chromatography: Theory of Interference*, Marcel Dekker, New York, 1970.

Jones, R. A., *An Introduction to Gas–Liquid Chromatography*, Academic Press, New York, 1970.

Leathard, D. A. and B. C. Shurlock, *Identification Techniques in Gas Chromatography*, Wiley-Interscience, New York, 1970.

Littlewood, A. B., *Gas Chromatography: Principles, Techniques and Applications*, 2nd ed., Academic Press, New York, 1970.

Purnell, H., *Gas Chromatography*, Wiley-Interscience, New York, 1968.

Zlatkis, A. and V. Pretorius, Eds., *Preparative Gas Chromatography*, Wiley-Interscience, New York, 1971.

LITERATURE CITED

1. Condal-Bosch, L., *J. Chem. Educ.*, **41**, A235 (1964).
2. Ettre, L. S., *Anal. Chem.*, **36** (8), 31A (1964).
3. Glueckauf, E., *Trans. Faraday Soc.* **51**, 34 (1955).
4. Kovats, E., *Helv. Chim. Acta*, **41**, 1915 (1958).
5. Rohrschneider, L., *J. Chromatogr.*, **22**, 6 (1966); *Advances in Chromatography*, Vol. IV, Marcel Dekker, New York, 1967.

19

Introduction to Electrometric Methods of Analysis

Electroanalytical chemistry encompasses a wide variety of techniques, each based on a particular phenomenon occurring within an electrochemical cell. Each basic electrical measurement—current, resistance, and voltage—has been used alone or in combination for analytical purposes. If these electrical properties are measured as a function of time, many other techniques are possible. Table 19-1 contains a brief summary of the various electrometric methods, classified according to the quantity measured. Electrometric methods can be grouped into two categories: *steady state* and *transient* or dynamic. In the former category the time variable is eliminated, leaving current, potential, and conductance. Time is effectively eliminated by stirring the solution with the electrode held static, or vice versa. In this way the concentration gradients of reacting species at an electrode surface are time independent, but related to bulk and surface concentration through the Nernst diffusion layer approximation. In transient methods, both the electrode and solution are static; then the concentration gradients at the electrode surface are time dependent. This category includes such methods as chronopotentiometry, chronoamperometry, and polarography. When either impressed potential or current are used, the dependent variable is a function of both concentration and time. If the data are taken in the form of current versus potential, time still enters the experiment because the value of the current observed for a given potential depends upon conductance and upon the elapsed time between application of the potential and the measurement.

The general relationships between current, potential, and composition of an electroactive system are depicted in the three-dimensional representation of Fig. 19-1. The various types of two-dimensional variations observed in the diverse techniques can be visualized from the intersection with the solid surface of a plane perpendicular to a particular axis. For example, at any particular concentration a polarographic curve can be obtained by changing the voltage and observing the current. Amperometric titration curves are obtained by observing the current at a constant voltage. A plane corresponding to zero current intersects the surface to yield the usual potentiometric titration curve.

Electrometric methods are characterized by a high degree of sensitivity, selectivity, and accuracy. Highly refined ways of making electrical measurements permit reliable determinations at the submicroampere and microvolt range. This means that analytical

TABLE 19-1 A Summary of Electrometric Methods of Analysis

Quantity Measured	Name or Description of the Method

Electromotive force vs. concentration of component of interest	1. At zero faradaic current
	  (a) Measurement of ionic activities or concentrations
	  (b) Potentiometric titration
	  (c) Null-point potentiometry
	2. With net faradaic current
	  (a) Potentiometric titration at constant current

Electromotive force vs. time (at constant current)	Chronopotentiometry
Weight of separated phase	Electrogravimetry
Resistance (conductance)	1. Conductometric titration 2. Concentrations measured by resistance after calibration with known mixtures
Current-voltage	Polarography, voltammetry
Current vs. concentration of component of interest	1. Amperometric titration 2. Amperometry at small constant applied potential
Coulombs (current × time)	1. Coulometry at constant electrode potential 2. Coulometric titration 3. Stripping analysis

sensitivity, limited largely by noise considerations, approaches and even exceeds the 10^{-10} molar level. Because fractions of a drop can be analyzed, reliable analyses at the subnanogram range are possible. Electrochemical selectivity minimizes preseparations, providing a further advantage. Remote control, in-line installations, and automation are possible with these techniques.

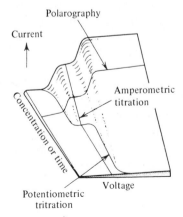

Fig. 19-1. Three-dimensional representation of the relationship between current, potential, and composition of a system. (After C. N. Reilley, W. D. Cooke, and N. H. Furman, *Anal. Chem.*, **23**, 1226 (1951). Courtesy of American Chemical Society.)

TYPES OF ELECTROCHEMICAL CELLS

There are two types of electrochemical cells: galvanic (or voltaic) and electrolytic. A *galvanic cell* consists of two electrodes and one or more solutions (i.e., two half-cells) and is capable of spontaneously converting chemical energy more or less completely into electrical energy and supplying this energy to an external source. In these cells, a chemical reaction involving an oxidation at one electrode and a reduction at the other electrode occurs. The electrons evolved in the oxidation step are transferred at the electrode surface, pass through the external circuit, and then return to the other electrode where reduction takes place. When one of the chemical components responsible for these reactions is depleted, the cell is no longer capable of supplying electrical energy to an external source and the cell is "dead."

If electrical energy is supplied from an external source, the cell through which the current is forced to flow is called an *electrolytic cell*. Electrochemical changes are produced at the electrode/solution interfaces, and concentration changes are produced in the bulk of the system. Actually, a galvanic cell is built up from the products of the electrolytic cell that accumulate at the electrodes. If the external current is turned off, the products tend to produce current in the opposite direction.

At the exact point where the galvanic emf is opposed by an equal applied emf, no current flows through the cell in either direction. In this condition of null balance, the potential generated at the interface of an indicator electrode will reflect the composition of the solution phase, provided that the indicator electrode is selected so that its potential is sensitive to the desired component in the solution phase. However, it is not possible to measure the potential of a single electrode, because any electrical contact between the bulk of the solution and an external circuit is itself another electrode/solution interface. It is only possible to measure the potential of one electrode relative to another—the reference electrode. Before we proceed, the matter of electrode potentials requires clarification.

ELECTRODE POTENTIALS

Two major factors determine the electrode potential relative to another electrode. First is the electrolytic solution pressure of the element, which is the tendency of an active element to send its ions into solution. At a given temperature this is a characteristic constant for a stable form of an element, but it varies if the electrode is strained mechanically or if a metastable crystalline form of the metal is present. Second is the activity of the dissolved ions of the element, which in turn varies with concentration at constant temperature.

Following the convention adopted by the International Union of Pure and Applied Chemistry at Stockholm in 1953,* electrode potentials are regarded as the emf of cells

---

*A complete account of the several sign conventions and the IUPAC recommendations can be found in a paper by T. S. Licht and A. J. deBéthune in *J. Chem. Educ.*, **34**, 433 (1957).

formed by the combination of an individual half-cell with a standard hydrogen electrode, any liquid junction potential which arises being set at zero. Thus, when the emf of each half-cell is mentioned, what is actually implied is the emf of the cell:

$$\text{Pt, } H_2(1 \text{ atm}) \mid H^+(m = 1.228) \parallel M^{n+}(a = 1), M^\circ$$

standard hydrogen electrode   individual half-cell
liquid junction

The emf is divided into two contributory electrode potentials, $E^\circ_{H^+, H_2}$ and $E^\circ_{M^{n+}, M^\circ}$, the cell emf being their difference. If all the substances participating in the reversible operation of the cell at a particular temperature are in their standard states, the free energy change of the cell reaction

$$\frac{n}{2} H_2 + M^{n+} = n H^+ + M^\circ \tag{19-1}$$

will have its standard value ΔG°, and the emf of the cell will be the standard cell emf, E°_{cell}. These are related by the expression

$$\Delta G^\circ = -nFE^\circ_{cell} \tag{19-2}$$

in which F is the value of the faraday. When the cell reaction is a spontaneous one, ΔG is negative; this requires the cell emf to be positive. The convention universally adopted is that the standard potential of the hydrogen electrode shall be taken as zero at all temperatures, thus setting up the normal hydrogen electrode (NHE) scale of electrode potentials. Returning to the expression for the emf of the cell, the electrode potential of the metal half-cell is equal in sign and magnitude to the electrical potential of the metallic conducting lead on the right when that of the similar lead on the left is taken as zero. The expression implies further that a reaction, as shown in Eq. 19-1, occurs when positive electricity flows through the cell from left to right. If this is the direction of the current when the cell is short-circuited, the emf of the half-cell (a reduction) will be positive, the reaction will proceed spontaneously (a galvanic cell), and the free energy change will be negative. Standard electrode potentials for a number of selected half-cell reactions are given in Appendix A.

The individual half-cell on the right is written to represent a metal/metal ion electrode reaction. It could equally well have been written to represent another gas/ion system or ion/ion system; each of these reactions take place at an inert metal electrode, such as gold or platinum.

When the electromotive forces of the half cells

$\text{Zn} \mid \text{Zn}^{2+}$

$\text{Ag, AgCl} \mid \text{Cl}^-$

$\text{Pt} \mid \text{Fe}^{2+}, \text{Fe}^{3+}$

are intended, these being oxidation reactions, the reactions implied are

$\text{Zn} \mid \text{Zn}^{2+} \parallel H^+ \mid H_2, \text{Pt}$	$\text{Zn}^\circ + 2H^+ \rightarrow \text{Zn}^{2+} + H_2$	(19-3)
$\text{Ag, AgCl} \mid \text{Cl}^- \parallel H^+ \mid H_2, \text{Pt}$	$\text{Ag}^\circ + \text{Cl}^- + H^+ \rightarrow \text{AgCl} + \frac{1}{2}H_2$	(19-4)
$\text{Pt} \mid \text{Fe}^{2+}, \text{Fe}^{3+} \parallel H^+ \mid H_2, \text{Pt}$	$\text{Fe}^{2+} + H^+ \rightarrow \text{Fe}^{3+} + \frac{1}{2}H_2$	(19-5)

These electromotive forces should not be called electrode potentials, although they may be denoted oxidation potentials.

To lessen the confusion that has arisen over the years concerning the two terms— electrode potential, an observed, invariant physical quantity, and the emf of a half-reaction, which may be defined as a reduction reaction or as an oxidation reaction—it should be remembered that the two terms are distinctly different. The sign of the emf of a half-reaction depends on the direction in which the reaction is written. Only when written as a reduction reaction will the sign of the emf of the half-reaction correspond to the sign of the electrode potential.

If direct electrical measurements prove impractical, the position of a couple in the standard electromotive series may be determined by thermochemical measurements, from equilibrium studies, or from kinetic experiments that show whether the half-cell is oxidizing or reducing relative to couples of known potentials.

Effect of Concentration on Electrode Potentials

The potential E of any electrode is given by the generalized form of the Nernst equation

$$E = E° - \frac{RT}{nF} \ln \frac{a_{red}}{a_{ox}} = E° - \frac{2.3206RT}{nF} \log \frac{a_{red}}{a_{ox}} \tag{19-6}$$

Where $E°$ is the standard electrode potential, R is the molar gas constant (8.316 VQ/degree), T is the absolute temperature, n is the number of electrons transferred in the electrode reaction, F is the faraday, and a_{ox} and a_{red} are the activities of the oxidized and reduced forms, respectively, of the electrode action. If concentrations are substituted for activities, common logarithms for natural logarithms, and numerical values inserted for the constants, assuming the temperature to be 25°C, the Nernst equation becomes

$$E = E° - \frac{0.05915}{n} \log \frac{[red]}{[ox]} \tag{19-7}$$

A change of one unit in the logarithmic term changes the value of E by $59.15/n$ mV. For many analytical purposes, a system is considered quantitatively converted when 0.1% or less of the original electroactive species remains. For a metallic ion-metal system, such as the $Ag^+/Ag°$ system,

$$E = E° + 0.0591 \log [Ag^+] \tag{19-8}$$

the value of the electrode need shift by only $3 \times 0.0591 = 0.177$ V, or in general, by $3 \times 0.0591/n$ V for a quantitative conversion. On the other hand, for an ion-ion system, such as Fe^{3+}/Fe^{2+},

$$E = E° - 0.0591 \log \frac{[Fe^{2+}]}{[Fe^{3+}]} \tag{19-9}$$

the shift would depend on the original concentration of both ions.

Effect of Complex Formation on Electrode Potentials

The effect of reagents that can react with one or both participants of an electrode process will be examined next. The simplest case involves a single ionic species formed over a range of concentrations of complexing agent. A typical example is the silver ion/ silver metal couple in the presence of aqueous ammonia, where the $Ag(NH_3)_2^+$ complex ion constitutes the major ionic species in the solution phase.

The formation of the silver diammine complex is represented by the equilibrium

$$Ag^+ + 2NH_3 \rightleftharpoons Ag(NH_3)_2^+ \tag{19-10}$$

for which the formation constant is written as

$$K_f = \frac{[Ag(NH_3)_2^+]}{[Ag^+][NH_3]^2} = 6 \times 10^8 \tag{19-11}$$

For the half-reaction involving the silver ion/silver couple,

$$Ag^+ + e^- = Ag^\circ \tag{19-12}$$

the Nernst equation is expressed by Eq. 19-8. Combining Eq. 19-11 with Eq. 19-8 yields the potential of a silver electrode in aqueous ammonia systems,

$$E = E^\circ + 0.0591 \log \frac{1}{K_f[NH_3]^2} + 0.0591 \log [Ag(NH_3)_2^+] \tag{19-13}$$

The shift in electrode potential caused by the complexing agent is contained in the second term of Eq. 19-13.

For a couple involving two oxidation states of a metal in solution, such as the aquo-cobalt species,

$$Co^{3+} + e^- = Co^{2+}; \quad E^\circ = 1.84 \text{ V} \tag{19-14}$$

in the presence of aqueous ammonia, both the cobalt(II) hexammine and the cobalt(III) hexammine species predominate. The respective formation constants are

$$\frac{[Co(NH_3)_6^{2+}]}{[Co^{2+}][NH_3]^6} = K_f' = 10^5$$

$$\frac{[Co(NH_3)_6^{3+}]}{[Co^{3+}][NH_3]^6} = K_f'' = 10^{34} \tag{19-15}$$

Substitution of these values into the Nernst equation for the cobalt system gives

$$E = E^\circ + 0.0591 \log \frac{K_f'}{K_f''} + 0.0591 \log \frac{[Co(NH_3)_6^{3+}]}{[Co(NH_3)_6^{2+}]} \tag{19-16}$$

Here the shift in potential is a function of the ratio of the formation constants for each electroactive species. Generally, the higher oxidation state will form the more stable complex and, if it does, the shift in electrode potential will be in the negative direction.

ELECTROCHEMICAL CELLS

For a complete electrochemical cell from which negligible current is drawn, the emf is given by:

$$E_{cell} = E_{ind} + E_{ref} + E_j \qquad (19\text{-}17)$$

where E_{ind}, E_{ref}, and E_j are the indicator electrode, the reference electrode, and the liquid junction potentials, respectively. The indicator electrode is the sensing or probe electrode, the reference electrode is independent of the sample solution composition, and the liquid junction is an interface between dissimilar solutions. In a properly designed system, E_{ref} is a constant and E_j is either constant or negligible. When these conditions are realized, the indicator electrode can supply information about ion activities.

REFERENCE ELECTRODES

Ideally, the reference electrode is of known and constant potential with negligible variation in the liquid-junction potential from one test or calibration solution to another. Reference electrodes contain these components (Fig. 19-2): (1) the actual reference internal half-cell, usually either silver chloride/silver or calomel; (2) the salt-bridge electrolyte; and (3) a small channel in the tip of the electrode through which the salt-bridge electrolyte flows very slowly and electrical contact is made with the other components of the electrochemical cell. The potential of the internal half-cell must not be significantly altered if a small current passes through it (approximately 10^{-8} A or less);

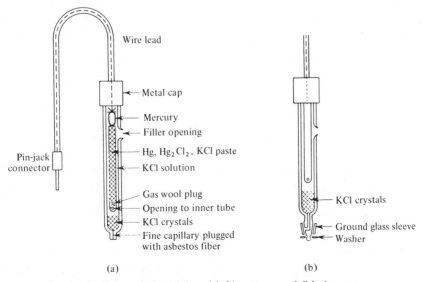

Fig. 19-2. Calomel electrodes: (a) fiber type and (b) sleeve type.

TABLE 19-2   Potentials of Reference Electrodes in Volts as a Function of Temperature (Liquid-Junction Potential Included)

Temperature °C	0.1 M KCl Calomel[a]	Sat'd. KCl Calomel[a]	1.0 M KCl Ag/AgCl[b]
0	0.3367	0.25918	0.23655
5			0.23413
10	0.3362	0.25387	0.23142
15	0.3361	0.2511	0.22857
20	0.3358	0.24775	0.22557
25	0.3356	0.24453	0.22234
30	0.3354	0.24118	0.21904
35	0.3351	0.2376	0.21565
38	0.3350	0.2355	
40	0.3345	0.23449	0.21208
45			0.20835
50	0.3315	0.22737	0.20449
55			0.20056
60	0.3248	0.2235	0.19649
70			0.18782
80		0.2083	0.1787
90			0.1695

[a]R. G. Bates et al., *J. Res. Natl. Bur. Std.*, **45**, 418 (1950).
[b]R. G. Bates and V. E. Bower, *J. Res. Natl. Bur. Std.*, **53**, 283 (1954).

the resistance of the entire electrode must not be too great; the electrode should be easily assembled, and the components should be stable in contact with the atmosphere and at the operating temperature. The standard potential of the reference electrode includes the liquid-junction potential of the electrochemical cell

$$\text{Pt} \mid \text{H}_2(g), \text{H}^+(a = 1) \parallel \text{KCl solution, reference half-cell}$$

Potentials of the common reference electrodes are given in Table 19-2; a more complete compilation appears in *Lange's Handbook of Chemistry*, 11th ed.

Hydrogen-Gas Electrode

The hydrogen-gas electrode consists essentially of a piece of clean platinum foil, coated with a thin layer of finely divided platinum to hasten establishment of the electrical potential. The platinum is capable of making the reaction

$$2\text{H}^+ + 2e^- = \text{H}_2(g)$$

at the platinum–solution interface proceed reversibly. The electrode is immersed in the solution under investigation or a known reference standard and electrolytic hydrogen gas (99.8% purity adequate) at 1 atm pressure is bubbled through the solution and over the

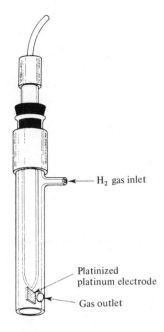

H$_2$ gas inlet

Platinized
platinum electrode

Gas outlet

Fig. 19-3. Hydrogen–gas electrode assembly.

electrode in such a way that the electrode surface and the adjacent solution will be saturated with the gas at all times. Electrode life is 7–20 days after which its response becomes sluggish.[2] One construction is illustrated in Fig. 19-3.

The essential purpose of a hydrogen-gas electrode in an analytical laboratory is to check the accuracy of other reference and indicator electrodes, in particular the errors of electrode combinations. However, the hydrogen-gas electrode is also used to test the magnitude of liquid junction potentials and the accuracy and stability of reference standard solutions. It should be recalled that the hydrogen-gas electrode is the primary standard against which all other electrode potentials are measured.

Calomel Electrodes

Calomel electrodes comprise a nonattackable element, such as platinum, in contact with mercury, mercury(I) chloride (calomel), and a neutral solution of potassium chloride of known concentration and saturated with calomel. The half-cell may be represented by

$$Hg \mid Hg_2 Cl_2 \text{ sat'd}, \quad KCl \, (xM)$$

where x represents the molar concentration of potassium chloride in the solution. The saturated calomel electrode (SCE), in which the solution is saturated with potassium chloride (4.2 M), is commonly used because it is easy to prepare and maintain. For accurate work the 0.1 M or 1.0 M electrodes are preferred because they reach their equilibrium potentials more quickly and their potential depends less on temperature than does the saturated type.

Construction of some commercial versions of calomel electrodes is illustrated in Fig. 19-2. A typical one consists of a tube 5–15 cm in length and 0.5–1.0 cm in diameter. The mercury-mercury(I) chloride paste is contained in an inner tube connected to the saturated potassium chloride solution in the outer tube by means of an asbestos fiber or ground glass seal in the end of the outer tubing. An electrode such as this has a relatively high resistance (2000–3000 Ω) and very limited current-carrying capacity before exhibiting severe polarization.

The saturated calomel electrode exhibits a perceptible hysteresis following temperature changes, due in part to the time required for solubility equilibrium to be established. Those designed for measurements at elevated temperatures have a large reservoir for potassium chloride crystals. Calomel electrodes become unstable at temperatures above $80°C$ and should be replaced with silver/silver chloride electrodes.

In measurements in which any chloride ion contamination must be avoided, the mercury(I) sulfate and potassium sulfate electrode may be used.

Silver/Silver Chloride Electrodes

The silver/silver chloride electrode consists of metallic silver (wire, rod, or gauze) coated with a layer of silver chloride and immersed in a chloride solution of known concentration that is also saturated with silver chloride. The cell formed is

$$Ag \,|\, AgCl_{sat'd}, \quad KCl \,(xM)$$

It is a small, compact electrode and can be used in any orientation. Electrode potentials are known up to $275°C$.

Preparation of the silver chloride coating can be more difficult than fabrication of a calomel electrode. Silver chloride is appreciably soluble in concentrated chloride solution, necessitating the addition of solid silver chloride to assure saturation in the bridge solution, yet entailing the risk that silver chloride may precipitate at the liquid junction when it is in contact with a solution of low chloride-ion content.

In nonaqueous titration studies this electrode occupied a preeminent position for many years, although the calomel electrode can and has been employed in virtually all types of solvent systems. Reproducibility of results vary from ±10 to ±20 mV in the more aqueous solvent mixtures, to ±50 mV in the nearly anhydrous media. Special salt bridges are often necessary.

Salt Bridges and Liquid Junctions

Connection between a separate reference and indicator electrode (or anode and cathode in electrolytic cells) is usually by a junction which allows the passage of ions but does not permit the solutions to mix. Various styles of electrolyte junction have been designed: a ground glass plug or tapered sleeve, a wick of asbestos fiber sealed into glass, an agar bridge rendered conductive by an electrolyte, a porous glass plug, a dual-junction glass rivet, and a flowing junction involving a palladium annulus or capillary drip. No single type can be used in all situations.

At the boundary between two dissimilar solutions or solids, there is always a fairly high resistance that involves an appreciable ohmic drop. A junction potential is always set up. It results from the fact that the mobilities of positive and negative ions diffusing across the boundary are unequal. Because of this difference, one side of the boundary accumulates an excess of positive ions. The junction potential adds to or subtracts from the potential of the reference electrode, depending on which side of the boundary becomes positive. In making electrode measurements it is very important that this potential be the same when the reference electrode is in the standardizing solution, as well as in the sample solution; otherwise the change in liquid junction potential will appear as an error in the measured electrode potential.

The junction potential is less when the ions of the electrolyte have nearly the same mobilities (see Table 26-1). In general a filling solution should be within 5% of being equitransferent. The ionic strength of the electrolyte bridge solution should be at least 5 to 10 times greater than the maximum ionic strength expected in the sample and standardizing solutions. The electrolytic filling solution need not be limited to only one salt. One equitransferent solution is a mixture of K^+, Na^+, NO_3^-, and a small amount of Cl^- in the appropriate ratios to satisfy the equitransferent condition. It gives a lower junction potential than saturated KCl in dilute (less than 10^{-3} M) ionic strength samples. Above 0.5 M (or μ = 0.5), KCl is preferable. For trace Cl^- determination, saturated KNO_3 should be used; likewise, saturated K_2SO_4 for trace nitrate determination, and 5 M lithium trichloroacetate for trace K^+ determination and solutions containing ClO_4^-.

When a saturated solution comprises one side of a boundary, the junction potential amounts to 1–2 mV under many aqueous solution conditions unless the second solution is strongly acidic or alkaline or the difference in concentration between the two sides of the junction is very large. A junction between 0.1 M HCl and 0.01 M HCl has a potential of 40 mV with the dilute side positive. For KCl, however, at the same concentrations, the junction potential is 1.0 mV with the dilute side negative because the mobility of K^+ is slightly less than that of Cl^-. Approximate liquid-junction potentials are given by Milazzo[1] for a number of boundary systems.

The salt-bridge electrolyte forms a continuous electrical conduction path from the internal cell to the lower tip of the electrode. It also serves to protect the internal half-cell from contamination that would cause changes in the reference electrode potential. The leakage rate of the bridge solution should be low, but satisfactory performance depends upon continuous, unimpeded, positive flow of the filling solution through the junction. The asbestos fiber and palladium annulus junction provide a small electrolyte flow, only 0.1–0.01 ml per day. Day-to-day stability of the junction potential is 2 mV for the asbestos fiber junction, 0.2 mV for the palladium annulus, and 0.06 mV for the ground glass sleeve. The asbestos fiber junction is good for general use, although it tends to clog in some media, especially colloids and suspensions. The palladium annulus is recommended for microtitrations and pH measurements for clinical solutions and applications under high pressure or vacuum. The ground glass sleeve is better for precipitation titrations, titrations in nonaqueous solvent systems, and the handling of colloids and suspensions. Double junction, sleeve type salt bridges overcome problems with leakage of undesirable ions into the sample solution or compatibility of filling and sample solutions (e.g., nonaqueous systems). The double-junction reference electrode contains inner and outer filling solution chambers connected by a ceramic plug. The inner chamber is inside the hollow body of the electrode, and the outer chamber is formed by the body of the electrode and the outer sleeve. Both chambers can be drained and refilled quickly. This type electrode must be used if the filling solution of the single-junction electrode contains the ion being measured or an ion that interferes with the electrode response to the ion being measured. The double-junction electrode must also be used if the filling solution of the single junction contains any ion that complexes or precipitates the ion being measured or that forms a precipitate with any ion in the sample.

BIBLIOGRAPHY

Clark, W. M., *Oxidation-Reduction Potentials of Organic Systems*, Williams and Wilkins, Baltimore, 1960.

Latimer, W. M., *Oxidation Potentials*, 2nd ed., Prentice-Hall, Englewood Cliffs, N.J., 1952.

Lingane, J. J., *Electroanalytical Chemistry*, 2nd ed., Wiley-Interscience, New York, 1958.

Ives, D. J. G. and G. J. Janz, *Reference Electrodes*, Academic Press, New York, 1961.

LITERATURE CITED

1. Milazzo, G., *Electrochemistry*, W. Schwabl, Trans., Springer-Verlag, Vienna, 1952, p. 98.
2. Perley, G. A., *J. Electrochem. Soc.*, **92**, 485 (1948).

20

pH and Ion Selective Potentiometry

Potentiometric methods embrace two major types of analyses: the direct measurement of an electrode potential from which the activity (or concentration) of an active ion may be derived, the topic of this chapter; and the changes in the electromotive force brought about through the addition of a titrant, which will be discussed in Chapter 21. The field of analytical potentiometry is experiencing a renewal of substantial activity, sparked by the development of novel ion-selective electrodes which are taking their place alongside the historical pH glass electrode.[7] The electrodes considered in this chapter fall into the following major categories:

1. Glass electrodes
2. Solid-state and precipitate electrodes
3. Liquid–liquid membrane electrodes
4. Enzyme and gas-sensing electrodes

All seem to involve an ion exchange process in the potential-determining mechanism. They yield potentials that can be adequately described by the classical Nernst equation or its expanded modifications.

Ion-selective electrodes measure ion activities, the thermodynamically effective free ion concentration, not concentrations. Activity measurements are valuable because the activities of ions determine rates of reactions and chemical equilibria. For example, ion activities are important parameters in predicting corrosion rates, extent of precipitation, formation of complexes, degree of acidity, solution conductivities, effectiveness of metal pickling baths and electroplating bath solutions, and physiological effects of ions in biological fluids.

GLASS-MEMBRANE ELECTRODES

The various types of ion-selective glass electrodes available are all members of a continuum of glass electrodes. To obtain desired response characteristics, a particular composition must be selected. Three subtypes of glass electrodes and their selectivity characteristics can be summarized as follows:

pH type, selectivity order: $H^+ >>> Na^+ > K^+, Rb^+, Cs^+ \cdots >> Ca^{2+}$

cation-sensitive type, selectivity order: $H^+ > K^+ > Na^+ > NH_4^+, Li^+ \cdots \gg Ca^{2+}$

sodium-sensitive type, selectivity order: $Ag^+ > H^+ > Na^+ \gg K^+, Li^+ \cdots \gg Ca^{2+}$

The second two subtypes also display considerable response to such univalent cations as Tl^+, Cu^+, and $R_nH_{4-n}N^+$, but are primarily responsive to univalent cations and generally are quite unresponsive to anions.

Not only the degree of electrode selectivity, but even the selectivity order can be changed with appropriate adjustment of the glass composition. As a general rule, cation selectivity (over hydrogen ion) can be achieved by adding elements whose coordination numbers are greater than their oxidation numbers [as in the substitution of aluminum(III) for silicon(IV)] to alkali metal-silicate glasses (20% Na_2O-10% CaO-70%-SiO_2). Such charge-deficient elements apparently leave the glass with an excess of negatively charged ion sites which attract cations having the proper charge-to-size ratio. Glasses containing less than about 1% Al_2O_3 yield good pH electrodes with little metal-ion response (see the section "Glass Electrodes for pH Measurement"). Glasses having a composition about 27% Na_2O-5% Al_2O_3-68% SiO_2 show a general cation response. Glasses of the composition 11% Na_2O-18% Al_2O_3-71% SiO_2 are highly sodium-selective with respect to other alkali metal ions. More complex glasses containing additives frequently yield electrodes with superior mechanical and electrical properties.

At present, according to Eisenman,[5] glass electrodes are the preferred electrodes for H^+, Na^+, Ag^+, and Li^+, because of their high specificity for these ions and their excellent stability characteristics. In addition, because these electrodes function well in organic solvents they can be used in nonaqueous media, as well as in the presence of lipid-soluble or surface-active molecules. Electrode response is relatively indifferent to the type of anion present, unless it chemically attacks the glass or reacts with the cation.

The glass electrode comprises a thin-walled bulb of cation-responsive glass sealed to a stem of noncation-responsive, high-resistance glass. In this manner the cation response is confined entirely to the area of the special glass membrane, eliminating any variance caused by the depth of immersion. A typical electrode is illustrated in Fig. 20-1. If intended for use outside a shielded electrode compartment, the stem of the electrode is shielded and a shielded lead is provided which should be grounded.

Both surfaces of the glass membrane are cation-responsive. Changes in the electrical potential of the outer membrane surface are measured by means of an external reference electrode and its associated salt bridge. An electrolyte of high buffer capacity and suitable chloride concentration fills the inside of the glass membrane; into this electrolyte dips an inner reference electrode. The complete electrochemical cell is

internal reference electrode		internal electrolyte		glass membrane		standard or unknown solution	external reference electrode

When immersed in an electrolyte, the glass electrode must be viewed as a structured continuum consisting of "dry" glass between a "swollen" or hydrated gel layer at the electrode–solution interface. A cross section through a glass membrane might be represented

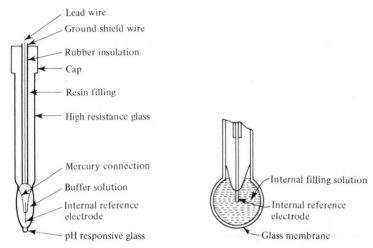

Fig. 20-1. Construction of glass membrane, pH-responsive electrode.

as

internal solution	hydrated gel layer	dry glass layer	hydrated gel layer	external solution

The dry glass layer constitutes the bulk of the membrane's thickness, about 50 μm; the hydrated layers vary in depth from 50 to 1000 Å, depending upon the hygroscopicity of the glass. When the dry glass electrode is first immersed in an aqueous medium, the formation of the hydrated (external) layer causes some swelling of the membrane. Thereafter, a constant dissolution of the hydrated layer takes place with accompanying further hydration of additional dry glass so as to maintain the thickness of the hydrated layer at some steady-state value. The rate of dissolution of the hydrated layer depends on the composition of the glass and also on the nature of the sample solution. The dissolution rate largely determines the practical lifetime of the electrode; lifetimes vary from a few weeks to several years.

The mechanism by which cations affect the potential of the glass membrane is not well understood.[4] According to the presently accepted concept, it is an ion-exchange process in the gel layer of the glass membrane producing a phase-boundary potential that determines the cation response of the electrode. Only the gel layer of the glass directly participates in the equilibirum. Cationic exchange occurs only in the external part of the gel layer, which does in fact act as a semipermeable membrane to cations. The inner regions of the glass have little effect on the potential formation. The concept of an actual penetration through the glass membrane by substantial amounts of cations has been definitely disproved.

SOLID-STATE SENSORS

The nonglass, solid-state sensors replace the glass membrane with an ionically conducting membrane. In the fluoride electrode, the active membrane portion is a single

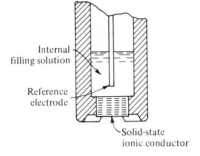

Internal filling solution

Reference electrode

Solid-state ionic conductor

Fig. 20-2. Cross-sectional view of solid-state sensor. (Courtesy of Orion Research, Inc.)

crystal of LaF_3 doped with europium(II) to lower its electrical resistance and facilitate ionic charge transport. The LaF_3 crystal, sealed into the end of a rigid plastic tube, is in contact with the internal and external solutions (Fig. 20-2). Typically, the internal solution is 0.1 M each in NaF and NaCl; the fluoride ion activity controls the potential of the inner surface of the LaF_3 membrane, and the chloride ion activity fixes the potential of the internal Ag/AgCl wire reference electrode. The electrochemical cell incorporating the LaF_3 membrane electrode is

$$Ag \mid AgCl, Cl^-(0.1\ M), F^-(0.1\ M) \mid LaF_3 \text{ crystal} \mid \text{test solution} \parallel \text{reference electrode}$$

It obeys a Nernst-type relation of the form

$$E = \text{constant} + \frac{RT}{F} \ln \frac{[F^-]_{int}}{[F^-]_{ext}} \tag{20-1}$$

which, because the $[F^-]_{int}$ is constant, simplifies to

$$E = \text{constant} + 0.05916\ \text{pF} \tag{20-2}$$

at 25°C. The activity calibration curve shows that the electrode follows Nernstian behavior to fluoride concentrations as low as 10^{-5} M and a useful, although non-Nernstian, response to at least 10^{-6} M fluoride ion.

The sulfide ion electrode has as its active element a polycrystalline Ag_2S membrane. If this membrane is altered from pure Ag_2S by dispersing within it another metal sulfide, such as CuS, CdS, or PbS, one obtains the corresponding metal-selective electrode. These electrodes transport charge by the movement of silver ions, but their potential is determined indirectly by the availability of the S^{2-} ion which, in turn, is fixed by the activity of the silver ion or divalent cation in contact with the membrane. Mixed crystals of AgX-Ag_2S compose the anion-selective electrodes of chloride, bromide, iodide, and thiocyanate, respectively. The solubility of the divalent metal sulfide, or the silver halide, must be greater than that of Ag_2S. Any of these electrodes also function as a silver-selective electrode.

Cast pellets of silver halides can also serve as the active membrane material for the respective halide-selective electrode. Selectivity is basically a function of the silver halide solubility.

The fine particles of a sparingly soluble salt can be immobilized in a coherent silicone rubber matrix.[6] The resulting membranes, prepared by the cold catalyzed polymerization of silicone rubber monomer, mixed with at least 50 weight percent of the appropriate precipitate (e.g., silver chloride for chloride response), have good mechanical properties and give reproducible potentials. The membrane is cemented to a glass body. Because the silicone rubber matrix has an immense resistance, the embedded precipitate must provide electrical conductivity across the membrane and this requires the precipitate particles to be in contact with each other.

A crystal membrane can be a highly selective device. Conduction in the crystalline phase proceeds by a lattice defect mechanism in which a mobile ion (usually the lattice ion with the smallest ionic radius and the smallest charge) adjacent to a vacancy defect moves into the vacancy. A vacancy for a particular ion is ideally tailored with respect to size, shape, and charge distribution to admit only the mobile ion; all other ions are unable to move and cannot contribute to the conduction process. For example, in the fluoride electrode the fluoride ion alone transports the electrical charge within the crystal, whereas in the mixed crystal-Ag_2S electrodes the mobile ion is the silver ion. Because no foreign ions can enter the crystalline phase, these electrodes always behave as a Nernstian device. Interferences can occur, but they arise from chemical reactions at the crystal–solution interface.

The main advantage of solid-state sensors over silver metal–silver halide electrodes of the second type is their insensitivity to redox interferences and surface poisoning. In the silver metal–silver halide electrodes, the potential is established via the couple Ag/Ag^+, which of course is sensitive to oxidants and reductants.

Solid-state electrodes have a typical life of 1–2 years in the laboratory and 1–3 months when used continuously at elevated temperatures or in continuously flowing systems containing abrasive materials. Although their operating temperature range for continuous use is $0°$ to $80°C$, they can be used intermittently in the $80°$ to $100°C$ range. Solid-state sensors behave in a Nernstian manner until the activities in solution approach the solubility of the membrane material, whereupon the potential value converges to a limit which gives the lower detection limit. For example, the silver halide electrodes exhibit useful pX ranges up to 5, 6, and 7, respectively, for chloride, bromide, and iodide ions.

LIQUID-MEMBRANE ELECTRODES

The liquid-membrane electrode consists of a double concentric tube arrangement in which the inner tube contains the aqueous reference solution and internal reference electrode (Fig. 20-3).   The outer compartment contains the organic liquid ion-exchanger reservoir which occupies the pores (100 μm) of a hydrophobic membrane. This membrane, a fluorocarbon body, is replaceable. The electrode can be taken apart to change the membrane or the ion exchanger, thus altering the electrode's ion selectivity. For example, the calcium-selective electrode uses the calcium salt of bis-(2-ethyl hexyl)phosphoric acid (d2EHP) dissolved in various straight-chain alcohols, or didecylphosphoric acid dissolved in di-n-octylphenyl phosphonate, as the liquid ion exchanger. The neutral,

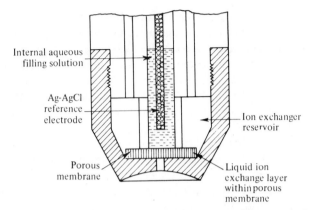

Fig. 20-3. Construction of liquid ion-exchange electrode. (Courtesy of Orion Research Inc.)

undissociated molecules of Ca(d2EHP)$_2$ diffuse easily in the solvent-saturated pores of the membrane but are insoluble in water. At the membrane–sample interface, these molecules can exchange their calcium ions for those in the sample solution. In this way the d2EHP groups can transport calcium ions back and forth across the membrane until equilibrium is established. The internal aqueous filling solution consists of a fixed concentration of calcium and chloride ions. This solution is in contact with the ion-exchanger reservoir (via the porous membrane) and with the inner Ag/AgCl reference electrode.

Site groups of the R—S—CH$_2$—COO$^-$ type, in which sulfur and the carboxylate groups are in position to form a chelate ring with cations, show good selectivity for copper(II) and lead(II) ions. For the nitrate and fluoroborate electrodes, a substituted nickel(II)-1,10-phenanthroline ion-pair site group is used. The corresponding iron(III) ion association complex is used in the perchlorate-selective electrode. With liquid ion exchangers the potential selectivity is largely dominated by their equilibrium ion-exchange selectivity toward cations or anions. Electrodes of this type will be only moderately selective for the ion of interest. However, considerable latitude exists for the development of numerous systems responsive either to cations or anions of varying charge.

Neutral extraction membranes are low dielectric liquids. A typical example is decane containing 10^{-4} to 10^{-7} M of an extractant such as valinomycin, nonactin, or monactin. These molecules are devoid of charged groups but contain an arrangement of ring oxygens energetically suitable through ion-dipole interaction to replace the hydration shell around cations. Thus, these lipid-soluble molecules are able to dissolve cations in organic solvents, forming mobile charged complexes with the cations therein, and in this way provide a mechanism for cation permeation across such normally insulating media. Neutral carriers are one of the ways used by biological membranes to discriminate among cations; and by using neutral carriers it may be possible to duplicate the exquisite selectivities characteristic of living cells in an artificial electrode. Valinomycin membranes show great potassium selectivity, about 3800 times that of sodium and much better than that observed (30 : 1) with the best available potassium-sensitive glass electrode. The actin-base membrane electrode is about 4 times more responsive to NH_4^+ than to K^+. Even more unusual is the 18,000 to 1 selectivity of the valinomycin membrane for potassium with respect to hydrogen ion; this means that the electrode is usable in strongly acidic media.

Liquid-membrane electrodes must be recharged every 1–3 months to replace the liquid ion exchanger and renew the porous membrane. It is desirable to have the electrode self-cleaning, and to effect this, the internal liquids of the electrode are put under mild hydrostatic pressure. The internal fluids then tend to flow out, rather than the sample fluid flowing into the electrode. This loss and the slight water-solubility of the exchanger limit the sensitivity of these electrodes to low concentrations of cations. The electrode will show a Nernstian response until the activity of the test ion is within a factor of 100 of the solubility of the liquid ion exchanger. Response time is about 30 sec in pure solutions; however, it is longer in mixtures because of the time required for the internal-membrane diffusion potential to reach steady state via adjustment in the liquid ion-exchanger concentration profile. In the presence of interfering ions, the ion-exchange process must establish new ion activities in the surface and this is not necessarily an instantaneous process. The temperature of operation must be kept within limits specified by the manufacturer, generally between $0°$ and $50°C$, so that water does not permeate the membrane and membrane liquids do not bleed excessively into the aqueous solution.

GAS-SENSING AND ENZYME ELECTRODES

Although the ammonia electrode is a gas-sensing electrode, it is used just as if it were a selective-ion electrode. Its construction is shown in Fig. 20-4. Dissolved ammonia from the sample diffuses through a gas-permeable fluorocarbon membrane until a reversible equilibrium is established between the ammonia level of the sample and the internal filling solution. Hydroxide ions are formed in the internal filling solution by the reaction of ammonia with water: $NH_3 + H_2O = NH_4^+ + OH^-$. The hydroxide level of the internal

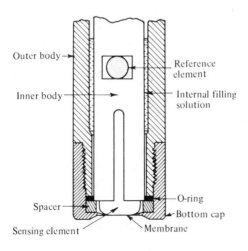

Outer body →

Inner body →

Reference element

Internal filling solution

Spacer

Sensing element

O-ring

Bottom cap

Membrane

Fig. 20-4. Construction of the ammonia gas-sensing electrode. (Courtesy of Orion Research, Inc.)

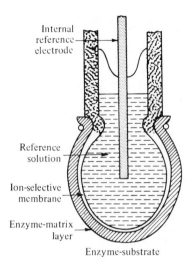

Internal
reference
electrode

Reference
solution

Ion-selective
membrane

Enzyme-matrix
layer

Enzyme-substrate

Fig. 20-5. Enzyme electrode.

filling solution is measured by the internal sensing element and is directly proportional to the level of ammonia in the sample. Samples and standards are adjusted to a fixed pH, or to pH > 11. The ammonia electrode, unlike glass and other ammonium ion electrodes, is almost totally free from interferences, although volatile amines may interfere. Anions, cations, and common gases, such as carbon dioxide, sulfide, cyanide, and sulfur dioxide do not interfere. Sensitivity extends from 10^{-6} M to 1 M, that is, from 0.017 to 17,000 $\mu g/ml$.

The basic arrangement of the enzyme electrode is shown in Fig. 20-5. The enzyme is immobilized in a gel layer which coats a conventional cation-sensitive glass electrode. For the ammonium ion-selective electrode, the enzyme urease is fixed in a layer of acrylamide gel held in place around the glass electrode bulb by porous nylon netting or a thin cellophane film. The urease acts specifically upon urea in the sample solution to yield ammonium ions which diffuse through the gel layer and are sensed by the electrode. The resulting potential is proportional to the substrate (urea) concentration in the sample solution. There are thousands of enzyme-substrate combinations that would yield products measurable with ion-selective electrodes. With some modifications, this electrode system can be reversed; that is, the substrate would surround the glass membrane, resulting in an enzyme-sensing electrode.

INTERFERENCES

Selective-ion electrodes are subject to two types of interferences: method interference and electrode interference. Method interferences occur when some characteristic of the sample prevents the probe from sensing the ion of interest. For example, a fluoride electrode can detect only fluoride ion. In acid solution, however, fluoride forms complexes with the hydrogen ion and is thereby masked from the fluoride detector.

Electrode interferences arise when the electrode responds to ions in the sample solution other than the ion being measured. The selectivity coefficient is an index of the ability of an electrode to measure a particular ion in the presence of another ion. In liquid and glass membrane electrodes, interference occurs when an interfering ion passes into the membrane just as does the ion being measured. As an example, barium ions will be transported across the liquid membrane of the calcium electrode if the activity of barium ion in the sample is sufficiently high and the activity of calcium ion is very low. Under these conditions the electrode potential will depend to some extent on the barium ion activity. The electrode response incorrectly attributed to the test ion's activity a, whose ionic charge is z, is related to the selectivity coefficient k_i by the equation

$$a = k_i a_i^{(z/z_i)} \tag{20-3}$$

where a_i is the background level of the interference whose ionic charge is z_i. Thus a selectivity coefficient of 0.01 means that the response attributable to Ca^{2+} alone at $a_{Ca^{2+}}$ would be matched in the absence of calcium by barium at an activity 100 times larger.

The interference mechanism that occurs in crystal electrodes is quite different and requires a different method for expressing the selectivity coefficient. Surface reactions can convert one of the components of the solid membrane to a second insoluble compound. As a result, the membrane loses sensitivity to the ion being measured. For example, thiocyanate ion can interfere with bromide ion measurements if the reaction

$$SCN^- + AgBr(s) \rightarrow AgSCN(s) + Br^- \tag{20-4}$$

takes place, which it will begin to do if the ratio of the thiocyanate ion activity to the bromide ion activity exceeds the value given by the ratio of the solubility products of silver thiocyanate to silver bromide, or

$$\frac{1}{k_i} = \frac{a_{SCN}}{a_{Br}} = \frac{1.00 \times 10^{-12}}{5.0 \times 10^{-13}} = 2.0 \tag{20-5}$$

Failure to give a Nernstian response also occurs in dilute solutions approximating the solubility of the membrane material; this is the limiting value of the cell emf regardless of dilution.

Solid electrodes can also give problems if they are used in samples containing a species that forms a very stable complex with one of the component ions of the crystal. For example, citrate ion forms with lanthanum ion a very stable citrate complex, which has the effect of increasing the solubility of the membrane and thereby increasing the lower limit of detection of the fluoride ion. More drastic is the effect of cyanide ion in contact with a mixed silver halide/silver sulfide membrane. In this case the reaction proceeds virtually to completion with the consumption of silver halide from the membrane. However, if an appropriate diffusion barrier is placed on the membrane surface, then a steady state is quickly established in which the cyanide level at the crystal surface is virtually zero, whereas the iodide (or other halide) level is very nearly one-half the sample cyanide concentration. The silver ion activity at the membrane surface is fixed by the iodide level via the solubility product of silver iodide. As a result the electrode depends in an almost Nernstian manner on the sample cyanide ion concentration. It has been found that the

Table 20-1 Chemical-Sensing Electrodes

Species Sensed	Type	Concentration Range, M	Preferred pH Range	Interferences
Ammonia	Gas	10^0-10^{-6}	11–13	Volatile amines
Bromide	Solid	$10^0-5\times10^{-6}$	2–12	$S^{2-} \leqslant 10^{-7}\,M$; I^- and CN^- may be present in trace amounts
Cadmium	Solid	10^0-10^{-7}	3–7	Ag^+, Hg^{2+}, $Cu^{2+} \leqslant 10^{-7}\,M$; high levels of Pb^{2+}, Fe^{3+}
Calcium	Liquid	10^0-10^{-5}	6–8	Selectivity coefficients: Zn^{2+}, 3.2; Fe^{2+}, 0.80; Pb^{2+}, 0.63; Mg^{2+}, 0.014; Na^+, 0.003; Cu^{2+}, 0.27; Ni^{2+}, 0.080; Ba^{2+}, 0.010; Sr^{2+}, 0.017
Chloride	Liquid	$10^{-1}-10^{-5}$	2–11	Selectivity coefficients: I^-, 17; NO_3^-, 4.2; Br^-, 1.6; ClO_4^-, 32; OH^-, 1.0; HCO_3^-, 0.19; SO_4^{2-}, 0.14; F^-, 0.10; OAc^-, 0.32
	Solid	$10^0-5\times10^{-5}$	0–14	$S^{2-} \leqslant 10^{-7}\,M$; $OH^- < 80\,Cl^-$; $Br^- < 300\,Cl^-$; $S_2O_3^{2-} < 100\,Cl^-$; $NH_3 < 8\,Cl^-$: I^-, CN^- may be present in trace amounts
Copper(II)	Solid	10^0-10^{-7}	3–7	S^{2-}, Ag^+, $Hg^{2+} \leqslant 10^{-7}\,M$; high levels of Cl^-, Br^-, Cd^{2+}, Fe^{3+} interfere; reagents which reduce Cu^{2+}
Cyanide	Solid	$10^{-2}-10^{-6}$	11–13	$S^{2-} \leqslant 10^{-7}\,M$; $I^- < 0.1\,CN^-$; $Br^- < 5000\,CN^-$; $Cl^- < 10^6\,CN^-$
Fluoride	Solid	10^0-10^{-6}	3–10	Maximum level: $OH^- < 0.1\,F^-$
Fluoroborate	Liquid	$10^{-1}-10^{-5}$		Selectivity coefficients: NO_3^-, 0.1; Br^-, 0.04; OAc^-, 0.004; HCO_3^-, 0.004; Cl^-, 0.001; SO_4^{2-}, 0.001; OH^-, 0.001; I^-; 20; F^-, 0.001
Iodide	Solid	$10^0-2\times10^{-7}$	3–12	$S^{2-} \leqslant 10^{-7}\,M$; $I^- \leqslant 10^6\,Cl^-$; $Br^- \leqslant 5000\,Cl^-$
Lead	Solid	10^0-10^{-7}	4–7	Ag^+, Cu^{2+}, $Hg^{2+} \leqslant 10^{-7}\,M$; Cd^{2+}, $Fe^{3+} \gg Pb^{2+}$
Nitrate	Liquid	$10^{-1}-10^{-5}$	3–10	Selectivity coefficients: I^-, 20; Br^-, 0.1; Cl^-, 0.004; NO_2^-, 0.04; SO_4^{2-}, 0.00003; CO_3^{2-}, 0.0002; ClO_4^-, 1000; F^-, 0.00006; ClO_3^-, 2; CN^-, 0.02; HCO_3^-, 0.02; OAc^-, 0.006; $H_2PO_4^-$, 0.003; HPO_4^{2-}, 0.00008

Perchlorate	Liquid	$10^{-1}-10^{-5}$	3–10	Selectivity coefficients: I^-, 0.012; NO_3^-, 0.0015; Br^-, 0.00056; F^-, 0.00025; Cl^-, 0.00022; OH^-, 1.0; OAc^-, 0.00051; HCO_3^-, 0.00035; SO_4^{2-}, 0.00016
Potassium	Liquid	$10^{0}-10^{-5}$	3–10	Selectivity coefficients: Cs^+, 1.0; NH_4^+, 0.03; H^+, 0.01, Ag^+, 0.001; Na^+, 0.002; Li^+, 0.001
Silver/Sulfide	Solid	$10^{0}-10^{-7}$ Ag^+ or S^{2-}	2–9 (Ag^+) 13–14 (S^{2-})	$Hg^{2+} \leqslant 10^{-7}$ M as silver sensor None as sulfide sensor
Sodium	Solid	$10^{0}-10^{-6}$	9–10	Selectivity coefficient: Li^+, 0.002; K^+, 0.001; Rb^+, 0.00003; H^+, 100; Cs^+, 0.0015; NH_4^+, 0.00003; Tl^+, 0.00003; Ag^+, 350; $(C_2H_5)_4N^+$, 0.0005
Sulfur dioxide	Gas	$10^{-2}-3 \times 10^{-6}$	0–2	Volatile acids
Thiocyanate	Solid	$10^{0}-5 \times 10^{-6}$	2–12	$OH^- < 100$ SCN^-; $Br^- < 0.003$ SCN^-; $Cl^- < 20$ SCN^-; $NH_3 < 0.13$ SCN^-; $S_2O_3^{2-} < 0.01$ SCN^-; $CN^- < 0.007$ SCN^-; I^-, $S^{2-} \leqslant 10^{-7}$ M

NOTE: All data courtesy of Orion Research, Inc.

porous silver sulfide matrix left behind as the silver iodide is consumed provides an almost ideal diffusion barrier; thus an interference can be turned to an advantage. The cyanide electrode has a lifetime of several months of continuous operation.

Interferences with chemical-sensing electrodes are summarized in Table 20-1.

ION-ACTIVITY EVALUATION METHODS

The direct-measurement technique requires a single potentiometric measurement on the sample solution. The sample's millivolt reading from an expanded scale pH/mV meter is compared to a previously prepared calibration curve, or to the sample concentration or activity read directly from the meter scale of a calibrated specific ion meter. Calibration procedures must use solutions of known activity or concentration, depending on which parameter is required. Below 10^{-3} to 10^{-4} M the two are practically indistinguishable. Direct-measurement techniques are useful where samples are essentially pure solutions of the ion sensed, or have a relatively high and constant total ionic strength. To swamp effects caused by variations in total ionic strength, the ionic strength of the sample solution and the calibrating solutions may be adjusted by adding a high level of a noninterfering ion, for example, 1 M KNO_3 when making halide determinations. Naturally the ion sensed must possess a large selectivity coefficient relative to possible interfering ions.

An approach, similar to the operational definition of pH values, can be applied to the problem of measuring the activities of other ions in solution. Based on procedures set forth by Bates and Alfenaar,[2] standard reference values for pNa and pCl in sodium chloride solutions, pCa in calcium chloride solutions, and pF in sodium fluoride solutions are suggested in Table 20-2. The uncertainty in activity of an ion from an emf measurement of ±0.2 mV amounts to approximately 0.75% for a monovalent ion and 1.5% for a divalent ion. Because of their freedom from junction potentials, cation-sensitive glass electrodes can yield a higher degree of precision with respect to emf measurements.

Table 20-2 Suggested Reference Standard Values, pI,[a] at 25°C

Material	Molality, mole kg^{-1}	pNa	pCa	pCl	pF
NaCl	0.001	3.015	–	3.015	–
	0.01	2.044	–	2.044	–
	0.1	1.108	–	1.110	–
	1.0	0.160	–	0.204	–
NaF	0.001	3.015	–	–	3.015
	0.01	2.044	–	–	2.048
	0.1	1.108	–	–	1.124
CaCl$_2$	0.000333	–	3.537	3.191	–
	0.00333	–	2.653	2.220	–
	0.0333	–	1.887	1.286	–
	0.333	–	1.105	0.381	–

[a] pI = –log [I].

In the *known increment* (or *decrement*) methods, another designation for the method of additions, the concentration of a specific ion sample is estimated by observing the change in electrode potential when a known incremental (or decremental) change is made in concentration of the ion in the sample. This approach requires neither preparation of a calibration curve nor calibration of logarithmic scales with standard solutions. The ion is added to the test solution in a known amount that changes the total concentration by a known amount ΔC, but does not change the total ionic strength appreciably or the fraction of the total concentration that is free. Thus, the initial reading is taken on a sample C_1, and the electrode response is

$$E_1 = \epsilon + S \log (\gamma_1 C_1) \qquad (20\text{-}6)$$

After the incremental addition

$$E_2 = \epsilon + S \log [\gamma_1 (C_1 + \Delta C)] \qquad (20\text{-}7)$$

Combining equations,

$$\Delta E = E_2 - E_1 = S \log \left(\frac{C_1 + \Delta C}{C_1} \right) \qquad (20\text{-}8)$$

where S is the Nernstian factor or emf/pC slope. Known addition and subtraction methods are particularly suitable for samples with high unknown total ionic strength. Where the species being measured is especially unstable, known subtraction is preferred over known addition.

THE MEASUREMENT OF pH

The pH scale is a series of numbers that express the degree of acidity (or alkalinity) of a solution, as contrasted with the total quantity of acid or base in some material as found by an alkalimetric (or acidimetric) titration. As defined by Sørensen, who introduced the term,

$$\text{pH} = -\log [H^+] \qquad (20\text{-}9)$$

Note that what is involved is the negative logarithm of the hydrogen-ion concentration expressed in molarity. However, it is the activity of the hydrogen ion that is formally consistent with the thermodynamics of the pH electromotive cell, and the activity definition is

$$pa\text{H} = -\log a_{H^+} \qquad (20\text{-}10)$$

Now $[H^+]$ and $f_\pm [H^+]$ are often the most useful units for expressing the acidity of aqueous solutions, where $f_\pm$ is the mean ionic activity coefficient. Unfortunately, the established experimental pH method cannot furnish either of these quantities. Consequently, the term pH is merely a mathematical symbol of convenience, widely accepted, but devoid of exact thermodynamic validity. For those interested in a detailed treatment of the historical development of the concept of pH, the treatise by Clark and the more recent work by Bates, both cited in the Bibliography, should be consulted.

The acidity of a solution will depend upon several factors: (a) the chemical nature of the acid, as expressed by the degree of dissociation (or association of a base), pK_a; (b) the relative concentrations of acid and its conjugate base, and the total ionic strength of the solution; and (c) the temperature of the solution as it affects the dissociation of water and the dissociation of the acid. For acids which are not dissociated completely, the expression for the pH of the solution is

$$pH = pK_a + \log \frac{[A^-] + [H^+]}{[HA] - [H^+]} \qquad (20\text{-}11)$$

Except when the hydrogen-ion concentration is comparable to the concentrations of HA or A^-, the expression can be simplified to

$$pH = pK_a + \log \frac{[A^-]}{[HA]} \qquad (20\text{-}12)$$

Buffer Solutions

A *buffer* may be defined as a solution which maintains a nearly constant pH value despite the addition of substantial quantities of acid or base. Generally it consists of a mixture of an incompletely dissociated acid and its conjugate base. In selecting a particular buffer, three characteristics should be considered: the buffer value β, the dilution value $pH_{1/2}$, and the change of pH with change in temperature, $\Delta pH/\Delta T$.

The Van Slyke buffer value β indicates the resistance of a buffer to change in pH upon addition of an acid or base. It is defined as

$$\beta = \frac{\Delta B}{\Delta pH}$$

where B is an increment of completely dissociated base (or acid) in gram-equivalents per liter that is required to produce unit change in pH within the solution. In the selection of a buffer system, pK_a should be as close as possible to the desired pH. Under this condition the ratio $[A^-]/[HA]$ in Eq. 20-12 is close to unity, and the buffer value will be large. For high buffering capacity, the concentrations of the buffering components should be high, yet consistent with considerations of ionic strength of the medium and the concomitant effect upon pH measurements.

The pH of the buffer solution should also be relatively insensitive to changes in the total concentration of the buffer components at a fixed ratio of $[A^-]/[HA]$. The dilution value is defined as the change of pH that results from a 1:1 dilution of the solution with pure water.

Solutions of specified composition for many pH values are compiled in the treatises by Clark and Bates. In the Clark and Lubs series which spans the range from pH 1.0 to 10.2, the ingredients are phthalic acid ($pK_1 = 2.90$, $pK_2 = 5.41$), potassium dihydrogen phosphate ($pK_2 = 7.13$), and boric acid ($pK = 9.24$), which are combined in suitable proportions with hydrochloric acid or sodium hydroxide. MacIlvaine's standard buffer spans the range from pH 2.2 to 8.0 and involves mixing citric acid ($pK_1 = 3.09$, $pK_2 =$

4.75, $pK_3 = 5.50$) and potassium dihydrogen phosphate solutions in certain proportions. In the last example, combining several acids of varying strength, but whose pK_a
values differ by less than two units, with the respective conjugate bases provides a universal buffer solution which covers a wider range of pH values than any single system, yet
a solution which exhibits considerable buffering capacity over the entire useful range.
Buffer tablets, available from chemical supply houses, eliminate the preparation, storage,
and mixing of buffering ingredients, and need only be dissolved in the specified volume of
pure water to obtain the pH value specified on the container.

Operational Definition of pH

The pH value is defined for an aqueous solution in an operational manner, according
to the Bates-Guggenheim convention, as

$$pH = pH_s + \frac{E - E_s}{2.302 \, RT/F} \tag{20-13}$$

In this definition, T is the temperature in degrees Kelvin, and E and E_s are, respectively,
the emf of an electrochemical cell of the usual design

electrode reversible to hydrogen ions	unknown or standard (s) buffer solution	salt bridge	reference electrode

which contains first the "unknown" solution, and secondly, a standard reference solution
of known pH, namely, pH_s.

The NBS pH standards were assigned pH_s values from measurements of the emf of
cells containing hydrogen gas and silver-silver chloride electrodes (i.e., without a liquid
junction):

$$Pt \mid H_2(1 \text{ atm}), H^+ Cl^- (\text{plus } K^+ Cl^-), AgCl \mid Ag$$

by the equation

$$E = E° - 0.000198T \log f_{H^+}f_{Cl^-}m_{H^+}m_{Cl^-} \tag{20-14}$$

where $E°$ is the standard potential of the cell[1,3]   Upon rearranging Eq. 20-14 in terms of
the acidity function, $p(a_{H^+}f_{Cl^-})$

$$p(a_{H^+}f_{Cl^-}) = -\log f_{H^+}f_{Cl^-}m_{H^+} = \frac{E - E°}{0.000198T} + \log m_{Cl^-} \tag{20-15}$$

The pH_s of the chloride-free buffer solution is computed from the equation

$$pH_s = p(a_{H^+}f_{Cl^-})° + \log f°_{Cl^-} \tag{20-16}$$

where $p(a_{H^+}f_{Cl^-})°$ is the value obtained by evaluation of $p(a_{H^+}f_{Cl^-})$ at several concentrations of chloride, and extrapolation to zero chloride concentration. The activity coeffi-

Table 20-3 National Bureau of Standards Reference pH$_s$ Buffer Solutions

Temp., °C	Secondary Standard, 0.05 M K tetroxalate	KH Tartrate (Sat'd. at 25°C)	0.05 M KH$_2$ Citrate	0.05 M KH Phthalate	0.025 M each KH$_2$PO$_4$ Na$_2$HPO$_4$	0.008695 M KH$_2$PO$_4$ 0.03043 M Na$_2$HPO$_4$	0.01 M Na$_2$B$_4$O$_7$	0.025 M each NaHCO$_3$ Na$_2$CO$_3$	Secondary Standard, Ca(OH)$_2$ (Satd. at 25°C)
0	1.666		3.863	4.003	6.984	7.534	9.464	10.317	13.423
5	1.668		3.840	3.999	6.951	7.500	9.395	10.245	13.207
10	1.670		3.820	3.998	6.923	7.472	9.332	10.179	13.003
15	1.672		3.802	3.999	6.900	7.448	9.276	10.118	12.810
20	1.675		3.788	4.002	6.881	7.429	9.225	10.062	12.627
25	1.679	3.557	3.776	4.008	6.865	7.413	9.180	10.012	12.454
30	1.683	3.552	3.766	4.015	6.853	7.400	9.139	9.966	12.289
35	1.688	3.549	3.759	4.024	6.844	7.389	9.102	9.925	12.133
40	1.694	3.547	3.753	4.035	6.838	7.380	9.068	9.889	11.984
50	1.707	3.549	3.749	4.060	6.833	7.367	9.011	9.828	11.705
60	1.723	3.560		4.091	6.836		8.962		11.449
70	1.743	3.580		4.126	6.845		8.921		
80	1.766	3.609		4.164	6.859		8.885		
90	1.792	3.650		4.205	6.877		8.850		
95	1.806	3.674		4.227	6.886		8.833		
Buffer value, β	0.070	0.027	0.034	0.016	0.029	0.016	0.020	0.029	0.09
Dilution value, $\Delta pH_{1/2}$	+0.186	+0.049	+0.052	+0.052	+0.080	+0.07	+0.01	+0.079	−0.28

SOURCE R. G. Bates, *J. Res. Natl. Bur. Std.*, **66A**, 179–183 (1962); B. R. Staples and R. G. Bates, ibid., **73A**, 37 (1969).
NOTE: Numbers given are "conventional" pH values. Properties of these buffer solutions are included at the foot of each column.

cient of chloride ion is given by the equation

$$-\log f_{Cl^-}^{\circ} = \frac{A\sqrt{\mu}}{1 + 1.5\sqrt{\mu}} \qquad (20\text{-}17)$$

where μ is the ionic strength, which should be maintained equal to, or less than, 0.1, and A is a parameter of the Debye-Hückel theory having a different value at each temperature. The recommended values of pH_s are summarized in Table 20-3. The total uncertainty in pH_s, exclusive of any liquid junction potentials introduced during calibration of pH equipment, is estimated as 0.005 pH unit (0°-60°C) and 0.008 pH unit (60°-95°C). The necessity of estimating the individual activity coefficients of chloride ion in each reference solution deprives the pH_s value of exact fundamental meaning. Nevertheless the operational definition of pH, chosen in part for its reasonableness but largely for its utility, agrees as closely as possible with the mathematical concepts evolved from the present state of solution theory.

Interpretation of Measured pH

The operational definition of pH emphasizes that the determination of pH is essentially a determination of a difference of emf as recorded in a pH cell containing first a reference buffer and then a test solution. The definition demands only that the electrode potential of the reference electrode remains constant while measurements of E and E_s are being made. Unfortunately, the definition makes no allowance for the presence of a liquid junction potential or a change in the value of the junction potential when the reference standard is replaced by an unknown solution. Hopefully, the liquid junction potential is assumed to remain constant from one measurement to another, and its value is combined with the value of the reference electrode. However, at pH values less than 2 or greater than 12, and for ionic strengths greater than 0.1, the reproducibility of the liquid junction potential is seriously impaired and errors as large as several tenths of a pH unit can result. To detect any serious impairment of the response of the measuring device and electrode assembly outside the pH range 2-12, the tetroxalate solution and the calcium hydroxide solution are included among the pH reference buffers but are designated secondary standards.

For pH measurements with an accuracy of 0.01 to 0.1 pH unit, the limiting factor is often the electrochemical system, that is, the characteristics of the electrodes and the solution in which they are immersed. Another source of error is due to temperature, for not only does the proportionality factor between cell emf and pH vary with temperature, but dissociation equilibria and junction potentials also have significant temperature coefficients. For accuracy of ±0.01 pH unit, the temperature should be known to ±2°C. Ideal solutions are those with compositions that match closely the primary standards of reference. Specifically, they are aqueous solutions of buffers and simple salts, of ionic strengths between 0.01 and 0.1, with only low concentrations of nonelectrolytes. In industrial processes, fortunately, a highly accurate knowledge of the pH of a solution is seldom required. Neither is it necessary to know exactly what a particular pH value means. It is sufficient to know that at a certain stage in an industrial process a particular pH value is maintained.

GLASS ELECTRODES FOR pH MEASUREMENT

Typical high-quality pH-sensitive glass membranes are chiefly lithium silicates with lanthanum and barium ions added. These ions act as lattice "tighteners" to retard silicate hydrolysis and lessen alkali ion, chiefly sodium ion, mobility. Lithium ions are the bulk mobile charge carriers under an applied electric field. After the membrane is soaked in water, the surface layer is depleted of Li^+, which is replaced by H^+. Virtually all surface silicate anion sites are neutralized by H^+ ions. Content of H^+ decreases in a complex way with increasing distance into the membrane, while Li^+ content increases is such a way that the sum of positive ions, charge carriers and other cations, balance the presumed uniform fixed-site concentration.

The activity of water in the solution appears to play an important role in the development of the pH response of the glass membrane. If the ionic strength is extremely high, or if a nonaqueous solvent is present, the measured potential deviates from the expected value. A direct relationship between hygroscopicity of the glass and pH response has been demonstrated. All glass electrodes must be conditioned for a time by soaking in water or in a dilute buffer solution, even though they may be used subsequently in media that are only partly aqueous.

The glass electrode displays an amazing versatility. Involving no electron exchange, it is the only hydrogen ion electrode uninfluenced by oxidizing and reducing agents. Nor is it disturbed by common electrode poisons. However, the glass membrane reaches equilibrium with the test solution slowly (response time is normally several seconds) and the surface of the glass is easily contaminated by adsorbed ions and particulate matter which delay the attainment of equilibrium between electrode and solution. The high electrical resistance (5–500 MΩ) necessitates measuring circuits with high input impedance, and is accompanied by a large temperature coefficient of resistance which changes exponentially with temperature. Chemical durability and electrical resistance are linked together. Electrodes durable against chemical attack at elevated temperatures have excessive electrical resistances when the temperature is lowered. Conversely, electrodes that are robust at low temperature will corrode rapidly in solutions at high temperatures. Consequently, electrodes are designed specifically for certain ranges of temperature and for certain ranges of pH. Frequently a general-purpose electrode is useful from $-5°$ to $60°C$ in acids and dilute alkalis with a negligible error to a pH of 11. In more alkaline solutions the observed values of pH are too small and must be corrected from nomographs supplied by the manufacturer (see Problem 20-1). The positive alkaline error is due to the partial exchange of cations, other than hydrogen ion, between the glass membrane and the solution. In general, the error will be large when the test solution contains a univalent cation in common with the glass membrane. For measurements above $60°$–$80°C$, a special glass is used that will withstand $100°C$ continuously, with intermittent use up to $130°C$; however, below $35°C$ these glasses are sluggish in response.

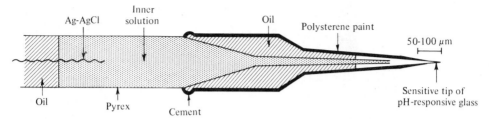

Fig. 20-6. Cross-sectional view of glass microelectrode.

The negative acid error is due to the change in the activity of the water in the gel layer. The gradual dissolution of the outermost layer of glass may also account for the error in strongly acid solution by preventing formation of a steady-state potential.

A glass electrode exhibits a reasonably rapid response to rapid and wide changes of pH in buffered solutions. However, valid readings are obtained more slowly in poorly buffered or unbuffered solutions, particularly when changing to these from buffered solutions, as after standardization. The electrodes should be thoroughly washed with distilled water after each measurement and then rinsed with several portions of the next test solution before making the final reading. Poorly buffered solutions should be vigorously stirred during measurement, otherwise the stagnant layer of solution at the glass–solution interface tends toward the composition of the particular kind of pH-responsive glass. Suspensions and colloidal material should be wiped from the glass surface with a soft tissue.

Commercial glass electrodes are fabricated in a wide variety of sizes and shapes and for many special applications. Syringe and capillary electrodes (Fig. 20-6) require only 1 or 2 drops of solution, even as little as 1 mm^3 volume in ultramicro work, whereas others will penetrate soft solids or are designed for measurements on smooth surfaces. The normal-size electrode operates with a volume of solution from 1 to 5 ml. Polyalcohols added to the solution in the reference electrode, and mercury inside the glass membrane to make direct contact with the glass, permit measurements of semifrozen materials at $-30°C$. Glass electrodes of special construction are available for operation under pressure conditions. Combination glass indicator and reference electrodes as a single unit are shown in Fig. 20-7. The outer cylinder contains the electrolyte for the salt bridge of the reference electrode and surrounds the usual glass-electrode assembly except for the pH-sensitive bulb.

ELECTROMETRIC MEASUREMENT OF pH AND pI

To achieve a reproducibility of ±0.005 pH or pI unit, the assigned limit of certainty of many reference buffer and ion activity standards and including unavoidable variations in liquid junction potentials, an instrument is needed that will be sensitive and reproducible to at least 0.2 mV. Negligible current must be drawn during measurement if changes in the ion concentration at the electrode surface are to be avoided, and no error is to arise from the voltage drop across the inherent resistance of the electrochemical cell. With

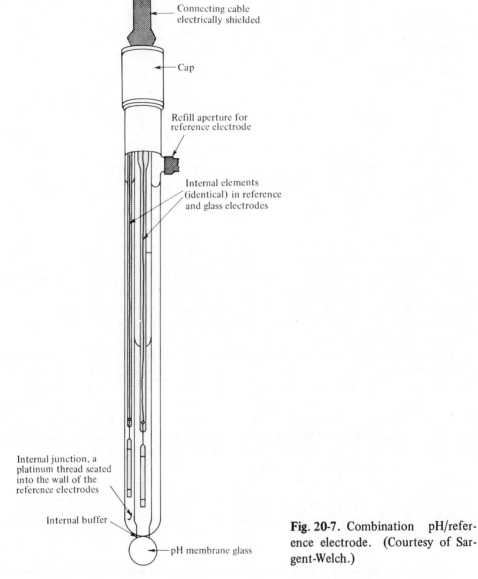

Connecting cable
electrically shielded

Cap

Refill aperture for
reference electrode

Internal elements
(identical) in reference
and glass electrodes

Internal junction, a
platinum thread seated
into the wall of the
reference electrodes

Internal buffer

pH membrane glass

Fig. 20-7. Combination pH/reference electrode. (Courtesy of Sargent-Welch.)

glass, solid-state, and membrane electrodes the current drawn should be 10^{-12} A or less. This restricts the choice of instrument to a high-impedance electronic voltmeter.

Null-Balance pH Meters

A schematic diagram of the first pH meter to be marketed, and still available, is shown in Fig. 20-8. It incorporates an ordinary potentiometric circuit with a null-balance amplifier circuit. The role of each of the adjustments and circuit components perhaps can best be appreciated in terms of the sequence of operations involved in the standardization and calibration of the instrument.

1. With the ganged switch in position 1, the grid of the input stage is connected to the contactor on slidewire R_1. The contactor is moved until the meter needle stands in its center position. Note that R_1 is the center portion of a potentiometer circuit which involves the same dry cells that supply the current through the filaments of the tubes. This operation selects an arbitrary "null" position on the meter, analogous to the mechanical adjustment of a D'Arsonval galvanometer, and compensates for aging of the batteries and changes in tube characteristics.

2. The rheostat marked *temperature adjust*, which shunts the *pH scale* slidewire, is set to the value of the solution temperature, thereby altering the denominator of the second term in Eq. 20-13. With the ganged switch in position 2, one side of the Weston standard cell is attached to the cathode of the input tube (through the contactor on R_1) and the grid of this tube is attached to the pH slidewire. Actually the Weston cell is in opposition to the potentiometer involving the pH slidewire, battery E_b, and rheostat R_2, with the amplifier serving as the current-measuring galvanometer. The rheostat R_2 is adjusted until the meter returns to its center position. Now the slide-wire scale is standardized in the correct number of millivolts per inscribed pH unit.

3. In the final step, the electrode assembly is immersed in a pH reference buffer and the pH slidewire is set at the value of the pH standard. With the ganged switch in position

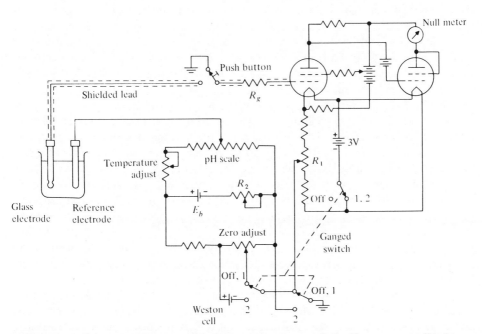

Fig. 20-8. Schematic circuit diagram of the Beckman model G pH meter, a null-balance amplifier circuit. (Courtesy of Beckman Instruments, Inc.)

1, a button is depressed which connects the glass electrode to the grid of the input tube. All this time the reference electrode has been connected to the contactor on the pH slidewire. Now the *zero adjust* rheostat (also called the *asymmetry control*) is adjusted until the meter returns to its center position. The inscribed pH scale is thereby brought into juxtaposition with the actual pH value by the zero adjust resistance which serves to lengthen one end of the pH scale while shortening by an equal amount the other end. Any changes in the asymmetry potential of the glass electrode are also compensated.

The slidewire in these instruments is also marked off in units of 100 mV and can be used in potentiometric methods other than pH measurements. The limiting factor in measurements with null-detector circuits is the slidewire accuracy, which is generally at least 0.1%. Because these circuits are subject to zero drift and require circuit readjustments frequently, they are unsuited to long-time unattended operation.

Direct-Reading pH and pI Meters

Meters with indicating scales in pH values are calibrated in voltage units for a glass-reference electrode pair on the basis of the relationship for the emf of a pH cell, described by

$$E = k - KT\,(\text{pH}) \tag{20-18}$$

the equation of a straight line with slope $-KT$ and a zero intercept of k. The electrode and its reference electrode must possess an isothermal point at 0 V. This is accomplished by making pH 7.000 correspond to 0 V at all temperatures by means of a glass internal solution with a buffer whose pH change with temperature exactly compensates the temperature changes of the internal and external reference electrodes. The proper emf/pH slope involves adjusting the KT factor (actually $2.3026RT/F$) to 59.16 mV per pH unit at 25°C by means of a slope control to rotate the emf/pH slope about the isothermal point. The temperature compensator, which is reserved to correct the slope for the actual temperature of the sample, varies the instrument definition of a pH unit from 54.20 mV at 0° to 66.10 mV at 60°C. Finally the meter scale is brought into juxtaposition with the pH value of a standard reference buffer whose pH value most closely approximates that of the test solution. The intercept (asymmetry) control effectively shifts the response curve laterally until it passes through the isothermal point. Figure 20-9 schematically illustrates these operations. A check with a second reference pH buffer, such that the two buffers bracket the pH of the test solution, should be done to ensure proper functioning of the cell assembly and measurement equipment and to verify conformity of the pH response with the theoretical Nernst slope. In effect, then, the pH value of the test solution is determined by interpolation.

A direct-reading instrument has few manipulative steps and is adaptable to continuous recording or control of industrial operations or processes. Temperature compensation can be provided by causing the feedback current to flow through a temperature-sensitive resistor (thermistor) immersed in the test solution. A useful feature found in some

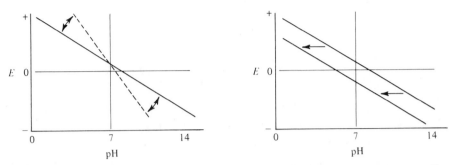

Fig. 20-9. Operation of the (a) *slope* and (b) *intercept* controls shown schematically.

meters is expanded ranges covering 0.5, 1, or 2 pH units over the full meter scale, thus permitting expanded range readings to 0.001 pH unit.

For ion-activity measurements, readout is required in terms of activity. This requires that the meter be equipped not only with slope and intercept calibrating controls but also with a logarithmic scale and, in the event of a divalent ion, the means for halving the slope ($n = 2$). Alternatively, an expanded millivolt scale and a calibration curve on semilog paper may be used. A scale with infinity at the center allows concentration to be read directly when using the known addition and subtraction techniques with selective ion electrodes.

PROBLEMS

1. The sodium ion correction nomograph for general-purpose glass electrodes manufactured by Beckman Instruments, Inc., is shown in the figure. (a) Estimate the

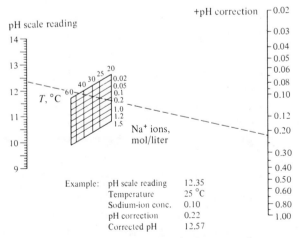

Problem 20-9. Sodium ion correction nomograph for general purpose Beckman glass electrodes. (Courtesy of Beckman Instruments, Inc.)

corrected pH for these measurements:

Solution number	A	B	C	D
pH scale reading	13.50	11.25	12.00	12.10
Temperature, °C	25	30	25	40
$[Na^+]$	0.05	0.2	0.1	0.02

(b) Below what pH value is the pH correction less than 0.02 pH unit at $[Na^+] = 0.2$? At $[Na^+] = 1.0$? (c) To render negligible the pH correction at pH 10.9, below what value must the sodium ion concentration be maintained?

2. Why is it necessary to protect carefully a potassium acid phthalate buffer solution from contamination with acids or alkalis?

3. Would the error in pH_s be significant if potassium acid tartrate solution were not completely saturated?

4. Why are alkaline reference buffer solutions subject to larger errors due to temperature changes than are acid buffers?

5. Why does the pH_s value of borax buffer solutions change so considerably with temperature? (*Hint:* Consider polymeric species of boric acid.)

6. Although no nitrite-selective electrode is available, suggest an indirect method to measure nitrite ion activity.

7. A fluoride solid-state electrode has a selectivity coefficient of 0.10 relative to hydroxide ion. At 10^{-2} M fluoride concentration, what hydroxide ion concentration could be tolerated?

8. What should be the lower pH limit when using the fluoride electrode if a 1% error is to be tolerated, and samples and standards are not adjusted to the same pH value?

9. Estimate the ratios at which the following ions can be present without impairing the response of the solid-state bromide electrode: chloride, iodide, hydroxide, and cyanide ions.

10. If calcium ion activity is to be measured with a liquid membrane electrode in samples containing up to 0.7 M sodium ion, estimate the minimum level of calcium which can be measured under these conditions. Assume a minimum level of 5% interference. The selectivity coefficient for sodium ion is 1.0×10^{-4}.

11. What is the maximum concentration of interfering anions that can be tolerated for a 1% interference level when measuring 10^{-5} M BF_4^- with a fluoroborate liquid ion-exchange membrane electrode? The interfering anions and their selectivity coefficients are: OH^-, 10^{-3}; I^-, 20; NO_3^-, 0.1; HCO_3^-, 4×10^{-3}; SO_4^{2-}, 1×10^{-3}.

12. What is the total sulfide concentration in 100 ml of sample that gives a potential reading of -845 mV before the addition of 1 ml of 0.1 M $AgNO_3$ and a reading -839 mV after the addition?

BIBLIOGRAPHY

Bates, R. G., *Determination of pH*, Wiley, New York, 1964.

Britton, H. T. S., *Hydrogen Ions*, 4th ed., Chapman & Hall, London, 1955.

Clark, W. M., *The Determination of Hydrogen Ions*, 3rd ed., Williams & Wilkins, Baltimore, 1928.

Durst, R. A., Ed., *Ion-Selective Electrodes*, National Bureau of Standards Spec. Publ. 314, Washington, D.C., 1969.

Eisenman, G., R. G. Bates, G. Mattock, and W. M. Friedman, *The Glass Electrode*, Wiley-Interscience Reprint, New York.

Kolthoff, I. M., *Acid-Base Indicators*, 4th ed., Macmillan, New York, 1937.

Lingane, J. J., *Electroanalytical Chemistry*, 2nd ed., Wiley-Interscience, New York, 1958.

Weissberger, A., and B. W. Rossiter, Eds., *Physical Methods of Chemistry*, Vol. 1, Part IIA, Wiley-Interscience, New York, 1971.

LITERATURE CITED

1. Bates, R. G., *J. Res. Natl. Bur. Std. (U.S.)*, **66A**, 179 (1962).

2. Bates, R. G. and M. Alfenaar, "Activity Standards for Ion-Selective Electrodes," in *Ion-Selective Electrodes*, R. A. Durst, Ed., Natl. Bur. Std. Spec. Publ. 314, Washington, 1969.

3. Bates, R. G. and V. E. Bowers, *J. Res. Natl. Bur. Std. (U.S.)*, **53**, 283 (1954).

4. Durst, R. A., *J. Chem. Educ.*, **44**, 175 (1967).

5. Eisenman, G. in *Ion-Selective Electrodes*, R. A. Durst, Ed., Natl. Bur. Std. Publ. 314, Washington, 1969.

6. Pungor, E., *Anal. Chem.*, **39** (13), 28A (1967).

7. Rechnitz, G. A., *Chem. Eng. News*, **45**, 146 (June 12, 1967); *Anal. Chem.*, **41** (12), 109A (1969).

21

Potentiometric Titrations

The measurement of the potential of an appropriate indicator electrode has been used for many years as a method of detecting the equivalence point in a variety of titrations. When a potentiometric titration is being performed, interest is focused upon changes in the emf of the electrochemical cell as a titrant of precisely known concentration is added to a solution of the test element. The method can be applied to any titrimetric reaction for which an indicator electrode is available to follow the activity of at least one of the substances involved. Reproducible equilibrium is of little concern here. Requirements for reference electrodes are greatly relaxed; it is only necessary that the response of one member of a pair of electrodes be substantially greater or faster than that of the other. In addition to the establishment of the equivalence point of a reaction, further information about the sample and its reactions may be obtained by the complete recording of a potentiometric titration curve.

The chief advantages of potentiometric titrations are applicability to turbid, fluorescent, opaque, or colored solutions, or when suitable visual indicators are unavailable or inapplicable. A succession of equivalence points in the titration of mixtures can be followed. Titrations in nonaqueous media are often dependent upon the method.

Equipment needed to carry out a classical potentiometric titration is illustrated in Fig. 21-1. In certain instances simplifications in equipment are possible.

In contrast to the direct potentiometric methods discussed in Chapter 20, potentiometric titrations generally offer an increase in accuracy and precision at the cost of increased time and difficulty. Accuracy is increased because measured potentials are used to detect rapid changes in activity that occur at the equivalence point of the titration, and this rate of emf change is usually considerably greater than the response slope which limits precision in direct potentiometry. Furthermore, it is the change in emf versus titrant volume rather than the absolute value of the emf which is of interest. Thus, the influences of liquid-junction potentials and activity coefficients have little or no effect.

CLASSIFICATION OF INDICATOR ELECTRODES

Electrodes of the First Kind

Electrodes of the first kind are reversible with respect to the ions of the metal phase. The electrode is a piece of metal in contact with a solution of its ions, for example, silver

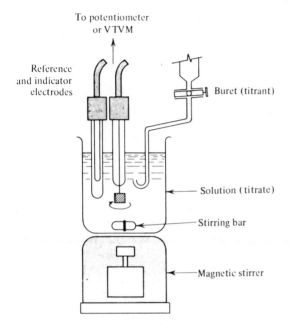

To potentiometer
or VTVM

Reference
and indicator
electrodes

Buret (titrant)

Solution (titrate)

Stirring bar

Magnetic stirrer

Fig. 21-1. Equipment for potentiometric titrations.

dipping into a silver nitrate solution. One interface is involved. For the half-cell

$$Ag^+ + e^- = Ag \qquad (21\text{-}1)$$

the Nernst expression is

$$E = 0.800 + 0.0592 \log [Ag^+] \qquad (21\text{-}2)$$

By convention, the activity of the pure massive metal (or any solid phase) is taken as unity.

The metal must be thermodynamically stable with respect to air oxidation, especially at low ion activities. In neutral solutions, suitable electrodes are restricted to Hg_2^{2+}/Hg and Ag^+/Ag. If oxygen is removed from the solution by deaeration, other electrodes become feasible: Cu^{2+}/Cu, Bi^{3+}/Bi, Pb^{2+}/Pb, Cd^{2+}/Cd, Sn^{2+}/Sn, Tl^+/Tl, and Zn^{2+}/Zn.

Simple amalgam electrodes are also electrodes of the first kind. For zinc, the reaction at the electrode is

$$Zn^{2+} + Hg + 2e^- = Zn(Hg) \qquad (21\text{-}3)$$

Electrodes of the Second Kind

Electrodes of the second kind involve two interfaces, such as metal coated with a layer of one of its sparingly soluble salts. The underlying electrode must be reversible. Consider a silver wire coated with a thin deposit of silver chloride. At the Ag/AgCl solution interface the electrochemical equilibrium is

$$AgCl(s) + e^- = Ag + Cl^- \qquad (21\text{-}4)$$

In addition, there is a chemical equilibrium

$$AgCl(s) = Ag^+ + Cl^-; \quad K_{sp} = 1.8 \times 10^{-10} \qquad (21\text{-}5)$$

Combining these two equations, we arrive at the Nernst expression for Eq. 21-4:

$$E = 0.800 + 0.0592 \log K_{sp} - 0.0592 \log [Cl^-] \qquad (21\text{-}6)$$

This simplifies to

$$E = 0.222 - 0.0592 \log [Cl^-] \qquad (21\text{-}7)$$

Electrodes of the second kind can be used for the direct determination of the activity of either the metal ion or the anion in the coating and also as an indicator electrode to follow titrations involving either. Limitations on these electrodes are severe. They can be used only over a range of anion activities such that the solution remains saturated with respect to the metal coating. Interferences from other anions can occur if they too form an insoluble salt with the cation of the underlying electrode.

Electrodes of the Third Kind

Reilley and coworkers[11,12] showed how to use an electrode of known reversibility to measure activities of ions for which no electrode of the first kind exists. They used a small mercury electrode (or gold amalgam wire) in contact with a solution containing metal ions to be titrated with a chelon Y, such as EDTA. A small added quantity of mercury(II) chelonate, HgY^{2-}, saturated the solution and established the half-cell:

$$Hg \mid HgY^{2-}, MY^{(n-4)+}, M^{n+}$$

where the electrode potential is given by

$$E = E^\circ + \frac{0.0592}{2} \log \frac{[M^{n+}]\ [HgY^{2-}]}{[MY^{(n-4)+}]} \qquad (21\text{-}8)$$

Because a fixed amount of HgY^{2-} is present, the potential is dependent upon the ratio $[M^{n+}]/[MY^{+n-4}]$. The species HgY^{2-} must be considerably more stable than MY^{+n-4}.

Redox Electrode

The redox electrode, usually gold, platinum, or carbon, immersed in a solution containing both the oxidized and reduced states of a homogeneous and reversible oxidation-reduction system, develops a potential proportional to the ratio of the two oxidation states. The only role of the redox electrode is to provide or accept electrons. An example is platinum in contact with a solution of iron(III) and iron(II) ions. For the half-reaction

$$Fe^{3+} + e^- = Fe^{2+}; \quad E^\circ = 0.76 \text{ V} \tag{21-9}$$

the Nernst expression is

$$E = 0.76 - \frac{0.0592}{1} \log \frac{[Fe^{2+}]}{[Fe^{3+}]} \tag{21-10}$$

Platinum electrodes are unsuitable for work with solutions containing powerful reducing agents, such as chromium(II), titanium(III), and vanadium(II) ions, because platinum catalyzes the reduction of hydrogen ion by these reductants at the platinum surface. Consequently, the interfacial electrode potential will not reflect the changes in the composition of the solution. In these cases a small pool of mercury can be used as the electrode because of the high overpotential associated with the deposition of hydrogen gas on a mercury surface.

In many redox titrations, the inert electrode is not reversible for one of the half-reactions, as in the case for thiosulfate in iodometric titrations. However, if the nonreversible system attains chemical equilibrium quickly with a reversible system (e.g., with I_2, I^-), the latter will serve as the potential-determining half-reaction. When chemical equilibrium is attained more slowly, mixed potentials may be involved; these bear no simple relationship to the activities of the reacting species although a stable equilibrium potential is rapidly attained. Sometimes the inert redox electrode behaves more or less like an oxygen electrode toward dissolved oxygen. If properly preconditioned, it may exhibit a memory for particular systems.

Membrane Electrodes

Ion-selective electrodes (Chapter 20) can also be used as indicating electrodes. Their usefulness can be extended to the determination of species for which ion-selective electrodes are not available by using the appropriate selective ion as the titrant. An example is the determination of lithium by precipitation of LiF in alcohol, the potentiometric titration being followed with the fluoride electrode.

LOCATION OF THE EQUIVALENCE POINT

The critical problem in a titration is the recognition of the point at which the quantities of reacting species are present in equivalent amounts—the *equivalence point*. The titration curve can be followed point by point, plotting as ordinate successive values of the cell emf versus the corresponding volume (or milliequivalents) of titrant added as abscissa. Additions of titrant should be the smallest accurately measured increments that provide an adequate density of points across the pH (or emf) range. Typical data are gathered in Table 21-1. Over most of the titration range the cell emf varies gradually, but near the end point the cell emf changes very abruptly as the logarithm of the concentra-

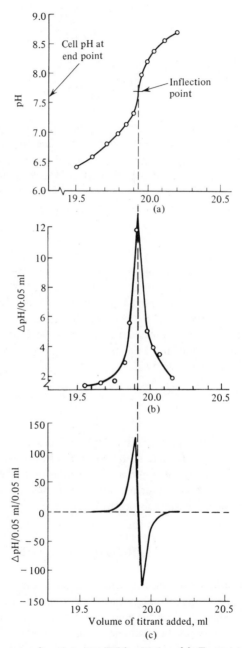

Fig. 21-2. Titration curves for data in Table 21-1. (a) Experimental titration curve, (b) first derivative curve, and (c) second derivative curve.

TABLE 21-1 Potentiometric Titration Data in Vicinity of an End Point[a]

Volume of Titrant, ml	Cell emf, pH units	$\Delta pH/\Delta V$	$\Delta^2 pH/\Delta V^2$
19.50	6.46		
		1.30	
19.60	6.59		+1.00
		1.40	
19.70	6.73		+2.00
		1.60	
19.80	6.89		+9.35
		3.00	
19.85	7.04		+52.0
		5.60	
19.90	7.32		+124.0
		11.80	
19.95	7.91		−136.0
		5.00	
20.00	8.16		−24.0
		3.80	
20.05	8.35		−5.00
		3.40	
20.10	8.52		
		1.60	
20.20	8.68		

[a] Volume at end point = 19.90 + 0.050 [124/(124 + 136)] = 19.92 ml.
pH at end point   = 7.32 + 0.59 [124/(124 + 136)] = 7.60.

tion(s) undergoes a rapid variation. The resulting titration curve will resemble Fig. 21-2a. The problem in general is to detect this sharp change in cell emf that occurs in the vicinity of the equivalence point. The equivalence point may be calculated, as will be outlined later. Usually the analyst must be content with finding a reproducible point, as close as possible to the equivalence point, at which the titration can be considered complete—the *end point*. By inspection the end point can be located from the inflection point of the titration curve: the point which corresponds to the maximum rate of change of cell emf per unit volume of titrant added. Distinctness of the end point increases as the reaction involved becomes more nearly quantitative. Once the cell emf has been established for a given titration, it can be used to indicate subsequent end points for the same chemical reaction.

In the immediate vicinity of the equivalence point the concentration of the original reactant becomes very small, and it usually becomes impossible for the ion or ions to control the electrode potential. The cell emf will become unstable and indefinite because the indicating electrode is no longer poised, that is, it is not bathed with sufficient quantities of each electroactive species of the desired redox couple. If the electroactive species are not too dilute, a drop or two of titrant will suffice to carry the titration through the equivalence point and into the region stabilized by the electroactive species of the titrant. However, solutions more dilute than 10^{-3} M generally do not give satisfactory titration end points unless special procedures are employed.

An end point may be located more precisely by plotting successive values of the rate of change of cell emf versus each increment of titrant in the vicinity of the inflection point. Increments need not be equal but should not be too large or too small. The position of the maximum on the first derivative curve, Fig. 21-2, corresponds to the inflection point on the normal titration curve. Once the end point volume is known, the corresponding cell emf at the end point can be obtained from the original titration curve. The end point can be even more precisely located from the second derivative curve, which is obtained by plotting the cell emf-volume acceleration versus the volume of titrant added. At the end point the second derivative becomes numerically equal to zero as the value of the ordinate rapidly changes from a positive to a negative number. Although either of these methods of selecting the end point is too laborious to do manually for each titration, they become feasible with appropriate electronic circuits (Fig. 21-10).

Oftentimes, tabulation of titration data will suffice to locate the end point by interpolation without the necessity of constructing derivative curves. In Table 21-1, the first two columns are original data. The third and fourth columns are the calculated values which would be used to plot the first and second derivative titration curves, respectively.

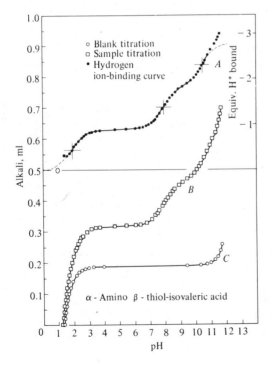

Fig. 21-3. Titration curves for α-amino-β-thiolisovaleric acid. (After T. V. Parke and W. W. Davis, *Anal. Chem.*, **26**, 642 (1954). Courtesy of American Chemical Society.)

A simple mathematical method for arriving at the pH (or emf) and volume of titrant at the end point is also outlined.

Particularly in acid-base titrations, a titration of a solution prepared like the sample solution, omitting only the sample itself (i.e., a blank), should also be run. In Fig. 21-3, curve *B* is a conventional titration curve with volume of titrant plotted against pH. Only the inflection at pH 8.0 is immediately apparent. The titration of the blank, curve *C*, is shown below. When the blank curve is subtracted volume-wise from the sample curve, the resulting curve *A* shows clearly the presence of two more inflections at pH values of 2.0 and 10.5. The preparation of blank curves is also desirable when handling relatively dilute solutions, when a minor constituent is suspected in the sample, or when impurities are present in the solvent.

Gran[4] proposed that plots of $\Delta ml/\Delta E$ versus ml of titrant be made to locate the equivalence point. Such plots are linear just before and after the end point, having a V-shape. Special graph paper is available on which the electrode potential (in mV) is plotted on the vertical (antilogarithmic) axis versus volume of titrant added on the horizontal linear axis. Only four or five points are necessary to define each segment of the curve; the intersection (extrapolated) of the two segments locates the end point. In some titrations the lines become curved some distance from the end point.

Use of Two Indicating Electrodes

Variation of the difference in electrode potential of two indicating electrodes can be followed sometimes during a titration. Generally only changes in the cell emf, but not the actual value, will be provided. Of course this is sufficient for many titration purposes. Elimination of the usual reference electrode and its attendant salt bridge eliminates leakage of the bridge electrolyte and minimizes the liquid-junction potential, which is desirable when solutions possessing high resistance (true for many nonaqueous systems) are involved.

A simple indicating electrode for reference purposes consists of a platinum (or other type) electrode inserted into the delivery tip from the buret, the buret electrode[15], or inside a capillary tube which contains a small portion of the original solution, the shielded electrode.[9] These electrodes, pictured in Fig. 21-4, assume a definite though not predictable potential and, because each is in contact with a solution of unchanging composition, maintain a constant potential which may be used for reference throughout the titration. Paired with a second indicating electrode dipping into the main solution, the usual S-shape titration curve is obtained. A platinum electrode serves as reference for most types of titrations. An antimony electrode in the buret paired with an antimony indicating electrode is useful for a strongly basic titrant such as sodium aminoethoxide in basic nonaqueous solvents.

The glass electrode may be employed as an electrode of reference in a medium of fixed, or buffered, pH or rather high hydrogen-ion concentration. Paired with a silver electrode, the combination finds use in argentometric titrations because leakage from the

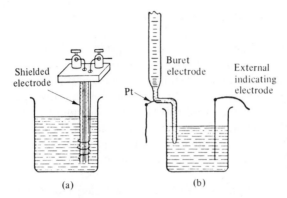

Shielded electrode

Buret electrode

External indicating electrode

Pt

(a)

(b)

Fig. 21-4. Systems consisting of two-indicating electrodes. (a) Shielded capillary reference electrode and (b) buret electrode.

salt-filling solution of a reference electrode is eliminated. In redox titrations, the glass electrode can be paired with platinum.

A graphite or tungsten electrode placed in the main solution serves well as a reference electrode, although these elements tend to reduce the amplitude of the change of cell emf by reason of their own response curves. This is insignificant in the case of a graphite/platinum electrode pair in neutralization reactions, and the tungsten/platinum pair in many redox systems. With two fast chemical half-reactions, a differential-shaped titration curve is obtained, but when one reaction is slow in establishing its equilibrium at the graphite or tungsten surface, a distorted S-shaped curve is obtained. An informative study of the tungsten electrode is available.[5]

An interesting system is provided by the pair: platinum and platinum/10% rhodium electrodes. Each acts as an indicating electrode of the second kind in neutralization reactions, presumably because of a thin layer of platinum oxide on each surface which renders each electrode responsive to hydrogen ions, in addition to the usual response to redox systems exhibited by an "inert" electrode. Whereas the platinum electrode responds rapidly to changes in solution composition, the platinum/10% rhodium electrode lags ever so slightly. Consequently, if the titrant is added rapidly and uniformly, the pair will exhibit a maximum difference in response at the equivalence point when the logarithmic term in the Nernst expression is changing most rapidly. A first-derivative curve is obtained. This pair of electrodes proves useful with automatic differential titrators.

Constant Current Potentiometry

Differential electrolytic potentiometry involves the observation of potentials across two indicator electrodes in a stirred solution during the passage of a minute, highly stabilized current. The apparatus required is a pair of wire electrodes, a resistor, a battery, and a meter (Fig. 21-5).

The electrode chosen must be appropriate to the reaction involved. Glass electrodes are not suitable. The electrode area should be kept small, 0.5 cm^2 or less, and should be of equal area for reversible reactions. For irreversible reactions there is some advantage in making the inactive electrode smaller. The current density required for optimum differ-

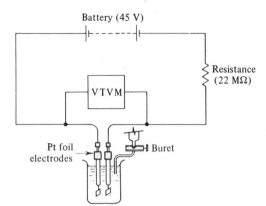

Battery (45 V)

VTVM

Resistance (22 MΩ)

Pt foil electrodes

Buret

Fig. 21-5. Equipment for conducting titrations at constant electrolysis current (approximately 2 μA with values of voltage and resistance indicated).

entiation depends on the equilibrium constant of the reaction and especially on the concentration of the titrant. For a 0.1 N titrant it is about 1 μA cm^{-2}, and it decreases with decreasing titrant concentration. Good stabilization of the current is required. The ballast load, the product of the ballast resistance in ohms and the source voltage in volts, should exceed 10^9 and preferably 10^{10}. At low values, potentials become unsteady and erratic, and current fluctuations also occur. At very high values, response times increase and the Johnson noise in the ballast resistor increases.

The shape of the titration curve can be predicted from current-voltage curves.[5] For example, consider the titration of copper(II) with EDTA,

$$Cu^{2+} + H_2Y^{2-} = CuY^{2-} + 2H^+ \qquad (21\text{-}11)$$

wherein the indicator electrodes are polarized with the small current indicated by the horizontal dashed line in Fig. 21-6, which contains the pertinent current–voltage curves. The initial potentials adopted by the cathode and anode will be the values at the intersections of the horizontal current line with the current–voltage curves; namely, E and F, respectively. The cell emf is then the difference, $F - E$. Curves 1 through 3 represent the Cu^{2+}/Cu° system at the beginning of the titration, when the titration is 91% complete, and when it is 99% complete. The difference in potential between each succeeding curve (from E to D, and from D to C) is 29.6 mV, as would be expected from the Nernst expression. The place where each curve intersects the zero-current axis (points E, D, and C) is the "null potential," as measured by the usual zero-current technique. However, in the vicinity of the equivalence point the mass transfer characteristics of copper(II) ions up to the cathode surface must be considered. A concentration gradient is established and the current carried through the cell by copper(II) ions becomes limited (curves 4 through 6). As soon as the current-carrying ability (the limiting current plateau) of the copper system falls below the electrolysis current being forced through the circuit, the potential of the cathode shifts quickly from point A to a value set by any other redox system which is able to poise the electrode, as, for example, the reduction of hydrogen ions at point R. All this time the other indicating electrode has maintained a constant potential because

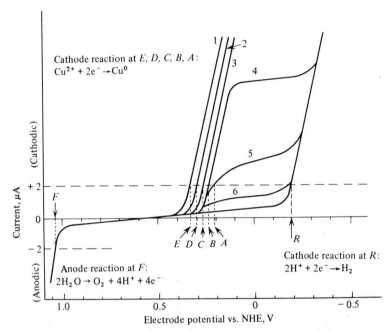

Fig. 21-6. Schematic current-voltage curves of copper(II)/(0) system in the presence of hydrogen ions. The potential scale is approximately real; the current scale is arbitrary.

the anode reaction at its surface is the oxidation of water (EDTA is not electroactive). The appearance of the titration curve is shown by curve 1 in Fig. 21-7.

When the titrant also possesses a set of current–voltage curves which lie within the boundary conditions imposed by the solvent and supporting electrolyte, the titration curve resembles curve 2 in Fig. 21-7. Such a titration would be iron(II) and cerium(IV). Prior to the equivalence point the cathode potential is established by the Fe^{3+}/Fe^{2+} system. After the equivalance point the cathode potential becomes stabilized and the anode becomes the indicating electrode for the Ce^{4+}/Ce^{3+} system. Titration curves take the form of the first derivative of the zero-current indicator electrode curve. The anode curve is displaced along the volume axis so that its potential rise occurs before that of the zero-current electrode, while the cathode curve is displaced in the opposite direction.

The vertical displacement of the two curves, the differential potential, then traces a sharp peak. The magnitudes of the anode lead and cathode lag are proportional to the differentiating current density. With electrodes of equal area, reversible reactions, and equal diffusion coefficients, the lead and lag are equal and a symmetrical differential curve is obtained. However, the end point and equivalence point usually do not coincide because of imperfection in reversibility of the electrode processes. Also electrolysis is occurring continuously at both electrodes; often the products are removed from the solution, for example, the copper(II) ions in the first example would be reduced to metallic copper at the cathode. The titration error will be a function of the electrolysis

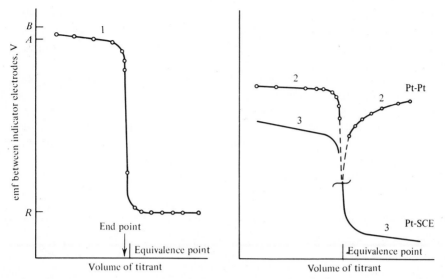

Fig. 21-7. Titration curves using constant electrolysis current.   Curve 1: Copper(II) titrated with EDTA (data from Fig. 21-6); Curve 2: Iron(II) titrated with cerium(IV) at finite current flow; and Curve 3: Iron(II) titrated with cerium(IV) under zero-current condition.

current and the time taken for the titration.   Furthermore, the end point will be premature by an amount that is a function of the magnitude of the electrolysis current in comparison with the concentration of unreacted test element, which gives rise to the limiting current of identical value.   The titration error can be minimized by setting the electrolysis current at as small a value as possible and providing rapid stirring to improve the rate of mass transfer to (and away from) the electrode surfaces.

A major advantage of titrations at constant current is that only one electroactive system needs to be present, either the titrant or the test element.   In fact, no advantage accrues in use of the method for two electroactive systems which establish steady potentials rapidly.   For the latter case the largest potential change occurs prematurely with polarized electrodes, whereas it occurs exactly at 100% of the equivalence-point volume when no electrolysis current is flowing.

Automatic Titrators

When performed manually so as to give a detailed titration curve or merely to locate precisely an end point, a potentiometric titration is a tedious and time-consuming operation.   For routine analyses the method does not have the speed and simplicity of comparable procedures employing visual indicators.   Automatic equipment for performing, and if desired, recording, the titration curve in its entirety provides a logical solution to the problem, albeit at some capital outlay.   An automatic titrator enables an operator to perform other tasks while the instrument delivers the requisite titrant and stops the delivery at a preset end point or, perhaps, continues beyond the end point when the entire curve is traced. Maximum benefit may be obtained from the equipment, particularly with slow reactions. The addition of the next increment will be delayed until the measured electrode potential falls below the value selected for the end point. The instrument will continue to repeat the final stages of the titration until a stable end point is obtained.

The basic features of commercial automatic titrators are alike. In the delivery unit, with no current passing through the solenoids, a short length of flexible tubing is squeezed shut in some manner. With the instrument set up and the buret level read, a switch is pressed to start the titration. The solenoid is energized, the pressure on the tubing is released, and titrant is allowed to flow through the delivery tip. The titration proceeds at a fast rate until a predetermined distance from the end point, when the anticipation control automatically slows the delivery of titrant. At the end point the delivery is stopped. The anticipation control is the key to highly precise automatic operation. It is set to anticipate the end point (preset on some instruments) by a chosen number of pH units or millivolts. The decreased rate of delivery precludes overstepping the preset end point while permitting a rapid delivery of titrant during the initial stages of the titration.

A schematic circuit diagram of an automatic titrator is shown in Fig. 21-8. The control unit includes a calibrated potentiometer, a null-sensing amplifier, and an anticipator circuit. To operate, the potentiometer is set at the pH or potential expected at the end point, the electrode assembly is immersed in the sample solution, and the operating switch is depressed. The difference signal arising between the cell emf and the preset voltage on the potentiometer is amplified, and the output from the amplifier energizes the solenoid value, or relay, in the delivery unit. As the end point is approached, the difference signal diminishes. When the two signals are matched, the delivery of titrant is stopped. If, upon additional mixing, the cell emf falls below the preset voltage, the controller relay will cause the delivery unit to dispense more titrant. This cycling repeats until a stable end point is reached without ever overshooting the end point.

An automatic recording titrator (Fig. 21-9) plots the complete titration curve. It is started and stopped manually. In this type of titrator the difference signal arises between the cell emf and an adjustable voltage from a calibrated potentiometer whose slidewire contact is positioned by the same motor that drives the recorder pen. The difference sig-

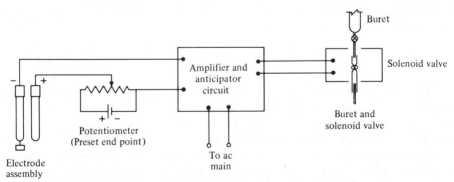

Fig. 21-8. Automatic titrator and its schematic circuit diagram.

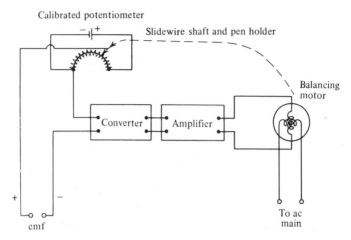

Calibrated potentiometer

Slidewire shaft and pen holder

Balancing motor

Converter   Amplifier

+   −
emf

To ac main

Fig. 21-9. Schematic circuit diagram of an automatic recording titrator.

nal, always very small, is converted to alternating current by a converter and then amplified. The output of the amplifier energizes one winding of the two-phase motor; the other winding is permanently connected to the 110-V main. Thus, the servosystem involves actuating the motor which drives the contactor on the slidewire of the potentiometer in a direction to match the cell emf and to preserve the null-balance. The pen traces the change in balancing voltage, and the corresponding cell emf, on the chart to provide a permanent record. No previous knowledge of the end point is required, and reevaluations can be made at a later date. This feature is a distinct advantage where completely unknown systems are run and where there may be successive inflection points. The chart-drive motor and syringe-delivery or constant buret-delivery unit must be synchronized to ensure a constant delivery rate throughout the entire operation.

In place of conventional burets, the delivery units can be designed around various types of syringes. These are operated by a motor-driven micrometer screw which actuates the plunger. Syringe-delivery units offer protection to the titrant from atmospheric oxidation, contamination, and loss of volatile solvent. To ensure rapid signal response, the delivery tip is placed close to the indicating electrode and in front, with respect to the direction of stirring, so that the indicating electrode is bathed by solution at a more advanced stage of titration.

A fully automatic unit will accept serially samples placed in a turntable. After each titration the turntable rotates, indexes the next sample solution beneath the electrode holder, lowers the electrode assembly, delivery tip, and stirring rod into the beaker, and actuates the titration switch to perform the next titration. Each time, the syringe is refilled with titrant and a printer prints out the amount of titrant delivered. This type of automatic instrument is ideal for performing multiple analyses in which the fundamental analytical procedure remains fixed over a period of time, as in a quality control situation.

The equivalent of plotting a second-derivative curve is accomplished automatically in another type of autotitrator,[7] whose schematic diagram is shown in Fig. 21-10. The cell

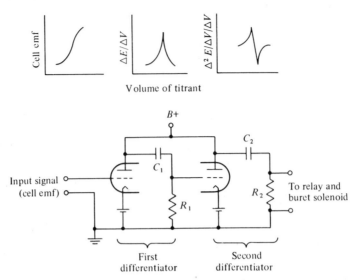

Fig. 21-10. Schematic diagram of a (Sargent-Malmstadt) derivative-autotitrator. (Courtesy of E. H. Sargent & Co.)

emf is fed directly to the control grid of a conventional amplifier. The amplified voltage is differentiated by a resistance-capacitance differentiator, R_1C_1, and the output is closely proportional to the first-derivative curve. Repeating the operation once again produces an output proportional to the second derivative of the titration curve, and a voltage ideally suited to trigger a relay system which closes the buret delivery unit, or terminates an external electrolysis signal, at the inflection point of the titration curve. All types of reactions are applicable if they possess a suitable reaction rate and the concentrations lie within 0.1–0.01 N. The derivative-autotitrator is definitely not applicable to titrations based on very slow reactions, whether these are the electrode reactions themselves, the fundamental chemical reaction, or intermediate secondary reactions. The method is invalidated when acceleration of cell emf becomes a function of some rate of change other than that of the rate of delivery of titrant, which is 1–6 ml/min. Too slow a delivery rate means too small a second-derivative signal, which may not actuate the thyratron relay. The relay must be set to reject spurious fluctuations to prevent a premature cutoff of the delivery unit; consequently, the actuating signal at the true end point must exceed any operating fluctuations.

Sensitivity

The sensitivity of a potentiometric titration is limited by the accuracy of the measurement of electrode potentials at low concentrations. Below 10^{-5} N, the residual current interferes with zero-current potentiometry. Similarly, the current in polarized titrations cannot be fixed at less than the residual current which is the order of the limiting current for a 10^{-5} N solution. A 10^{-2} N solution can therefore be titrated with an accuracy of 0.1%, but a 10^{-3} N solution can be titrated with an accuracy of only 1%. Other titration methods are needed for solutions more dilute than 10^{-3} N, the limiting concentration in potentiometric titration methods.

NULL-POINT POTENTIOMETRY

In principle, null-point potentiometry is a concentration cell technique that compares the solution to be analyzed with a solution of known composition.[8] The usual procedure consists of adjusting the composition of one of the half-cell solutions to match the other as evidenced by zero cell potential when measured between identical indicator electrodes selective for the species being determined. The dependence upon an absolute measurement of emf is eliminated. Very small volumes of solution, ranging from 5 to 100 μl, may be handled.

For a cell of this type, the emf is given by

$$E_{cell} = \frac{RT}{nF} \ln \frac{\gamma_1 C_1}{\gamma_2 C_2} + E_j \qquad (21\text{-}12)$$

where the subscripts 1 and 2 indicate the test and variable known solutions. If a sufficiently large excess of an inert electrolyte is used in both half-cells, E_j will become negligible. Also, the activity coefficients of the ionic species ($\gamma_1 \gamma_2$) in the two half-cells will be approximately equal due to the constant high ionic strength maintained by the inert electrolyte.

The null-point technique can be applied in two ways. In the *dead-stop* method, diluent or a standard solution of the species of interest is added until the null point is reached; the test element concentration is evaluated from the known concentration in the variable half-cell. In the *linear-interpolation titration* technique, aliquots of the diluent or standard solution are added incrementally, and the emf measurements are made prior to and after the null point. These emf data are then plotted (E_{cell} vs. log C_2); at the $E_{cell} = 0$ intercept, $C_1 = C_2$. Ideally the slope is 59.16/n mV per decade change in concentration; the slope serves as a check on the electrode behavior.

CLASSES OF CHEMICAL TITRATIONS

The principles governing the major types of potentiometric titrations will be examined briefly as a foundation on which potentiometric methods can be built. For more detailed discussion, refer to Lingane, Charlot, Badoz-Lambling, and Tremillon, or Kolthoff and Elving.

Oxidation–Reduction Reactions

Oxidation–reduction reactions can be followed by an inert indicator electrode. The electrode assumes a potential proportional to the logarithm of the concentration ratio of the two oxidation states of the reactant or the titrant, whichever is capable of properly poising the electrode. Let us assume that the reactant is the principal system, for example, the iron(III)/(II) system in the titration with cerium(IV). At the start of the titration the minute amount of one oxidation form iron(III) leaves the system without a definite electrode potential. However, as soon as a drop or two of cerium(IV) has been added, the concentration ratio of iron(III)/(II) assumes a definite value and, likewise, the electrode potential of the indicator electrode. During the major portion of the titration the electrode potential changes gradually. Only as the equivalence point is approached does the concentration ratio change rapidly again. Past the equivalence point, the indicator electrode ceases to be affected by the iron(III)/(II) system and will assume a potential dictated by the cerium(IV)/(III) system. For various ratios of iron(III)/(II) system, the corresponding electrode potential is

Ratio, Fe^{3+}/Fe^{2+}	10^{-3}	10^{-2}	10^{-1}	1	10	100	1000
Electrode potential,	0.594	0.653	0.712	0.771	0.830	0.889	0.948

At the equivalence point in an oxidation-reduction reaction,

$$a_{ox_1} + b_{red_2} = a_{red_1} + b_{ox_2} \tag{21-13}$$

the electrode potential is the weighted mean of the standard electrode potentials of reactant and titrant,

$$E_{\text{equiv pt}} = \frac{bE_1^\circ + aE_2^\circ}{a + b} \tag{21-14}$$

When $a = b$, the titration curve is symmetrical around the equivalence point, but when $a \neq b$, the titration curve will be markedly asymmetrical and the point of inflection will not coincide with the equivalence point. The difference will depend upon the ratio a/b. If $a > b$, the inflection point will occur when excess oxidant$_1$ is present in solution, that is, before the equivalence point. The inverse is true when $b > a$.

Chemical reactions (acid-base changes, complexation) can displace the electrode potentials in a way that often lends itself to a quantitative treatment. For example, cobalt salts may be titrated potentiometrically by making use of the reaction with standard potassium ferricyanide solution in the presence of a high concentration of ammonium citrate and of aqueous ammonia.[2] Under these conditions, the aquo-cobaltous ion is converted to the corresponding ammine complex which can be oxidized to the cobalt(III) hexammine complex by the ferricyanide. Although the aquo-cobalt species possess a very high reduction potential,

$$Co^{3+} + e^- = Co^{2+}; \quad E^\circ = 1.84 \text{ V} \tag{21-15}$$

as compared with the ferri/ferrocyanide couple,

$$Fe(CN)_6^{3-} + e^- = Fe(CN)_6^{4-}; \quad E^\circ = 0.36 \text{ V} \tag{21-16}$$

in the presence of aqueous ammonia, the ammine complexes of cobalt predominate. The formation constants for the cobalt(III) hexammine and cobalt(II) hexammine are

$$\frac{[Co(NH_3)_6^{3+}]}{[Co^{3+}][NH_3]^6} = 10^{34} \qquad \frac{[Co(NH_3)_6^{2+}]}{[Co^{2+}][NH_3]^6} = 10^5$$

Substituting the values for cobalt(III) and cobalt(II) ions into the Nernst equation for the cobalt equilibrium gives

$$E = 1.84 + 0.059 \log \frac{[Co(NH_3)_6^{3+}]}{[Co(NH_3)_6^{2+}]} + 0.059 \log \frac{10^5}{10^{34}} \qquad (21\text{-}17)$$

Although the difference between the reduction potential of the titrant and the cobalt system is slightly less than the minimum of 0.36 V, the difference of 0.26 V does permit the reaction to proceed within about 1% of completion. Air must be rigorously excluded and the temperature held at 0°–5°C. Usually the standard ferricyanide solution is added in excess and the unused amount is back-titrated with standard aquo-cobalt(II) solution. Replacement of ammonia by a stronger complexing agent, such as ethylenediammine, results in a greater difference in cell emf and a sharper end point. Unfortunately, additional interferences are introduced, whereas the titration in ammoniacal medium is virtually specific for cobalt.

Complexation, coupled with control of pH, plays an important role in the titration of manganese(II) ion with permanganate ion in pyrophosphate solution:

$$4Mn^{2+} + MnO_4^- + 8H^+ + 15H_2P_2O_7^{2-} = 5Mn(H_2P_2O_7)_3^{3-} + 4H_2O \qquad (21\text{-}18)$$

The change in potential at the equivalence point at a pH between 6 and 7 is large (about 300 mV); the equivalence-point potential is 0.47 V versus SCE. The method is subject to few interferences. Vanadium causes difficulties when the amount is equal to or larger

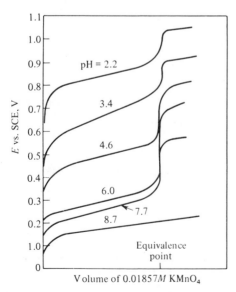

Fig. 21-11. Titration of manganese(II) with permanganate ion in pyrophosphate solution. (After J. J. Lingane and R. Karplus, *Ind. Eng. Chem., Anal. Ed.*, 18, 191 (1946). Courtesy of American Chemical Society.)

than the amount of manganese because of its sluggish reoxidation by permanganate, unless the titration is performed at a pH of 3.5. The titration curves are shown in Fig. 21-11.

In general, the reduction potential of a metal complex, or of a metal ion in equilibrium with a complexing agent, is decreased by complex formation. Three effects are involved[6]: The *coordinating effect* is the combination of a metal ion with an electron donor. The *charge effect* is simply the charge on the resulting complex. The *electronic effect* relates to the degree of stability of the electron configuration in the metal complex. The first two always tend to increase the tendency toward oxidation to a higher valence state; the third effect may work in either direction.

Ion Combination Reactions

Reactions in this category involve the formation of a sparingly soluble compound or a slightly dissociated material. Argentometric titrations are illustrative. For the chloride, and similar precipitation systems, the titration curve can be calculated from Eq. 21-7 or equivalent equations. At the equivalence point,

$$[Ag^+] = [Cl^-] = \sqrt{K_{sp}} \tag{21-19}$$

Typical halide titration curves are shown in Fig. 21-12.

For an ion combination reaction involving a soluble complex, such as

$$Ag^+ + 2CN^- = Ag(CN)_2^- \tag{21-20}$$

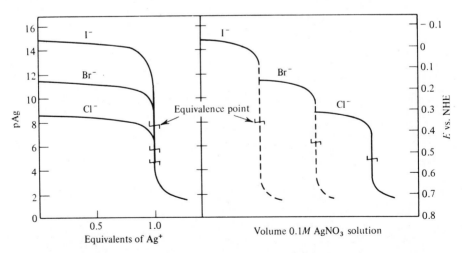

Fig. 21-12. Theoretical titration curves of halide ions (0.1 *M* each) with silver nitrate solution and a silver indicator electrode. The dashed segments are the separate curves for iodide and bromide ions.

the electrode potential is given by the expression

$$E = 0.80 + 0.059 \log \frac{1}{K_f[CN^-]^2} + 0.059 \log [Ag(CN)_2^-] \qquad (21\text{-}21)$$

where K_f is the formation constant of the dicyanoargentic ion. At the equivalence point

$$[Ag^+] = \tfrac{1}{2}[CN^-] = \sqrt[3]{\frac{[Ag(CN)_2^-]}{4K_f}} \qquad (21\text{-}22)$$

The magnitude of the inflection point on the titration curve depends upon the degree of insolubility of a precipitate, or the extent of dissociation of a complex. Successive titrations are feasible when one compound is markedly less soluble, or dissociated, than another, that is, $K_1/K_2 \geqslant 10^6$ and $K_1 > 10^8$, where the constants symbolize solubility product constants or instability constants. Application of potentiometric titration methods to precipitation reactions is limited by factors adversely affecting the character

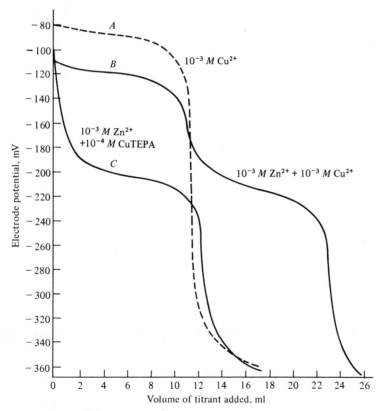

Fig. 21-13. Complexometric titration using a copper-selective electrode and tetraethylene pentamine (TEPA) titrant at pH 8 (curve *A*) and at pH 10 (curves *B* and *C*). (Courtesy of Orion Research, Inc.)

of precipitates, and applicability to this category is restricted by unavailability of indicator electrodes.

Ion-selective electrodes can be utilized for many complexometric titrations. Best results can be obtained with reagents such as EDTA which form only one complex with the metal ion and, consequently, give a single end-point break. An important type of complexometric titration involves the titration of a solution containing two metal ions, both of which form a complex with a reagent L. If an electrode is available that senses the ion M_1, which forms the more stable complex, then a titration curve similar to Fig. 21-13 results. If only M_1 were present in the sample, curve A would be obtained. If M_2 is also present, curve B results. In this case, after all the M_1 has been complexed by L at the first end point, the concentration of L increases up to a point where the second complex involving M_2 begins to form. This type of titration is very useful, as it allows an electrode sensitive to M_1 to be used to titrate any metal M_2 that forms a less stable complex with a given reagent than does M_1. In practice, the sample containing M_2 is spiked with an indicator concentration of the complex M_1L (10^{-5} to 10^{-4} M), which need not be accurately measured. The initial solution is therefore equivalent to the solution in Fig. 21-13 at a point corresponding to the first end point. The rest of the titration with L proceeds exactly as described.

Figure 21-14 explains the significance and value of pH-potential plots, and shows

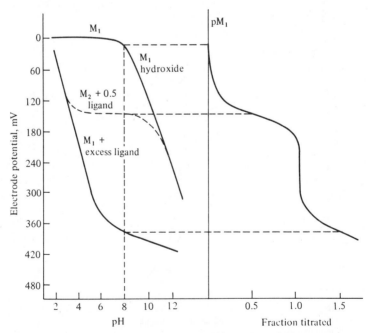

Fig. 21-14. Metal-ion electrode potential vs. pH in presence of excess ligand (*lower left curve*) and absence of ligand (*upper left*). *Right*, titration curve of M_2 and M_1 as an indicator and at pH 8. (Courtesy of Orion Research, Inc.)

how easily and quickly suitable titration conditions can be established.[10,13] The upper-most line is the electrode response at a fixed ion activity level as a function of pH. In general, there is no change until the precipitation of the hydroxide occurs. In the absence of complexing agents, the precipitation of hydroxide establishes the upper limit of metal ion which can be present at any given pH during the titration. Similarly, the electrode response is determined in the presence of a 50% excess of complexing agent, which is the level of free metal ion that will be present after the end point is reached. These two curves define the pH and metal ion levels within which useful titrations can be performed. It is also convenient to measure the potential of the equimolar mixture of metal ion and complex as a function of pH; this defines the midpoint of the titration curve for the system.

Acid–Base Reactions

Titrations of acids and their conjugate bases can be broken down conveniently into several categories, including consideration of nonaqueous solvent systems. Indicating electrodes are discussed in Chapter 20.

In the titration of a completely dissociated acid or base, the pH at the equivalence point is that of pure water (in the absence of dissolved CO_2), namely, 7. For a reaction to be complete within 0.1%, the initial concentration must not be less than 10^{-4} N.

For an incompletely dissociated acid, the hydrogen-ion concentration at the equivalence point is given by the expression

$$[H^+] = \sqrt{\frac{K_w K_a}{C_{salt}}} \tag{21-23}$$

As the acid becomes progressively weaker, the distinctness of the inflection point diminishes, and the pH at the equivalence point shifts to higher values. Feasibility of a particular titration is determined by the product, $K_a[HA]$. For an uncertainty of 0.1 percent or less, and in aqueous solution, the product should exceed 10^{-8}, assuming that the titrant is completely dissociated and 0.1 N in strength.

The accuracy with which two successive equivalence points may be located will depend on the absolute and relative strengths of the two acid groups and their concentrations. Difficulty is encountered in locating the break in the titration curve at a ratio of the first to the second dissociation constant of 100, and the ratio must be greater than 10^5 to 1 to give a sharp inflection point. However, by using transparent masks with theoretical curves, it is possible to estimate pK_a values at the extremes of the pH scale where only a portion of the complete theoretical curve can be distinguished (see Fig. 21-3). Groups whose pK_a values differ by as little as 1 pK unit can be resolved, provided at least half of each hydrogen ion-binding curve is free from overlap and permits matching with the masks. The chart in Fig. 21-15 relates pK_a values with a number of dissociating groups of organic acids. In addition, it is possible to distinguish further between various types of acidic and basic groups by performing the titration at several temperatures or in a nonaqueous solvent. The heat of dissociation for a monocarboxylic acid is usually less than 2 kcal mole^{-1}, and for a phenolic group it is about 6 kcal mole^{-1}, whereas for basic groups it is generally larger than 5 kcal mole^{-1}.

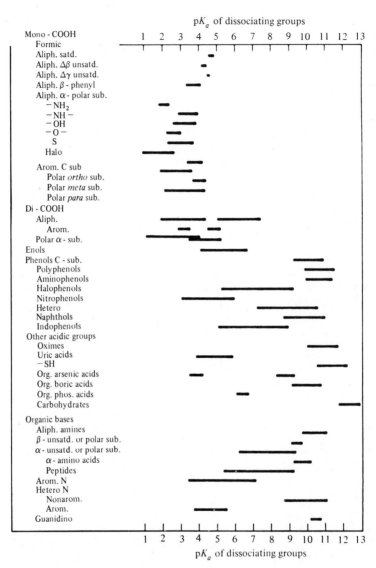

Fig. 21-15. Range of pK_a of dissociating groups from literature data. (After T. V. Parke and W. W. Davis, *Anal. Chem.*, **26**, 642 (1954). Courtesy of American Chemical Society.)

Acid–Base Titrations in Nonaqueous Solvents

Many acids or bases that are too weak for determination in water become susceptible to titration in appropriate nonaqueous solvents. The resolution of mixtures, particularly dibasic acids, may be improved.

The major considerations in the choice of a solvent for acidimetric reactions are its acidity and basicity, its dielectric constant, and the physical solubility of a solute. Acidity is important because it determines to a large extent whether or not a weak acid can be titrated in the presence of a relatively high concentration of solvent molecules. Phenol, for example, cannot be titrated as an acid in aqueous solution because water is too acid and present in too high a concentration to permit the phenolate ion to be formed stoichiometrically by titration with a base. In other words the intrinsic basic strength of the phenolate and hydroxide ions are not sufficiently different for the reaction

$$\phi-OH + OH^- \rightleftharpoons \phi-O^- + H_2O \tag{21-24}$$

to proceed quantitatively to completion. In less acidic solvents, such as dimethylformamide or pyridine, this titration can be carried out readily with a stronger basic titrant, the alkoxide ion,

$$\phi-OH + RO^- \longrightarrow \phi-O^- + ROH \tag{21-25}$$

The solvent must not be strongly basic if resolution of the strong and moderately strong acids is to be achieved, because of the "leveling effect" of a basic solvent on the stronger acids. In ethylenediamine, sulfonic and carboxylic acids are both leveled to the ammonium type ion, thus,

$$\left.\begin{array}{l} RSO_3H \\ RCOOH \end{array}\right\} + H_2N-C_2H_2-NH_2 \longrightarrow \left.\begin{array}{l} RSO_3^- \\ RCOO^- \end{array}\right\} + H_2N-C_2H_2-NH_3^+ \tag{21-26}$$

whereas in dimethylformamide only the sulfonic acid is leveled. An ideal solvent for the titration of an acidic mixture should be sufficiently weak in acidity to permit titration of the most weakly acid component and sufficiently weak in basicity to permit resolution of the strongest components.

Most of the acids can be classified as uncharged, positively charged, or negatively charged. The members of any one class may vary in relative strength to a certain extent as the dielectric constant of the solvent is changed, but in general they behave in a similar manner. However, acids of different charge type change greatly in relative strength as the solvent is changed. The positively charged acids, such as the ammonium ion, become stronger relative to an uncharged acid, such as acetic acid, as the dielectric constant is reduced. The protolysis reaction involving the solvent, SH,

$$NH_4^+ + SH \rightleftharpoons SH_2^+ + NH_3 \tag{21-27}$$

does not result in the formation of new charged species as does the reaction

$$HOAc + SH \rightleftharpoons SH_2^+ + OAc^- \tag{21-28}$$

Negatively charged acids, such as the bisuccinate ion, tend to become weaker relative to an uncharged acid.

The apparent strength of an acid in a solvent may be expressed empirically in terms of its midpoint potential or half-neutralization point. The difference in the midpoint potentials of two acids in the same solvent can serve as a measure of the resolution achieved;

it should be roughly 200–300 mV, depending on the slope of the plateaus in the titration curves.

Solvents may be divided into several classes. Amphiprotic solvents are those that possess both acidic and basic properties. They undergo self-dissociation, or autoprotolysis, that is,

$$SH + SH \rightleftharpoons SH_2^+ + S^- \tag{21-29}$$

to produce a solvonium ion SH_2^+ and a solvate ion S^-. Representative amphiprotic solvents include water, the lower alcohols, and glacial acetic acid. The product of the ion concentrations gives the autoprotolysis constant

$$K_{auto} = [SH_2^+][S^-] \tag{21-30}$$

It is 14 in water, varies from 15 to 19 in alcohols, and is about 14 in glacial acetic acid. Aprotic solvents have no acidic or basic properties, and if their dielectric constant is low, they have low ionizing power. These include aromatic and aliphatic hydrocarbons and carbon tetrachloride. Finally, there are a number of solvents with basic properties but essentially no acidic tendencies. These include amines, dimethylformamide, ketones, and ethers.

When a solute is dissolved in an amphiprotic solvent, the position of equilibrium depends on the relative acidic or basic strengths of the solute and solvonium ion (or solvate). The position of the autoprotolysis ranges, relative to the intrinsic strength of index acids, is indicated schematically in Fig. 21-16. Although a base must have a dissociation con-

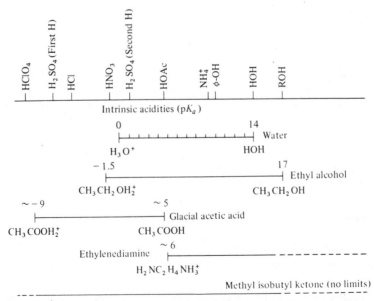

Fig. 21-16. Schematic representation of autoprotolysis ranges of selected solvents, in relation to the intrinsic strength of certain index acids. Influence of dielectric constant is not included.

stant greater than about 10^{-8} for successful titration in water, the accurate determination of compounds with basic dissociation constants (in water) of 10^{-12} is possible with a less basic solvent such as glacial acetic acid. Amino acids yield sharp end points because the carboxylic acid group is swamped, thus removing the zwitterion equilibrium. Commonly, the titrant is a solution of perchloric acid in dioxane or glacial acetic acid which has been standardized with either potassium acid phthalate (phthalic acid is the product) or sodium carbonate. In an analogous fashion, basic solvents enhance the properties of weak acids. Phenols and carboxylic acids produce distinctive end points in butylamine or ethylenediamine, and these plus sulfonic acids can be differentiated from each other in dimethylformamide. The titrant is usually sodium aminoethoxide or a quaternary base.

The effect of the dielectric constant coupled with the decrease in solvent basicity is well illustrated for the titration of oxalic acid, succinic acid, or sulfuric acid singly in isopropyl alcohol. In each instance two well-resolved end points are present, whereas in water a single inflection is obtained for sulfuric and succinic acid and only a poorly defined inflection indicates the existence of the first replaceable hydrogen in oxalic acid. Similarly, solutions containing perchloric and acetic acid exhibit two distinct inflections, and the degree of resolution is comparable with that of the dibasic acids. The reason is the difference in charge type between perchloric and acetic acid (or more correctly the solvonium ion and acetic acid), which makes the change in relative strength so large. Likewise, dibasic acids are a mixture of a positively charged acid (the solvonium ion) or an uncharged acid (succinic acid, first hydrogen) and a negatively charged acid (HSO_4^-, $HC_2O_4^-$, etc.). In addition, the weaker basicity of isopropyl alcohol lessens the extent of "leveling" of the first hydrogen of sulfuric acid and the perchloric acid, which increases the difference between the midpoint potentials of the titration curves.

Aprotic solvents have certain advantages over the other classes of solvents. By not interacting with dissolved solutes, no leveling action is exerted. Except for the influence of the dielectric constant, each solute will exhibit its intrinsic acidic or basic strength. Having no autoprotolysis limits, the range of applicability is limited only by the strength of the acid or base titrant. The former is usually perchloric acid or p-toluene sulfonic acid, the latter a quaternary ammonium base. If suitably spaced, a series of successive end points can be achieved in a solvent such as methyl isobutyl ketone (Fig. 21-17).

The electrode systems vary with the solvent employed. The glass-calomel electrode system is suitable where the solvent is either acetonitrile, an alcohol, or a ketone, or for differentiating titrations in dimethylformamide, provided that the titrant consists of either potassium hydroxide or alkoxide or tetraalkyl-ammonium hydroxide (i.e., no sodium compounds). It is advisable to replace the aqueous salt bridge of the calomel electrode by either a saturated solution of potassium chloride in methanol or N-tetraalkyl ammonium chloride solution. A glass electrode does not function as an indicator electrode in the more strongly basic solvents if the titrant contains sodium compounds. In this situation a pair of antimony electrodes forms a satisfactory combination, one dipping into the titrant and the other into the solution. A platinum electrode sealed into the buret and a glass indicator electrode give stable and reproducible potentials when solutions in methyl isobutyl ketone are titrated with quaternary ammonium hydroxide in benzene-methanol. The chloranil indicator electrode has been used in glacial acetic acid. Solvents with low dielectric constant exhibit such high internal resistance that it is diffi-

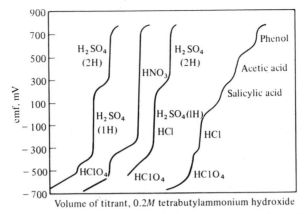

Fig. 21-17. Resolution of acid mixtures in methyl isobutyl ketone. Glass-calomel electrodes. (After D. B. Bruss and G. E. A. Wyld, *Anal. Chem.*, **29**, 232 (1957). Courtesy of American Chemical Society.)

cult to find electrodes which function satisfactorily. When the dielectric constant is 5 or less, potentiometric methods are unsuitable. A review of potentiometric electrode systems in nonaqueous titrimetry has been published.[14] Brief surveys of titrations in nonaqueous media are available and should be consulted for operational details.[1,3]

Other Nonaqueous Titrations

Other types of reactions are also feasible in nonaqueous solvents, particularly when solubility is a strong consideration. In petroleum products, the titration of hydrogen sulfide and mercaptans, either singly or in combination, is done in a 1:1 mixture of methanol-benzene (plus dissolved sodium acetate) with a methanolic solution of silver nitrate. The electrode system consists of a silver indicator electrode coupled with a calomel reference electrode connected to the sample solution by means of an agar bridge containing 3% potassium nitrate in the gel.

The Karl Fischer method for the determination of water is an excellent example of a well-known redox reaction conducted in methanol solution. Iron(II) perchlorate is useful as a reductant in glacial acetic acid.

LABORATORY WORK

General Instructions for Potentiometric Titrations

Assemble the titration equipment as shown in Fig. 21-1, or consult the manufacturer's bulletin supplied with a particular commercial instrument. The reference electrode may be a calomel or silver–silver chloride half-cell. The indicator electrode is assigned by the instructor or else is selected from among those discussed in the text for the appropriate class of titration.

Stir the solution in the beaker gently without producing a vortex. Measure or read the potential difference between the electrodes before any titrant is added. Record the reading and the volume of titrant in the buret. Add 2–3 ml of titrant from a 50-ml buret (or appropriate quantities from other size burets or microsyringes), stir for about 30 sec or until the potential difference becomes constant, measure the cell emf, and record the reading of the titrant volume in the buret. Additions of titrant should be of such volume to provide an adequate density of points across the pH or emf range. For example, about 5–10 points per pH unit (or $59/n$ mV) are adequate in the region where a group (pK_a or $E°$) is being titrated and in the vicinity of an end point. Continue the additions until the equivalence point has been exceeded by 1–2 pH units (or 50–100 mV).

Plot the cell emf as ordinates vs. the volume of titrant added as abscissa. Draw a smooth curve through the points. Locate the equivalence point by calculating the anticipated cell emf. Compare this value with the point on the curve which corresponds to the steepest portion of the titration graph. Also locate the end point by plotting $\Delta E/\Delta ml$ for small increments of titrant in the vicinity of the end point, or by estimating the volume of titrant by the method outlined in Table 21-1.

Experiment 21-1 Acid–Base Titrations

A. Aqueous Systems

Each student will be assigned a titration system; the sample size and titrant strength will be specified. Suggested systems are:

1. Acetic acid, 0.1 N, with 0.1 N NaOH.
2. Sodium carbonate, 0.05 M, with 0.1 N HCl.
3. Phosphoric acid, 0.05 M, with 0.1 N NaOH.
4. Boric acid, 0.1 N, in the presence of mannitol (4 g/50 ml volume) with 0.1 N NaOH.

B. Glacial Acetic Acid

Perchloric acid, 0.1 N, is prepared by adding slowly with stirring 8.5 ml of 72% perchloric acid to 900 ml of glacial acetic acid (or purified dioxane) followed by 30 ml of acetic anhydride. Allow the mixture to stand for 24 hr before use. Standardize against primary grade potassium hydrogen phthalate.

Suggested basic solutes:

1. Potassium hydrogen phthalate. Dissolve approximately 0.1 g (weighed accurately) in 25 ml of glacial acetic acid with gentle boiling to effect solution. Insert a pair of small glass and calomel electrodes. Stir the solution and titrate with 0.1 M perchloric acid solution, using a 10-ml buret.
2. Sodium carbonate. Dissolve approximately 0.05 g of the anhydrous salt, weighed accurately.

3. Aniline, or chloroaniline. Dissolve about 0.1 ml, weighed accurately by difference from a small beaker.

C. Methanol or Ethanol

Hydrochloric acid, 0.1 N, is prepared by adding 9.0 ml of reagent grade hydrochloric acid to 1 liter of absolute methanol. It is standardized against sodium carbonate as follows: Dissolve approximately 0.1 g of freshly dried sodium carbonate (weighed accurately) in the smallest possible amount of glacial acetic acid, taking care to avoid loss during the effervescence. Evaporate to dryness and dissolve the residue in 20 ml of methanol. Insert a glass-calomel electrode system. The calomel electrode should be the sleeve-type salt bridge or some type of salt bridge with a relatively large area of solution contact. Titrate with 0.1 N hydrochloric acid.

Potassium hydroxide, 0.1 M, is prepared by dissolving approximately 6 g of pellets in a small volume of methanol in the absence of air. Dilute to 1 liter with additional methanol and protect the solution from absorption of carbon dioxide. Standardize against the 0.1 M hydrochloric acid solution.

Suggested titration systems:

1. Mixture of hydrochloric acid and acetic acid. Place in a 150-ml beaker 10 ml of 0.1 M hydrochloric acid in methanol and 0.1 ml of glacial acetic acid. Dilute to 50 ml with methanol. Titrate with 0.1 M potassium hydroxide in methanol.

   Repeat the titration with the inclusion of 10 ml of water in the solvent mixture. Note the change in position of the titration curve for hydrochloric acid and the indistinctness of the first inflection point.

2. Oxalic acid. Dissolve 0.2 g (weighed accurately) in 50 ml of isopropyl alcohol and titrate with 0.1 M potassium hydroxide solution. Be sure to titrate both replaceable hydrogen ions.

Experiment 21-2 Oxidation–Reduction Titrations

A. Determination of Cobalt with Ferricyanide

Into a 250-ml beaker place 10 ml of 0.05 M potassium ferricyanide solution, 10 ml of 5% ammonium citrate solution, and 100 ml of 5 N aqueous ammonia solution. Immerse the beaker in crushed ice (or add crushed ice directly) to lower the temperature to 3°–5°C. Insert a smooth platinum foil electrode and a calomel or other reference electrode. Stir the solution and titrate with 0.1 N cobaltous solution (prepared from uneffluoresced crystals of the reagent grade cobalt sulfate heptahydrate). The cell emf falls from about 300 mV initially to about 100 mV in the vicinity of the end point when the reference electrode is a SCE.

An unknown cobalt solution may be determined by adding an aliquot to an excess of standard potassium ferricyanide solution, and titrating the unused ferricyanide solution with standard cobalt solution. For samples that may be high in iron, use 30% ammonium citrate solution.

B. Determination of Manganese with Permanganate

Place 150 ml of a 0.27 M sodium pyrophosphate solution (freshly prepared) in a 400-ml beaker, and adjust the pH to 6–7 by the addition of concentrated sulfuric acid from a graduated pipet. Add 25 ml of 0.05 M manganese solution and, if necessary, readjust the pH. Insert a bright platinum electrode and a calomel electrode. Stir the solution and titrate with 0.01 M potassium permanganate solution. Remember that the permanganate ion undergoes only a 4-electron change to the 3+ state. If the vanadium content exceeds the manganese content in a steel sample, adjust the pH to 3.4–4.0.

C. Other Suggested Systems

Titrate an iron(II) sulfate solution, 0.05–0.1 M, with 0.1 M cerium(IV), 0.02 M potassium permanganate, or 0.0167 M potassium dichromate in a solution that is approximately 1 M in sulfuric acid. Repeat the titration with the addition of 5 ml of 85% phosphoric acid to complex the iron(III) as it is produced.

Experiment 21-3   Argentometric Titration of Halides

A. Determination of a Single Halide

Place 25 ml of a 0.1 M sodium chloride solution in a 250-ml beaker and dilute to about 100 ml with water. Insert a bright silver wire plus a reference electrode with a fiber-type connection or with an intermediate salt bridge containing a nonhalide filling solution. Titrate with 0.1 M silver nitrate solution.

Repeat the titration with a 0.1 M potassium iodide or a 0.1 M potassium bromide solution. Note the differences in cell emf at the midpoint of each titration curve; compare with the calculated values.

Repeat the titration with a 0.1 M potassium iodide solution in an aqueous solution that is 1 M in ammonia. Note the difference in the position of the titration curve after the equivalence point.

B. Mixture of Halides

Place 10 ml of a 0.1 M sodium chloride solution and 15 ml of a 0.1 M potassium iodide solution in a 250-ml beaker and dilute to about 100 ml with water. Titrate with 0.1 M silver nitrate solution.

Repeat the titration with this change: titrate the iodide ion in an ammoniacal solution, and after the end point is reached, acidify the solution with nitric acid and titrate the chloride ion.

C. Determination of Iodide with Redox Indicating System

Place 25 ml of a 0.1 M potassium iodide solution in a 250-ml beaker and dilute to about 100 ml with water. Add about 0.5 ml of a saturated solution of iodine in alcohol, freshly prepared. Insert a bright platinum electrode and a calomel reference electrode. Titrate with 0.1 M silver nitrate solution.

Experiment 21-4   Complexometric Titrations with EDTA

Mercury indicator electrode Lightly amalgamate a gold wire electrode with mercury, or use a mercury pool emanating from a J-tube or other small cup-like container. Wash the mercury with dilute nitric acid and then rinse thoroughly with distilled water. After each titration, rinse the mercury thoroughly.

Procedure Place 25 ml of the metal-ion solution (approximately 0.05 M) in a 250-ml beaker, add 25 ml of the appropriate buffer solution (about 0.5 M in each component), and 1 drop of 0.0025 M mercury(II) EDTA solution. Insert the amalgam electrode and a calomel reference electrode.   Titrate with 0.05 M disodium dihydrogen ethylenediaminetetraacetate.

Conditions for the titration of selected metal ions are:

At pH 2 in chloroacetic acid system—thorium, mercury, or bismuth.
At pH 4.7 in a sodium acetate–acetic acid buffer—copper, zinc.
At pH 10 in an ammonium chloride–aqueous ammonia buffer—cobalt.

Experiment 21-5   Determination of Zinc with Ferrocyanide

Place 25 ml of 0.1 M zinc chloride solution in a 250-ml beaker. Add approximately 10 g of ammonium chloride and 1 ml of 0.001 M potassium ferricyanide solution. Insert a bright platinum electrode and calomel reference electrode. Titrate with a 0.067 M potassium ferrocyanide solution.

Experiment 21-6   Determination of Apparent Dissociation Constants

Dissolve about 10 mg of the base (or acid) in exactly 5 ml of water. Insert a set of micro glass and calomel electrodes. Titrate with a 2 M HCl (or KOH) solution delivered by a microsyringe buret having a total capacity of 0.2 ml and reading to 0.0002 ml. Obtain readings over the interval of pH from the initial point up to pH 12.

Plot the conventional titration curve. Repeat the titration on a blank solution prepared like the sample solution, omitting only the sample itself. Subtract the blank curve volumewise from the sample curve, and plot the resulting data. Estimate the pK_a value from each inflection within ±0.05 pK_a unit. For overlapping inflection points, or inflection points near the extremities of the pH scale, it is desirable that the curves be replotted as equivalents of bound hydrogen ion vs. pH on a uniform scale—perhaps 1 in. per equivalent and 0.5 in. per pH unit. Compare with a theoretical curve calculated from the expression

$$pH = pK_a + \log\left(\frac{a}{1-a}\right)$$

where a is the fraction of the sample in the dissociated state for acids or the associated state for bases.

Make a tentative assignment of structure of the dissociating group from the chart in Fig. 21-15 if your compound is an organic acid or base.

Experiment 21-7   Titration with Constant Electrolysis Current

Connect one terminal of a 45-V dry cell through a 22-MΩ resistor to a platinum foil electrode, and connect the other terminal of the dry cell directly to a second platinum foil electrode. Also connect a high-impedance VTVM across the terminals of the platinum electrodes. The circuit should resemble Fig. 21-5.

Place 10 ml of a 0.1 M iron(II) sulfate solution in a 150-ml beaker. Add 40 ml of water and 3 ml of concentrated sulfuric acid. Insert the pair of platinum electrodes. Titrate with 0.1 M cerium(IV) solution.

Repeat the titration with 0.0167 M potassium dichromate solution as the titrant.

If time is available, the titration can be performed with various values of the electrolysis current flowing through the solution other than the 2 μA suggested in the directions for assembly of the circuit components.

Experiment 21-8   Study of the Response of a Platinum–Tungsten Electrode Combination

Place 25 ml of 0.1 N iron(II) ammonium sulfate solution into a 250-ml beaker. Add 5 ml of 36 N sulfuric acid and, if desired, 5 ml of 85% phosphoric acid. Dilute to 100 ml.

Insert into the solution a platinum electrode, a tungsten electrode, and a standard reference electrode. The tungsten electrode should be polished immediately before use or else immersed for a few seconds in molten sodium nitrite and then washed thoroughly with distilled water. The fusion mixture should be barely molten or the tungsten will quickly dissolve.

Connect the leads from the platinum and tungsten electrodes to the opposite ends of a single-pole double-throw switch. Connect the lead from the center tap of the switch and the lead from the reference electrode to a potentiometer or vacuum tube voltmeter.

Proceed with the titration using 0.1 N potassium dichromate as titrant. After the addition of each increment of titrant, measure the potential of the platinum and tungsten electrodes against the reference electrode.

Plot the indicator electrode potential against the volume of titrant for the platinum–calomel and the tungsten–calomel pair of electrodes separately. Also plot the difference in potential between the platinum and tungsten electrodes.

In a separate titration, measure the difference in potential between the platinum and tungsten electrodes now connected directly to the potentiometer or vacuum tube voltmeter. Connect the tungsten electrode in place of the calomel reference electrode.

PROBLEMS

1. Sketch the titration curves you would expect from the titration of each of the following aqueous systems: (a) 0.1 M solution of H_3PO_4 titrated with 0.2 M NaOH, (b) a solution 0.05 M in Na_3PO_4 and 0.1 M in Na_2HPO_4 titrated with 0.1 M HCl, (c) a 0.1 M solution of ammonia titrated with 0.1 M HCl.

2. Construct the complete curve for the titration of 50 ml of 0.1 M titanous chloride ($TiCl_3$), $E° = 0.10$ V, with a 0.1 M solution of methylene blue ($E° = 0.52$) in 1 M HCl.

3. Construct the titration curve for the titration of the solution resulting from passage of 0.01 M vanadium solution through an amalgamated zinc reductor; titrant is 0.1 M cerium(IV). $E° = -0.255$ for V^{3+}/V^{2+}, $E° = 0.361$ for VO^{2+}/V^{3+}, and $E° = 1.000$ for VO_2^+/VO^{2+}.

4. Sketch the titration curves you would expect from the titration of each of the following nonaqueous systems. Express the ordinate values in units of 0.059 V.
   (a) A mixture of an alkyl sulfonic acid ($pK_a = -7$) and an alkyl carboxylic acid ($pK = 4$) dissolved in methyl isobutyl ketone and titrated with tetra-n-butylammonium hydroxide.
   (b) A solution of aniline in glacial acetic acid titrated with $HClO_4$.
   (c) A mixture of HCl and acetic acid dissolved in isopropyl alcohol and titrated with sodium isopropoxide.
   (d) A mixture of acetic acid and phenol in n-butylamine titrated with sodium aminoethoxide.

5. Calculate the potential at the equivalence point in the potentiometric titration of each of these systems; assume the reference electrode is saturated calomel: (a) Titration of tin(II) with cerium(IV) ions; (b) titration of uranium(IV) with iron(III); (c) titration of VO^{2+} with cerium(IV) in 1 M H_2SO_4; (d) titration of arsenic(III) with bromate in 5 M HCl.

6. From the information in Fig. 21-12, estimate the error in the location of the end point, as contrasted with the equivalence point, for (a) iodide and (b) bromide when all three halides are present in mixtures. Disregard any error attributable to mixed salt formation.

7. The following pH readings were obtained for corresponding volumes of 0.100 N NaOH in the potentiometric titration of a weak monobasic acid:

0.00 ml = 2.90	14.00 ml = 6.60	16.00 ml = 10.61
1.00 ml = 4.00	15.00 ml = 7.04	17.00 ml = 11.30
2.00 ml = 4.50	15.50 ml = 7.70	18.00 ml = 11.60
4.00 ml = 5.05	15.60 ml = 8.24	20.00 ml = 11.96
7.00 ml = 5.47	15.70 ml = 9.43	24.00 ml = 12.39
10.00 ml = 5.85	15.80 ml = 10.03	28.00 ml = 12.57
12.00 ml = 6.11		

(a) Plot the above values of pH against milliliters of NaOH solution. (b) What is the pH value at the equivalence point? (c) What volume of NaOH corresponds to the equivalence point? (d) What is the ionization constant of the acid?

8. In the titration of iron(II) using the differential potentiometric method with a constant electrolysis current, the following readings were obtained for corresponding volumes of 0.1 M cerium(IV):

2.00 ml = 50 mV	9.60 ml = 410 mV	10.75 ml = 400 mV
4.00 ml = 50 mV	9.80 ml = 740 mV	11.00 ml = 365 mV
6.00 ml = 50 mV	10.00 ml = 705 mV	12.00 ml = 300 mV
8.00 ml = 100 mV	10.25 ml = 515 mV	14.00 ml = 250 mV
9.00 ml = 155 mV	10.50 ml = 460 mV	16.00 ml = 205 mV
9.40 ml = 205 mV		

(a) Plot the millivolt readings against the volume of cerium solution. (b) What volume of cerium corresponds to the end point?

BIBLIOGRAPHY

Beckett, A. H. and E. H. Tinley, *Titrations in Non-Aqueous Solvents*, 3rd ed., British Drug Houses, Ltd., Poole, England, 1962.

Browning, D. R., Ed., *Electrometric Methods*, McGraw-Hill, Maidenhead, England, 1969.

Charlot, G., J. Badoz-Lambling, and G. Tremillon, *Electrochemical Reactions*, American Elsevier, New York, 1962.

Gyenes, I., *Titration in Nonaqueous Media*, Van Nostrand Reinhold, New York, 1967.

Kolthoff, I. M. and P. J. Elving, Eds., *Treatise on Analytical Chemistry*, Vol. 1, Part I, Wiley-Interscience, New York, 1959; Chapters 11–14 and 16.

Lingane, J. J., *Electroanalytical Chemistry*, 2nd ed., Wiley-Interscience, New York, 1958.

Ringbom, A., *Complexation in Analytical Chemistry*, Wiley, New York, 1963.

Rossotti, H., *Chemical Applications of Potentiometry*, Van Nostrand Reinhold, New York, 1969.

LITERATURE CITED

1. Beckett, A. H. and E. H. Tinley, *Titrations in Non-Aqueous Solvents*, 3rd ed., British Drug Houses, Ltd., Poole, England, 1962.
2. Chirnside, R. C., H. J. Cluley, and P. M. C. Proffitt, *Analyst*, 72, 354 (1947).
3. Fritz, J. S., *Acid–Base Titrations in Nonaqueous Solvents*, G. Frederick Smith Chemical Co., Columbus, Ohio, 1952.
4. Gran, G., *Acta Chem. Scand.*, 4, 559 (1950); *Analyst*, 77, 661 (1952).
5. Kolthoff, I. M., *Anal. Chem.*, 26, 1685 (1954).
6. Laitinen, H. A., *Chemical Analysis*, McGraw-Hill, New York, 1960.

7. Malmstadt, H. V. and E. R. Fett, *Anal. Chem.*, **26**, 1348 (1954).
8. Malmstadt, H. V. and J. D. Winefordner, *Anal. Chim. Acta*, **20**, 283 (1959).
9. Müller, E., *Z. Physik. Chem.*, **135**, 102 (1928).
10. Reilley, C. N., *Anal. Chem.*, **30**, 947 (1958).
11. Reilley, C. N. and R. W. Schmid, *Anal. Chem.*, **30**, 947 (1958).
12. Reilley, C. N., R. W. Schmid, and D. W. Lamson, *Anal. Chem.*, **30**, 953 (1958).
13. Ross, J. W. and M. S. Frant, *Anal. Chem.*, **41**, 1900 (1969).
14. Stock, J. T. and W. C. Purdy, *Chem. Rev.*, **57**, 1159 (1957).
15. Willard, H. H. and A. W. Boldyreff, *J. Am. Chem. Soc.*, **51**, 471 (1929).

22

Voltammetry, Polarography, and Related Techniques

The polarographic method of analysis is based on the current-voltage curves arising at a microelectrode when diffusion is the rate-determining step in the electrochemical reaction. The development of polarography, commencing with the work of Heyrovsky in 1922, marked a significant advance in electrochemical methodology because it introduced the element of selectivity through control of electrode potential, an element which was largely lacking in the older electrochemical methods of potentiometry and conductimetry. However, the fundamental dc polarographic technique suffered from a number of difficulties that made it less than ideal for routine analytical purposes and made the results obtained somewhat difficult to interpret. With the advent of low-cost, fast, stable, operational amplifiers in the early 1960s, various problems began to be overcome. Investigations of such techniques as ac polarography, pulse polarography, and derivative pulse polarography demonstrated the utility and desirability of the *new* polarography for fingerprint purposes and analytical applications. Today multipurpose instruments in an all-electronic form provide sensitivity to the parts-per-billion level for many electroactive substances. All the techniques discussed in this chapter are united in their applicability to electrodes other than the standard dropping mercury electrode employed in classical polarography; these include glassy carbon and wax-impregnated graphite electrodes that permit voltammetry in the anodic region. The reemergence of voltammetry is timely, since it has the greatest sensitivity to environmentally important elements such as lead and cadmium. It is being used in forensic work to determine drugs and arsenic in urine.

CURRENT–VOLTAGE RELATIONSHIPS

An electrode is considered to be *polarized* when it adopts a potential impressed upon it with little or no change of the current. Take, for example, a platinum electrode dipping into a solution of copper(II) ions which is also 0.1 M in sulfuric acid. When short-circuited with a calomel reference electrode, the platinum electrode will assume the potential of the calomel electrode with no flow of current. The platinum electrode is polarized, and it will remain polarized until an emf is impressed across the two electrodes

633

that is sufficient to exceed the decomposition potential of the copper(II) ions. When the impressed voltage does exceed the decomposition potential, copper will deposit on the platinum. Until this potential is attained, there is no reversible electrode reaction. After some copper has plated out, the electrode becomes depolarized and its potential is determined by the Nernst equation,

$$E = E° + \frac{0.0592}{2} \log [Cu^{2+}] \tag{22-1}$$

As long as the electrode is ideally depolarized, passage of current does not cause the potential to deviate from its reversible value.

Figure 22-1 gives idealized current–voltage curves of copper(II) ion solutions with varying concentrations. Starting out with a well-stirred solution of 0.05 M copper(II) ions and impressing a voltage across the platinum electrode and reference electrode, the current–voltage curve will be traced by curve OAB. No current will be observed to flow until the applied emf exceeds the decomposition potential of a solution of 0.05 M copper(II) ions. At point A, copper commences to plate out on the electrode and current starts to flow. As the voltage is increased further, the current increases linearly in accordance with Ohm's law. On the other hand, if the voltage is reduced from B to A, the current will diminish gradually to zero as the deposit of copper dissolves from the platinum electrode. Similarly, for smaller concentrations of copper the current–voltage traces will be given by curves OCD and OEF. In each case, the decomposition potential will be shifted along the voltage axis to a more negative value by 29.5 mV per decadic change in concentration.

The picture changes if the experiment is repeated without stirring the solution and with a microelectrode, that is, an electrode with a small area of contact with the test solution. Consider a concentration of 0.005 M copper(II) ions at an impressed emf corresponding to D in Fig. 22-1. If the voltage is held constant, the current will start to flow

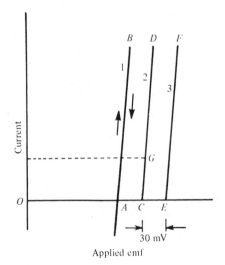

Fig. 22-1. Current–voltage curves without polarization. Curve 1: 0.05 M copper; Curve 2: 0.005 M copper; Curve 3: 0.005 M copper.

at a value corresponding to D but will decrease rapidly to some value G as the concentration of copper(II) ions at the microelectrode surface becomes depleted by deposition. The current, represented by G, is characteristic of the rate at which fresh copper(II) ions are supplied to the microelectrode. The microelectrode is polarized under these conditions.

In general, there are three mass-transfer processes by which a reacting species may be brought to an electrode surface. These are (1) diffusion under the influence of a concentration gradient; (2) migration of charged ions in an electric field; and (3) convection due to motion of the solution or the electrode. In voltammetry the effect of migration is usually eliminated by adding a 50- or 100-fold excess of an inert "supporting electrolyte." The ions of this electrolyte migrate to relieve the electric fields but do not undergo an electrochemical reaction at the electrode. A potassium salt is often employed. Because the potassium ions cannot be discharged at the cathode until the impressed voltage becomes rather large, large numbers of them remain as a cloud around the cathode. This positively charged cloud restricts the potential gradient to a region so very close to the electrode surface that there is no longer an electrostatic attraction operative to attract other reducible ions from the bulk of the solution. Convection can be minimized by using unstirred vibration-free solutions. Under such conditions, the limiting current is controlled solely by diffusion of the reacting species through the concentration gradient adjacent to the electrode. The latter gradient arises because of the relative slowness of diffusion.

According to Fick's law, the net rate of diffusion of a species to a unit area of electrode surface A at any time t is proportional to the magnitude of the concentration gradient, that is,

$$\text{flux} = -D\left(\frac{dC}{dx}\right)_{x=0} = \frac{-D\,(C_{\text{bulk}} - C_0)}{\delta} \tag{22-2}$$

where D is the diffusion coefficient of the species and δ is the thickness of the hypothetical diffusion layer about the microelectrode (Fig. 22-2). As the region around the microelectrode becomes depleted of electroactive species, that is, as C_0 approaches zero, the rate of diffusion becomes proportional to the concentration in the bulk of the solution, C_{bulk}.

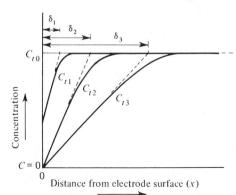

Fig. 22-2. Concentration profiles for reacting species at various times after the start of electrolysis. $C_{t,0} = C_{\text{bulk}}$; $t_3 > t_2 > t_1$; $t_0 =$ before electrolysis; δ_i refers to thickness of the Nernst diffusion layer at time t_i.

When equilibrium is established at the microelectrode, the rate of discharge of the ions will be equal to the rate of diffusion to the electrode. If i is the faradaic current, then the rate of discharge of ions is equal to i/nFA, where n is the number of electrons involved in the discharge process and, F, the faraday, is the quantity of electricity carried by one equivalent of electroactive species. Expressing the flux in terms of electrical current density, i/A, the diffusion limited current is given by

$$i_{\text{lim}} = \frac{nFADC_{\text{bulk}}}{\delta} \tag{22-3}$$

showing that the limiting current is proportional to concentration and inversely proportional to the thickness of the diffusion layer. A diffusion-limited current decreases with time because of the buildup of the diffusion layer.

In its simplest form, classical or conventional polarography involves applying a

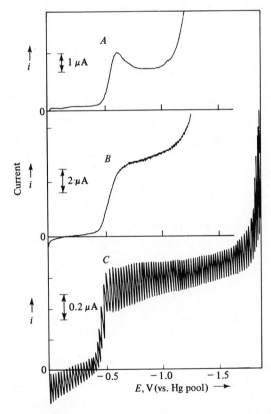

Fig. 22-3. Qualitative comparison of current–voltage curves recorded for 10^{-4} M Tl^+ in $1\,M$ KCl with various types of electrodes. Curve A, gold wire electrode (mercury coated), scanning rate = 500 mV/min; curve B, rotating platinum wire electrode (600 rpm); curve C, dropping mercury electrode.

linearly varying dc potential between two electrodes, one small and easily polarizable and the other large and relatively immune to polarization. The current between these electrodes is recorded as a function of the applied potential. A characteristic step-like current–voltage curve is obtained for each electroactive species in the solution (see Fig. 22-4). The potential at the midpoint of the rising portion of each step, the half-wave potential, $E_{1/2}$, is characteristic of the particular active species causing the transition in the solvent system. In addition, the difference in current i_d between the baseline before the step and the plateau after it is proportional to the concentration of the species in question.

Current–voltage curves with several types of microelectrodes are shown in Fig. 22-3. For the curves shown, the voltage scanning rate was between 2 and 8 mV/sec. Curve A was obtained with a stationary gold wire microelectrode. The electrochemical reaction begins gradually and a hump-shaped current–voltage curve is observed. The height of the hump is proportional to the square root of the scanning rate. Curve B was recorded using a rotating platinum microelectrode. The electrode is rotated rapidly through the solution so that the diffusion layer cannot increase in thickness, but remains constant and very thin. A steady current is obtained immediately. Rotating microelectrodes find considerable use in amperometric titrations. Curve C was obtained with a dropping mercury electrode. With this type of microelectrode the growth of the drops offsets the effect of a widening diffusion layer. The current oscillates between a near-zero value, just after a drop falls, to a maximum value, just before the next drop falls. Usually no attempt is made to follow or to record the entire current excursion, but rather a somewhat damped system is used to indicate the average current flowing during the life of the drop.

CHARACTERISTICS OF THE DROPPING MERCURY ELECTRODE

The most commonly used type of microelectrode is the *dropping mercury electrode*, the electrode of classical polarography. This small polarizable electrode is produced by passing a stream of mercury through a very fine bore (0.05–0.08 mm i.d.) glass capillary. A steady flow of droplets issue from the capillary at a rate of one drop every 3–5 sec. A dropping mercury electrode has several advantages: (1) Its surface area is reproducible with any given capillary. (2) The constant renewal of the electrode surface eliminates passivity or poisoning effects. (3) The high overpotential of hydrogen on mercury renders the electrode useful for electroactive species whose reduction potential is considerably more negative than the reversible potential of hydrogen discharge. (4) Mercury forms amalgams with many metals and thereby lowers their reduction potential. (5) The diffusion current assumes a steady value immediately and is reproducible.

The dropping mercury electrode is useful over the range +0.3 to −2.8 V versus the SCE. At potentials more positive than 0.3 V, mercury dissolves and gives an anodic wave. The most positive potentials may be attained in the presence of noncomplexing anions that form soluble mercury (I and II) salts, for example, nitrate or perchlorate ions. Anions that form insoluble mercury salts or stable complexes shift the anodic dissolution potential to more negative values. At potentials more negative than −1.2 V, visible hydrogen evolution occurs in 1 M HCl solutions, and at -2 V the usual supporting electrolytes of alkali salts begin to discharge. The most negative potentials may be attained in solutions in which a quaternary ammonium hydroxide is used as supporting electrolyte. With tetra-n-butyl-ammonium hydroxide the limit is -2.7 V.

The Charging or Capacitative Current

Even when no reducible species is present in solution, an appreciable current must flow to charge the double layer capacitance at the surface of each growing drop up to the new applied potential. At the solution–electrode interface, a separation of charge takes place which makes the interface look like a large capacitor to the external circuitry. Current is required to charge this capacitor, in addition to the current required by any reacting species. Because the electrode surface repeatedly grows to a maximum, then suddenly falls to zero as the drop detaches, the current flowing in the system fluctuates in the same fashion. More importantly, this charging current appears as a surge at the beginning of each drop when a new capacitor must be charged. The magnitude of this surge increases with applied potential, because the capacitor must be charged to a higher potential, thereby producing a highly sloping baseline.[*] This charging current is the principal factor limiting the sensitivity of polarography and its accuracy at low concentrations. At concentrations of electroactive species of 10^{-3} M or greater, the charging current is negligible compared to the faradaic current and may be ignored. At concentrations of 10^{-4} M the charging current is an appreciable fraction of the total current and a correction must be made for it. At concentrations around 10^{-5} M, the charging current is usually larger than the faradaic current and the precision of the polarographic determination depends principally on how precisely the contribution of the charging current can be estimated, compensated, or eliminated at the potential at which the diffusion current is measured.

Diffusion Current

The theoretical equation, which has become known as the Ilkovic equation, for the faradaic diffusion current is[†]

$$(i_d)_{av} = 607 \, n \, C D^{1/2} \, m^{2/3} \, t^{1/6} \tag{22-4}$$

where i_d is the average current (in microamperes) flowing during the life of a drop, n is the number of equivalents per mole of the electrode reaction, D is the diffusion coefficient of the electroactive substance in square centimeters per second, C is the concentration of the electroactive material in millimoles per liter, m is the mass rate of flow of mercury through the capillary in milligrams per second, and t is the drop time in seconds.

---

[*]The capacity of a mercury surface in dilute aqueous chloride or nitrate solution is about 44 μF/cm^2 on the positive side of the electrocapillary maximum and about half this value on the negative side.

[†]With instantaneous current, the constant is 708.

For a typical dropping mercury electrode with the following characteristics: $m = 2$ mg/sec, $t = 4$ sec, and taking D as 1×10^{-5} cm^2/sec, the response would be $i_d/nC = 3.8$ μA/mequiv/liter.

A more dimensionally correct modification of the diffusion equation, which recognizes that the diffusion is toward a curved surface, is

$$i_d = 607 \, n \, C D^{1/2} \, m^{2/3} \, t^{1/6} \left(1 + 34 \frac{D^{1/2} \, t^{1/6}}{m^{1/3}}\right) \qquad (22\text{-}5)$$

For typical values of D, m, and t, the last term is roughly 5 to 10% of the first term. Equation 22-5 should be used whenever maximum precision is desired for comparing results from different capillaries and for calculating the diffusion coefficient. For analytical applications of polarography, the original Ilkovic equation is adequate and much more convenient; its errors tend to cancel in practical use.

Factors Affecting the Diffusion Current

The Ilkovic equation indicates that the diffusion current should increase directly with the sixth root of the drop lifetime. Galvanometers usually employed have a 3- to 6-sec time period and therefore are unable to follow the periodic growth and fall of the current with each individual drop. The saw-toothed waves actually observed [Fig. 22-3(c)] correspond to the oscillations about the true average current to which the Ilkovic equation refers. In measuring the diffusion current, therefore, one should measure the average of the oscillations.

To obtain the true diffusion current of a substance, a correction must be made for the residual current i_r. The most reliable method for making this correction is to evaluate in a separate polarogram the residual current of the supporting electrolyte alone. The value of the residual current at any particular potential of the dropping electrode is then subtracted from the total current observed. In practice, an adequate correction can be obtained by extrapolating the residual current portion of the polarogram immediately preceding the rising part of the polarogram, and taking as the diffusion current the difference between this extrapolated line and the current–voltage plateau. Both methods are illustrated in Fig. 22-4. Because the slope of the charging current curve is not linear with changing potential and because a change in drop time affects the charging current and the faradaic current differently, estimation of residual currents by extrapolation techniques are inaccurate and questionable at low concentrations. Even with introduction of linear compensation for residual current and a "curve follower" to permit subtraction of the residual current run on a blank solution, the lower limit with sophisticated instrumentation and an ideal reductant lies in the concentration range of 0.2 to 2×10^{-6} M.

The Ilkovic equation points out two facts: (1) The observed diffusion current is directly proportional to the concentration of electroactive material. This relationship is the foundation of quantitative polarography. (2) The diffusion current is proportional to the product $m^{2/3} t^{1/6}$. The quantities m and t depend on the dimensions of the dropping capillary microelectrode and on the pressure exerted on the capillary orifice due to the

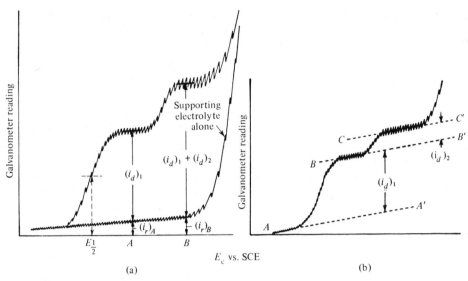

Fig. 22-4. Measuring a diffusion current: (a) exact method and (b) extrapolation method.

height of the mercury column attached to the electrode. An increase in pressure will not alter the size of the individual drops, which is a function of the capillary bore, but it will increase the number of drops forming in a given time and consequently the total electrode area exposed to the solution. A mercury reservoir of large area is customarily attached to the mercury column to prevent any change in height of the column during a series of analyses.

The drop time varies as a function of the emf impressed across the polarographic cell. Actually the drop time follows very closely the electrocapillary curve of mercury,* as shown in Fig. 22-5. As the emf is increased, the drop time first increases, then passes through a maximum at about -0.52 V, and decreases rapidly with increasing negative cathodic potential. The product $m^{2/3}t^{1/6}$ is less affected because it is influenced by the sixth root of t only, and for practical purposes, may be assumed constant over the range of cathode potential from zero to -1.0 V. At more negative potentials, however, its decrease is more rapid and must be taken into account.

The influence of temperature on the diffusion current is quite marked, particularly as the diffusion coefficient of many ions changes 1 to 2% per degree in the vicinity of 25.0°C, the standard temperature chosen for polarographic work. This implies that the temperature of the solution in the polarographic cell must be controlled to within 0.5°C.

Gelatin or some other maximum suppressor has a very pronounced effect on the critical drop time below which the Ilkovic equation fails. Without gelatin the Ilkovic equation fails with drop times less than 4 or 5 sec. As gelatin is added, the critical drop time decreases to the neighborhood of 1.5 sec. At faster drop rates there is appreciable stirring

*The electrocapillary curve expresses the relation between the potential of mercury and the surface tension at a mercury–electrolyte solution interface.

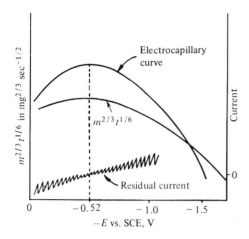

Fig. 22-5. Comparison of the capillary characteristics and the electrocapillary curve for mercury with increasing negative potential. The magnitude of the residual current is shown on the lower curve.

of the solution and a significant variation in the thickness of the diffusion layer which produces an abnormally large current. With drop times between 2 and 5 sec, 0.005–0.01% of gelatin present and 0.5 M or larger concentration of supporting electrolyte, the diffusion current will be directly proportional to concentration.

The nature and viscosity of the solvent medium also influence the diffusion current. The diffusion coefficient varies inversely with the viscosity coefficient of the solution. Ionic species vary in size and consequently in their rate of diffusion, depending on whether they are present as aquo complexes or some other type. The effect of complex formation is shown by the data in Appendix B. In some cases the nature of the complex species determines whether or not a satisfactory polarographic wave will be obtained. With tin(IV) ions, for example, no reduction is obtained in nitrate or perchlorate media in which only an aquo complex exists, whereas well-defined waves are found in chloride solutions in which the predominant species is $SnCl_6^{2-}$.

Polarographic Maxima

Current–voltage curves obtained with the dropping mercury electrode are frequently distorted by more or less pronounced maxima. These maxima vary in shape from sharp peaks to rounded humps. In all cases the current rises sharply, but instead of developing into a normal diffusion current, it increases abnormally until a critical value is reached and then rapidly decreases to the normal diffusion-current plateau. No exact explanation has been proposed. Maxima are especially prevalent when the decomposition potential is considerably removed from the electrocapillary zero of mercury.

Whatever the cause, maxima must be eliminated in order to obtain the true diffusion-current plateau. They can usually be suppressed by surface-active agents. Gelatin is often used, but the amount present in the solution must be carefully controlled and should lie between 0.005 and 0.01%. Less is useless, and more will suppress the diffusion current. Agar and methyl cellulose are also employed. Generally the proper amount of suppressor is added to every polarographic solution during the preparative step as a precautionary measure.

THE HALF-WAVE POTENTIAL

The electroactive material in polarography is characterized by its half-wave potential, $E_{1/2}$. This is the potential at the point of inflection of the current–voltage curve, one-half the distance between the residual current and the final limiting current plateau, as shown in Fig. 22-4. The significance of the half-wave potential is demonstrated by an oxidation-reduction system

$$ox + ne^- \rightleftharpoons red \tag{22-6}$$

The reversible potential of the system as it exists at the electrode-solution interface will be recorded on the polarogram. This electrochemical equilibrium may be represented as follows:

$$E = E^\circ - \frac{0.0592}{n} \log \frac{[red]_i}{[ox]_i} \tag{22-7}$$

where the subscripts denote concentrations at the electrode–solution interface.

Assume, for example, that the solution at the electrode surface consists entirely of the oxidized form before the commencement of the current–voltage scan. As soon as the applied emf is made large enough to reduce some of the oxidant, the concentration of oxidant at the electrode surface begins to decrease. Some ions will move in from the bulk of the solution as the concentration gradient builds up between the electrode surface and the bulk of the solution. The observed current depends upon the rate of diffusion established by the concentration gradient

$$i = K ([ox] - [ox]_i) D_{ox}^{1/2} \tag{22-8}$$

where K includes capillary characteristics and other terms from the Ilkovic equation. When the current attains the limiting value represented by the diffusion–current plateau, the concentration of oxidant at the electrode–solution interface will be essentially zero, and

$$i_d = K [ox] D_{ox}^{1/2} \tag{22-9}$$

Solving Eq. 22-8 for $[ox]_i$, and then combining with Eq. 22-9,

$$[ox]_i = \frac{i_d - i}{K D_{ox}^{1/2}} \tag{22-10}$$

For metals that form amalgams with the dropping electrode, the concentration of metal amalgam is directly proportional to the current on the current–voltage curve, and generally the concentration of reductant formed is proportional to the observed current, so

$$i = K [red]_i D_{red}^{1/2} \tag{22-11}$$

Solving for $[red]_i$ and substituting the result into Eq. 22-7 along with Eq. 22-10, the potential of an oxidation–reduction system can be expressed as

$$E = E° - \frac{0.0592}{n} \log \frac{i}{i_d - 1} + \frac{0.0592}{n} \log \left(\frac{D_{red}}{D_{ox}} \right)^{1/2} \tag{22-12}$$

By definition, the half-wave potential is the point where

$$i = i_d - i \quad \text{or} \quad i = \frac{i_d}{2}$$

Thus,

$$E_{1/2} = E° + \frac{0.0592}{n} \log \left(\frac{D_{red}}{D_{ox}} \right)^{1/2} \tag{22-13}$$

This equation represents the current–potential relation for a reversible voltammetric wave. The half-wave potential is a quantity characteristic of a given electroactive species in a given medium, often being very close to the value of the standard potential of the couple.

When both the oxidized and reduced form of the electroactive species are present initially in the bulk of the solution, the current–potential relation will assume the form

$$E = E_{1/2} + \frac{0.0592}{n} \log \frac{(i_d)_c - i}{i - (i_d)_a} \tag{22-14}$$

where $(i_d)_c$ and $(i_d)_a$ correspond to the cathodic and anodic diffusion-limited currents of the oxidized and reduced species, respectively. There is no characteristic half-wave potential for a reversible reaction that yields an insoluble product on the electrode surface, such as the deposition of silver on a platinum microelectrode. The potential at $i = 0.5$ i_d is dependent on the bulk concentration of the metal and the stirring rate.

For the current–potential relations just described, the rate for the electron transfer at the electrode surface was assumed to proceed fast enough to maintain the surface concentrations of reactants and products very close to their equilibrium values. When this is true, the reaction is termed *reversible*. If the energy of activation for the electron-transfer reaction is large, and the rate of electron transfer correspondingly slow, the concentrations at the electrode surface will not be equilibrium values and the Nernst equation is not applicable. Voltammetric reactions of this type are termed *irreversible*. To determine whether an electrode reaction is reversible, one should prepare a plot of E versus $\log i/(i_d - i)$. For a reversible electrode reaction, a straight line should result whose slope is $0.0592/n$. Furthermore, $E_{1/2}$ for the cathodic reduction should coincide with $E_{1/2}$ in the anodic oxidation reaction, after a correction is made for the iR drop across the electrode–solution interface. When the electrode reaction is irreversible, a straight line may still result but the slope of the log plot will differ from the theoretical value.

In essence, there is no such thing as a perfectly reversible electrode reaction. In a practical sense a reaction is termed reversible if, within the limits of error of experimental measurement, its behavior follows the Nernst equation under the given experimental conditions. Consequently, a reaction that appears to be reversible under one set of experimental conditions may not behave reversibly under a different set of experimental conditions. For example, the reaction may become irreversible if it is required to proceed at a much faster pace, or if it is carried out in the presence of a masking agent which forms a complex with unfavorable electron-transfer reaction-rate characteristics.

INSTRUMENTATION

Conventional polarography uses a dropping mercury electrode as the microelectrode and a layer of mercury as the counter electrode. A typical electrical circuit and polarographic cell are shown in Fig. 22-6. A suitable slidewire is a linear 10-turn Helipot potentiometer. The resistance of the potentiometer should not exceed 100 Ω in order that the current flowing through the cell always remains a negligible fraction of the current flowing through the slidewire and does not affect the iR drop at any point along the slidewire. The potential drop across the slidewire is adjusted to any desired value by inserting two dry cells (3.0 V) and regulating the voltage by means of a radio potentiometer (about 500 Ω) placed in series. The voltage span selected is indicated by the voltmeter, V. To measure the current a galvanometer with sensitivity of 0.005 μA/mm scale deflection is

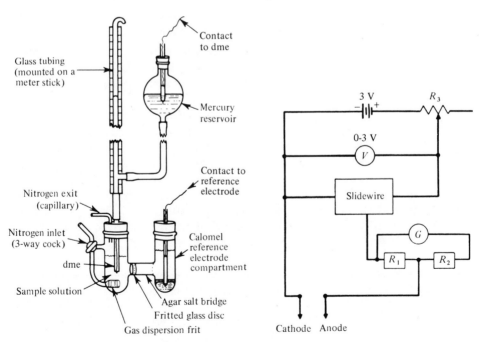

Fig. 22-6. *Left*, H-cell and reservoir arrangement for polarographic cell. *Right*, basic circuit for obtaining classical dc polarograms.

satisfactory. The shunt for the galvanometer can be an Ayrton shunt or two 1000-Ω resistance boxes connected as shown. Keeping the total resistance at some fixed value, for example, $R_1 + R_2 = 1000$ Ω, the sensitivity of the galvanometer circuit will be given by $S_g R_1 / (R_1 + R_2)$, where S_g is the sensitivity of the galvanometer alone.

The circuit shown in Fig. 22-6 can be made the basis of a simple automatic recording polarograph if the slidewire is driven at 1 rpm with a synchronous motor, and if the galvanometer is removed and a recording millivolt potentiometer connected across R_1. Capacitive current and drop-induced fluctuations limit the use of simple battery-potentiometer polarographs to systems in which test element concentrations are high enough to yield polarograms undisturbed by these difficulties.

A mercury pool at the bottom of the polarographic cell is the simplest form of a counter electrode. Because the layer of mercury has a large area and the current is generally very small, the concentration overpotential at this electrode is negligible and its potential may be regarded as constant. In chloride solutions it maintains approximately the potential of the calomel electrode of the particular chloride ion concentration. While convenient, the mercury pool never possesses a definite, known potential. In the absence of chloride ions or another depolarizing electrolyte, and particularly in nonaqueous solutions, the potential does not attain a steady value. Moreover, in the presence of substances capable of forming complexes with mercury ions, the dissolution potential of mercury will be shifted to more negative reduction potentials, thus compressing the useful range of the dropping mercury electrode. In order to eliminate the possibility of unknown or nonreproducible anode potentials, it is necessary to replace the mercury pool with a separate reference (external) electrode connected to the polarographic cell through a salt bridge. The SCE is commonly employed, and it is almost universal practice in polarography to express half-wave potentials with reference to this electrode.

At potentials where mercury oxidizes, the dropping mercury electrode can be replaced by various electrode materials including pyrolytic graphite, carbon paste, wax-impregnated carbon rods, boron carbide, and platinum.

Drop Detachment

Mercury drop synchronization is essential for reproducible measurements in single-sweep polarography and desirable in other types of polarography. The detachment of a drop of mercury from the capillary is accomplished by an electromechanical drop dislodger. One type moves the capillary away from the drop at a fixed time interval; others deliver a sharp knock to the capillary. At the same moment a trigger signal is sent to the time-base. Each drop is permitted to grow until its area changes the least, usually after the first 1.5–2.0 sec of drop life. Then the current in the cell is sampled just prior to the dislodgement of the drop. The sampled current is stored in a memory and read out to the recorder until the next measurement is taken. In this way the recorder plots a curve representing the peak current flowing during each drop's life cycle, and is devoid of drop-induced fluctuations.

Three-Electrode Potentiostat

A characteristic of modern polarographic instrumentation is potentiostatic control of the working electrode potential. In classical polarography the dc ramp voltage was applied across the entire cell, rather than across the working electrode–solution interface. Current flowing through a polarographic cell of high resistance causes an appreciable voltage drop. Thus, the potential at the dropping mercury electrode differs from that at the other end of the polarographic cell. This causes several deleterious effects: (1) The half-wave potential (or peak potential in derivative methods) is shifted to more negative values. (2) The total reduction current is less. (3) The distortion of the shape of the wave and slope of the polarographic step is often severe. Despite these shortcomings, it is desirable to be able to use nonaqueous solvents of high resistance and aqueous electrolytes of millimolar concentrations. Three-electrode potentiostatic control makes this possible.

A third electrode, a reference electrode of constant potential, can be considered a probe that is positioned as close as possible to the mercury drop, and can sense the potential at that point. It is connected to the polarograph (point C in Fig. 22-7) through a circuit which draws essentially no current; the input impedance is in excess of 10^{14} Ω. If the voltage sensed by the reference electrode is less than the dc ramp voltage provided to the scan amplifier, the feedback to the scan amplifier from the operational amplifier control loop will provide a corrective voltage that will change the applied emf enough to compensate for the resistance of the cell and electrolyte. Thus, the voltage measured at point C should always be the same as that applied to point A and, if momentarily it is not, the voltage at point B will automatically increase to maintain point C equal to point A. Present-day analog potentiostats operate in the same manner but they are extremely fast, with a rise time on the order of microseconds when driving a resistive load. A set control potential can be maintained to within better than 1 mV for long periods, and the reference electrode is loaded only to the extent of picoamperes or less.

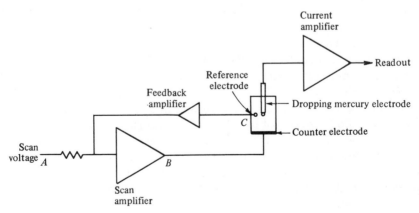

Fig. 22-7. Schematic diagram of three-electrode potentiostat.

MODERN VOLTAMMETRIC TECHNIQUES

For the remainder of this chapter it will be assumed that all the instrumentation discussed possesses potentiostatic capabilities. Only those techniques that have found the greatest degree of analytical application will be treated. Commercial instrumentation for all these techniques is available, and all techniques are applicable to electrodes other than the standard dropping mercury electrode.

In any electrochemical system, an equilibrated diffusion layer is established between the electrode and the bulk of the solution. Once the reaction potentials have been reached, the concentration of the electroactive species of interest changes from the bulk concentration value at positions far removed from the electrode surface to essentially zero at the electrode surface. When the classical slow scan rates are employed, the slope of the concentration gradient within the diffusion layer is determined primarily by the rate of depletion of the electroactive species at the electrode surface. It varies from essentially zero at potentials significantly more positive than the reduction potential, to a value governed by the concentration and diffusion coefficient of the electroactive species at potentials well past the reduction potential. However, when the polarographic method uses rapid changes in potential, either because the scan rate is fast or because a pulse modulation of some type is employed, the slope of the concentration gradient at any particular potential will be greater than in the slow-scan instance. The bulk concentration will be closer to the electrode surface, the number of electroactive particles arriving at the surface per unit time will be greater, and larger current signals will result.

Derivative Polarography

Derivative methods of polarographic analysis depend on differentiation of the current–voltage curve while it is being recorded. The resulting polarogram consists of a series of peaks superimposed upon the background. Measurement of a peak rather than a plateau is inherently more sensitive, particularly when the nonlinearity of the baseline and the oscillations accompanying drop formation are considered. The most elegant experimental answer to the problem of derivative polarography is the apparatus of Kelley et al.,[5] manufactured by Indiana Instrument and Chemical Corporation, in which a three-electrode system using potentiostatic control of the dropping electrode potential is employed to yield first derivatives of theoretical slope. A special RC filter network is used to smooth the curves. For the case of general interest, that is, negligible solution resistance and appreciable residual current,

$$\left[\frac{d(\Sigma i)}{dt} \right]_{max} = - \left(\frac{dE}{dt} \right) \left(\frac{n i_d}{4RT/F - d i_r/dE} \right) \tag{22-15}$$

where Σi is the total current (faradaic plus residual), $-dE/dt$ is the rate of potential scan, and $d i_r/dE$ is the slope of the residual current curve at the half-wave potential. According to Eq. 22-15 the sensitivity of derivative polarography (about $10^{-7}\ M$) is proportional to the scan rate up to the maximum of about $40/n$ mV/min. As the scan rate increases, the potential at the electrode–solution interface, through the capacitive nature of the interface, penetrates deeper into the solution and causes a larger number of ions to be reduced. The maximum scan rate at which undistorted peaks may be observed is set by time lags introduced by the *RC* filter network. Important also is the slope of the residual current curve, which varies with the resistance of the solution. The main advantage of this technique is the minimization of the effect of diffusion currents due to the presence of more easily reducible constituents. Polarographic waves differing in half-wave potential by $90/n$ mV are completely resolved, whereas smaller differences lead to peaks that are sufficiently separated to be used analytically (Fig. 22-8). Because the derivative signal returns to the baseline, each wave can be recorded at its maximum sensitivity. The peak potential is typically $28/n$ mV more negative than the half-wave potential reported for classical polarography.

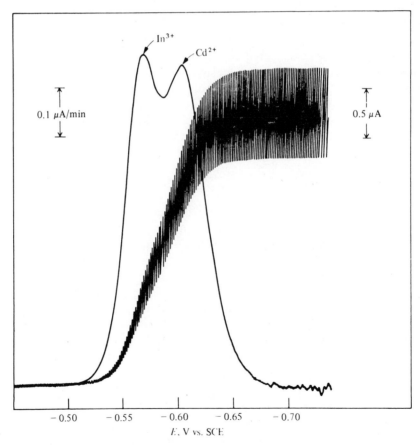

Fig. 22-8. Comparison of resolution of a regular and a derivative polarogram recorded with a solution containing 1×10^{-4} M In^{3+} and 2×10^{-4} M Cd^{2+} in 0.1 M KCl.

Pulse Polarography[2]

Pulse polarography takes advantage of the fact that following a sudden change in applied potential, the capacitive current surge decays much more rapidly than does the faradaic current. In this technique a small amplitude voltage pulse, in addition to the linearly increasing dc ramp (about 1 mV/sec) normally used for dc polarography, is applied to the polarographic cell. As each mercury drop forms it is allowed to grow for a period of time, perhaps 1.9 sec at the dc ramp potential, after which a sudden voltage pulse of perhaps 50 msec duration is applied. The pulse is synchronized with the maximum growth of the mercury drop, if a dropping mercury electrode is used. The current is measured 40 msec after the application of the pulse, to allow time for the charging current to decay to a very low value. The capacitive current actually decays exponentially at a rate governed by the magnitude of the capacitance and series resistance of the system. During this time interval the faradaic current also decays somewhat but does not reach the diffusion-controlled level because the concentration gradient at the instant of current measurement is considerably larger. Each succeeding drop is polarized with a somewhat larger pulse. This method gives a current–voltage curve similar to that obtained in dc polarography except for the cancellation of the capacitive components. The measured signal is the faradaic current that flows at the pulse potential minus any faradaic current flowing due to the fixed dc potential (Fig. 22-9). The limiting current is given by the Cottrell equation,

$$i_{\text{lim}} = nFCA \sqrt{\frac{D}{\pi t}} \qquad (22\text{-}16)$$

In comparison with classical dc polarography, the sensitivity is about 6.5 times better. A much larger gain in real sensitivity is achieved through the virtual elimination of the charging current.

Derivative pulse polarography employs two current sampling intervals of equal time periods. The first sample period occurs just before the pulse application; the second sample period occurs at the end of the pulse. The current samples are stored on memory capacitors and the difference is displayed on a recorder operated in a display-and-hold manner. Differences in the two stored samples occur only in the region of the half-wave potential where the current is changing rapidly with the potential. The voltammogram recorded in this way is a peak-shaped incremental derivative of a conventional dc polarogram. Because the applied pulse is held for an appreciable length of time, the system response is not strongly dependent on the electrode kinetics, at least not to the extent expected of ac polarography. This implies that pulse polarographic techniques may be readily used for organic and other electrochemically irreversible systems. For pulses of small amplitude, ΔE, the peak current is given by

$$\Delta i_{\text{max}} = \frac{n^2 F^2}{4RT} A C (\Delta E) \sqrt{\frac{D}{\pi t}} \qquad (22\text{-}17)$$

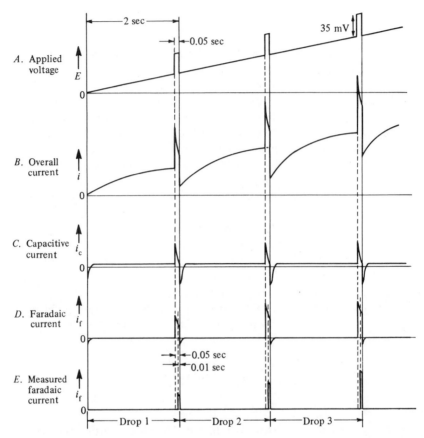

Fig. 22-9. Current and voltage diagrams illustrating the operation of a pulse polarograph. Curve A, a linearly increasing scan voltage upon which a 35-mV pulse is superimposed during the last 50 msec of the 2-sec drop time; B, the overall current flowing through the cell as a result of the applied voltage; C, the capacitative component of cell current; D, the faradaic component of the cell current measured above the "dc" background; E, the net current signal measured during the last 10 msec of the life of the drop after the capacitative current has decayed to near zero.

whereas for pulses of large amplitude, it is given by

$$\Delta i_{max} = nFAC \sqrt{\frac{D}{\pi t}} \left(\frac{\sigma - 1}{\sigma + 1} \right) \tag{22-18}$$

where $\sigma = \exp(\Delta E\, nF/2RT)$. In the limit, for $\Delta E \gg RT/nF$, $(\sigma - 1)/(\sigma + 1)$ approaches unity and i_{max} becomes the limiting current as given by the Cottrell equation. The use of a large pulse amplitude allows one to obtain large signals from extremely dilute solutions, albeit with some distortion in the curve shape, because the pulse amplitude is an appreciable portion of the total polarographic wave. In principle the derivative mode is less sensitive than the normal mode, but the resolution is better, about $50/n$ mV. It is capable of detecting 10^{-8} M, a 100-fold increase over classical polarography and a 10-fold increase over derivative dc polarography.

ac Polarography[6]

In the ac technique a potential periodic in time, such as a sine wave, and of relatively small amplitude (1–35 mV) and low frequency (10–60 Hz), is superimposed upon the slow linear voltage sweep of dc polarography. This sinusoidal variation in potential is similar to that employed in cyclic voltammetry with the exception that the potential excursions are of much smaller magnitude, typically 10-mV peak-to-peak. The direct current component of the total current is blocked out and only the rectified and damped alternating component is displayed as a function of dc potential. By looking only at the alternating portion of the current that flows and detecting its amplitude, one is in effect looking at the difference in current that flows between the minimum and maximum applied potentials during the modulation period. The current is sampled just before the mercury drop is dislodged. A peak output signal with its maximum amplitude at the half-wave potential, rather than a step-shaped wave, is produced, because the reversible wave attains its maximum slope at the half-wave potential; therefore, a given ac signal causes the periodic changes in concentration of the electroactive species to be maximal at this potential. At other dc potentials along the wave, the sinusoidal potential variation causes less of a perturbation of the dc surface concentration and the ac current is correspondingly less. An important characteristic of ac polarography is that it responds only to reversible electrode reactions. This limits its applicability, but in some cases can be advantageous in avoiding interferences.

Detection of only the ac component allows one to separate the faradaic and capacitive currents because of the phase difference between them. By employing a phase-sensitive lock-in amplifier, one can select either the faradaic (phase-shifted $45°$ from the applied potential) or the capacitive current (phase-shifted $90°$), while rejecting the other. The capacitive current is important in studies of kinetics and adsorption. The maximum height of the faradaic alternating current is given by

$$\Delta i_{max} = \frac{n2F^2 AVC\omega^{1/2}D_{ox}^{1/2}}{4RT} \tag{22-19}$$

where A is the electrode area, V is the amplitude of the voltage signal, and ω is the angular frequency. Because even a moderate cell resistance serves to mix the phases, successful application of the phase discriminator requires very low cell resistances.

Another successful approach to the separation of faradaic and capacitive components of the cell current is through the measurement of second (or higher) harmonics of the alternating current. The method is based on the fact that the capacitive current varies essentially linearly with voltage, whereas the faradaic process varies nonlinearly. Thus, although the first derivative of the capacitive current may be appreciable, the second derivative will be near zero. Whereas the baseline current may be 75% of the peak current in the case of normal ac polarography, it is only about 5% or less when the second harmonics are recorded. The second harmonic, because its signal crosses through zero at the half-wave potential, is very useful for resolution problems, especially in complex mixtures.

Cyclic or Fast Linear Sweep Voltammetry

Cyclic voltammetry involves the measurement of current–voltage curves under diffusion-controlled, mass transfer conditions at a stationary electrode, utilizing symmetrical triangular scan rates ranging from a few millivolts per second to hundreds of volts per second. The triangle returns at the same speed and permits the display of a complete polarogram with cathodic (reduction) and anodic (oxidation) waveforms one above the other (Fig. 22-10). Two seconds or less is required to record a complete polarogram following deaeration of the solution. This technique yields information about reaction reversibilities and also offers a very rapid means of analysis for suitable systems. The method is particularly valuable for the investigation of stepwise reactions, and in many cases direct investigation of reactive intermediates is possible. By varying the scan rate, systems exhibiting a wide range of rate constants can be studied, and transient species with half-lives of the order of milliseconds are readily detected. The method can be applied to stationary electrodes as well as to a single mercury drop, and to reactions for which stripping analysis is inapplicable due to highly irreversible electrode processes or the formation of solution-soluble reaction products.

In cyclic voltammetry the typical concentration gradient prevails at potentials more positive than the reduction potential. Once this point has been passed, however, the rate of variation of potential is too rapid for diffusional processes to maintain equilibrium with the bulk of the solution. More and more material is consumed and the diffusion layer extends further and further into the solution. Unlike the case with dropping

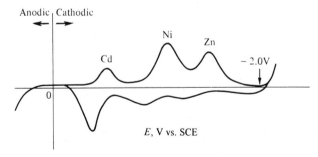

Fig. 22-10. A cyclic voltammetric plot of Cd, Ni, and Zn in 0.1 M KCl. The initial scan in the cathodic direction from zero to -2.0 V causes the reduction of Cd, Ni, and Zn. The reverse scan from -2.0 V to zero causes the oxidation of the ions and produces a peak moving in a downward direction whose height is related to both concentration and reversibility of the reaction.

electrodes, this process is not periodically reversed by the stirring associated with drop fall, so that the signal decay continues and a peak-shaped curve is obtained. Furthermore, in the course of the cathodic variation in potential, the reduced form of the reactant is produced in the vicinity of the electrode while a depletion of the oxidized form occurs. Given sufficient time, the reduced form would diffuse into the bulk of the solution, but the potential is taken back to the initial value at a rate such that some of the reduced form is still present at the electrode surface and undergoes a process of oxidation back to the form of the couple initially present in the solution. A maximum occurs at $(E_{dc} - E_{1/2})_n = -0.0285$ V. Thus the peak potential is

$$E_{peak} = E_{1/2} - 0.0285 \tag{22-20}$$

for a reduction reaction. For an oxidation reaction the sign of the numerical term is reversed. On the reverse scan, the position of the peak depends on the switching potential. As this potential moves more negative, the position of the anodic peak becomes constant at $29.5/n$ mV anodic of the half-wave potential. With the switching potential more than $100/n$ mV cathodic of the reduction peak, the separation of the two peaks will be $59/n$ mV and independent of the rate of potential scan. This is a commonly used criterion of reversibility. Reversibility can also be ascertained by plotting $(E_{cath} - E_{anod})$ as a function of $\sqrt{velocity}$, which should be a straight line if reversible.

To do qualitative analysis with cyclic voltammetry, one observes four characteristics of an electroactive substance: peak potential, wave slope, reversibility, and the effect of changing the supporting electrolyte. To illustrate: An unknown sample in 0.1 M KCl exhibits peaks at -0.40 (slope 28 mV) and at -0.60 V (slope 25 mV); on the reverse scan there is an anodic peak at -0.60 V. In considering the first peak at -0.40 V, likely choices would be lead(II) and thallium(I). However, lead has a slope of 28 mV for a two-electron reduction, whereas thallium(I) has a slope of 56 mV. The wave at -0.60 V could be cadmium(II), chromium(III), indium(III), or europium(III). Cadmium has a slope of 28 mV and is reversible, indium has a slope of 19 mV and is reversible, chromium is not reversible, and europium has a peak potential near -0.7 V with a slope of 64 mV. The tentative conclusion is that the sample contains lead and cadmium. Further validation would involve the use of a different electrolyte.

Chronopotentiometry

Chronopotentiometry is based on the observation of the change in potential of a working electrode as a function of time during electrolysis. Usually the electrolysis is performed with a constant current in a quiescent solution. Ultimately, the exhaustion of electroactive substance at the surface of the electrode causes a more or less rapid change of potential. The time necessary for the potential to go from one level to another is measured. This transition time is a measure of the rate at which the concentration of the electroactive species at the electrode surface is reduced to the point where it can no longer sustain the required current. A schematic chronopotentiogram is shown in Fig. 22-11. For a symmetrical, reversible reaction the potential of the electrode as a function of the

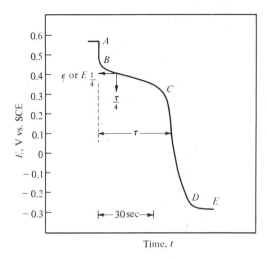

Fig. 22-11. Chronopotentiogram for the reduction of iron(III) at a platinum cathode. The concentration of iron(III) was 0.0250 M, the supporting electrolyte solution was 1 M HCl, and the temperature was 25°C. A cylindrical wire electrode of area 0.281 cm² and radius 0.0255 cm was used, and the constant current was 2.79 × 10⁻⁴ A.

time t obeys the relationship (at 25°C):

$$E = \epsilon - \frac{0.0592}{n} \log \frac{t^{1/2}}{\tau^{1/2} - t^{1/2}} \tag{22-21}$$

where τ is the transition time, and ϵ, which is identical with the polarographic half-wave potential, is very nearly equal to the standard potential of the electrode reaction. The potential becomes equal to ϵ when $t = \tau/4$. With linear diffusion to a plane electrode and when diffusion is the controlling factor, the transition time (in seconds) is given by the Sand equation

$$\tau^{1/2} = \frac{\pi^{1/2} nFAD^{1/2}C}{2i} \tag{22-22}$$

where i is the constant current (in amperes) imposed on the system. The constant current-density is selected so that the transition time will be less than about 30 sec, as otherwise the strong convective stirring produced by the density gradient at the electrode disrupts the diffusion layer at unshielded electrodes. The Sand equation is seldom used directly because D usually is not known with sufficient accuracy in a variety of supporting electrolytes. Instead the electrode is calibrated empirically under a given set of conditions with known concentrations of the test element. Lingane[3] has discussed the analytical aspects of chronopotentiometry.

When several electroactive species are present in the solution, the chronopotentiogram will show a succession of transition times. However, transition times are nonadditive. After the first inflection, when reaction of the second or subsequent substance begins, the first (or prior) substance still continues to diffuse to, and react at, the electrode. Consequently, the current results from both reactions at points on the second wave. With a constant total current, the current from the second reaction is smaller than it would be in the absence of the first substance, and the transition time is increased.

In the two-stage reduction or oxidation of a single substance, only the two n-values differ. Since $C_1 = C_2$, and $D_1 = D_2$, it follows that the ratio of the individual transition times is

$$\frac{\tau_2}{\tau_1} = \frac{2n_2}{n_1} + \left(\frac{n_2}{n_1}\right)^2 \qquad (22\text{-}23)$$

where n_1 and n_2 are the separate number of electrons for each stage.

Both chronopotentiometry and dc polarography are based on similar principles. Because chronopotentiometry is more sensitive to the rates of the various steps that may be involved in an overall electrode process, chronopotentiograms often tend to be less well defined than polarograms, especially at small concentrations. In general, chrono-potentiometry is disappointing at concentrations below about $5 \times 10^{-4} M$. It is a powerful tool for the study of electrode processes at higher concentration. Thin-layer chronopotentiometry is most important when run under conditions of quantitative electrolysis; it is discussed under coulometry.

Stripping Analysis[1]

The technique of stripping analysis, also called linear-potential sweep stripping chronoamperometry, involves two steps: a concentration or pre-electrolysis step in which the desired component is deposited cathodically or anodically, followed by a reverse electrolysis in which the component is determined. In the anodic variant of stripping analysis, the metal (or metals) concerned is reduced at a controlled potential for a definite time under fixed conditions of geometry and stirring. The working electrode may be a hanging mercury drop or a mercury film on a wax-impregnated graphite electrode, or in some cases an inert solid electrode such as platinum or carbon. The final anodic dissolution, or stripping process, involves a linear anodic scan in which the metal is oxidized. The resulting stripping voltammogram shows peaks, the heights (or areas) of which are generally proportional to the concentrations of the corresponding electroactive metal ions, and the potentials of which have the same qualitative interpretation as their half-wave potentials in polarography. Standards are carried through identical pre-electrolysis and stripping steps. A typical anodic scan is shown in Fig. 22-12 for a 0.100 ml sample of whole blood prepared for analysis by digestion with perchloric acid and brought to 5.0 ml volume with sodium acetate. As little as $10^{-8} M$ cadmium (corresponding to about 10^{-6} weight percent) has been determined with a precision of $\pm 3\%$ using 15 min of pre-electrolysis and linear scan rate of 21 mV/sec. Because the standard addition method is usually used for evaluating the unknown concentration, the reproducibility will be the same as the precision. By extending the pre-electrolysis time to 60 min, the sensitivity is extended to $10^{-9} M$ but deviations will be about 10–20% at this concentration level.

Cathodic stripping analysis consists of forming an insoluble layer at the electrode surface during an anodic pre-electrolysis, and stripping it off by reverse electrolysis. Unlike anodic stripping analysis, the deposition potential depends on the concentration to be determined and shifts about $60/n$ mV to more positive values for each order-of-magnitude decrease in the anion concentration. It is not possible to carry the deposition to completion because of the limitation imposed by finite solubility of the layer; usually salts of

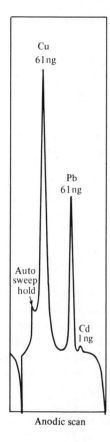

Anodic scan

Fig. 22-12. Stripping analysis: anodic scan. Experimental data: plating time, 30 min; plating potential, −1.0 V; sweep rate, 60 mV/ sec; chart speed, 5 in/min; blood sample, 0.2 ml; current range, 200 μA full scale; stripping time, ∼ 16 sec. (Courtesy of Environmental Sciences Associates, Inc.)

mercury(I) precipitate. Such a method has been used for determination of the halides, tungstate, and molybdate. Lower concentration limits are about $5 \times 10^{-6}\ M$.

Complete exhaustion of the desired component in the solution during the concentration step is not necessary and not even desirable, as long as this step is carried out under reproducible conditions. Thus an enormous saving in time is effected with only a moderate sacrifice in sensitivity. The plating potential should surpass the half-wave potential by some hundreds of millivolts until the current reaches its limiting value. Then after a short rest period to allow the amalgam concentration to become homogeneous and to ensure that convection in the solution has ceased (perhaps 30–60 sec), the anodic redissolution procedure is initiated. If desired, the electrolyte may be changed to one better suited for the stripping process. Using the hanging mercury drop electrode, the peak current during anodic scan is proportional to the square root of the scan rate; with thin-film mercury electrodes, a direct proportionality is observed. Compared with other highly sensitive electroanalytical methods, stripping analysis gives comparable performance at lower cost.

APPLICATIONS

Determinations can be performed of inorganic or organic species that are either molecular or ionic, if they undergo oxidation or reduction at a mercury electrode in the region of potential bounded at the positive limit by the potential of oxidation of mercury in the medium employed, and at the negative limit by the potential at which the supporting electrolyte or the solvent is reduced. If materials other than mercury are used for the microelectrode, the positive limit can be extended considerably and will be limited only by the oxidation of the solvent, the anion present, or the electrode material. Nonaqueous solvents can be used for organic substances that are insoluble in water, the only limitation being that if the resistance of the medium is high, the polarographic wave may be severely distorted. A precision of ±3% to 5% is readily attainable at concentrations between 10^{-2} and 10^{-4} M. With additional effort and care to maintain the experimental conditions constant, a precision approaching ±1% can be obtained.

Reversible organic systems are confined largely to quinones and a few other functional systems such as the phenylene diamines, which resemble quinones in forming resonating systems. However, polarography offers the possibility of characterizing oxidation–reduction properties of numerous irreversible systems. These systems involve a step with a high activation energy, and the half-wave potential is a function of the rate constant of the electrode process. Among structural factors which affect half-wave potentials are the nature of the electroactive group, that is, the group where cleavage or bond formation occurs during electrolysis, stereochemistry, and the nature of substituents. Also important is the molecular frame to which the electroactive group is attached and the groups situated in the immediate vicinity of the electroactive group.

The types of bonds that can be reduced at a dropping mercury electrode are enumerated in Table 22-1. Often the presence of a single group is insufficient to bring the wave of the substance studied to an accessible potential range. For these compounds it is necessary for the molecule to contain, in addition to the electroactive group, another activating group that affects the electron distribution in the substrate and shifts the half-wave potential of the reduction wave. In a system of conjugated multiple bonds, the whole system of conjugated bonds represents a single electroactive group. Both inductive and resonance effects can ease the reduction of groups attached to aromatic systems or to double bonds. For example, the reduction potential of the disulfide group linked to a

TABLE 22-1 Reducible Organic Functional Groups

$>C{=}O$	Ketone	$-C{\equiv}N$	Nitrile	$-NO_2$	Nitro
$-CHO$	Aldehyde	$-N{=}N-$	Azo	$-NO$	Nitroso
$>C{=}C<$	Alkene	$-NO{=}N-$	Azoxy	$-NHOH$	Hydroxylamine
$\phi-C{\equiv}C-$	Aryl alkyne	$-O-O-$	Peroxy	$-ONO$	Nitrite
$>C{=}N-$	Azomethine	$-S-S-$	Disulfide	$-ONO_2$	Nitrate

Also dibromides, aryl halide, alpha-halogenated ketone or aryl methane, conjugated alkenes and ketones, polynuclear aromatic ring systems, and heterocyclic double bond.

phenyl group is -0.5 V versus SCE and to an alkyl group is -1.25 V. As the number of condensed aromatic rings increases, the reduction is made easier. Single C—X (where X is halogen) bonds are usually reduced at more negative potentials than C=C—C—X, but at more positive potentials than C—C=C—X. The ease of reduction of the C—X bond increases with the increasing polarizability of the halogen, that is, C—F < C—Cl < C—Br < C—I. One should be aware of the extent to which organic functional groups can be converted to an active polarographic group because such conversion can markedly extend the method. A number of examples are enumerated in Table 22-2. These are only a few of the many facets of organic polarography. The subject is thoroughly discussed by Zuman.[7]

TABLE 22-2 Organic Functional Group Analysis of Nonpolarographic Active Groups

Functional Group	Reagent	Active Polarographic Group
Carbonyl	Girard T and D	Azomethine
	Semicarbazide	Carbazide
	Hydroxylamine	Hydroxylamine
Primary amine	Piperonal	Azomethine
	CS_2	Dithiocarbonate (anodic)
	$Cu_3(PO_4)_2$ suspension	Copper(II) amine
Secondary amine	HNO_2	Nitrosoamine
Alcohols	Chromic acid	Aldehyde
1,2-Diols	Periodic acid	Aldehyde
Carboxyl	(Transform to thiouronium salts)	—SH (anodic)
Phenyl	Nitration	—NO_2

Oxygen Determinations

Several compact portable units are available for the determination of dissolved oxygen. The oxygen-sensing probe is an electrolytic cell with gold (or platinum) cathode separated from a tabular silver anode by an epoxy casting. The anode is electrically connected to the cathode by electrolytic gel, and the entire chemical system is isolated from the environment by a thin gas-permeable membrane (often Teflon). A potential of approximately 0.8 V (from a solid-state power supply) is applied between the electrodes. The oxygen in the sample diffuses through the membrane, is reduced at the cathode with the formation of the oxidation product, silver oxide, at the silver anode. The resultant current is proportional to the amount of oxygen reduced. To counteract temperature effects, a thermistor is built into the sensor. The analyzer unit operates over the range from 0.2 to 50 ppm of dissolved oxygen. Gases that reduce at -0.8 V will interfere; these include the halogens and SO_2. H_2S contaminates the electrodes.

EVALUATION METHODS

Direct Comparison

The direct comparison method calls for recording the current–voltage curves of a standard solution of the test ion under the same conditions as the unknown. Then, using the Ilkovic equation in the simplified form, the diffusion current quotient, i_d/C, can be computed. When divided into the height of the unknown wave, it yields the concentration of test ion in the unknown. The unknown will be most accurately determined when the concentration of the comparison standard is about the same as that of the unknown, particularly if a nonlinear relation exists or is suspected to exist between wave height and concentration. The quantity of standard can be estimated by remembering that the diffusion current of simple ions in neutral or acid solution is about 4 μA per milliequivalent of reducible ion.

Relative measurements of this type do not demand knowledge of the exact capillary characteristics, only that they remain constant during the comparison. Likewise, temperature need not be controlled at any fixed value, merely maintained the same for all solutions. Immersion of the solution cells in a large container of water is adequate. However, it is important that the composition of the supporting electrolyte and the amount of maximum suppressor added be identical for the unknown and the comparison standard.

Standard Addition

If a single analysis is to be performed, it is possible to dispense with the preparation of a known solution that is the exact duplicate of the test solution. The polarogram of the unknown solution is recorded, then a known volume of a standard solution of the test ion is added and the polarogram repeated. From the increase in the diffusion current, the original concentration can be computed by interpolation. For the unknown solution,

$$i_d = KC_x = h \qquad (22\text{-}24)$$

and after the addition of v ml of a standard solution, whose concentration of test ion is C_s, to V ml of unknown,

$$KC_x\left(\frac{V}{V+v}\right) + KC_s\left(\frac{v}{V+v}\right) = H \qquad (22\text{-}25)$$

Solving for the concentration of the unknown,

$$C_x = \frac{-vC_sh}{hV - H(V+v)} \qquad (22\text{-}26)$$

For the maximum precision the amount of standard solution added should be sufficient to about double the original wave height. (See also page 379 in Chapter 12).

Internal Standard Method

The internal standard method, also called the "pilot ion" method, is based on the fact that the relative wave heights of two electroactive substances in a particular supporting electrolyte are constant for equal concentrations and independent of capillary characteristics. Even small temperature differences between analyses can be tolerated. In practice, one element is used to standardize the dropping assembly and all the other diffusion-current constants for other elements are measured relative to this same ion. Thus, it is only necessary to add a known concentration of the reference ion to an unknown; and from the wave heights of the unknown and the reference ion, the concentration of the unknown can be computed

$$C_x = \frac{i_x(I_d)_s}{i_s(I_d)_x} \cdot C_s \tag{22-27}$$

This method simplifies work when different capillary systems must be used. Only a single standard solution is required for a series of test substances once the internal standard ratio, $(I_d)_s/(I_d)_x$, has been established. Whenever the nature or the concentration of the supporting electrolyte is altered in any manner, the ratio must be determined anew. The method has only limited application because only a small number of ions give sufficiently well-defined waves for use as internal standards. In multicomponent mixtures there may not be sufficient difference among existing half-wave potentials to introduce another wave.

LABORATORY WORK

Assembling the Equipment and Solutions

Prepare a dropping mercury electrode assembly similar to the one shown in Fig. 22-6. Fill the leveling bulb half-full of clean mercury. Insert a 7- to 10-cm length of capillary and open the stopcock or release the screw hose-clamp. Raise the leveling bulb until drops of mercury begin to form at the tip of the capillary.

Allow the drops to form continuously during the entire laboratory period. When a laboratory period is completed, rinse the capillary with distilled water and then dry by blotting with filter paper. Insert the capillary through an inverted cone of filter paper and clamp vertically over a small beaker. Lower the leveling bulb until the drops cease to form.

MERCURY IS POISONOUS. Use a large tray under the dropping electrode assembly and be sure to clean up all spilled mercury.

Cover the bottom of the electrolysis cell with a $\frac{1}{8}$-in. layer of mercury. Pour in enough solution to fill the cell within a $\frac{1}{2}$-in. from the top, and place the cell in position so that the capillary is dipping into the solution. Adjust the height of the mercury reservoir until the drops detach themselves from the capillary every 2–4 sec.

In place of the mercury pool, an external reference electrode may be connected to the cell by means of a salt bridge. The area of the mercury surface should be at least 10 cm^2.

Removal of dissolved oxygen   Connect the gas-inlet tube to a tank of nitrogen, and bubble gas through the solution for 10 min. Removal of oxygen is complete within 1 to 2 min when a filter stick is used to disperse the nitrogen. When necessary or desired, traces of oxygen which may be present in tank gases can be removed by passing the gas through a gas-washing bottle filled with a solution of chromous chloride, or alkaline pyrogallol, or ammoniacal cuprous chloride and a second washing bottle filled with distilled water to remove any spray.

Gelatin solution, 0.2%   Dissolve 0.2 g of gelatin in 100 ml of freshly boiled water that has been cooled to 60°C. Add a small crystal of thymol and stopper firmly with a rubber stopper.

Adjustment of the galvanometer index   Flip the toggle switch marked LAMP to ON and bring the galvanometer index to the desired position by rotating the COMPENSATOR control UPSCALE or DOWNSCALE. See Fig. 22-13.

Use of sensitivity switch   The markings on the CURRENT MULTIPLIER are factors by which the galvanometer deflections must be multiplied to obtain the reading for full sensitivity, 1. If the galvanometer index goes off-scale as the current–voltage curve is recorded, decrease the galvanometer sensitivity by turning the CURRENT MULTIPLIER switch to a larger number.

Selection of the applied emf   Rotate the EMF switch clockwise to connect the batteries in the bridge circuit. Continue the rotation clockwise until the voltage desired is applied across the bridge. If a VOLTMETER is located on the front panel, it indicates the total voltage applied across the slide wire.

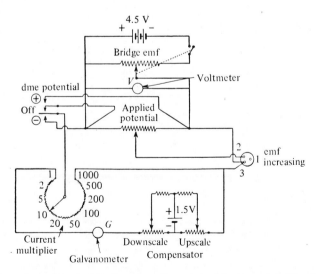

Fig. 22-13. The schematic circuit diagram of a manual instrument: The Sargent Polarograph, model III. (Courtesy of Sargent-Welch.)

Plotting the current-voltage curve Set the CURRENT MULTIPLIER to 10, the APPLIED POTENTIAL dial to 0, the VOLTMETER reading to 1.0 V, the toggle switch marked DME to "–", and the toggle switch marked GALV to "+", Increase the applied emf by rotating the APPLIED POTENTIAL dial in steps of 25 or 50 (0.025 or 0.050 V). Record the maximum (or preferably the average) galvanometer index reading each time.

When the APPLIED POTENTIAL dial reaches 1000 (1.000 V), return it to 500 and advance the BRIDGE EMF control until the VOLTMETER reads 2.00 V. Again increase the applied emf by rotating the APPLIED POTENTIAL dial from 500 (now 1.000 V) to 1000 (2.000 V). The span of the APPLIED POTENTIAL dial is now 0–2 V.

Use of the compensator The COMPENSATOR control may be used to balance out a large diffusion current when it precedes the current-voltage curve of an ion of interest (Fig. 22-14). Adjust the BRIDGE EMF and APPLIED POTENTIAL dial to a value at which the interfering current-voltage curve is completely developed, then rotate the COMPEN-SATOR control DOWNSCALE until the galvanometer index is brought back to the reading that corresponded to the residual current. Decrease the CURRENT MULTIPLIER setting and plot the current-voltage curve at the increased galvanometer sensitivity. To avoid oscillations of excessively large amplitude, a condenser (1000- to 2000-μf, electrolytic type) can be connected across the galvanometer terminals.

Note In these generalized instructions names of operational controls may differ among the various instruments. Compare the manufacturer's directions with his circuit diagram and the diagram in Fig. 22-13.

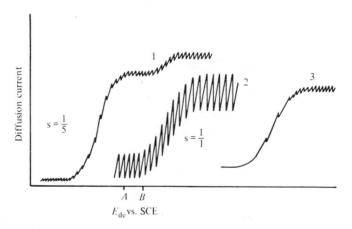

Fig. 22-14. Compensation method and condenser damping. Curve 1 is the wave of the predominant constituent preceding the wave due to a small concentration of less easily reducible ion. Curve 2 is the second wave after compensation and at an increased galvanometer sensitivity. Curve 3 is the same, but with the insertion of an electrolytic condenser across the terminals of the galvanometer.

Experiment 22-1   Evaluation Methods: Direct Comparison

1. Prepare a series of standard solutions, each containing 20 ml of $1\,M$ potassium chloride, 2.5 ml of 0.2% gelatin, and these quantities of $0.01\,M$ cadmium sulfate solution: 25, 20, 15, 10, 5, 2, and 0 ml. Dilute to the mark in 100-ml volumetric flasks and mix well.
2. Pour the strongest solution into the electrolysis cell and run the polarogram from 0.4 to 0.8 V negative to the pool or reference electrode. For the remainder of the standards, determine only the current flowing before the cadmium wave develops and the diffusion current of the fully developed wave.
3. Transfer aliquots of the unknown solution containing 0.10 to 0.25 mM of cadmium into 100 ml volumetric flasks. Add 20 ml of $1\,M$ potassium chloride and 2.5 ml of 0.2% gelatin and dilute to volume. Measure the residual current and the diffusion current as before.
4. Plot a calibration curve on a sheet of graph paper. The ordinate will be the diffusion current in microamperes or arbitrary scale divisions and the abscissa the corresponding cadmium concentration. Correct all observed diffusion current values by subtracting the value of the residual current found for the blank or by the extrapolation method. (Both are illustrated in Fig. 22-4.) From your calibration curve determine the concentration of the unknown.

Note   At the discretion of the instructor, other metal salt solutions may be substituted for cadmium sulfate.

Experiment 22-2   Evaluation Methods: Standard Addition

1. Pipet into a 100-ml volumetric flask 10 ml of the unknown cadmium solution. Add 20 ml of $1\,M$ potassium chloride solution, 2.5 ml of 0.2% gelatin solution, and dilute to the mark. Mix well.
2. Transfer a known volume of the solution to the electrolysis cell. Run the current–voltage curve from 0.4–0.8 V negative with respect to the reference electrode.
3. Estimate the concentration of the unknown from the fact that the diffusion current of aquo ions in neutral or acid solution is about 4 μA per milliequivalent of reducible ion. Pipet into the electrolysis cell a volume of standard cadmium solution which contains approximately the same amount of cadmium that is estimated to be present in the original solution but multiplied by the fraction $\frac{25}{100}$ or $\frac{50}{100}$, whichever numerator corresponds to the volume transferred to the electrolysis cell in the preceding step.
4. Run the current–voltage curve as before. Calculate the concentration of the unknown solution.

Experiment 22-3   Evaluation Methods: Internal Standard Method

1. Pipet into a 100-ml volumetric flask 10 ml of a cadmium stock solution and 10 ml of a zinc stock solution, each about $0.01\,M$. Add 20 ml of $1\,M$ potassium chloride solution, 2.5 ml of 0.2% gelatin solution, and dilute to the mark. Mix well.

2. Transfer a portion of the solution to the electrolysis cell. Run the current–voltage curves between 0.4 and 1.5 V negative with respect to the reference electrode. Determine the diffusion current for the cadmium and the zinc wave, respectively.

3. Pipet into a separate 100-ml volumetric flask 10 ml of the unknown cadmium solution and 10 ml of the same zinc stock solution. Add 20 ml of 1 M potassium chloride solution, 2.5 ml of 0.2% gelatin solution, and dilute to the mark. Mix well.

4. Transfer a portion of the second solution to an electrolysis cell. Run the current–voltage curves between −0.4 and −1.5 V as before. Determine the diffusion current for the cadmium and zinc waves. Calculate the concentration of the unknown solution from Eq. 22-27 and with the information obtained in steps 2 and 4.

Note If the experiment seems somewhat artificial the drop rate of the mercury issuing from the capillary may be altered between steps 2 and 4. Also the concentration of the unknown solution may be calculated by substituting the values for the diffusion current constant of the cadmium and zinc ions into Eq. 22-7. Appendix B lists these constants.

Experiment 22-4 Cathodic, Anodic, and Mixed Current–Voltage Curves

1. To a series of three 150-ml beakers add 50 ml of 1 M sodium citrate solution and 2.5 ml of 0.2% gelatin solution. Remove dissolved oxygen from the second and third solution before proceeding.

2. To the first beaker add 10 ml of 0.01 M iron(III) ammonium sulfate solution; to the second add 10 ml of 0.01 M iron(II) ammonium sulfate solution; and to the third add 5 ml each of the iron(II) and iron(III) solutions. Adjust each solution to pH 5.6. Transfer the contents of each beaker to 100-ml volumetric flasks and dilute to the mark. Mix well.

3. Run the current–voltage curves from +0.2 to −0.5 V with respect to the reference electrode. To plot the anodic portions of each curve reverse the connections to the galvanometer or initially adjust the galvanometer index to the midpoint of its scale when only the supporting electrolyte is in the electrolysis cell. To employ positive values of applied emf reverse the leads to the dropping electrode and pool or reference electrode.

Notes The iron(II) ammonium sulfate solution must be prepared fresh each time using water that has been deaerated. Even with these precautions a small cathodic wave due to iron(III) may appear on the current–voltage curve.

The half-wave potentials of the three curves will not be exactly the same unless a correction is made for the iR drop within the electrolysis cell. If this correction is attempted, remember the signs of the diffusion currents: positive for cathodic currents and negative for anodic currents.

Experiment 22-5 Analysis of a Copper-Base Alloy[4]

Lead, tin, nickel, and zinc are to be determined by the standard addition method. The major portion of the copper is removed from the dissolved sample, and lead is determined in one aliquot of the residual solution and tin in a second aliquot. Nickel and zinc are determined on a second sample from which copper, tin, and lead have been removed. Suitable solutions will be available if copper and tin plus lead were deposited electrolytically at controlled cathode potentials (see Experiments 23-1A and 23-1B). Synthetic solutions that simulate the metal contents of the alloy may be employed if the electrolytic separations were not carried out. In the latter event a pseudo-alloy solution is prepared that contains 4–20 mg each of lead and tin, 2–4 mg of nickel, and 4–100 mg of zinc, in a total volume of 100 ml.

Lead   Transfer a 25-ml aliquot of the synthetic solution to a 100-ml volumetric flask.   Add 4.8 g of sodium hydroxide pellets, 2.5 ml of 0.2 percent gelatin, and dilute to the mark. Mix well. Transfer exactly 25 ml of the solution to the electrolysis cell. Run the current–voltage curve from 0.6 to 0.9 V negative with respect to a SCE (approximately 0.5–0.8 V negative with respect to a mercury pool in contact with 1 M sodium hydroxide solution).

Estimate the concentration of lead in the 25-ml aliquot from the fact that the diffusion current is about 4 μA per milliequivalent of lead. Evaluate the amount of lead present by the standard addition method.

Tin   Transfer a 25-ml aliquot of the synthetic solution to a 100-ml volumetric flask. Add 21 g of ammonium chloride, 6.6 ml of 12 M hydrochloric acid, and dilute to about 90 ml. Shake until all the salt has dissolved, warming if necessary. Add 2.5 ml of 0.2% gelatin solution and dilute to the mark. Mix well. Transfer exactly 25 ml of the solution to the electrolysis cell. Run the current–voltage curve from 0 to 0.7 V negative with respect to a SCE.

Estimate the amount of tin in the 25-ml aliquot from the diffusion current of the second wave. Subtract the diffusion current previously found for lead alone, because the half-wave potential of lead is virtually coincident with that of the second tin wave in the 4 M ammonium chloride plus 1 M hydrochloric acid-supporting electrolyte. Evaluate the amount of tin present by the standard addition method.

To correct the diffusion current of the second tin wave for the contribution of the coincident lead wave, multiply the diffusion current of lead found in the sodium hydroxide-supporting electrolyte by 1.035 and subtract the result from the total diffusion current due to tin plus lead. The factor 1.035 is the ratio of the diffusion current constants of lead in the ammonium chloride-hydrochloric acid medium and the sodium hydroxide medium, namely 3.52/3.40.

Nickel and Zinc   Transfer a 25-ml aliquot of the synthetic solution from which tin and lead are absent to a 100-ml volumetric flask. Add 25 ml of a supporting electrolyte stock solution, which contains 43 g of ammonium chloride and 270 ml of concentrated aqueous ammonia, made up to 1 liter. Add 1 g of sodium sulfite, 2.5 ml of 0.2% gelatin solution, and dilute to the mark. Mix well and allow to stand 10 min to let the sulfite react with the dissolved oxygen. Transfer exactly 25 ml of the solution to the electrolysis cell. Run the current–voltage curves from 0.8 to 1.6 V negative with respect to the SCE.

If the height of the first wave due to nickel ions is much smaller than that of the second wave due to zinc ions, run a second current–voltage curve from 0.8 to 1.3 V negative with respect to the SCE at an increased galvanometer sensitivity.

Estimate the amounts of nickel and zinc in the 25-ml aliquot from the respective diffusion currents, and evaluate the amount of nickel and zinc present by the standard addition method.

Notes When working with the alkaline-supporting electrolytes, one observes a large diffusion current that begins at zero applied emf. Proceed as follows: Adjust the applied emf to 0.6 V negative to the reference electrode and return the galvanometer index to zero by means of the diffusion current compensator (or bias control).

If the residual solutions from the electrodeposition experiments are employed, transfer the residual solutions to 250-ml volumetric flasks and dilute to the mark. Use 50-ml aliquots in each of the preceding steps.

PROBLEMS

1. Compare the evaluation methods with respect to (a) applicability to routine analyses of similar type samples, (b) applicability to routine analyses of samples of widely varying composition but for the analysis of a few constituents by polarography, and (c) applicability to occasional analyses of a wide variety of samples and for a wide variety of constituents.

2. Discuss the factors that may contribute to the observed polarographic limiting current of a single ion in addition to the diffusion current. How may all the undesirable factors be eliminated?

3. Polarographic curves resemble potentiometric titration curves. When might polarography yield useful data not obtainable by potentiometric methods?

4. The data below were obtained in 25°C with cadmium ion in a supporting electrolyte composed of 0.1 M potassium chloride and 0.005% gelatin. Galvanometer deflections were measured at −1.00 V vs. SCE. The galvanometer sensitivity was stated as 0.0055 μA/mm; the current multiplier was set at 50. $t = 2.47$ sec; $m = 3.30$ mg sec^{-1}.

Cd^{2+}, mM	i_d, mm
0.00	4.5
0.20	11.0
0.50	21.0
1.00	37.5
1.50	54.0
2.00	70.5
2.50	86.5

Plot the calibration curve on a sheet of graph paper with the ordinate the diffusion current in microamperes (or arbitrary scale divisions) and the abscissa the corresponding cadmium concentration. From the calibration curve determine the concentration of an unknown solution, prepared similarly, which has a diffusion current of 39.5 mm after correction for the residual current.

5. Exactly 25.0 ml of the unknown solution, which gave a diffusion current of 39.5 mm in Problem 4, has been transferred to the polarographic cell. To this solution is added exactly 5.0 ml of 0.0120 M cadmium solution. The corrected diffusion current is now 88 mm. Calculate the concentration of the unknown cadmium solution.

6. At the time the unknown cadmium solution was prepared in Problem 4 a second solution was also prepared, identical in all respects except that it also contained zinc ions, 10.0 ml of 0.0100 M zinc in a total volume of 100 ml. The corrected diffusion current was 32 mm for the zinc wave. Calculate the concentration of the unknown cadmium solution by the internal standard method. Obtain the necessary diffusion current constants from Appendix B.

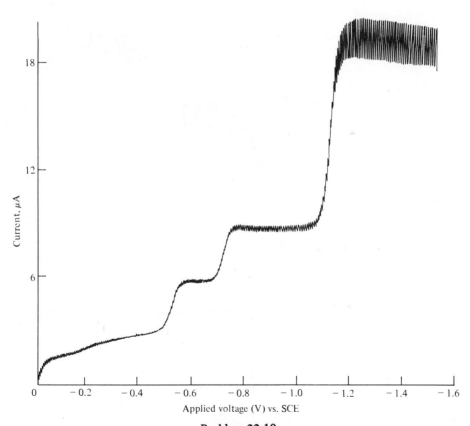

Problem 22-10

7. An unknown amount of copper(II) ions produces 12.3 μA on a dc polarogram. By adding 0.100 ml of 1.00×10^{-3} M copper(II) ions to the original volume of 5.00 ml, the new current is 28.2 μA. Calculate the original amount of copper.

8. From each of the following sets of transition times obtained from chronopotentiograms, deduce the successive electron reactions: (a) $\tau_2 = 3\tau_1$ for the reduction of oxygen at a mercury cathode; (b) $\tau_2 = 8\tau_1$ for the reduction of uranium; and (c) $\tau_2 = 35\tau_1$ for the oxidation of iodide ion (< 0.0025 M) at a platinum anode in dilute sulfuric acid.

9. To what extent will the charging current decay under the following operating conditions in normal pulse polarography: $R = 1000$ Ω (the electrolytic resistance of the solution); C (the double layer capacitance) $= 20$ μF/cm^2 for a dropping mercury electrode of area 0.03 cm^2; measurements made 40 msec after the application of each pulse.

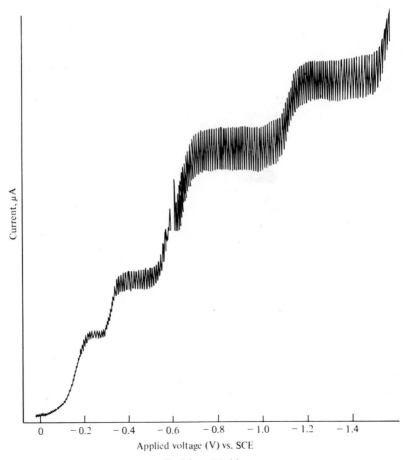

Problem 22-11

10. The polarogram of a sample solution is shown. Before the solution was made up to volume in $0.1\,M$ KCl, sufficient cadmium was added to bring its concentration to $0.03\,mM$. Identify and estimate the concentrations of the other two metals present in the nonferrous alloy sample. *(Illustration for Problem 10 on page 667.)*

11. On the conventional dc polarogram shown, identify the five different metals present in the $0.1\,M$ KCl solution. *(Illustration for Problem 11 on page 668.)*

12. Identify the six metals present in the derivative polarogram shown which was obtained in $0.1\,M$ HCl.

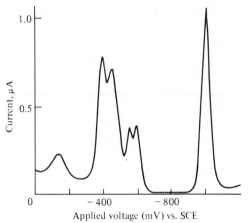

Applied voltage (mV) vs. SCE

Problem 22-12

13. The polarogram of a sample is shown. Sufficient cadmium was added to the sample before dilution to volume in $0.1\,M$ HCl to bring its concentration to $2.00 \times 10^{-4}\,M$. Calculate the concentration of the other two metals present.

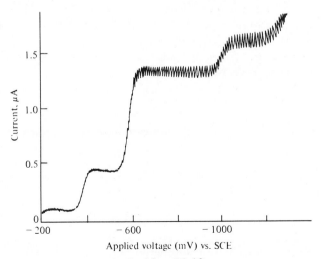

Applied voltage (mV) vs. SCE

Problem 22-13

BIBLIOGRAPHY

Adams, R. N., *Electrochemistry of Solid Electrodes*, Marcel Dekker, New York, 1969.

Ellis, W. D., "Anodic Stripping Voltammetry," *J. Chem. Educ.*, **50**, A131 (1973).

Flato, J. B., "The Renaissance in Polarographic and Voltammetric Analysis," *Anal. Chem.*, **44** (11), 75A (1972).

Flato, J. B., "Two New Electrochemical Instruments," *Am. Laboratory*, **1** (2), 10 (February 1969).

Heyrovsky, J. and J. Kuta, *Principles of Polarography*, Academic Press, New York, 1965.

Kolthoff, I. M. and J. J. Lingane, *Polarography*, 2nd ed;. Wiley-Interscience, New York, 1952, in 2 volumes.

Meites, L., *Polarographic Techniques*, 2nd ed., Wiley-Interscience, New York, 1966.

Milner, G. W. C., *The Principles and Application of Polarography*, Wiley, New York, 1957.

Müller, O. H. in *Physical Methods of Chemistry*, A. Weissberger and B. W. Rossiter, Eds., Vol. I, Part IIA, Wiley-Interscience, New York, 1971, Chapter 5.

Schaap, W. B. in *Standard Methods of Chemical Analysis*, F. J. Welcher, Ed., 6th ed., Vol. 3, Part A, Van Nostrand Reinhold, New York, 1966, Chapter 19.

Schmidt, H. and M. von Stackelberg, *Modern Polarographic Methods*, Academic Press, New York, 1963.

LITERATURE CITED

1. Barendrecht, E., "Stripping Voltammetry," in *Electroanalytical Chemistry*, A. J. Bard Ed., Vol. 2, Marcel Dekker, New York, 1967, p. 53.
2. Burge, D. E., *J. Chem. Educ.*, **47**, A81 (1970).
3. Lingane, J. J., *Analyst*, **91**, 1 (1966).
4. Lingane, J. J., *Ind. Eng. Chem.*, *Anal. Ed.*, **18**, 429 (1946).
5. Kelley, M. T., H. C. Jones, and D. J. Fisher, *Anal. Chem.*, **31**, 1475 (1959).
6. Smith, D. E. in *Electroanalytical Chemistry*, A. J. Bard, Ed., Vol. 1, Marcel Dekker, New York, 1966, Chapter 1.
7. Zuman, P., *Chem. Eng. News*, p. 94 (March 18, 1968); *Substituent Effects in Organic Polarography*, Plenum Press, New York, 1967.

23

Separations by
Electrolysis

Electroseparation is electrolysis in which a quantitative reaction or, at the very least, an appreciable amount of reaction of analytical interest takes place at an electrode. Electrooxidation or reduction may be accomplished. Electroreductions for analytical purposes have long been carried out with the current kept more or less constant by adjusting the voltage applied to the cell. Probably the most widely used type of electroreduction involves a mercury cathode to which a constant potential is applied. Although selectivity is poor, this method finds extensive use for removal of interfering elements. Better selectivity can be achieved by controlling the potential of the working electrode with a potentiostat such that constituents may be sequentially removed from a solution either by reduction or oxidation. Electroorganic synthesis is a useful tool for synthetic organic chemists.

Other uses of electroseparations include the dissolution of refractory metals; the metal sample is made the anode and a platinum, tantalum, or graphite cathode is employed. Electrolytic methods are useful in the preparation of very pure inorganic substances or in the preparation of unusual oxidation states of metals. As a preliminary step in the preparation of samples for spectroscopic techniques, elements are deposited on a suitable electrode material and determined by emission spectroscopy or by spark source mass spectrometry.

BASIC PRINCIPLES

Let us assume a typical electrogravimetric cell. A pair of relatively large platinum electrodes is immersed in the electrolyte, and a voltage is applied across them. At first, as the applied voltage is gradually increased, virtually no current (except a small residual current) is observed to flow through the cell. However, as a particular point on the voltage axis is reached (Fig. 23-1), a noticeable reaction will be observed and a current begins to flow through the cell. This particular value of applied voltage is called the *decomposition potential* with respect to the particular electrode reaction. As soon as the decomposition potential is exceeded, continuous electrolysis of the solution is sustained. With further increase in applied voltage the current increases linearly in accordance with Ohm's law. Ideally, the slope of the curve is the reciprocal of the resistance between the terminals of the electrolytic cell. Actually, the current is soon limited by the rate of mass

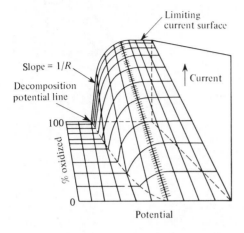

Fig. 23-1. Three-dimensional representation of a reversible oxidation–reduction system as it applies to electroanalysis. (After W. H. Reinmuth, *J. Chem. Educ.*, **38**, 149 (1961).)

transfer of electroactive material to the electrode surface. The mass transfer coefficient depends on the rate of stirring, cell and electrode geometries, diffusion coefficient of the electroactive species, and the like. As the concentration of electroactive species in solution decreases, so also does the limiting current plateau. The decomposition potential simultaneously shifts to more cathodic (if reducible species are involved) or anodic (if oxidizable species are involved) values.

To determine what voltage must be applied to a cell to cause electrolysis, it is necessary to know first what reactions will occur at the two electrodes. If these reactions are known, it is possible to calculate the potential of each electrode and thereby determine the emf of the galvanic cell which exerts its potential in opposition to the applied voltage. For example, in the electrolysis of 0.100 M $CuSO_4$ in 1 N H_2SO_4 with platinum electrodes, the reaction at the cathode will be

$$Cu^{2+} + 2e^- \rightarrow Cu^\circ \tag{23-1}$$

and at the anode

$$2H_2O \rightarrow O_2 + 4H^+ + 4e^- \tag{23-2}$$

Using the Nernst equation to calculate the potential of the copper electrode,

$$E = 0.337 + 0.0296 \log (0.100) = 0.307 \text{ V} \tag{23-3}$$

For the other electrode at which oxygen is evolved,

$$E = 1.229 + \frac{0.0591}{4} \log \frac{[O_2][H^+]^4}{[H_2O]^2} \tag{23-4}$$

This expression can be simplified by realizing that the term $[O_2]$ is the same as the partial pressure of oxygen gas in the atmosphere (approximately 0.21), and that the concentration of water is essentially constant. Then,

$$E = 1.229 + 0.0148 \log p_{O_2} [H^+]^4 \tag{23-5}$$

In 1 N H_2SO_4, the potential is about 1.22 V.

The cell emf is the algebraic difference of the electrode potentials of the two half-cells comprising the electrolytic cell. The spontaneous reaction which will occur in this cell is

$$2Cu^{\circ} + O_2 + 4H^+ = 2Cu^{2+} + 2H_2O \qquad (23\text{-}6)$$

This corresponds to subtracting Eq. 23-3 from Eq. 23-5; therefore, the emf of the galvanic cell is $1.22 - 0.31 = 0.91$ V. Correspondingly, the emf of the electrolytic cell is -0.91 V. When the applied voltage is exactly 0.91 V, no current will flow through the cell in either direction, which is, of course, the principle of potentiometric measurements. As soon as the applied voltage is in excess of the galvanic cell emf, the iR drop, and the sum of any overpotential effects,

$$E_{\text{applied}} = E_{\text{cell}} + iR + \omega_{\text{anod}} + \omega_{\text{cath}} \qquad (23\text{-}7)$$

current is forced to flow through the electrolytic cell and the reaction expressed by Eq. 23-6 proceeds from right to left. The maximum value of the current is governed by the cell resistance and mass transfer conditions.

Completeness of Depositions

The emf that is applied to an electrolytic cell must be sufficient to ensure the removal of the desired electroactive species to an extent adequate for the purpose of the experiment. Take the deposition of silver for example. As the silver deposits, the concentration of silver ions in solution decreases, and, according to the Nernst equation,

$$E = 0.80 + 0.0591 \log [Ag^+] \qquad (23\text{-}8)$$

the potential at which silver deposits becomes more negative. The values in Table 23-1 represent the changes in cathode potential for various fractions of the original silver concentration, which was 0.1 M in our example. The cathode potential changes very nearly 59 mV for each tenfold decrease in the concentration of the silver ion remaining undeposited (and, in general, $59/n$ mV for metallic ions). Assuming that the anode process proceeds at a constant level and that the iR drop is essentially constant, the decomposition potential of the solution should vary by a like amount.

TABLE 23-1   Electrode Potential as a Function of Silver-Ion Concentration

Concentration of Silver Ion, M	Fraction of Original Remaining, %	Cathode Potential, V
1×10^{-1}	100	0.74
1×10^{-2}	10	0.68
1×10^{-3}	1	0.62
1×10^{-4}	0.1	0.56
1×10^{-5}	0.01	0.50

If only a very minute amount of a metal is to be deposited on platinum, perhaps a fraction of a microgram, then the amount of deposit may not be enough to form a mono-layer of atoms, and hence the activity of the metal phase cannot be assumed to be constant. Rogers and Stehney[15] discuss the theory of depositions in such situations.

Overpotentials

When a current flows across an electrode–solution interface, it is normally found that the electrode potential changes from the reversible value it possesses before the passage of current. The difference between the measured potential (or cell emf) and its reversible value is the *overpotential* (also called overvoltage). The electrodes are said to be polarized. Both cathodic and anodic processes exhibit overpotential, and it is affected by many factors. When an anodic process shows an overpotential effect, the applied potential necessary to cause electrolysis will always be a more positive value than the calculated potential and, for cathodic processes, overpotential causes the applied potential to be more negative than the calculated value. Although overpotential phenomena complicate the calculation of the applied voltage necessary for electrolysis to occur, its effect makes feasible certain separations that would not be expected from standard electrode potentials.

Various types of overpotential may be distinguished. In some electrode processes a film of oxide or some other substance forms on the electrode surface and sets up a resistance to the passage of current across it. What may be termed an ohmic-pseudo-overpotential is also observed when the capillary tip used in measuring the potential of an electrode is at an appreciable distance from the electrode surface. The latter effect only becomes appreciable at high current densities or low concentrations.

A second type of overpotential is due to concentration changes in the vicinity of the electrode and is consequently referred to as *concentration overpotential*. Whenever a finite current flows across an electrode–solution interface, the concentration of the electrode surface is somewhat altered from its concentration in the bulk of the solution. The reason is that the species at the electrode surface is not replenished at a rate commensurate with the current demand. Consequently, the potential of the cathode will exhibit a more negative value as the voltage applied to the electrolytic cell is increased. Generally, the metal ion concentration at the electrode interface is only 1% of the bulk concentration; thus, the concentration overpotential is approximately $0.118/n$ V. In practice, concentration overpotential is minimized by the use of electrodes with large surface areas and by keeping electrolysis currents small, although the latter stipulation is not conducive to rapid electrolysis. The rate of mass transfer is aided by mechanical stirring and increase in temperature of the solution.

Although concentration changes are probably the most important source of overpotential accompanying the deposition of a metal, small overpotentials arise from other causes. As a general rule, the overpotential of a metal upon itself is not large (about 0.01 V) at low current densities. This is not true, however, for such hard metals as cobalt, nickel, iron, chromium, and molybdenum. In an ammoniacal solution the deposition potential of copper on a platinum electrode is considerably lower than on a silver electrode. Consequently, in the separation of silver from copper in an ammoniacal solution, if the silver is plated on a platinum electrode and if toward the end of the electrolysis the level of the solution is raised, copper will plate on the bare platinum surface exposed.

Complexation may affect the overpotential in either direction, because the rate of exchange of electrons between the electrode and complex species may be greater or less than the rate of exchange with the aquated ion. For example, aquo nickel ions show an overpotential of about 0.6 V at a mercury surface, whereas complexes of nickel with thiocyanate or pyridine, although actually shifting the equilibrium electrode potential more negative, show a decrease in overpotential which more than compensates for the shift in the equilibrium potential.

The evolution of gases at an electrode is usually associated with an overpotential significantly larger than concentration overpotentials. It is particularly marked in the evolution of hydrogen and oxygen, and is called the *gas overpotential*. Some values of the overpotential of hydrogen on various surfaces are given in Table 23-2. The anodic overpotential of oxygen on smooth platinum in acid solutions is approximately 0.4 V.

Gas overpotential depends on several factors: (a) *Electrode material.* At a given current density, overpotential for many metal surfaces seems to decrease roughly in a parallel manner to the thermionic work function of the electrode material. (b) *Current density.* An increase in current density invariably increases the overpotential up to a limiting value. (c) *Electrode condition.* Whether smooth or rough, bright or platinized, the overpotential at a given current density decreases if the electrode surface is roughened.

TABLE 23-2 Hydrogen Overpotential on Various Cathodes

Electrolyte is $1\ M\ H_2SO_4$. Overpotential given in volts.

Cathode	First Visible Gas Bubbles	Current Density $0.01\ A\ cm^{-2}$	$0.1\ A\ cm^{-2}$
Antimony	0.23	0.4	—
Bismuth	0.39	0.4	—
Cadmium	0.39	~0.4[a]	—
		—	1.2
Copper	0.19	0.4	0.8
Gold	0.017	0.4	1.0
Lead	0.40	0.4	1.2
Mercury	0.80	1.2	1.3
Platinum (bright)	~0	0.09	0.16
Silver	0.097	0.3	0.9
Tin	0.40	0.5	1.2
Zinc	0.48	0.7[b]	—

SOURCE: J. J. Lingane, *Electroanalytical Chemistry*, 2nd ed., Wiley-Interscience, New York, 1958, p. 209. Reproduced by permission.

[a] $0.005\ M\ H_2SO_4$.
[b] $0.01\ M\ Zn\ (C_2H_3O_2)_2$.

This is due partly, if not entirely, to an increase in the effective area and the consequent decrease in the actual current density. (d) *Temperature*. As the temperature is raised, overpotential diminishes. For most electrodes the temperature coefficient is about 2 mV/degree. (e) *The pH*. At low current densities the overpotential is independent of the pH. At high acid concentrations and with some metals there appears to be some dependence on pH.

The hydrogen overpotential is large on metals such as bismuth, cadmium, lead, tin, zinc, and especially mercury. With a mercury cathode, a number of useful separations become possible, as will be discussed in a later section. Overpotential of hydrogen on cadmium and zinc is important in electrolysis because it permits their determination in an aqueous solution.

Processes at the Anode

The behavior at an anode is, in general, analogous to that at a cathode. The process associated with the smallest oxidation potential, whether it be dissolution of the metallic anode to form cations or the discharge of anions, will take place first. Subsequent anodic processes will follow in order of increasing oxidation potential. The discharge of anions involves a consideration of sulfide, halide, and hydroxyl ions only. For other anions, the hydroxyl ion, derived from the water at the surface of the anode, or water itself, will be preferentially discharged, leading to the evolution of oxygen. Consequently, for solutions of metal salts other than halides or sulfides, the decomposition potential will depend primarily upon the metal ion. Equations 23-4 and 23-5 will express the anodic reaction.

Under suitable conditions, PbO_2, MnO_2, and Tl_2O_3 can be deposited at the anode and thereby separated from nearly all other metallic ions. Halide ions can be deposited on a silver anode—selectively if the anode potential is controlled.

EQUIPMENT FOR ELECTROLYTIC SEPARATIONS[10]

In order to make electrolytic separations, it is necessary to have a source of direct current, an adjustable resistance, a cell for electrolysis, including the electrodes, and usually some means for stirring the solution. In order to measure the current and applied voltage, an appropriate ammeter and voltmeter are needed. The schematic arrangement of the equipment is shown in Fig. 23-2. The direct current is most conveniently supplied from storage batteries because they give a steady voltage. However, compact commercial power supplies are available that operate from alternating current to supply the direct current. A schematic diagram of one unit is shown in Fig. 23-3. A fixed transformer steps the voltage down to 6 or 10 V, and the current is then passed through a selenium rectifier bridge of the full- or half-wave type, and finally through a filter circuit. The latter, a combination of an inductance or choke and a capacitor, converts the pulse of raw dc output from the rectifiers into a more or less smooth flow of direct current. A filter circuit which leaves the ripple remaining at 1% or less is satisfactory. The variable autotransformer is used to control the voltage applied to the stepdown transformer, and thereby the rectified output voltage.

The electrolysis cell is frequently a tall-form beaker, covered with a split watch glass to exclude dirt and to minimize loss of solution through spray during the electrolysis. The cathode is generally a cylinder (perhaps corrugated) of platinum gauze or a perforated platinum foil. Anodes may be a coiled wire, a platinum paddle, or a second gauze electrode smaller in diameter than the cathode. A gauze construction presents the largest surface area consistent with adequate mechanical strength. The effective area will be the total foil area or the length of wire, calculated from the number of meshes and dimensions of the electrode, multiplied by πd, where d is the diameter of the wire. Effective stirring of the electrolyte is usually essential. This can be accomplished by a motor stirrer and a glass impellor, a magnetic stirrer and rotating bar magnet, or with the anode rotating within the cathode. The last method is commonly used where commercial equipment is employed.

After the electrolysis is complete, the deposited metal must be removed from the solution without contaminating the solution if further analyses are to be made on the solution, and without loss of the deposited metal if this deposit is to be weighed or analyzed. If the deposit has been made on a platinum electrode and is to be weighed, the electrode must be washed thoroughly as it is removed from the solution. Furthermore, because of the voltaic cell which is present and which would cause dissolution of the deposited metal if the applied voltage were interrupted, the electrode should be washed without breaking the electric current. This is best done by slowly lowering the electrolysis cell from the electrodes while washing the electrodes with a stream of water from a wash bottle. The electrodes are then rinsed with alcohol or acetone prior to

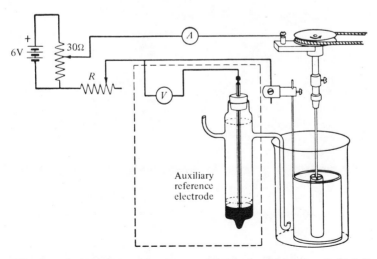

Fig. 23-2. Equipment for electrodeposition. Enclosed within the dashed lines is the additional equipment required for measuring the electrode potential.

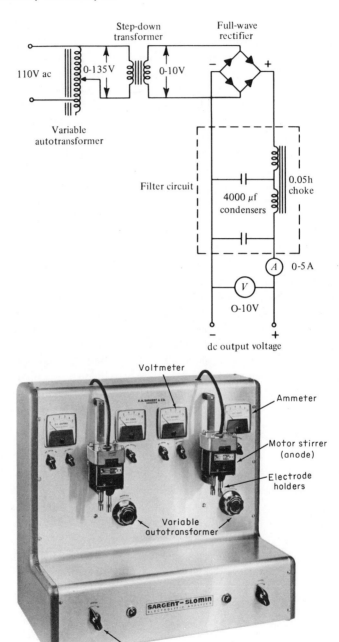

Fig. 23-3. Direct-current power supply for electrodeposition. *Upper*, the electrical circuit; *lower*, exterior view showing the controls. (Courtesy of Sargent-Welch.)

drying them at an elevated temperature. The weight of the deposit is obtained by weighing the electrode before and after deposition.

ELECTROGRAVIMETRY

Constant-Current Electrolysis

Electrolysis has long been carried out with the current kept more or less constant with time by adjusting the voltage applied to the cell. The technique can be represented by the intersection of the system surface with a plane parallel to the base, or zero-current plane, of Fig. 23-1. No control is exerted over the cathode potential; rather, a predetermined current is forced through the electrolytic cells regardless of mass transport conditions. The electrochemical process with the most positive reduction potential will occur first at the cathode, then the next most positive electrochemical process, and so on. Thus, if a current is passed through a solution containing cupric, hydrogen, and zinc ions, copper will be deposited first at the cathode. As the copper deposits, the reduction potential of the cupric ions becomes more negative, requiring periodic changes in the applied emf to more negative values as the electrolysis proceeds. More significantly, the rate at which cupric ions can be brought to the electrode surface will eventually fall under the rate required by the current forced through the cell, that is, the reduction of cupric ions alone will not hold the current at the desired level. Further increases in the applied emf will then result in a rapid change of cathode potential to a point where it equals that of hydrogen ions, and liberation of hydrogen gas begins. From this point on an increasing fraction of the current is devoted to the evolution of hydrogen, although the cathode potential will become relatively stable at a level fixed by the electrode potential and the overpotential for the evolution of hydrogen gas. This second process would be intolerable were it to involve another metal deposit; even so, the continual evolution of gas at the electrode is unsuited for adherent deposits.

Since, in the foregoing example, the hydrogen-ion concentration remains virtually constant in a solution during the evolution of hydrogen at the cathode and oxygen at the anode, the potential of the cathode cannot become sufficiently negative for the deposition of the zinc ions to commence. It should be evident, then, that metallic ions with a positive reduction potential may be separated, without external control of the cathode potential, from metallic ions having negative reduction potentials. However, for this separation to be successful, the hydrogen overpotential on the cathode plus the reversible reduction potential of the hydrogen ions must be less than the negative reduction potential of any of the metallic ions that are to remain in solution. For example, cupric ions in a solution containing $1\,M$ hydrogen ions may be separated from all metallic ions whose reduction potentials are more negative than about -0.4 V, the hydrogen overpotential on a copper electrode for relatively large current densities. Additional selectivity can be achieved through use of masking agents or potential buffers, or control of pH.

Example 23-1

Under what conditions would it be possible to initiate the deposition of zinc onto a copper-clad electrode from a solution that is 0.01 M in zinc ions? Also, what conditions are necessary for quantitative removal of zinc?

Answer   From the Nernst equation, the deposition potential for zinc is

$$E = -0.76 + 0.03 \log (0.01) = -0.82 \text{ V}$$

and this value increases to -0.94 V when 0.01% remains in solution. Turning to the expression for the evolution of hydrogen, we can evaluate the minimum pH necessary to allow the deposition of zinc to commence, assuming the overpotential of hydrogen on copper to be 0.4 V:

$$E = 0.0 + 0.059 \log [H^+] + (-0.4) = -0.82$$

from which the pH is calculated to be 7. Although it might be expected that the pH would have to be raised to about 8.5 to remove the zinc quantitatively, this is true only if the amount of zinc is insufficient to coat completely the electrode surface exposed to the solution. As soon as the electrode becomes coated with zinc metal, the overpotential of hydrogen rises to the value on a zinc surface, and, consequently, the deposition of zinc proceeds to completion at pH 7 approximately. In practice, an ammoniacal buffer is employed, partly to take advantage of the superior deposit from zinc ammine ions.

Separations with Controlled Electrode Potentials

To carry out the electrolytic separation of two metals whose deposition potentials differ by an adequate amount, yet which lie on the same side of hydrogen, provision must be made to control the electrode potential. This is achieved by introducing an auxiliary reference electrode and placing the tip of the salt bridge adjacent to the working electrode. The potential of the working electrode is determined by measuring the emf of the cell established by the electrode and the reference electrode, from which the potential of the working electrode (herein assumed to be the cathode) can be calculated:

$$E_{cath} = E_{cell} - E_{ref} \tag{23-9}$$

A potentiometer or vacuum-tube voltmeter serves to measure the cell emf. The extra items of equipment, enclosed within the dashed lines of Fig. 23-2, are the only changes required in the conventional apparatus for electrodeposition by the constant-current method.

With the isolation of the term: $E_{cath} + \omega_c$, from the cell emf, controlled-potential electrolysis may cleanly separate two elements, put an element into a particular oxidation state, or synthesize an organic compound. The potential of the cathode (or anode in oxidation reactions) is controlled so that it never becomes sufficiently negative to allow the deposition of the next element. In practice, the controlled potential should be $0.118/n$ V more negative than the final equilibrium electrode potential to correspond to the hundred-fold difference that usually prevails in concentration between the electrode surface and the bulk concentration. This control of the electrode potential is achieved by adjusting the voltage applied to the electrolysis cell—manually with a battery and a variable resistor, by manual adjustment of the autotransformer at the input of commercial electroanalyzers, or electronically with a potentiostat. The current steadily decreases as the metallic ions are removed, but the maximum current permissible is used at all times and thus the electrolysis proceeds at its maximum rate. The electrolysis is discontinued when the current has fallen to a constant low value, usually 10 or 20 mA. The intersection of the system surface with a plane parallel to the zero-potential plane in Fig. 23-1 shows the course of the current and the percent reduced (or oxidized) during the electrolysis.

A variety of electronic circuits are available which are capable of decreasing or increasing automatically the applied emf in order to maintain a constant electrode potential. The schematic circuit of one instrument is shown in Fig. 23-4. It can be used in conjunction with the power supply shown in Fig. 23-3. In use, the electrode potential is set at any desired value from 0 to 3 V versus the reference electrode. The reference half-cell is then balanced against the working electrode through a potentiometer. When the voltage of the working electrode changes from the preset potential, a current flows in the potentiometer circuit. This current, amplified by a dc amplifier, activates one of two relays that control the reversible motor that adjusts the contactor of the autotransformer (Fig. 23-3). If the electrode potential is low, one of the relays operates the autotransformer to increase the electrode potential. If it is high, the other relay turns the motor in reverse to decrease it until the difference signal has been reduced to zero.

An approximate value of the limiting electrode potential can be calculated from the Nernst equation, but lack of knowledge concerning the overpotential term for a system severely limits its usefulness. A more reliable method involves the determination of the limiting potential empirically from current–potential curves. The current–potential curve is determined for each reaction under exactly the same conditions that will prevail in the actual analysis. The potential of the working electrode is increased in regular increments by increasing the voltage applied to the cell. The current is observed at each value of the electrode potential. To minimize any change in the concentration around the electrode, the cell circuit should be closed only long enough to secure the current measurement. Schematic current–cathode potential curves for the reduction of copper(II) are shown in Fig. 21-6. Ordinarily polarograms obtained with the dropping mercury electrode serve excellently to define the conditions for electrolysis with a large mercury cath-

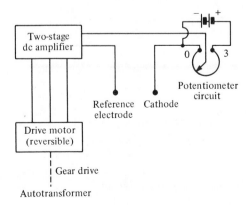

Fig. 23-4. Schematic circuit diagram of an automatic potentiostat.

ode. Usually these will be about 0.1–0.15 V more negative than the polarographic half-wave potential.

Consider, for example, a solution which is 0.1 M in copper(II) and 0.01 M in lead and tin ions. A diagram of the course of the electrolysis of these metals in hydrochloric acid medium and using platinum electrodes is shown in Fig. 23-5.[3] If hydrazine is present, the potential at which the oxidation of hydrazine will occur will be approximately 0.4–0.6 V versus SCE. This anode potential, shown as the lower solid line in Fig. 23-5(b), will remain fairly constant because the factors that affect this reaction do not change appreciably. Copper will be deposited first on the cathode at a potential of about 0.0 V versus SCE from the reduction of the $CuCl_3^{2-}$ and $CuCl_2^{-}$ ions. (The overpotential term is roughly 0.10–0.15 V for the currents initially employed.) Thus, the emf required for electrolysis will be about 0.6 V plus any ohmic drop. As copper is deposited, the concentration of the chlorocuprate(I) ions decreases and the potential of the cathode becomes more negative. When the copper concentration is lowered to about 10^{-5} M, the potential of the cathode is -0.40 V versus SCE. Actually, the copper concentration need only be lowered to about 10^{-4} M for the removal to be quantitative and, consequently, the cathode potential is usually controlled at -0.35 V versus SCE. At -0.40 V versus SCE the lead and tin will start to deposit. Thus, if the potential of the cathode can be controlled so that it never becomes more negative than -0.35 V versus SCE, no lead or tin will be deposited, whereas the deposition of copper from chloride medium is virtually complete.

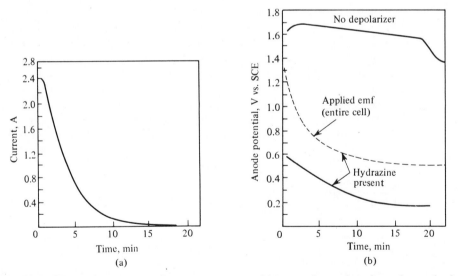

Fig. 23-5. Electrodeposition of copper onto a platinum electrode when the cathode potential was maintained at -0.35 V vs. SCE. (a) Current decay with time. (b) Anode potential in absence and presence of hydrazine; also the change in total applied emf with time when hydrazine was present. (Adapted from J. J. Lingane, *Anal. Chim. Acta,* 2, 589 (1948) and J. J. Lingane and S. Jones, *Anal. Chem.,* 23, 1804 (1951).)

After the copper is separated, the tin plus lead can be separated on the copper-clad electrode by controlling the potential of the cathode at another value which is slightly less negative than that required for the liberation of hydrogen gas. In the presence of tartrate ions and at pH 4.5–5.0, only lead ions will be deposited, because the tartrate complex of tin(IV) is sufficiently stable to prevent the codeposition of tin.[9]

Separations with controlled electrode potentials are very satisfactorily done with a mercury cathode. By including a silver or some other coulometer in series with the electrolysis cell, it is comparatively simple to perform a series of separations and analyses without replacing the mercury cathode between successive separations and, perhaps, without weighing any electrodes.

Halides may be determined by electrolyzing their solutions between a platinum cathode and a silver anode. By controlling the anode potential, the gain in weight of the anode is equal to the weight of the particular halide in the original solution.

The use of controlled electrode potentials is now quite prevalent in preparative organic chemistry. If a certain organic compound can undergo a series of reductions (or oxidations), each at a definite potential, it is then possible to reduce the starting material selectively and efficiently to some desired compound by controlling the potential of the cathode during the reduction. Because this procedure for organic compounds produces essentially only the desired product, it is much more economical than most reductions or oxidations performed chemically, where side reactions producing undesired products usually occur.

Constant-Voltage Electrolysis

If we consider all the terms embodied in Eq. 23-7 for the separation of copper and lead ions at the cathode, with oxygen liberated at the anode, the copper is practically completely deposited at an applied voltage of 1.91 V:

$$E_{applied} = (1.23 + 0.40) - (0.21 - 0.01) + 0.50 \qquad (23\text{-}10)$$

where the ohmic drop is assumed to be 0.50 V. The lead will not begin to deposit until the applied voltage exceeds 2.28 V. Consequently, the separation of copper from lead can be accomplished by an electrolysis at any constant applied voltage between these two values.

This technique assumes that the ohmic drop and the overpotential terms in Eq. 23-7 remain constant. However, as a consequence of the cell reaction, the copper concentration diminishes and the hydrogen-ion concentration increases. This results in both E_{anod} and E_{cath} becoming more negative. From the Nernst equation the theoretical cathode potential will change from 0.31 to 0.19 V if the corresponding copper(II) concentration changes from 0.1 to 10^{-5} M. On the other hand, the anode potential will have decreased only slightly. The greater conductance of the hydrogen ion decreases the cell resistance and affects the iR term. Finally, the overpotential terms are dependent on the current density and cannot be expected to remain constant when the current tends to vary. These variations render it difficult to control the cathode potential within as close limits as is possible by the controlled potential method.

The electrolysis process can be represented by the intersection of the system surface in Fig. 23-1 with a slanted plane. When the current is zero, the cell emf is equal to the applied voltage. During the electrolysis, as the cathode potential becomes more negative, the copper(II) concentration becomes depleted to the point where the supply of ions at the cathode surface is insufficient to meet the current demand. When this occurs, the current must of necessity decrease, ultimately falling to zero. Thus, the deposition of lead does not occur, but this advantage is achieved at the cost of diminished current at any time. Hence a longer time is required for complete electrolysis in comparison with the controlled potential method.

Composition of the Electrolyte

It is clearly possible to separate one metal from another if the respective deposition potentials are sufficiently far apart and, in constant-current electrolysis, if one potential is more negative than that required for the evolution of hydrogen. When two metals have similar discharge potentials, sometimes the electrolyte composition can be altered sufficiently for separation to be possible. By addition of a masking agent which forms a complex with one of the metal ions, the discharge potential for this ion usually becomes more negative. Take, for example, an alkaline solution of copper and bismuth ions to which cyanide ions are added. Copper(II) ions are reduced to the monovalent state and form the tricyanocuprate(I) ion:

$$Cu^+ + 3CN^- \rightleftharpoons Cu(CN)_3^{2-}; \quad K_f = 10^{27} \tag{23-11}$$

In this environment the discharge potential of the copper complex is -1.05 V when $1\,M$ cyanide ion is present. By contrast, the addition of cyanide hardly affects the deposition potential for bismuth, and if tartrate ions are present to keep the bismuth in solution, quantitative separation from copper is possible.

The temperature dependence of a series of homologous metal complexes sometimes provides a means for separating discharge potentials. For example, in the case of nickel and zinc in ammoniacal solution, the deposition potentials are similar at $20°C$, being -0.90 and -1.14 V, respectively, but differ markedly at $90°C$, now being -0.60 and -1.05 V. In general the less dissociated the complex at room temperature, the greater the change in the degree of dissociation as the temperature rises.

Potential Buffers

Suitable oxidation–reduction systems that are preferentially reduced at the cathode, or oxidized at the anode, may be employed to limit and maintain a constant potential at the electrode. The name *potential buffer* is applied to this type of electrolyte because of its functional resemblance to pH buffers. For example, the uranium(III)/(IV) system has been used to prevent the cathode potential from exceeding approximately -0.5 V, the reduction potential of the uranium system. As the cathode potential attains this value, there is increased competition of the uranium(IV) reduction reaction with the deposition process. The reduced uranium(III) ions are reoxidized at the anode and remain in the electrolyte to continue the cyclic process. Eventually the entire current flow through the electrolytic cell is consumed in the reduction of uranium(IV) at the cathode and the oxidation of uranium(III) at the anode. One application of the uranium couple has been to prevent the deposition of chromium and manganese at a mercury cathode while permitting copper, tin, lead, and nickel to deposit in the normal manner.

Nitrate ions have long been employed in the constant-current deposition of copper and lead dioxide to prevent the formation of metallic lead at the cathode. Nitrate ions are less easily reduced than copper ions, and sufficiently so that the copper can be quantitatively removed; yet nitrate ions are more easily reduced than lead ions. At the anode, on the other hand, lead(II) ions are oxidized to lead dioxide more easily than hydroxyl ions to oxygen, thus,

$$Pb^{2+} + 2H_2O \rightarrow PbO_2 + 4H^+ + 2e^- \tag{23-12}$$

for which the electrode potential is 1.46 V, whereas the liberation of oxygen occurs at about 1.70 V. Chloride ions must be absent for several reasons: Chloride ions are oxidized more easily (E° = 1.36 V), the overpotential of chlorine gas is less than that for the deposition of PbO_2, and formation of $PbCl_3^-$ is appreciable.

The successful deposition of copper from chlorocuprate(I) ions, and the prevention of the competing oxidation of these ions to copper(II) at the anode, are due to the buffering action of hydrazine. The oxidation of hydrazine,

$$N_2H_5^+ \rightarrow N_2 + 5H^+ + 4e^- \tag{23-13}$$

for which the electrode potential is 0.17 V, takes place in preference to chlorocuprate(I) ions,

$$CuCl_3^{2-} \rightarrow Cu^{2+} + 3Cl^- + e^- \tag{23-14}$$

for which the electrode potential is 0.51 V. As long as hydrazine is present in excess, the anodic oxidation of chloride ions and copper(I) will be prevented. Without any hydrazine, the anode potential remains at a relatively high value between the chlorocuprate(I)–copper and the water–oxygen couples, as shown in Fig. 23-5.

Physical Characteristics of Metal Deposits

It is important to conduct the electrolysis under conditions which ensure that the deposit is pure, adherent to the electrode, and quantitative. This requires consideration of the factors that influence the nature of the deposit: current density, the chemical nature of the ion in solution, that is, complexed or as an aquated ion, the rate of stirring, the temperature, and the presence of depolarizers to minimize the evolution of gases.

The optimum conditions for achieving the best deposit vary from one metal to another. Adherence to the electrode is the most important physical characteristic of a deposit. Generally a smooth deposit and adherence are congruent. Flaky, spongy, or powdery deposits adhere only loosely to an electrode. Simultaneous evolution of a gas is often detrimental. Continual evolution of bubbles on the electrode surface disturbs the orderly growth of the crystal structure of a metal deposit, and porous and spongy deposits may be obtained. Once under way, these continue because the current density is high at points on an otherwise smooth surface, and this is conducive to irregular, tree-like growths. The discharge of hydrogen frequently causes the film of solution in the vicinity of the cathode to become alkaline, with the consequent formation of hydrous oxides or basic salts.

The chemical nature of the ion in solution often has an important influence on the physical form of the deposited metal. For example, a pure, bright, and adherent deposit of copper can be obtained by electrolyzing a nitric acid solution of Cu^{2+} ions. By contrast, a coarse, tree-like deposit of silver is obtained under similar conditions. If a suitable deposit of silver is to be obtained, the electrolysis must be carried out from a solution in which the silver ions are complexed as $Ag(CN)_2^-$. Similarly, the best deposits of iron are obtained from an oxalate complex and those of nickel from an ammonia complex. Halide ions facilitate the deposition of some metals, probably because the overpotential is lower for metal halide ions than for aquated metal ions. Complex ions also exhibit what is known as "throwing power" to a considerable degree—that is, the property of a solution by virtue of which a relatively uniform deposit of metal may be obtained on irregular surfaces.

The time required for electrodeposition is shortened if the electrode is rotated or the electrolyte stirred vigorously. As an adjunct to the normal diffusion process and mass transfer conditions, it lowers the concentration overpotential and enables a higher current density to be employed without deleterious results.

Increase in temperature has two effects which oppose each other. On one hand, diffusion is favored. On the other hand, hydrogen overpotential is decreased and the stability of many complex ions is decreased.

Factors Governing Current

When the electrode potential is controlled so that only a single reaction can occur, the decrease of the current is a reliable criterion of the progress of the electrolysis. The current–time curve usually obeys the relation

$$i_t = i_0 e^{-kt} \quad \text{or} \quad 2.3 \log\left(\frac{i_0}{i_t}\right) = kt \tag{23-15}$$

where i_0 is the initial current and i_t the current at time t. If $t = t_{1/2}$ when $i/i_0 = \frac{1}{2}$, then

$$t_{1/2} = \frac{2.3}{k} \log 2 = \frac{0.69}{k} \tag{23-16}$$

Deposition is quantitative after 10 half-lives, or when $i = 0.001\, i_0$.

A plot of Eq. 23-15 on semilog coordinates produces a straight line, from which the value of k can be evaluated. Lingane[7] has shown that the constant k is given by

$$k \text{ (in min}^{-1}) = \frac{0.43\, DA}{\delta V} \tag{23-17}$$

where A is the electrode area in centimeters squared, V is the electrolyte volume in milliliters, D is the rate of diffusion of the reacting ion expressed in square centimeters divided by minutes, and δ is the thickness of the diffusion layer surrounding the electrode expressed in centimeters. Equation 23-17 provides a logical basis for the selection of experimental conditions to achieve rapid electrolysis. The electrode area should be as large as possible and the solution volume as small as possible. This explains the advantage of tall-form beakers and large, cylindrical, gauze electrodes. Stirring and elevated temperatures reduce the thickness of the diffusion layer, and an increase in temperature also increases the diffusion coefficient. The current at concentration C is given by

$$i = \frac{nFACD}{\delta} \tag{23-18}$$

THE MERCURY CATHODE

Cathodes comprising a pool of mercury or an amalgamated platinum or brass gauze electrode warrant special consideration. The mercury cathode is not generally used to determine any of the metals plated out because of the difficulties involved in weighing and drying the mercury before and after the determination. However, it is one of the most useful aids for the removal of certain base metals, even in considerable quantities, that interfere in the determination of elements high in the electromotive series. Two factors set mercury apart from other electrode materials. Many of the metals depositing on mercury can form an alloy (amalgam) with the mercury. Owing to the alloy formation the deposition potentials of these metals on mercury are displaced from their normal value in the positive direction with respect to reduction potentials. Their deposition is also aided by the fact that the hydrogen overpotential on mercury is particularly large. As a result the deposition from a fairly acid solution is possible for such metals as iron, nickel, chromium, zinc, and even manganese under certain conditions.

In its simplest form the mercury cathode cell consists of a shallow pool of mercury covering the bottom of a beaker. Electrical contact to the mercury is made by a glass-enclosed platinum wire, either immersed in the mercury pool or sealed through the base of the container. The cell designed by Melaven,[13] shown in Fig. 23-6, is a slightly more refined form and is in common use. The cathode consists of 35 to 50 ml of pure mercury in a modified separatory funnel. The apparatus has a conical base fitted with a three-way stopcock. One arm of the stopcock is connected to a leveling bulb that controls the level of the mercury in the cell; the other permits removal of the electrolyte. With a beaker, this removal is accomplished by siphoning. The anode is a platinum wire in the form of a spiral. Agitation is accomplished by a mechanical stirrer or a stream of air.

Simple, unitized cells have been designed.[11] A sturdy, compact, self-contained immersion electrode is shown in Fig. 23-7. This cell is a glass dish about 30 mm in diameter by 15 mm high, from the side of which extends a glass tube carrying the wire for electrical contact. A flat, spiral anode completes the cell. The unitized electrode is easily removed from the electrolyte and washed with a stream of wash solution quickly enough to prevent appreciable dissolution of the deposited metals. The consumption of mercury is a mini-

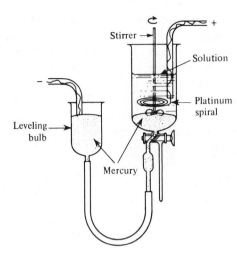

Fig. 23-6. Mercury cathode cell. (After A. D. Melaven, *Ind. Eng. Chem., Anal. Ed.*, **2**, 180 (1930).)

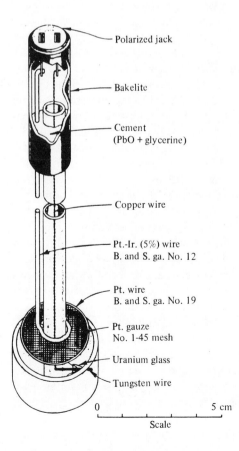

Fig. 23-7. Unitized mercury cathode cell. (After H. O. Johnson, J. R. Weaver, and L. Lykken, *Anal. Chem.*, **19**, 481 (1947).)

mum, usually 5 ml per electrolysis, and the simplicity with which duplicate assemblies are interchanged encourages frequent substitution of fresh mercury. This increases the efficiency of a separation and decreases the time for electrolysis. Finally, the difficulty from loss of mercury in handling and from dispersion during electrolysis is minimized.

Specially designed mercury cathode equipment patterned after the preceding designs is available commercially and can be used in conjunction with ordinary electrolysis apparatus.

Vigorous stirring materially shortens the electrolysis. Agitation can be accomplished by any type of mechanical stirrer or by a stream of air. Rapid countercurrent stirring of the mercury and the electrolyte at the deposition interface favors a more efficient deposition and constantly exposes fresh mercury to the electrolyte. This can be provided by a magnetic stirring bar floating on the mercury surface, or by letting the impeller blades be only partially immersed in the mercury. A commercial unit has been devised in which a magnetic circuit provides the stirring, the electrolyte and the mercury becoming the two independent rotors of a dc motor.[1] In addition, the magnetic field immediately removes deposited ferromagnetic materials from the mercury–solution interface and retains them beneath the surface of the mercury. The instrument is pictured in Fig. 23-8.

The electrolyte is usually a 0.1–0.5 M solution of sulfuric acid or perchloric acid. Nitric and hydrochloric acids are avoided; the reduction of nitrate lowers the current efficiency for the reaction of interest, and the anode may be attacked in the presence of chloride.

A current density of 0.1–0.2 A cm^{-2} is common, but substantially higher current densities have been used in cells with appropriate cooling devices to remove heat developed by the resistance of the electrolyte, as is done in the Dyna-Cath (Fig. 23-8). The amount of metal removed is proportional to the current and the area of the mercury surface.

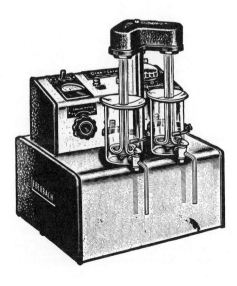

Fig. 23-8. The Dyna-Cath, a commercial mercury cathode instrument. (Courtesy of Eberbach & Son.)

Table 23-3 Electrolysis with a Mercury Cathode in 0.3 N Sulfuric Acid Solution

H																	He
Li	Be											B	C	N	O	F	Ne
Na	Mg											Al	Si	P	S	Cl	Ar
K	Ca	Sc	Ti	V	Cr	Mn	Fe	Co	Ni	Cu	Zn	Ga	Ge	As	Se	Br	Kr
Rb	Sr	Y	Zr	Nb	Mo	Tc	Ru	Rh	Pd	Ag	Cd	In	Sn	Sb	Te	I	Xe
Cs	Ba	La	Hf	Ta	W	Re	Os	Ir	Pt	Au	Hg	Tl	Pb	Bi	Po	At	Rn
Fr	Ra	Ac															
			Ce	Pr	Nd	Pm	Sm	Eu	Gd	Tb	Dy	Ho	Er	Tm	Yb	Lu	
			Th	Pa	U	Np	Pu	Am	Cm	Bk	Cf	Es	Fm	Md	No	Lw	

SOURCE: G. E. F. Lundell and J. I. Hoffman, *Outlines of Methods of Chemical Analysis*, John Wiley & Sons, Inc., New York, 1938, p. 94.

NOTE: On the periodic chart of the atoms the theoretical separation possibilities have been indicated for the mercury cathode. The elements enclosed by solid line (———) are quantitatively deposited in the mercury. Those surrounded by a dotted line (·····) are quantitatively separated from the electrolyte but not quantitatively deposited in the mercury. Elements enclosed by a wavy line (∿∿∿) are incompletely separated from the electrolyte.

In most cases the constant-current technique serves excellently as the separation method for the elements shown in Table 23-3. Most uses of the mercury cathode concern the removal of an interfering element or elements before the determination of a substance that remains in the electrolyte. In this respect it has been extensively applied to facilitate the determination of aluminum, titanium, vanadium, and magnesium in a wide variety of materials. The element most commonly deposited is iron. The mercury cathode has also been used to effect the reduction of an element or compound to a lower oxidation state in solution. Other applications are enumerated in review articles.[12,14]

INTERNAL ELECTROLYSIS

Internal electrolysis is the term applied by Sand[16] to electrogravimetric analyses which employ an attackable anode. The latter is connected directly to the cathode. In reality the arrangement is nothing but a short-circuited galvanic cell. It is convenient in some applications because the electrolysis proceeds spontaneously without the application of an external voltage and the choice of an attackable anode limits the cathode potential without elaborate instrumentation or the operator's attention. However, the driving force—that is, the difference between the potential of the system plating at the cathode and the dissolution of the anode—is small, and in consequence, the cell resistance is a critical factor in determining the rate of metal deposition. The application of the method is restricted to small amounts of material if the time of electrolysis is not to be excessively long.

The selection of an anode is made with a knowledge of the reversible potentials of the various metal ion–metal couples. A typical application is the removal of small amounts of copper and bismuth from pig lead.[2] Because the reduction potential of lead is sufficiently far apart from the reduction potentials of the copper and bismuth systems, the anodes can be constructed from helices of pure lead wire. The arrangement of equipment is shown in Fig. 23-9. Dual anodes are often used to provide a larger electrode area. These are inserted within a porous membrane (Alundum shell) in order to isolate them from the sample and forestall any direct plating on the lead itself. A platinum gauze electrode is placed between the anode compartments. The electrolysis is begun by short-circuiting the cathode to the anode.

To keep the ohmic resistance small, the anode solution must have a high concentration of electrolyte. In addition it must contain a higher concentration of the ions formed from the dissolution of the anode (i.e., lead ions in the example) than does the catholyte containing the dissolved sample. The anode reaction will be the dissolution of the lead; the cathode reaction the deposition of copper. The cell can be represented as

$$-Pb^\circ \mid Pb^{2+} \parallel Cu^{2+} \mid Cu^\circ(Pt) +$$

Because the cell operates spontaneously, the cathode is the positive electrode. As the cathode and anode are shortcircuited, the only dissipation of energy is in the form of the

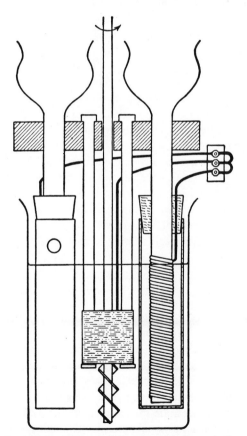

Fig. 23-9. Apparatus for internal electrolysis.

ohmic resistance, which in turn limits the maximum current flow through the cell:

$$iR = -E_{Pb^{2+},Pb^o} + E_{Cu^{2+},Cu^o} + \frac{0.0591}{2} \log \frac{[Cu^{2+}]}{[Pb^{2+}]} \qquad (23\text{-}19)$$

For an anolyte solution that is 1 M in Pb^{2+} ion, and inserting standard electrode potentials, Eq. 23-19 becomes

$$iR = 0.22 + 0.0296 \log [Cu^{2+}] \qquad (23\text{-}20)$$

With this arrangement the electrode potential of the cathode cannot exceed -0.12 V. Only those metal ions will deposit whose electrode potentials are more positive than this value. In the example taken, as the electrolysis progresses, the concentration of copper(II) ions diminishes and the electrode potential of the cathode becomes more negative until it becomes equal to the anode potential (or the decomposition potential of another substance is exceeded). At no time will the decomposition potential of lead at the cathode be exceeded. There is no danger of lead contamination due to the concentration-overpotential factor because the rate of cathodic deposition is controlled by the rate of anodic dissolution.

The anode need not always be constructed of the material that constitutes the matrix of the sample. For selective reduction of several trace constituents in zinc, for example, four separate samples would be dissolved for the separation of traces of silver, copper, lead, and cadmium. In the first, an attackable anode of copper would permit the complete removal of silver but control the cathode potential below the deposition potentials of the others. Similarly, a lead anode would make it possible to remove silver plus copper; a cadmium electrode would remove silver, copper, and lead; and with a zinc anode, all four elements would be removed.

The amount of deposit is generally limited to quantities not exceeding 25 mg. Although larger quantities have been handled, the deposit is apt to be spongy and some of the metal ions may diffuse to the anode during the longer time required for complete electrolysis. Little attention is required during an analysis except to flush the anolyte compartments once or twice. Halide solutions may be employed without removing the halide ion and without adding an anodic depolarizer. Average running time is 30 min per sample.

ELECTROGRAPHY

The electrographic method,[6] developed by Glazunov[5] and Fritz,[4] is a useful microanalytical tool for accurately identifying and determining substances. This method consists in anodically dissolving a minute amount of the test substance onto a piece of bibulous paper or, for more accurate rendition, gelatin-coated paper which has been soaked in a suitable electrolyte. The test sheet is held under pressure between the sample surface, the anode, and a suitable cathode surface. The latter may be a flat square electrode for flat surfaces, a long narrow electrode for use on metal ribbons, or sponge rubber covered with aluminum foil for uneven sample surfaces. The unit is connected to a battery of dry cells and a current is allowed to flow for several seconds. A general laboratory circuit

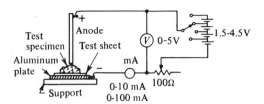

Fig. 23-10. Schematic arrangement of equipment and electrical circuit for electrographic analysis.

is shown in Fig. 23-10. While the current is flowing, ions leave the surface of the specimen and migrate into the permeable test sheet. Their presence can be made manifest, if they are colorless, by treating the test sheet with selective reagents. Distinctive identifying colors result and appear in an exact chemical and physical image of the surface.

The magnitude of the current required and its duration can be approximated from the second law of Faraday

$$it = \frac{96,487\ Adn}{W} \tag{23-21}$$

where i = current in amperes

t = duration of the current in seconds

A = area of the specimen surface in centimeters squared

W = atomic weight of the element dissolving and forming n equivalents per gram-atom

d = the minimum weight of material needed for detection by the method chosen in grams per square centimeter.

In general, 50 μg of most metals will produce brilliantly colored products when the reaction is confined to an area of 1 cm^2. These conditions would require a current of 15 mA and an exposure time of 10 sec. For multicomponent samples the current may be carried almost entirely by ions of highest mobility, with the result that the pattern will be under-exposed with respect to the poorly conducting constituents. This difficulty may be partially remedied by moistening the specimen with a mineral acid instead of a neutral electrolyte and by eluting the interfering ions prior to spot testing.

The test sheet may be moistened with only a neutral electrolyte, such as sodium nitrate or sodium chloride, or it may be impregnated with a reagent for the metal or metals to be detected, such as potassium ferricyanide for iron and ammonium sulfide for copper and silver. With a neutral electrolyte the print must be further developed by immersion in a developing reagent that forms a reaction product of distinctive color. Individual patterns can be secured by developing successive prints with different selective reagents, for example, α-benzoinoxime for copper, dimethylglyoxime for nickel, and α-nitroso-β-naphthol for cobalt from a sulfide mineral surface which has been electrographed with an ammonia solution. The test sheet should be fine-grained and held in close contact with the surface to be tested. By this method, in contrast with contact printing, lateral diffusion or "bleeding" is minimized. Prints are sharp and permit many fine features to be detected.

The electrographic method is applicable only to materials that are conductors of the electric current. It can be applied for the inspection of lacquer coating and of plated metals for pinholes and cracks in their surface. It can be used for many alloy identifications, such as the differentiation of lead-containing brass from ordinary brass, nickel in steel, and the distribution of metal constituents within an alloy. In the biological field the method is applicable to the localization of those constituents which are normally present within the tissue in an ionic state. One important advantage of this method, besides those of simplicity and rapidity, is the fact that so little of the sample is consumed; the sample remains essentially unaltered. Portable field kits have found extensive use in inspection and sorting work, in the laboratory as well as in the stockroom and in mineralogical field work.

Analogous to the anodic oxidation transfer is the cathodic reduction of certain anions of tarnish or corrosion films on metals. These are often tied up as basic insoluble salts and are not detectable in simple contact printing. Electrolytic reduction will free these ions.

By controlling the time, pressure, and current in a series of transfers, quantitative determinations can be made by comparing the color intensity of the pattern of an unknown sample with a series of patterns produced by known amounts of the metallic ion.

Electrolytic Purification

Electrolysis is of great practical value in the commercial refining of certain metals. Purification of copper by electrolytic refining is a large-scale operation. The relatively impure copper metal (about 99% copper) is made the anode in a sulfuric acid bath and the copper is electrolytically dissolved, forming copper sulfate solution. The copper ions are redeposited at the cathode to form the purified copper metal (about 99.99%). Total impurities are decreased by a factor of 100. The high efficiency of the method depends on the fact that most of the impurities present at this particular stage in the refining of metallic copper are not soluble in the electrolyte.

In the refining of gold, impure gold is made the anode in a hydrochloric acid bath, and the purified gold is electroplated out at the cathode. The small amounts of silver impurity will form insoluble silver chloride. However, when too much silver is present, as in the reclaiming of jewelry gold, a nitric acid bath is used. The gold falls to the bottom of the bath as purified metal powder, and the silver is deposited at the cathode.

LABORATORY WORK

General Instructions

Platinum electrodes are expensive and, if of gauze construction, somewhat delicate and must be handled carefully. They are cleaned by heating in 5 M nitric acid, to which a reductant (such as a little hydrogen peroxide) is added when dealing with a deposit of lead dioxide. Before weighing, the electrode is passed through the oxidizing flame of a burner to remove any grease picked up from hands.

At the end of an electrolysis, lower the solution away from the electrodes without interrupting the current, and wash the electrode surface as it becomes exposed with a stream of wash water. Remove the electrode containing the deposit from its holder and rinse with ethanol or acetone, then dry for several minutes in an oven at 80°C. Weigh the electrode plus deposit.

The electrolyte is generally composed of nitrates or sulfates because halides in acid solution give active halogens which attack the anode unless a proper depolarizer is present, for example, hydrazine in chloride solutions. Tin, silver, zinc, and bismuth damage the platinum and, in such separations, the cathode is copper-clad before depositing these metals.

Experiment 23-1   Controlled Cathode Separations

A. Determination of Copper in Brass[3,8]

Transfer a 0.5-g sample of brass (weighed accurately) to a 200-ml tall-form beaker. Dissolve the brass in 10 ml of concentrated HCl plus 5 ml of water to which concentrated nitric acid is added dropwise until dissolution is complete. In an alternative method, add 10 ml of 1:1 HCl and, in small portions, 5 ml of 30% hydrogen peroxide. Boil the solution to expel the oxides of nitrogen and chlorine (or excess hydrogen peroxide).

Dilute the solution to 25 ml, and add 4 g of hydrazine hydrochloride. Heat the solution to 95°C and maintain it at this temperature until the dark green color changes to a light olive-green, indicating considerable reduction to the chlorocuprate(I) ion.

Place the beaker in position in the electrolysis apparatus, and add water until the electrodes are covered completely. Place the tip of the salt bridge from the reference electrode on the outside and near the middle of the cathode.

Turn on the electrolysis current and adjust the applied emf to maintain the cathode potential at –0.35 V vs. SCE. The initial current should range from 2 to 4 A. Copper may not deposit for several minutes or until all the copper(II) ions have been reduced to chlorocuprate(I) and, indeed, the reference electrode may at first be negative to the cathode. Copper will commence to plate out when the cathode potential is about –0.2 V. As the deposition of copper proceeds, continuously adjust the applied emf to maintain the cathode potential at the limiting value. Continue the electrolysis until the current decreases to 10–20 mA.

B. Determination of Tin Plus Lead in Brass

To the solution from which copper has been removed, add 1 g of hydrazine hydrochloride, and insert a copper-clad electrode (weighed accurately). Adjust the applied emf to –0.60 V vs. SCE initially, but after the electrode becomes coated with a deposit of tin and lead, raise the applied emf to –0.70 V. Continue the electrolysis until the current decreases to a constant value and remains steady for 10 min. Before removing the electrodes, carefully neutralize the solution to pH 5–6 with aqueous ammonia.

C. Successive Determination of Copper, Bismuth, Lead, and Tin in Brass[9]

This separation is conducted in a tartrate solution. Dissolve the sample as described in part A. Add 100 ml of 0.1 M sodium tartrate solution and sufficient 5 M sodium hydroxide solution to adjust the pH at 5.0. Add 4 g of hydrazine hydrochloride and recheck the pH. Warm the solution to 70°C and proceed with the electrolysis.

Copper is deposited at −0.30 V vs. SCE, bismuth at −0.40 V (or copper plus bismuth at −0.40 V), and lead at −0.55 V initially and raising to −0.60 V after the electrode becomes coated with lead. The solution remaining from the lead deposition is acidified to destroy the tartrate complex of tin(IV), and tin is determined at −0.60 V initially and raising to −0.65 V after the electrode becomes coated with tin.

The amount of lead must not exceed 100 mg per 200 ml at a pH of 5 because larger amounts precipitate as the tartrate.

Experiment 23-2   Separation of Copper by Internal Electrolysis

Assemble the apparatus as shown in Fig. 23-9. For anodes, wind pure lead wire, 10 to 12 gauge, around the stem of a thistle tube which has been shortened to 6 in. in length, to form a compact helix. Leave a sufficient length of wire to connect to the common binding post. Each anode compartment is a porous alundum shell (extraction thimble obtained from the Norton Company, RA 84 or 360, 19 × 90 mm). The solution is stirred by a glass corkscrew stirrer or with a magnetic stirrer and bar.

Transfer aliquots containing 10–30 mg of copper into 400-ml beakers. To each, add 3 ml of concentrated nitric acid and 3 ml of concentrated sulfuric acid. Dilute to 200 ml. Warm the solution to 70°C.

Fill the anode compartments with a solution composed of nitric acid (3% v/v) and lead nitrate (5% w/v), and insert the anode compartments and the platinum cathode into the sample. Add 0.3 g of urea and short-circuit the electrodes.

Electrolyze for 15 min and then flush out the anode compartments with lead nitrate–nitric acid solution. Continue the electrolysis until the copper is completely deposited, as indicated by failure to plate on a fresh surface when the solution level is raised. Lower the beaker, and rinse the electrode with a stream of distilled water. Remove the cathode, dry, and weigh.

Experiment 23-3   Removal of Metals at the Mercury Cathode

Transfer aliquots containing approximately 0.5 g of iron (or copper, nickel, or other heavy-metal ion) to a 180-ml tall-form beaker. Dilute to about 80 ml. Add 1 ml of concentrated sulfuric acid.

Fill the cathode compartment with fresh mercury and insert the unitized assembly into the beaker. Adjust the current to 5 A. After 15 min raise the unitized electrode assembly and rinse with a stream of distilled water. Replace the mercury with fresh mercury and electrolyze for another interval. Repeat the cycle at 15-min intervals the first 3 times and then at 30-min intervals until the removal of the metal ions is considered complete.

Notes An auxiliary motor stirrer rotating slowly at the level of the mercury surface, or a magnetic stirrer bar floating on the surface, will speed the rate of deposition.

It is not necessary to change the mercury when using the Melaven cell.

The instructor may wish to have the student follow the removal of the metal ions by means of spot tests or other methods.

Experiment 23-4   Electrographic Spot Testing

A schematic diagram of an electrograph is given in Fig. 23-10. The sample is made the anode, in contact with a sheet of hardened filter paper, such as Schleicher & Schull Nos. 575 or 576 or Whatman No. 50, and backed by a thick, soft, backing paper such as S & S No. 601 or blotting paper.

General directions Moisten the pad of the printing medium and backing paper with electrolyte, blot lightly, and place on the aluminum base plate with the printing surface upward. Place the specimen on the paper and make contact by bringing the other electrode onto the specimen by hand pressure or by a clamp. Close the electrical circuit for the length of time calculated by Eq. 23-21. Generally an exposure of 10 sec with a current of 15 mA/cm^2 of surface area is sufficient.

Examination of a print of pure metal Moisten the pad with an electrolyte consisting of 0.5 M sodium carbonate solution plus 0.1 M sodium chloride solution. After printing, remove the upper sheet and cut it into 4 parts. Hold one part over a beaker of warm, concentrated aqueous ammonia, hold the second part over a beaker of warm, concentrated hydrochloric acid, and hold the third part over a warm surface until dry. Compare the original color and the colors of the treated portions of the test sheet with those listed in Table 23-4.

Table 23-4 Colors of Transfer Products of Certain Metals

Metal	Electrolyte: 0.5 M Na$_2$CO$_3$ + 0.1 M NaCl	Fuming Over		Exposure to Heat and Light
		NH$_3$	HCl	
Copper	Greenish blue	Deep blue	Green-yellow	Green-blue
Silver	Colorless	—	—	Black
Iron	Brown	Brown	Orange-yellow	Brown
Nickel	Light green	Light violet	Green	Light green
Cobalt	Dirty brown	Brown	Blue	Deep blue
Molybdenum	Deep blue-violet	Gray	Gray	Gray
Chromium	Yellow	Yellow	Yellow	Yellow

SOURCE: H. W. Hermance and H. V. Wadlow, "Electro Spot Testing and Electrography," in *Am. Soc. Testing Materials, Spec. Tech. Pub.*, **98**, 25 (1950).

Use of color-producing reagents

1. Solder lugs may be lead- or tin-dipped. A test sheet impregnated with a 0.5 M ammonium molybdate solution yields a blue color with tin; a zinc sulfide test paper, subsequently treated with warm, yellow ammonium polysulfide, reveals leads as a black stain.

2. A specimen of steel may be examined for nickel and chromium. A test sheet moistened with a saturated solution of barium hydroxide plus a 1% alcoholic dimethylglyoxime solution yields yellow barium chromate and red nickel dimethylglyoxime when these metals are present.

3. A flat laboratory spatula, labeled stainless, may be tested for pinholes in the coating over the underlying iron. A test sheet impregnated with 0.5 M potassium ferricyanide solution yields blue dots where holes exist in the coating unless the chromium plate has an underlying layer of nickel.

PROBLEMS

1. At what value should the cathode potential be controlled if one desires to separate silver from a 0.005 M solution of Cu^{2+} ions? If the initial silver concentration is 0.05 M, how long should the deposition take, assuming $\delta = 2 \times 10^{-3}$ cm; $D = 7 \times 10^{-5}$ cm^2 sec^{-1}; $V = 200$ ml; and $A = 150$ cm^2?

2. Copper and nickel can be separated by a constant current procedure provided the pH is carefully controlled. Calculate the minimum pH necessary to initiate the deposition of nickel from a 0.005 M solution. Assume that the current is 1 A, area = 150 cm^2, and remember that the electrode will be covered with the copper deposit.

3. Under the conditions of Problem 2, how completely will the copper have been removed up to the point when hydrogen gas is initially liberated?

4. What is the initial cathode potential when cadmium deposits from a solution 0.01 M in cadmium?

5. What weights of each of the following would be deposited by 3378 coulombs? (a) $Cu°$ from Cu^{2+}, (b) PbO_2 from Pb^{2+}, (c) Cl^- as AgCl at a silver anode, (d) $Sn°$ from $SnCl_4$.

6. A solution is initially 0.01 M in silver ion and 0.5 M in copper(II) ions. (a) What cathode potential is needed theoretically for the complete deposition of silver? (b) What cathode potential may be required considering concentration polarization? (c) How much silver remains in the solution when the cathode potential has been brought to 0.45 V vs. NHE?

7. Under a given set of electrolysis conditions, 0.500 g of silver was deposited at the cathode and oxygen liberated simultaneously at the anode. Calculate the number of millimoles of hydrogen ion added to the solution. If the solution volume were 200 ml, what would be the change in pH, assuming initially that the solution was neutral and unbuffered?

8. In an electrolytic determination of bromide ion from 100 ml of solution, the silver anode, after electrolysis was completed, was found to have gained 0.8735 g. (a) Calculate the molarity of bromide in the original solution. (b) Calculate the potential of the silver electrode at the beginning of the electrolysis, assuming the solubility product of AgBr is 4×10^{-13}.

9. A solution which is $0.01\ M$ in zinc sulfate, and buffered at pH 4 with an acetate buffer, is to be electrolyzed using a copper-clad cathode. If the overpotential of hydrogen on copper is 0.75 V at the current density to be used, and that of oxygen on the platinum anode is 0.50 V, (a) calculate the decomposition potential of the solution, assuming that the iR drop is 0.5 V. (b) Will the decomposition potential change as the electrolysis proceeds? (c) How much zinc will remain in solution at the point when hydrogen gas begins to be liberated?

10. At a current density of $0.01\ \text{A cm}^{-2}$ the overpotential of hydrogen gas on cadmium is 0.4 V. Would it be possible to deposit cadmium quantitatively in a solution buffered at pH = 2?

11. The cathode potential is controlled 0.05 V less negative than the value at which tin would be deposited from a $0.005\ M$ solution. (a) Calculate the molarity of copper ions remaining in a sulfate solution. (b) Estimate the quantity of undeposited copper in a solution $1.0\ M$ in HCl and containing hydrazine hydrochloride.

12. If lead is used as the soluble anode in the internal electrolysis of a lead solution containing a small amount of copper, what would be the final concentration of copper in solution if the lead concentration is $0.2\ M$?

13. By means of suitable calculations, show why zinc can be successfully plated onto a copper-clad electrode from a solution buffered at pH = 10.5, whereas the deposition would not occur without concurrent evolution of hydrogen gas if smooth platinum electrodes were substituted. Assume a current density equivalent to the appearance of first visible gas bubbles.

14. Suggest an electrographic method for detecting the presence of copper filings on an ax blade suspected of being used to cut telephone cables. The method should be adaptable to courtroom demonstration before a jury of laymen.

15. Suggest a field method for the detection of fool's gold (FeS_2).

16. For a typical laboratory deposition of 0.200 g copper onto a platinum electrode (area $160\ \text{cm}^2$) from 200 ml of $0.5\ M$ tartrate solution adjusted to pH 4.5 and containing hydrazine, calculate the time required to reduce the copper concentration (a) to 1% of its original value and (b) to 0.1%. Starting at an initial value of 2.6 A the current decreased to 1.3 A after 2 min, to 0.65 A after 4 min, and to 0.33 A after 6 min.

17. In Problem 16, assuming that the diffusion coefficient of the tartrate complex is approximately $2 \times 10^{-5}\ \text{cm}^2\ \text{sec}^{-1}$, estimate the thickness of the diffusion layer about the electrode.

18. A solution is 0.1 M in cadmium ion and 0.01 M in hydrogen ion. (a) What is the difference between the two cathode potentials? (b) To this solution is added sufficient ammonia to convert the cadmium to $Cd(NH_3)_4^{2+}$ and to make the solution 0.57 M in free NH_3 and 0.10 M in NH_4^+. Assuming that the volume of the solution is unchanged, what is the difference between the two cathode potentials?

19. In an ammoniacal solution, 0.10 M in both NH_3 and NH_4^+, would the deposition of copper be complete if the cathode potential were limited at -0.40 V vs. SCE?

20. Outline a procedure for the successive determination of lead, cadmium, and zinc in metallurgical materials, such as flue dust from zinc refineries, composed chiefly of the oxides of these three metals.

21. During the determination of silver in silver solder, the $Cu(NH_3)_4^{2+}$ complex undergoes reduction to the $Cu(NH_3)_2^+$ complex. What difficulty is thereby introduced into the electrodeposition of silver from the silver ammine complex?

22. For the determination of cyanide ion, Baker and Morrison [*Anal. Chem.*, 27, 1306 (1955)] employed a silver and platinum electrode pair, connected together through a microammeter, immersed in 1 M NaOH solution to which the cyanide sample is added. (a) Write the cell representing the reaction and (b) the electrode reactions involved. (c) Why does a reaction, nevertheless, proceed spontaneously?

23. During the deposition of copper from a chloride medium and in the presence of hydrazine hydrochloride as anodic depolarizer, the following current readings were obtained for corresponding times:

3.00 A = 1.00 min	1.2 A = 7.0 min	0.075 A = 13.0 min
3.00 A = 2.00 min	0.8 A = 8.0 min	0.052 A = 14.0 min
2.85 A = 3.00 min	0.50 A = 9.0 min	0.036 A = 15.0 min
2.70 A = 4.00 min	0.30 A = 10.0 min	0.027 A = 16.0 min
2.2 A = 5.0 min	0.18 A = 11.0 min	0.020 A = 17.0 min
1.8 A = 6.0 min	0.12 A = 12.0 min	0.016 A = 18.0 min
		0.016 A = 20.0 min

(a) Graph the results on semilog paper. (b) From the descending slope of the graph, determine the value of the constant k (see Eq. 23-15). (c) Estimate the time required to reduce the copper concentration to 0.1% of its original value [after all the residual copper(II) ions are reduced to the chlorocuprate(I) ion].

BIBLIOGRAPHY

Bard, A. J., Ed., *Electroanalytical Chemistry—A Series of Advances*, Marcel Dekker, New York, Vol. 1 (1966), Vol. 2 (1967), Vol. 3 (1969), Vol. 4 (1970), Vol. 5 (1972).

Charlot, G., J. Badoz-Lambling, and B. Tremillon, *Electrochemical Reactions*, American Elsevier, New York, 1962.

Eberson, L. E. and N. L. Weinberg, "Electroorganic Synthesis," *Chem. Eng. News*, p. 40 (Jan. 25, 1971).

Lingane, J. J., *Electroanalytical Chemistry*, 2nd ed., Wiley-Interscience, New York, 1958.

Rechnitz, G. A., *Controlled Potential Analysis*, Pergamon, London, 1963.

Shanefield, D., "Electrolysis as a Purification Tool," *Ann. N.Y. Acad. Sci.*, **137**, 135 (1966).

Wawzonek, S., "Synthetic Electroorganic Chemistry," *Science*, **155**, 39 (1967).

LITERATURE CITED

1. Center, E. J., R. C. Overbeck, and D. L. Chase, *Anal. Chem.*, **23**, 1134 (1951).
2. Clarke, B. L., L. A. Wooten, and C. L. Luke, *Trans. Electrochem. Soc.*, **76**, 63 (1939).
3. Diehl, H., *Electrochemical Analysis with Graded Cathode Potential Control*, G. Frederick Smith Chemical Co., Columbus, Ohio, 1948.
4. Fritz, H., *Z. Anal. Chem.*, **78**, 418 (1929).
5. Glazunov, A., *Chim. Ind.*, Special Number, p. 425 (Feb. 1929).
6. Hermance, H. W. and H. V. Wadlow, "Electrography and Electrospot Testing," in *Standard Methods of Chemical Analysis*, 6th ed., F. J. Welcher, Ed., Vol. 3, Part A, pp. 500-520, D. Van Nostrand, New York, 1966.
7. Lingane, J. J., *Anal. Chim. Acta*, **2**, 591 (1948).
8. Lingane, J. J., *Ind. Eng. Chem., Anal. Ed.* **17**, 640 (1945).
9. Lingane, J. J. and S. Jones, *Anal. Chem.*, **23**, 1804 (1951).
10. P. F. Lott, *J. Chem. Educ.*, **42**, A261 (1965).
11. Johnson, H. O., J. R. Weaver, and L. Lykken, *Anal. Chem.* **19**, 481 (1950).
12. Maxwell, J. A. and R. P. Graham, *Chem. Rev.*, **46**, 471 (1950).
13. Melaven, A. D., *Ind. Eng. Chem., Anal. Ed.*, **2**, 180 (1930).
14. Page, J. A., J. A. Maxwell, and R. P. Graham, *Analyst*, **87**, 245 (1962).
15. Rogers, L. B. and A. F. Stehney, *J. Electrochem. Soc.*, **95**, 25 (1949).
16. Sand, H. J. S., *Analyst*, **55**, 309 (1930).

24
Coulometric Methods

Coulometric methods of analysis measure the quantity of electricity, that is, the number of coulombs, required to carry out a chemical reaction. Reactions may be carried out either directly by oxidation or reduction at the proper electrode (primary coulometric analysis), or indirectly by quantitative reaction in the solution with a primary reactant produced at one of the electrodes (secondary coulometric analysis). In any case, the fundamental requirement of coulometric analysis is that only one overall reaction must occur and that the electrode reaction used for the determination proceed with 100% current efficiency.

Coulometric methods eliminate the need for burets and balances, and the preparation, storage, and standardization of standard solutions. These methods can be automated readily. In a sense the electron becomes the primary standard. Coulometric methods can be used to produce reagents in solution that would otherwise be difficult to employ: volatile reactants such as chlorine, bromine, or iodine, or unstable reactants such as titanium(III), chromium(II), copper(I), or silver(II). The method is particularly useful and accurate in the range from milligram quantities down to microgram quantities and, therefore, in trace analyses. In addition, coulometric techniques are especially adaptable to remote control and operation and are presently very useful in the analyses of radioactive materials.

There are two general techniques used in coulometry. One, the *controlled-potential method*, maintains a constant electrode potential by continuously monitoring the potential of the working electrode as compared to a reference electrode. The current is adjusted continuously to maintain the desired potential. The other method, known as *constant-current coulometry*, maintains a constant current throughout the reaction period. In this method an excess of a redox buffer substance must be added so that the potential does not rise to a value that will cause some unwanted reaction to take place. Furthermore, the product of electrolysis of the redox buffer must react quantitatively with the substance to be determined; that is, it serves as an intermediate in the reaction. Examples of both types of coulometry will be discussed later in this chapter.

FUNDAMENTAL PRINCIPLES

If a reaction is 100% current efficient, the passage of one faraday of electricity, 96,487 coulombs, will cause the reaction of one equivalent weight of substance. The relationship between the weight in grams, W, the number of coulombs, Q, the molecular weight, M, the number of faradays involved in the reaction of one mole, n, and the value of the faraday in coulombs, F, is given by the relationship

$$W = \frac{QM}{Fn} \tag{24-1}$$

If the current, i, remains constant as in constant-current coulometry, then the total number of coulombs is given by the product of the current times the time; thus

$$Q = it \tag{24-2}$$

If, as in controlled-potential coulometry, the current changes continuously, then Q is given by the integration of time versus current as in Eq. 24-3.

$$Q = \int_0^\infty i \, dt \tag{24-3}$$

Most controlled-potential coulometric titrations are carried out under conditions in which the current is diffusion controlled and the relationship between current, i_t, at any time, t, and concentration, C_t, is given by the equation

$$i_t = \frac{nFADC_t}{\delta} \tag{24-4}$$

where A is the area of the electrode, D is the diffusion coefficient, and δ is the thickness of the Nernst diffusion layer. From Faraday's law, the rate of change of concentration with time is given by the relationship

$$\frac{dC_t}{dt} = -\frac{i_t}{nFV} \tag{24-5}$$

where V is the volume of the solution. Substitution of Eq. (24-4) into (24-5) and integration yields the following results for concentration as a function of initial concentration, C_0, and current as a function of initial current, i_0,

$$C_t = C_0 e^{-kt} \tag{24-6}$$

$$i_t = i_0 e^{-kt} \tag{24-7}$$

where $k = DA/V\delta$.

The number of coulombs, Q_t, passing up to time t is given by integration.

$$Q_t = \int_0^t i \, dt = \int_0^t i_0 e^{-kt} \, dt = \frac{i_0}{k} - \frac{i_t}{k} \tag{24-8}$$

Such a relationship is useful in estimating the total number of coulombs required for complete reaction before the reaction is actually completed. Q_t is read at several values of t, preferably in the range of 90 to 99% completion, and then these values of Q_t are

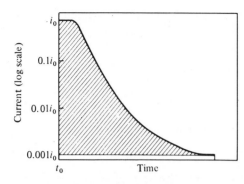

Fig. 24-1. Typical current–time relationship in controlled-potential coulometry. Shaded area indicates graphical method of integration.

plotted versus i_t. Q_∞ is determined by extrapolation of the straight line so obtained to the coulomb axis. The limiting value of Q is, obviously, i_0/k.

If the extrapolation method is not used, then the electrolysis is usually carried out until i_t has diminished to 0.1% or less of its original value or until the current becomes equal to the residual current as measured on a sample of supporting electrolyte alone.

The behavior of current versus time in an actual controlled-potential electrolysis is represented in Fig. 24-1. The initial horizontal portion of the curve arises because most coulometers have an upper limit to the current they can furnish and therefore may not be able to carry out the electrolysis at a rate sufficiently high to reach the limiting current set by Eq. 24-4 until the concentration, C, is reduced somewhat. The tail at the end of the curve is due to residual current in the supporting electrolyte.

Example 24-1

If a constant current of 10.00 mA passes through a chloride solution for 200 sec, what weight of chloride reacts with the silver anode?

Answer The net charge involved is

$$Q = it = (10 \times 10^{-3}\,\text{A})\,(200\,\text{sec}) = 2.00\ \text{coulombs}$$

or

$$\frac{2.00}{96{,}487} = 2.075 \times 10^{-5}\ \text{equivalents of chloride ion}$$

Since $n = 1$ for the reaction

$$\text{Ag}^\circ + \text{Cl}^- \rightarrow \text{AgCl} + e^-$$

2.075×10^{-5} equivalents equal 2.075×10^{-5} mole of chloride ion. In weight, the amount of chloride ion, now present as AgCl, is

$$(2.075 \times 10^{-5})\,(35.45) = 0.735 \times 10^{-3}\,\text{g (or 0.735 mg)}$$

INSTRUMENTS USED IN CONSTANT-CURRENT METHODS

Constant-current procedures require only a knowledge of the current and elapsed time to determine the number of coulombs. Since both current and time can be measured with high accuracy and with relatively simple equipment, this method of coulometry is both accurate and simple.

The schematic diagram of a coulometric setup for constant-current methods is illustrated in Fig. 24-2. The major problem is adequate stabilization of the constant-current supply in the range 1 to 200 mA. A true constant-current source as described below is, of course, desirable, but fairly constant current can be obtained from batteries with a series-regulating resistance. Either standard radio B batteries with voltages of 45 to 300 V or a line-operated, constant-voltage power supply with large series resistors (ballast resistor) can be used to maintain a constant current. If the voltage and the resistor are sufficiently large, changes in the resistance of the cell and the cell potential will have little effect on the current. This resistor is also varied to adjust the cell current to the desired level. Usually a current is selected that allows the electrolysis to be completed within 10 to 200 sec. To maintain the series resistance in thermal equilibrium and minimize adjustments of the cell current, it is advisable to employ a switching arrangement whereby the electrolytic cell is replaced by a dummy resistance (high-wattage type, approximately 20 Ω) during the intervals between analyses.

The current can be indicated approximately by a calibrated milliammeter, and measured precisely by means of the voltage drop across a precision resistor incorporated

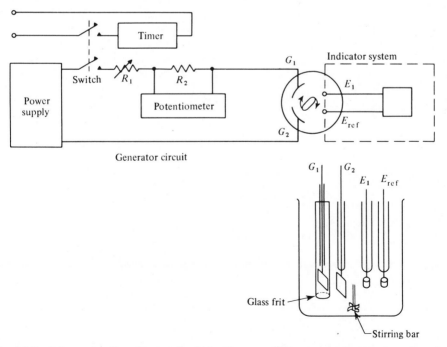

Fig. 24-2. Schematic of equipment (and titration vessel) for constant-current coulometry. R_1 is series (ballast) resistor; R_2 is precision resistor; G_1 and G_2 are generator electrodes (one isolated behind a porous frit barrier); E_1 and E_{ref} are electrodes for end-point detector system.

directly in series with the electrolytic cell. The voltage drop across the resistor can be measured very precisely with a manual or a recording potentiometer when the voltage drop is about 1 V. Under these conditions the error in the current measurement is about 0.002%.

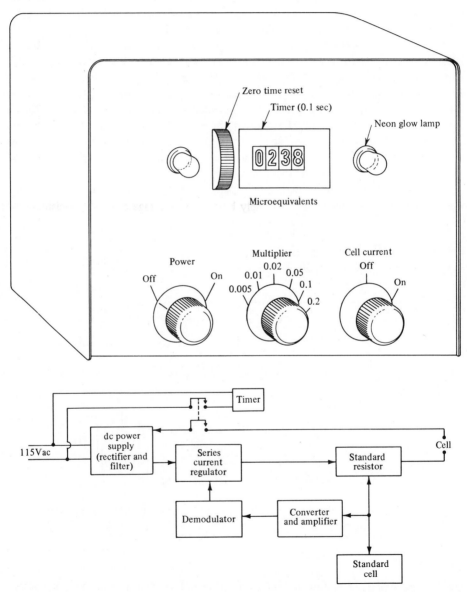

Fig. 24-3. Sargent coulometric current source block diagram (and front panel). (Courtesy of Sargent-Welch.)

Time measurements are normally made with a precision electric stopclock. A single switch control actuates both the timer and the electrolysis current. Times accurate to 0.01 sec are possible with modern electric chronometers.

One commercial source of equipment (Sargent-Welch) that maintains a constant current in any of several selected ranges is shown in Fig. 24-3. It comprises basically a power supply of the conventional ac rectifier and filter type to provide a dc voltage with a maximum of about 300 V, sufficient for work with high-resistance electrolytes. Current that is drawn from this power supply by connection to the cell electrode system passes through a series-regulating tube and a precision resistor, which is one of several selected by a current-selector switch. The size of this resistor is so chosen that, at the specified current level, an iR drop or voltage is developed which is equal to the potential of a standard cadmium cell. Any instantaneous error or difference resulting from a change in current due to line voltage or cell resistance variation is converted to an ac signal, amplified many times, and reconverted to a dc signal which is applied to the series regulator. Other commercial units are available from Fisher Scientific Co., Allied Electronics, Ltd., and A.E.I. (Woolwich), Ltd.

Generating electrodes must be of sufficient area to permit a low enough current density to keep electrode polarization within the limits necessary for 100% current efficiency. Currents normally employed require a substantial electrode surface area (10 cm^2 or larger), and often utilize a half-cylinder of sheet or gauze platinum. The nonworking electrode must be isolated in most cases by a salt bridge and frit barrier. The latter arrangement, however, increases the internal resistance of the cell and entails larger energy losses, so that a potentiostat may have difficulty in stabilizing the potential properly.

INSTRUMENTS USED IN CONTROLLED-POTENTIAL METHODS

In controlled-potential methods, the current is continuously changing and some sort of integrating device, a coulometer, is needed. In addition a potentiostat is necessary to control the potential of the working electrode at the desired value.

Chemical Coulometers

The electrolytic method for determining the current-time integral employs a standard chemical coulometer in series with the electrolysis cell. With electrogravimetric coulometers, the change in weight of one electrode is a measure of the charge transferred. The coulometer is simply a second electrolytic cell in which an electrochemical reaction is known to proceed with 100% current efficiency, as, for example, the deposition of silver or copper. A modern version of the electrogravimetric coulometer is the coulometric coulometer.[5,13] After completion of the coulometric step, the coulometer is included in another circuit, by means of which a perfectly constant current is passed through the coulometer in the opposite direction. In this way the electrodes are returned to their original condition. The time required for this, multiplied by the current, gives the num-

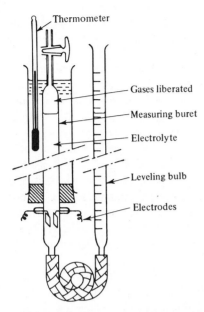

Thermometer

Gases liberated

Measuring buret

Electrolyte

Leveling bulb

Electrodes

Fig. 24-4. Gas coulometer. (After J. J. Lingane, *J. Am. Chem. Soc.*, **67**, 1916 (1945).)

ber of coulombs consumed in the actual determination. For example, if the reaction involved reduction of copper(II) to metallic copper, the deposit is redissolved anodically and the end point is indicated by a sharp change in electrode potential from the value for copper to that for the discharge of oxygen (see Electrolytic Stripping). This method is particularly suited for analyses on a microscale.

Lingane has described two gas coulometers. In one the total volume of hydrogen and oxygen liberated in the electrolysis of an aqueous solution of 0.5 M potassium sulfate is collected in a thermostatted gas-measuring buret.[8] Its lower limit of accuracy is 10 Q. For the range from 5 to 20 Q, 0.1 M hydrazine sulfate is used as electrolyte. Nitrogen is evolved at the anode and hydrogen at the cathode. The net coulometer reaction is

$$N_2H_5^+ = N_2 + 2H_2 + H^+ \tag{24-9}$$

The total volume of nitrogen and hydrogen evolved is measured in a 5-ml buret.[10] A typical arrangement of the gas coulometer is shown in Fig. 24-4. In using a gas coulometer, one must adjust liquid levels before, during, and after the electrolysis, but the coulometer thereby furnishes a semicontinuous indication of the progress of the reaction. After correction of the gas volume to standard conditions of pressure and temperature, 16,810 ml of gas corresponds to one faraday (96,487 Q) theoretically; the actual volume per coulomb is 0.1739 ml at standard conditions.

Electromechanical Coulometers

A simple electromechanical integrator is constructed by connecting the ends of a series resistor (in the current circuit) to one of the coil windings of a low-inertia integrating motor—essentially a dc motor built for 0.25-1.5 V operation in which friction and heat losses have been reduced to a minimum. The speed of shaft rotation is a linear function of the applied voltage—derived from the current flowing through the series resistor. Rotation of the armature shaft is followed by a mechanical counter.[1]

Electronic Coulometers

Electronic integration, although requiring complicated equipment, enables one to integrate even small charge transfers and is extremely accurate. The main limitation (0.01%) with electronic equipment arises from background signal. Voltage-to-frequency converters measure the voltage drop over a standard resistor and feed the output to a scaler, from which the current-time integral is obtained as a number of counts. For example, an input signal of 1 V may be converted to an output signal of 10,000 counts per second. The operational amplifier-capacitor integrator is similar to those used in analog computers. Response time to current changes is as fast as 10 μsec. Usable range extends from 10 μA to 10 mA or greater. A typical circuit is included in Fig. 24-5.

Potentiostats

In controlled-potential coulometry, four instrumental units are involved: a coulometer, a dc current supply, a potentiostat, and an electrolytic cell. The test material is reduced (or oxidized) directly at the working electrode, and the charge transfer during this process is integrated by a coulometer. In order that only the desired reaction may take place, the potential of the working electrode is controlled within 1-5 mV of the limiting electrode potential with the aid of a potentiostat. As the desired constituent reacts at the working electrode, the current decreases from a relatively large value at the beginning to essentially zero at the completion of the reaction (Fig. 24-1). Potentiostats have been mentioned in Chapter 23. An electronic potentiostat is included in the complete instrument described below.

Electronic Coulometric Titrator

In the apparatus of Kelley, Jones, and Fisher (marketed by Indiana Instruments and Chemical Corp.), the potential of the working electrode is controlled by a stabilized difference amplifier combined with a transistor current amplifier.[7] The electrolysis current is integrated by a stabilized amplifier and the integral is read out as a voltage. The block diagram, switched for reduction, is shown in Fig. 24-5. The command signal to the control amplifier is the algebraic sum of the control potential from the control potential source and the potential of the controlled electrode with respect to the solution, as seen through the reference electrode. The control potential is a selected fraction of the constant potential across a silicon voltage (Zener) diode. The current integrator is an analog computer circuit.

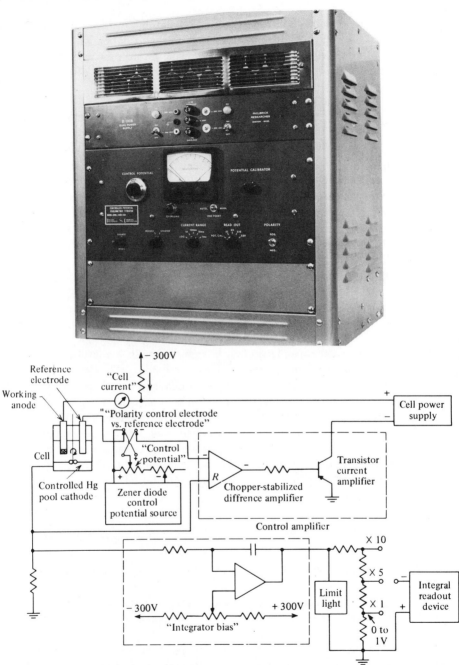

Fig. 24-5. *Upper*, electronic controlled-potential coulometric titrator. (Courtesy of Indiana Instrument and Chemical Corp.) *Lower*, block diagram, switched for reduction. (After M. T. Kelley, H. C. Jones, and D. J. Fisher, *Anal. Chem.*, **31**, 488 (1959). Courtesy of American Chemical Society.)

Titrations are made in an inert atmosphere provided by a nitrogen blanket. For controlled anode-potential oxidations, such as those of iodide or iron(II), platinum electrodes are used, and the cathode is isolated by a salt bridge and frit barrier. For controlled cathode-potential reductions, such as those of uranium(VI) or copper(II), the anode is a platinum wire that is isolated by a sulfuric acid salt bridge and frit barrier, and the cathode is a mercury pool. Thorough agitation of the mercury-solution interface is necessary to obtain a high initial current by providing adequate mixing of the solution to replenish ions depleted by electrolysis. A rotated platinum cell with fast sparging characteristics, low sample volumes (2 ml), and high electrolysis rates has also been developed.[2] A standard reference electrode, positioned as close to the working electrode as possible, monitors the potential of the working electrode. The titration is terminated when the current drops to a predetermined fraction of the initial current or to the residual current from the supporting electrolyte.

APPLICATIONS OF CONTROLLED-POTENTIAL METHODS

To apply controlled-potential coulometry, current-potential diagrams must be available for the oxidation-reduction system to be determined and also for any other system capable of reaction at the working electrode. As discussed in earlier chapters, in any sample system where two or more ions are capable of oxidation (or reduction) at a working electrode, that which requires the least free energy for transformation will determine the electrode process. For this to be compatible with the requirement of 100% current efficiency in generation, it is necessary to control the potential of the working (generating) electrode within specified limits.

These limits can be understood better from an example.[4]   Consider a mixture of antimony(V) and antimony(III) in a supporting electrolyte containing 6 M HCl plus 0.4 M tartaric acid (Fig. 24-6).[4] Plots of Q versus cathode potential show plateaus centered at -0.21 and -0.35 V versus SCE. First, one prereduces the supporting electrolyte at -0.35 V. Then the sample is introduced and the system deaerated. Finally the reduction is started, at -0.21 V for $Sb^{5+} \rightarrow Sb^{3+}$, followed at -0.35 V for $Sb^{3+} \rightarrow Sb^{\circ}$. Initially, electrolysis proceeds at a constant rate (initial current = i_0) until the potential of the working electrode reaches the limited value, in this case -0.21 V versus SCE. At this point the

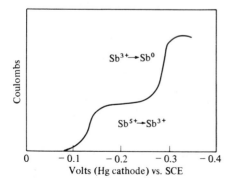

Fig. 24-6. Electrolytic reduction of antimony(V) by two-step process in 6 M HCl plus 0.4 M tartaric acid. (After L. B. Dunlap and W. D. Shults, *Anal. Chem.*, **34**, 499 (1962). Courtesy of American Chemical Society.)

potentiostat takes over, and the current through the cell gradually decreases until all antimony(V) has been reduced to antimony(III). This pattern is repeated at -0.35 V.

Current-potential diagrams (also denoted "coulograms") are obtained by plotting current against cathode-reference electrode potential (rather than cathode-anode potential which would include the large and variable iR drop in the cell). The necessary data can be obtained by setting the potentiostat to one cathode-reference electrode potential after another in sequence, allowing only enough time at each voltage setting for the current indicator to balance. Alternatively, the reduction (or oxidation) is performed in the usual manner except that periodically throughout the electrolysis the potential is adjusted to a value that causes cessation of current flow. The net charge transferred up to this point and the electrode potential are noted and the electrolysis then continued. Curves plotted from a series of points establish optimum electrode potentials because they relate extent of reaction with electrode potentials under actual titration conditions and electrode material.

By controlling the potential of the electrode at a suitable value, it is possible to reduce a metal completely to a lower valency state, and then, by controlling at a more positive potential, the metal can be oxidized quantitatively to a higher valency state on allowing the current to reach its background value. For example, at -0.15 V with a mercury electrode, reduction of uranium(IV to III) and chromium(III to II) occur simultaneously. If a pre-electrolysis is carried out at -0.55 V, only uranium(III) is oxidized. Although the reaction does not occur with 100% current efficiency, it is complete. When all uranium (III) has been removed from the solution, chromium is determined by oxidation to chromium(III) with a 100% yield at -0.15 V.

Indirect methods are possible. In the determination of plutonium in the presence of iron, the first step is the reduction of plutonium(VI) to plutonium(III) and partial reduction of iron(III) to iron(II) at a platinum electrode in a sulfuric acid electrolyte. When this is followed by oxidation of the mixture to plutonium(IV) and iron(III), the net reaction is the reduction of plutonium(VI) to plutonium(IV). Interference caused by the presence of uranium is thereby avoided.[12] Also, the reaction which occurs at a mercury electrode,

$$Hg^\circ + Y^{4-} = HgY^{2-} + 2\bar{e} \qquad (24\text{-}10)$$

may be used to follow a number of electrochemical reactions wherein a metal, M^{n+}, is not electroactive

$$M^{n+} + HgY^{2-} + 2\bar{e} = Hg^\circ + MY^{(n-4)+} \qquad (24\text{-}11)$$

An excess of HgY^{2-} is added to the solution of M^{n+} (Y^{4-} is the symbol for the anion of EDTA). The current is limited by the diffusion of M^{n+} to the electrode, and becomes zero at the end point.

Controlled-potential methods are finding ever-increasing numbers of applications, especially since the advent of the modern, electronic coulometric titrators. A recent example[6] is the determination of nitrite, which is difficult by ordinary chemical means. The nitrite is oxidized directly to nitrate at a platinum electrode in pH 4.7 acetate buffer solution. Errors of about 0.05% were recorded for the determination of 1 mg of NO_2^-.

Methods for a few organic substances, particularly halogen and nitro compounds, have also been developed. Controlled-potential methods can be used in preparative work to oxidize or reduce compounds at controlled potentials. They can also be used to determine the probable result of redox reactions because, if the weight of material and the coulombs required in reaction are known, then n, the number of electrons required per mole, can be computed.

Controlled-potential coulometry suffers from the disadvantages of requiring relatively long electrolysis times and expensive equipment, although it proceeds virtually unattended with automatic coulometers. However, direct indication of optimum conditions for successive reactions are easily obtained. No indicator electrode system is necessary, since the magnitude of the final current is sufficient indication of the degree of completion of the reaction. Although the concentration limits vary for each individual case, the upper limit is about 2 milliequivalents and the lower limit is about 0.05 microequivalent. The latter limit is largely set by the magnitude of the residual current and the many factors that affect it.

CONSTANT-CURRENT METHODS

In constant-current coulometric methods some external means must be used to determine the end point of the reaction because, at constant current, the potential will rise at the completion of each reaction to a value which will permit some other reaction to take place and thus maintain the current.

Detection of the End Point

Various methods are used for detection of the end point. It can be found by means of normal colored indicators, provided the indicator itself is not electroactive, or by instrumental methods—potentiometry, amperometry, and photometry. No correction for volume changes is necessary when plotting the results if internal generation of titrant is employed. Potentiometric and photometric indication find use in acid-base and redox titrations, while amperometric procedures are applicable to redox and precipitation reactions and, in particular, for these systems as the solutions become more dilute.

Electrolytic Stripping

The removal (stripping) of deposits has been used to measure the thickness of plated metals and of corrosion or tarnish films. In the case of oxide tarnish on the surface of metallic copper, the specimen is made the cathode and the copper oxide is reduced slowly with a small, but constant, known current to metallic copper. When the oxide film has been quantitatively reduced, the potential of the cathode changes rapidly to the discharge potential of hydrogen. The equivalence point is taken as the point of inflection of the voltage-time curve, as illustrated in Fig. 24-7. From the known current i, expressed in

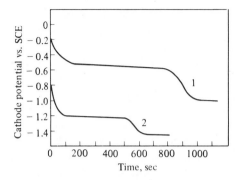

Fig. 24-7. Cathodic reduction of tarnish films on copper: curve 1, copper(I) oxide; curve 2, copper(I) sulfide.

milliamperes, and the elapsed time t, in seconds, the film thickness T, in angstrom units, can be calculated from the known film area A in centimeters squared, and the film density ρ according to the equation

$$T = \frac{10^5 Mit}{AnF\rho} \tag{24-12}$$

where M is the gram-molecular weight of the oxide comprising the film. From mixed films of oxide and sulfide on a metal, two inflection points are obtained. Similar methods have been described for the determination of the relative amounts of tin(IV) and tin(II) oxides on a tinplate surface.

Analogous anodic dissolution is used to determine the successive coatings on a metal surface. Iron is sometimes clad with a tin undercoating for adhesion and a copper-tin surface layer for protection from corrosion. The two coatings will exhibit individual potential breaks. In a similar manner, the thickness of chromium plate on iron, copper, or nickel, and of zinc or nickel plate on either copper or iron can be determined. The method may also be used whenever the substance to be determined can be deposited beforehand so as to adhere to a solid electrode or to form an amalgam with a mercury electrode. Sensitivity is high. Accuracy is limited by the residual current and by the fact that the last traces of deposit do not dissolve uniformly from the surface.

Primary Coulometric Titrations

In primary coulometric titrations at constant current the substance to be determined reacts directly at the electrode. Consequently, no other substance should be able to be electrolyzed at the working electrode until much higher potentials are attained, usually at least 0.5 V from the desired value. Since the potential of the working electrode is not controlled, this class of titrations is limited generally to reactants which are nondiffusible.

One major area of application involves the electrode material itself participating in an anodic process as, for example, the reaction of mercaptans, sulfhydryl groups, and ionic halide ions with silver ions generated at a silver anode. For chloride samples the initial reaction may be

$$Cl^- + Ag^\circ \rightarrow AgCl + e^- \qquad (24\text{-}13)$$

followed by

$$Ag^\circ \rightarrow Ag^+ + e^- \qquad (24\text{-}14)$$

as soon as the limiting current (supply of chloride ions to the anode) has become smaller than the current forced through the electrolytic cell. At this point the silver ion generated anodically diffuses into the solution, and precipitation occurs with the chloride ions left in solution. Of course, the result of the two reactions is identical. The end point of the titration is ascertained amperometrically. Commercial titrators for biological and industrial samples based on this method are available (Aminco-Cotlove, Buchler Instruments). Combustion by the oxygen flask method precedes the titration step for nonionic halides in organic compounds. Mercaptan samples are dissolved in a mixture of aqueous methanol and benzene to which aqueous ammonia and ammonium nitrate are added to buffer the solution and to supply sufficient electrolyte to lower the solution resistance.

Secondary Coulometric Titrations

In secondary coulometric titrations an active intermediate is first generated quantitatively by the electrode process, and this then reacts directly with the substance to be determined. The standard potential of the auxiliary system has to lie between the potential of the system to which the substance to be determined belongs and the potential at which the supporting electrolyte or a second electroactive system undergoes an electrode reaction.

A knowledge of current-potential curves aids the analyst in choosing the auxiliary system. Current-potential curves for systems pertinent to the coulometric determination of iron(II) at constant current are illustrated in Fig. 24-8. To complete the titration within a reasonable period of time, usually 10-200 sec, a finite current must be selected, say i_0. The necessary applied emf will result in a voltage drop across the cell given by V_0 for the initial concentration of iron(II) present in the solution. At the beginning, iron(II) is oxidized directly at the anode

$$Fe^{2+} \rightarrow Fe^{3+} + e^- \qquad (24\text{-}15)$$

As the concentration of iron(II) decreases with the progress of the oxidation, the current will tend to decrease. However, since the current is being maintained constant, the voltage must be increased continually until ultimately the decomposition potential of water is exceeded. If i_0 is selected sufficiently small to delay the other anodic oxidation, the time required for a determination will become too long for practical consideration. One alternative, of course, is to control the anode potential at a value below the decomposition potential of the undesired reactant, as was discussed in an earlier section.

Interposition of an auxiliary system between the potentials at which iron(II) and water are oxidized is the basis of secondary coulometric titrations. No interfering elec-

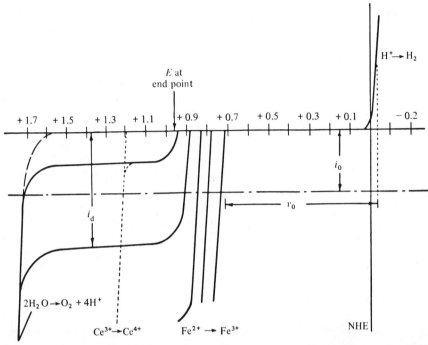

Fig. 24-8. Current–potential curves pertinent to the coulometric titration of iron(II) with cerium(III) as the auxiliary system.

trode reaction can occur if the potential of the working electrode, in this case the anode, is prevented from reaching the value that would occasion initiating the decomposition of water. Limitation of such potential drift is achieved by having a precursor of the secondary titrating agent present in relatively high concentration. In our example a large excess of cerium(III) is added to the solution. Now as soon as the limiting current of iron(II) falls below the value of the current forced through the cell, that is, $i_0 > i_{\text{limiting}}$, the cerium(III) commences to undergo oxidation at the anode in increasing amounts until it may be the preponderant anode reactant. Since the cerium(IV) formed reacts instantly and stoichiometrically with the iron(II),

$$Ce^{4+} + Fe^{2+} \to Ce^{3+} + Fe^{3+} \qquad (24\text{-}16)$$

the total current ultimately employed in attaining the oxidation of iron(II) is the same as would have been required for the direct oxidation. Because there is a relatively inexhaustible supply of cerium(III), the anode potential is stabilized at a value less than the decomposition potential of water. The end point is signaled by the first persistence of excess cerium(IV) in the solution and may be detected the ordinary way with a platinum-reference electrode pair, or photometrically at a wavelength at which cerium(IV) absorbs strongly.

Errors due to impurities in the supporting electrolyte or in the auxiliary substance can be avoided by performing a pretitration, then performing the sample titration, or a succession of titrations, in the same supporting electrolyte.

Exploitation of titrants that for one reason or another are difficult to use in conventional titrimetry are among the virtues of secondary coulometric methods. Electrolytic generation of hydroxyl ion has some advantages over conventional methods. Very small amounts of titrant can be prepared, and in a carbonate-free condition. To analyze dilute acid solutions, such as would result from adsorption of acidic gases, the cathode reaction

$$2H_2O + 2e^- \rightarrow H_2 + 2OH^- \tag{24-17}$$

generates the hydroxyl ion. Of course, in the initial stages it is also possible for the hydrogen ion to react directly at the cathode,

$$2H_3O^+ + 2e^- \rightarrow 2H_2O + H_2 \tag{24-18}$$

but in the vicinity of the end point, the secondary generation predominates. The anode reaction must also be considered. If a platinum anode is used, it must be isolated in a separate compartment, for hydrogen ions would be liberated at its surface. Alternatively, a silver anode may be used within the electrolytic cell in the presence of bromide ions, for then the anode reaction is

$$Ag^\circ + Br^- \rightarrow AgBr + e^- \tag{24-19}$$

and the silver and bromide ions are fixed as a coating of silver bromide on the electrode surface.

Halogens generated internally, and particularly bromine, have found widespread application, especially in organic analysis. In contrast with certain difficulties encountered in the use of bromate-bromide mixtures by conventional volumetric procedures, coulometry is much simpler.[14] Bromates are not soluble in many organic solvents, and many organic samples are not soluble in water. However, sodium and lithium bromides are quite soluble in various organic solvents in which brominations can be conducted.

The complexing ability of EDTA has been exploited in the coulometric titration of metal ions. The method depends on the reduction of the mercury(II) or cadmium chelate of EDTA and the titration, by the anion of EDTA that is released, of the metal ion to be determined. If the direct reaction of metal with EDTA is too slow, excess of the EDTA anion is generated and then the excess is back-titrated by cadmium generated at a cadmium-amalgam electrode.

Dual intermediates can be used whenever the substance to be titrated does not react rapidly with the auxiliary system or, at least, when the reaction rate is not as rapid as the generation rate. An excess of titrant is generated and permitted to react for the necessary time. Then the polarity of the working electrode is reversed and a back-titration is conducted with a second auxiliary system. In this manner an excess of bromine can be titrated with electrically generated chlorocuprate(I) ion,

$$Cu^{2+} + 3Cl^- + e^- \rightarrow CuCl_3^{2-} \tag{24-20}$$

$$Br_2 \text{ (in excess)} + 2CuCl_3^{2-} \rightarrow 2Cu^{2+} + 2Br^- + 6Cl^- \tag{24-21}$$

External Generation

Internal generation methods possess limitations. Often conditions conducive to optimum generation of reactant and to rapid reaction with the substance to be titrated are not compatible. Or the sample may contain two or more substances that are capable of undergoing reactions at the electrode and that are not different sufficiently in electrode potential to permit use of an auxiliary system. For example, the titration of acids by electrically generated hydroxyl ion is precluded in the presence of certain other reducible substances. These limitations are circumvented when the reagent is generated in an electrolytic cell that is isolated from the solution to be titrated, and the desired electrolytic product is allowed to flow via a capillary tube into the test solution.

Cross-sectional views of a double-arm[3] and a single-arm generator cell[11] are illustrated in Figs. 24-9 and 24-10. The supporting electrolyte is fed continuously from a reservoir into the top of the generator cell. The incoming solution is then divided at the T-joint, in the two-arm design, so that about equal quantities flow through each of the arms of the cell. Platinum electrodes are sealed on either side of the T-joint. The products of electrolysis are swept along by the flow of solution through the arms and emerge from the delivery tips on either side. A beaker containing the sample to be titrated is placed beneath the appropriate delivery tip. Thus, determinations performed with external generation of titrant hardly differ from normal volumetric methods in essentials; the only difference is that the titer is referred to unit of time and not unit of volume. Naturally, one-half of the liquid continuously discharged is conducted to waste. When a solution of Na_2SO_4 is supplied, H_2SO_4 is formed at the anode and $NaOH$ at the cathode. For bromination, a solution of KBr is used.

The single-arm generator cell is useful for the generation of reagents in those cases in which mixing of the cathode and anode electrolysis products can be tolerated. The working electrode can be made of platinum or it can be a mercury pool. The other electrode compartment is isolated by a frit barrier. The flow of supporting electrolyte is usually 6 ml/min, or larger.

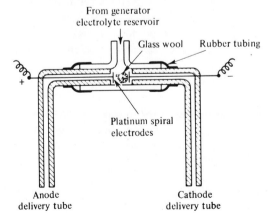

From generator
electrolyte reservoir

Glass wool   Rubber tubing

Platinum spiral
electrodes

Anode   Cathode
delivery tube   delivery tube

Fig. 24-9. Double-arm electrolytic cell for external generation of titrant. After D. D. Deford, J. N. Pitts, and C. J. Johns, *Anal. Chem.*, **23**, 938 (1951). (Courtesy of American Chemical Society.)

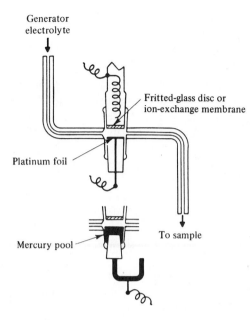

Generator
electrolyte

Fritted-glass disc or
ion-exchange membrane

Platinum foil

Mercury pool

To sample

Fig. 24-10. Single-arm generator cell with working electrode either of platinum or a mercury pool. After J. N. Pitts et al., *Anal. Chem.*, **26**, 628 (1954). (Courtesy of American Chemical Society.)

The titration of azo dyes with titanium(III) illustrates the advantage of external generation. At room temperature the rate of reaction of titanium(III) with the dye is slow. Yet on raising the temperature, hydrolysis of titanium(IV) and bubble formation at the electrode surface lead to low current efficiencies. However, if the titanium(III) is generated at room temperature and then delivered to the hot dye solution, optimum conditions prevail for each step. A mercury-pool cathode or an amalgamated working electrode is used to take advantage of the favorable hydrogen overpotential on mercury.

PRECISION CONSTANT-CURRENT COULOMETRY

Constant-current methods with careful attention to eliminating or decreasing all sources of error can be extremely precise. Marinenko and Taylor[9] have measured the electrochemical equivalents of benzoic and oxalic acid and, knowing the purity of the acids, the atomic weights and the precise values of the current and time required to convert all of the hydrogen ion to hydrogen at a platinum electrode, have calculated the value of the faraday. They obtained values of the faraday equal to 96,486.7 $\pm$ 2.5 coulombs/gram-equivalent and 96,485.4 $\pm$ 3.4 C/g-equivalent, respectively. These compare with 96,487.0 $\pm$ 1.6 C/g-equivalent presently recommended by the National Academy of Science-National Research Council. They propose that, when applicable, the conformance of a given material to its theoretical electrochemical equivalent should define the absolute purity. The faraday, then, would in fact be the primary standard.

LABORATORY WORK

General Instructions for Constant-Current Experiments

Assemble the titration apparatus as shown in Fig. 24-2. Connect a 45-V B battery through a 5000-Ω potentiometer (or bank of fixed resistors of different values), a 3000-Ω limiting resistor, the current-measuring device, the generator electrodes, and an ON-OFF switch. Measure the current with a calibrated 0-10 milliammeter or determine the iR drop across a precision 100-Ω resistor (for currents not exceeding 10 mA) with a student potentiometer. Time measurements made with a stopwatch or stopclock will provide results of moderate accuracy.

Arrange the generator electrodes and indicator system as shown in Fig. 24-2. Positioning the generator electrode from which the reactant is derived adjacent to the indicator electrode, in the direction of stirring, gives a more rapid warning of the approach of the end point. For photometric indication, the electrolysis cell is positioned in the photometer in place of the usual cuvette.

When using amperometric indication for the end point, plot the amperometric signal vs. time. The end point will be signalled by an abrupt change in the amperometric current, which may be taken as the end point, or the coulometric titration may be terminated momentarily, the amperometric current and generation time noted, and then the generation continued for perhaps an additional 5-10 sec. This series of steps is repeated until 4 or 5 readings are obtained beyond the end point, which is then established by extrapolation of the two branches of the plot.

Pretitration of the supporting electrolyte should be done to remove impurities and to familiarize oneself with the end point signal. Then the sample is added and the titration continued until the end point signal reappears.

Experiment 24-1   Electrically Generated Hydroxyl Ion

Place 100 ml of 0.05 M KBr solution (6 g/liter) in a 200-ml tall-form beaker. Add several drops of an appropriate indicator, or insert glass-calomel electrodes for potentiometric indication. The generator electrodes may be a platinum foil cathode (10 cm^2) and a helix of silver (No. 6 gauge) wire as anode (or a second platinum foil electrode isolated by a frit barrier).

Turn on the stirrer and generator current. Adjust the current by means of the potentiometer (or bank of resistors) to 10 mA or less. Titrate to the theoretical pH at the end point (adding a trace of acid, if necessary), then discontinue the current. Add the sample, then turn on the current and timer simultaneously. Adjust the variable rheostat whenever necessary to maintain the current at the selected value throughout the titration. Select an aliquot of the sample that will require about 200 sec (for example, 10-20 ml of 0.001 N acid, transferred with a pipet or microsyringe if more concentrated acids are employed).

After every titration clean the anode with emery cloth or by dipping it into concentrated aqueous ammonia or potassium cyanide solution. Several consecutive samples may be titrated without renewing the supporting electrolyte.

Experiment 24-2   Electrically Generated Bromine

Place 100 ml of 0.2 M KBr solution (24 g/liter) and 3 ml of 18 M H_2SO_4 in a 200-ml tall-form beaker. The generator electrodes are two platinum foil electrodes; 10 cm^2 or larger in area. If amperometric indication is employed, insert a small platinum electrode and a large-area calomel electrode and apply 0.2 V positive with respect to the SCE. Pretitrate to an end-point signal, then add 1.00 ml of 0.005 M As_2O_3 solution (0.987 g/liter) and titrate until the same signal is repeated.

Experiment 24-3   Electrolytically Generated Cerium(IV) Ion

Place 40 ml of 0.1 M cerous ammonium sulfate and 10 ml of 9 M H_2SO_4 in the electrolysis cell. Insert a platinum foil working electrode in the cell and isolate a second platinum foil electrode (the cathode) inside a tube with a fritted glass end and filled with 1.5 M H_2SO_4. Purge the supporting electrolyte with nitrogen gas for 10 min, and maintain a stream of gas through the electrolyte during the titration.

Add a sample of iron(II) ammonium sulfate and titrate at 50 mA. The end point is ascertained potentiometrically with a platinum-calomel electrode pair or amperometrically with a platinum indicator electrode. A pretitration is recommended.

Experiment 24-4   Electrically Generated Iodine

Place 50 ml of 0.1 M KI solution (16.6 g/liter) and 20 ml of 0.25 M Na_2HPO_4 solution (36 g/liter) in the electrolysis cell. Add a few drops of 0.005 M As_2O_3 solution (0.987 g/liter) and pretitrate to the end point (starch-iodide color, amperometric indication, or potentiometric indication).

Add 1.00 ml of 0.005 M As_2O_3 solution and titrate once more to the end point signal.

Experiment 24-5   Electrically Generated Silver Ion

For the titration of bromide and iodide, the supporting electrolyte is 0.5 M KNO_3 (51 g/liter). The working electrode (anode) can be a clean silver foil (10 cm^2) or a silver rod*; the cathode is a platinum foil electrode. For the titration of chloride, the supporting electrolyte is a nitric-acetic acid system (38 ml concentrated HNO_3 and 200 ml glacial acetic acid per liter) plus 0.05% gelatin.

---

*Any previous coating of silver halide must be completely removed (see Exp. 24-1).

Use 5.00-15.00 ml of 0.025 M KBr (2.975 g/liter), or 0.025 M KCl (1.864 g/liter), solution when the generating current is 30 mA. For amperometric indication, a large-area SCE may be short-circuited to the platinum indicator electrode through a suitable galvanometer.

Experiment 24-6   External Titration

Assemble an external generator, double-arm cell as shown in Fig. 24-10.

Connect a source of direct current, approximately 200 mA with 0.1% or less ripple, through a precision resistor, the pair of generator electrodes, and, if necessary, a 1250-Ω (100-W) rheostat and a 125-Ω (20-W) rheostat.

Feed a solution of the supporting electrolyte continuously into the generator cell to provide a delivery rate of 6 ml/min from each delivery tip. Use a 600- or 800-ml beaker for the titration.

Suggested supporting electrolytes: 1.0 M sodium sulfate solution (adjusted to pH 7) for generation of hydroxyl or hydrogen ions; 0.05 M potassium iodide solution in 0.1 M boric acid (to neutralize the hydroxyl ion produced at the cathode) for generation of iodine; and 0.05 M potassium bromide in 0.1 N sulfuric acid for generation of bromine.

General Instructions for Controlled-Potential Experiments

Set up the controlled-potential coulometer and run through the checkout procedure as specified in the manual of instructions furnished by the manufacturer.

Experiment 24-7   Determination of Silver and Copper

Dissolve a weighed sample of silver or copper metal in nitric acid, and remove the nitric acid by fuming with perchloric acid. Dilute to volume in a volumetric flask and take an aliquot containing several milligrams of alloy for analysis. To the aliquot in the analysis cell, add enough distilled water to cover the electrodes, set the potentiostat at +0.13 V with respect to a saturated calomel electrode, and electrolyze until completion. Read the number of coulombs. Reset the potentiostat to −0.12 versus the SCE and electrolyze again. Read the number of coulombs. Calculate the percentage of silver and of copper in the alloy. Should the alloy contain other, more easily reducible metals, try pre-electrolyzing the sample at a potential insufficient to plate out either copper or silver.

PROBLEMS

1. The initial current is 90.0 mA and decreases exponentially with $k = 0.0058$ sec^{-1}; the titration time is 714 sec. How many milligrams of uranium(VI) are reduced to uranium(IV)?

2. When an integrating motor was calibrated, these results were obtained:

Current, mA	Shunt Resistance, Ω	Time, sec	Counts, N
10.02	2220	600	9102
20.03	1110	600	9180
30.00	770	600	9458
50.00	475	600	9773

Calculate the microequivalents per count.

3. The calibration factor of an integrating motor is 0.00267 microequivalents per count. Calculate the normality of an acid solution, 10.0 ml of which produced 40.72 counts during a titration.

4. These results were obtained during the titration of three successive 1.00-ml aliquots of As_2O_3 solution with electrically generated iodine at pH 8 and using amperometric indication of the end point. Graph the results and determine the normality of the As_2O_3 solution. The microequivalents of iodine generated are followed by the amperometric signal in microamperes: Pretitration—0.00 microequivalents = 0.4 μA; 5.10 = 0.7; 9.90 = 1.3; and 15.0 = 1.7. First aliquot—15.0 = 0.4; 50.0 = 0.4; 100.0 = 0.4; 149.5 = 1.3; 154.5 = 2.0; 160.0 = 2.6; 164.9 = 3.0. Second aliquot—164.9 = 0.4; 200.0 = 0.4; 250.0 = 0.4; 273.8 = 1.0; 277.4 = 2.2; 280.8 = 2.7; 286.1 = 3.2. Third aliquot—286.1 = 0.4; 350.0 = 0.4; 400.0 = 0.4; 402.0 = 0.8; 406.3 = 2.6; 411.0 = 3.1; 416.1 = 3.8.

5. In coulometric titrations, a milliampere-second corresponds to how many grams of (a) hydroxyl ions, (b) antimony (III to V), (c) chloride ions, (d) copper (II to 0), (e) arsenious oxide (III to V)?

6. A 0.5 M K_2SO_4 solution in a gas coulometer gave 22.33 ml of hydrogen plus oxygen at the end of an electrolysis; temperature was 24.0°C in the water jacket and the pressure was 740 mm. The partial pressure of water over the electrolyte is 22 mm at 24°C. How many coulombs were involved in the electrolysis?

7. Calculate the concentration of acid in a 10.0-ml aliquot that required a generation time of 165 sec for the appearance of the pink color of phenolphthalein. The voltage drop across a 100-Ω resistor was 0.849 V.

8. Sketch the current-potential curves that would pertain to each of these coulometric systems: (a) The titration of acids with electrically generated hydroxyl ion in a potassium bromide electrolyte and using a silver anode. (b) The generation of excess bromine in a potassium bromide electrolyte, followed by the generation of copper(I) to react with the unused bromine. (c) The titration of zinc with generated ferrocyanide ions.

9. In the coulometric determination of permanganate ion by generating iron(II) from iron(III), the permanganate was all reduced to manganese(II) by a constant current of 2.50 mA acting for 10.37 min. Calculate the molarity of the permanganate if the initial volume was 25.00 ml.

10. In an electrolytic determination of bromide from 100.0 ml of solution, the quantity of electricity, as read on a mechanical current–time integrator, was 105.2 Q. Calculate the weight of bromide ions in the original solution. Calculate the potential of the silver electrode that should be employed throughout the electrolysis. $K_{sp\ AgBr} = 4 \times 10^{-13}$.

11. The following measurements were made in a coulometric titration of arsenic(III) ions with generated bromine.

$$\text{Generation time: 132.6 sec}$$
$$\text{Calibrated resistance: 100 } \Omega$$
$$iR \text{ drop across resistance: 0.620 V}$$

Calculate the amount of arsenic present in the sample.

12. Using generating currents of 1 to 10 mA, and corresponding titration times of about 300 to 100 sec, what range in weights of mercaptans may be present in a solution volume of 50 ml? The reaction is $Ag^\circ + RSH \rightarrow AgSR + H^+ + e^-$.

13. In Fig. 24-11 are shown the coulograms of iron(III), manganese(VII), and vanadium (V) for their reduction in 1 M phosphate medium at pH 2 with a platinum cathode. Outline a procedure for the determination of each element in a mixture of the others by controlling the cathode potential.

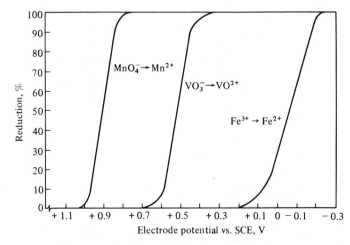

Fig. 24-11. Coulograms of iron(III), vanadium(V), and manganese(VII) in 1 M phosphate medium at pH 2 with a platinum cathode. (Courtesy of B. W. Conroy and O. Menis, NUMEC, Apollo, Pa.)

14. Assuming that the coulograms for iron and vanadium, shown in Fig. 24-11, are reversible, outline a constant-current procedure for the determination of iron(II) in the presence of vanadium(IV).

15. From the information contained in Fig. 24-11, outline a procedure for the determination of the amounts of vanadium(V) and vanadium(IV) in a mixture containing the two oxidation states.

16. How long should a constant current of 100.0 mA be passed through a solution to prepare 100 ml of a solution of 0.0100 M Ni^{2+} using an anode of pure nickel?

17. Determine the equivalent weight of an organic acid if 0.0400 g in alcohol-water mixture required a constant current of 50 mA for 500 sec to generate sufficient hydroxyl ion to reach a phenolphthalein end point.

BIBLIOGRAPHY

Abresch, K., and I. Classen, *Die coulometrische Analyse*, Verlag Chemie, Weinheim, 1961.

Adams, R. N., *Electrochemistry at Solid Electrodes*, Dekker, New York, 1969.

Bard, A. J. (ed.), *Electroanalytical Chemistry, A Series of Advances*, Dekker, New York, Vol. 1, 1966; Vol. 2, 1967; Vol. 3, 1969; Vol. 4, 1970; Vol. 5, 1971; Vol. 6, 1973.

Charlot, G., J. Badoz-Lambling, and B. Tremillion, *Electrochemical Reactions*, Elsevier, Amsterdam, 1962.

DeFord, D. D., and J. W. Miller in I. M. Kolthoff and P. J. Elving, Eds., *Treatise on Analytical Chemistry*, Part I, Vol. 4, Chap. 49, Wiley-Interscience, New York, 1963.

Kies, H. L., *J. Electroanal. Chem.*, **4**, 257 (1962).

Lewis, D. T., *Analyst*, **86**, 494 (1961).

Lingane, J. J., *Electroanalytical Chemistry*, 2nd ed., Wiley-Interscience, New York, 1958.

Milner, G. W. C. and G. Phillips, *Coulometry in Analytical Chemistry*, Pergamon, London, 1968.

Shain, I. in I. M. Kolthoff and P. J. Elving, Eds., *Treatise on Analytical Chemistry*, Part I, Vol. 4, Chap. 50, Wiley-Interscience, New York, 1963.

Tutundzic, P. S., *Anal. Chem. Acta*, **18**, 60 (1958).

LITERATURE CITED

1. Bett, N., W. Nock, and G. Morris, *Analyst*, **79**, 607 (1954).
2. Clem, R. G., *Anal. Chem.* **43**, 1853 (1971).
3. DeFord, D. D., J. N. Pitts, and C. J. Johns, *Anal. Chem.*, **23**, 938 (1951).
4. Dunlap, L. B. and W. D. Schults, *Anal. Chem.*, **34**, 499 (1962).
5. Ehlers, V. B. and J. W. Sease, *Anal. Chem.*, **26**, 513 (1954).
6. Harrar, J. E., *Anal. Chem.*, **43**, 143 (1971).
7. Kelley, M. T., H. C. Jones, and D. J. Fisher, *Anal. Chem.*, **31**, 489 (1959).

8. Lingane, J. J., *J. Am. Chem. Soc.*, 67, 1916 (1945).
9. Marinenko, G. and J. K. Taylor, *Anal. Chem.*, **40**, 1645–51 (1968).
10. Page, J. A. and J. J. Lingane, *Anal. Chim. Acta*, **16**, 175 (1957).
11. Pitts, J. N., et al., *Anal. Chem.*, **26**, 628 (1954).
12. Shults, W. D., *Anal. Chem.*, **33**, 15 (1961).
13. Smith, S. W. and J. K. Taylor, *J. Res. Natl. Bur. Std.*, (U.S.) **63C**, 65 (1959).
14. Swift, E. H. and co-workers, *Anal. Chem.*, **19**, 197 (1947); **22**, 332 (1950); **24**, 1195 (1952); **25**, 591 (1953); and *J. Am. Chem. Soc.*, **70**, 1047 (1948); **71**, 1457, 2717 (1949).

Amperometric Titration Methods

The polarographic method can be used as the basis of an electrometric titration method comparable with the potentiometric, the conductometric, and the photometric methods. In this case the voltage applied across the indicator electrode and reference electrode is kept constant, and the current passing through the cell is measured and plotted against the volume of reagent added. Hence the name *amperometric titration*.

The current is measured, in general, on a diffusion-current region of a current–voltage curve. On such a region the current is independent of the potential of the indicator electrode because of an extreme state of concentration polarization at the electrode. Because at the electrode surface the concentration of material undergoing electrode reaction is maintained at a value practically equal to zero, the current is limited by the supply of fresh material to the electrode surface by diffusion. The rate of diffusion, and hence the current, is proportional to the concentration of diffusing substance in the bulk of the solution.

This technique can best be described by an example—the titration of a reducible substance, lead ion, with a nonreducible reagent, sulfate ion. A polarogram of a solution containing lead ions is represented by curve A in Fig. 25-1. If the voltage is held at any value on the diffusion current plateau, the current will be represented by i_0. The titrant exhibits no diffusion current at the applied emf. Increments of titrant remove some of the electroactive lead ions. As the concentration of lead ions decreases, the current decreases to i_1, i_2, i_3, and finally i_r, at which point the lead ions have completely reacted and the only current flowing is a residual current characteristic of the supporting electrolyte.

If successive values of the diffusion current are plotted against the volume of titrant added, the result is a straight line which levels off at the end point (Fig. 25-2). The intersection of the extrapolated branches of the titration curve gives the end point.

When both titrant and unknown give diffusion currents at the applied voltage chosen, the current will drop to the end point, then increase again to give a V-shaped titration curve, as seen in Fig. 25-3. If the original material does not react electrolytically, but the titrant does, a horizontal line, rising at the end point, results. This is shown in Fig. 25-4.

METHODOLOGY

If the optimum value needed to maintain the titration voltage is not known, the polarograms are determined for the materials involved, an appropriate voltage is selected,

727

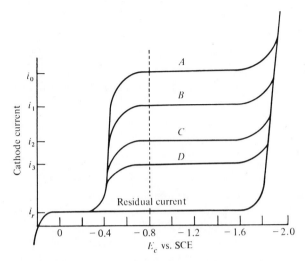

Fig. 25-1. Successive current–voltage curves of lead ion made after increments of sulfate ion were added.

and the titration is carried out. As sometimes happens, a choice between two applied emf values can be made. In the titration of lead with dichromate, the titration can be conducted by choosing as the voltage a value E_1 at which dichromate ions are reduced, but not lead ions. The current–voltage curves are shown in Fig. 25-5. The titration curve is a horizontal line, rising at the end point, resembling Fig. 25-4. By shifting the cathode potential to E_2, both dichromate and lead ions are reduced. The current will drop to the end point, then increase again to give a V-shaped titration curve.

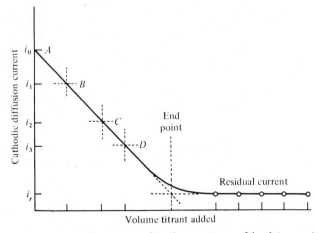

Fig. 25-2. Amperometric titration curve for the reaction of lead ions with sulfate ions. See Fig. 25-1 for corresponding current–voltage curves. Performed at -0.8 V vs. SCE.

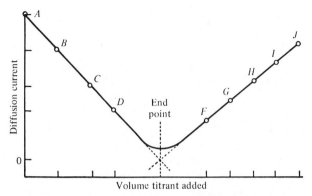

Fig. 25-3. Type of amperometric titration curve when both reactant and titrant give diffusion currents; e.g., the titration of lead ions with dichromate ions at −0.8 V vs. SCE. See also Fig. 25-5.

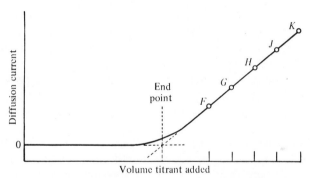

Fig. 25-4. Type of amperometric titration curve when only titrant gives a diffusion current; e.g., the titration of lead ions with dichromate ion performed at 0.0 V vs. SCE in an acetate buffer of pH 4.2. See also Fig. 25-5.

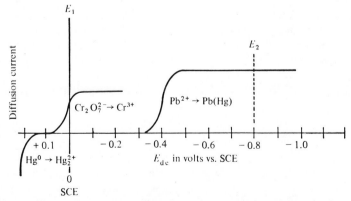

Fig. 25-5. Current–voltage curves of dichromate and lead ions shown schematically.

In practice, the reversed L-shaped type of curve, illustrated by Fig. 25-4, is preferred. Because the titrant in this case produces no current, it can be added continuously at a moderate rate until the end point is passed. This will be noted by a permanent increase in the diffusion current. Then three or four additional readings, taken after successive increments of excess titrant have been added, will establish the rising branch of the curve.

Strictly speaking, a correction for dilution is necessary to attain a linear relation between current and volume of titrant, but by working with a reagent which is tenfold more concentrated than the solution being titrated, the correction becomes negligibly small. Incompleteness of reaction in the vicinity of the end point usually will not detract from the results provided reaction equilibrium is attained rapidly during the titration. Points can be selected between 0 and 50% and 150 and 200% of the end-point volume for the construction of the two branches of the titration curve. In these regions the common ion effect will repress dissociation and solubility of precipitates.

Apparatus

The equipment for conducting amperometric titrations is simple. Although it may be the same as for polarography, several simplifications are possible. The potential of the indicator electrode need only be selected within 0.1 V if it lies on a limiting current region of a current–voltage curve. Often the potential of a reference electrode will lie in the permissible range, so that it is necessary only to short-circuit the indicator electrode through a suitable current-measuring instrument to a reference electrode of relatively large area.

No thermostat is necessary. The temperature of a solution will seldom vary appreciably during the short time, 10 min or less, necessary to conduct a titration.

The indicator electrode may be a dropping mercury electrode or a rotating metal microelectrode. The latter is simple to construct. It consists of a short length of wire, usually platinum, protruding 5–10 mm from the wall of a piece of glass tubing. The latter is bent at right angles a short distance from the end of the stem so as to sweep an area of the solution with the wire. It is illustrated in Fig. 25-6. The electrode is mounted in the shaft of a motor and rotated at a constant speed of about 600 rpm. By using a rotating electrode, the diffusion layer thickness is decreased, thereby increasing the sensitivity and the rate of attainment of a steady diffusion state. The limiting current may be up to 18 or 20 times larger than that with a dropping electrode; it is proportional to the $\frac{1}{3}$ power of the number of revolutions per minute above 200 rpm. A stationary electrode with a magnetic stirrer to pass the solution by the electrode exhibits the same response as a rotated electrode.

The larger currents attained with a rotating electrode allow correspondingly smaller concentrations to be measured without loss of accuracy. Also the absence of drops disengaging themselves at regular intervals eliminates the charging current observed with a dropping electrode, which in turn permits the use of ordinary rugged microammeters. However, many systems whose oxidation potentials or reduction potentials lie in the range of the platinum microelectrode do not give limiting currents with a rotating elec-

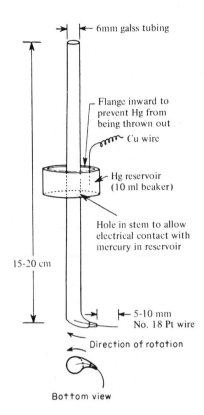

6mm galss tubing

Flange inward to
prevent Hg from
being thrown out

Cu wire

Hg reservoir
(10 ml beaker)

Hole in stem to allow
electrical contact with
mercury in reservoir

15-20 cm

5-10 mm
No. 18 Pt wire

Direction of rotation

Bottom view

Fig. 25-6. Rotating platinum microelectrode.

trode. And where the discharge of hydrogen interferes, a dropping mercury electrode, with its larger value of hydrogen overpotential, must be used.

The removal of oxygen is generally mandatory over most of the useful range of the dropping electrode. Nitrogen or hydrogen gas must be bubbled through the solution preceding the titration and for a minute or two after the addition of each increment of titrant. The oxidation of mercury limits the anodic range of the dropping electrode, but when applicable, the rotating electrode extends the useful range to about +0.9 V versus the SCE, at which point the oxidation of water to oxygen commences.

SUCCESSIVE TITRATIONS

Iodide, bromide, and chloride can be successively titrated in mixtures with silver, using the rotating electrode.[3] In a 0.1–0.3 N solution of ammonia only silver iodide will precipitate when a silver solution is added. The indicator reaction is the reduction of the complex diammine silver ion. Consequently, the potential of the rotating electrode must be made negative enough to plate out silver, but must not be negative enough to give an appreciable current due to the reduction of dissolved oxygen. The range of permissible potential is strictly limited, as is evident from an examination of the current–voltage curves

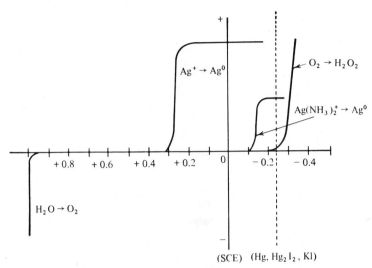

Fig. 25-7. Schematic current–voltage curves of silver and oxygen obtained with a rotating microelectrode.

of diammine silver reduction and oxygen reduction using a silver-plated microelectrode in an ammoniacal solution (Fig. 25-7). Fortunately the mercury/mercuric iodide/potassium iodide reference electrode happens to lie in the permissible range (-0.23 V vs. SCE) and can be short-circuited through the current-measuring device to the rotating electrode. During the titration of iodide the current remains constant at zero, or nearly so, until the iodide ions are consumed, and then rises. After three or four points have been recorded past the end point, the solution is acidified to make it 0.8 N in nitric acid. Immediately the silver ions added in excess and now released from the ammine complex combine with the bromide ions and precipitate as silver bromide, and the current drops to zero.

The titration of bromide and chloride is carried out at a less negative potential, for in these titrations the indicator reaction is the deposition of silver from aquo-silver ions.

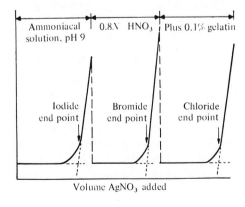

Fig. 25-8. A composite of the consecutive titration curves for a mixture of iodide, bromide, and chloride ions.

Because the potential of the saturated calomel electrode lies in the limiting current region, it also may be short-circuited to the rotating electrode. Chloride does not interfere with the titration of bromide because silver chloride particles cause a cathodic current even in the presence of a large excess of chloride. Therefore a second rise in the current indicates the end point of the bromide titration. A chloride end point can be obtained by adding gelatin, which suppresses the current due to silver chloride, and continuing the titration until the current again rises after the chloride end point. A composite of these titration curves is shown schematically in Fig. 25-8.

TITRATIONS TO ZERO CURRENT

With systems in which both the oxidant and reductant yield a diffusion current, the titration curve obtained is of the type shown in Fig. 25-9. In such systems as, for example, the titration of iron(III) with titanium(III), a voltage E is impressed upon the indicator electrode, so that the diffusion current for the reduction of iron(III) is set up at the start of the titration. As the iron(III) concentration is decreased linearly, the current decreases in a similar fashion and reaches zero at the end point. When the end point is passed, a diffusion current caused by the oxidation of the titanium(III) is set up. A change in slope caused by the difference in diffusion coefficients is usually evident as the lines cross the zero axis. Actually the zero axis is the value of the residual current for the supporting electrolyte and usually will be different from zero.

Titrations of this type without any chemical reaction are also possible.[4] Copper(II) and tin(II) in a tartrate medium at pH 4 possess the current–voltage curves schematically represented in Fig. 25-10. At an applied potential midway between the half-wave potential

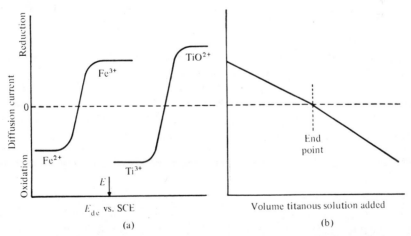

Fig. 25-9. Current–voltage curves of (a) iron(II/III) and titanium(IV/III) systems and (b) amperometric titration system for iron(III) titrated with titanium(III) solution at E_{de} = point E on the graph.

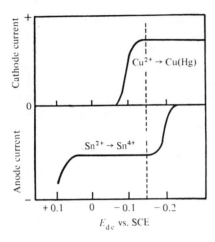

Fig. 25-10. Current–voltage curves for cathodic reduction of copper(II) ions and anodic oxidation of tin(II) ions, both in tartrate medium at pH = 4.

of the cathodic copper(II to 0) wave and the anodic tin(II to IV) wave, the diffusion currents of both waves are fully developed. Titrating with a solution of copper(II), the anodic diffusion current of the tin(II) wave will be compensated by the increasing cathodic current of the copper(II) wave. The net diffusion current will be zero at the end point.

TWO INDICATOR ELECTRODES

In a modification of the usual or classical amperometric system, two similar platinum electrodes can be immersed in the titration cell. A small and constant voltage is applied to these electrodes as in the classical method. For the method to be applicable, the only requirement is that a reversible oxidation–reduction system be present either before or after the end point.

In a titration with two indicator electrodes, and when the reactant involves a reversible system, a small amount of electrolysis takes place. The amount of oxidized form reduced at the cathode is equal to that formed by oxidation of the reduced form at the anode. Both electrodes are depolarized until either the oxidized or the reduced member of the system has been consumed by a titrant. After the end point, only one electrode remains depolarized if the titrant does not involve a reversible system. The solution at this juncture resembles a one-electrode method connected to a depolarized (reference) electrode. Current flows until the end point. At and after the end point the current is zero or close to zero.

The method was introduced years ago under the name "dead-stop end point."[1] The reverse of this type of end point, and the more desirable in practice, might be called "kick off" and resembles a reversed L-shaped amperometric curve. When both the system titrated and the reagent are reversible oxidation–reduction systems, the current is zero or close to zero only at the end point, and a V-shaped titration curve results.

Three regions appear in a titration curve when two indicator electrodes are employed.[2]

Take for example the iodine–iodide system being titrated with thiosulfate. If a considerable quantity of iodide is in solution, the system will be well poised and the current will maintain a steady value (Region 1 on Fig. 25-11). As the titration progresses and the concentration of iodine gets smaller, the concentration overpotential (polarization) at the cathode begins to play a role, and the current tends to become diffusion controlled. Now the current tends to vary in proportion to the concentration of iodine remaining in the bulk of the solution. The characteristics of the line giving the change of current from the point where the system is well poised to the vicinity of the end point is represented as Region 2. Near the end point the line becomes straight, as in a titration with one indicator electrode at constant applied emf. No current flows after the end point, because the thiosulfate–tetrathionate system is not a reversible couple and insufficient emf is applied to cause the oxidation of iodide ions at the anode and the discharge of hydrogen (or dissolved oxygen) at the cathode.

If the system involving the reactant is poorly poised, as it is when no iodide ions are initially present in the solution, the current at the start of the titration will be essentially zero. The current will rise as iodide ions are formed and will reach a maximum when the degree of completion of the titration is 50%. This portion of the curve is represented by the dashed lines in Region 1 of Fig. 25-11. The remainder of the titration curve will follow the descending branch.

Both one- and two-indicator electrode methods become identical if the applied emf in the two-indicator method is made large enough to yield a diffusion-controlled current early in the titration. Whereas some workers recommend only sufficient applied emf to balance the back emf, approximately 20 mV in the iodine titration, the diffusion current is not completely developed until the applied emf is 100 mV or greater, as shown in Fig. 25-12. With larger values of applied emf the value for the iR term remains negligibly small over a larger region, and thus the titration line remains straight over a longer distance in the vicinity of the end point. In effect, this places the system under a diffusion-controlled condition well in advance of the end point. A second effect is to increase the current sensitivity, but this can be varied at will by varying the size of the electrodes and the speed of stirring.

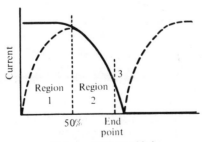

Fig. 25-11. Amperometric titration lines using two indicator electrodes. The dashed line in Region 1 is followed when the reactant is poorly poised initially; after the end point, when the titrant also forms a reversible oxidation–reduction system.

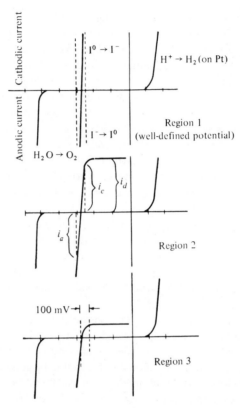

Fig. 25-12. Schematic current–voltage curves for two indicator electrodes corresponding to regions on Fig. 25-11; iodine titrated with thiosulfate with excess iodide present.

COMPARISON WITH OTHER TITRATION METHODS

Several advantages of amperometric methods are immediately apparent. The equipment is simple. Because it is a relative method, there are fewer disturbing variables, as contrasted with polarography. Electrode characteristics are unimportant. There is no need to determine the capillary characteristics of a dropping electrode. Use of rotating electrodes is possible, and in fact, desirable when applicable. The lack of current oscillations when using rotating electrodes makes it possible to use rugged microammeters for measurement of the current. Accuracy is higher than in polarography because each branch of the titration curve effectively is the average of the recorded points. An error in the end point is primarily determined by the accuracy of the titrant delivery.

The method possesses greater sensitivity than conductometric and potentiometric titrations. In fact, amperometric methods are best for determining traces with good precision. Concentrations from 0.1 to 0.0001 M, and even in favorable cases to 0.000001 M can be measured with ease and accuracy.

Applications of amperometric methods are more general than classical potentiometric methods and polarography. Many systems do not possess a measurable equilibrium potential but can be electrolyzed under an applied emf. However, even if one reactant is not oxidizable or reducible, the titration can be conducted by utilizing the oxidation-reduction characteristics of the other reactant. This method is one of the few generally applicable to precipitation reactions.

LABORATORY WORK

Experiment 25-1   Titration of Lead with Dichromate

1. Transfer 25 ml of $0.02\ M$ lead nitrate solution to a polarographic cell. Add 25 ml of a supporting electrolyte, which is approximately $0.1\ M$ in potassium nitrate, $0.17\ M$ in acetic acid, and $0.06\ M$ in sodium acetate. The pH should be about 4.2. (The weights of the three ingredients in the supporting electrolyte are 10, 10, and 5 g per liter, respectively.) Add 2.5 ml of 0.2% gelatin solution. Remove dissolved oxygen.
2. Determine the current–voltage curve of lead from 0 to 1.0 V negative, using a dropping mercury electrode and an external saturated calomel electrode of large area ($> 10\ cm^2$).
3. Plot the current–voltage curve on a sheet of graph paper.
4. Apply 0 V across the electrodes; that is, short the dropping electrode directly through galvanometer and shunt. Titrate with $0.05\ M$ potassium dichromate solution from a 10-ml buret. Take readings every 0.5 ml until the galvanometer registers a definite deflection, and then take readings every 0.25 ml until several points have been obtained beyond the equivalence point.
5. Run a current–voltage curve on the solution containing excess potassium dichromate. Plot the curve on the sheet of graph paper used in step 3.
6. Repeat step 4 with the cathode at 1.0 V negative with respect to the SCE.
7. Repeat steps 4 and 6 with $0.002\ M$ lead nitrate solution and $0.005\ M$ potassium dichromate solution.
8. On a second graph, plot the two pairs of curves of current vs. volume of titrant for steps 4, 6, and 7.

Note See Fig. 25-5 for the current–voltage curves of lead and chromium, and Figs. 25-3 and 25-4 for the general shape of the titration curves.

Experiment 25-2   Titration of Arsenic with Bromate

1. Transfer 25 ml of $0.001\ N$ arsenious oxide to the polarographic cell. Add 5 ml of $12\ M$ hydrochloric acid and 20 ml of $0.125\ M$ potassium bromide (15 g/liter). Center a button-type platinum electrode in the cell (a 150-ml beaker) and insert the arm of the salt bridge from a saturated calomel electrode whose surface area exceeds $10\ cm^2$. Insert a stirring bar and adjust the stirring rate between 200 and 600 rpm without producing a vortex.

2. Apply 0.2 V positive to the rotating platinum electrode (which is the anode in this titration). Titrate with 0.05 N potassium bromate solution from a 10-ml buret. At first the galvanometer will be deflected only slightly, and 1-ml increments may be added. When a definite galvanometer deflection occurs, decrease the size of the increments to 0.25 ml and secure several readings after the equivalence point.
3. Run the current–voltage curve of bromine from 0.3 V negative to 0.9 V positive vs. the SCE. The excess bromate reacts with the bromide ions in the presence of hydrogen ions to form free bromine, the electroactive species undergoing reduction at the rotating platinum electrode.
4. Repeat step 2 with duplicate samples to ascertain the precision attainable by this titration method.

Experiment 25-3   Use of Two-Indicator Electrode System (Dead-Stop Method)

1. Transfer 5 ml of 0.01 N iodine solution to a 150-ml beaker. Add 0.1 g of potassium iodide and dilute to about 60 ml.
2. Apply 0.1 V across two similar platinum wire (or foil) electrodes. Pass the solution by the electrodes with a magnetic stirrer. Titrate with a 0.01 N sodium thiosulfate solution from a 10-ml buret. Take readings every 0.5 ml until the galvanometer index remains constant through 4 or 5 additions. Plot the results on graph paper.

In place of a conventional instrument, connect a battery of 1.5 V through 0.1-MΩ and 7000-Ω resistors in series. Connect two similar platinum-wire electrodes to the terminals of the 7000-Ω resistor. In series with the cathode, insert a galvanometer with a sensitivity of about 0.1 μA/mm.

The titrant may be sodium arsenite; if it is, the pH of the solution must be adjusted between 4 and 9 with a suitable buffer.

PROBLEMS

1. Compare polarographic determinations with amperometric titrations with respect to (a) relative accuracy, (b) permissible range of applied potential usable, and (c) applicability of each method in regard to types of ions and range of concentrations.

2. Cupric ions form a precipitate with alpha-benzoinoxime in an ammoniacal solution. The $Cu(NH_3)_4^{2+}$ present in a supporting electrolyte consisting of 0.05 M ammonia and 0.1 M ammonium chloride is reduced stepwise, giving polarographic waves at -0.2 and -0.5 V vs. SCE. Alpha-benzoinoxime gives a polarographic wave with $E_{1/2} = -1.6$ V vs. SCE. Deduce the shape of the amperometric titration curve that will be obtained at applied voltage of -0.8 V vs. SCE, and also the sketch of the titration curve obtained at -1.7 V. Which potential would be preferred under normal circumstances? When

nickel and zinc ions are also present in the titrating solution, which potential would be preferred?

3. Contrast amperometric titration methods with potentiometric titration methods.

BIBLIOGRAPHY

Dolezal, J. and J. Zyka in *Standard Methods of Chemical Analysis*, 6th ed., F. J. Welcher, Ed., Vol. IIIA, D. Van Nostrand, New York, 1966; Chapter 20.

Kolthoff, I. M. and J. J. Lingane, *Polarography*, 2nd ed., Wiley-Interscience, New York, 1952.

Lingane, J. J., *Electroanalytical Chemistry*, 2nd ed., Wiley-Interscience, New York, 1958.

Stock, J. T., *Amperometric Titrations*, Wiley, New York, 1965.

LITERATURE CITED

1. Foulk, C. W. and A. T. Bawden, *J. Am. Chem. Soc.*, **48**, 2045 (1926).
2. Kolthoff, I. M., *Anal. Chem.*, **26**, 1685 (1954).
3. Laitinen, H. A., W. P. Jennings, and T. D. Parks, *Ind. Eng. Chem.*, *Anal. Ed.*, **18**, 355, 358 (1946).
4. Lingane, J. J., *J. Am. Chem. Soc.*, **65**, 866 (1943).

26

Conductance Methods

Conductance measurements were among the first to be used for determining solubility products, dissociation constants, and other properties of electrolyte solutions. Conductance is an additive property of a solution depending on all the ions present. Solution conductance measurements, therefore, are nonspecific. This nonspecificity restricts the quantitative analytical use of this technique to situations where only a single electrolyte is present or where the total ionic species needs to be ascertained. In these special situations, however, conductance measurements are capable of extreme sensitivity.

The possibility of using conductance to locate end points in titrations was also recognized early in the development of instrumental methods. Changes in the slope of conductance versus titrant volume occur because ionic mobilities vary and also because of the formation of insoluble or nonionized materials. Accordingly, conductometric indication of end points is possible. A related technique, high-frequency conductometric titration, was developed in recent years. High-frequency measurements permit the determination of changes in conductance, or dielectric constant, without the introduction of electrodes into direct contact with the solution.

ELECTROLYTIC CONDUCTIVITY

Electrolytic conductivity is a measure of the ability of a solution to carry an electric current. Solutions of electrolytes conduct an electric current by the migration of ions under the influence of an electric field. Like a metallic conductor, they obey Ohm's law. Exceptions to this law occur only under abnormal conditions, for example, very high voltages or high-frequency currents. Thus, for an applied electromotive force E, maintained constant but at a value that exceeds the decomposition voltage of the electrolyte, the current i flowing between the electrodes immersed in the electrolyte will vary inversely with the resistance of the electrolytic solution R. The reciprocal of the resistance $1/R$ is called the *conductance*, and is expressed in reciprocal ohms, or mhos.

The standard unit of conductance is specific conductance κ, which is defined as the reciprocal of the resistance in ohms of a 1-cm cube of liquid at a specified temperature. The units of specific conductance are the reciprocal ohm-cm (or mho/cm). The observed conductance of a solution depends inversely on the distance d between the electrodes and directly upon their area A

$$\frac{1}{R} = \kappa \frac{A}{d} \qquad (26\text{-}1)$$

The electrical conductance of a solution is a summation of contributions from all the ions present. It depends upon the number of ions per unit volume of the solution and upon the velocities with which these ions move under the influence of the applied electromotive force. As a solution of an electrolyte is diluted, the specific conductance in Eq. 26-1 will decrease. Fewer ions to carry the electric current are present in each cubic centimeter of solution. However, in order to express the ability of individual ions to conduct, a function called the *equivalent conductance* is employed. It may be derived from Eq. 26-1, where A is equal to the area of two large parallel electrodes set 1 cm apart and holding between them a solution containing one equivalent of solute. If C_s is the concentration of the solution in gram-equivalents per liter, then the volume of solution in cubic centimeters per equivalent is equal to $1000/C_s$, so that Eq. 26-1 becomes

$$\Lambda = 1000 \frac{\kappa}{C_s} \tag{26-2}$$

At infinite dilution the ions theoretically are independent of each other and each ion contributes its part to the total conductance, thus

TABLE 26-1   Limiting Equivalent Ionic Conductances in Aqueous Solution at 25°C

Cations	λ_+	Anions	λ_-
H^+	350	OH^-	198
Li^+	39	F^-	54
Na^+	50	Cl^-	76
K^+	74	Br^-	78
NH_4^+	73	I^-	77
Ag^+	62	NO_3^-	71
Mg^{2+}	53	IO_4^-	55
Ca^{2+}	60	HCO_3^-	45
Sr^{2+}	59	Formate	55
Ba^{2+}	64	Acetate	41
Zn^{2+}	53	Propionate	36
Hg^{2+}	53	Butyrate	33
Cu^{2+}	55	Benzoate	32
Pb^{2+}	71	Picrate	30
Co^{2+}	53	SCN^-	66
Fe^{2+}	54	SO_4^{2-}	80
Fe^{3+}	68	CO_3^{2-}	72
La^{3+}	70	$C_2O_4^{2-}$	74
Ce^{3+}	70	CrO_4^{2-}	85
$CH_3NH_3^+$	58	PO_4^{3-}	69
$N(Et)_4^+$	33	$Fe(CN)_6^{3-}$	101
$N(Bu)_4^+$	19	$Fe(CN)_6^{4-}$	111

SOURCE: J. A. Dean, Ed., *Lange's Handbook of Chemistry*, 11th ed., McGraw-Hill Book Company, New York, 1973, Table 6-7, p. 6–30.
NOTE: All values rounded to nearest unit.

$$\Lambda_\infty = \Sigma(\lambda_+) + \Sigma(\lambda_-) \tag{26-3}$$

where λ_+ and λ_- are the ionic conductances of cations and anions, respectively, at infinite dilution. Values for the limiting ionic conductances for selected ions in water at 25°C are given in Table 26-1. The ionic conductance is a definite constant for each ion in a given solvent, its value depending only on the temperature. Since these are actually equivalent conductances, symbols such as $\frac{1}{2}Ba^{2+}$ are sometimes employed. At finite concentrations interionic forces generally lower the ionic mobilities.

Example 26-1

The equivalent conductance at infinite dilution of H_2SO_4 is

$$\Lambda_\infty = 350 + 80 = 430 \ \Omega^{-1} \ cm^2$$

The molar conductance is given by

$$(2)(350) + (2)(80) = 860 \ \Omega^{-1} \ cm^2$$

The conductivity of solutions is quite temperature-dependent. An increase of temperature invariably results in an increase of ionic conductance, and for most ions this amounts to about 2% per degree. For precise work, conductance cells must be immersed in a constant-temperature bath. It is customary to select 25°C for measurements in the United States, although generally 18°C is preferred in Europe. For relative measurements, as in titrations, the conductance cell need only attain thermal equilibrium with its surroundings before proceeding with conductance measurements.

MEASUREMENT OF ELECTROLYTIC CONDUCTANCE

Electrolytic conductance measurements usually involve determination of the resistance of a segment of solution between two parallel electrodes by means of Ohm's law. These electrodes are platinum metal that has been coated with a deposit of platinum black to increase the surface area and reduce the polarization resistance. Some of the more important phenomena associated with the application of a voltage between electrodes immersed in a liquid electrolyte are indicated in Fig. 26-1 for an idealized system. To eliminate the effects of processes associated with the electrodes, such as those discussed in Chapter 23, measurements are made with an alternating current at 60, 100, 1000, or 3000 Hz. Some variation of the Wheatstone bridge is generally employed. The evaluation involves a comparative procedure. A conductance cell is calibrated by determining its cell constant, using a solution of known conductivity as will be described later.

Generally the bridge circuit must contain not only resistance, but also capacitance (or inductance) to balance the capacitive effects in the conductance cell. The latter arises from the electric double-layer at the electrode–electrolyte interface at applied voltages below the decomposition voltage, and from the frequency-dependent resistance (impedance) associated with the Faradaic processes at voltages above the decomposition voltage. For resistances of less than $10^4 \ \Omega$ the model of a conductance cell as the electrolyte resistance in series with the double-layer capacitance is a reasonable physical approximation.[1]

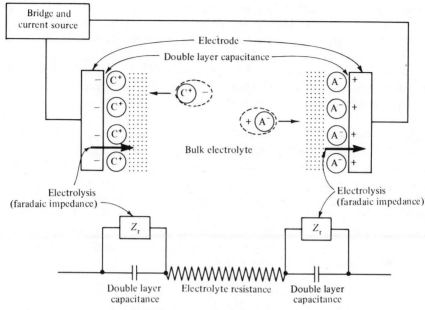

Fig. 26-1. Electrolytic conductance cell—a simplified representation of the double-layer at the electrodes, faradaic processes, and migration of ions through the bulk electrolyte. (From J. Braunstein and G. D. Robbins, *J. Chem. Educ.*, **48**, 52 (1971). Courtesy of American Chemical Society.)

The magnitude of impedance at 1000 Hz is of the order of

$$\frac{1}{2\pi f C} = 1.6 \text{ to } 16 \ \Omega \tag{26-4}$$

for capacitance values of 10–100 $\mu f/cm^2$ of electrode surface. Introduction of a variable capacitance (or inductance) into the bridge circuit permits compensation of the phase shift between current and voltage caused by the capacitance in the electrolytic cell. Parallel RC balancing arms are used more frequently than a series arrangement because smaller capacitance values are needed, and small capacitances can be obtained with higher accuracy and less frequency dependence than large ones. For example, the parallel capacitance required to compensate a series capacitance of 100 μf at 1000 Hz with a resistance of 1000 Ω is only 300 pf.

Instrumentation

A typical commercial conductivity bridge, shown in Fig. 26-2, is designed to measure electrolytic conductance in micromhos and resistance in ohms. The reading for either is indicated directly on a digital readout dial. A built-in generator provides bridge current at frequencies of 100 and 1000 Hz. Generally the lower frequency should be used when the measured resistance is high, and the higher frequency when the measured resistance is low. The instrument is balanced with the aid of a phase-sensitive detector and a null meter. Inexpensive process control instruments use an electron-ray (magic-eye) tube. With precision-class instruments, oscilloscopes are used for balance indication. The null condition is indicated as a straight horizontal line on the oscilloscope, whereas resistive and reactive imbalance are shown respectively by tilting of the line and by widening to an ellipse.

The operation of the instrument can be understood from Fig. 26-3(a). The generator supplies a sinusoidal drive voltage to the bridge arm as well as a reference voltage for the phase detector. In the resistance mode, the bridge is an equal-arm bridge with cell and decade resistance in adjacent arms. At balance the decade resistance equals the cell resistance. The bridge is balanced by adjusting the readout dial resistor, a 10-turn potentiometer coupled to a mechanical counter which constitutes the readout device. The range of the instrument is changed by switching in different multiplier resistors for each range. The cell constant of the conductivity cell should be selected to maintain the measured resistance between 100 Ω and 1.1 MΩ.

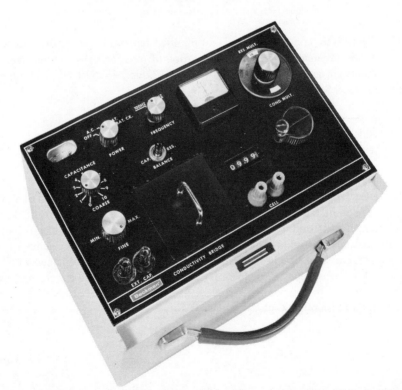

Fig. 26-2. A laboratory and field conductivity bridge. (Courtesy of Beckman Instruments, Inc.)

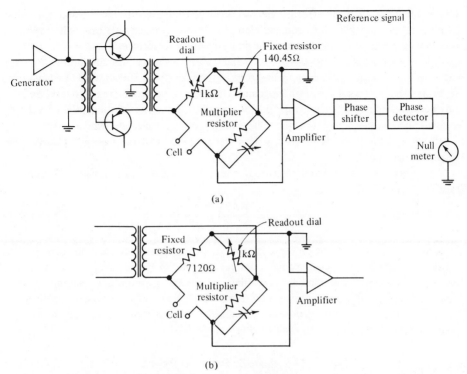

Fig. 26-3. Simplified schematic of a conductivity bridge: (a) resistance mode and (b) conductance mode. (Courtesy of Beckman Instruments, Inc.)

The small output signal of the bridge at balance is amplified and applied to a phase-shifting circuit where in the resistive position the signal is shifted slightly to make up for any phase change that occurs in the amplifier and transformers. In the capacitive position the phase is shifted 90° so that the detector will be sensitive only to capacitive imbalances in the bridge circuit. To facilitate reactive balance, a continuously variable 8 to 200 pf capacitor, plus 10 steps of 200 pf capacitance, is placed in the arm of the bridge adjacent to the cell. The output of this circuit is then applied to the detector where it is added to the reference signal and the resultant signal is read on the meter.

Figure 26-3(b) shows the bridge in the conductance mode. Now the decade resistance is in the arm opposite the cell. All other circuits remain the same and perform the same functions as they do when measuring resistance.

Conductance Cells

In the design of conductance cells for precision measurements a number of factors must be taken into consideration. However, for many purposes two parallel sheets of platinum fixed in position by sealing the connecting tubes into the sides of the measuring cell are adequate. Also satisfactory are two sheet-platinum electrodes or wands immersed in the solution and held by ordinary clamps. These arrangements make the measured resistance independent of sample volume and proximity to surface.

There are practical limits of measured electrolytic resistance for any desired accuracy and sensitivity. The optimum appears to be in the vicinity of 500–10,000 Ω when errors are to be ±0.1%. In solutions of low conductance, the electrode area A should be large and the plates spaced (ℓ) close together; for highly conducting solutions, the area should be small and the electrodes far apart. The platinum electrodes are almost always lightly plated with platinum black to reduce the polarizing effect of the passage of current between the electrodes.

For a given cell with fixed electrodes, the ratio ℓ/A is a constant, called the cell constant Θ. It follows that

$$\kappa = \frac{1}{R}\left(\frac{\ell}{A}\right) = \frac{\Theta}{R} \tag{26-5}$$

For conductance measurements a cell is calibrated by measuring R when the cell contains a standard solution of known specific conductance, and Θ is then computed by means of Eq. 26-5. The electrolyte almost invariably used for this purpose is potassium chloride. Values of the specific conductance of potassium chloride solutions are given in Table 26-2. For conductometric titrations the absolute conductance need not be known, merely relative conductances as the titration progresses.

TABLE 26-2 Specific Conductances of
Potassium Chloride Solutions

Grams KCl/kg of Solution	κ in Ω^{-1} cm^{-1}	
	18°C	25°C
71.1352	0.09784	0.11134
7.4191	0.01117	0.01286
0.7453[a]	0.001221	0.001409

SOURCE: G. Jones and B. C. Bradshaw, *J. Am. Chem. Soc.*, **55**, 1780 (1933).
[a]Virtually 0.0100 *M*.

Example 26-2

When a certain conductance cell was filled with 0.0100 *M* solution of KCl, it had a resistance of 161.8 Ω at 25°C, and when filled with 0.005 *M* NaOH it had a resistance of 190 Ω.

The cell constant is

$$\Theta = (0.001409)(161.8) = 0.2280 \text{ cm}^{-1}$$

The specific conductance of the sodium hydroxide solution is

$$\kappa = \frac{\Theta}{R} = \frac{0.2281}{190} = 0.00120 \ \Omega^{-1} \text{ cm}^{-1}$$

and the equivalent conductance is

$$\Lambda = \frac{(1000)\,(0.00120)}{0.005} = 240 \text{ cm}^2 \text{ equiv}^{-1} \ \Omega^{-1}$$

Various types of conductivity cells are commercially available. The dip cell (Fig. 26-4) is the simplest to use whenever the liquid to be tested is in an open container. It is merely immersed in the solution to a depth sufficient to cover the electrodes and the vent holes. Liquid volumes of 5 ml or less suffice for small-diameter dip cells. A pair of individual platinum electrodes on glass wands is useful in conductance titrations. Epoxy cells are used for high-temperature work in corrosive solutions except concentrated

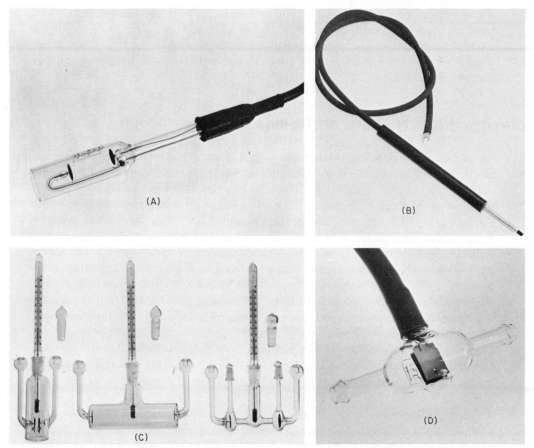

Fig. 26-4. Conductivity cells. (A) Dip-type cell for medium conductance solutions; cell constants from 0.5 to 2. (B) Individual wands for conductometric titrations. (C) Fill-type cell for laboratory work to contain the sample under test and to be immersed in a temperature bath. (D) Flow-through cell; cell constants from 0.01 to 0.2. (Courtesy of Beckman Instruments, Inc.)

oxidizing acids. Pipet cells permit measurements with small volumes of solution, as little as 0.01 ml in some designs.

Temperature Compensation

Conductivity varies with temperature as well as with electrolyte concentration. The temperature coefficient of conductance of electrolyte solutions in water is almost always positive and of a magnitude from about 0.5 to 3% per degree Centigrade. A practical means of providing temperature compensation is to introduce into the bridge circuit a resistive element that will change with temperature at the same rate as the solution under test. In different forms, this temperature compensator arm of the bridge can be a rheostat calibrated in temperature and requiring manual adjustment or a thermistor and fixed resistive network in thermal contact with the test solution, to provide automatic compensation. Regardless of the means employed, accurate compensation for temperature changes requires that the temperature coefficient of resistance of the compensator match that of the test solution.

DIRECT CONCENTRATION DETERMINATIONS

Although the electrical conductance of a solution is a general property and is not specific for any particular ion, a number of analyses can be made by means of a measurement of conductivity. In general, the success of a measurement depends upon relating the property of the sample that is to be estimated to the conductance of some highly conducting ion. For example, free caustic remaining in scrubbing-tower solutions can be estimated by observing the decrease in conductance of the solution. This is possible even in the presence of the salts formed upon neutralization of the alkali, because the conductance of the hydroxyl ion is approximately fivefold greater than that of any other anion. Similarly, the unusually high conductance of the hydrogen ion permits an estimation of the free acid content in acid pickling baths. The changing conductivity resulting from the absorption of gaseous combustion products in suitable solutions is frequently employed for the determination of carbon, hydrogen, oxygen, and sulfur individually, in organic and in inorganic compounds. The change in conductivity is always measured in relation to an identical solution not in contact with the combustion products. On the other hand, when checking the purity of distilled or deionized water, steam distillates, rinse waters, boiler waters, or in regeneration of ion exchangers, it is the total salt content which is sought.

For all these purposes very compact and inexpensive conductance bridges are available with scales calibrated directly in pounds per gallon, parts per million, grains per gallon, or percent. For example, instruments can be supplied for direct indication such as 1–12 lb Na_2CO_3 per 100 gal., 0–40 parts per thousand salinity, 0.4–10% H_2SO_4, 0.4–12% NaOH, and 96–99.5% H_2SO_4. These units are intended for industrial monitoring as well as for following an industrial process.

Example 26-3

The scale of a conductivity bridge is inscribed from 0.005 to 2.0% H_2SO_4 in approximately a logarithmic manner. For these solutions the specific conductance ranges from about 0.00044 to 0.176 Ω^{-1} cm^{-1}. What range of resistances are involved and what cell constant is compatible?

Answer The resistance values will range from

$$R = \frac{\Theta}{0.00044} = 22{,}800$$

to

$$R = \frac{\Theta}{0.176} = 56.9$$

A suitable cell constant is 10.0 cm^{-1}; the resistance ranging from 57 to 22,800 Ω. A cell constant of 20 cm^{-1} would also be suitable. A smaller cell constant would provide too low a resistance for the stronger acid solutions.

A cell with constant 10 cm^{-1} would have electrodes of moderate area and some distance apart, perhaps electrodes 0.5 cm^2 in area and spaced 5 cm apart.

Direct-reading conductivity meters are also available. These instruments apply a stabilized ac voltage to the conductivity cell and a series resistor, rectify the voltage drop across the series resistor, and measure the resultant dc signal. As long as the resistor in series with the cell is smaller in resistance, the dc signal will be directly related to the cell resistance. Continuous indication is provided on a linear meter scale.

CONDUCTOMETRIC TITRATIONS

In this method the variation of the electrical conductivity of a solution during the course of a titration is followed. It is not necessary to know the actual specific conductance of the solution; any quantity proportional to it is satisfactory. This may result in considerable simplification of equipment. The titrant is introduced by means of a buret, and the conductance readings corresponding to various increments of titrant are plotted against the latter. Figure 26-5 illustrates the conductometric titration of hydrochloric acid with sodium hydroxide. As seen from Eq. 26-3, the measured conductance is a linear function of the concentration of ions present. In the example, the falling branch represents the conductance of the hydrochloric acid still present in the solution, together with that of the sodium chloride already formed. The rising branch represents the conductance of the excess base present after neutralization, together with that of the sodium chloride. Since the variation of conductance is linear, it is sufficient to obtain six or eight readings, covering the range before and after the end point, and draw two straight lines through them. The point of intersection of the two branches gives the end point.

If the reaction is not quantitative, there is curvature in the vicinity of the end point. Hydrolysis, dissociation of the reaction product, or appreciable solubility in the case of

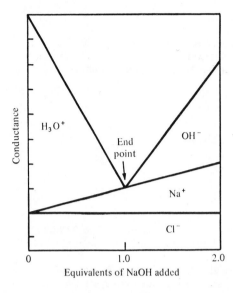

Conductance

H$_3$O$^+$

End
point

OH$^-$

Na$^+$

Cl$^-$

0   1.0   2.0
Equivalents of NaOH added

Fig. 26-5. Titration of hydrochloric acid with sodium hydroxide.

precipitation reactions will give rise to this type of curvature. At a sufficient distance on either side away from the end point, from 0 to 50% and between 150 and 200% of the equivalent volume of titrant, sufficient common ion is present to repress these effects, and the branches are straight lines. By extrapolating these portions of the lines, the position of the end point can be determined.

The acuteness of the angle at the point of intersection of the two branches will be a function of the individual ionic conductances of the reactants. In Fig. 26-5 the falling branch was steep because it involved the replacement of hydrogen ion ($\lambda_+ = 350$) by sodium ions ($\lambda_+ = 50$), and a large difference exists between the two conductances. Similarly, the rising branch on this curve is relatively steep also, but not as steep as the falling branch, because the conductance of the hydroxyl ion ($\lambda_- = 198$) is considerably smaller than the corresponding value for the hydrogen ion.

The titrant should be at least ten times as concentrated as the solution being titrated in order to keep the volume change small. If necessary, a correction may be applied:

$$\left(\frac{1}{R}\right)_{actual} = \left(\frac{V+v}{V}\right)\left(\frac{1}{R}\right)_{obs} \tag{26-6}$$

where V is the initial volume and v is the volume of titrant added up to the particular conductance reading.

In principle, all types of reactions can be employed. The method can be used with very dilute solutions, about 0.0001 M. On the other hand, because every ion present contributes to the electrolytic conductivity, large amounts of extraneous electrolytes should be absent. In the presence of large amounts of such electrolytes, the change in conductance accompanying a reaction would be a very small part of the total conductance and would be difficult to measure with accuracy. For this reason oxidation–reduction reactions usually cannot be performed because the solutions are generally well buffered or strongly acidic.

Under optimum conditions the end point in conductometric titrations can be located with a relative error of approximately 0.5%. A single titration requires about 10 min.

Acid–Base Titrations

The conductometric titration of an acid with a base, each completely dissociated, is illustrated by Fig. 26-5 for the titration of hydrochloric acid with sodium hydroxide:

$$(H^+ + Cl^-) + (Na^+ + OH^-) \longrightarrow (Na^+ + Cl^-) + H_2O \qquad (26\text{-}7)$$

The highly conducting hydrogen ions initially present in the solution are replaced by sodium ions having a much smaller ionic conductance, while the concentration of chloride ions remains constant except for the small dilution by the titrant. The conductance of the solution at any point on the descending branch of the titration curve is given by the expression

$$\frac{1}{R} = \frac{1}{1000\Theta} (C_H\lambda_H + C_{Na}\lambda_{Na} + C_{Cl}\lambda_{Cl}) \qquad (26\text{-}8)$$

In terms of the initial concentration of hydrochloric acid C_i and the fraction of the acid titrated f,

$$C_H = C_i(1 - f), \qquad C_{Na} = C_i f, \qquad \text{and} \qquad C_{Cl} = C_i$$

Substituting these values into Eq. 27-8,

$$\frac{1}{R} = \frac{C_i}{1000\Theta} [\lambda_H + \lambda_{Cl} + f(\lambda_{Na} - \lambda_H)] \qquad (26\text{-}9)$$

As a result of the term within the parentheses in Eq. 26-9, the conductance of the solution diminishes up to the equivalence point. Beyond the equivalence point the conductance increases in direct proportion to the excess base added.

Example 26-4

Let us assume that one is titrating 100 ml of 0.01 N HCl solution with 0.1 N NaOH in a cell whose constant is 1.0 cm^{-1}. Under these conditions the initial conductance is

$$\frac{1}{R} = \frac{0.01}{(1000)\,(1)} (350 + 76) = 0.00426 \ \Omega^{-1}$$

In this dilute solution, little error is introduced by assuming that the actual ionic conductances are essentially those at infinite dilution when dealing with completely ionized materials.

When the titration is 0.9 complete,

$$\frac{1}{R} = \frac{0.01}{(1000)\,(1)}\,[350 + 76 + (0.9)\,(50 - 350)]$$

$$= 0.00156\ \Omega^{-1}$$

and, correcting for the dilution caused by the titrant,

$$\frac{1}{R} = \frac{(100 + 9)}{100}\,(0.00156) = 0.00170\ \Omega^{-1}$$

At the equivalence point, there exists a solution of sodium chloride whose conductance is

$$\frac{1}{R} = \frac{0.01}{(1000)\,(1)}\,(76 + 50)\left(\frac{110}{100}\right) = 0.00139\ \Omega^{-1}$$

When the equivalence point has been exceeded by 10%

$$\frac{1}{R} = \frac{0.01}{(1000)\,(1)}\,[126 + (0.1)\,(248)]\left(\frac{111}{100}\right) = 0.00167\ \Omega^{-1}$$

If the strong acid is titrated with a weak base, for example, an aqueous solution of ammonia, the first part of the conductance titration curve, representing the removal of hydrogen ion and replacement by ammonium ion, will be very similar to the descending branch of Fig. 26-5, because most cations have similar ionic conductances. After the equivalence point is passed, however, the conductance will remain almost constant, because a solution of ammonia has a very small conductance compared with that of ammonium chloride (Fig. 26-6).

$$(H^+ + Cl^-) + NH_3 \longrightarrow NH_4^+ + Cl^- \tag{26-10}$$

Titrations of weak acids or weak bases are somewhat more difficult. Acetic acid, for example, is present partly in the form of H^+ and CH_3COO^-, but largely as nonionized molecules. The proportions of each are regulated by the law of mass action. Initially the solution has a low conductance. As neutralization proceeds, the common ion formed, that is, the acetate ion, represses the dissociation of the acetic acid so that an initial fall in conductance may occur. With further addition of sodium hydroxide the conductance of the sodium and acetate ions soon exceeds that of the acetic acid which they replace, and so the curve passes through a minimum and thereafter the conductance of the solution increases. The shape of the initial portion of these conductance curves will vary with the strength of the weak acid and its concentration, as indicated in Figs. 26-6 and 26-7.

When a weak acid is titrated with a weak base, the initial portion of the conductance titration curve follows the pattern described above. Beyond the equivalence point, there is no change in the conductance because of the very small conductance of the excess free base. The intersection of the two branches is sharper than for a corresponding titration of a weak acid with a strong base.

Pronounced hydrolysis in the vicinity of the equivalence point makes it necessary to select the experimental points for the construction of the two branches considerably re-

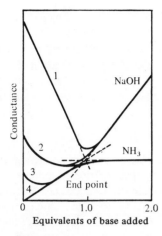
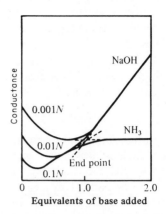

Fig. 26-6. Conductometric titration curves of various acids of sodium hydroxide and by aqueous ammonia. The numbered curves are (1) HCl, (2) acid, pK_a of 3, (3) acid, pK_a of 5, and (4) acids, $pK_a > 7$.

Fig. 26-7. Conductometric titration curves of acetic acid $(pK_a = 4.8)$ at various concentrations.

moved from the equivalence point. It will be observed in Fig. 26-7 that for dilute solutions of weak acids no linear region is obtained preceding the equivalence point, and therefore no reliable point of intersection seems possible. Addition of ethanol or other solvent with a smaller dielectric constant than water often reduces the dissociation of the weak acid sufficiently to yield a region of linear conductance preceding the equivalence point. When this does not suffice, two other experimental modifications may aid in the location of the equivalence point. In one procedure, titrations are conducted with duplicate portions of test solution, using an aqueous solution of ammonia as one titrant and a strong base as the other titrant, both titrants equivalent in normality. The conductance curves preceding the equivalence point are practically identical in shape, but beyond the equivalence point the curves will have quite different slopes. After superimposing the fore-portions, the intersection of the two branches beyond the equivalence point establishes the end point, as shown in Fig. 26-7. In the second procedure, sufficient ammonia is added to neutralize about 80% of the weak acid; then the titration is carried out with standard sodium hydroxide. If the conductance were plotted during the addition of the aqueous solution of ammonia, the curve would resemble those discussed. As long as any of the original acid is present the curve will continue to parallel the shape shown in Fig. 26-6, but when all the acid has been consumed, the ammonium ion formed commences to react with the hydroxyl ion, thus

$$(NH_4^+ + CH_3COO^-) + (Na^+ + OH^-) \longrightarrow NH_3 + H_2O + Na^+ + CH_3COO^- \qquad (26\text{-}11)$$

and the conductance falls owing to the replacement of the ammonium ion $(\lambda_+ = 73)$ by the sodium ion $(\lambda_+ = 50)$. When the replacement is complete, the conductance abruptly increases, as shown in Fig. 26-8, and is then parallel to the corresponding part of Fig. 26-5 after the equivalence point.

The determination of a very weak acid or a very weak base is merely an extension of the titration of weak acids or weak bases. Take, for example, the titration of boric acid ($pK_a = 9.2$) with sodium hydroxide, as shown in Fig. 26-6. Initially the conductance is very small, since boric acid is dissociated to a negligible extent in aqueous solution. During the titration the reaction

$$HBO_2 + (Na^+ + OH^-) \longrightarrow Na^+ + BO_2^- + H_2O \qquad (26\text{-}12)$$

occurs, and the conductance increases linearly with the formation of borate and the addition of sodium ions. Beyond the equivalence point, further addition of sodium hydroxide introduces hydroxyl ions and there is a further increase of conductance. The inflection point will not be particularly sharp; in fact, it is often impossible to locate accurately the intersection of the two branches. This is true when the acid becomes so weak that extensive hydrolysis occurs throughout the reaction and the nonlinear portion extends from the end point to the initial point. Only if the product of the ionization constant and the acid concentration exceeds 10^{-11} can the titration be performed.

One of the valuable features of the conductance method of titration is that it permits the analysis of a mixture of a strong and a weak acid in one titration. Figure 26-9 illustrates the neutralization of oxalic acid ($pK_1 = 1.2$; $pK_2 = 4.3$), representative of a strong acid and a moderately strong acid present in equivalent amounts. The initial decrease in conductance is due to the removal of hydrogen ions supplied by the relatively complete dissociation of $H_2C_2O_4$. This is followed by an increase in conductance as the weak acid, $HC_2O_4^-$, is consumed and replaced by $C_2O_4^{2-}$ ions and the cation of the titrant. A rounded

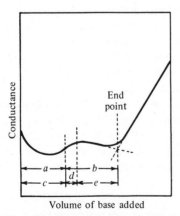

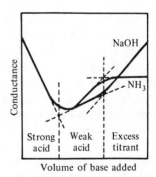

Fig. 26-8. Conductometric titration of a weak acid employing preliminary addition of aqueous ammonia followed by sodium hydroxide: (a) volume of aqueous ammonia added; (b) volume of NaOH added up to the end point; (c) amount of acid neutralized by the NH$_3$; (d) amount of acid neutralized by the NaOH; and (e) displacement of NH$_3$ from NH$_4^+$ formed in (c).

Fig. 26-9. Conductometric titration of a mixture of a strong acid and a weak acid. Example: oxalic acid.

section joins these two branches because the pK_a values are too close to each other and consequently the neutralization of the second acid begins while that of the first is being completed. When the neutralization of the second acid is complete, there is an increase in conductance when sodium hydroxide is the titrant due to the sodium and hydroxyl ions. Substitution of ammonia as the titrant results in a greater rise in conductance on the middle portion of the titration curve and little further change in conductance after the second end point. The first point of intersection gives the amount of strong acid in the mixture, and the difference between the first and second is equivalent to the amount of weak acid. Practical applications include the determination of mineral acids in vinegar and the titration of sulfonic acid groups followed by either carboxylic or phenolic groups in mixtures of organic acids.

The conductance method is also useful in the titration of the conjugate base of a weakly ionized acid, and vice versa. For example, organic acid salts such as acetates, benzoates, and nicotinates can be titrated with a standard solution of a completely ionized acid. As long as the ionization constant of the displaced acid or base divided by the original salt concentration does not exceed approximately 5×10^{-3}, the displaced acid will not contribute to the total conductance. The descending branch in the middle portion of Fig. 26-8 illustrates the nature of the titration curve that will be obtained.

Determinations by Precipitations and Through Formation of Complexes

Mercuric nitrate and perchlorate have found use as reagents for complexometric re-actions. These salts exist almost entirely in the form of free ions. If a solution of a cyanide is added, the reaction

$$(Hg^{2+} + 2ClO_4^-) + 2(K^+ + CN^-) \longrightarrow Hg(CN)_2 + 2(K^+ + ClO_4^-) \qquad (26\text{-}13)$$

occurs. Before the equivalence point, 1 mercuric ion is replaced by 2 potassium ions. The conductance varies only slightly. Beyond the end point the addition of potassium and cyanide ions causes the conductance to increase. In this class of reactions the slopes of the branches of the curve are determined both by the change in the ionic conductances of the ions present and by any change in the total number of electrical charges carried by the ions in solution. For these reasons, an acetate salt is preferable when titrating an anion, and a lithium salt when titrating a cation.

Even in favorable cases, results obtained through electrical conductivity tend to be less accurate than in acid–base systems. In precipitations, all the factors influencing the formation of the precipitate and the nature of the product must be considered just as in ordinary gravimetric methods. A slow rate of precipitation, excessive solubility of the insoluble materials, and all types of adsorption or occlusion difficulties make it difficult if not impossible to locate the equivalence point with any degree of accuracy.

HIGH-FREQUENCY METHODS

A current alternating at frequencies exceeding 1 MHz is affected by the conductance and capacitance of a solution within the field. The vessel containing the solution is placed in the field of an inductance coil or between the plates of a capacitor carrying the high-frequency current. Because the inductance coil or the capacitor is part of the high-frequency oscillator circuit, any changes in the composition of the solution will be reflected as changes in the oscillator, or changes in the plate and grid currents and voltages. Capacitance balance is the major factor in high-frequency measurements, whereas resistive balance was the more important in the determination of low-frequency conductivity. The unique advantage of methods based on high-frequency alternating currents is the possibility of placing the electrodes outside the vessel and out of direct contact with the solution. Significant measurements can be made without danger of electrolysis or electrode polarization, and without altering or consuming any solution.

Oscillator Circuit

The fundamental circuit of an oscillator is shown in Fig. 26-10. It consists of a capacitance C and an inductance L, plus the resistive components attributed to both. Upon adjustment of the circuit capacitance (called "tuning"), a condition will be attained which is known as resonance. When the condenser is discharged through the inductance, current will surge back and forth from the inductor to the capacitor at a frequency given by

$$f = \frac{1}{2\pi\sqrt{LC}} \tag{26-14}$$

Since the circuit always has some resistance, the current will decay rapidly. If, however, the resonant circuit of Fig. 26-10 is coupled to the input of an electronic amplifier, as shown in Fig. 26-11, the voltage impressed on the grid will alternate at the frequency given by Eq. 26-14. As a result the output voltage at the plate, which is μe_g, will also alternate at the same frequency. Now if an inductance L_2 is inserted in the plate circuit and coupled to L_1 in the grid circuit by means of a common core, the amplified plate voltage will appear across L_2. With the windings of the coils properly arranged, coil L_2 can serve to convey the output voltage oscillations back to the input grid circuit. Then if L_1 is selected so that the magnitude of the feedback voltage delivered from it exactly

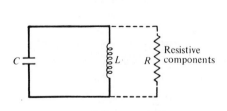
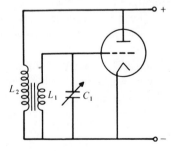

Fig. 26-10. Fundamental oscillator circuit.
Fig. 26-11. Schematic diagram of a tuned grid oscillator circuit.

equals the original input grid voltage, then it can replace it. Furthermore, the electronic circuit will continue to amplify the returned signal and oscillations will be maintained. The resonant circuit can be placed in either the grid or plate circuit, or in both circuits and coupled through a common inductance.

High-Frequency Titrimeters

The test solution is made part of the high-frequency oscillating circuit, either in the capacity or the induction branch. The cell may be the tube upon which the coil is wound, with corresponding alterations in capacitance between the windings and in inductance due to the change in permeability of the material making up the cell, or it may be the dielectric material between the plates of a condenser. Changes in the solution's conductivity and dielectric properties will affect the value of the particular component in the oscillator circuit which contains the cell and consequently alter the high-frequency conductance and capacitance of the cell. Depending upon the particular oscillator circuit, in some instruments this effect can be observed as a change in the plate or grid currents, the voltages at the plate or grid, or the frequency at which resonance occurs; in others the effect of changes in the electrolyte upon the inductance or capacitance is offset by retuning the oscillator to the original resonant frequency. As shown in Fig. 26-12, the sample cell is in parallel with a calibrated capacitor. In order to return to the resonant frequency it is necessary to remove the exact amount of capacitance that was added by the sample.

Another convenient method is to measure the output frequencies of two identical oscillator circuits, one of which contains the sample cell as part of the oscillator capacitance (f), the other serving merely as a reference unit (f_0). The outputs from the two units are fed into a mixer unit from which is obtained a low-frequency difference $(f - f_0)$ directly proportional to the change in the high-frequency capacitance of the cells. Either the difference frequency is measured directly, or after each change in the test solution the value of the variable precision condenser is altered to make up for the capacitance change in the cell. Instruments of a beat-frequency type or those involving compensatory alteration in circuit components are more stable in practice than direct-reading instruments. An instrument of the compensatory type which is commercially available is illustrated in Fig. 26-13.

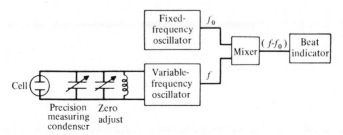

Fig. 26-12. Schematic diagram of a high-frequency titrimeter.

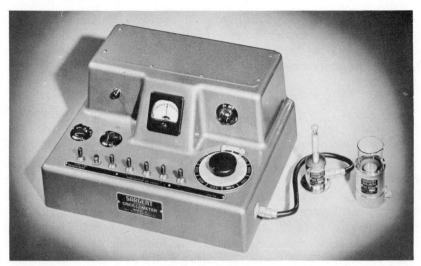

Fig. 26-13. The chemical oscillometer. Test-tube sample holder and titration cell are shown at the right. Switches along the panel front are calibrated capacitors in parallel with a variable capacitor (main dial). (Courtesy of E. H. Sargent & Co.)

Equivalent Circuit of Cell and Solution

The vertical cross section of a typical cell is shown in Fig. 26-14(a). The outside and inside surfaces of the annular space are plated with metal. When the annular space between the metal plates is filled with liquid, the high-frequency behavior of the system will be similar to that of a system in which the metal plates of a condenser are separated by a dielectric, comprising in this case the glass walls and the sample. The fundamental equivalent circuit of the cell may be treated as shown in Fig. 26-14(c). C_s represents the capaci-

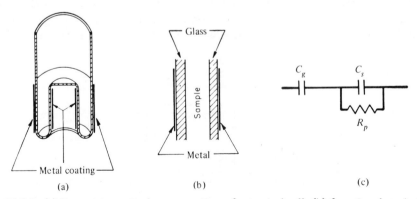

Fig. 26-14. (a) Isometric vertical cross section of a typical cell; (b) functional equivalent of the cell; (c) equivalent circuit of cell and solution.

tance across the liquid sample, and C_g is the capacitance of the glass walls of the cell. In case the solution within the cell has appreciable conductance, the approximate equivalent circuit may be represented by placing a resistance, R_p, in parallel with C_s. The resistive component of the container walls is so high as to be a negligible contribution.

High-Frequency Conductance Term

When the frequency at which the field between the two electrodes alternates becomes greater than 1 MHz, the conductance of the solution begins to include capacitance terms in addition to the ordinary low-frequency conductance (Fig. 26-15). Consider what happens to the individual ions of an electrolyte and to polar molecules when exposed to a rapidly alternating field. Each ion or dipolar molecule tends to move or align itself in the direction of the electrode of opposite polarity. Once every cycle the electrode polarity changes, and the ion or dipole must reverse its motion or orientation. The conductivity of the solution is the result of movement of positive and negative ions relative to their neighbors and to the solvent molecules. Each ion tends to move ahead of its ionic atmosphere, with the result that an unsymmetrical charge distribution forms around each central ion and exerts a retardation force on the ion in a direction opposite to its motion. However, at alternating frequencies greater than 1 MHz, the central ion changes its direction of motion so quickly with every cycle of the applied field that there is little chance for the dissymmetry of the ionic atmosphere to arise. As a result the conductance increases.

The relationship between conventional low-frequency conductance measurements

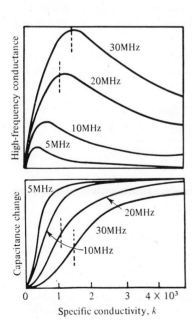

Fig. 26-15. High-frequency conductance and capacitance change as a function of low-frequency conductivity. Zero refers to a cell filled with pure solvents.

and high-frequency conductance, G, or the capacitance, C_p, is given by[2]

$$G = \frac{1}{R_p} = \frac{\kappa(2\pi f)^2 C_g^2}{\kappa^2 + (2\pi f)^2 (C_g + C_s)^2} \tag{26-15}$$

$$C_p = \frac{C_g\kappa^2 + 2\pi f(C_g C_s^2 + C_g^2 C_s)}{\kappa^2 + (2\pi f)^2 (C_g + C_s)^2} \tag{26-16}$$

where κ is the low-frequency specific conductance and the other terms have the meanings expressed by Eq. 26-14 and in Fig. 26-14. As C_g increases, that is, thinner cell walls, the value of G increases, and when C_g approaches infinite capacitance, the high-frequency conductance approaches the low-frequency conductance as a limit. When κ approaches a very small or very large value, G approaches zero so that there is a peak for a given frequency. The position of the peak is important from the standpoint of reversals and curvature in the titration curves based on high-frequency conductance. Upon differentiating Eq. 26-15, setting the result equal to zero, and solving,

$$\kappa_{peak} = 2\pi f(C_g + C_s) \tag{26-17}$$

When C_g becomes negligibly small in comparison with C_s, the value of the high-frequency conductance at the maximum increases with frequency, and

$$f_{peak} = \frac{1.8 \times 10^{12}\kappa}{D} \tag{26-18}$$

where D is the dielectric constant of the solution. From this relation it would seem that very high frequencies must be employed if solution concentrations in the range of 0.1 M are to be measured. Table 26-3 gives the concentration of electrolyte solutions giving maximum sensitivity for various oscillator frequencies.

It is difficult to construct oscillators which will operate at frequencies much above 30 MHz and provide trouble-free service in routine laboratory use. Thus the method is

TABLE 26-3 Relation of Electrolyte Concentration and Frequency for Maximum Sensitivity

Oscillator Frequency, MHz	Concentration of Maximum Sensitivity		
	NaCl	CaCl$_2$	HCl
5	0.0025 M	0.0013 M	0.0006 M
30	0.014	0.008	0.003
57	0.032	0.015	0.005
100	0.05	0.027	0.01
375	0.2	0.1	0.036

SOURCE: W. J. Blaedel and H. V. Malmstadt, *Anal. Chem.*, **22**, 734 (1950); J. Forman and D. J. Crisp, *Faraday Soc.* **42A**, 186 (1946).

severely restricted in respect to the range of concentrations of electrolytes that can be employed with a reasonable degree of instrument response.

Capacitance Changes

The change in equivalent parallel capacitance is actually the quantity measured by titrimeters of the beat-frequency type or those involving compensatory alteration in circuit components to make up for the capacitance change in the test solution. Reference to Fig. 26-15 shows that the capacitance, for dilute ionic solutions, follows an S-shaped curve when plotted as a function of the low-frequency conductivity of the solution. The peak high-frequency conductance and the inflection point on the capacitance curve occur at the same low-frequency conductivity value. Consequently, Eq. 26-17 applies when speaking of the midpoint of a capacitance curve; similarly, the discussion also applies to the relationship for the concentration of electrolyte solutions giving maximum sensitivity for various oscillator frequencies.

Lowering the dielectric constant of the solvent decreases C_s and results in an increased capacitance change, and at the same time shifts the inflection point to lower values of conductivity. This is advantageous for acid–base titrations in certain non-aqueous solvents.

Titration Curves

Before attempting to carry out high-frequency titrations it is advisable to determine the response of the instrument as a function of the electrolyte concentration. Each instrument with associated cell has an optimum range for which changes in the test solution will produce the maximum response. Typical high-frequency conductance and capacitance curves are shown in Fig. 26-15. The relationship between the forms of ordinary low-frequency titration curves and high-frequency curves are best shown by means of transfer plots. In Fig. 26-16, upper right and lower right, are two plots of specific conductivity at low frequency versus the corresponding response at 10 and 3 MHz, respectively. To ascertain the corresponding high-frequency titration curve from an ordinary titration curve, the high-frequency response corresponding to the specific conductivity for each volume of titrant added is found from either the 10- or the 3-MHz transfer plot and then plotted versus the respective volume of titrant. The shapes of the high-frequency titration curves are shown in the upper left and lower left of Fig. 26-16.

At 10 MHz under the conditions employed, V-shaped high-frequency conductance and capacitance curves are obtained. Although similar in shape to the low-frequency curves, considerable curvature exists in the curves both near to and considerably removed from the vicinity of the end point. This curvature is a distinct drawback and renders extrapolations uncertain.

Further difficulties arise when measuring high-frequency conductance during a titration at lower frequencies, for example, 3 MHz. The high-frequency readings lie on both

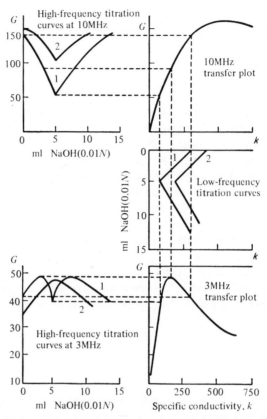

Fig. 26-16. Correlation of high-frequency titration curves to low-frequency titration curves. (1) 5 ml of 0.01 M HCl in total volume of 60 ml. (2) 5 ml of 0.01 M HCl plus 55 ml 0.001 M KCl. (After C. N. Reilley and W. H. McCurdy, Jr., *Anal. Chem.*, **25**, 86 (1953). Courtesy of American Chemical Society.)

sides of the peak value, and it is possible to obtain an M-shaped titration curve or an inverted V-shaped curve. The complexity of these curves could cause errors of interpretation.

Acid–base, precipitation, and ion-combination reactions can be followed by high-frequency methods in the same manner as discussed for low-frequency titrations. In fact, ordinary frequencies will provide as good or better titration curves except in those few instances where it is advantageous to remove the electrodes from direct contact with the test solution or when titrations are conducted in nonaqueous solvents.

Measurement of Dielectric Constant

When a nonconducting liquid is placed between the metal plates of Fig. 26-14, the system behaves as two capacitors, C_g and C_s, in series. The capacitance C_s will vary from

TABLE 26-4 Dielectric Constants

Formamide	109.5	Acetic anhydride	20.7	Acetic acid	6.15
Water	81	1-Propanol	20.1	Ethyl acetate	6.02
Formic acid	58.5	2-Propanol	18.3	1-Butylamine	5.3
Acetonitrile	37.5	1-Butanol	17.1	Chloroform	4.806
Nitrobenzene	34.8	Ethylenediamine	14.2	Benzene	2.379
Methanol	32.6	Benzyl alcohol	13.1	Carbon tetrachloride	2.238
Ethanol	24.3	Phenol	9.78	1,4-Dioxane	2.209
Acetone	20.7	Aniline	6.89		

a fixed value C_0, determined by the geometry of the cell when filled with air, to a value D dependent upon the dielectric constant of the sample; that is,

$$C_s = C_0 D \qquad (26\text{-}19)$$

The value C_0 depends upon the thickness of the glass walls, the effective plate area, and the dielectric constant of the container itself. The capacitance of the cell is not a linear function of the dielectric constant, but will approach the value of C_g as a limit. Figure 26-17 is a plot of cell response with solutions of various dielectric constants. The region of linear response may be extended by inserting a variable inductance between the measuring cell and the oscillator circuit.

Table 26-4 lists the dielectric constants of some common materials. The high value for water renders it possible to determine extremely small amounts of water in organic liquids or moisture in granular or powdered substances. For many years a dielectric type of moisture meter for use with grains, cereals, and other powdered substances has been available commercially. The test material is uniformly packed between two parallel plates which serve as the measuring condenser.

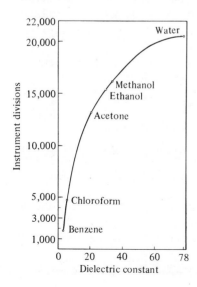

Fig. 26-17. Instrument response as a function of dielectric constant; Sargent chemical oscillometer, model V.

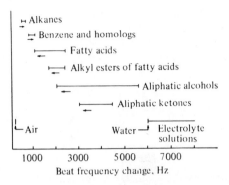

Fig. 26-18. Frequency change induced by organic compounds. Arrow indicates direction of increasing molecular weight. (After P. W. West, T. S. Burkhalter, and L. Broussard, *Anal. Chem.*, 22, 469 (1950). Courtesy of American Chemical Society.)

High-frequency methods have been applied to the discrimination of organic mixtures. Organic compounds lie in rather definite groups in the range between the dielectric constant of air and water, as shown in Fig. 26-18.

LABORATORY WORK

Experiment 26-1   Titration of a Completely Ionized Acid

1. Transfer 50 ml of 0.01 *M* hydrochloric acid to a 250-ml beaker and dilute to about 100 ml. Measure the conductance. Titrate with 0.1 *N* sodium hydroxide added in 0.5-ml increments from a 10-ml buret. Measure the conductance (or resistance) after the addition of each increment.
2. Repeat the titration, using 0.001 *M* and 0.0001 *M* solution of hydrochloric acid and titrating with 0.01 *N* and 0.001 *N* sodium hydroxide, respectively.

Experiment 26-2   Titration of an Incompletely Ionized Acid

1. Transfer 50 ml of 0.1 *N* acetic acid to a 250-ml beaker and dilute to about 100 ml. Measure the conductance. Titrate with 1 *N* sodium hydroxide added in 0.5-ml increments from a 10-ml buret.
2. Repeat the titration, using 0.01 *N* and 0.001 *N* solutions of acetic acid and titrating with 0.1 *N* and 0.01 *N* sodium hydroxide, respectively.
3. Repeat the entire series of titrations, substituting aqueous solutions of ammonia as the titrant.
4. Plot all results on one piece of graph paper. The curves should resemble Fig. 26-7.

Experiment 26-3   Titration of Incompletely Ionized Acid: Partial Neutralization with Ammonia Followed by Titration with Sodium Hydroxide

1. Transfer 50 ml of 0.1 N acetic acid to a 250-ml beaker. Dilute to about 100 ml with distilled water. Measure the conductance.
2. Add, in 0.5-ml increments, a total of 8 ml of 0.5 N aqueous ammonia. Complete the titration with 1 N sodium hydroxide added from a 10-ml buret.

   Note The aqueous ammonia need not be standardized. The titration curve should resemble Fig. 26-8.

Experiment 26-4   Titration of a Mixture of Acids

1. Transfer 40 ml of 0.01 N acetic acid and 10 ml of 0.01 N hydrochloric acid to a 250-ml beaker. Dilute to about 100 ml with distilled water. Measure the conductance.
2. Titrate with 0.1 N sodium hydroxide added from a 10-ml buret. Use 0.25 ml-increments of titrant.
3. Repeat the titration, using 10 ml of 0.01 N acetic acid and 40 ml of 0.01 N hydrochloric acid.

   Note If desired, a 0.1 N aqueous ammonia solution may be used as titrant as was done in Experiment 26-2.

Experiment 26-5   Precipitation Titrations

1. Transfer 50 ml of 0.01 N sodium chloride to a 250-ml beaker and dilute to about 100 ml. Measure the conductance.
2. Titrate with 0.1 N silver nitrate added from a 10-ml buret.
3. Repeat the titration with 0.1 N silver acetate as titrant.
4. Plot the results on the same graph and note the slope of the two branches of the titration curve for the different titrants.

Experiment 26-6   Ion Combination Titrations

1. Transfer 5 ml of 0.1 N potassium iodide to the conductivity cell. Dilute with distilled water to about 100 ml. Measure the conductance.
2. Titrate with 0.1 N mercuric perchlorate. A small break in the titration curve occurs when the titration is 50% complete due to the formation of K_2HgI_4. A much larger

break occurs when the reaction

$$K_2HgI_4 + Hg(ClO_4)_2 \longrightarrow 2HgI_2 + 2KClO_4$$

is complete.

Note Prepare mercuric perchlorate by saturating a 0.1 N perchloric acid solution with red mercuric oxide.

Experiment 26-7   Response Curves for a High-Frequency Titrimeter

1. Fill the cell with exactly 80 ml of distilled water (or any known amount of water sufficient to cover the condenser plates or the top of an inductance coil). Start the stirrer and determine the instrument reading.
2. Add, in 1-ml increments, a total of 9 ml of 0.001 N hydrochloric acid. Determine the instrument reading after each addition.
3. Repeat step 2 using a total of 9 ml of 0.01 N acid.
4. Repeat step 2 using a total of 10 ml of 0.1 N acid.
5. Repeat steps 1 through 4 using a series of sodium hydroxide solutions.
6. Repeat steps 1 through 4 using a series of sodium chloride solutions.
7. Repeat steps 1, 3, and 4 using a series of acetic acid solutions.
8. Plot the respective instrument readings vs. the normality of the solution in the titration cell after each increment was added to obtain a series of response curves characteristic of the particular high-frequency instrument employed. If a Sargent oscillometer was employed, the response curves will resemble the lower portion of Fig. 26-15.

Experiment 26-8   Titrations with a High-Frequency Titrimeter

From the response curves obtained in Experiment 26-7, select the range of optimum concentrations for conducting titrations with your particular titrimeter. Fill the titration cell with distilled water until the condenser plates are covered. Transfer an appropriate volume of solution to the cell and, with the stirrer running, add 0.5-ml increments of the titrant. Continue the addition of titrant until the end point is reached and seven or eight readings are obtained beyond it. Plot the condenser readings (or respective instrument readings) vs. volume of titrant added; the intersection of the extrapolated straight-line portions of the two branches of the curve is taken as the end point of the titration.

Suggested titrations: (1) Experiment 26-1, steps 1 and 2; (2) Experiment 26-2, steps 1 and 2; (3) Experiment 26-4; and (4) Experiment 26-5.

Experiment 26-9   Measurement of Dielectric Constant

1. Connect the test-tube cell holder to a high-frequency instrument, such as the Sargent oscillometer. Balance the instrument with the cell empty.

2. Prepare a calibration curve of instrument response as a function of dielectric constant by successively measuring pure solutions of benzene, chloroform, acetone, ethanol, methanol, and water.
3. Determine the instrument reading of an unknown solution and report the dielectric constant of the unknown solution by reference to your calibration curve.

Note The instructor may wish to have the student prepare a calibration curve for a binary mixture—for example, nitrobenzene and benzene or aniline and nitrobenzene—and then determine the composition of an unknown.

PROBLEMS

1. An aqueous 20% HCl solution has a specific conductance of about 0.85 Ω^{-1} cm^{-1} at 25°C. What is the measured resistance with a cell of constant (a) 100, (b) 20, (c) 10, (d) 1, (e) 0.2 cm^{-1}? Are these resistance values feasible to measure with standard conductivity bridges?

2. A cell constant of 20.0 cm^{-1} is recommended for a commercial conductivity bridge designed to span the range from 1 to 18% HCl. The corresponding conductance ranges from 0.0630 to about 0.750 Ω^{-1}. What range of resistance values are involved?

3. A meter scale is to be inscribed from 2 to 1000 ppm Na_2SO_4, and the midpoint of the logarithmic scale shall correspond to 40 ppm. Suggest a compatible set of instrument parameters; i.e., resistance range and cell constant.

4. Similar to Problem 3, individual instruments are to be designed for each of these systems. Compute the resistance range and a compatible cell constant. Use handbooks to locate necessary conductance values, and assume average distilled water has a specific conductance of 2 × 10^{-6} Ω^{-1} cm^{-1}. (a) 0%–5% HCl; (b) 0.5%–5% NH_3; (c) 0–60 ppm sodium formate; (d) 0–40 ppm salinity (as NaCl); (e) 96%–99.5% H_2SO_4; (f) 0.1%–10% CrO_3.

5. The equivalent conductance of a 0.002414 N acetic acid solution is found to be 32.22 at 25°C. Calculate the degree of dissociation of acetic acid at this concentration, and calculate the ionization constant.

6. The specific conductance at 25°C of a saturated solution of barium sulfate was 4.58 × 10^{-6} Ω^{-1} cm^{-1}, and that of the water used was 1.52 × 10^{-6}. What is the solubility of $BaSO_4$ at 25°C in moles per liter and in grams per liter? Calculate the solubility product constant.

7. The solubility product of silver iodate at 25°C is 3.1 × 10^{-8}. What would be the resistance of a saturated solution of silver iodate, measured with a cell whose cell constant was 0.2 cm^{-1}? (Neglect the solvent correction.)

8. A very dilute solution of NaOH (100 ml) is titrated with 1.00 N HCl. The following resistance readings (in ohms) were obtained at the indicated buret readings: 0.00 ml,

3175; 1.00 ml 3850; 2.00 ml, 4900; 3.00 ml, 6500; 4.00 ml, 5080; 5.00 ml, 3495; 6.00 ml, 2733. Determine the normality of the solution and the weight of NaOH present.

9. In the titration of 100 ml of a dilute solution of acetic acid with 0.500 N aqueous ammonia, the following resistance readings (in ohms) were obtained at the indicated buret readings: 8.00 ml, 750; 9.00 ml, 680; 10.00 ml, 620; 11.00 ml, 570; 12.00 ml, 530; 13.00 ml, 508; 15.00 ml, 515; 17.00 ml, 521. What is the normality of the acetic acid solution?

10. In the titration of 100 ml of H_2SO_4 in glacial acetic acid with 0.500 M sodium acetate in the same solvent, the following specific conductance ($\times 10^6$) data were obtained at the indicated buret readings:

0.50 ml = 2.95	3.50 ml = 4.78	7.00 ml = 3.20
1.00 ml = 3.30	4.00 ml = 4.73	7.50 ml = 3.20
1.50 ml = 3.65	4.50 ml = 4.40	8.00 ml = 3.47
2.00 ml = 4.00	5.00 ml = 4.04	8.50 ml = 3.82
2.50 ml = 4.35	5.50 ml = 3.76	9.00 ml = 4.18
3.00 ml = 4.65	6.00 ml = 3.43	9.50 ml = 4.50

What is the molarity of the sulfuric acid solution?

11. The following relative conductance readings were obtained during the titration of a mixture containing an aliphatic carboxylic acid and an aromatic sulfonic acid. The titrant was 0.200 N NH_3. Readings have been corrected for titrant volume.

0.00 ml = 2.01	3.20 ml = 1.19	5.00 ml = 1.51
1.00 ml = 1.75	3.50 ml = 1.26	6.00 ml = 1.52
2.00 ml = 1.47	4.00 ml = 1.41	8.00 ml = 1.53
2.50 ml = 1.33	4.20 ml = 1.47	
3.00 ml = 1.19	4.50 ml = 1.51	

Calculate the number of equivalents of each acid present in the mixture.

12. Using the equivalent conductance values obtained from Table 27-1, sketch the general form of the titration curve in each of the following cases: (a) titration of $Ba(OH)_2$ with HCl; (b) titration of NH_4Cl with NaOH; (c) titration of silver nitrate with potassium chloride; (d) titration of silver acetate with lithium chloride; (e) titration of sodium acetate with HCl; (f) titration of barium carbonate with H_2SO_4 [hint: check solubility of intermediate products]; (g) titration of a mixture of a sulfonic acid and a carboxylic acid with NaOH; (h) titration of $KH_3(C_2O_4)_2$ with NH_3; titration of a mixture containing Na_3PO_4 and Na_2HPO_4 with HCl.

13. A commercial liquor contains nicotinic acid, ammonium nicotinate, and nicotinamide. Device a conductometric titration for the determination of the free acid and the ammonium salt.

14. Exactly 50 ml of a 0.001 N solution of HCl is titrated with 0.01 N KOH. Calculate the conductance and resistance observed after the addition of 0, 25, 50, 90, 100,

110, 150, 175, and 200% of the equivalent amount of titrant. Assume the cell constant is 0.2 cm^{-1}. Plot the results.

15. Exactly 100 ml of a 0.1 N solution of NH$_4$NO$_3$ is titrated with 1.0 N KOH. As in Problem 14, calculate the conductance observed after the addition of the stated increments of titrant. Assume the cell constant is 0.5 cm^{-1}. Plot the results.

16. The response of the Sargent oscillometer to varying concentrations of hydrochloric acid was obtained by adding known increments of acid to exactly 100 ml of distilled water. The following instrument readings were obtained for the indicated acid additions: (A) using 0.001 N HCl.   0.00 ml = 21,297, 1.00 ml = 21,277, 5.00 ml = 21,208, 8.00 ml = 21,262, 10.00 ml = 21,327; (B) using 0.01 N HCl and continuing with the solution from set A: 1.00 ml = 21,821, 3.00 ml = 23,135, 5.00 ml = 24,429, 7.00 ml = 25,430, 9.00 ml = 26,192; (C) continuation with 0.1 N HCl: 1.00 ml = 27,330, 2.00 ml = 28,357, 4.00 ml = 28,600, 8.00 ml = 28,690, 10.00 ml = 28,700. Plot the instrument response vs molarity of HCl. Estimate the concentration of maximum sensitivity and compare with the value given in Table 26-3. The Sargent oscillometer operates at 4.89 MHz.

17. On the graph prepared in Problem 16, sketch the response curve for NaCl using the information contained in Table 26-3 and assuming in all other respects the curve parallels the one for HCl. Sketch the general form of the titration curve in each of the following cases, assuming the response curve for NaOH coincides with the curve for HCl: (a) titration of 0.01 N HCl with 0.1 N NaOH, (b) titration of 0.0005 N HCl with 0.005 N NaOH, (c) titration of 0.001 N HCl with 0.01 N NaOH.

18. Exactly 100 ml of a dilute HCl solution was titrated with 0.0200 N NaOH. The following oscillometer readings were obtained at the indicated buret readings:

0.00 ml = 26,552	4.00 ml = 23,538	6.00 ml = 22,940
1.00 ml = 26,036	4.50 ml = 23,047	7.00 ml = 23,631
2.00 ml = 25,328	5.00 ml = 22,585	7.75 ml = 24,174
3.00 ml = 24,506	5.50 ml = 22,631	9.00 ml = 24,930
		10.00 ml = 25,445

What is the molarity of the HCl solution?

19. Ammonia in gas streams has been determined by bubbling the gas through exactly 100 ml of 0.0400 M boric acid at a rate of 10 liters/min and measuring the change in conductance. The following data was obtained for a series of standard solutions of ammonia added to boric acid:

NH$_3$ Added,	Conductance, Ω^{-1}, ($\times 10^4$)	
mequiv	HBO$_2$	HBO$_2$ + NH$_3$
3.665	0.55	45.55
0.715	0.55	10.65
0.0715	0.55	1.61
0.0071	0.55	0.66

Plot the calibration curve. Calculate the ammonia concentration present in each of these gas streams for which these conductance readings ($\times 10^4$) were obtained: sample A, 12.75; sample B, 6.35; sample C, 1.15; sample D, 3.45.

20. In Problem 19 the cell constant was 0.069 cm^{-1}. Assuming that the equivalent conductance of the borate ion at infinite dilution is about 30 cm^2 equivalent^{-1} ohm^{-1}, and that the specific conductance of the water employed was 2×10^{-6} Ω^{-1} cm^{-1}, estimate the degree of ionization of boric acid at the concentration employed and calculate the ionization constant.

BIBLIOGRAPHY

Britton, H. T. S., "Conductometric Analysis" in *Physical Methods in Chemical Analysis*, Vol. II, W. G. Berl, Ed., Academic Press, New York, 1951.

Braunstein, J. and G. D. Robbins, "Electrolytic Conductance Measurements and Capacitive Balance," *J. Chem. Educ.*, **48**, 52 (1971).

Lingane, J. J., *Electroanalytical Chemistry*, 2nd ed., Wiley-Interscience, New York, 1958.

Ross, J. W., "Conductometric Titrations" in *Standard Methods of Chemical Analysis*, Vol. 3, Part A, F. J. Welcher, Ed., Van Nostrand Reinhold, New York, 1966, Chapter 18.

Shedlovsky, T. in *Physical Methods of Organic Chemistry*, 3rd ed., Vol. I, Part 4, A. Weissberger, Ed., Wiley-Interscience, New York, 1960, Chapter 45.

LITERATURE CITED

1. Braunstein, J. and G. D. Robbins, *J. Chem. Educ.*, **48**, 52 (1971).
2. Reilley, C. N. and W. N. McCurdy, Jr., *Anal. Chem.*, **25**, 86 (1953).

27

Process Instruments and Automatic Analysis

Methods for continuous analysis and control of on-stream process systems and automatic analyzers for routine laboratory methods will be considered in this final chapter. While it is true that specific types of automatic equipment have been used for many years to control the process stream environment—temperature, pressure, flow rate, and others— only recently have automatic measurements of stream composition become more wide-spread in use. In control analyses, generally, chemical constituents have been determined on grab samples which were subsequently analyzed by conventional manual techniques with obvious shortcomings in time, economy, and human-error possibilities. More and more industrial processes, however, require constant surveillance and control at each step in the process. Instruments and methods are needed to provide a dynamic rather than historical analysis. This requires either continuous analysis, or at least rapid repetitive analysis, of starting materials, intermediate products, and the end products for the desired component. Often monitored also are various possible contaminants that could be deleterious at various stages in the process. Only by means of analyses of these types can production facilities reduce off-specification materials to a minimum. Remedial action can be taken within a few minutes (for discontinuous repetitive analyses) or immediately with continuous on-stream analyzers. By suitable feedback circuitry, the latter instruments will make necessary adjustments in a process stream to maintain the specific variable or component at a predetermined value.

Although instrument costs are high, they still represent a small fraction of plant or laboratory costs. Because the limits of accuracy of both process and laboratory work can be no better than the reliability of instruments used in making measurements, high costs are easily justified. Other advantages to be gained include greater safety of operation, greater operating economy, and new evaluations. Applicability of a method that initially appears to be borderline may actually prove to be acceptable in service because continuous sampling tends to improve precision. Flowing samples are easier to protect from contamination.

Not every process or analytical method lends itself to automation. Application to solid samples is difficult for many types of measuring techniques now used in continuous analysis. In fact, there will probably always remain a hard core of complex chemical analyses that will totally defy automation or be too costly to automate.

AUTOMATIC ANALYZERS

Instruments for automatically determining specific chemical variables or process parameters will be discussed first. Some of these have received attention in preceding chapters.

Continuous Analysis

Continuous analysis adopts many of the instrumental techniques which have been developed to a high degree in laboratory use and applies them to characteristic measurements on flowing samples in chemical processes. The features of continuous analysis are illustrated schematically in Fig. 27-1. A sample at a controlled rate is obtained from the main process stream or other source of material. When necessary, provision is made for preparing the initial sample for analysis by further operations such as the addition of reagents or filtration. The actual measurement is made by a suitable sensor or transducer. The results of the analysis are indicated in terms of the concentration of the desired component in the sample, usually by a chart recorder or other readout device.

One class of primary sensors consists of devices that measure physical or chemical variables such as temperature, pressure, fluid flow, and liquid level. The other broad category of primary sensors, often referred to as *on-stream* or *on-line analyzers*, measures chemical composition. In many cases they achieve this by measuring some physical property related to chemical composition, such as infrared or ultraviolet-visible absorption, thermal or electrical conductivity, dielectric constant, paramagnetic susceptibility, density, refractive index, and the like.

Most significant among the factors retarding development of on-stream analyzers for chemical process control is the difficulty in obtaining samples and preparing them

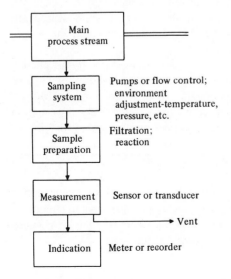

Fig. 27-1. Features of continuous analysis.

properly for analysis. Sampling may represent as much as 90% of the total analysis problem. Chemical process streams may be hot and under pressure, supersaturated with respect to certain dissolved salts, highly corrosive, or laden with fibrous or particulate matter. Even when acceptable samples can be obtained, they then may need to be cooled, filtered, diluted, or otherwise processed before they are ready for analysis. All of these operations must be performed either continuously or on an automatic semicontinuous basis. Transport lag must be eliminated and dead volume reduced if events within the analyzer are to be representative of the stream being monitored. Special bypass pumping devices are often needed to keep fresh sample rapidly supplied to the input of the analyzer. Care must be taken to prevent fractionation of liquid samples or the adsorption of gaseous constituents onto the tubing or fittings ahead of the analyzer. Extreme care is required in continuous analysis of trace quantities. Leakage of sample at a fitting or connection can cause serious back diffusion of the atmosphere into the stream; examples include dissolved oxygen in power plant condensate streams or the analysis for water in dry gas streams. Corrosive samples or materials under extremes of pressure or temperature may restrict latitude in the design of sample systems, especially with regard to sample cells for the various optical methods of analysis. Gas handling equipment should always include a small protective filter, preceded by a major filter if the gas contains suspended matter requiring removal. If a gas sample has a water vapor concentration high enough to cause condensation within the analyzer, or if moisture constitutes an interferent, stream drying equipment must be installed.

If several separate process streams are reasonably identical in major constituents and the concentration of the desired component does not vary widely, all streams can be connected through solenoid-operated, three-way valves to a common manifold which leads to the instrument. Each stream can then be selected sequentially by a multipoint recorder at intervals of 3–5 min, giving almost continuous analysis of each stream.

Methods Based on Nonselective Properties

A number of instrumental methods have been available for some time for continuous measurement of nonselective properties such as pH, density, viscosity, conductance, capacitance, and combustibility. Selected examples will serve to illustrate these types of analysis.

The composition of binary systems where each component has a somewhat different physical parameter can be determined. A calibration curve of the parameter signal versus composition yields the correlation. For example, water in many organic materials is readily determined, because water has a dielectric constant of 80 whereas most organic materials have a dielectric constant between 1 and 10. Thus, the moisture content of the paper web during manufacture at speeds of up to 3000 ft/min can be measured as shown in Fig. 27-2. The measuring head is an electrical capacitor which uses the paper web as part of the measuring circuit. The dielectric constant of dry paper is about 3.

In many process streams a pseudo-binary situation exists wherein the components of the stream fall into two responsive groups. If the signal difference between the groups is

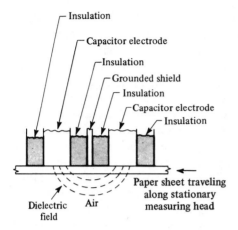

Fig. 27-2. Schematic diagram of measuring head for control of moisture in paper web. (Courtesy of Foxboro Co.)

large compared with the difference between the individual members of each group, a successful "group type" analysis is possible. In this manner, the concentration of aromatics in a reformate stream consisting of naphthenes, paraffins, and aromatics can be determined by refractometric methods. Other examples have been discussed in the chapter on conductance methods.

Determination of the pH value of process streams is handled by heavy-duty, industrial pH analyzers which feature solid-state electronics for longer life and plug-in circuit components for quick replacement when necessary. It is important that the conditions to which the pH-responsive electrodes are subjected do not impair the performance of the electrodes, or at least affect them to such a degree that excessive maintenance is required in order to keep the system dependably operative. Extraneous material must never be

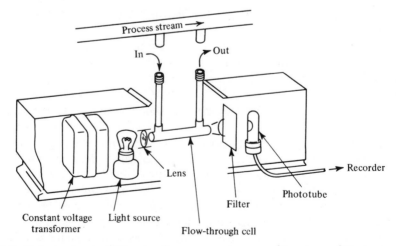

Fig. 27-3. Colorimeter–turbidimeter for process streams. (Courtesy of Beckman Instruments, Inc.)

allowed to scratch or to coat the pH-sensitive portion of the glass surface. Such materials can be oils, tars, suspended solids, or precipitated chemicals.

Simple, but rugged, flow colorimeters are used for continuous monitoring of process streams. A diagram of a typical installation is shown in Fig. 27-3. Components consist of a well-regulated light source, a proper interference filter, a flow-through sample compartment, and a photodetector. A cylindrical shutter surrounds the source and facilitates checking and adjustment of zero and 100% span in plant service. Units of this type find application in measurement of color in a final product or imparted to a process stream, decoloration of a product with bleaching earth, measurement of clarity of a product in filtration or sedimentation control, and continuous detection of particles, bubbles, immiscible droplets, and other suspended matter. Group analysis for aromatics, diolefins, ketones, and aldehydes is handled with ultraviolet absorption instruments employing a mercury vapor lamp and suitable filters.

Absorptiometry with X rays provides a unique method for studying the fluidization of a finely divided solid by a gas, as in beds of silicon fluidized by air.

Differential Analyzers

Differential instruments permit measurements to more decimal places, even with plant equipment, than are possible with instruments that take absolute measurements. Signal specificity can also be improved while lessening the extraneous influence of other parameters such as temperature variations, alteration in sample flows, and slight mismatch of electronic components. Differential methods tend also to compensate for suspended particulate matter and stray colors in the sample stream.

Differential refractometers achieve sensitivities of 4×10^{-6} refractive index unit, which is equivalent to 40 ppm of water in glacial acetic acid. Other measurements include concentration of sugar solutions and high molecular weight organic compounds, and the composition of mixed Freons for aerosol propellants.

Flow colorimeters can be improved by measuring the signal ratio from two phototubes. Zero drift is checked by purging a standard sample through the sample cell occasionally and making appropriate optical adjustments. These instruments are ideal for low levels of color and turbidities in liquids, that is, concentrations down to 1 ppm.

Differential conductivity analysis provides another method for on-line control. For example, the conductivity of a boiler water sample is measured and recorded before and after neutralization by carbon dioxide. The difference between these two signals is used to control chemicals fed to the boiler. In the mining industry acid samples are neutralized with amines; the resulting differential conductivity is a measure of free acid in leach liquor, electrowinning process liquor, and similar acid solutions.

METHODS FOR THE ANALYSIS OF GAS STREAMS

Several methods for the analysis of gas streams will be discussed in this section. These are only a few of the possibilities. Ewing,[3] for example, has discussed more than 25 distinct analytical methods for moisture measurement.

Thermal Conductivity

All gases possess the ability to conduct heat, but in varying degrees. This difference can be used to determine quantitatively the composition of binary gaseous mixtures or more complex mixtures if all the components of the mixture have about the same thermal conductivity except for the one component being measured.

The *thermal conductivity* of a gas is defined as the quantity of heat (in calories) transferred in unit time (seconds) in a gas between two surfaces 1 cm^2 in area, and 1 cm apart when the temperature difference between the surfaces is 1°C. Absolute values of the thermal conductivity are seldom needed for analytical purposes, but a number of values are gathered in Table 27-1.

The hot-wire type of thermal conductivity cell has been discussed in Chapter 18. The hot wires may be replaced with thermistors. These are electronic semiconductors of fused metal oxides whose electrical resistance varies with temperature and lies between that of conductors and insulators. Thermistors are extremely sensitive to relatively minute temperature changes because they possess a large negative temperature coefficient. A thermistor bead encapsulated in glass or Teflon may replace the hot wire in thermal conductivity cells. The sensitivities of a 4-W-2 hot-wire assembly (see Table 27-1, footnotes) and a thermistor cell are comparable when the cell block is near room temperature. The thermistor has the advantage of smaller cell volume but this is offset by the thermal inertia of the thermistor cell and decreased sensitivity when the cell block is maintained at elevated temperatures.

TABLE 27-1 Thermal Conductivities of Gases and Vapors

Gas	$\lambda \times 10^5$ (0°C)	λ/λ AIR (100°C)	Approximate Bridge Output Relative to Air, mV[a]		
			4-W	4-W-2	2-GBT
Acetone	2.37	0.557	−3.4	−10.2	−12.0
Air	5.83	1.000	−	−	−
Ammonia	5.22	1.04	2.5	7.5	8.8
Argon	−	0.696	−2.0	−6.0	−7.0
Carbon dioxide	3.52	0.700	−3.0	−9.0	−10.5
Carbon monoxide	−	0.960	−0.4	−1.2	−1.4
Chlorine	−	0.323	−10	−30	−35
Ether	−	0.747	−3.6	−10.8	−
Helium	34.80	5.53	17.2	51.6	60
Hydrogen	41.60	7.10	28	84.0	98
Methane	7.21	1.45	3.5	10.5	12.2
Nitrogen	5.81	0.996	−0.25	−0.75	−0.9
Oxygen	−	1.014	0.45	1.35	1.6
Sulfur dioxide	−	0.350	−37.5	−37.5	−44

SOURCE: Courtesy of Gow-Mac Instrument Company.
[a]4-W filaments are helical bare tungsten wires, four cells in parallel; 4-W-2 filaments consist of two helical wires in series in each cell; 2-GBT are two nominal 8,000-Ω glass bead thermistors on platinum support.

The thermal conductivity method is easily applied to the determination of the composition of a binary mixture, provided that the two gases have different thermal conductivities. An example is oxygen in electrolytic hydrogen or nitrogen in helium in the latter stages of separation from natural gas. A multicomponent mixture can be treated as a binary mixture if all components but one remain constant, as, for example, the determination of hydrogen in uncarburated water gas from which carbon dioxide and steam have been removed, leaving only nitrogen and carbon monoxide. Similarly, if all components of the mixture other than the one being measured vary in the same ratio to each other, an analysis is possible. Helium, methane, or carbon dioxide in air are examples. However, temperature coefficients are important in some mixtures because the difference between certain pairs of gases may reverse its sign as the temperature changes, for example, ammonia with air or nitrogen, water with air, and butane with air. Sometimes the proportion of constituents other than the test substance are related to an equilibrium. An example is the relationship among hydrogen, carbon dioxide, carbon monoxide and air in exhaust gases of combustion engines which permits the air/fuel ratio to be determined.

Example 27-1

The feasibility of the thermal conductivity method for a particular system can be estimated from the information in Table 27-1, columns 4–6, and by reference to Fig. 27-4. In many cases a small segment of the curve, such as 0–1% H_2 in O_2, is virtually linear.

The approximate bridge output (for four tungsten, W-2, filaments which consist of two helical bare wires in series in each cell) is 1.35 mV for pure oxygen and 84.0 mV for pure hydrogen. Assuming for the moment that over the 0–100% range the response is linear, contrary to information in Fig. 27-4, the bridge output will vary from 1.35 mV to

$$(0.01)(84.0) + (0.99)(1.35) = 2.18 \text{ mV}$$

for 1% H_2 in O_2, a span of 0.81 mV. On a 2-mV potentiometric recorder, with zero suppression to place the reading for pure oxygen at zero, the sensitivity would be $\pm 0.012\%$ H_2 per 1% full-scale deflection, equivalent to a sensitivity of ± 120 ppm. Actually, the slope of the hydrogen response curve from 0–10% is approximately 10-fold greater, which would make the sensitivity $\pm 0.0012\%$ H_2 (± 12 ppm) per 1% full scale deflection. A slight improvement would result from the use of a thermistor bridge.

In the foregoing examples, the reference gas is often sealed in the reference arms of the Wheatstone bridge. Usually only a limited range over which linearity can be assumed is needed for the particular process control.

Frequently a constituent may be removed by some chemical means and the thermal conductivity of the unaltered gas sample is compared with the conductivity of the treated gas sample. The unaltered gas sample is passed through the sample cells, through a chemical reagent or adsorbent, and then through the reference arms of the bridge. The procedure can be continued with another pair of thermal conductivity cells, ad infinitum, until all the components are determined. For example, the use of copper oxide at elevated temperatures removes carbon monoxide and hydrogen, whereas passage over Hopcalite

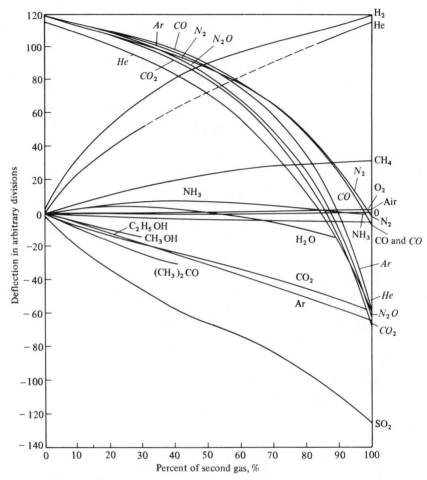

Fig. 27-4. Response of thermal conductivity cell as function of percentage of second gas in air (Roman symbols) or in hydrogen (italicized symbols). (Courtesy of Gow-Mac Instrument Co.)

catalyst* removes only carbon monoxide. Methane can be burned to carbon dioxide and water, and the products removed by adsorption.

Gas Density

At constant temperature and pressure one mole of any pure, ideal gas will occupy the same volume as one mole of any other ideal gas. Consequently, the density of an

*A mixture of Ag_2O, Co_3O_4, and MnO_2.

ideal gas is a direct linear function of the molecular weight of that gas. Although this is strictly true only for ideal gases, nearly all real gases behave as ideal gases at temperatures near room temperature and pressures near atmospheric. Two designs of the gas-density detector are in commercial use: the original Martin[9] design and the recent Nerheim[10] design. Both function on the same basic operational principles but differ in sensing elements, configuration, and simplicity.

The Nerheim configuration, Fig. 27-5, illustrates the principle of operation: With the conduit network mounted vertically, the reference (carrier) gas enters at A, splits into two streams, and exits at D. Two flowmeters, B_1, B_2, are installed, one in each stream, and are wired in a Wheatstone bridge. When the flow is balanced, the detector elements, which are a matched pair, are equally cooled and the bridge is balanced, thus giving a base-line (zero) trace. The detector elements may be either hot wires or thermistors, depending upon the desired operating temperature. These are connected via an electrical bridge to a recording potentiometer (Fig. 18-4). The sample gas (or effluent from a chromatographic column) enters at C, splits into two streams, mixes with the reference gas in the horizontal conduits, and exits at D. The sample gas never comes into contact with the detector element, thus avoiding problems caused by corrosion or carbonization.

If the sample gas is of the same density as the reference, there will be no unbalance of reference streams or of the detector elements. When the sample gas carries transient trace impurities which are heavier than the reference gas, the density of the heavier molecules will cause a net downward flow, partially obstructing the flow A—B_2—D, with a temperature rise of element B_2, and permitting a corresponding increase in the flow A—B_1—D, with a temperature decrease of element B_1. In a similar manner, lighter molecules will cause a net upward flow and the reverse will be true, namely, a temperature rise of element B_1 and decrease of element B_2 with a signal of opposite polarity from the first case. Bridge unbalance is linear over a broad range and directly related to the gas density difference. Calibration for individual components is eliminated because the response depends on a predictable relationship, the difference in molecular weight of component and reference gas:

$$\Delta\rho = k\,\frac{n_s(M_s - M_r)P}{RT}\tag{27-1}$$

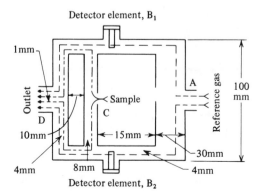

Fig. 27-5. Nerheim gas-density balance. Schematic view as mounted in a vertical plane. (Courtesy of Gow-Mac Instrument Co.)

where k is a constant whose value depends on cell geometry, viscosity, the flow measuring system, and the thermal conductivity of the gases; M_s and M_r are the molecular weights of the sample and reference gases; and n_s is the mole fraction of component in the sample.

The sensitivity of the gas density detector with thermistor sensor is comparable to that of a thermistor type of thermal-conductivity cell and requires no amplification at room temperature. The hot-wire sensor, which has one-sixth the sensitivity of thermistors at 25°C, may require low-level amplification at 100°C. The low noise level of the detectors permits effective use to 300°C, although the sensitivity decreases rapidly with increasing temperature for thermistors, being at 250° one-fifth that at 50°C. The Nerheim design has an effective sample volume of approximately 5 ml.

The Ranarex specific-gravity indicator operates on an interesting principle. The gas is given a rotating motion by means of an impeller fan. This fan drives the gas against the

Follow the 5 points for explanation of principle

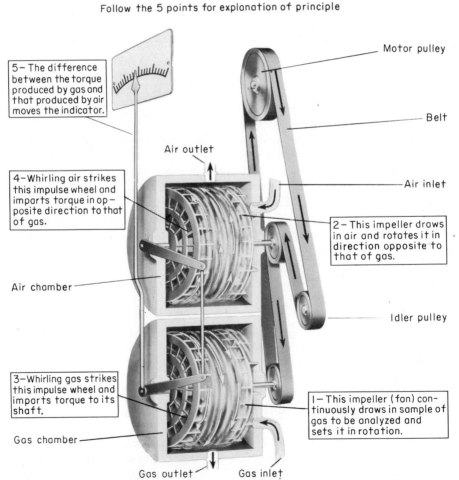

5 — The difference between the torque produced by gas and that produced by air moves the indicator.

Motor pulley

Belt

Air outlet

4 — Whirling air strikes this impulse wheel and imparts torque in op — posite direction to that of gas.

Air inlet

2 — This impeller draws in air and rotates it in direction opposite to that of gas.

Air chamber

Idler pulley

3 — Whirling gas strikes this impulse wheel and imparts torque to its shaft.

1 — This impeller (fan) continuously draws in sample of gas to be analyzed and sets it in rotation.

Gas chamber

Gas outlet   Gas inlet

Fig. 27-6. Principle of the Ranarex specific gravity indicator. (Courtesy of Permutit Co.)

blades of an impulse wheel, producing a torque. The greater the density of the gas, the greater the coupling between the two wheels and the greater the torque produced on the second wheel. In order to eliminate changes in fan speed, temperature, humidity, and atmospheric pressure, a comparison set of wheels is used with air as the reference gas. The two fan wheels are run by the same motor but in opposite directions. The two torque wheels are mechanically coupled to each other, and the difference in torque is registered on a dial. The dials may be calibrated directly in percentage of constituent sought or in specific gravity units. Automatic recording and control can be provided for, if desired. The principle and design of this device are further illustrated in Fig. 27-6.

Heat of Combustion

The heat of reaction evolved by a gas when it burns at a filament or in the presence of a catalyst can be used to determine combustible gases in a mixture. One device used for the detection of explosive gas mixtures is quite similar in construction to the thermal-conductivity apparatus. The gas is passed through a cell containing a heated filament. The combustion of the gas raises the temperature and the resistance of the filament as compared to that of a reference filament. The reference and active filaments form two arms of a Wheatstone bridge circuit. The degree of imbalance of the bridge furnishes a measure of the concentration of combustible gas. The scales can be calibrated directly in percentage of alcohol, ether, methane, carbon monoxide, or other combustible gas.

Another type of instrument measures the increase in temperature as the combustible gas is burned in contact with a catalyst. The gas is drawn in by a motor through a flow-meter to ensure a constant rate of flow. The sample passes through a dehydrating agent to remove water vapor and finally through a bed of "Hopcalite" catalyst, which promotes the oxidation of the carbon monoxide to carbon dioxide. The heat liberated by this oxidation is proportional to the concentration of carbon monoxide. Two thermocouples (one in the incoming gas stream and one imbedded in the catalyst) are used to measure the heat evolved. Such instruments are very sensitive. The range for carbon monoxide is 0–0.15%. The recorder scale can be calibrated to 0.005% and can be estimated to 0.001%.

Velocity of Sound in Gases

The velocity of sound in a gas is given by the following equation:

$$v = \sqrt{\frac{\gamma P}{\rho}} \tag{27-2}$$

where v is velocity, P is pressure, ρ is density, and $\gamma = C_p/C_v$, the ratio of the specific heats at constant pressure and constant volume. The ratio of the specific heats, γ, depends somewhat on the nature of the gas, being 1.67 for a monatomic gas, 1.40 for diatomic gases, 1.33 for triatomic gases, and approaches 1.0 for polyatomic gases. This is not a great variation, and for most gases or gas mixture γ can be assumed to be essentially a constant. If we substitute in Eq. 27-2

$$\frac{nM}{V} = \rho \qquad (27\text{-}3)$$

where n is number of moles, M is molecular weight, and V is volume, we obtain the following relationships:

$$v = \sqrt{\frac{\gamma PV}{nM}} \qquad (27\text{-}4)$$

or since

$$PV = nRT = nK$$

where K = a constant if T remains constant,

$$v = \sqrt{\frac{\gamma K}{M}} \qquad (27\text{-}5)$$

Thus the velocity of sound in a gas is practically proportional, at constant temperature, to the reciprocal of the square root of the molecular weight. The molecular weight must be interpreted as the average molecular weight for a gaseous mixture. Ewing[4] has discussed the production, detection, and application of sonic waves.

Magnetic Susceptibility

Oxygen, nitric oxide, and nitrogen dioxide are unique among the ordinary gases in that they are paramagnetic; that is, they are attracted into a magnetic field. Most gases are slightly diamagnetic—repelled out of a magnetic field. Oxygen is several times more paramagnetic than nitric oxide or nitrogen dioxide. The values of the volume suscepti- bilities are $146.6, 65.2$, and 4.3×10^{-9}, respectively. Advantage is taken of this property of oxygen in gaseous oxygen analyzers.

Gaseous oxygen is measured on the basis of change in magnetic force acting on a test body suspended in a nonuniform magnetic field when the test body is surrounded by the gas sample. The Beckman paramagnetic oxygen analyzer (Fig. 27-7) incorporates a small glass dumbbell suspended on a taut quartz fiber in a nonuniform magnetic field. When no oxygen is present, the magnetic forces exactly balance the torque of the fiber. However, when a sample containing oxygen is drawn into the chamber surrounding the dumbbell, the magnetic force is altered, causing the dumbbell spheres to rotate away from the region of maximum magnetic flux density. The degree of rotation is proportional to the partial pres- sure of oxygen in the sample. A small mirror attached to, and rotating with, the dumbbell throws a beam of light on a translucent scale of the instrument. The scale is calibrated in concentration units of oxygen present. Instruments are capable of sampling static or flowing gas samples, free of solids or liquids. With a span of 5% oxygen full scale, an accuracy of ±0.05% oxygen can be achieved. Standard cell volume is 8–10 ml, although for static samples a 3-ml cell volume is possible. Response time is about 10 sec.

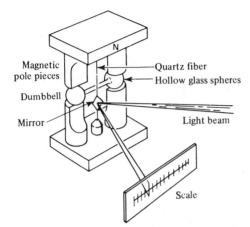

Magnetic pole pieces

Quartz fiber

Hollow glass spheres

Dumbbell

Mirror

Light beam

N

Scale

Fig. 27-7. Schematic of Beckman paramagnetic oxygen analyzer. (Courtesy of Beckman Instruments, Inc.)

The instrument shown in Fig. 27-8 utilizes the magnetic properties of oxygen along with thermal conductivity for the measurement of oxygen in a gas. The gas sample is passed across the bottom of a gas cell containing an electrically heated hot wire. A strong magnetic flux from a permanent magnet is directed across the wire. The oxygen is pulled into the region around the hot wire by the magnetic flux and is heated by the wire. Oxygen loses its magnetic susceptibility in inverse proportion to the square of the absolute temperature, and therefore the heated, relatively nonparamagnetic gas is continually displaced by the cooler, more paramagnetic oxygen moving in from below. A flow of gas proportional to the amount of oxygen present is set up around the hot wire. The

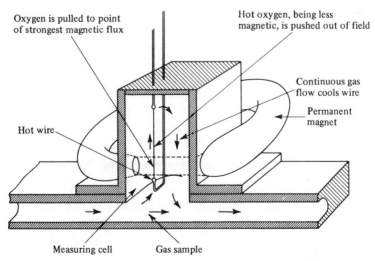

Oxygen is pulled to point of strongest magnetic flux

Hot oxygen, being less magnetic, is pushed out of field

Continuous gas flow cools wire

Permanent magnet

Hot wire

Measuring cell

Gas sample

Fig. 27-8. Schematic of Hays Magno-Therm oxygen recorder: schematic operation of analyzing cell. (Courtesy of Hays Corporation.)

hot wire is cooled, and its resistance is thereby decreased. The resistance of the wire in the analysis cell is compared with the resistance of a similar wire in a comparison cell by means of a Wheatstone bridge circuit. The comparison cell contains the sample of gas as the measuring cell but does not have a magnetic flux around the wire. Thus all variables except the cooling due to the oxygen present are canceled. The zero setting of the instrument is checked by swinging the magnet away from the measuring cell without interrupting the gas flow. The overall accuracy of the instrument is claimed to be ±0.25% oxygen up to 20% and ±2.5% of range up to 100%.

Electroanalytical Methods

The electrolytic hygrometer provides a specific method for the determination of water down to 1 ppm in gaseous samples. It involves continuous passage of the gas stream through an electrolytic cell—a capillary tube of Teflon, 0.12 mm in diameter, containing two intertwined platinum electrodes isolated from each other by a Teflon coil. A thin film of a hygroscopic electrolyte (phosphoric acid) coats the electrodes. A dc potential applied between the electrodes electrolyzes any moisture absorbed into hydrogen and oxygen. The electrolysis current is directly proportional to the concentration of water in the sample. This current becomes the linear signal which drives the indicating meter. Instruments of this type are used to monitor the moisture content of hydrogen chloride, protective atmospheres such as argon or nitrogen blankets in atomic reactors, and natural-gas pipelines to prevent formation of explosive hydrocarbon hydrates.

The Hersch galvanic cell provides an electrolytic method for the determination of low concentrations of oxygen. The gas stream passes through an electrolytic cell which consists of a silver cathode and an anode of active lead or cadmium. The electrodes are separated by a porous tube saturated with an electrolyte of potassium hydroxide. Oxygen in the gas sample is absorbed on the silver cathode and dissolves in the electrolyte as hydroxyl ions. The metallic lead or cadmium in turn is oxidized to plumbite ions or cadmium hydroxide. The magnitude of the cell current is a measure of the oxygen in the sample—a sensitivity of one part per million is attainable. Acidic substances are removed in advance by scrubbing the gas stream with caustic. Calibration is achieved by periodically generating known amounts of oxygen in a separate electrolysis cell.

Reactive sulfur compounds in a gas stream have been determined by a coulometric method. The gas stream is continuously bubbled through a solution of potassium bromide and sulfuric acid contained in the reaction cell. The reactive sulfur compounds are absorbed through reaction with electrolytic bromine present. The residual bromine concentration is automatically maintained at a constant value by means of sensing electrodes that match the rate of electrolytic generation of bromine to the absorption by reactive compound. The bromine-generating current, which is recorded, varies directly with the amount of sulfur compounds present in the sample. With timed cycles and selective absorbents preceding the reaction cell, results for hydrogen sulfide, mercaptans, and sulfur dioxide, in succession, are obtainable.

Absorption Spectroscopy

Among the most popular on-stream analyzers are infrared analyzers, available since the late 1940s. These are described in Chapter 6. The nondispersive instrument uses an integrated total of the energy absorbed by a sample over a broad spectral region and must be sensitized for a particular gas by means of special cells. On the other hand, the dispersive instrument can be set at any wavelength within its range and absorbance measured. Thus, the dispersive instrument is a general-purpose tool, able to detect and quantitatively measure any gas having infrared absorption rather than being limited to a single, preselected gas as is the case with the nondispersive instrument. Components frequently analyzed by infrared analyzers in industrial processes are listed in Table 27-2; these include gases and vapors of substances that are liquids at room temperature. By contrast, infrared instruments cannot selectively measure similar materials, such as butane in propane. Considering only the industrial processes where the measurement of carbon dioxide is important, one finds such applications as the control of combustion, the manufacture of cement, and the production of ethylene oxide, phthalic anhydride and ammonia.

Ultraviolet analyzers are used to measure chlorine, bromine, sulfur dioxide, hydrogen sulfide, hydrogen peroxide, phenol, phosgene, and toluene in process streams. Compared to infrared instruments, ultraviolet analyzers are generally able to measure substances more selectively.

TABLE 27-2 Applications of Infrared Spectroscopy to Process Streams[a]

Gas	Analytical Wavelength, μm	Minimum Detectable Concentration (ppm) 20-m Path	Maximum Concentration 1.0 Absorbance (ppm or %) $\frac{3}{4}$-m Path
Cyclohexane	3.4	0.04	4000
CO_2	4.25	0.08	7500
N_2O	4.5	0.03	2100
CO	4.65	1.2	8.3%
COS	4.85	0.02	1100
NO	5.3	1.5	8.2%
CH_3COCH_3	5.75	0.1	4000
SO_2	7.4	0.1	3000
Vinyl acetate	8.2	0.06	1000
Dioxane	8.8	0.2	2100
CH_3CH_2OH	9.4	0.4	5000
NH_3	10.75	2.2	2.1%
Freon 11	11.8	0.06	300
CCl_4	12.6	0.05	250
CH_2Cl_2	13.3	0.4	1900

SOURCE: Courtesy of Wilks Scientific Corp.

[a]Data for Miran (miniature infrared analyzers) dispersive gas analyzer with variable path gas cell (0.75 to 20 m).

LIQUID PROCESS STREAMS

A refractometer is the instrument of choice whenever the sample to be analyzed is a simple binary liquid mixture.[8] Density-measuring analyzers are applicable when the range of compositions is broad. But when the range is narrow, and when an analysis of the liquid phase of a suspension or slurry is required, refractometry is again the preferred method. Streams too dark or turbid to be analyzed by the differential refractometer (Chapter 14) are easily handled with the critical-angle reflectance-type refractometer. The prism is mounted in the pipeline itself and is therefore in intimate contact with the process fluid because its surface is flush with the wall of the pipeline. Because only the refractive index at the interface is measured, care must be exercised to make the material at the interface representative of the bulk solution. An automatic jet washing system is provided in the pipeline head to prevent film buildup on the prism surface. A thermistor probe inserted in the pipe measures the temperature of the process line and handles modest temperature changes (about $10°C$) and automatically corrects the refractive index reading if linked with appropriate computer software.

Process refractometers have been used to control the blending of styrene and butadiene streams in the manufacture of GR-S rubber, and to monitor the purity of each component stream. Because the refractive indices of styrene and butadiene are 1.5434 and 1.4120, respectively, a change of 0.1% in composition is easily detectable.

The proper consistency of ketchup is determined by refractometry, and the heat is turned off when the mixture of tomatoes, spices, and water has reached the right consistency and enough water has been evaporated. Because making ketchup is a batch process, a sample is pumped from the bottom of the kettle through the in-line refractometer and back into the finishing kettle. When the product reaches the desired concentration the kettle is drained and refilled with another batch.

Refractometry works well for solutions of sugar and water. Many carbonated beverages have their sugar–water base controlled to within 0.05 percent of the desired concentration. By employing a small analog computer in the production of crystalline sugar, syrup concentration is brought to the proper point and seed crystals are automatically added to begin crystallization. The result is high-quality crystalline sugar produced in about a third of the time needed for conventional processes.

AUTOMATIC CHEMICAL ANALYZERS

Most routine laboratory analyses require the close attention of a trained scientist or technician. With the development of automatic, wet-chemical instruments, many of these measurements can be made just as effectively and with just as much accuracy by far less skilled technicians, or by no one at all. Almost any repetitive analytical determination can be automated.

Automatic laboratory analyzers operating on two fundamentally different principles are available. In one type, standard laboratory tests are used and each sample is handled as a discrete entity. Only the reagents are added and the measurements made by the device itself. From a sample turntable the sample is introduced, diluted, and reacted with a color-producing reagent; then the resulting color is measured and the answer is printed on tape. Finally, the solution from the sample cell is removed to a waste container and the cell rinsed prior to the next determination.

In the second type of automatic chemical analyzer, the samples are treated in a continuously flowing system. The analytical system of the AutoAnalyzer consists of a series of individual modules, each a separate component performing one specific function in a programmed sequence (Fig. 27-9). Modules can be interchanged and rearranged for different analytical purposes. The system automatically prepares the sample, introduces it, purifies, isolates, senses, reads out, and records the concentration. As a sample-processing turntable rotates, a sampling capillary dips into each container in succession, aspirating its contents for a timed interval and feeding them into the analytical system. Between samples the capillary is raised and sucks in air for another timed interval.

The heart of the analyzer is the proportioning pump. It can handle 12 or more separate fluids (reagents, diluents, air, etc.) simultaneously while varying their delivered output in any ratio up to 79 to 1. The pump consists of two parallel stainless-steel roller chains with spaced roller thwarts that bear continuously against a spring-loaded platen. Flexible resistant tubing with different inside diameter is employed.

At each point in the system where two different liquids come together there is a mixing coil of suitable length. Digestive procedures are accomplished at $95° \pm 0.1°C$ in a continuous glass or plastic helix which provides a thin layer of fluids which quickly reach the bath temperature.

After the sample is processed, the end product is quantitated by a suitable sensing device—dual-beam spectrophotometer, fluorimeter, flame spectrometer, or titrator—and the results are printed on a continuous strip chart or on a paper tape next to an identifying sample number. Known standards are interspersed periodically to assure correctness of calibration.

Cross contamination between samples is an ever-present hazard in continuous flow systems. Air bubbles both between the individual samples and within the sample volume minimize contamination. The bubbles act as a wiping influence in the tubular system and as an unbreachable barrier between samples. Wash liquids can also be introduced.

One important difference between automatic and manual systems is that reactions do not have to be carried to completion. Because in analytical processing systems conditions are constant and unvarying within the unit, and known standards are also subjected to identical treatment, sensing of incomplete reactions provides an accurate, reproducible determination. Answers are supplied for most analyses from 10 to 100 times faster than by manually operated methods, that is, in minutes rather than in hours or even days. Furthermore, results are not affected as they would be if the operator happened to be mentally fatigued or otherwise momentarily inattentive.

On one eight-channel instrument designed for clinical laboratories, the determination of sodium, potassium, chloride, carbon dioxide, glucose, urea nitrogen, albumin, and total protein is performed on a single 1.2-ml serum sample. In an eight-hour day, one instrument can run 120 complete analyses or 960 individual tests, not including blanks and control samples. By comparison with ordinary manual methods, this particular instrument handles the output of one technician over a period of three weeks when working

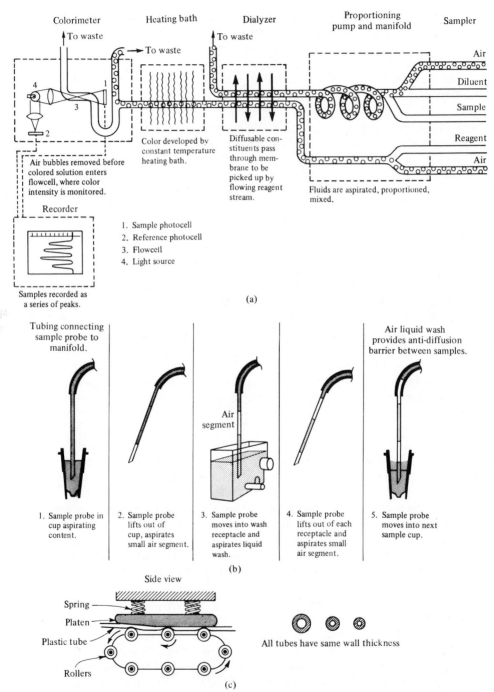

Colorimeter   Heating bath   Dialyzer   Proportioning pump and manifold   Sampler

To waste

To waste

To waste

Air

Diluent

Sample

Reagent

Air

Color developed by constant temperature heating bath.

Diffusable constituents pass through membrane to be picked up by flowing reagent stream.

Air bubbles removed before colored solution enters flowcell, where color intensity is monitored.

Fluids are aspirated, proportioned, mixed.

Recorder

1. Sample photocell
2. Reference photocell
3. Flowcell
4. Light source

Samples recorded as a series of peaks.

(a)

Tubing connecting sample probe to manifold.

Air liquid wash provides anti-diffusion barrier between samples.

Air segment

1. Sample probe in cup aspirating content.

2. Sample probe lifts out of cup, aspirates small air segment.

3. Sample probe moves into wash receptacle and aspirates liquid wash.

4. Sample probe lifts out of each receptacle and aspirates small air segment.

5. Sample probe moves into next sample cup.

(b)

Side view

Spring

Platen

Plastic tube

Rollers

All tubes have same wall thickness

(c)

Fig. 27-9. (a) Typical single channel flow schematic, (b) details of sampler, and (c) proportioning pump of the AutoAnalyzer. (Courtesy of Technicon Corporation.)

manually. For automation to justify itself for a particular test method, at least 20–30 relatively simple analyses of one type must be run a day. Before some applications become feasible, analytical methods for specific products will have to be developed with automation in mind.

Example 27-2

The methodology of phosphate determination (in water analysis) is sketched in the flow diagram (Fig. 27-10), using the reduction of the phosphomolybdate complex to molybdenum blue by 1-amino-2-naphthol-4-sulfonic acid in sulfuric acid medium. The sample is diluted with water and sent into the dialyzer. Here a portion of the phosphate dialyzes into an air-segmented stream of the reductant. This enriched stream then meets the sulfuric acid-ammonium molybdate reagent. In the heating bath, phosphomolybdate formation and reduction take place, the color intensity then being read at 650 nm. The sample and reagent volumes shown may be compared with the amounts used in the conventional colorimetric procedure as a guide in designing procedures for the AutoAnalyzer.

A number of additional flow diagrams will be found in the book by Welcher.[11]

The basic problem with slow analyzers (about 20 sec to 2 min between data points), such as the preceding automated system and the sequential AutoAnalyzer, is that data is stored in the form of a strip-chart record, or of punched cards or magnetic tape. Baseline drift of mechanical, chemical, or electronic origin is indigenous to such slow-output systems. If the data signals for 15 or more analyses can be produced in a small fraction of a second, electronic or other drift will be minimized within that set. The possibility then exists of repeating all readings on each sample many times, averaging them, doing the necessary statistical analyses, all in much less than a second. A computer-interfaced fast analyzer accomplishes these objectives.[1] In this type of analyzer centrifugation provides the driving force for mixing the sample and reagent, and transferring the reaction mixture

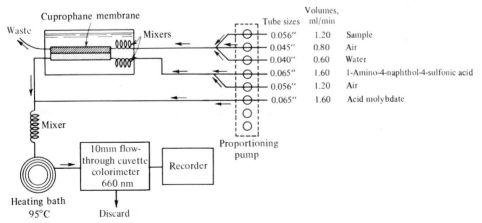

Fig. 27-10. AutoAnalyzer methodology: flow diagram for phosphate determination. (Courtesy of Technicon Corporation.)

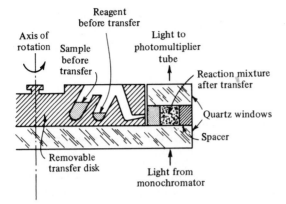

Reagent
before transfer

Axis of
rotation   Sample
before
transfer

Light to
photomultiplier
tube

Reaction mixture
after transfer

Quartz windows

Spacer

Removable
transfer disk

Light from
monochromator

Fig. 27-11. Cross section of sample disk and centrifugal analyzer section.

to the cuvette. Figure 27-11 shows the cross section of the sample disk and centrifugal analyzer section of one version of this type instrument.[2] In use, samples and reagents are placed in cups arranged in concentric circles in the transfer disk. The cups in the various circles are aligned radially and arranged in such a manner that samples and reagents are thrown through a transfer cavity (when the rotor attains a speed of about 350 rpm) and finally into the corresponding cell, where the transmittance or luminescence of the analysis mixture is measured. Figure 27-12 shows the pattern obtained when the sweep rate of an oscilloscope is keyed to the rotation rate of the analyzer section (about 600 rpm). The peak (lower portion of signal) is flat with a width of about 650 μsec; the top portion of the signal is obtained when the light beam is blocked by the partitions dividing the cells and therefore constitutes a dark current value. The computed data collection function (Digital Equipment Company PDP-8/I computer) observes the detector signal 16 times on the flat portion of the transmittance peak and averages the 16 readings. The observations can be repeated and averaged on successive passes of the cell (16 for most applications), yielding a final result which is the average of 256 observations. This result is then corrected for dark current and individual cell differences, with the resulting value being stored in memory. An important advantage of this approach to analysis is that samples and standards are observed virtually simultaneously and under the same conditions. For

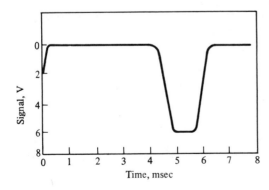

Signal, V

Time, msec

Fig. 27-12. Drawing of oscilloscopic readout showing signal from a single cell.

systems in which the rate of change is slow with respect to the rotor speed, the analysis time can be optimized without waiting for maximum and/or equilibrium color development. Procedures can be changed simply by changing the reagents in the loading disk and the monochromator setting.

AUTOMATIC ELEMENTAL ANALYZERS[5]

Present instrumentation for carbon, hydrogen, and nitrogen (CHN) analyzers is based upon one of two general procedures. One involves the separation of carbon dioxide, nitrogen, and water by a gas–liquid chromatographic column. The other involves separation by means of specific absorbants for water and for carbon dioxide with the resulting change in composition of the gas mixture being measured. Thermal conductivity is the detection method in both techniques. Results are calculated by analyzing standard samples and an occasional blank.

The sample (0.1 to 3 mg range) is either burned under static conditions in a pure oxygen atmosphere at 900°C, whereby the sample boat can subsequently be removed for weighing any residue, or mixed with cobaltic oxide [or a mixture of the oxides of manganese(IV) and tungsten(VI)] to provide the oxygen and heated to 900°C. Combustion converts the carbon in the sample to carbon dioxide and carbon monoxide, and hydrogen to water. Nitrogen is released as the free gas, along with some oxides. A stream of helium carries the combustion gases into a reaction furnace operated at 750°C. Here a chemical change completes the simultaneous oxidation and reduction of the sample gases. Hot copper reduces the nitrogen oxides to nitrogen and removes the oxygen. Copper oxide converts the carbon monoxide to carbon dioxide. If needed, a magnesium oxide layer in the middle of the furnace removes fluorine, and a silver-wool plug at the exit removes chlorine, iodine, and bromine, and also any sulfur or phosphorus compounds resulting from the combustion of the sample. In CH analyzers, oxides of nitrogen are removed a little later in the train with manganese dioxide.

Separation of the combustion gases by gas chromatography involves the following sequence: The gases pass through a charge of calcium carbide where water vapor is converted to acetylene. A nitrogen cold-trap freezes the sample gases and isolates them in a loop of tubing. A valve seals off the combustion train, which is then ready for another sample. The chromatographic stage is begun by lowering the cold-trap and heating the injection loop. Another stream of dry helium gas carries the gases (as a plug) into the chromatographic column where the three gases—N_2, CO_2, and C_2H_2—are completely separated. Most organic samples can be completely burned in 10–12 min, and the chromatographic separation requires another 10 min. Figure 27-13 gives the schematic diagram of a CHN analyzer of this type.

The other general procedure involves separation by means of specific absorbents for water and carbon dioxide. Three pairs of thermal conductivity cells are used in series for detection: one pair each for water, carbon dioxide, and nitrogen. A magnesium perchlorate trap between the first pair of cells absorbs any water from the gas mixture before it enters the second pair of cells. Therefore, the signal is proportional to the amount of

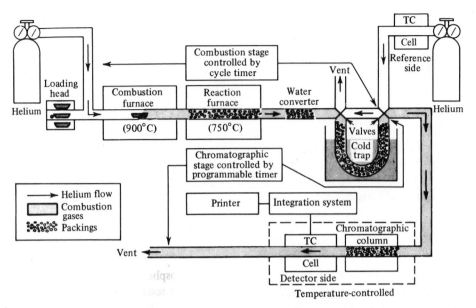

Fig. 27-13. Schematic diagram of the Fisher Carbon-Hydrogen-Nitrogen Analyzer. (Courtesy of Fisher Scientific Co.)

water removed. Similarly, a soda–asbestos trap between the second pair of cells results in a signal in proportion to the carbon dioxide removed from the sample. The last pair of cells detects nitrogen by comparing the helium–nitrogen mixture with pure helium. A modified sequence handles the three gases as follows: Water is absorbed on silica gel, then carbon dioxide is absorbed in a length of molecular sieve. The nitrogen–helium mixture is compared with pure helium. The carbon dioxide is desorbed from the molecular sieve, mixed with helium, and passed through the detector. Finally, the silica gel section is heated and the desorbed water is swept through the pair of detectors (bypassing the molecular sieve). These programs take about 13 min per sample.

In the total carbon analyzer for water samples, depending on the range of carbon contents, a 20–200 μl water sample is injected via syringe into a flowing stream of air and swept into a catalytic combustion tube containing a cobaltic oxide-impregnated asbestos packing and maintained at 950°C. The water is vaporized, and all carbonaceous material is oxidized to carbon dioxide and steam. The air flow carries the cloud of steam and carbon dioxide out of the furnace, the steam is condensed and removed, and the carbon dioxide is swept into a nondispersive infrared analyzer, sensitized to measure carbon dioxide. The transient signal is recorded; it includes organic carbon, inorganic carbon, and carbon dioxide dissolved in the sample. In a second operation a similar-sized sample is injected into the stream of air and swept into a second reaction tube at 150°C which contains quartz chips wetted with 85% phosphoric acid. Only carbon dioxide from inorganic carbonates is released. The difference in the two signal peaks yields an analysis of total organic carbon present. Time of a single analysis is 2–5 min.

The entire procedure for sulfur in metals has been automated. In a programmed induction furnace the metal sample is burned in a stream of purified oxygen. The sulfur dioxide is absorbed in hydrogen peroxide and titrated to a pH 4.8 end point with an automatic titrator. All the operator does is weigh out the sample, pop it into the induction furnace, start the timed combustion cycle and, in 3–6 min, read the digital results on printed tape.

CONTINUOUS ON-STREAM CONTROL ANALYSIS

The final step in automatic process instrumentation involves transfer of the instrumentation from the control laboratory to the process stream. A continuous analyzer is attached to a sampling line and thereafter obtains automatically and continuously a reading which is proportional to the instantaneous concentration of a particular component in the flowing stream. The information provided is then used to set automatically the environmental controllers and to take any corrective action necessary in the process stream. In essence, continuous stream analyzers take over the function of the control laboratory with a considerable increase in speed and efficiency.

In some processes (e.g., acetylene-from-hydrocarbon processes) certain steps are so critical that the close control available only by continuous on-stream analysis is a necessity. In other instances, continuous on-stream analysis leads to increased throughput through the ability to operate closer to safety limits and through better material utilization. Accurate continuous analysis permits the holding of a narrow specification range. It can also keep a close check on waste streams for excessive loss of valuable components or warn of the development of pollution levels in effluent streams. Analyzers can provide accurate records for accounting purposes and can be important from a legal standpoint.

A number of steps are involved in setting up on-stream control facilities. The analyst, in close collaboration with the project engineer, must determine the analytical task or tasks to be done in order to follow a process as effectively as possible. The number of constituents monitored and the number and location of checkpoints must be decided. Never deluge the operator with a mass of information, much of which may be useless. The guiding principle should be to provide the operator with the least amount of data needed to produce the desired product. Economic considerations and the manpower requirements for installation, calibration, and maintenance must not be overlooked. Different analytical methods may be desirable at each checkpoint. As process control improves, production costs decrease; however, analysis costs rise. The selection of an optimum analytical method depends on minimizing the sum total of these costs. Leemans[7] discusses the selection procedure in detail.

Design Features

When an analyzer is selected for a given problem, the first step is to make certain that the particular analysis can be made by the instrument and that it has sufficient sensitivity to determine the component of interest in the range of concentrations expected. Although similar in operating principles to their laboratory cousins, process stream analyzers differ in a number of important respects. Moving the automatic instrument sensors from the laboratory to the plant confronts instrument designers with major problems. Design criteria must incorporate these features: (1) reliability, (2) operational simplicity, (3) readout as foolproof as possible, (4) ease of maintenance, and (5) flexibility for future growth.

First and foremost is reliability. Instrument downtime represents plant process downtime, or operation without control, which is very costly. Hence, long-term stability and reliability are essential characteristics. Availability of modular plug-in type construction shortens necessary repairs and goes a long way toward making the instrumentation as reliable as any of the links in the process. Operational simplicity implies a minimum of controls and infrequent attention by the operator. Preferably once every shift a cursory check is run. Thorough overhaul and inspection is carried out only during normal process downtime.

Readout and control functions must be made as foolproof as possible. Digital readout devices are utilized extensively. The environment in which the analyzer is used differs from the calm of the laboratory. Analyzers must withstand wide ambient temperature fluctuations and heavy vibrations, and they must not create explosion hazards. Often the units are completely sealed so as to operate independent of outside conditions and to withstand the onslaught of monkey-wrench mechanics.

Closed Control Loop

An automatic control system generally consists of the sensing element, the amplifier and recorder, the controller, and the final control element. These elements react upon each other and form the closed loop, as illustrated in Fig. 27-14. All are of equal importance.

Automatic control systems depend on deviations for their operation. Before the controller can act, a deviation in the process variable must occur. The controller, which receives the signal from the sensing element, compares it with the set point signal and

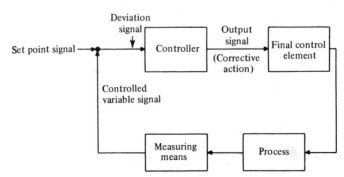

Fig. 27-14. Control loop diagram. (After W. G. Holzbock, *Instruments for Measurement and Control*, 2nd ed., Van Nostrand Reinhold, New York, 1962, p. 246.)

responds to it by sending an output signal to the final control element, which produces a correction in the process. The corrective cycle terminates when the modified signal from the sensing element equals the set point signal. It is this feedback that actuates the correction.

A time lag is involved in each of the steps in the control loop; this lag is composed of the dead time and time lapse before corrective action can be initiated. A thorough treatment of this subject can be found elsewhere.[6]

The ultimate goal is to tie on-stream process analyzers or sensors to computers. This goal requires studying the controllability of processes in order to understand the reactions that are taking place, determine how flows respond dynamically to changes in valve positions, and work out the mathematics of processes. Equations are written that express the desired amount of influence from a given variable. In predictive control the desired correction is made not on the basis of set points previously determined by the operator or process engineer, but on the basis of continuing calculations made by the computer. If a computer completes the closed control loop, when the operating conditions suddenly vary, the computer can rapidly make the necessary calculations so that several of the set points are properly changed to bring about the optimum yield, the desired product quality, and so forth. Computers can also be used to evaluate measurement results and search for useful correlations not obviously discernible.

Examples of Process Control

A typical use of continuous analyzers in ethylene purification is shown in Fig. 27-15. Purity and recovery are monitored at five checkpoints. Other stream components during

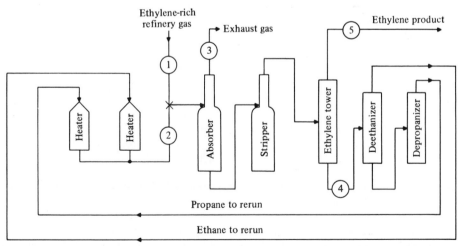

Fig. 27-15. Use of continuous infrared analyzers in ethylene purification: (1) ethylene analysis on feed stock for accounting purposes; (2) ethylene analysis beyond crackers; (3) ethylene analysis of absorber off-gas for absorber efficiency; (4) ethylene analysis in ethylene tower bottoms for fractionation tower efficiency; (5) end point analysis for ethylene purity. (After P. A. Wilks, Jr., *Chem. Eng. Progr.*, **51**, 358 (1955).)

ethylene manufacture and initial purification might be any mixture of these: CH_4, C_2H_6, C_3H_8, C_2H_2, CO, CO_2, and N_2. In the final purification, impurities are likely to be CH_4, C_2H_6, and C_2H_2, whose total concentration will be less than 2%. Instrument readability (1% of scale) should correspond to 0.05% ethylene in the manufacture, 0.02% in the initial purification, and 0.005% in the final purification. Long-term stability (8 hours or longer) should be within 2% of scale.

On-stream analyzers specifically designed for water treatment of cooling tower waters are based on conductivity and pH measurements. A sample of recirculating water flows to an analyzer, where the conductivity and pH are continuously measured and recorded. The pH measurement provides a control signal for the addition of pH control chemicals. The signal for complete water replacement is based on the conductivity measurements, which are directly related to the total amount of dissolved solids in the system. In addition, the makeup water flow rate is measured and recorded, and provides a control signal for feed of corrosion inhibitor, biocide, and so forth.

Nondestructive methods for gauging the thickness of a homogenous material are often based on the absorption or diminution in initial intensity of an X-ray, gamma-ray,

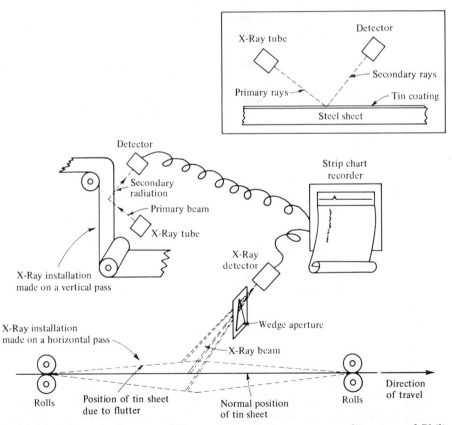

Fig. 27-16. Schematic diagram of X-ray plating thickness gauge. (Courtesy of Philips Electronic Instruments.)

or beta-ray beam in passage through a layer. The unabsorbed radiation passes into the the detector, in which the current is amplified and opposed by the signal received from a beam that passes through a stationary sample of standard required gauge. Deviations from balance activate a servo-mechanism which automatically readjusts the mill rolls more or less tightly. This technique of gauging is now established practice for a wide variety of rolled metal sheets travelling at 800 ft/min on the rolling mill. In addition to X rays, sources of radiation are radioisotopes mounted in shielded containers. For example, the normal gauging range for ^{90}Sr is 300–10,000 mg cm^{-2}. Beta gauges are employed for lower gauging ranges from 1 to 1200 mg cm^{-2} and utilize ^{85}Kr, ^{106}Ru, or ^{204}Tl.

Operation of the X-ray tin gauge, outlined in Fig. 27-16 is based on the principle of absorption by the tin coating of fluorescent iron X rays arising from the excitation of the iron base. The iron radiation is absorbed in direct proportion to the thickness of the tin plate. Dual units simultaneously measure the tin coating on both sides of the steel strip, moving at speeds of 500–650 ft/min. Signals, reflecting the coating weight, regulate either the speed of the strip through the plating bath or the amount of current applied to the tin anodes. The wedge aperture compensates for flutter of the sheet during its travel between rollers. Usually the X-ray equipment is positioned at the point where the plated steel is ready for coiling, and flutter is a minimum.

PROBLEMS

1. Design a thermal conductance circuit for handling the following product gases:
   (a) Automobile carburetors: 10%–15% CO_2 in air (plus CO).
   (b) Ammonia synthesis: 70%–80% H_2 in N_2 plus NH_3.
   (c) Electrolytic H_2 with O_2 plus N_2 impurities (0%–0.4%).
   (d) Carbon dioxide (0%–10%) in air.
   (e) Hydrogen (0%–0.2%) impurity in electrolytic oxygen.
   (f) Methane (0%–5%) in air.

2. For each mixture in Problem 1, compute the bridge output when using four tungsten filaments and a 2-mV recorder. Estimate the sensitivity per 1% full-chart deflection.

3. Design a system for the continuous analysis of CO_2 and H_2 in a stream consisting of 34% CO_2, 15% CO, 40% H_2, 10% CH_4, and 1% N_2.

4. Suggest a method for handling the following on-stream process situations: (a) fractionating tower control in the separation of cyclohexane and n-hexane; (b) fractionating column for the determination of butane in isobutane; (c) control of the end point in making shortenings and margarine using materials such as soybean and cottonseed oils; (d) changes in the concentration of individual solutions of inorganic salts, HCl, or H_2SO_4.

5. Design a method for monitoring boiler stack gas in pulp mills for soda, both to aid in minimizing soda losses and to help control air pollution. Keep in mind that light transmission or scattering are handicapped by rigid sample-handling requirements and high emission rates. In addition, highly conductive dissolved gases that are also present offset the advantages of measuring the electrical conductivity in scrubbed gas samples.

6. Discuss several limitations inherent in the gas-chromatographic system of analysis employed in some CHN analyzers.

7. Design an on-line analyzer to close the loop in a hot process lime–soda softener.

8. In the separation of aromatics from saturates, using refractive index to follow a process, what precision can be expected? The aromatics have a refractive index of about 1.50, and the saturates about 1.40.

9. A CHN analyzer was calibrated by burning 1.657 mg of dimethylglyoxime. The total signal for C = 17,500 units, for H = 1062 units, and for N = 4128 units. The unknown compound, a 2.021-mg sample, gave these signals: C = 40,760 units, H = 1078 units, and N = 3195 units. Calculate the percent carbon, hydrogen, and nitrogen.

BIBLIOGRAPHY

Frazer, J. W., "Design Procedures for Chemical Automation," *Am. Laboratory*, **5**, 21 (February 1973).

Holzbock, W. G., *Instruments for Measurement and Control*, 2nd ed., Van Nostrand Reinhold, New York, 1962.

Horton, W. S. and C. C. Carson, "Gas Analysis: Determination of Gases in Metals" in *Treatise on Analytical Chemistry*, Vol. 10, Part 1, I. M. Kolthoff and P. J. Elving, Eds., Wiley-Interscience, New York, 1972, Chapter 103.

Kiesebach, R. and J. A. Schmit, "Gas Analysis: Thermal Conductivity" in *Treatise on Analytical Chemistry*, Vol. 10, Part 1, I. M. Kolthoff and P. J. Elving, Eds., Wiley-Interscience, New York, 1972, Chapter 102.

Kramer, E., "The Future of Laboratory Automation," *Am. Laboratory*, **2**, 32 (February 1970).

Lewis, C. D., "Continuous Automatic Instrumentation for Process Applications" in *Treatise on Analytical Chemistry*, Vol. 10, Part 1, I. M. Kolthoff and P. J. Elving, Eds., Wiley-Interscience, New York, 1972, Chapter 105.

Nueberger, E. D., "On Target with On-Line," *Ind. Res.*, p. 42 (Nov. 1970).

Perone, S. P., K. Ernst, and J. W. Frazer, "A Systematic Approach to Instrument Automation," *Am. Laboratory*, **5**, 39 (Feb. 1973).

Sanders, H. J., "Process Instruments," *Chem. Eng. News*, p. 33 (Oct. 13, 1969).

Siggia, S., *Continuous Analysis of Chemical Process Systems*, Wiley, New York, 1959.

LITERATURE CITED

1. Anderson, N. G., *Science*, **166**, 317 (1969).
2. Coleman, R. L., W. D. Shults, M. T. Kelley, and J. A. Dean, *Am. Laboratory*, **3**, 26 (July 1971).
3. Ewing, G. W., *J. Chem. Educ.*, **45**, A377 (1968).
4. Ewing, G. W., *J. Chem. Educ.*, **43**, A1037 (1966).

5. Fish, V. B., *J. Chem. Educ.*, **46**, A323 (1969).
6. Holzbock, W. G., *Automatic Control: Principles and Practice*, Van Nostrand Reinhold, New York, 1958.
7. Leemans, F. A., *Anal. Chem.*, **43**, (11), 36A (1971).
8. Maley, L. E., *J. Chem. Educ.*, **45**, A467 (1968).
9. Martin, A. J. P. and A. T. Jones, *Biochem. J.*, **63**, 138 (1956).
10. Nerheim, A. G., *Anal. Chem.*, **35**, 1640 (1963).
11. Welcher, F. J. Ed., *Standard Methods of Chemical Analysis*, Vol. III, Part B, Van Nostrand Reinhold, New York, 1966.

Answers to Problems

1. (a) 2 2/9 W in R_1; 4 4/9 W in R_2. **(b)** 20 W in R_1; 10 W in R_2. **(c)** 6.6 W in R_1; 17.5 W in R_2; 0.44 W in R_3. **2.** $i_l = 0.001$ A; $i_g = 5 \times 10^{-7}$ A. **3.** $Z = 4796\ \Omega$; $\theta \simeq 78°$. **4.** Approximately 504 Hz. **5.** 0.7 μf.

1. e_c = about -4.4 V. **2.** 60 V, 130 V. **3.** 18.3 and 1610.

1. $\sin \theta = 3(3.8 \times 10^{-5}$ cm$)/d < \sin \theta = 2(7 \times 10^{-5}$ cm$)/d$ whatever the grating spacing. **2.** From $17°14'$ (violet) to $33°4'$ (red). **3. (a)** 1773 grooves/mm; **(b)** 2387 Å; **(c)** $10.2°$; **(d)** 1180 grooves/mm; **(e)** 5000 Å; **(f)** $20.5°$; **(g)** 590 grooves/mm; **(h)** 1.0 μm; **(i)** 295 grooves/mm; **(j)** $14.8°$; **(k)** 148 grooves/mm; **(l)** $21.6°$. **4.** Resolution is 179,000 (first grating) and 45,000 (second grating) when both are used at the same angle in the first order. A line imaged by first grating in the first order will be produced by second grating in the fourth order at the same dispersion and resolution. **5. (a)** M_2 has 17 Å/mm; M_1 has 34 Å/mm. For same slit height and width, $(F_T)_1/(F_T)_2 = 23.9/6.0 = 4.0$. M_1 has $f/17.2$ and M_2 has $f/8.6$. **(b)** Both monochromators have f/number $= f/8.6$. M_2 has 8.5 Å/mm. $(F_T)_1/(F_T)_2 = 23.9/12 = 2.0$. See G. L. Clark, Ed., *Encyclopedia of Spectroscopy*, Van Nostrand Reinhold, New York, 1960; p. 249. **6. (a)** Resolution $= 19,200$ in first order. **(b)** $\Delta\lambda = 0.05$ mm. **(c)** Yes. **7.** $\Delta\lambda = 0.035$ Å. **8. (a)** 982 grooves (first order). **(b)** 3 Å bandpass. **(c)** 0.186 mm. **9. (a)** Mn 4033.07 Å and Ga 4032.98 Å are separated by 0.09 Å; 44,811 grooves (first order) required. **(b)** 3 μm (for base-line separation). **(c)** 10 μm (to distinguish existence of two lines); third order needed for base-line separation. **10. (a)** 0.153 μm. **(b)** 0.165 μm. **(c)** 0.214 μm. **(d)** 0.277 μm. **11.** No; at base-line an overlap of 17 Å exists. **12. (a)** $f/8.6$. **(b)** Rs $= 61,400$. **(c)** 17 Å/mm. **13.** Gain is 500.

Chapter 4

1. 2.42 mg/50 ml. 2. $T = 0.617$. 3. $\epsilon = 3120$. 4. 6.07×10^{-5} M. 5. % Ti = 0.540 A_{400} - 0.338 A_{460} and % V = 1.595 A_{460} - 0.795 A_{400}.
Sample 1, 0.054 % Ti and 0.050 % V; sample 2, 0.052 % Ti and 0.397 % V; sample 3, 0.099 % Ti and 0.242 % V; sample 4, 0.197 % Ti and 0.186 % V; sample 5, 0.294 % Ti and 0.193 % V; sample 6, 0.101 % Ti and 0.58 % V; sample 7, 0.140 % Ti and 0.031 % V; sample 8, 0.067 % Ti and 0.043 % V; sample 9, 0.116 % Ti and 0.026 % V. 6. $C_A = A_{600}$. $C_B = (3 A_{400} - A_{500})/5.95$. $C_C = (40 A_{500} - A_{400})/71.4$. 7. Tyrosine at 294 nm and tryptophan at 280 nm. 8. *Ortho* at 276 nm and *para* at 256 nm. 9. Sodium acetate, 0.0681 M, and o-chloroaniline, 0.0308 M. 10. Aniline, 0.97 ml; 2-naphthylamine, 1.98 ml. 11. First rise, extrapolated back to zero absorbance gives system blank. Net titrant volume: 2.70 - 0.23 = 2.47 ml; 78.0 μg of magnesium present.
12. C=C—C—C—Br; double bond and bromine are not in conjugation.
13. CH_2=C(CH_3)—CH_2—CO—CH_3 shows no high-intensity band beyond 220 nm.
14. Structure has isolated conjugated systems (C=C—C=C) and (C=C—C=O), one in the ring and one in the side chain, leading to different molar absorptivities. 15. Structure II is α-isomer; it possesses only two conjugated units. 16. A high-intensity, three conjugated system is indicated which would require the carbonyl to be adjacent to the benzene ring. 17. The calculated λ_{max} values permit identification. (a) 254 nm. (b) 219 nm. (c) 268 nm. (d) 298 nm. 18. (a) 254.5 nm band and most intense. (b) 250.5 nm; *meta*-substituent lower molar absorptivity. (c) 259.5 nm (low intensity) and 228 nm (forced into nonplanar configuration by *ortho*-substituents). 19. $pK_a = 9.39$.
20. $pK_a = 6.45$. 21. p-Nitrophenol: $pK_a = 6.99$. Papaverine (protonated cation): $pK_a = 6.40$ (average). 22. Bromophenol blue: $pK_a = 4.05$. Methyl red: $pK_a = 4.93$. Bromocresol purple: $pK_a = 6.29$.

23. $\Delta C/C$	Concentration, M	24. $\Delta C/C$	Concentration, M
-0.0330	0.0050	-0.00255	0.0600
-0.0190	0.0100	-0.00237	0.0700
-0.0145	0.0150	-0.00232	0.0800
-0.0124	0.0200	-0.00230	0.0900
-0.0110	0.0300	-0.00230	0.100
-0.0110	0.0400	-0.00234	0.110
-0.0124	0.0600	-0.00240	0.120
-0.0164	0.0800	-0.00246	0.130
-0.0224	0.100	-0.00250	0.140
-0.0268	0.110	-0.00255	0.150

25. $\Delta C/C = -0.0114$ for 0.0200 M solution. 26. $\Delta C/C = -0.00212$ for 0.0700 M solution. 27. (a) $\epsilon = 133$ and $a = 0.023$ (mg/100 ml)$^{-1}$. (b) Increase in precision is $(1 - T_r)$; $\Delta C = 0.345$ times the value using the ordinary method. (c) Normal blood: 2.96 mg/100 ml; ketotic blood: 8.22 mg/100 ml; normal urine: 4.22 mg/100 ml; ketotic urine: 215 mg/100 ml. 28. (a) 12-fold increase in precision based on the expanded scale.

(b) $\Delta C/C = 0.173 \Delta T$. **29.** (a) FeL_3. (b) Conditional formation constant is 3.2×10^{17}
30. (a) FeL_3. (b) $\epsilon = 12{,}600$. **31.** (a) FeL_3. **32.** (a) 0.333. See *Anal. Chem.*, **29**,
1426 (1957). (b) $K_f = 8.6 \times 10^9$.
33. Concentration Error, %, for Stray Light at

0.5%	5.0%	Absorbance
0	0	0.000
0.7	7.0	0.301
1.2	10.0	0.602
1.4	14.0	0.903
2.5	20.0	1.200

34. One method is differential spectrophotometry using the wavelength shift of phenol
with pH change. The difference in absorption at 289 nm between a strongly basic sample
and a slightly acid sample is proportional to the phenol concentration. Incidentally, a
platinum hollow cathode emission lamp (Chapter 12) could supply the proper wavelength, thus eliminating the need for a monochromator.

Chapter 5

1. 1.2 $\mu g/g$. **2.** Eq. 5-5 is valid up to 8 $\mu g/50$ ml. **3.** For anthracine: $\lambda_{ex} = 2550$ or
3600 Å; $\lambda_F = 3980$ Å. For quinine: $\lambda_{ex} = 2500$ or 3500 Å; $\lambda_F = 4600$ Å. For riboflavin:
$\lambda_{ex} = 2700$, 3700, or 4450 Å; $\lambda_F = 5200$ Å. **4.** Filter must possess a bandpass sufficiently narrow to exclude the unwanted emission. **5.** Excitation at 3400 Å produces
only phenanthrene emissions. Excitation at 2750 Å maximizes the naphthalene fluorescence which is spectrally quite distinct. **6.** See *Anal. Chem.*, **29**, 202 (1957). **7.** For
aluminum, fluorescent peaks are at 4171, 4400, and 4676 Å, giving $^1S^* \rightarrow {}^1S_0$ transitions
to $\nu = 0$, 2, 4 ground-state vibrational levels (23,975 cm^{-1}, 22,727 cm^{-1}, and 21,386
cm^{-1}, respectively). Phosphorescent peaks occur for $\nu = 0, 1, 2, 3, 4, 5$ and involve transitions from the triplet state of 20,790, 20,202, 19,531, 18,857, 18,182, and 17,482 cm^{-1},
respectively. See also *J. Chem. Phys.*, **17**, 1182 (1949). **8.** Fluorescent transitions occur
to $\nu = 0, 2, 3, 4$ of ground state; phosphorescent transitions involve $\nu = 0, 1, 2$ (or 3), 4.

Chapter 6

1. 910 cm^{-1}. **2.** First overtone at 1108 cm^{-1}; second overtone at 1662 cm^{-1}. NaCl.
3. (a) 3080 cm^{-1}. (b) 3290 cm^{-1}. (c) 1126 cm^{-1}. (d) 1465 cm^{-1}. (e) 2150 cm^{-1}.
(f) 1750 cm^{-1}. **4.** (a) 2.22 $\mu g/ml$. (b) 1.52 $\mu g/ml$. (c) 0.41 $\mu g/ml$. (d) 0.328 $\mu g/ml$.
(e) 0.455 $\mu g/ml$. (f) 1.98 $\mu g/ml$. **5.** 1.72%. **6.** Bands shift about 1 μm to longer wave-
lengths due to slowing down of allyl group vibration by a heavy atom. **7.** (a) 2.00×10^{-4} *M*. (b) 5.00×10^{-4} *M*. (c) 1.70×10^{-3} *M*. (d) 1.24×10^{-4} *M*. (e) 5.90×10^{-5}
M. **8.** (a) Cell 1, 0.085 cm; cell 2, 0.13 cm; cell 3, 0.044 cm; cell 4, 0.022 cm. (b)
0.032 mm. **9.** (a) *meta*. (b) *para*. (c) *ortho*. **10.** *p*-Bromotoluene. **11.** Hexachloro-

benzene. **12.** OH str (intermolecular H-bond) at 3380 cm^{-1}; CH str at 2940 cm^{-1}; CH$_2$ in-plane bend at 1460 cm^{-1}; nonsymmetrical breathing mode of phenyl ring at 690 cm^{-1}; out-of-plane bending of 5 adjacent hydrogen atoms at 740 cm^{-1}; C—O str at 1060 cm^{-1} of primary alcohol. Absent: C=O str at 1735 cm^{-1}; CHO str of aldehydes at 2720 cm^{-1}; C—CH$_3$ bend at 1380 cm^{-1}; C—O—C asymmetric stretching of ethers. Compound is benzyl alcohol. **13.** 3-Methyl pentane. **14.** Structure II. **15.** Allyl cyanide. **16.** Unusually strong overtone at 1848 cm^{-1} plus C=C group at 1650 cm^{-1} and vinyl group at 990 and 910 cm^{-1}; no methyl bending at 1390 cm^{-1} and no methylene rocking at 720 cm^{-1}. Thus, the limiting formula would have two terminal vinyl groups with no more than 3 methylene groups in between, and no chain branching. Compound is C=C—C—C—C=C. **17.** p-Chlorobenzaldehyde. **18.** (CH$_3$)$_3$C—CO—CH$_3$. **19.** (CH$_3$)$_3$C—COOCH$_3$. **20.** (CH$_3$)$_2$CH—CH$_2$—NH$_2$. **21.** p-Cyanobenzaldehyde.

22.

23.

Chapter 7

1. There is a failure to appreciate the magnitude of the decrease in the efficiency of the photomultiplier tube coupled with the decrease in grating efficiency over the Raman range of 2960 cm^{-1} when the spectrum is excited by a He–Ne laser. **2. (a)** 20 cm^{-1} per sec or 1200 cm^{-1} per min. **(b)** Scanning speed can be 13.4 cm^{-1}/mm so the slit can be opened to 0.60 mm. **3.** It is reduced by a factor of approximately 22.5. **4.** $(15,798/20,487)^4 = 0.35$.

5.

Benzene, Δ cm^{-1}	Wavelength of Raman Lines (Å) for Exciting Lines				
	6328 Å	4880 Å	5145 Å	5682 Å	6471 Å
606	6580	5029	5311	5885	6735
850	6688	5091	5380	5970	6848
991	6751	5128	5421	6021	6914
1176	6837	5177	5476	6089	7004
1584	7033	5289	5602	6244	7210
1605	7043	5295	5608	6252	7221
3047	7839	5732	6102	6872	8060
3063	7849	5738	6107	6879	8071

6. The C—H stretching frequencies at 3047 and 3063 cm^{-1} would overlap. **7.** N$_2$O has no center of symmetry while CS$_2$ does have one. The structures must be N—N—O and S—C—S (type of bonds not intended to be indicated). **8.** The molecule has a center of symmetry. If planar the molecule would have three unpolarized Raman lines; if tetra-hedral it would have but one. The spatial configuration is tetrahedral.

9.

Unknown	Volume Percent, %		
	1,2,3-	1,2,4-	1,3,5-
A	33.3	33.3	33.3
B	40.0	26.0	34.0
C	25.0	37.0	38.0

10. Benzoyl chloride. 11. Dimethyl acetylene. 12. Nitrobenzene.

Chapter 8

1. Bearing in mind the $1:2$ relationship of the enolic hydrogen to the keto methylene group: % enol = $[37.0/(37.0 + \frac{1}{2} \times 19.5)] \times 100 = 79.1\%$. 2. As benzophenone contains 5.53% hydrogen, the sample contains: $(0.8023/0.3055)(184/228)(5.53\%) = 11.72\%$. 3. The average number of methylene bridges per chain is $(n + 1)$, so that the average numbers of aromatic and methylene protons per molecule are $(2n + 6)$ and $(2n + 2)$, respectively. Thus, $(2n + 6)/(2n + 2) = 30/18$. Hence, $n = 2.0$, indicating that the average chain length is four phenolic nuclei and the average molecular weight is 860. 4. The three signals are attributable to the methyl group in o-cyanotoluene, the methylene group in o-cyanobenzyl chloride, and the methine proton in o-cyanobenzal chloride. The relative molar proportions are 13/3, 20/2, and 10/1; and the proportions by weight are $1.0:3.0:3.7$. 5. $CH_2=C-CH_2$ 6. In compound I the CH_3 resonance would be

$$O-C=O$$

a doublet with $J = 6$ Hz, showing that there was a proton attached to the adjacent carbon atom. In compound II the CH_3 resonance, in addition to being further downfield, would be coupled very weakly with $J \approx 1$ Hz. 7. $HPO(OH)_2$; $H_2PO(OH)$. 8. First structure. 9. Second structure. 10. Structure III with intensity ratios of $4:3:1$. Note that the methylene protons are not precisely equivalent, since two of them are cis to the methyl group and two are $trans$. 11. Peak at 7.3 δ assigned to olefinic proton in structure A and the peak at 6.7 δ to the olefinic proton in structure B. Isomer A: $(50/72) \times 100 = 69.4$ mole %; and isomer B: $(22/72) \times 100 = 30.5$ mole %.

12.

$$O_2N-\langle\text{benzene}\rangle-\overset{\overset{CH_2CH_3}{|}}{CH}-COOH$$

13. Ditolyl disulfide. 14.

$$\langle\text{benzene}\rangle-CH_2-\overset{\overset{CH_3}{|}}{\underset{\underset{CH_3}{|}}{C}}-Cl$$

15.

$$CH_3O-\langle\text{benzene}\rangle-\langle S\rangle$$

16. $CH_3-CH_2-CH(Br)-COO-CH_2-CH_3$

17. Ethyl chloroacetate.   18. $CH_3-CH_2-O-\underset{\underset{O}{\|}}{C}-CH_2-CH_2-\underset{\underset{O}{\|}}{C}-O-CH_2-CH_3$

19. $HC{\equiv}C-CH_3$ 21.

CH_3-CH_2-O-⬡$-N(H)-\underset{\underset{O}{\|}}{C}-CH_3$

22.

⬡$-CH_2-CH_2-O-\underset{\underset{O}{\|}}{C}-CH_3$

23. $\begin{array}{c} CH_2-CH_2 \\ | \qquad\quad | \\ CH_2-O \end{array}$

Chapter 9

1. $H_0 = 12{,}500$ G; $\Delta E = 2.3 \times 10^{-16}$ erg/molecule. 2. $\Delta t < 10^{-6}$ sec. 3. Copper (I) and silver(II). 4. Spectrum 1 from dihydrofumaric acid; spectrum 2 from ascorbic acid; and spectrum 3 from reductic acid. 5. Spectrum A from semiquinone itself; B, trichloro-; C, monochloro-; D, 2,3-dichloro-; and E, tetrachloro-derivative. 6. (a) $\dot{C}H_3$, four peaks in $1:3:3:1$ intensity ratio; (b) C_6H_6, seven lines; $1:6:15:20:15:6:1$; (c) $(CH_3)_3\dot{C}$, ten lines, intensity ratio $1:9:36:81:126:126:81:36:9:1$; (d) a quartet from CH_3 with coupling constant of 65 G which is further split into triplets with coupling constant of 50 G due to CH_2—the $CH_3CH_2 \cdot$ radical; (e) $\dot{C}H(OH)COOH$ radical with $a_{CH}^H = 15.3$ G and $a_{OH}^H = 2.5$ G; (f) $X = CH_2Y$. 7. Lines 2 and 4 result from proton splitting; each line in turn is split into three components due to interaction with nucleus X; line 3 results from overlap of triplet wings. 8. The predominant radical is $\dot{C}H(COOH)_2$ with a smaller amount of $\dot{C}H_2COOH$. 9. (a) SO_2^- in equilibrium with $S_2O_4^{2-}$. (b) SO_2^- contains no nuclei with magnetic moments. (c) Enrichment with ^{33}S ($I = \frac{3}{2}$) would give a quartet $(1:3:3:1)$ with SO_2^-. 10. Three lines arise from hyperfine splitting by ^{14}N; the additional lines arise from splitting by ^{15}N and their intensity corresponds to natural abundance of ^{15}N. 11. The principal intermediate is 2,6-di-t-butyl-4-methylphenoxy radical. The primary quartet arises from the three protons of the 4-methyl group $(a_i = 10.7)$; the closely spaced triplets $(a_i = 1.8)$ result from weaker coupling of electron with two *meta*-protons on the ring. 12. $-CH_2-\dot{C}H-CH_2-$; four equivalent protons are being split by the methine proton. The difference between the alpha proton and the four beta protons becomes observable. 13. Larger triplet arises from electron coupling with protons 2 and 3 $(a_i = 3)$; each triplet is split into five lines due to protons 5, 6, 7, and 8 $(a_i = 0.3)$. 14. A quintet of quintets with the coupling of the alpha protons about three times larger than the coupling of the beta protons. 15. The radical $(CH_3)_2\dot{C}-OH$. The small hydroxyl–hydrogen splitting, likely to have considerable anisotropy, is not resolved but is responsible for the line widths. 16. Spectrum A is from 2-methylcyclohexanone; spectrum B is from 4-methylcyclohexanone; spectrum C, a mixture of approximately three parts of B to one part of A, is from 3-methylcyclohexanone. 17. Structure I would give a quartet of triplets whereas structure II would give a triplet of triplets.

Coupling of methyl protons is 10.7 G, slightly larger than the 9.0 G observed for the methylene protons of the ethyl group. **18.** Compound A is a triplet of triplets; compound B is a triplet; compound C is a septet of triplets; and compound D is a quartet of triplets. **19.** In compound A, the *p*-methyl group was oxidized to COOH or CHO; the proton is in the plane of the ring and spin polarization will not be induced. In compound B the 4-ethyl group is oxidized to acetyl. In compounds C and D the 4-methyl group is oxidized to COOH or CHO.

Chapter 10

1. 0.1241 Å; $Z = 87$. **2.** (a) L_{II} to K, (b) M_{III} to K, (c) M_V to L_{III}.

3. and 4.

Element	K Edge, eV	L_{III} Edge, eV	$K\alpha_1$, Å
Al	1.559	0.073	8.34
Cr	5.990	0.599	2.30
Zr	18.0	2.22	0.786
Nd	43.51	6.22	0.333
W	69.66	10.20	0.209
U	115.9	17.17	0.126

5. For L_{III} spectra, $Z < 92$; for K spectra, $Z < 39$. **6.** K edges of iodine and argon, respectively.

7.

Crystal	Al	S	Ca	Cr	Mn	Co	Br	Sr	Ag	Mo	W
LiF	—	—	112.6°	69.4°	63.0°	52.6°	29.8°	25.1°	16.0°	20.4°	6.0°
CaF$_2$	—	115.8°	64.2°	42.4°	38.9°	32.8°	18.9°	16.1°	10.2°	12.9°	3.8°
EDDT	142°	71.8°	44.8°	30.2°	27.6°	26.5°	13.6°	11.4°	7.3°	9.7°	3.0°

8. Use of pulse height analyzer, regulation of voltage applied to source, use of appropriate filter, and incorporation of monochromator into the system.

9.

U, wt %	Slope, counts sec^{-1} %$^{-1}$ U	Time (1.96 σ), min
2	165	2.57
5	110	1.77
10	67	1.80
15	45	2.08
20	31	2.87

10. NaCl crystal; $2\theta = 143°36'$. Tube voltage: 3.5 kV. Counting times: background, 0.5 min; sample, 95 sec for 0.4% S. **11.** ADP, 16.6°; mica, 8.9°; LiF, 45.1°; EDDT, 20.2°; lead palmitate, 1.94°. **12. (a)** Set base line at 6.5 V and window for 6 V. **(b)** Set base line at 15.5 V and window for 7.5 V. **(c)** Set base line at 17.5 V and window for 4.0 V. **(d)** Set window for greater than 27 V. **(e)** Include the interval from 9.0 to 19.5 V. **13. (c)** Insert a nickel filter whose K edge at 1.48 Å would selectively absorb the Nb $K\alpha_1$

line. **14. (b)** Action of a selective filter; see Table 10-3. **15. (a)** Sr: 16.10 kV. Y: 17.05 kV. **(b)** Sr: $K\alpha_1 = 25.2°$ and $K\beta = 22.4°$. Y: $K\alpha_1 = 23.9°$ and $K\beta = 21.2°$. **(c)** A: 0.250%. B: 0.177%. C: 0.060%. **16.** $\Delta\lambda = 0.002$ Å. $\Delta\theta = 2$ min. Use filter to reduce Zr $K\alpha_1$ intensity or, better, use a proportional counter with pulse height discrimination.

17.

	Top Layer			Bottom Layer	
2θ	λ, Å	Line	2θ	λ, Å	Line
27.6°	2.54	Ti $K\beta$	34.0°	3.14	Ca $K\beta$
30.0°	2.74	Ti $K\alpha_1$	36.0°	3.32	Ca $K\alpha$
58.0°	5.30	Ti $K\beta$(2nd)	71.0°	6.36	Ca $K\alpha$(2nd)
63.0°	5.76	Ti $K\alpha_1$(2nd)			

18. (a) 17 cm²/g. **(b)** 46 cm²/g. **(c)** 47 cm²/g. **19.** Decrease in intensity is 63% per 1 cm cell length vs. air, or 35% decrease vs. an octane blank. **20.** Decrease in intensity is 97.5% per 1 cm cell length vs. air, or 59% decrease vs. an octane blank. **21.** Sample 1, Ni $K\alpha_1$ on Mo $K\alpha_1$. Sample 2, Cu $K\alpha_1$ on Mo $K\alpha_1$. Sample 3, Au $L\beta_1$ on Ni $K\alpha_1$. Sample 4, Pd $K\alpha_1$ on Ni $K\alpha_1$.

22.

2θ:	111.0°	100.2°	57.8°	48.8°	45.1°	44.0°	38.0°
λ, Å:	3.320	3.085	2.01	1.671	1.562	1.517	1.319
Element:	Ni $K\alpha_1$(2nd)	Cu $K\alpha_1$(2nd)	Mn $K\alpha_1$	Ni $K\alpha_1$	Cu $K\alpha_1$	Ni $K\beta$	Zn $K\beta$

23. See G. L. Clark, Ed., *Encyclopedia of Spectroscopy*, Van Nostrand Reinhold, New York, 1960; pp. 704–711. **24.** Elements with $Z < 25$, and especially good for Cl, P, and S in hydrocarbon matrices. See *Appl. Spectrosc.*, **24**, 557 (1970) and *Anal. Chem.*, **44** (14) 57A (1972). **25.** 1.66 Å, Ni $K\alpha_1$; 2.1 Å, Mn $K\alpha_1$; 3.32 Å, Ni $K\alpha_1$(2nd); 4.20 Å, Mn $K\alpha_1$(2nd); 6.64 Å, Ni $K\alpha_1$(3rd). **27.** A fully extended planar zigzag carbon chain is indicated since the C—C distance is essentially this value.

Chapter 11

1. Operating voltage: 1300 to 1700 V. **2.** Operating voltage: 1470 to 1545 V. **3.** Shelf 1, 35%; shelf 2, 8%; shelf 3, 3.55%; shelf 4, 1.5%. A sample of 50 mg U_3O_8 gives 31,800 disintegrations per min. **4. (a)** 3.5 µg. **(b)** 30 mg. **(c)** 40.5 ng. **(d)** 4.1 pg. **(e)** 8.85 µg. **5.** 0.00145 µCi. **6.** 5.5×10^9 disintegrations/min.

7.

		N/N_0 Fraction	
Nuclide	14 days	30 days	60 days
^{32}P	~0.5	~0.23	0.052
^{131}I	0.30	0.0755	0.0058
^{198}Au	0.0274	4.48×10^{-4}	2.18×10^{-7}

8.

Number of Counts	P.E.(%)	$1\,\sigma(\%)$	$2\sigma(\%)$
3200	1.19	1.77	3.54
6400	0.84	1.25	2.50
8000	0.75	1.12	2.24
25,600	0.42	0.63	1.26
102,400	0.21	0.31	0.62

9. Dead time is 200 μsec; loss is 704 counts/min. **10. (a)** 50%. **(b)** 80%. **(c)** 96%. **(d)** 99.9% efficiency. **11.** For 99% counting efficiency, counting rate could not exceed 1% of reciprocal value of dead time. **(a)** 40,000 counts/sec. **(b)** 10,000 counts/sec. **(c)** 2000 counts/sec. **(d)** 37 counts/sec. **12.** Less than 6.6%. **13. (a)** Accumulate 800 counts for sample plus background, and 40 counts for background. Counting time would be 4 min for sample plus background, and 1 min for background alone. **(b)** 5000 and 300 counts for sample plus background and background alone, respectively. Counting time: 25 min and 7.5 min, respectively. **14.** Increase in counting rate: Al, 3.4%; Cu, 4.3%; and Pb, 16.5%. **15.** $A_0 = 870$ counts/min; $t_{1/2} = 11$ sec. $B_0 = 93$ counts/min; $t_{1/2} = 89$ sec. **16.** $A_0 = 2250$ counts/min; $t_{1/2} = 18$ min; ^{80}Br. $B_0 = 440$ counts/min; $t_{1/2} = 264$ min; ^{80m}Br. **17.** $A_0 = 9000$ counts/min; $t_{1/2} = 15$ min. $B_0 = 1000$ counts/min; $t_{1/2} = 122$ min. **18.** $E_{max} = 0.34$ MeV; 9 mg/cm^2. **19.** Range is approximately 500 mg/cm^2; energy is 1.20 MeV. **20.** 1.67 MeV. **21.** $\mu = 0.0248$ cm^2/g; range (for 99.9% absorption) is 280 mg/cm^2; energy is 0.714 MeV. **22.** 200 mg. **23.** 6.5 mg. **24.** Fraction of ^{35}S: 2.36×10^{-10}. **25.** In air: 162 cm for ^{90}Sr and 910 cm for ^{90}Y. In iron: 0.25 mm for ^{90}Sr and 1.39 mm for ^{90}Y. **26.** 0.0379 N. **27.** To absorb sodium activity, use aluminum, 634 mg/cm^2. **28.** 69 min. **29.** 0.26 μg. **30.** 2.7 mg. **31.** Cu: 71 sec; Mn: 2 sec; Ni: 4.6 min; Co: 12.9 days. **32.** It will take 50 hr for manganese to decay to 1 count/min. An aluminum absorber, 0.52 mg/cm^2 will remove all manganese beta radiation which is the most energetic of the beta radiations. **33.** For an irradiation period equal to half-life of radionuclide, ^{128}I would give 9.0 counts min^{-1} mg^{-1}; ^{80}Br, 9.7 counts min^{-1} mg^{-1}; ^{122}Sb, 5.1 counts min^{-1} mg^{-1}; and ^{56}Mn, 42.9 counts min^{-1} mg^{-1}. **34.** Nuclides observed: ^{22}Na, ^{47}Ca, ^{59}Fe, ^{60}Co, ^{65}Zn, ^{85}Sr, ^{86}Rb, ^{95}Zr, ^{95}Nb, ^{103}Ru, ^{124}Sb, ^{134}Cs, ^{140}Ba, and ^{140}La. **35.** Arsenic and antimony are present, typical of lead bullets. Check against a bullet seized from suspect's gun. **36.** Half-width is 3.46×10^{-8} eV; velocity is 0.44 mm/sec. **37.** Iron(III) possesses a spherically symmetrical charge distribution and does not give rise to interaction with the electrical quadrupole moment whereas iron(II) salts, with an additional d electron, do interact. In the octahedral complex compounds of iron, two of the $3d$ orbitals of the iron are used for the formation of the six hybrid orbitals, and the $3d$ electrons therefore occupy the remaining three $3d$ orbitals. These are fully occupied in the case of a complex like ferrocyanide, and this leads to a spherically symmetrical charge distribution. Ferricyanide ion lacks one electron for completing the remaining $3d$ orbital, and splitting occurs.

Chapter 12

1. Cement A: 0.14% Na_2O and 0.62% K_2O. Cement B: 0.37% Na_2O and 0.43% K_2O. Cement C: 0.23% Na_2O and 0.55% K_2O. 2. Emission from molecular band systems of CaO and CaOH. Blank would be larger because reading would increase approximately with the square of the bandwidth. 3. $350/5.5 = 63.6$ at 0.02 mm slit width and $1400/35 = 40.0$ at 0.05 mm slit width. Background increases with square of slit aperture: $35/5.5 = 6.35 \cong (0.05/0.02)^2$. Line emission increased with 3/2 power of slit aperture: $1400/350 = 4 \cong (0.05/0.02)^{3/2}$. 4. (a) 55 $\mu g/ml$ of boron. (b) 170 $\mu g/ml$. (c) 130 $\mu g/ml$. See *Anal. Chem.*, 27, 42 (1955). 5. (b) Ionization of strontium atoms in flame. (c) The ionization in (b) is repressed by large excess of calcium atoms whose own ionizations supply a large number of free electrons. 6. Ionization at low concentrations (slope > 1); self-absorption (slope ≈ 0.5) at concentrations above 10 $\mu g/ml$. 7. 117 $\mu g/ml$ of manganese. 8. 62.8 $\mu g/ml$ of manganese. 9. 52.8 $\mu g/ml$ of potassium. 10. 29.7 ± 0.3 $\mu g/ml$ of barium. 11. Water: 26.9 μm. 50% methanol-water: 18.6 μm. 40% glycerol-water: 29.4 μm.

12.

	Droplet Diameter, μm		
Flow Rate, ml/min	Water	50% EtOH–H_2O	MIBK
1.0	16.1	12.0	10.9
2.0	18.3	16.3	13.0
3.0	21.0	21.8	15.7
5.0	28.0	35.9	22.5

See also *Applied Optics*, 7, 1353 (1968).
13. (a) 0.0094. (b) 0.094. (c) 0.29. 14. (a) 0.36. (b) 0.77. (c) 0.97.
15. To suppress the ionization of 0.23 $\mu g/ml$ of sodium to 0.01 (1%), add:

Element	At 2500°K	At 2800°K
Cs	0.44 $\mu g/ml$	0.74 $\mu g/ml$
Rb	1.05 $\mu g/ml$	1.6 $\mu g/ml$
K	1.04 $\mu g/ml$	1.4 $\mu g/ml$
Li	22.50 $\mu g/ml$	19.5 $\mu g/ml$

16. 0.71 $\mu g/200$ ml. 17. 13.4°K for sodium, and 15°K for potassium. 18. (a) $S/N = 13.3$ (or 6.7 peak-to-peak) for 2 $\mu g/ml$. (b) Sensitivity = 0.22 $\mu g/ml$. (c) Detection limit = 0.30 $\mu g/ml$.

Chapter 13

1. 3267.61 Å. 2. Resolving power = 37,500; second order. 3. Cadmium, magnesium, and zinc. 4. (a) 0.26 mg/ml. (b) 0.34 mg/ml. (c) 0.41 mg/ml. 5. Gamma factor = 1.51; F is proportional to 0.16. 6. 0.83%. 7. Between 0.05 and 0.1%.

Chapter 14

1. Allyl alcohol. 2. 26.51. 3. 1.516 (calculated).

Compound	Specific Refraction	Molar Refraction
Benzene	0.3334	26.05
Ethanol	0.2794	12.87
Ethyl acetate	0.2511	22.12
Toluene	0.3356	30.92
Nitrobenzene	0.2642	32.53
Water	0.2061	3.712

4.

5. 66.7% D_2O and 33.3% H_2O. 6. 1.5415. 7. 12.11% by volume.

Chapter 15

1. +66.50°. 2. 96.79%. 4. 50%. 5. $[\theta] = \dfrac{3305}{bC} \log \dfrac{P_d}{P_l}$. 6. $N_r - N_l = -1.75 \times 10^{-7}$

8. $[\theta] = -0.70$.

1. 2478 V for mass 18 and 223 V for mass 200. 2. 8.87 μsec for mass 44 and 8.75 μsec for mass 43. 3. 1.63 MHz. 4. For the pair, tridecyl benzene and phenyl undecyl ketone, resolution required is 7140; for 1,2-dimethyl-4-benzoyl naphthalene and 2,2-naphthyl benzothiophene, resolution required is 9320. 5. (a) 10,000. (b) 6,500. (c) 10,000. 6. Mass = 237.1473 ± 0.0005; $C_{12}H_{19}N_3O_2$. 7. (a) $C_9H_8O_3$. (b) C_8H_8O. (c) $C_{14}H_{12}$. (d) $C_5H_6N_2$. (e) C_6H_7NO. (f) $C_6H_3Cl_2NO_2$. (g) $C_{13}H_{11}N$. (h) $C_{17}H_{18}O_5S$. (i) $C_9H_7NO_3$. 8. (a) C_3H_6NS; C_4H_8S; $C_3H_6NO_2$; $^{13}CC_3H_7S$. (b) The hydrogen 3 carbon atoms from the carbonyl group is transferred to the carbonyl oxygen (McLafferty rearrangement) with simultaneous cleavage of the C-2, C-3 bond. m/e 75.0267 is $CH_3-S-CH_2-CH_2$. 9. CH_3I; only one carbon is indicated plus a heavy monoisotopic element. 10. (a) 3 chlorine atoms. (b) 5 chlorine atoms. (c) 4 bromine atoms. (d) 1 chlorine and 1 bromine atom. (e) 2 chlorine and 1 bromine atoms. 11. 90(P): 1 sulfur atom plus mass 58; therefore, $C_4H_{10}S$ because 5.61 - 0.78 = 4.58 and 4.58/1.08 = 4C's, or 58/14 = 4CH_2 groups plus 2H's. Compound is a dialkyl sulfide.   89(P): 1 chlorine, 1 nitrogen, plus residual mass 40. Since $P + 1$ indicates not over 2 carbon atoms, probable formula is C_2ClNO.   206(P): 2 sulfur atoms plus mass 142. $P + 1$ is 12.5% which indicates not over 10 carbon atoms [12.5 - 2(0.78) = 10.9]. Formula is $C_{10}H_{22}S_2$.   230(P): 2 bromine atoms plus mass 72; 72/14 = 5CH_2 groups plus 2H's. Formula is $C_5H_{12}Br_2$. 140(P): 1 sulfur atom. $P + 1$ is 9.54 - 0.78 = 8.76; therefore, not over 7 carbon atoms.   151(P): 1 chlorine and 1 nitrogen atom; probably aromatic compound since

base peak is parent peak. Residual mass is 102, and 102/12 = 8C's plus 6H's. Formula is C_8H_6ClN.   12. Appropriate masses: 32, 39, 46, and 59.

13. Unknown	MeOH, %	EtOH, %	n-PrOH, %	iso-PrOH, %
A	0.9	8.2	11.7	79.0
B	2.5	13.5	6.9	77.0
C	7.1	7.8	52.7	31.4

14. Metastable at 45.0 indicates the parent ion decomposes to mass 91 with loss of mass 93 neutral fragment(s). Metastable at 46.5 indicates that mass 91 decomposes further to mass 65 plus a neutral fragment of mass 26. This data, coupled with the peak intensities, indicates that we are dealing with a ring compound possessing the structure C_6H_5—CH_2 plus mass 93. Mass 93 is probably C_6H_5O. Molecule is tolylphenyl ether.   15. Metastable at 69.4 indicates the decomposition route is from mass 122 to mass 92 plus a neutral fragment of mass 30. Metastable at 46.5 indicates the decomposition of mass 91 to mass 65 plus a fragment of mass 26. Ion peak at mass 92, an even mass arising from an even mass molecular ion, suggests a rearrangement reaction. Coupled with the possible transition from mass 122 to mass 91, involving a loss of mass 31, this suggests the presence of a CH_2OH group which could participate in a McLafferty rearrangement to account for loss of CH_2O. Base peak at mass 91 suggest a tolyl group. Compound is 2-phenyl ethanol.   16. Metastable peaks indicate these transitions.   147.9: $200^+ \to 172^+ + 28$ (probable loss of CH_2=CH_2). 121.7: $200^+ \to 156^+ + 44$ (probable loss of CH_2=CH—OH). 106.3: $108^+ \to 107^+ + 1$ (loss of H). 67.9: $173^+ \to 108^+ + 64$ (probable loss of SO_2). 53.5: $155^+ \to 91^+ + 64$ (probable loss of SO_2). 46.5: $91^+ \to 65^+ + 26$ (loss of HC≡CH from tolyl). One sulfur atom is indicated by the $P + 2$ peak. The loss of CH_2=CH_2 and CH_2=$CHOH$, both as a result of rearrangement processes, plus the loss of mass 45 in the transition from mass 200 to mass 155, shows the presence of an ethoxy group which is linked to a sulfoxide group as shown by the subsequent loss of SO_2 from mass 155. Base peak at mass 91 suggests a tolyl group. These deductions account for the molecular weight of 200 and the presence of not more than 9 carbon atoms ($P + 1$ abundance). The compound is C_2H_5—OSO_2—C_6H_5—CH_3; ring substitution cannot be ascertained.   17. 88(P): butyric acid. 86(P): methyl acrylate. 134(P): 3-phenyl propanaldehyde. 152(P): Formula is $C_8H_8O_3$; methyl salicylate.   18. Metastable peaks indicate these transitions.   102.2: $153^+ \to 125^+ + 28$. 100.2: $151^+ \to 123^+ + 28$. 54.7: $103^+ \to 75^+ + 28$. All mass 28 losses involve ethylene fragment from an ethoxy group.

$$CH_3-CH_2-O-CH-O-CH_2-CH_3$$

19. CH_3CCl_3. Apparent loss of 20 mass units (117 − 97) cannot be correlated with any known fragment except HF. However, mass 117 contains 3 chlorine atoms, and 117 − 3(35) leaves only 12 mass units, or 1 carbon atom. Thus, HF is ruled out and mass 117 is CCl_3. Now mass 97 contains 2 chlorine atoms, yet cannot arise from mass 117 by loss of

1 chlorine atom. Mass 97 is CH_3CCl_2. Thus, compound is CH_3CCl_3; no molecular ion peak is present. **20.** From empirical formula, two unsaturated bonds and/or rings are indicated. Mass 114, a rearrangement peak, arises from loss of water, indicating a COOH group. Mass 101 indicates loss of CH_3O or $HOCH_2$. Mass 59 suggests an ester, our second double bond. Pieces are: $CH_3O + CO$; COOH; and remainder of C_2H_4. Compound is methyl hydrogen succinate.

Chapter 17

1. $2NaHCO_3 \rightarrow Na_2CO_3 + CO_2 + H_2O$.   **2.** $\Delta H = -(\text{slope})\,(2.303)\,(1.987) = 39.0$ kcal/mole, where slope is found from plotting $\log p_{CO_2}$ vs. $1/T$ (in °K).   **3.** Plot specific heat vs. temperature; T_g, given by intersection of two linear segments, is 415°K. **4.** (a) 24.0%. (b) 80.0%. (c) 49.0%. (d) 40.0%. (e) 76.0%.   **5.** Chain rotation, 13.7 cal/g; fusion, 38.0 cal/g.   **6.** 89.1 cal/g of UO_2.   **7.** (a) $w_{CaO} = (w_{600°} - w_{900°})\,(56/44)$; $w_{MgO} = 1.5(w_{300°} - w_{600°})\,(40.6/44)$. (b) $3MgCO_3 \rightarrow MgO \cdot 2MgCO_3 + CO_2$. $MgO \cdot 2MgCO_3 \rightarrow 3MgO + 2CO_2$. (c) 40.0% CaO and 13.5% MgO.   **8.** In all cases, the first loss is the two moles of hydrated water. In an oxidizing atmosphere the final product is NiO; in CO and N_2, it is nickel metal.   **9.** $K_{sp}^{158°} = 3.0 \times 10^{-8}$.   **10.** $\Delta H(\text{neutralization}) = -13.5 + 3.3 = -10.5$ kcal/mole. $\Delta H(\text{ionization}) = -RT\ln K_a + T\Delta S = 12.6 - 9.3 = 3.3$ kcal/mole. See *Chimia*, **17**, 102 (1963).   **11.** Titration curve exhibits a rising portion (Ca^{2+} reacting) followed by a descending portion (Mg^{2+} reacting). For Ca: $\Delta H = -5.7$ kcal/mole; for Mg: $\Delta H = 5.6$ kcal/mole.   **12.** In the first step, NH_3 is lost and HVO_3 remains. Then a mole of water is lost and residue is V_2O_5.   **13.** (a) $MnCO_3 \xrightarrow{400°} MnO_2 \xrightarrow{550°} Mn_2O_3 \xrightarrow{900°} Mn_3O_4$. (b) In CO_2 atmosphere, decomposition is delayed; Mn_2O_3 is formed at 600°. Formation of MnO_2 requires presence of oxygen.   **14.** $CaSiF_6 \cdot 2H_2O \xrightarrow{100°} CaSiF_6 \xrightarrow{300°} CaF_2 + SiF_4$. $ZnSiF_6 \cdot 6H_2O \xrightarrow{100°-200°} ZnF_2 + 6H_2O + SiF_4$.   **15.** Successive weight losses lead to: $NaHSO_4 \cdot 1/2H_2O$, $NaHSO_4$, $Na_2S_2O_7$, $2Na_2SO_4 \cdot SO_3$ and Na_2SO_4. Melting of Na_2SO_4 occurs at 884°C.   **16.** Glass transition at 40°C; melting of coating at 125°C. **17.** (a) Glass transition at $-43°$C; melting at 40°C. (b) 90×10^{-6} °C^{-1} up to transition point; 200×10^{-6} °C^{-1} between -43 and $+40$°C.   **18.** The sharp endothermic peak at 70°C is superimposed upon a glass transition at 61°C, followed by the exothermic curing reaction which appears to be complete near 260°C. Cured sample manifests a glass transition at 116°C.   **19.** At a mole ratio of 1:1 there occurs an intermolecular substance; 110°C is transition point of hexamethylbenzene.   **20.** Thoriated nickel, 0.0925 cal/g; Zytel 61, 0.487 cal/g; and gold, 0.0305 cal/g.

Chapter 18

1.

Compound	V_R', ml	V_N, ml	V_g, ml/g	K_d
Benzene	981	739	247	287
Cyclohexene	856	637	214	239
Cyclohexane	642	475	159	176

2. Average values: $N = 525$ and $H = 1.75$ mm. **3.** For cyclohexane/cyclohexene: $\alpha = 1.33$; at 4σ, $N = 200$ and $L = 35$ cm; at 6σ, $N = 400$ and $L = 70$ cm. For cyclohexene/benzene: $\alpha = 1.15$; at 4σ, $N = 1450$ and $L = 252$ cm; at 6σ, $N = 2500$ and $L = 490$ cm.

4. (a–b)

Column	Flow Rate, liters/hr	Linear Velocity, cm/sec	Plate Height, cm	Plate Number
31% substrate	1	1.65	0.340	1060
	2	3.10	0.310	1155
	4	6.20	0.405	890
	6	8.68	0.500	720
	10	13.46	0.700	515
23% substrate	1	1.43	0.335	1075
	2	2.77	0.245	1480
	3	4.03	0.216	1670
	4	5.20	0.217	1645
	5	6.24	0.230	1585
	10	10.60	0.302	1190
13% substrate	2	2.49	0.246	1460
	4	4.78	0.170	2135
	6	7.15	0.160	2260
	10	9.32	0.167	2150

(d) For 31% substrate: $A = 0.09$, $B = 0.26$, $E = 0.042$. For 23% substrate: $A = 0.099$, $B = 0.35$, $E = 0.010$. **(e)** For 31% substrate: $H_{opt} = 0.29$ cm at $v = 2.65$ cm/sec. For 23% substrate, values are 0.218 cm and 5.8 cm/sec. For 13% substrate, values are 0.160 cm and 7.25 cm/sec. **(f)** On 31% column: $K_d = 15.5$, $k = 5.75$. On 23% column: $K_d = 16.2$, $k = 3.62$. On 13% column: $K_d = 18.8$, $k = 2.67$.

5.

Column		1	2	3	4
K_d	1-MeN	5660	3100	536	154
	2-MeN	5030	2760	480	140
k	1-MeN	36.5	61.2	48.6	66.5
	2-MeN	32.4	54.5	44.0	60.6
N	1-MeN	27,500	16,200	12,650	13,200
	2-MeN	19,400	15,500	12,900	15,000
Phase ratio		155	50.6	11.1	2.3
Resolution		4.5	3.7	3.06	3.3

6. In all cases, resolution = 1.5. For case 1, N is 4360 to 6240 (limit). For case 2, $N = 39,200$. For case 3, $N = 7740$. **7.** **(b)** An isothermal column temperature in the interval 144–$162°C$ (or in the low-temperature region around $50°C$). About 8500 plates required since $\alpha = 1.045$ for the least satisfactory separation. **(c)** $V_g = 45$ ml/g for all the n-alkanes. **(d)** $\Delta H_s = 6$ kcal/mole. **8.** **(A)** n-Hexane. **(B)** n-Butylbenzene. **(C)** Toluene.

(D) n-Octane. (E) n-Pentane. (F) Ethylbenzene. (G) n-Heptane. (H) n-Propylbenzene.
9. (c) For n-butyl acetate, $V_R' = 190$ ml (Carbowax) and 320 ml (Nujol). For n-amyl alcohol, $V_R' = 800$ and 210 ml. **10.** Present: C-12, C-16, C-18, C-20, and C-22.
11. (a) ΔH values, kcal/mole:

Column	CH_2Cl_2	$CHCl_2CH_3$	$CHCl_3$	CCl_3CH_3	CCl_4	$CCl_2{=}CHCl$	$CCl_2{=}CCl_2$	C_6H_5Cl	C_6H_6
Paraffin	7.12	6.8	7.2	7.4	7.4	7.9	9.1	9.3	9.5
Tricresyl phosphate	7.57	7.5	8.2	7.4	7.4	8.3	9.0	9.3	11.4
Carbowax 4000	8.80	8.5	9.4	7.8	7.5	8.5	9.1	10.0	10.4

(b) $K_d = V_g\rho$. Specific retention volumes, given at column temperature per gram of liquid phase, need only be multiplied by density of liquid phase at given temperature to obtain K_d. **12. (a)** $\Delta H = 12.5$ kcal/mole. **(b)** $t_R' = 4.95$ min at 150°C, 1.01 min at 200°C, and 0.52 min at 225°C. **13. (c)** For $\eta = 1\%$ at 138°C (and 175°C), N_{req} are the following: 80 (100) for 1,5-, 1,6- and 1,7- from each other; 2400 (1800) for DETA/1,6-; 2400 (4700) for TEDA/1,5-; 45 (70) for DETA/TEDA. **(d)** $\alpha = 1.17$ ($\eta = 1\%$), TEDA separated from 1,5- and 1,4- between 80°–130°C; DETA from 1,7- and 1,6- at 220°C.
14. A family of parallel straight lines, one for each homologous series, is obtained.
15. A, 14:0; B, 16:1; C, 12:0; D, 14:1; E, 16:2; F, 22:0; G, 16:3; H, 17:0; I, 16:4; J, 24:0; K, 10:0. **16.** A (0.73, 0.65); B (3.8, 1.9) these are estimated from a line drawn parallel to X:3 family and spaced apart the average distance of each homologous family; C (6.3, 3.6); D (6.0, 13.4); E (6.6, 11). **17. (b)** At 25°C, $N = 1620$, 1920, 1990, and 2300. At 16°C, $N = 1720$, 2040, 2120, and 2400. **(c)** At 25°C, resolution = 3.13, 1.62, and 2.65. At 16°C, resolution = 3.33, 1.75, and 2.79. **18. (a–b)** C-5: 8.3% and 178 plates: C-6: 13.2% and 140; C-7: 12.8% and 280; C-8: 16.7% and 290; C-9: 21.2% and 240; C-10: 28.0% and 192. **(c)** Separation of C-5/C-6 would require about 150 plates (3%) and about 400 plates (0.1%). These requirements take account of unequal peak area; the factor is 1.11. **19.** Case I. Packed column: $k = 1.00$; $N = 17,400$; $t = 348$ sec. WCOT: $k = 0.17$; $N = 206,000$; $t = 1200$ sec. SCOT: $k = 0.33$; $N = 70,000$; $t = 311$ sec. Case II. For packed column: $k = 6.00$; $N = 5910$; $t = 354$ sec. WCOT: $k = 1.00$; $N = 17,400$; $t = 174$ sec. SCOT: $k = 2.00$; $N = 9800$; $t = 97$ sec. Case III. For packed column: $k = 30.0$; $N = 4630$; $t = 1434$ sec. WCOT: $k = 5.0$; $N = 6260$; $t = 188$ sec. SCOT: $k = 10.0$; $N = 5270$; $t = 193$ sec. **20.** n-Butane, 400; n-pentane, 500; n-hexane, 600; n-heptane, 700; n-octane, 800; 2-methyl butane, 433; butene-1, 375; hexene-1, 571; benzene, 608; n-butanol, 729. **21.** ΔI values on OV-1, TCP, and Carbowax 4000 columns, respectively. cyclohexane: 8, 42, 82; toluene: 13, 172, 315; acetone: 55, 270, 409; chloroform: 23, 227, 449; dioxane: 40, 263, 442. **22.** Peaks 2, 4, 6, and 9 are n-C_1 to C_4 formate esters; peaks 3, 5, 8, and 12 are the n-C_1 to C_4 acetates; peaks 7 and 11 are the iso-C_3 and iso-C_4 acetates; and peak 10 is sec-butyl acetate. **23.** C-14, 2.4%; C-16, 4.4%; C-18, 0.8%; C-20, 6.5%; C-22, 4.0%; and methyl ricinoleate (hydroxy oleate), 81.7%.

Chapter 20

1. (a) A: 14.08; B: 11.43; C: 12.15; D: 12.26. (b) pH $<$ 9.6 at $[Na^+]$ = 0.2; pH $<$ 9.3 at $[Na^+]$ = 1.0. (c) Below 0.02 M. 2. Not a true buffer, only a stoichiometric mixture of K^+ and hydrogen phthalate ions. 3. A 10% error in the concentration would change the pH by only 0.01 unit. 4. Because of the increased dissociation of water at higher temperatures and because of the change in the liquid-junction potential caused by the presence of increasing numbers of highly mobile hydroxyl ions in the solution. 6. Bromine is added to the sample to oxidize nitrite to nitrate; the amount of bromide ion produced is a measure of the nitrite originally present. 7. Electrode is 10 times more sensitive to fluoride ion than to hydroxide ion; a 0.001 M hydroxide ion concentration could be tolerated. 8. pH $>$ 5.2 to avoid complexation as HF; one could tolerate about pH 4.2 if standards and samples are adjusted to this pH beforehand. 9. Ratios of interferant to bromide ion: chloride, 330; iodide, 1.7×10^{-4}; hydroxide, 40,000; cyanide, 4×10^{-4}. 10. Minimum calcium activity = $(1.0 \times 10^{-4}) (0.7)^2/0.05 = 9.8 \times 10^{-4}$ M. 11. Hydroxide, 10^{-4} M; iodide, 5×10^{-9} M; nitrate, 10^{-6} M; hydrogen carbonate, 2.5×10^{-5} M; and sulfate, 10^{-8} M. 12. Total sulfide = 1.33×10^{-3} M.

Chapter 21

5. (a) 0.393 V. (b) 0.236 V. (c) 1.06 V. (d) 1.04 V. 6. (a) 0.017%. (b) 0.30%. 7. (b) 8.85. (c) 15.65 ml. (d) pK_a = 5.57. 8. (b) 10.0 ml.

Chapter 22

4. 1.2×10^{-3} M. 5. 0.0012 M. 6. 0.0012 M. 7. 1.49×10^{-5} M. 8. (a) Successive 2-electron steps for: $O_2 \rightarrow H_2O_2 \rightarrow 2H_2O$. (b) Uranium: VI to V, and finally to III. (c) Successive 1-electron and 5-electron reactions: $I^- \rightarrow \frac{1}{2}I_2 \rightarrow IO_3^-$. 9. To about 10^{-29} of its initial value. 10. Lead, 0.25 mM, and zinc, 1.08 mM. 11. Copper or bismuth; lead; indium; cadmium; nickel or zinc. 12. Copper (or bismuth); lead; thallium; indium; cadmium; zinc (or nickel). 13. Lead: 0.77×10^{-4} M; zinc: 0.67×10^{-4} M.

Chapter 23

1. (a) Cathode potential: $\geqslant$ 0.26 V vs. NHE. (b) 10.1 min to remove 99.9% of the silver. 2. pH $\geqslant$ 2.2. 3. Residual copper concentration is 10^{-22} M. 4. -0.46 V vs. NHE. 5. (a) 1.11 g. (b) 4.20 g. (c) 1.24 g. (d) 1.04 g. 6. (a) 0.50 V vs. NHE. (b) 0.38 V vs. NHE. (c) pAg = 5.93. 7. (a) 4.64 mmol. (b) ΔpH = 5.37. 8. (a) 0.109 M. (b) 0.124 V vs. NHE. 9. (a) 3.29 V. (b) Yes. (c) 10^{-12} M. 10. Probably not completely since the cadmium ion concentration is lowered only to 10^{-4} M. 11. (a) 5×10^{-17} M. (b) 1.8×10^{-6} M. 12. 5×10^{-17} M. 13. The overpotential term for hydrogen gas on copper shifts the point of incipient evolution of hydrogen more negative than

the potential for initial deposition of zinc from 0.1 M solution, after which the still higher overpotential term for hydrogen gas on zinc takes over. **14.** One possibility is to use a Na_2CO_3–$NaNO_3$ electrolyte, wash with dilute acetic acid, and develop green color with 1% alcoholic solution of α-benzoinoxime. **15.** For sulfide, lead carbonate paper can be used with a sodium carbonate electrolyte. Specimen is cathodic. **16. (a)** 13.3 min. **(b)** 19.9 min. **17.** 0.028 mm. **18. (a)** Cd: -0.433 V and H: -0.118 V. **(b)** Cd: -0.61 V and H: -0.59 V. **19.** No, a cathode potential of -0.78 V vs. SCE would be required to lower the copper(I) concentration to 1×10^{-5} M. **20.** From a weakly acid solution, remove lead while controlling the cathode potential at -0.35 V vs. NHE, then cadmium at -0.70 V, and finally zinc at -0.91 V. The electrode is returned to the electrolysis cell after each plated metal is weighed without removing the accumulated deposits. **21.** A cyclic process is set up involving the oxidation of the copper(I) ammine complex at the anode and the reduction of the copper(II) ammine complex at the cathode. Also the copper(I) ammine reduces the silver ammine to metallic silver. **22. (a)** $Ag \mid Ag(CN)_2^-$, CN^-, $NaOH(1\ M) \mid H_2(Pt)$. **(b)** Anode: $Ag + 2CN^- = Ag(CN)_2^- + e^-$; $E° = -0.31$ V. Cathode: $2H_2O + 2e^- = H_2 + 2OH^-$; $E° = -0.828$ V. **(c)** Although the spontaneous reaction is: $Ag(CN)_2^- + H_2 + 2OH^- \rightarrow Ag° + 2CN^- + 2H_2O$, when the electrodes are first placed in the solution the potential of the silver electrode actually is more negative than that of the platinum electrode because the solution originally contains no $Ag(CN)_2^-$ and no hydrogen. Consequently, the above reaction proceeds from right to left although the spontaneous current decreases exponentially with time and approaches zero as equilibrium is established. **23. (b)** 0.46 min^{-1}. **(c)** Theoretically 15 additional min.

Chapter 24

1. 19.13 mg. **2.** 1.98×10^{-5} μequiv/count at 30 mA. **3.** 0.01087 N. **4.** 0.133 ± 0.003 N. **5.** 1 mA · sec corresponds to 1.04×10^{-8} equiv. **(a)** 176 ng. **(b)** 632 ng. **(c)** 368 ng. **(d)** 331 ng. **(e)** 513 ng. **6.** 111.7 Q. **7.** 0.00145 N. **9.** 1.290×10^{-4} M. **10. (a)** 87.1 mg. **(b)** $E \leqslant 0.362$ V. **11.** 3.192 mg. **12.** From 0.033 to about 1 mg. **13.** At $E = 0.75$ V, manganese is reduced (VII → II); at 0.3 V, vanadium is reduced (V → IV); and at -0.3 V the iron is reduced (III → II). **16.** 193 sec. **17.** 154.4 g.

Chapter 25

2. (a) At -0.8 V, the titration curve is L-shaped; at -1.7 V, the curve is V-shaped. **(b)** -0.8 V. **(c)** -0.8 V to avoid the nickel ammine and zinc ammine waves.

Chapter 26

1. (a) 118 Ω. **(b)** 11.8 Ω. **(c)** 2.35 Ω. **(d)** 1.18 Ω. Only (a) and (b). **2.** 26.7 to 318 Ω. **3.** Resistance should range from 1,000 Θ to 500,000 Θ. If $\Theta = 0.2$, the resistance readings will range from 200 to 10,000 Ω with 5000 Ω at midscale. A commercial instrument covers this range. **4. (a)** 4 to 17,000 Ω; cell constant = 2.00. **(b)** 180 to 330 Ω; cell constant = 0.200. **(c) and (d)** Cell constant = 0.100; 1,000 to 46,000 Ω. **(e)** Cell constant = 50.0. **(f)** Cell constant = 3.1. **5.** $\alpha = 0.0825$ and $K_a = 1.79 \times 10^{-5}$. **6.** Solubility is 1.07×10^{-5} mol/liter or 2.50×10^{-3} g/liter. $K_{sp} = 1.15 \times 10^{-10}$. **7.** Specific conductance is 1.80×10^{-5} Ω^{-1} cm^{-1}; resistance of saturated solution would be 110,000 Ω. **8.** 0.0325 N; 0.130 g NaOH. **9.** 0.0625 N. **10.** 0.0176 M; individual end points for each replaceable hydrogen. **11.** 0.630 mequiv of the aromatic sulfonic acid and 0.230 mequiv of the aliphatic carboxylic acid. **14.** Conductance (in units of 10^{-4} Ω^{-1} cm^{-1}) after dilution by titrant: 0%, 21.26; 25%, 17.40; 50%, 13.8; 90%, 8.15; 100%, 6.81%; 110%, 7.97; 150%, 12.39; 175%, 14.99; 200%, 17.50. **18.** 0.00104 M. **19.** Use a double logarithmic plot. (A) 0.9 mequiv; (B) 0.4 mequiv; (C) 0.04 mequiv; (D) 0.2 mequiv. **20.** $\alpha = 0.000118$; $K_a = 5.6 \times 10^{-10}$.

Chapter 27

4. All can be handled by refractometric methods. See *J. Chem. Educ.*, **45**, A470 (1968). **5.** Gas sample is drawn through the probe by means of a steam-operated aspirator. Steam–gas mixture is condensed in a cooling chamber and the resulting condensate, containing all the sodium ion entrained in the original gas sample, is separated from the gas and passed through a rotameter ahead of the sodium analyzer (selective-ion electrode). Gas flow measured in second rotameter provides a multiplier, which may easily be corrected for absolute humidity, whereby the recorded sodium data can be reported on a dry-gas basis. **6.** All the parameters of the gas chromatograph must be optimized and controlled. The method depends upon the use of small samples of about 0.6 mg which allows for rapid combustion and the introduction of the products into the chromatographic column without excessive dilution by the carrier gas. **7.** At regular programmed intervals, a sample of treated water is drawn into the analyzer cell system. After the initial pH is measured and recorded, the sample automatically is titrated to pH 8.3, with the results fed to the analyzer computer and stored. Then the titration is continued to pH 4.3, the total alkalinity end point. This signal is fed to the computer and also the soda ash feed controller, and is combined with a raw water flow signal. The combined signal controls addition of soda ash to the unit. The residual hydroxide and/or bicarbonate ion value is computed, recorded, and transmitted to the lime feed controller, where it also is combined with a raw water flow signal. This combined signal controls the addition of lime to the softener, or lime and soda ash are fed in a fixed ratio with residual hydroxide and/or bicarbonate ion value providing the control signal. **8.** Measurements to the fourth decimal place provide a precision of about 0.1%. **9.** 78.99% C; 5.78% H; 15.29% N.

Appendices

Potentials of Selected Half-Reactions at $25°C$

A summary of reduction-oxidation half-reactions arranged in order of decreasing oxidation strength and useful for selecting reagent systems.

Half-Reaction		$E°$, V
$F_2(g) + 2H^+ + 2e^-$	$= 2HF$	3.06
$O_3 + 2H^+ + 2e^-$	$= O_2 + H_2O$	2.07
$S_2O_8^{2-} + 2e^-$	$= 2SO_4^{2-}$	2.01
$Ag^{2+} + e^-$	$= Ag^+$	2.00
$H_2O_2 + 2H^+ + 2e^-$	$= 2H_2O$	1.77
$MnO_4^- + 4H^+ + 3e^-$	$= MnO_2(s) + 2H_2O$	1.70
$Ce(IV) + e^-$	$= Ce(III)$ (in 1 M HClO$_4$)	1.61
$H_5IO_6 + H^+ + 2e^-$	$= IO_3^- + 3H_2O$	1.6
Bi_2O_4 (bismuthate) $+ 4H^+ + 2e^-$	$= 2BiO^+ + 2H_2O$	1.59
$BrO_3^- + 6H^+ + 5e^-$	$= \frac{1}{2}Br_2 + 3H_2O$	1.52
$MnO_4^- + 8H^+ + 5e^-$	$= Mn^{2+} + 4H_2O$	1.51
$PbO_2 + 4H^+ + 2e^-$	$= Pb^{2+} + 2H_2O$	1.455
$Cl_2 + 2e^-$	$= 2Cl^-$	1.36
$Cr_2O_7^{2-} + 14H^+ + 6e^-$	$= 2Cr^{3+} + 7H_2O$	1.33
$MnO_2(s) + 4H^+ + 2e^-$	$= Mn^{2+} + 2H_2O$	1.23
$O_2(g) + 4H^+ + 4e^-$	$= 2H_2O$	1.229
$IO_3^- + 6H^+ + 5e^-$	$= \frac{1}{2}I_2 + 3H_2O$	1.20
$Br_2(l) + 2e^-$	$= 2Br^-$	1.065
$ICl_2^- + e^-$	$= \frac{1}{2}I_2 + 2Cl^-$	1.06
$VO_2^+ + 2H^+ + e^-$	$= VO^{2+} + H_2O$	1.00
$HNO_2 + H^+ + e^-$	$= NO(g) + H_2O$	1.00
$NO_3^- + 3H^+ + 2e^-$	$= HNO_2 + H_2O$	0.94
$2Hg^{2+} + 2e^-$	$= Hg_2^{2+}$	0.92
$Cu^{2+} + I^- + e^-$	$= CuI$	0.86
$Ag^+ + e^-$	$= Ag$	0.799
$Hg_2^{2+} + 2e^-$	$= 2Hg$	0.79
$Fe(III) + e^-$	$= Fe^{2+}$	0.771
$O_2(g) + 2H^+ + 2e^-$	$= H_2O_2$	0.682
$2HgCl_2 + 2e^-$	$= Hg_2Cl_2(s) + 2Cl^-$	0.63
$Hg_2SO_4(s) + 2e^-$	$= 2Hg + SO_4^{2-}$	0.615
$Sb_2O_5 + 6H^+ + 4e^-$	$= 2SbO^+ + 3H_2O$	0.581
$H_3AsO_4 + 2H^+ + 2e^-$	$= HAsO_2 + 2H_2O$	0.559

APPENDIX A *Continued*

Half-Reaction		$E°, V$
$I_3^- + 2e^-$	$= 3I^-$	0.545
$Cu^+ + e^-$	$= Cu$	0.52
$VO^{2+} + 2H^+ + e^-$	$= V^{3+} + H_2O$	0.337
$Fe(CN)_6^{3-} + e^-$	$= Fe(CN)_6^{4-}$	0.36
$Cu^{2+} + 2e^-$	$= Cu$	0.337
$UO_2^{2+} + 4H^+ + 2e^-$	$= U^{4+} + 2H_2O$	0.334
$Hg_2Cl_2(s) + 2e^-$	$= 2Hg + 2Cl^-$	0.2676
$BiO^+ + 2H^+ + 3e^-$	$= Bi + H_2O$	0.32
$AgCl(s) + e^-$	$= Ag + Cl^-$	0.2222
$SbO^+ + 2H^+ + 3e^-$	$= Sb + H_2O$	0.212
$CuCl_3^{2-} + e^-$	$= Cu + 3Cl^-$	0.178
$SO_4^{2-} + 4H^+ + 2e^-$	$= SO_2(aq) + 2H_2O$	0.17
$Sn^{2+} + 2e^-$	$= Sn$	0.15
$S + 2H^+ + 2e^-$	$= H_2S(g)$	0.14
$TiO^{2+} + 2H^+ + e^-$	$= Ti^{3+} + H_2O$	0.10
$S_4O_6^{2-} + 2e^-$	$= 2S_2O_3^{2-}$	0.08
$AgBr(s) + e^-$	$= Ag + Br^-$	0.071
$2H^+ + 2e^-$	$= H_2$	0.0000
$Pb^{2+} + 2e^-$	$= Pb$	-0.126
$Sn^{2+} + 2e^-$	$= Sn$	-0.136
$AgI(s) + e^-$	$= Ag + I^-$	-0.152
$Mo^{3+} + 3e^-$	$= Mo$	*approx.* -0.2
$N_2 + 5H^+ + 4e^-$	$= H_2NNH_3^+$	-0.23
$Ni^{2+} + 2e^-$	$= Ni$	-0.246
$V^{3+} + e^-$	$= V^{2+}$	-0.255
$Co^{2+} + 2e^-$	$= Co$	-0.277
$Ag(CN)_2^- + e^-$	$= Ag + 2CN^-$	-0.31
$Cd^{2+} + 2e^-$	$= Cd$	-0.403
$Cr^{3+} + e^-$	$= Cr^{2+}$	-0.41
$Fe^{2+} + 2e^-$	$= Fe$	-0.440
$2CO_2 + 2H^+ + 2e^-$	$= H_2C_2O_4$	-0.49
$H_3PO_3 + 2H^+ + 2e^-$	$= H_3PO_2 + H_2O$	-0.50
$U^{4+} + e^-$	$= U^{3+}$	-0.61
$Zn^{2+} + 2e^-$	$= Zn$	-0.763
$Cr^{2+} + 2e^-$	$= Cr$	-0.91
$Mn^{2+} + 2e^-$	$= Mn$	-1.18
$Zr^{4+} + 4e^-$	$= Zr$	-1.53
$Ti^{3+} + 3e^-$	$= Ti$	-1.63
$Al^{3+} + 3e^-$	$= Al$	-1.66
$Th^{4+} + 4e^-$	$= Th$	-1.90
$Mg^{2+} + 2e^-$	$= Mg$	-2.37
$La^{3+} + 3e^-$	$= La$	-2.52
$Na^+ + e^-$	$= Na$	-2.714
$Ca^{2+} + 2e^-$	$= Ca$	-2.87
$Sr^{2+} + 2e^-$	$= Sr$	-2.89
$K^+ + e^-$	$= K$	-2.925
$Li^+ + e^-$	$= Li$	-3.045

APPENDIX B Polarographic Half-Wave Potentials and Diffusion-Current Constants[a]

Generally, solutions contained 0.01% gelatin, and the data pertain to a temperature of 25°C. Half-wave potentials are referred to the saturated calomel electrode, and values of $i_d/Cm^{2/3}t^{1/6}$ are based on i_d in microamperes, C in millimoles per liter, m in mg sec^{-1} and t in seconds.[a]

Ion	Supporting Electrolyte	$E_{1/2}$	I_d
Bi^{3+}	1 M HCl	−0.09	
	0.5 M tartrate (pH 4.5)	−0.29	
	0.5 M tartrate + 0.1 M NaOH	−1.0	
Cd^{2+}	0.1 M KCl	−0.60	3.51
	1 M NH$_3$ + 1 M NH$_4^+$	−0.81	3.68
Co^{2+}	1 M KSCN	−1.03	
	0.1 M KCl	−1.20	
	0.1 M pyridine + 0.1 M pyridinium ion	−1.07	
CrO_4^{2-}	0.1 M KCl (basic chromic chromate)	−0.3	
	$(CrO_4^{2-} \rightarrow Cr^{3+})$	−1.0	
	$(Cr^{3+} \rightarrow Cr^{2+})$	−1.5	
	$(Cr^{2+} \rightarrow Cr^{\circ})$	−1.7	
Cu^{2+}	0.1 M KCl (HCl)	+0.04	3.23
	0.5 M tartrate, pH = 4.5	−0.09	2.37
	1 M NH$_3$ + 1 M NH$_4^+$ (1st wave)	−0.24	(Total
	(2d wave)	−0.50	3.75)
Fe^{3+}	0.5 M citrate, pH = 5.8 (1st wave)	−0.17	0.90
	(2d wave)	−1.50	
	0.1 M EDTA + 2 M NaAc (1st wave)	−0.13	
	(2d wave)	−1.3	
Fe^{2+}	0.05 M BaCl$_2$	−1.3	
In^{3+}	0.1 M KCl	−0.561	
Mn^{2+}	1 M KCl	−1.51	
	1 M KSCN	−1.55	
Ni^{2+}	0.01 M KCl	−1.1	
	1 M NH$_3$ + 1 M NH$_4^+$	−1.09	3.56
	1 M KSCN	−0.70	
	0.5 M pyridine + 1 M KCl	−0.78	
O_2	pH 1 to 10 ($O_2 \rightarrow H_2O_2$)	−0.05	(Total
	($H_2O_2 \rightarrow H_2O$)	−0.94	12.3)
Pb^{2+}	0.1 M KCl	−0.40	3.80
	1 M HNO$_3$	−0.40	3.67
	1 M NaOH	−0.75	3.39
	0.5 M tartrate + 0.1 M NaOH	−0.75	2.39
Sb^{3+}	1 M HCl	−0.15	
	0.5 M tartrate + 0.1 M NaOH	−1.32	
Sn^{2+}	1 M HCl	−0.47	4.07
	0.5 M tartrate + 0.1 M NaOH (anodic)	−0.71	2.86
	(cathodic)	−1.16	2.86
Sn^{4+}	1 M HCl + 4 M NH$_4^+$ (1st wave)	−0.25	2.84
	(2d wave)	−0.52	3.49

APPENDIX B *Continued*

Ion	Supporting Electrolyte	$E_{1/2}$	I_d
Zn^{2+}	0.1 M KCl	−1.00	3.42
	1 M NaOH	−1.50	3.14
	1 M NH$_3$ + 1 M NH$_4^+$	−1.33	3.82
	0.5 M tartrate, pH = 9	−1.15	2.30

[a]Reproduced, by permission, from *Polarography*, by I. M. Kolthoff and J. J. Lingane, 2nd ed. Copyright 1952 by Interscience Publishers, Inc.

APPENDIX C Proton-Transfer Reactions of Materials in Water at 25°C

Substance	pK_1	pK_2	pK_3	pK_4
Acetic acid	4.76			
Ammonium ion	9.24			
Anilinium ion	4.60			
Arsenic acid	2.20	6.98	11.5	
Arsenous acid	9.22			
Ascorbic acid	4.30	11.82		
Benzoic acid	4.21			
Boric acid: meta-	9.24			
tetra-	4	9		
Bromocresol green	4.68			
Bromocresol purple	6.3			
p-Bromophenol	9.24			
Bromophenol blue	3.86			
Bromothymol blue	7.1			
Carbonic acid (CO_2 + H_2O)	6.38	10.25		
Chloroacetic acid	2.86			
Chlorophenol red	6.0			
Chromic acid		6.50		
Citric acid	3.13	4.76	6.40	
Cresol purple (acid range)	1.51			
(base range)	8.32			
Cresol red	8.2			
Dichloroacetic acid	1.30			
Ethanolammonium ion	9.50			

APPENDIX C *Continued*

Substance	pK_1	pK_2	pK_3	pK_4
Ethylammonium ion	10.63			
Ethylenediaminetetraacetic acid (EDTA)	2.0	2.67	6.27	10.95
Ethylenediammonium ion	6.85	9.93		
Ferrocyanic acid			2.22	4.17
Formic acid	3.75			
Glycine (protonated cation)	2.35	9.78		
Hydrazinium ion	−0.88	7.99		
Hydrocyanic acid	9.21			
Hydrofluoric acid	3.18			
Hydrogen peroxide	11.65			
Hydrogen sulfide	6.88	14.15		
Hydroquinone	10.0	12.0		
Hydroxyammonium ion	5.96			
N,N-bis (2-Hydroxyethyl) glycine (Bicine) (protonated cation)	8.35			
tris (Hydroxymethyl) aminomethane (TRIS) (protonated cation)	8.08			
N-2-Hydroxyethylpiperazine-N'-2-ethanesulfonic acid (HEPES)	7.55			
N-tris (Hydroxymethyl) methylglycine (TRIS) (protonated cation)	8.08			
Hypochlorous acid	7.50			
Methyl orange	3.40			
Methyl red	4.95			
2-(N-Morpholino) ethanesulfonic acid (MES)	6.15			
Nitrous acid	3.35			
Oxalic acid	1.27	4.27		
1,10-Phenanthrolinium ion	4.86			
Phenol	9.99			
Phenol red	7.9			
Phenolphthalein	9.4			
Phenylacetic acid	4.31			
Phosphoric acid: ortho	2.15	7.20	12.36	
pyro	1.52	2.36	6.60	9.25
o-Phthalic acid	2.95	5.41		
Pyridinium ion	5.21			
Salicylic acid	3.00	12.38		
Succinic acid	4.21	5.64		
Sulfamic acid	0.988			
Sulfuric acid		1.92		
Sulfurous acid ($SO_2 + H_2O$)	1.90	7.20		
Tartaric acid: meso-	3.22	4.81		
Thymol blue	8.9			
Thymolphthalein	10.0			
Triethanolammonium ion	7.76			
Vanillin	7.40			
Veronal	7.43			

APPENDIX D Cumulative Formation Constants for Metal Complexes at 25°C

	$\log K_1$	$\log K_2$	$\log K_3$	$\log K_4$	$\log K_5$	$\log K_6$
AMMONIA						
Cadmium	2.65	4.75	6.19	7.12	6.80	5.14
Cobalt(II)	2.11	3.74	4.79	5.55	5.73	5.11
Cobalt(III)	6.7	14.0	20.1	25.7	30.8	35.2
Copper(I)	5.93	.10.86				
Copper(II)	4.31	7.98	11.02	13.32	12.86	
Nickel	2.80	5.04	6.77	7.96	8.71	8.74
Silver(I)	3.24	7.05				
Zinc	2.37	4.81	7.31	9.46		
CHLORIDE						
Copper(I)		5.5	5.7			
Copper(II)	0.1	−0.6				
Tin(II)	1.51	2.24	2.03	1.48		
Tin(IV)						4
CITRATE (L^{3-} anion)						
Cadmium	11.3					
Cobalt(II)	12.5					
Copper(II)	14.2					
Iron(II)	15.5					
Iron(III)	25.0					
Nickel	14.3					
Zinc	11.4					
CYANIDE						
Cadmium	5.48	10.60	15.23	18.78		
Copper(I)		24.0	28.59	30.30		
Nickel				31.3		
Silver(I)		21.1	21.7	20.6		
Zinc				16.7		
ETHYLENEDIAMINE-N,N,N',N'-TETRAACETIC ACID						
Calcium	11.0					
Copper(II)	18.7					
Iron(II)	14.33					
Iron(III)	24.23					
Magnesium	8.64					
Mercury(II)	21.80					
Zinc	16.4					
1,10-PHENANTHROLINE						
Iron(II)	5.85	11.45	21.3			
Iron(III)	6.5	11.4	23.5			

APPENDIX E Flame Emission and Atomic Absorption Spectra

Element	Wavelength, Å	Emission Detection Limit, $\mu g/ml/0.1$ mV	Absorption Sensitivity, $\mu g/ml/1\%$ Abs
Aluminum	3961.5	0.5 OAn	3.0 NAr
Antimony	2175.8		0.6 AA
	2598.1	0.6 OAn	
Arsenic	1972.0		1.3 AA
	2349.8	2.2 OAn	
Barium	5535.5	0.03 OA	2.6 NA
Bismuth	2230.6		0.7 AA
	2276.6	6.4 OAn	
Boron	2497.7	7.0 OAnr	10.0 NA
(as BO$_2$)	5180	3.0 OAn	
Cadmium	2288.0	4.0 OAn	0.03 AA
	3261.1	2.0 OAn	20.0 AA
Calcium	4226.7	0.07 OA	0.08 AA
Cesium	4555.4	0.01 OH, OA	10.0 AP
	8521.1	0.02 OH, OA	0.16 AP
Chromium	3578.7	0.2 OAn	0.22 AAr, NA
	4254.3	0.1 OAn	0.6 AAr, NA
Cobalt	2407.2		0.02 AA
	2424.9	1.7 OAr	0.2 AA
Copper	3247.5	0.6 OA	0.1 AA
	3274.0	0.01 NA	0.2 AA
Gallium	2874.2		2.3 AA
	4172.1	0.02 NA	3.7 AA
Gold	2428.0	5.0 OH	0.3 AA, NA
Indium	3256.1	2.2 OH	1.0 AA
	4511.3	0.002 NA	2.8 AA
Iron	2483.3		0.15 AAr
	3719.9	0.12 OAn	1.0 AAr
Lanthanum (as LaO)	7410	0.005 OAn	
Lead	2170.0		0.23 AA
	2833.1	6.0 OHn	
Lithium	6707.8	0.0002 OH	0.02 AA
Magnesium	2852.1	0.2 OAr	0.008 AA
Manganese	2794.8	0.02 AA	0.06 AA
	4030.8	0.005 NA	0.6 AA
Mercury	2536.5	2.5 OAn	2.0 AA
Molybdenum	3132.6	0.2 NA	0.8 NA
	3798.3	0.5 OAn	2.0 NA
Nickel	2320.0		0.15 AA
	3524.5	0.2 OAn	0.6 AA
Niobium	4058.9	1.0 NA	28.0 NA
Palladium	2476.4		0.3 AA
	3634.7	0.1 OAn	
Phosphorus (as HPO)	5249	6.0 AH (reversed)	
Platinum	2659.4	15.0 OAn	2.2 AA

APPENDIX E *Continued*

Element	Wavelength, Å	Emission Detection Limit, $\mu g/ml/0.1$ mV	Absorption Sensitivity, $\mu g/ml/1\%$ Abs
Potassium	4044.1	1.7 OH	3.7 AP
	7664.9	0.0005 AA	0.01 AP
Rubidium	7800.2	0.001 AA	0.04 AP
Silicon	2516.1	4.0 OAnr	2.0 NA
Silver	3280.7	0.2 OA	0.13 AA
	3382.9	0.2 OA	0.22 AA
Sodium	5890.0	0.0005 AA	0.004 AP
Strontium	4607.3	0.06 OA	0.06 AA, NA
Sulfur (as S_2)	3940	3.0 AH (reversed and shielded)	
Tellurium	2142.8	7.0 OAn	0.5 AA
	2385.8	2.0 OAn	43.0 AA
Thallium	2767.9		0.1 AA
	3775.7	0.02 NA	0.03 AA
	5350.5	0.05 NA	
Tin	2840.0	0.3 NA	
	2863.3		10.0 AH
Vanadium	3184.0	0.4 NA	0.4 NA
Zinc	2138.6	50.0 OAn	0.025 AA

NOTE: The symbols used in this table and their meanings are as follows: AA, air–acetylene flame; AH, air–hydrogen flame; AP, air–propane flame; NA, nitrous oxide–acetylene flame; OA, oxygen–acetylene flame; OH, oxygen–hydrogen flame; n, organic solvent containing solute aspirated directly into flame, solvent often is methyl isobutyl ketone; r, reaction zone of a fuel-rich flame.

APPENDIX F Values of Absorbance for Percent Absorption

%A	.0	.1	.2	.3	.4	.5	.6	.7	.8	.9
0.0	.0000	.0004	.0009	.0013	.0017	.0022	.0026	.0031	.0035	.0039
1.0	.0044	.0048	.0052	.0057	.0061	.0066	.0070	.0074	.0079	.0083
2.0	.0088	.0092	.0097	.0101	.0106	.0110	.0114	.0119	.0123	.0128
3.0	.0132	.0137	.0141	.0146	.0150	.0155	.0159	.0164	.0168	.0173
4.0	.0177	.0182	.0186	.0191	.0195	.0200	.0205	.0209	.0214	.0218
5.0	.0223	.0227	.0232	.0236	.0241	.0246	.0250	.0255	.0259	.0264
6.0	.0269	.0273	.0278	.0283	.0287	.0292	.0297	.0301	.0306	.0311
7.0	.0315	.0320	.0325	.0329	.0334	.0339	.0343	.0348	.0353	.0357
8.0	.0362	.0367	.0372	.0376	.0381	.0386	.0391	.0395	.0400	.0405
9.0	.0410	.0414	.0419	.0424	.0429	.0434	.0438	.0443	.0448	.0453
10.0	.0458	.0462	.0467	.0472	.0477	.0482	.0487	.0491	.0496	.0501
11.0	.0506	.0511	.0516	.0521	.0526	.0531	.0535	.0540	.0545	.0550

APPENDIX F *Continued*

%A	.0	.1	.2	.3	.4	.5	.6	.7	.8	.9
12.0	.0555	.0560	.0565	.0570	.0575	.0580	.0585	.0590	.0595	.0600
13.0	.0605	.0610	.0615	.0620	.0625	.0630	.0635	.0640	.0645	.0650
14.0	.0655	.0660	.0665	.0670	.0675	.0680	.0685	.0691	.0696	.0701
15.0	.0706	.0711	.0716	.0721	.0726	.0731	.0737	.0742	.0747	.0752
16.0	.0757	.0762	.0768	.0773	.0778	.0783	.0788	.0794	.0799	.0804
17.0	.0809	.0814	.0820	.0825	.0830	.0835	.0841	.0846	.0851	.0857
18.0	.0862	.0867	.0872	.0878	.0883	.0888	.0894	.0899	.0904	.0910
19.0	.0915	.0921	.0926	.0931	.0937	.0942	.0947	.0953	.0958	.0964
20.0	.0969	.0975	.0980	.0985	.0991	.0996	.1002	.1007	.1013	.1018
21.0	.1024	.1029	.1035	.1040	.1046	.1051	.1057	.1062	.1068	.1073
22.0	.1079	.1085	.1090	.1096	.1101	.1107	.1113	.1118	.1124	.1129
23.0	.1135	.1141	.1146	.1152	.1158	.1163	.1169	.1175	.1180	.1186
24.0	.1192	.1198	.1203	.1209	.1215	.1221	.1226	.1232	.1238	.1244
25.0	.1249	.1255	.1261	.1267	.1273	.1278	.1284	.1290	.1296	.1302
26.0	.1308	.1314	.1319	.1325	.1331	.1337	.1343	.1349	.1355	.1361
27.0	.1367	.1373	.1379	.1385	.1391	.1397	.1403	.1409	.1415	.1421
28.0	.1427	.1433	.1439	.1445	.1451	.1457	.1463	.1469	.1475	.1481
29.0	.1487	.1494	.1500	.1506	.1512	.1518	.1524	.1530	.1537	.1543
30.0	.1549	.1555	.1561	.1568	.1574	.1580	.1586	.1593	.1599	.1605
31.0	.1612	.1618	.1624	.1630	.1637	.1643	.1649	.1656	.1662	.1669
32.0	.1675	.1681	.1688	.1694	.1701	.1707	.1713	.1720	.1726	.1733
33.0	.1739	.1746	.1752	.1759	.1765	.1772	.1778	.1785	.1791	.1798
34.0	.1805	.1811	.1818	.1824	.1831	.1838	.1844	.1851	.1858	.1864
35.0	.1871	.1878	.1884	.1891	.1898	.1904	.1911	.1918	.1925	.1931
36.0	.1938	.1945	.1952	.1959	.1965	.1972	.1979	.1986	.1993	.2000
37.0	.2007	.2013	.2020	.2027	.2034	.2041	.2048	.2055	.2062	.2069
38.0	.2076	.2083	.2090	.2097	.2104	.2111	.2118	.2125	.2132	.2140
39.0	.2147	.2154	.2161	.2168	.2175	.2182	.2190	.2197	.2204	.2211
40.0	.2218	.2226	.2233	.2240	.2248	.2255	.2262	.2269	.2277	.2284
41.0	.2291	.2299	.2306	.2314	.2321	.2328	.2336	.2343	.2351	.2358
42.0	.2366	.2373	.2381	.2388	.2396	.2403	.2411	.2418	.2426	.2434
43.0	.2441	.2449	.2457	.2464	.2472	.2480	.2487	.2495	.2503	.2510
44.0	.2518	.2526	.2534	.2541	.2549	.2557	.2565	.2573	.2581	.2588
45.0	.2596	.2604	.2612	.2620	.2628	.2636	.2644	.2652	.2660	.2668
46.0	.2676	.2684	.2692	.2700	.2708	.2716	.2725	.2733	.2741	.2749
47.0	.2757	.2765	.2774	.2782	.2790	.2798	.2807	.2815	.2823	.2832
48.0	.2840	.2848	.2857	.2865	.2874	.2882	.2890	.2899	.2907	.2916
49.0	.2924	.2933	.2941	.2950	.2958	.2967	.2976	.2984	.2993	.3002
50.0	.3010	.3019	.3028	.3036	.3045	.3054	.3063	.3072	.3080	.3089
51.0	.3098	.3107	.3116	.3125	.3134	.3143	.3152	.3161	.3170	.3179
52.0	.3188	.3197	.3206	.3215	.3224	.3233	.3242	.3251	.3261	.3270
53.0	.3279	.3288	.3298	.3307	.3316	.3325	.3335	.3344	.3354	.3363

APPENDIX G Four-Place Table of Common Logarithms

N	0	1	2	3	4	5	6	7	8	9
10	0000	0043	0086	0128	0170	0212	0253	0294	0334	0374
11	0414	0453	0492	0531	0569	0607	0645	0682	0719	0755
12	0792	0828	0864	0899	0934	0969	1004	1038	1072	1106
13	1139	1173	1206	1239	1271	1303	1335	1367	1399	1430
14	1461	1492	1523	1553	1584	1614	1644	1673	1703	1732
15	1761	1790	1818	1847	1875	1903	1931	1959	1987	2014
16	2041	2068	2095	2122	2148	2175	2201	2227	2253	2279
17	2304	2330	2355	2380	2405	2430	2455	2480	2504	2529
18	2553	2577	2601	2625	2648	2672	2695	2718	2742	2765
19	2788	2810	2833	2856	2878	2900	2923	2945	2967	2989
20	3010	3032	3054	3075	3096	3118	3139	3160	3181	3201
21	3222	3243	3263	3284	3304	3324	3345	3365	3385	3404
22	3424	3444	3463	3483	3502	3522	3541	3560	3579	3598
23	3617	3636	3655	3674	3692	3711	3729	3747	3766	3784
24	3802	3820	3838	3856	3874	3892	3909	3927	3945	3962
25	3979	3997	4014	4031	4048	4065	4082	4099	4116	4133
26	4150	4166	4183	4200	4216	4232	4249	4265	4281	4298
27	4314	4330	4346	4362	4378	4393	4409	4425	4440	4456
28	4472	4487	4502	4518	4533	4548	4564	4579	4594	4609
29	4624	4639	4654	4669	4683	4698	4713	4728	4742	4757
30	4771	4786	4800	4814	4829	4843	4857	4871	4886	4900
31	4914	4928	4942	4955	4969	4983	4997	5011	5024	5038
32	5051	5065	5079	5092	5105	5119	5132	5145	5159	5172
33	5185	5198	5211	5224	5237	5250	5263	5276	5289	5302
34	5315	5328	5340	5353	5366	5378	5391	5403	5416	5428
35	5441	5453	5465	5478	5490	5502	5514	5527	5539	5551
36	5563	5575	5587	5599	5611	5623	5635	5647	5658	5670
37	5682	5694	5705	5717	5729	5740	5752	5763	5775	5786
38	5798	5809	5821	5832	5843	5855	5866	5877	5888	5899
39	5911	5922	5933	5944	5955	5966	5977	5988	5999	6010
40	6021	6031	6042	6053	6064	6075	6085	6096	6107	6117
41	6128	6138	6149	6160	6170	6180	6191	6201	6212	6222
42	6232	6243	6253	6263	6274	6284	6294	6304	6314	6325
43	6335	6345	6355	6365	6375	6385	6395	6405	6415	6425
44	6435	6444	6454	6464	6474	6484	6493	6503	6513	6522
45	6532	6542	6551	6561	6571	6580	6590	6599	6609	6618
46	6628	6637	6646	6656	6665	6675	6684	6693	6702	6712
47	6721	6730	6739	6749	6758	6767	6776	6785	6794	6803
48	6812	6821	6830	6839	6848	6857	6866	6875	6884	6893
49	6902	6911	6920	6928	6937	6946	6955	6964	6972	6981
N	0	1	2	3	4	5	6	7	8	9

APPENDIX G *Continued*

N	0	1	2	3	4	5	6	7	8	9
90	9542	9547	9552	9557	9562	9566	9571	9576	9581	9586
91	9590	9595	9600	9605	9609	9614	9619	9624	9628	9633
92	9638	9643	9647	9652	9657	9661	9666	9671	9675	9680
93	9685	9689	9694	9699	9703	9708	9713	9717	9722	9727
94	9731	9736	9741	9745	9750	9754	9759	9763	9768	9773
95	9777	9782	9786	9791	9795	9800	9805	9809	9814	9818
96	9823	9827	9832	9836	9841	9845	9850	9854	9859	9863
97	9868	9872	9877	9881	9886	9890	9894	9899	9903	9908
98	9912	9917	9921	9926	9930	9934	9939	9943	9948	9952
99	9956	9961	9965	9969	9974	9978	9983	9987	9991	9996
N	0	1	2	3	4	5	6	7	8	9

Index

Periodic Chart

IA	IIA	IIIB	IVB	VB	VIB	VIIB	VIIB		VIII	
1 **H** 1.0079										
3 **Li** 6.941	**4** **Be** 9.01218									
11 **Na** 22.98977	**12** **Mg** 24.305									
19 **K** 39.098	**20** **Ca** 40.08	**21** **Sc** 44.9559	**22** **Ti** 47.90	**23** **V** 50.9414	**24** **Cr** 51.996	**25** **Mn** 54.9380	**26** **Fe** 55.847	**27** **Co** 58.9332	**28** **Ni** 58.71	
37 **Rb** 85.4678	**38** **Sr** 87.62	**39** **Y** 88.9059	**40** **Zr** 91.22	**41** **Nb** 92.9064	**42** **Mo** 95.94	**43** **Tc** 98.9062	**44** **Ru** 101.07	**45** **Rh** 102.9055	**46** **Pd** 106.4	
55 **Cs** 132.9054	**56** **Ba** 137.34	**57** ***La** 138.9055	**72** **Hf** 178.49	**73** **Ta** 180.9479	**74** **W** 183.85	**75** **Re** 186.2	**76** **Os** 190.2	**77** **Ir** 192.22	**78** **Pt** 195.09	
87 **Fr** (223)	**88** **Ra** 226.0254	**89** **†Ac** (227)	**104** **§** (260)	**105** **§** (260)						

( ) Numbers in parentheses are mass numbers of most stable or most common isotope.

Atomic weights corrected to conform to the 1971 values of the Commission on Atomic Weights.

© by Fisher Scientific Company. Used by permission.

*Lanthanum Series

58 **Ce** 140.12	59 **Pr** 140.9077	60 **Nd** 144.2	61 **Pm** (147)	62 **Sm** 150.4	63 **Eu** 151.96

'Actinium Series

90 **Th** 232.0381	91 **Pa** 231.0359	92 **U** 238.029	93 **Np** 237.0482	94 **Pu** (244)	95 **Am** (243)